Advanced Functional Analysis

T0174593

Advanced Functional Analysis

Eberhard Malkowsky
Vladimir Rakočević

CRC Press
Taylor & Francis Group
Boca Raton London New York

CRC Press is an imprint of the
Taylor & Francis Group, an **informa** business

CRC Press
Taylor & Francis Group
6000 Broken Sound Parkway NW, Suite 300
Boca Raton, FL 33487-2742

First issued in paperback 2020

ISBN 13: 978-0-367-65656-0 (pbk)
ISBN 13: 978-1-1383-3715-2 (hbk)

Library of Congress Cataloging-in-Publication Data

Names: Malkowsky, Eberhard, author. | Vladimir Rakočević, author.
Title: Advanced functional analysis / Eberhard Malkowsky and Vladimir Rakočević.
Other titles: Functional analysis
Description: Boca Raton, Florida : CRC Press, [2019] | Includes bibliographical references and index.
Identifiers: LCCN 2018048992| ISBN 9781138337152 (hardback : alk. paper) | ISBN 9780429442599 (e-book)
Subjects: LCSH: Functional analysis.
Classification: LCC QA320 .M2845 2019 | DDC 515/.7--dc23
LC record available at https://lccn.loc.gov/2018048992

**Visit the Taylor & Francis Web site at
http://www.taylorandfrancis.com**

**and the CRC Press Web site at
http://www.crcpress.com**

Contents

10 Fixed Point Theory   377

Preface

Several books have been devoted to functional analysis and its applications. This volume, however, is special in comparison to the previous ones. It contains topics and recent research results in the fields of linear and nonlinear functional analysis and operator theory with a special emphasis on Fredholm theory, measures of noncompactness, summability, sequence spaces, and fixed point theory.

In order to make the book self–contained and comprehensive and of interest for a larger mathematical community, it also presents the necessary concepts and results for the covered advanced research topics. This book is based on several of the authors' lectures at international conferences and courses at universities in Australia, Germany, India, Jordan, Mexico, Serbia, Turkey, the USA, and South Africa.

In view of its original approach, the book differs from other existing ones in those areas.

The book contains ten chapters with a great number of illustrating examples and remarks concerning related results. It can be used as a textbook for graduate and post–graduate courses in advanced functional analysis, and also as a basis for research in the fields above, and is intended to address students, teachers, and researchers alike.

Chapter 1 provides a survey of useful notations, important inequalities, and fundamental concepts and results concerning linear and topological spaces which are essential throughout the book. It also deals with the concepts of semimetric and seminormed spaces. Detailed studies are dedicated to net convergence, subnets, and compactness in topological spaces, and the proofs are given for those results which are regarded as less familiar by the authors. Preference was given to net convergence over filter convergence, since net convergence seems to be the more natural extension to topological spaces of convergence of sequences in semimetric spaces.

Chapter 2 is dedicated to the study of linear topological spaces. Concerning purely algebraic concepts, thorough treatments are presented of convex and affine sets, balloons and cones, and quotient spaces. The studies of purely topological concepts concentrate on supremum, weak, and product topologies. The notion of linear topological spaces arises from combining the concepts of linearity and topology in a natural way by demanding that the algebraic operations of linear spaces are continuous in the topology. Various important properties of linear topological spaces are established and a characterization of linear topologies is given in terms of their open sets. Furthermore, the chapter deals with properties of closed maps, and presents the closed graph lemma in linear topological spaces, and Baire's category theorem in complete metric spaces.

Chapter 3 deals with linear metric and linear semimetric spaces, the concepts of paranormed spaces and Schauder bases, and their most important properties. The highlights of the chapter are the open mapping theorem, the closed graph theorem, the uniform boundedness principle and the Banach–Steinhaus theorem. Furthermore, the chapter contains studies of useful properties of seminorms, the concept of local convexity, and the Minkowski functional and its role in defining a seminorm or norm on a linear space. Finally, a sufficient condition is established for the metrizability of a linear topology, and also a criterion is given for a linear topology to be generated by a seminorm.

Chapter 4 deals mainly with the studies of Banach spaces. It contains the most important examples of Banach spaces and results on the bounded linear operators between normed and Banach spaces. The first highlight is the fundamental Hahn–Banach extension theorem which is of vital importance in the proof of many results in functional analysis; also many of its corollaries are given. Furthermore, the versions in Banach spaces are given of the open mapping and closed graph theorems, Banach's theorem of the bounded inverse, and the Banach–Steinhaus theorem, an alternative proof of which is added that does not use Baire's theorem. Two important applications of the closed graph theorem are the criterion for a closed subspace in a Banach space to have a topological complement, and the Eni–Karauš theorem. Further applications of the Hahn–Banach and Banach–Steinhaus theorems are the classical representation theorems for the continuous linear functionals on the classical sequence spaces, including the representation theorem in the case of the space of bounded real sequences, and the Riesz representation theorem for the continuous linear functionals on the space of continuous real functions on the unit interval. The other topics focus on the reflexivity of spaces, the studies of adjoint operators, quotient spaces, Cauchy nets, the equivalence of norms, compactness and the Riesz lemma, compact operators and operators with closed range.

Chapter 5 contains the study of Hilbert spaces. Apart from the standard examples and more familiar results such as the P. Jordan–J. von Neumann theorem, the concepts of orthogonality and orthonormality, the theorem of the elements of minimal norm, the theorem on the orthogonal decomposition, the Riesz representation theorem, Bessel's inequality, Fourier coefficients, Parseval's equality and the Gram–Schmidt orthogonalization process, the topics of Hilbert adjoint and Hermitian, normal, positive and unitary operators are also discussed. Finally, the chapter contains a detailed study of projectors and orthogonal projectors, and formulas and results on the norm of idempotent operators.

Chapter 6 is dedicated to the study of Banach algebras. The introduction of the concept of a Banach algebra is followed by a great number of examples and a study of the invertibility of elements in a complex Banach algebra with identity, and of the spectrum, resolvent and spectral radius. Also the concept of the topological divisor of zero is introduced and several results related to this concept are established. Further highlights include the study of subalgebras, one theorem by Hochwald–Morell and two theorems by Harte for regular elements in Banach algebras, conditions for the invertibility of operators in the Banach algebra of bounded linear operators from a Banach space into itself, and detailed studies of the spectra of adjoint, normal, and compact operators. Finally, the concept of C^*–algebras is introduced, their most important properties are established and conditions for the invertibility of the difference of projections are given.

Chapter 7 deals with measures of noncompactness, their properties and some applications. Measures of noncompactness are very useful tools in functional analysis—for instance in metric fixed point theory and the theory of operator equations in Banach spaces, and in the characterizations of compact operators between Banach spaces. After a motivation, results from fixed point theory and the discussion of compact operators and the Hausdorff distance, an axiomatic introduction of measures of noncompactness in complete metric spaces is given, and their most important properties are established, such as monotonicity and the generalized Cantor intersection property. There also are detailed studies of the most important properties of the Kuratowski and Hausdorff measures of noncompactness on bounded sets of Banach spaces related to the linear structure of Banach spaces—in particular, the invariance under the passage to the convex hull, which is crucial in the proof of Darbo's fixed point theorem. Furthermore, the famous Goldenštein–Goh'berg–Markus theorem is proved which gives a very useful estimate for the Hausdorff measure of noncompactness of bounded sets in Banach spaces with a Schauder basis. Finally, the notion of measures of noncompactness of operators and some of their properties are considered, and Fredholm's alternative is proved.

Chapter 8 is focused on the Fredholm theory and Fredholm operators, which are generalizations of operators that are the difference of the identity and a compact linear operator on a Banach space. Fredholm operators play a very important role in the spectral theory of operators. The chapter presents a study of Fredholm and semi–Fredholm operators, the index and Atkinson's theorems and Yood's results for all upper semi–Fredholm operators with nonpositive index, lower semi–Fredholm operators with nonnegative index, and properties of right and left Fredholm operators. It also establishes the openness of the set of proper semi–Fredholm operators in the space of bounded linear operators between Banach spaces, and gives the proofs of the punctured neighborhood theorem and the Kato decomposition theorem. Finally, it provides detailed studies of the ascent and descent of operators, the properties of Browder and semi–Browder operators, essential spectra, and essential type subsets of the spectrum.

Chapter 9 presents useful and interesting applications of the results of the preceding chapters to modern topics in the large field of summability—in particular, in the characterizations of (infinite) matrix transformations and of compact linear operators between sequence spaces. The chapter contains an introduction to the theory of FK and BK spaces, which is a very powerful tool for the characterization of classes of matrix transformations because of the fundamental result that matrix maps between FK spaces are continuous. Further topics are the studies and determinations of multiplier spaces and β–, γ–, functional and continuous duals, and relations between them, of properties of transposes of matrices and the complete known characterizations of matrix transformation between the classical sequence spaces of bounded, convergent, and null sequences and of p–summable series for $1 \leq p < \infty$ in terms of conditions on the entries of the matrices. Finally, the focus is on the study of compact matrix operators, the representation of compact operators on the space of convergent sequences, an estimate of the Hausdorff measure of noncompactness and the characterization of such operators, and the characterizations of compact matrix operators between the classical sequence spaces by the application of the Hausdorff measure of noncompactness.

Chapter 10 deals with a great number of the most important results in fixed point theory, starting with Banach's classical contraction principle for which several different proofs are presented, as well as various corollaries and examples. Furthermore, it presents results by Edelstein and Rakotch, the concept of nonlinear contraction by Boyd and Wong, and theorems by Meir–Keeler, Kannan, Chatterjee, and Zamfirescu. It also deals with results related to the concepts of Ćirić's generalized contractions and quasi–contractions, and establishes the theorems by Reich, Hardy–Rogers, Caristi, and Bollenbacher and Hicks. The chapter closes with studies of the Mann iteration and fixed point theorems by Mann, Reinermann, Franks, and Marzec for real functions on compact intervals of the real line.

Finally, the authors express their sincere gratitude to A. Aasma, M. Abbas, J. Banaś, M. Cvetković, A. Dajić, I. Djolović, R. Harte, A. Jeribi, E. Karapınar, M. A. Khamsi, J. J. Koliha, R. T. Mısırlıoğlu, R. P. Pant, V. Pavlović, M. Petrović, and M. A. Sofi for carefully reading the manuscript and for their valuable comments and suggestions. The authors are particularly grateful to J. Banaś for his encouragement and support to finish work on the book.

Authors

Eberhard Malkowsky is a Full Professor at the Department of Mathematics of the State University in Novi Pazar, Serbia. His research interests include functional analysis, summability theory, matrix transformations, measures of noncompactness and operator theory.

Vladimir Rakočević is a Full Professor at the Department of Mathematics of the Faculty of Sciences and Mathematics of the University of Niš, Serbia, and a Corresponding Member of the Serbian Academy of Sciences and Arts (SANU) in Belgrade, Serbia. His research interests include functional analysis, fixed point theory, operator theory, linear algebra and summability theory.

Symbol Description

$\mathbb{N}$	set of natural numbers	$\mathcal{L}(X,Y)$	set of all linear functions from X to Y
$\mathbb{Z}$	set of integers		
$\mathbb{Q}$	set of rational numbers	$X^\#$	$= \mathcal{L}(X,\mathbb{C})$, set of all linear functionals on X
$\mathbb{R}$	set of real numbers		
$\mathbb{C}$	set of complex numbers	$X_\mathbb{R}^\#$	$= \mathcal{L}(X,\mathbb{R})$, set of all real linear functionals on X
$\mathcal{P}(A)$	power set of a set A		
S^c	complement of a subset S in a set	d	semimetric or metric
		(X,d)	semimetric or metric space
$\mathcal{F}[A]$	class of all complex functions on A	$B_r(x)$	open ball of radius r and centre in x in a semimetric or metric space
$\mathcal{F}_\mathbb{R}[A]$	class of all real functions on A		
ω	set of all complex sequences	$\bar{B}_r(x)$	closed ball of radius r and centre in x in a semimetric or metric space
c_0	set of all null sequences		
c	set of all convergent sequences	$S_r(x)$	sphere of radius r and centre in x in a semimetric or metric space
ℓ_∞	set of all bounded sequences		
ℓ_p	set of all absolutely p summable series		
		$\lim x_n$	limit of the convergent sequence (x_n) in a metric space
ϕ	set of all finite sequences		
$\prod A_\beta$	(Cartesian) product of the sets A_β	$\mathrm{diam}(S)$	diameter of the subset S of a semimetric space
$e = (e_k)$	sequence with $e_k = 1$ for all k	$\mathrm{dist}(S,S')$	$= \inf\{d(s,s') : s \in S, s' \in S'\}$, distance of the subsets S and S' of the semimetric space (X,d)
$e^{(n)} = (e_k^{(n)})$	n–th unit sequence, that is, $e_n^{(n)} = 1$ and $e_k^{(n)} = 0$ for $k \neq n$		
P_t	projection map	$\mathrm{dist}(x,S)$	$=\mathrm{dist}(\{x\},S)$ distance of the point x and the subset S of a semimetric space
P_n	n–th co–ordinate map (projection)		
$\mathbb{F}$	a field, always $\mathbb{R}$ or $\mathbb{C}$	p	seminorm
$\mathrm{span}(S)$	span or linear hull of a subset S of a linear space	(X,p)	seminormed space
		$\|\cdot\|$	norm
$\mathrm{span}(S_1,S_2)$	span of the union of the subsets S_1 and S_2 of a linear space	$(X,\|\cdot\|)$	normed space
		$\mathcal{T}$	topology
$\mathrm{span}(x,S)$	$=\mathrm{span}(\{x\},S)$	$(X,\mathcal{T})$	topological space
$S_1 \oplus S_2$	direct sum of the subspaces S_1 and S_2 of a linear space	d_∞	natural metric on c_0, c, and ℓ_∞
		d_p	natural metric on ℓ_p
$\dim X$	algebraic dimension of the linear space X	d_ω	natural metric on ω
		$\|\cdot\|_\infty$	natural or supremum norm on c_0, c, and ℓ_∞
$N(f)$	null space or kernel of the function f		
		$\|\cdot\|_p$	natural norm on ℓ_p for $1 \leq p < \infty$
$R(f)$	range of the function f		

$\bar{S}$	closure of a subset S in a topological space
$\mathcal{T}_S$	relative topology on a subset S of a topological space $(X, \mathcal{T})$
$\bigvee \Phi$	topology generated by the union of the class Φ of topologies on a set
$\mathcal{T}_1 \vee \mathcal{T}_2$	topology generated by the union of the class $\Phi = \{\mathcal{T}_1, \mathcal{T}_2\}$ of topologies of a set
$d_1 \vee d_2$	topology $\mathcal{T}_1 \vee \mathcal{T}_2$ when the topologies $\mathcal{T}_1$ and $\mathcal{T}_2$ are given by the metrics d_1 and d_2
$(x_\delta)_{\delta \in D}$, (x_δ)	net
x_δ	terms of a net
$x_\delta \to x$	convergence of a net $(x_\delta)_{\delta \in D}$ in a topological space to a limit x
$\lim x_\delta$	limit of a convergent of a net $(x_\delta)_{\delta \in D}$, if the limit is unique
$T(\delta_0)$	tail of a net
$A + B$	sum of subsets A and B of a linear space
$a + B$	$= \{a\} + B$
ΛA	multiplication of a subset Λ of the scalars and a subset A of a linear space
λA	$= \{\lambda\} A$
$\mathrm{cvx}(A)$	set of all convex combinations of a subset A of a linear space
$\mathrm{co}(A)$	convex hull of a subset A of a linear space
$\mathrm{span}_{\mathbb{R}}(A)$	real span of a subset A of a linear space
X/S	quotient space X modulo S
$\mathrm{codim}_X(S)$	$= \dim X/S$
$w(A, f)$	weak topology on a set A by f
$w(A, \Phi)$	weak topology on a set A by a class Φ of functions from A into a class of topological spaces (Definition 2.6.6)
$\sigma(X, \Phi)$	weak topology on a linear space X by a class Φ of seminorms on X, or linear maps from X to a collection of linear topological spaces (Definition 2.8.10)
$\sigma(\Phi)$	short for $\sigma(X, \Phi)$
$\mathcal{Q}_f$	quotient topology
$\{f < \lambda\}$	$= \{x : f(x) < \lambda\}$ when $R(f) \subset \mathbb{R}$
$\mathrm{graph}(f)$	graph of a function f
p	paranorm
$\{p < \lambda\}$	$= p^{-1}([0, \lambda))$ for the paranrom p
$\{p \leq \lambda\}$	$= p^{-1}([0, \lambda])$ for the paranorm p
$\sum\limits_{n=1}^{\infty} a_n$	series in a semimetric space
$\ell^{(n)}$	n–th coordinate with respect to the Schauder basis (b_n)
$\mathcal{B}[X]$	space of all bounded complex–valued functions on a set X
$\|p\|$	norm of the seminorm p
$\|p\|_S$	restriction of $\|p\|$ on the subset S of a seminormed space (X, p)
$\|f\|$	norm of the linear map f between seminormed spaces, in particular, norm of the linear functional f on a normed space
μ_B	Minkowski functional or gauge of the balloon B in a linear space
$\mathrm{C}[a, b]$	space of all continuous real- or complex-valued functions on the interval $[a, b] \subset \mathbb{R}$
$\mathrm{BV}[a, b]$	space of all real- or complex-valued functions of bounded variation on the interval $[a, b] \subset \mathbb{R}$
$V_a^b(f)$	total variation of $f \in \mathrm{BV}[a, b]$ on the interval $[a, b] \subset \mathbb{R}$
$\mathrm{B}[a, b]$	space of all bounded real- or complex-valued functions on the interval $[a, b] \subset \mathbb{R}$
$\mathcal{B}(X, Y)$	space of all bounded linear operators from the normed space X into the normed space Y
$\mathcal{B}(X)$	$= \mathcal{B}(X, X)$
X^*	$= \mathcal{B}(X, \mathbb{C})$, space of all continuous linear functionals on the normed space X (continuous dual of X)
$\mathrm{sgn}(z)$	sign of the complex number z
$M \oplus N$	topological direct sum of the closed subspaces M and N of a normed space
$\mathcal{B}[H]$	set of all real–valued bounded functions on a set H
$\sigma(f; \Pi, (h_k))$	Riemann sums of the function f with respect to the partition Π and the points h_k

χ_E	characteristic function of the set E	$\mathrm{Re}A$	$= (A + A^*)/2$, the real part of the bounded linear operator A from a Hilbert space into itself
$\mathcal{M}[H]$	set of all additive set functions μ on H for which $\|\mu\| < \infty$	$\mathrm{Im}A$	$= (A - A^*)/(2i)$, the imaginary part of the bounded linear operator A from a Hilbert space into itself
X^{**}	$= (X^*)^*$ bidual of the normed space X		
J	canonical map from a normed space into its bidual	$\mathcal{A}_l^{-1}$	the set of all left invertible elements in the algebra $\mathcal{A}$
L'	adjoint operator of the operator $L \in \mathcal{B}(X, Y)$	$\mathcal{A}_r^{-1}$	the set of all right invertible elements in the algebra $\mathcal{A}$
U^0	set of all annihilators of a subset U of a normed space	$\mathcal{A}^{-1}$	the set of all invertible elements in the algebra $\mathcal{A}$
^0W	set of all annihilators of a subset W of the continuous dual of a normed space	$f * g$	convolution of the functions $f, g \in L_1$
z_F	$= \sum_{\delta \in \mathcal{F}} x_\delta$ where the sum is taken the finite subset $\mathcal{F}$ of the directed set D of the net x_δ	$\sigma(a)$	spectrum of a
		$\sigma_l(a)$	left spectrum of a
		$\sigma_r(a)$	right spectrum of a
$\mathcal{K}(X, Y)$	set of all compact operators in $\mathcal{B}(X, Y)$	$\rho(a)$	resolvent set of a
		R_a	resolvent function of a
$\mathcal{K}(X)$	$= \mathcal{K}(X, X)$	$R_a(\lambda)$	$= (\lambda - a)^{-1}$ for $\lambda \in \rho(a)$
$\mathcal{F}(X, Y)$	set of all finite rank operators in $\mathcal{B}(X, Y)$	$r(a)$	spectral radius if a
		$\mathcal{Z}^l$	the set of all left topological divisors of zero in a Banach algebra with identity
$\mathcal{F}(X)$	$= \mathcal{F}(X, X)$		
$j(A)$	minimum modulus or modulus of injectivity of $A \in \mathcal{B}(X, Y)$	$\mathcal{Z}^r$	the set of all right topological divisors of zero in a Banach algebra with identity
$\mathcal{J}(X, Y)$	set of all bounded-below operators in $\mathcal{B}(X, Y)$		
$\gamma(A)$	reduced minimum modulus of the operator $A \in \mathcal{B}(X, Y)$	$\mathcal{Z}$	$= \mathcal{Z}^l \cup \mathcal{Z}^r$
		$\mathcal{H}^l$	$= \mathcal{A} \setminus \mathcal{Z}^l$
$q(A)$	modulus or coefficient of surjectivity of the operator $A \in \mathcal{B}(X, Y)$	$\mathcal{H}^r$	$= \mathcal{A} \setminus \mathcal{Z}^r$
		$\mathcal{H}$	$= \mathcal{A} \setminus \mathcal{Z}$
		$\sigma_\mathcal{B}(a)$	spectrum of a with respect to the subalgebra $\mathcal{B}$
$\mathcal{Q}(X, Y)$	set of all $A \in \mathcal{B}(X, Y)$ with $q(A) > 0$	$r_\mathcal{B}(a)$	spectral radius of a with respect to the subalgebra $\mathcal{B}$
$\langle \cdot, \cdot \rangle$	inner product		
$L_2^a(G)$	Bergmann space	$\mathcal{A}^\sqcap$	the set of all regular elements in the Banach algebra $\mathcal{A}$ with identity
$x \perp y$	x and y are orthogonal (in an inner product space)		
$E^\perp$	orthogonal complement of a subset E of an inner product space	$\sigma_p(A)$	point spectrum of $A \in \mathcal{B}(X)$
		$\sigma_a(A)$	approximate spectrum of $A \in \mathcal{B}(X)$
$X = M \oplus M^\perp$	orthogonal sum (in the Hilbert space X)	$\sigma_l(A)$	left spectrum of $A \in \mathcal{B}(X)$
		$\sigma_d(A)$	defect spectrum of $A \in \mathcal{B}(X)$
A^*	Hilbert adjoint operator of the operator A	$\sigma_\delta(A)$	approximate defect spectrum of $A \in \mathcal{B}(X)$
$m(A)$	lower bound of the self–adjoint operator A	$\sigma_r(A)$	right spectrum of $A \in \mathcal{B}(X)$
		$\sigma_{res}(A)$	residual spectrum of $A \in \mathcal{B}(X)$
$M(A)$	upper bound of the self-adjoint operator A	$\sigma_c(A)$	continuous spectrum of $A \in \mathcal{B}(X)$

$A^\dagger$	Moore–Penrose pseudo–inverse of $A \in \mathcal{B}(X)$, where X is a Hilbert space	$\|L\|_{(\phi,\psi)}$	(ϕ,ψ)–norm or (ϕ,ψ)–measure of noncompactness of the operator $L \in \mathcal{B}(X,Y)$
$\mathcal{F}(f)$	set of all fixed points of the function f	$\mathrm{asc}(A)$	ascent of the operator $A \in \mathcal{L}(X)$
$\mathcal{F}(X,Y)$	set of all finite rank operators in $\mathcal{B}(X,Y)$	$\mathrm{dsc}(A)$	descent of the operator $A \in \mathcal{L}(X)$
$\mathcal{F}(X)$	$= \mathcal{F}(X,X)$	$\alpha(A)$	nullity of the operator $A \in \mathcal{B}(X)$
$\mathcal{M}_X$	class of all bounded subsets of the metric space X	$\beta(A)$	defect of the operator $A \in \mathcal{B}(X)$
$\mathcal{M}_X^c$	class of all closed bounded subsets of the metric space X	$i(A)$	$= \alpha(A) - \beta(A)$ index of the operator $A \in \mathcal{B}(X)$
d_H	Hausdorff distance		
$d_H(S,\tilde{S})$	the Hausdorff distance of the bounded subsets S and $\tilde{S}$ of a metric space	$\alpha(A)$	nullity of the operator $A \in \mathcal{L}(X,Y)$
$B_r(S)$	open ball with radius $r > 0$ and centre in the subset S of a metric space	$\beta(A)$	defect of the operator $A \in \mathcal{L}(X,Y)$
		$N^\infty(A)$	$= \bigcup_{n=1}^{\infty} N(A^n)$, the generalized kernel of $A \in \mathcal{L}(X)$
α	Kuratowski measure of noncompactness	$R^\infty(A)$	$= \bigcap_{n=1}^{\infty} R(A^n)$, the generalized range of $A \in \mathcal{L}(X)$
$\alpha(Q)$	Kuratowski measure of noncompactness of the bounded subset Q of a complete metric space	$k(A)$	$= \dim[N(A)/(N(A) \cap R^\infty(A))]$
		$\Phi_+(X,Y)$	the class of all operators $A \in \mathcal{B}(X,Y)$ with $\alpha(A) < \infty$ and closed range, the class of the upper semi–Fredholm operators in $\mathcal{B}(X,Y)$
$\overline{\mathrm{co}}(Q)$	convex closure of the nonempty bounded subset Q of a normed space	$\Phi_-(X,Y)$	$= \{A \in \mathcal{B}(X,Y) : \beta(A) < \infty\}$, the class of lower semi–Fredholm operators in $\mathcal{B}(X,Y)$
χ	Hausdorff or ball measure of noncompactness		
$\chi(Q)$	the Hausdorff or ball measure of noncompactness of the bounded subset Q of a complete metric space	$\Phi_\pm(X,Y)$	$= \Phi_+(X,Y) \cup \Phi_-(X,Y)$, the class of semi–Fredholm operators in $\mathcal{B}(X,Y)$
$\mathcal{N}_X^c$	the class of all nonempty, bounded and compact subsets of the metric space X	$\Phi(X,Y)$	$= \Phi_+(X,Y) \cap \Phi_-(X,Y)$, the class of Fredholm operators in $\mathcal{B}(X,Y)$
χ_i	inner Hausdorff or inner ball measure of noncompactness	$\Phi(X)$	$= \Phi(X,X)$
		$i(A)$	$= \alpha(A) - \beta(A)$, index of a Fredholm operator A
$\chi_i(Q)$	inner Hausdorff or inner ball measure of noncompactness of the bounded subset Q of a complete metric space	$\Phi_0(X,Y)$	$= \{A \in \Phi(X,Y) : i(A) = 0\}$, the class of Weyl operators
		$\Phi_+^-(X,Y)$	$= \{A \in \Phi_+(X,Y) : i(A) \leq 0\}$
		$\Phi_-^+(X,Y)$	$= \{A \in \Phi_-(X,Y) : i(A) \geq 0\}$
β	separation or Istrăţescu measure of noncompactness	$\Phi_r(X,Y)$	class of right Fredholm operators, or right essentially invertible operators
$\beta(Q)$	separation or Istrăţescu measure of noncompactness of the bounded subset Q of a complete metric space	$\Phi_l(X,Y)$	class of left Fredholm operators, or left essentially invertible operators
$\|\cdot\|_{(\phi,\psi)}$	(ϕ,ψ)–norm or (ϕ,ψ)–measure of noncompactness	$\mathcal{C}(X)$	$= \mathcal{B}(X)/\mathcal{K}(X)$, Calkin algebra

$r_e(A)$ — essential spectral radius of $\pi(A)$, where $\pi : \mathcal{B}(X) \to \mathcal{C}(X)$ is the natural homomorphism

$a_n(A)$ — $= \dim(N(A)/[N(A) \cap R(A^n)])$

$b_n(A)$ — $= \dim([R(A) + N(A^n)]/R(A))$

$\Phi_{\pm}(A)$ — $= \{\lambda \in \mathbb{C} : A - \lambda \in \Phi_+(X) \cup \Phi_-(X)\}$, semi–Fredholm domain of $A \in \mathcal{B}(X)$

$\Phi(A)$ — $= \{\lambda \in \mathbb{C} : A - \lambda \in \Phi(X)\}$, Fredholm domain of $A \in \mathcal{B}(X)$

$ju(A)$ — the jump of a semi–Fredholm operator A

$\mathfrak{B}(X)$ — the class of Browder operators in $\mathcal{B}(X)$

$\mathfrak{B}_+(X)$ — the class of all upper semi–Browder operators

$\mathfrak{B}_-(X)$ — the class of all lower semi–Browder operators

$\sigma_{ef}(A)$ — the Fredholm (Wolf, or Calkin) essential spectrum of the operator $A \in \mathcal{B}(X)$

$\sigma_{ew}(A)$ — the Weyl (Schechter) essential spectrum of the operator $A \in \mathcal{B}(X)$

$\sigma_{eb}(A)$ — the Browder essential spectrum of the operator $A \in \mathcal{B}(X)$

$\sigma_{ek}(A)$ — the Kato essential spectrum of the operator $A \in \mathcal{B}(X)$

$\sigma_{e\alpha}(A)$ — the upper semi–Fredholm or Gustafson essential spectrum of the operator $A \in \mathcal{B}(X)$

$\sigma_{e\beta}(A)$ — the lower semi–Fredholm or Gustafson essential spectrum of the operator $A \in \mathcal{B}(X)$

$\sigma_{ea}(A)$ — the essential approximate point spectrum of the operator $A \in \mathcal{B}(X)$

$\sigma_{ed}(A)$ — the essential approximate defect spectrum of the operator $A \in \mathcal{B}(X)$

$\sigma_{ab}(A)$ — the Browder essential approximate point spectrum of the operator $A \in \mathcal{B}(X)$

$\sigma_{db}(A)$ — the Browder defect spectrum of the operator $A \in \mathcal{B}(X)$

$\mathrm{cl}_Y(E)$ — the closure of a subset E in a topological space Y

$x^{[m]}$ — $= \sum_{k=0}^m x_k e^{(k)}$, the m–section of a sequence $x = (x_k)_{k=0}^\infty$

A_n — $= (a_{nk})_{k=0}^\infty$, the sequence in the nth row of the infinite matrix $A = (a_{nk})_{n,k=0}^\infty$

$A_n x$ — $= \sum_{k=0}^\infty a_{nk} x_k$, the nth A transform of the sequence x

Ax — $= (A_n x)_{n=0}^\infty$, the A transform of the sequence x

X_A — the matrix domain of A in X

c_A — the convergence domain of A

(X,Y) — the class of all matrix transformations from X into Y

cs — the set of all convergent series

bs — the set of all bounded series

$M(X,Y)$ — the multiplier space of X in Y

X^α — the α–dual of the set X of sequences

X^β — the β–dual of the set X of sequences

X^γ — the γ–dual of the set X of sequences

X^f — the functional dual of the set X of sequences

A^T — transpose of the infinite matrix A

Chapter 1

Fundamentals of Linear and Topological Spaces

Chapter 1 provides a survey of useful fundamental notations and concepts such as partially ordered and directed sets, and Zorn's lemma, as well as important inequalities, such as the Hölder, Minkowski and Jensen inequalities. It also contains a review of fundamental well–known concepts and results from linear algebra concerning linear spaces, linear maps, which are essential throughout the book. Furthermore, it revises the concepts of semimetric, seminormed and topological spaces, open and closed sets and neighborhoods, continuity of functions between topological spaces, bases and subbases of topological spaces, and separation axioms. Finally, detailed studies are dedicated to net convergence, subnets and compactness in topological spaces, and the proofs are given for those results which the authors regard as being less familiar. Preference was given to net convergence over filter convergence, since net convergence seems to be the more natural extension to topological spaces of convergence of sequences in semimetric, metric and normed spaces.

1.1   Introduction

In this chapter, we present and revise the fundamental concepts, results and notations that are needed throughout the book, such as partially ordered and directed sets, and Zorn's lemma, as well as important inequalities, such as the Hölder, Minkowski and Jensen inequalities. It also contains a review of fundamental well–known concepts and results from linear algebra concerning linear spaces, linear maps, which are essential throughout the book. Furthermore it revises the concepts of semimetric, seminormed and topological spaces, open and closed sets and neighborhoods, continuity of functions between topological spaces, bases and subbases of topological spaces, and separation axioms. Finally, detailed studies are dedicated to net convergence, subnets and compactness in topological spaces, and the proofs are given for those results which the authors regard as being less familiar. Preference was given to net convergence over filter convergence, since net convergence seems to be the more natural extension to topological spaces of convergence of sequences in semimetric, metric and normed spaces.

The material of this chapter can be found in many books on linear algebra and topology.

We use the standard notations $\mathbb{N}$, $\mathbb{Z}$, $\mathbb{Q}$, $\mathbb{R}$ and $\mathbb{C}$ for the sets of all natural numbers, integers, rational, real and complex numbers.

The symbol $\emptyset$ denotes the empty set. We write $A \subset B$, if A is a subset of B, and $A \subsetneq B$, if A is a proper subset of B. For any set A, we write $\mathcal{P}(A) = \{B : B \subset A\}$ for the *power set of A*. If S is a subset of A, then $S^c = A \setminus S$ denotes the *complement of S (in A)*.

1.2   Partially ordered sets and Zorn's lemma

In this section, we recall the definitions of *partially ordered* and *directed sets*, which are essential in the concept of net convergence in topological spaces studied in Section 1.10. We also state *Zorn's lemma* and the equivalent *axiom of transfinite induction*. Zorn's lemma will be applied, for instance, in the proof of the famous Hahn–Banach extension theorem, Theorem 4.5.2.

Definition 1.2.1 (a) Let A be a nonempty set. A binary relation $\preceq$ on A is said to be a *partial order* if it satisfies the following conditions for all $a, b, c \in X$

$$a \preceq a \qquad \qquad \textit{(reflexivity)}$$
$$a \preceq b \text{ and } b \preceq c \text{ imply } a \preceq c \quad \textit{(transitivity)}.$$

A set A with a partial order $\preceq$ is called a *partially ordered set*, or *poset*, for short. Elements a and b of a partially ordered set are called *comparable*, if $a \preceq b$ or $b \preceq a$; otherwise they are said to be *incomparable*. We write $a \succeq b$ to mean $b \preceq a$.
(b) A *directed set D* is a partially ordered set which satisfies the following condition:

$$\text{for all } a, b \in D \text{ there exists } c \in D \text{ such that } a \preceq c \text{ and } b \preceq c.$$

(c) A partially ordered set is called *totally ordered*, if all elements are comparable and the ordering $\preceq$ also satisfies the condition that

$$a \preceq b \text{ and } b \preceq a \text{ imply } a = b, \quad \textit{(antisymmetry)}.$$

(d) A *chain* is a totally ordered subset of a partially ordered set. A *maximal chain* is a chain which is not properly included in any chain.

The most familiar example for a totally ordered set is $(\mathbb{R}, \leq)$ where $\leq$ denotes the usual ordering on $\mathbb{R}$.

Example 1.2.2 (Ordering by inclusion, ordering by containment) *Let A be a set and $\mathcal{A}$ be a subclass of $\mathcal{P}(A)$.*

(a)   For all $B, C \in \mathcal{A}$, let $B \preceq C$ mean $B \subset C$.

(b)   For all $B, C \in \mathcal{A}$, let $B \preceq C$ mean $B \supset C$.

The orderings defined by (a) or (b) are called ordering by inclusion *or* ordering by containment, *respectively.*

Example 1.2.3 *Let S be a non–empty finite subset of a directed set D. Then there is an element $d \in D$ such that $d \geq s$ for all $s \in S$.*

Axiom 1 (Axiom of transfinite induction)
Every partially ordered set includes a maximal chain.

Remark 1.2.4 *(a) The axiom of transfinite induction is also referred to as Zorn's lemma. (b) Clearly a finite partially ordered set includes a maximal chain. One picks an element at random, then, if possible, another one comparable with the first, another comparable with both, and so on until no further choice is possible. In the case of an infinite partially ordered set, the result is postulated.*

The axiom of transfinite induction can be expressed in the following form.

Theorem 1.2.5 *Every chain is included in a maximal chain.*

Proof. Let A be a partially ordered set and C be a nonempty chain in A. Let B be the subset of A consisting of all elements comparable with all the elements of C; clearly $C \subset B$. Also B is a partially ordered set with the ordering it inherits from A. Hence B includes a maximal chain. This includes C, for if it would not, we could enlarge it by adjoining C to it. It is maximal in A, since any larger chain would, by definition of B, be included in B. $\square$

1.3   Important notations

Here, we list the important fundamental notations we are going to use in the sequel. Let A be a set. We write

$$\mathcal{F}[A] = \{f : f : A \to \mathbb{C}\} \text{ and } \mathcal{F}_{\mathbb{R}}[A] = \{f : f : A \to \mathbb{R}\}$$

for the classes of all complex– and real–valued functions on A.

The class $\omega = \mathcal{F}[\mathbb{N}]$ is the set of all complex sequences $x = (x_k)_{k=1}^{\infty}$. Important subsets of ω are the sets

$$c_0 = \{x \in \omega : x_k \to 0 \ (k \to \infty)\} \qquad \text{of all } null \ sequences,$$

$$c = \{x \in \omega : x_k \to \xi \ (k \to \infty) \text{ for some } \xi \in \mathbb{C}\} \quad \text{of all } convergent \ sequences,$$

$$\ell_{\infty} = \left\{x \in \omega : \sup_k |x_k| < \infty\right\} \qquad \text{of all } bounded \ sequences,$$

$$\ell_p = \left\{x \in \omega : \sum_{k=1}^{\infty} |x_k|^p < \infty\right\} \ (0 < p < \infty) \qquad \text{of all } p \ summable \ series,$$

$$\phi \qquad \text{of all } finite \ sequences;$$

thus $x \in \phi$ if and only if there exists a $k_0 = k_0(x) \in \mathbb{N}$ such that $x_k = 0$ for all $k > k_0$. Some authors use the notation c_{00} for the set of all finite sequences. Occasionally we identify the n–tuples $x = (x_1, x_2, \ldots, x_n) \in \mathbb{C}^n$ with the finite sequence $x = (x_1, x_2, \ldots, x_n, 0, \ldots)$, enabling us to write $\mathbb{C}^4 \subset \mathbb{C}^7 \subset \phi$.

Let B be a set and A_β be a set for each $\beta \in B$. Then the set

$$\prod A_\beta = \prod \{A_\beta : \beta \in B\} = \left\{f : B \to \bigcup_{\beta \in B} A_\beta \ : \ f(\beta) \in A_\beta \text{ for each } \beta \in B\right\}$$

is called the *(Cartesian) product* of the sets A_β. The sets B and A_β are called the *indexing set* and the *factors of the product*. In the case of finitely or countably many sets A_β, we have $B = \{1, 2, \ldots, n\}$ or $B = \mathbb{N}$, and may write

$$A_1 \times A_2 \times \cdots \times A_n \text{ or } A_1 \times A_2 \times \cdots.$$

If $A_\beta = A$ for all $\beta \in B$, then we write

$$A^B = \prod \{A_\beta : \beta \in B\},$$

hence A^B is the set of all functions $f : B \to A$; for instance, $\omega = \mathbb{C}^\mathbb{N}$.

Now we consider some important special functions.

We write e for the sequence with $e_k = 1$ $(k = 1, 2, \ldots)$, and $e^{(n)}$ $(n \in \mathbb{N})$ for the sequence with $e_n^{(n)} = 1$ and $e_k^{(n)} = 0$ for $k \neq n$.

If B is a set and A_β is a set for each $\beta \in B$ then, for every $t \in B$, the map

$$P_t : \prod \{A_\beta : \beta \in B\} \to A_t \text{ with } P_t(f) = f(t) \text{ for all } f \in \prod \{A_\beta : \beta \in B\}$$

is called the *projection map*. If A and B are sets and $t \in B$, then the projection map $P_t : A^B \to A$ is called the *evaluation map at* t. In particular, if $A = \mathbb{C}$ and $B = \mathbb{N}$, for each $n \in \mathbb{N}$, the evaluation map $P_n : \omega \to \mathbb{C}$ is called the *n–th co–ordinate map* or *n–th projection*. The following example shows that projection map is a suitable term.

Example 1.3.1 *Let $A = \mathbb{R}$ and $B = \{1, 2, 3\}$. Then we have $\mathbb{R}^3 = \mathbb{R} \times \mathbb{R} \times \mathbb{R}$ and $P_2(x, y, z) = y$ for all $(x, y, z) \in \mathbb{R}^3$ is the projection on the y–axis.*

1.4   Inequalities

Here we list a few important inequalities, among them the *Hölder*, *Minkowski* and *Jensen inequalities*.

For $1 \leq p \leq \infty$, let q denote the *conjugate number*, that is, $q = \infty$ for $p = 1$, $q = p/(p-1)$ for $1 < p < \infty$ and $q = 1$ for $p = \infty$.

The first inequality can be used to prove Hölder's inequality. Let $1 < p < \infty$, and a and b be nonnegative real numbers. Then we have

$$ab \leq \frac{a^p}{p} + \frac{b^q}{q} \text{ with equality if and only if } a^p = b^q.$$

If $1 < p < \infty$, and $x = (x_k)_{k=1}^\infty \in \ell_p$ and $y = (y_k)_{k=1}^\infty \in \ell_q$, then we have $xy = (x_k y_k)_{k=1}^\infty \in \ell_1$ and

$$\sum_{k=1}^\infty |x_k y_k| \leq \left(\sum_{k=1}^\infty |x_k|^p \right)^{1/p} \cdot \left(\sum_{k=1}^\infty |y_k|^q \right)^{1/q} \quad \text{(Hölder's inequality).} \quad (1.1)$$

If $1 \leq p < \infty$ and $x = (x_k)_{k=1}^\infty, y = (y_k)_{k=1}^\infty \in \ell_p$, then we have $x + y = (x_k + y_k)_{k=1}^\infty \in \ell_p$ and

$$\left(\sum_{k=1}^\infty |x_k + y_k|^p \right)^{1/p} \leq \left(\sum_{k=1}^\infty |x_k|^p \right)^{1/p} + \left(\sum_{k=1}^\infty |y_k|^p \right)^{1/p} \quad \text{(Minkowski's inequality).} \quad (1.2)$$

If $0 < p \leq p' < \infty$ and $x \in \ell_p$, then we have $x \in \ell_{p'}$ and

$$\left(\sum_{k=1}^\infty |x_k|^p \right)^{1/p} \geq \left(\sum_{k=1}^\infty |x_k|^{p'} \right)^{1/p'} \quad \text{(Jensen's inequality).}$$

Finally, we state the analogue of Minkowski's inequality for $0 < p < 1$. If $0 < p < 1$ and $x, y \in \ell_p$, then we have $x + y \in \ell_p$ and

$$\sum_{k=1}^{\infty} |x_k + y_k|^p \leq \sum_{k=1}^{\infty} |x_k|^p + \sum_{k=1}^{\infty} |y_k|^p.$$

1.5 Linear spaces, algebraic bases and dimension

The concept of linear spaces is fundamental in functional analysis. It involves algebraic structures on sets. Here we recall the elementary definitions, concepts and results.

Definition 1.5.1 Let $\mathbb{F}$ be a field.
(a) A *linear space over* $\mathbb{F}$, or *vector space over* $\mathbb{F}$ is a non–empty set X with two maps $+ : X \times X \to X$ and $\cdot : \mathbb{F} \times X \to X$ such that $(X, +)$ is an Abelian group and the following distributive laws hold for all $\lambda, \mu \in \mathbb{F}$ and for all $x, y \in X$
(D.1) $\lambda(x + y) = \lambda x + \lambda y,$
(D.2) $(\lambda + \mu)x = \lambda x + \mu x,$
(D.3) $(\lambda \cdot \mu)x = \lambda(\mu x),$
(D.4) $1 \cdot x = x$ for the unit element $1 \in \mathbb{F}.$
The elements of X are called *vectors*; the elements of $\mathbb{F}$ are called *scalars*. The inverse element of $x \in X$ with respect to addition is denoted by $-x$. We write

$$x - y = x + (-y) \text{ and } x + \lambda y = x + (\lambda y).$$

(b) A *subspace* of a linear space over $\mathbb{F}$ is a subset which is a linear space with the same operations over $\mathbb{F}$.
(c) Throughout, unless explicitly stated otherwise, $\mathbb{F}$ denotes the fields $\mathbb{R}$ or $\mathbb{C}$ with the usual addition, multiplication and absolute value $|\cdot|$. The term *linear space* or *vector space* will always mean a linear space over $\mathbb{C}$, whereas the term *real linear space* or *real vector space* will always stand for a linear space over $\mathbb{R}$.

So a nonempty subset S of a linear space X is a subspace if and only if

$$\lambda s + \tilde{s} \in S \text{ for all } s, \tilde{s} \in S \text{ and all scalars } \lambda,$$

or equivalently, if and only if for all $s, \tilde{s} \in S$ and all scalars λ

$$s + \tilde{s} \in S \text{ and } \lambda s \in S.$$

The following result is well known.

Theorem 1.5.2 *The intersection I of any family Φ of subspaces of a linear space X is a subspace of X.*

We recall the important and well–known concept of *linear dependence* and *independence* of sets of vectors.
Let S be a subset of a linear space.
A *linear combination of S* is an element $\sum_{k=1}^{n} \lambda_k s_k$, where $\lambda_1, \ldots, \lambda_n$ are scalars and $s_1, s_2 \ldots, s_n \in S$ are distinct vectors.
Every linear combination that contains at least one non–zero scalar coefficient is called *non–trivial*, otherwise *trivial*.
A subset S of a linear space is called *linearly dependent*, if 0 is a non–trivial linear combination of S; otherwise S is called *linearly independent*.

Remark 1.5.3 *Obviously a subset S of a linear space is linearly independent if and only if every finite subset of S is linearly independent. This means,*

$$\text{if } s_1, s_2, \ldots, s_n \in S \text{ are distinct and } \sum_{k=1}^{n} \lambda_k s_k = 0 \text{ then } \lambda_1 = \lambda_2 = \cdots = \lambda_n = 0.$$

The set of all linear combinations of a subset S of a linear space is called the *span of S*, or *linear hull of S*, denoted by span(S). We say that S *spans* T if $T \subset$span(S).

The following results are well known from linear algebra.

Theorem 1.5.4 *If two linear combinations of a linearly independent set are equal then they have to contain the same vectors and scalars.*

No element of a linearly independent set is a linear combination of the rest of the set.

If S is a subset of a linear space X, then span(S) *is the smallest subspace of X that contains S.*

In general, the union of subspaces of a linear space is not a subspace. Therefore it is useful to define the *sum of S_1 and S_2* of subsets S_1 and S_2 of a linear space X by

$$\text{span}(S_1, S_2) = \text{span}(S_1 \cup S_2) \text{ and } \text{span}(x, S_2) = \text{span}(\{x\}, S_2) \text{ for all } x \in X.$$

Remark 1.5.5 *If S_1 and S_2 are subspaces of a linear space, then we have*

$$\text{span}(S_1, S_2) = S_1 + S_2 = \{s_1 + s_2 : s_1 \in S_1, s_2 \in S_2\};$$

furthermore, if $S_1 \cap S_2 = \{0\}$, then we write $S_1 \oplus S_2 = S_1 + S_2$ for the direct sum *of S_1 and S_2.*

Example 1.5.6 *Let S be a set. Then $\mathcal{F}[S]$ becomes a linear space with addition $f + g$ and multiplication by scalars $\lambda \cdot f$ for all $f, g \in \mathcal{F}[S]$ and all $\lambda \in \mathbb{C}$ defined by*

$$(f + g)(s) = f(s) + g(s) \text{ and } (\lambda \cdot f)(s) = \lambda f(s) \text{ for all } s \in S.$$

In particular, $\omega = \mathcal{F}[\mathbb{N}]$ is a linear space with

$$x + y = (x_k + y_k)_{k=1}^{\infty} \text{ and } \lambda x = (\lambda x_k)_{k=1}^{\infty} \text{ for all } x, y \in \omega \text{ and all } \lambda \in \mathbb{C}.$$

Obviously, we have

$$\phi \subset \bigcap_{p>0} \ell_p \subset \bigcup_{p>0} \ell_p \subset c_0 \subset c \subset \ell_\infty \text{ and } \ell_p \subset \ell_{p'} \text{ for all } 0 < p < p';$$

each of the inclusions is strict. Each of the sets is a subspace of ω.

The concept of *algebraic bases* is of vital importance in the theory of linear spaces, since the whole space can be described in terms of a basis.

Definition 1.5.7 *An algebraic basis, or Hamel basis H for a linear space X, is a linearly independent set with $X =$span(H).*

Example 1.5.8 *The set $H = \{e^{(n)} : n \in \mathbb{N}\}$ is an algebraic basis for ϕ, but not for c_0 since* span(H) $= \phi$ *and ϕ is a proper subset of c_0.*

As a consequence of Zorn's lemma, we have:

Theorem 1.5.9 *Every linear space has an algebraic basis.*

The next result is well known from linear algebra.

Theorem 1.5.10 *Any two algebraic bases of a linear space are in one–to–one correspondence.*

In view of Theorem 1.5.10, it makes sense to define the dimension of a linear space. The *algebraic dimension*, or *Hamel dimension of a linear space* is the cardinality of its basis. The *algebraic dimension*, or *Hamel dimension of a subset* of a linear space is the dimension of its span.
If X is a linear space, then we write $\dim X$ for the algebraic dimension of X.

Example 1.5.11 *The algebraic dimension of c_0 is equal to the cardinality of the continuum, that is, $\dim c_0 = |\mathbb{R}|$.*

Theorem 1.5.12 *Every linearly independent set in a linear space can be extended to an algebraic basis of the linear space.*

1.6 Linear maps

Now we consider the important class of functions that preserve algebraic structures, namely the class of *linear functions*.

Definition 1.6.1 Let X and Y be linear spaces over the same field $\mathbb{F}$, and $f : X \to Y$ be a map. Then f is said to be
(a) *additive* if $f(x + \tilde{x}) = f(x) + f(\tilde{x})$ for all $x, \tilde{x} \in X$;
(b) *homogeneous* if $f(\lambda x) = \lambda f(x)$ for all $x \in X$ and all $\lambda \in \mathbb{F}$;
(c) *linear* if f is additive and homogeneous; a linear function is also called a *homomorphism*, we denote the set of all linear functions from X to Y by $\mathcal{L}(X, Y)$;
(d) an *isomorphism* if f is one to one and linear; two linear spaces are said to be *isomorphic* if there is an isomorphism from one space onto the other.
(e) An element of $\mathcal{L}(X, \mathbb{F})$ is called a *linear functional on* X; if $\mathbb{F} = \mathbb{R}$, then an element of $\mathcal{L}(X, \mathbb{R})$ is called a *real linear functional*. We denote the set of all linear functionals on X by $X^{\#}$, and the set of all real linear functionals on A by $X_{\mathbb{R}}^{\#}$.
(f) The sets $N(f) = f^{-1}(\{0\}) = \{x \in X : f(x) = 0\}$ and $R(f) = f(X) = \{y \in Y : f(x) = y \text{ for some } x \in X\}$ are called the *kernel* or *null space* and the *range of* f.

Warning 1 Many authors refer to a linear map as an isomorphism if it is both one to one and onto.

Remark 1.6.2 *(a) It is clear that $N(f)$ and $R(f)$ in Definition 1.6.1 are linear spaces.*
(b) The range $R(f)$ can be defined for any map $f : X \to Y$ and any sets X and Y. The null space $N(f)$ can be defined for any map $f : X \to Y$, any set X and any subspace Y of a linear space. Some authors also write $f^{\perp} = N(f)$.

Example 1.6.3 *(a) Let $\mathcal{P}$ be the linear space of all polynomials on $\mathbb{R}$ and $D : \mathcal{P} \to \mathcal{P}$ be the differential operator. Then D is a linear map.*
(b) We define the map $f : \ell_\infty \to c_0$ by

$$f(x) = \left(\frac{x_k}{k}\right)_{k=1}^{\infty} \quad \text{for all } x = (x_k)_{k=1}^{\infty} \in \ell_\infty.$$

*Then f is a linear map. The map $g : c_0 \to c_0$ with $g(x) = (x_k^2)_{k=1}^\infty$ for all $x = (x_k)_{k=1}^\infty \in c_0$
is not a linear map.*
(c) The coordinate maps $P_n : \omega \to \mathbb{C}$ of Section 1.3 are linear functionals.
*(d) The map $\lim : c \to \mathbb{C}$ with $\lim x = \lim_{k \to \infty} x_k$ for all $x = (x_k)_{k=1}^\infty \in c$ is a linear
functional.*

A map f is obviously linear if and only if

$$f(\lambda x + \mu \tilde{x}) = \lambda f(x) + \mu f(\tilde{x}) \text{ for all vectors } x \text{ and } \tilde{x} \text{ and all scalars } \lambda \text{ and } \mu.$$

The following result holds. Its proof is an easy exercise for the reader.

Theorem 1.6.4 *An additive function is* rational homogeneous, *that is,*

$$f(\lambda x) = \lambda f(x) \text{ for all } \lambda \in \mathbb{Q} \text{ and all vectors } x,$$

in particular, we have for all vectors x and all $m \in \mathbb{Z}$

$$f(0) = 0, \ f(-x) = -f(x) \text{ and } f(mx) = mf(x).$$

The following results are well known from elementary linear algebra.

Theorem 1.6.5 *Let S be a proper subspace of a linear space X and $x \in X \setminus S$. Then there
exists $f \in X^\#$ such that*

$$f(x) = 1 \text{ and } f|_S \equiv 0. \tag{1.3}$$

Example 1.6.6 *We have $\lim \in c^\#$ with $\lim(e) = 1$ and $\lim|_{c_0} \equiv 0$. with the linear functional
$\lim$ from Example 1.6.3 (d).*

Theorem 1.6.7 *Let X and Y be linear spaces, S be a subspace of X and $f \in \mathcal{L}(S, Y)$.
Then f can be extended to all of X, that is, there is a function $F \in \mathcal{L}(X, Y)$ with $F|_S = f$.*

Theorem 1.6.8 *Two linear spaces with the same algebraic dimension are isomorphic.*

Definition 1.6.9 (a) A set Φ of functions defined on a set S is said to be *separating (over
S)* if for all $s, \tilde{s} \in S$ with $s \neq \tilde{s}$ there exists a function $f \in \Phi$ such that $f(s) \neq f(\tilde{s})$.
(b) A set Φ of functions from a linear space X into a linear space is said to be *total (over
X)* if for every $x \in X \setminus \{0\}$ there exists a function $f \in \Phi$ such that $f(x) \neq 0$.

Example 1.6.10 *(a) Let $\Phi = \{P_t : t \in [0, 1]\}$ where each $P_t : C[0, 1] \to \mathbb{C}$ is the evaluation
map defined by $P_t(f) = f(t)$ for all $f \in C[0, 1]$. Then Φ is separating over $C[0, 1]$. For if
$f, g \in C[0, 1]$ are given with $f \neq g$, then there is $t_0 \in [0, 1]$ with $f(t_0) \neq g(t_0)$, and so*

$$P_{t_0}(f) = f(t_0) \neq g(t_0) = P_{t_0}(g).$$

*(b) Let $\Phi = \{f_t : t \in \mathbb{R}\}$ where each $f_t : \mathbb{R} \to \mathbb{R}$ is defined by $f_t(x) = t(x - 1)$ for all $x \in \mathbb{R}$.
Then Φ is not total over $\mathbb{R}$, since $f_t(1) = 0$ for all $t \in \mathbb{R}$.*

Remark 1.6.11 *(a) The condition in Definition 1.6.9 (b) can be phrased as: If $f(x) = 0$
for all $f \in \Phi$ then $x = 0$.*
*(b) Let Φ be a family of maps between linear spaces. Then Φ may be separating but not total;
also Φ may be total but not separating.*

 (i) *Let $\Phi = \{f\}$ where $f : \mathbb{R} \to \mathbb{R}$ is defined by $f(x) = x - 1$ for all $x \in \mathbb{R}$. Then Φ is
      separating but not total over $\mathbb{R}$.*

(ii) Let $\Phi = \{f\}$ where $f : \mathbb{R}^2 \to \mathbb{R}$ is defined by $f(x, y) = \sqrt{x^2 + y^2}$ for all $(x, y) \in \mathbb{R}^2$. Then Φ is total over $\mathbb{R}^2$, since $f(x, y) \neq 0$ for all $(x, y) \neq (0, 0)$. But Φ is not separating over $\mathbb{R}^2$, since if $(x, y) \neq (0, 0)$ then $(x, y) \neq (-x, -y)$ and $f(x, y) = f(-x, -y)$.

(c) The concepts are equivalent for families of linear functions, since $f(x) \neq f(\tilde{x})$ if and only if $f(x - \tilde{x}) \neq 0$.

The next result follows from Theorem 1.6.5 with $S = \{0\}$.

Corollary 1.6.12 *Let X be a linear space. Then $X^{\#}$ is total over X.*

Remark 1.6.13 *It follows from Remark 1.6.11 (c) and Corollary 1.6.12 that $X^{\#}$ is also separating over X, since the elements of $X^{\#}$ are linear functions.*

1.7   Semimetric and metric spaces

Whereas linear spaces involve an algebraic structure, *semimetric* and *metric spaces* are sets endowed with a function that measures distances which enables us to introduce notions such as the convergence of sequences and continuity of functions. No linear structure is needed, in general.

Definition 1.7.1 (a) A *semimetric* for a non–empty set X is a function $d : X \times X \to \mathbb{R}$ which satisfies the following conditions for all $x, y, z \in X$
(SM.1) $d(x, y) = d(y, x) \geq 0$,
(SM.2) $d(x, x) = 0$,
(SM.3) $d(x, z) \leq d(x, y) + d(y, z)$ *(triangle inequality)*.
The set X with the semimetric d is called a *semimetric space*, denoted by (X, d), or simply by X.
(b) A *metric* is a semimetric which satisfies
(SM.4) $d(x, y) > 0$ if $x \neq y$.

Example 1.7.2 *The natural or Euclidean metric d on $\mathbb{R}^n$ or $\mathbb{C}^n$ for $n \in \mathbb{N}$ is defined by*

$$d(x, y) = \left(\sum_{k=1}^{n} |x_k - y_k|^2 \right)^{1/2} \quad \text{for all } n \text{ tuples } x = (x_1, \dots, x_n) \text{ and } y = (y_1, \dots, y_n).$$

Of course, in the special case of $n = 1$, the Euclidean metric reduces to the absolute value of the difference of two complex or real numbers.
If not explicitly stated otherwise, the metrics of $\mathbb{R}^n$ or $\mathbb{C}^n$ for $n \in \mathbb{N}$ are always assumed to be the natural ones.

We use the following notations throughout.
Let (X, d) be a semimetric space, and $r > 0$ and $x_0 \in X$. Then the sets

$$B_r(x_0) = \{x \in X : d(x, x_0) < r\}$$
$$\overline{B}_r(x_0) = \{x \in X : d(x, x_0) \leq r\}$$
$$S_r(x_0) = \{x \in X : d(x, x_0) = r\}$$

are called the *open ball*, *closed ball*, and *sphere* of radius r and centre in x_0.

Let S and $\tilde{S}$ be subsets of a semimetric space (X, d), then

$$\begin{cases} \operatorname{diam}(S) = \sup\{d(s_1, s_2) : s_1, s_2 \in S\}, \ \operatorname{dist}(S, \tilde{S}) = \inf\{d(s, \tilde{s}) : s \in S, \ \tilde{s} \in \tilde{S}\} \\ \text{and } \operatorname{dist}(x, S) = \operatorname{dist}(\{x\}, S) \end{cases} \tag{1.4}$$

are called the *diameter of S*, the distance of S and $\tilde{S}$, and the distance of the point x and the set S.

We recall the following concepts in semimetric spaces.

A point x in a semimetric space (X, d) is said to be an *interior point* of a subset S of X, if there exists $r > 0$ such that $B_r(x) \subset S$.

A subset S of X is said to be a *neighborhood of x*, if x is an interior point of S. A subset S of X is said to be *open*, if X is the empty set, or is a neighborhood of each of its points; S is said to be *closed*, if its complement S^c is open.

The following results are well known.

Remark 1.7.3 *Let (X, d) be a semimetric space.*
(a) Then the sets $B_r(x)$ are open; the sets $\overline{B}_r(x)$ and $S_r(x)$ are closed for eacj $x \in X$.
(b) We have

$$|d(x, y) - d(x, z)| \leq d(y, z) \text{ for all } x, y, z \in X.$$

(c) Every union and each finite intersection of open sets is open. (By a finite intersection of sets we mean the intersection of finitely many sets.)

Now we recall the concepts of convergence of sequences and continuity of functions in semimetric spaces.

Definition 1.7.4 (a) A sequence (x_n) in a semimetric space (X, d) is said to be *convergent to a limit x*, if, for every neighborhood G of x, we have $x_n \in G$ for almost all n, that is, $x_n \in G$ for all but finitely many n; we denote this by $x_n \to x$.
(b) Let X and Y be semimetric spaces. A map $f : X \to Y$ is said to be *continuous at $x \in X$*, if, for each neighborhood G of $f(x)$, the set $f^{-1}(G)$ is a neighborhood of x; f is said to be *continuous on $S \subset X$*, if f is continuous at each point of S; f is said to be *continuous*, if f is continuous on X.

Remark 1.7.5 *Let X be a semimetric space.*
(a) Then $x_n \to x$ if and only if $d(x_n, x) \to 0$. We can state this equivalently as follows: for every $\varepsilon > 0$ there exists $n_0 \in \mathbb{N}$ such that $d(x_n, x) < \varepsilon$ for all $n \geq n_0$.
(b) A sequence in a semimetric space may converge to distinct limits. If, however, X is a metric space then the limit of a convergent sequence (x_n) is unique. In this case, we write $x = \lim x_n = \lim_{n \to \infty} x_n$ for its limit.
(c) Let X and Y be semimetric spaces and $f : X \to Y$ be a map. Then f is continuous at $x \in X$ if and only if $f(x_x) \to f(x)$ for every sequence (x_n) with $x_n \to x$. We can also express this equivalently as follows: for every $\varepsilon > 0$ there exist $\delta = \delta(\varepsilon, x)$ such that $f(B_\delta(x)) \subset B_\varepsilon(f(x))$.

We recall the concepts of a Cauchy sequence and of completeness.

A sequence (x_n) in a semimetric space (X, d) is said to be a *Cauchy sequence*, if, for every $\varepsilon > 0$, there exists $n_0 \in \mathbb{N}$ such that $d(x_m, x_n) < \varepsilon$ for all $m, n \geq n_0$. Obviously every convergent sequence is a Cauchy sequence, but the converse implication is not true, in general. A semimetric space X with the property that every Cauchy sequence converges to an element $x \in X$ is said to be *complete*.

1.8 Seminormed and normed spaces

Seminormed spaces are spaces that have both a linear and semimetric structure. The importance of seminorms p in functional analysis arises from the fact that the set $\{p < 1\} = \{x : p(x) < 1\}$ is *convex*, as we will see later.

We recall the concepts of seminorms and norms.

Definition 1.8.1 A *seminorm* p is a real–valued function on a linear space satisfying the following conditions for all vectors x and y and for all scalars λ

(SN.1) $\qquad p(x) \geq 0$,

(SN.2) $\qquad p(x + y) \leq p(x) + p(y)$ *(triangle inequality)*,

(SN.3) $\qquad p(\lambda x) = |\lambda| p(x)$ *(homogeneity)*.

A linear space X with a seminorm p is called a *seminormed space*, denoted by (X, p), or simply by X.

If p is a total seminorm, that is, if the family $\{p\}$ is total, then p is said to be a *norm*, and we write $\| \cdot \| = p$. A linear space X with a norm $\| \cdot \|$ is called a *normed space*, denoted by $(X, \| \cdot \|)$.

Remark 1.8.2 *It is clear from Definition 1.6.9 (b) that a norm satisfies*

(SN.4) $\qquad \|x\| > 0$ *whenever* $x \neq 0$.

Obviously, every seminorm p (norm $\| \cdot \|$) on a linear space X defines a semimetric (metric) on X by

$$d(x, x') = p(x - x') \ (d(x, x') = \|x - x'\|) \text{ for all } x, x' \in X; \qquad (1.5)$$

the converse implication obviously does not hold true, in general.

Since every seminorm defines a semimetric, the concepts of convergence of sequences, Cauchy sequences and continuity of functions of Section 1.7 are meaningful in seminormed spaces and formulated by using (1.5).

A complete normed space is called a *Banach space*.

1.9 Topological spaces

Topological spaces enable what might be considered as generalized analysis. This involves the introduction of some structures on sets such that the definition of concepts such as the continuity of functions makes sense. Such general structures are generated by means of so–called *topologies*, which are certain collections of subsets of a given sets, the so–called *open sets*.

Open sets are also of vital importance in the context of continuity of functions between metric spaces. The openness of a set in a metric space, however, depends on the underlying metric. If we want to be independent of the reference to a metric then it makes sense to use the following well–known properties of open sets, namely that any union and any finite intersection of open sets are open. Those properties are used in the well–known definition of topologies.

Definition 1.9.1 Let $X \neq \emptyset$ be a set. A subclass $\mathcal{T}$ of the power set $\mathcal{P}(X)$ of X is said to

be a *topology on* X, if
(T.1) $\emptyset, X \in \mathcal{T}$,
(T.2) $\bigcup \mathcal{O} \in \mathcal{T}$ for all subclasses $\mathcal{O}$ of $\mathcal{T}$,
(T.3) $O_1 \cap O_2 \in \mathcal{T}$ for all $O_1, O_2 \in \mathcal{T}$.
The set X with the topology $\mathcal{T}$ is called a *topological space*, denoted by $(X, \mathcal{T})$, or simply
by X. The sets in the topology $\mathcal{T}$ are called *open sets*.

Example 1.9.2 *(a) Let $X \neq \emptyset$. Then the class $\mathcal{T} = \mathcal{P}(X)$ obviously is a topology, the so–called* discrete topology.
The class $\mathcal{T} = \{\emptyset, X\}$ is obviously also a topology, the so–called indiscrete topology.
(b) Every metric generates a topology in the following way. If X is a metric space then the class of all open sets with respect to the metric is a topology on X, referred to as the topology given by the metric.
The natural topology for the sets c_0, c and ℓ_∞ is given by the so–called natural metric d_∞ with $d_\infty(x, y) = \sup_k |x_k - y_k|$. The natural topology for the sets ℓ_p $(1 \leq p < \infty)$ is given by the so–called natural metric d_p with

$$d_p(x, y) = \left(\sum_{k=0}^{\infty} |x_k|^p \right)^{1/p} \quad for \ all \ x = (x_k)_{k=0}^{\infty}, y = (y_k)_{k=0}^{\infty} \in \ell_p.$$

The natural topology of ω is given by the so–called natural metric d_ω with

$$d_\omega(x, y) = \sum_{k=0}^{\infty} \frac{1}{2^k} \cdot \frac{|x_k - y_k|}{1 + |x_k - y_k|}.$$

Clearly $\|x\|_p = d_p(x, 0)$ for $1 \leq p \leq \infty$ define norms on c_0, c and ℓ_p for $p = \infty$, and on ℓ_p for $1 \leq p < \infty$; these norms are referred to as the natural norms.
(c) A topology need not be given by a metric, in general. For instance, the indiscrete topology of a set of at least two elements is not given by any metric, since, in metric spaces, for any two distinct points, there exist disjoint open sets that contain the points.
If, however, for a given topology there exists a metric such that the topology is given by that metric, then the topology is said to be metrizable.

The well–known concepts in metric spaces of *closed sets*, *interior points* of sets and *neighborhoods* can easily be extended to topological spaces.

Definition 1.9.3 Let $(X, \mathcal{T})$ be a topological space.
(a) A subset F of X is said to be *closed*, if its complement F^c is open. The *closure* $\overline{S}$ of a subset S of X is the intersection of all closed sets that contain S.
(b) A point x is called an *interior point of $S \subset X$*, if there exists an open set O such that

$$x \in O \subset S;$$

then S is called a *neighborhood of x*.
(c) If S is a subset of X, then $S_0 \subset S$ is said to be *S–open*, if there exists an open set O in $\mathcal{T}$ such that

$$S_0 = S \cap O.$$

The class of all S–open sets is called the *(relative) topology of X on S*, denoted by $\mathcal{T}_S$.
(d) Let $\mathcal{T}$ and $\mathcal{T}'$ be topologies on a set X. Then $\mathcal{T}'$ is said to be *stronger* or *finer* than $\mathcal{T}$, if $\mathcal{T}' \supset \mathcal{T}$; and then $\mathcal{T}$ is said to be *weaker* or *coarser* than $\mathcal{T}'$.

Example 1.9.4 *The indiscrete topology is always the weakest possible topology on a set, whereas the discrete topology is always the strongest possible topology on a set.*

Remark 1.9.5 *A set G in a topological space is open if and only if it is a neighborhood of each of its points.*

Definition 1.9.6 Let $(X, \mathcal{T})$ and $(Y, \mathcal{T}')$ be topological spaces.
(a) A sequence (x_n) in a topological space is called *convergent to a limit x*, if, for every neighborhood G of x, there exists an $n_0 \in \mathbb{N}$ such that

$$x_n \in G \text{ for all } n \geq n_0.$$

(b) A map $f : (X, \mathcal{T}) \to (Y, \mathcal{T}')$ is said to be *continuous at $x \in X$*, if $f^{-1}(G)$ is a neighborhood of x for every neighborhood G of $f(x)$; f is said to be *continuous on $S \subset X$*, if it is continuous at every point of S; f is said to be *continuous*, if it is continuous on X.
(c) A map $f : (X, \mathcal{T}) \to (Y, \mathcal{T}')$ is called a *homeomorphism*, if it is one to one and continuous, and if the inverse function $f^{-1} : f(X) \to X$ is continuous; f is said to be *open*, if the set $f(G)$ is open for all open sets $G \subset X$.
(d) Let S_1 and S_2 be subsets of X. Then S_1 is said to be *dense in S_2*, if $\overline{S_1} \supset S_2$; S_1 is said to be *dense* if S_1 is dense in X.

Warning 2 A similar warning applies here as in the case of isomorphisms in Warning 1, namely that many authors include the property of onto in the definition of a homeomorphism.

Remark 1.9.7 *(a) Obviously a topology $\mathcal{T}'$ is stronger than a topology $\mathcal{T}$ if and only if every $\mathcal{T}$-neighborhood of a point x is also a $\mathcal{T}'$-neighborhood of x.*
(b) If $\mathcal{T}$ and $\mathcal{T}'$ are topologies on a set X then $\mathcal{T}'$ is stronger than $\mathcal{T}$ if and only if the identity map $\mathrm{id}: (X, \mathcal{T}') \to (X, \mathcal{T})$ is continuous.
(c) It is easy to see that a map $f : (X, \mathcal{T}_X) \to \mathcal{T}_Y$ is continuous if and only if the pre-image $f^{-1}(G)$ of every $\mathcal{T}_Y$-open set G is $\mathcal{T}_X$-open; or equivalently, if the pre-image $f^{-1}(B)$ of every $\mathcal{T}_Y$-closed set B is $\mathcal{T}_X$-closed.

Definition 1.9.8 Let $(X, \mathcal{T})$ be a topological space.
(a) A *local basis at a point x for $\mathcal{T}$* is a class Σ of neighborhoods of x such that, for every neighborhood N of x there exists an element $G \in \Sigma$ such that $G \subset N$.
(b) The space $(X, \mathcal{T})$ is said to be *first countable at $x \in A$*, if it has a countable local basis at x; it is said to be *first countable*, if it is first countable at all of its points.
(c) A *subbasis for $\mathcal{T}$* is a subclass Σ of $\mathcal{T}$ such that, for every $x \in X$ and every neighborhood G of x, there exists a finite class $\{G_1, \ldots, G_n\} \subset \Sigma$ such that

$$x \in \bigcap_{k=1}^{n} G_k \subset G.$$

Example 1.9.9 *(a) Every metric space (X, d) has a local basis at each of its points x, namely $\Sigma_x = \{B_{1/n}(x) : n \in \mathbb{N}\}$ for $r > 0$. Thus every metric space is first countable.*
(b) The topology given by the familiar Euclidean metric on $\mathbb{R}^2$ has a subbasis the elements of which are vertical and horizontal strips of the form $(a, b) \times \mathbb{R}$ for $a < b$ and $\mathbb{R} \times (c, d)$ for $c < d$. For if G is a neighborhood of a point $P = (r, s) \in \mathbb{R}^2$, then there exists $\delta > 0$ such that $U = B_\delta(P) \subset G$. Let $H = \{(x, y) \in \mathbb{R}^2 : |s - y| < \delta/2\}$ and $V = \{(x, y) \in \mathbb{R}^2 : |r - x| < \delta/2\} \in \Sigma$. Then we have

$$P = (r, s) \in H \cap V \subset U \subset G.$$

Every subclass of the power set of a nonempty set can be uniquely extended to a smallest topology that contains the subclass.

Theorem 1.9.10 *Let X be a nonempty set and Σ be a class of subsets of X such that $\bigcup \Sigma = X$. Then there is a unique topology $\mathcal{T}$ which has Σ as a subbasis and $\mathcal{T}$ is the weakest topology that contains Σ; $\mathcal{T}$ is called the* topology generated by Σ.

Proof.

(i) First we show the uniqueness.
Let Σ be a subbasis of $\mathcal{T}$ and $\mathcal{T}'$, $x \in X$, and G be a $\mathcal{T}'$–neighborhood of x. Then there exist $G_1, G_2, \ldots, G_n \in \Sigma$ such that

$$x \in I = \bigcap_{k=1}^{n} G_k \subset G.$$

Since $\Sigma \subset \mathcal{T}$, Σ consists of $\mathcal{T}$–open sets. Therefore I is also a $\mathcal{T}$–open set, hence G is also a $\mathcal{T}$–neighborhood of a. Thus Remark 1.9.7 (a) implies $\mathcal{T}' \subset \mathcal{T}$. Similiarly it can be shown that $\mathcal{T} \subset \mathcal{T}'$.

(ii) Now we construct $\mathcal{T}$.
Let the class $\mathcal{T}$ consist of X, $\emptyset$ and all unions of finite intersections of sets in Σ. Obviously $\mathcal{T}$ is a topology.

(α) We have to show that Σ is a subbasis of $\mathcal{T}$.
Indeed, $\Sigma \subset \mathcal{T}$, and if $x \in X$ and G is a neighborhood of x, then there exists an open set $O \in \mathcal{T}$ such that
$$x \in O \subset G.$$

Since $x \in O$, and O, by the definition of $\mathcal{T}$, is a union of finite intersections of sets of Σ, x is an element of an intersection of sets $S_1, S_2, \ldots, S_n \in \Sigma$, hence

$$x \in \bigcap_{k=1}^{n} S_k \subset G.$$

Thus we have shown that Σ is a subbasis of $\mathcal{T}$.

(iii) Finally, we show that $\mathcal{T}$ is the weakest topology that contains Σ.
Let $\mathcal{T}' \supset \Sigma$ be a topology and $O \in \mathcal{T}$. Then we have by construction $O = \emptyset$, or $O = X$, or O is the union of finite intersections of sets from Σ. Since $\Sigma \subset \mathcal{T}'$ and $\mathcal{T}'$ is a topology, it follows that $O \in \mathcal{T}'$. Thus we have shown that $\mathcal{T} \subset \mathcal{T}'$. $\square$

Definition 1.9.11 *Let X be a set and Φ be a class of topologies on X. The topology generated by $\bigcup \Phi$ is called the* supremum of Φ, *denoted by* $\bigvee \Phi$.
If $\Phi = \{\mathcal{T}_1, \mathcal{T}_2\}$ then we write $\mathcal{T}_1 \vee \mathcal{T}_2$, and similarly, for $\Phi = \{\mathcal{T}_1, \mathcal{T}_2, \ldots, \mathcal{T}_n\}$,

$$\mathcal{T}_1 \vee \mathcal{T}_2 \vee \cdots \vee \mathcal{T}_n.$$

If the topologies $\mathcal{T}_1$ and $\mathcal{T}_2$ are given by the metrics d_1 and d_2, we also write $d_1 \vee d_2$ to mean the topology $\mathcal{T}_1 \vee \mathcal{T}_2$.

Example **1.9.12** *On $\mathbb{R}^2$, let $\mathcal{H}$ be the topology generated by the horizontal strips, and $\mathcal{V}$ be the topology generated by the vertical strips. Then $\mathcal{H} \vee \mathcal{V}$ is the Euclidean topology.*

We also recall the important concept of a topological *Hausdorff space*.

A topological space is called a *Hausdorff space* or T_2–*space*, if any two distinct points are separated by open sets; that is, if $x_1 \neq x_2$ are points then there are open sets O_1 and O_2 such that
$$x_1 \in O_1, \ x_2 \in O_2 \text{ and } O_1 \cap O_2 = \emptyset.$$

Some interesting topologies, apart from Hausdorff topologies, have additional properties which are referred to as *separation axioms*.

Definition 1.9.13 A topological space X is said to be a
(a) T_0 *space*, if of any two distinct points, at least one has a neighborhood that does not contain the other;
(b) T_1 *space*, if each of two distinct points has a neighborhood that does not contain the other;
(c) *regular space*, if any closed set and any point not in it can be separated by open sets, that is, for any closed subset S and any $x \in X \setminus S$ there exist open sets $O_S \supset S$ and $O_x \supset \{x\}$ such that $O_S \cap O_x = \emptyset$;
(d) T_3 *space*, if it is a regular T_1 space.

Remark 1.9.14 *Obviously the T_n property of a topological space implies the T_{n-1} property for $n = 1, 2, 3.$*

Example **1.9.15** *A T_0 semimetric is a metric.*

Proof. Let d denote the semimetric that generates the T_0 topology. If $d(x, y) = 0$, then every neighborhood of x contains y. Since d generates the T_0 topology, it follows that $a = b$, for otherwise, the neighborhood of one of the points would contain the other point. □

1.10   Net convergence

The notion of sequences is not general enough and appropriate to study convergence in topological spaces. A more general concept is needed.

There are basically two ways to study convergence in topological spaces, namely by the use of *nets* or by the use of *filters*. The choice is a matter of taste. We are going to use nets, since they somehow are generalizations of sequences.

A sequence is a map defined on the set $\mathbb{N}$ of natural numbers, but unfortunately the special properties of natural numbers result in a lack of flexibility. The generalization to nets is achieved by replacing the set $\mathbb{N}$ by arbitrary directed sets (Definition 1.2.1 (b)).

We recall the definition of a net.

Definition 1.10.1 (a) A function defined on a directed set D is called a *net*. If the range of a net is contained in a set A then we refer to it as a *net in A*. We denote nets by

$$(x_\delta)_{\delta \in D} \text{ or } (x_\delta).$$

The points x_δ are called the *terms of a net*. A net $(x_n)_{n \in \mathbb{N}}$ is a sequence; the notations x_n and x_k refer to sequences.
(b) A net $(x_\delta)_{\delta \in D}$ is said to be *eventually in a set S*, if there exists $\delta_0 \in D$ such that $x_\delta \in S$ for all $\delta \geq \delta_0$; we write $x_\delta \in S$ *eventually*.
(c) A net $(x_\delta)_{\delta \in D}$ is said to be *frequently in a set S*, if, for every $\delta_0 \in D$, there exists a $\delta \geq \delta_0$ such that $x_\delta \in S$; we write $x_\delta \in S$ *frequently*.

We recall a few basic results.

Lemma 1.10.2 *Let $S_1, S_2, \ldots, S_n$ be sets and $(x_\delta)_{\delta \in D}$ be a net such that $x_\delta \in S_k$ eventually for each $k = 1, 2, \ldots, n$. Then we have $x_\delta \in \bigcap_{k=1}^{n} S_k$ eventually.*

Proof. For each $k = 1, 2, \ldots, n$ there exists $\delta_k \in D$ such that $x_\delta \in S_k$ for all $\delta \geq \delta_k$. Since $\{\delta_1, \delta_2, \cdots, \delta_n\}$ is a finite subset of the directed set D, it follows from Example 1.2.3 that there exists $\delta_0 \in D$ such that $\delta_0 \geq \delta_k$ for all $k = 1, 2, \ldots, n$. Then $\delta \geq \delta_0$ implies $\delta \geq \delta_k$ for all $k = 1, 2, \ldots, n$, and so $x_\delta \in S_k$ for all $k = 1, 2, \ldots, n$, hence $x_\delta \in \bigcap_{k=1}^{n} S_k$. □

Next we recall the definition of net convergence in a topological space.

Definition 1.10.3 A net (x_δ) in a topological space X is said to be *convergent to $x \in X$*, if $x_\delta \in G$ eventually for every neighborhood G of x; we denote this by $x_\delta \to x$. If the net (x_δ) converges to a uniquely determined point x, we write $x = \lim x_\delta$.

The following standard examples are needed.

*Example **1.10.4** (a) Let $D = (0, 1) \subset \mathbb{R}$ have the usual order of $\mathbb{R}$, and X be a topological space. A net (x_δ) in X is a function $f : (0, 1) \to X$, and $x_\delta \to x$ then means $\lim_{\delta \to 1-} f(\delta) = x$ in the usual sense; this means, for every neighborhood G of x, there is $\delta_0 \in (0, 1)$ such that $f(\delta) \in G$ for all δ with $\delta_0 \leq \delta < 1$.*
(b) Let D be as in Part (a) and $X = \mathbb{R}$. A net in $\mathbb{R}$ is now a real–valued function f defined on $(0, 1)$. In such a case it is particularly easy to determine if a net converges. For instance, if $x_\delta = f(\delta) = \sin(1/(1 - \delta))$ then (x_δ) does not converge to any limit.
(c) Let $D = [1, 3] \setminus \{2\} \subset \mathbb{R}$. We define for $\alpha, \beta \in D$

$$\alpha \geq \beta \text{ to mean } |\alpha - 2| \leq |\beta - 2|,$$

where $\leq$ denotes the usual order on $\mathbb{R}$. Then $x_\delta \to x$ for a net (x_δ) means $x_\delta \to x$ for $\delta \to 2$ in the usual sense.
(d) A net (x_δ) in a metric space (X, d) converges to a point x if and only if $d(x_\delta, x) \to 0$. (Of course, $(d(x_\delta, x))$ is a net in $\mathbb{R}$.)

(i) To show this, we first assume $x_\delta \to x$. Let $G \subset \mathbb{R}$ be a neighborhood of 0. Then there exists $\varepsilon > 0$ such that $(-\varepsilon, \varepsilon) \subset G$. It follows from $x_\delta \to x$, that $x_\delta \in B_\varepsilon(x)$ eventually, hence $d(x_\delta, x) < \varepsilon$ eventually, that is, $d(x_\delta, x) \subset G$ eventually and consequently $d(x_\delta, x) \to 0$.

(ii) Conversely, we assume $d(x_\delta, x) \to 0$. Let G be an arbitrary neighborhood of x. Then there is $\varepsilon > 0$ such that $B_\varepsilon(x) \subset G$. Now $d(x_\delta, x) < \varepsilon$ eventually implies $x_\delta \in B_\varepsilon(x) \subset G$ eventually.

(e) Let X be a topological space and $x \in X$. Also let D be the set of all neighborhoods of x ordered by inclusion, that is,

$$\alpha \geq \beta \text{ if and only if } \alpha \subset \beta.$$

Then D is a directed set.
For each $\delta \in D$ let x_δ be a point of δ. Then (x_δ) is a net in X. It follows that $x_\delta \to x$. To see this, let G be a given neighborhood of x. Then we have $G \in D$ and we choose $\delta_0 = G$. Then it follows for all $\delta \geq \delta_0$, that is, for all δ with $\delta \subset \delta_0 = G$, that $x_\delta \in \delta \subset G$, hence $x_\delta \in G$ eventually.

(f) Let X be a topological space, $x \in X$, and D be the set of all pairs (y, G) where G is a neighborhood of x and $y \in G$. We define

$$(y_2, G_2) \geq (y_1, G_1) \ by \ G_2 \subset G_1.$$

Then D is a directed set. For $\delta \in D$, that is, for $\delta = (y, G)$, we put $y_\delta = y$. Then (y_δ) is a net in X, and $y_\delta \to x$. For if G_0 is a neighborhood of x, then we put $\delta_0 = (x, G_0)$, and we have $y_\delta = y \in G$ for all $\delta = (y, G) \geq \delta_0$ by the definition of D, and $G \subset G_0$ by the definition of $\geq$, hence $y_\delta \in G_0$ for all $\delta \geq \delta_0$.

The following statements are generalizations of well–known facts for sequences in metric or semimetric spaces.

Theorem 1.10.5 *(a) Let S be a set and x be a point in a topological space. Then we have $x \in \overline{S}$ if and only if there exists a net (x_δ) in S such that $x_\delta \to x$.*
(b) A map $f : X \to Y$ between topological spaces X and Y is continuous at a point $x \in X$ if and only if, for every net (x_δ) in X with $x_\delta \to x$, we have $f(x_\delta) \to f(x)$.
(c) The map $f : X \to Y$ is continuous if and only if

$$f(\overline{S}) \subset \overline{f(S)} \ \text{ for all subsets } S \text{ of the domain of } f. \tag{1.6}$$

(d) Let $\mathcal{T}$ and $\mathcal{T}'$ be topologies on a set. Then $\mathcal{T}'$ is stronger than $\mathcal{T}$ if, for any net (x_δ), $x_\delta \to x$ in $\mathcal{T}'$ implies $x_\delta \to x$ in $\mathcal{T}$.
(e) Let S be a subset of a topological space $(X, \mathcal{T})$, (x_δ) be a net in S and $x \in S$. Then $x_\delta \to x$ in the relative topology $\mathcal{T}_S$ of X on S if and only if $x_\delta \to x$ in $\mathcal{T}$.
(f) Every convergent net in a topological space has a unique limit if and only if the space is a Hausdorff space.

Proof.

(a) (i) We show that if there exists a net (x_δ) in S with $x_\delta \to x$ then we have $x \in \overline{S}$. We assume $x \notin \overline{S}$. Then the complement $(\overline{S})^c$ of $\overline{S}$ is a neighborhood of x with $(\overline{S})^c \cap S = \emptyset$. Since every net in S must lie entirely outside $(\overline{S})^c$, it cannot converge to x.

   (ii) Now we show that if $x \in \overline{S}$ then there exists a net (x_δ) in S with $x_\delta \to x$. Let $x \in \overline{S}$. Then $G \cap S \neq \emptyset$ for every neighborhood G of x. We choose the directed set D as in Example 1.10.4 (e), and choose an $x_\delta \in \delta \cap S$ for every $\delta \in D$. Then (x_δ) is a net in S with $x_\delta \to x$ by Example 1.10.4 (e).

(b) If (x_δ) is a net in X, then obviously $f(x_\delta)$ is a net in Y.

   (i) Let f be continuous at $x \in X$, (x_δ) be a net in X with $x_\delta \to x$, and G be a neighborhood of $f(x)$. Then $f^{-1}(G)$ is a neighborhood of x by Definition 1.9.6 (b), since f is continuous at x. This implies $x_\delta \in f^{-1}(G)$ eventually, hence $f(x_\delta) \in G$ eventually, and so $f(x_\delta) \to f(x)$.

   (ii) We assume that f is not continuous at $x \in X$. Then, by Definition 1.9.6 (b), there exists a neighborhood G of $f(x)$ for which $f^{-1}(G)$ is not a neighborhood of x. We choose a net (x_δ) in X with $x_\delta \to x$ and $f(x_\delta) \not\to f(x)$ as follows. Let D be the directed set of Example 1.10.4 (e). For any $\delta \in D$, we choose a point $x_\delta \in \delta \setminus f^{-1}(G)$; this is possible, since $f^{-1}(G)$ is not a neighborhood of x_δ, and so $\delta \not\subset f^{-1}(G)$. Then we have $x_\delta \to x$ in X by Example 1.10.4 (e), but $x_\delta \notin f^{-1}(G)$ implies $f(x_\delta) \notin G$, and so $f(x_\delta) \not\to f(a)$.

(c) (i) First we show that the continuity of f at x implies that the condition in (1.6) holds.

Let $f : X \to Y$ be continuous and $y \in f(\overline{S})$. Then there exists a point $x \in \overline{S}$ with $f(x) = y$. By Part (a), there exists a net (x_δ) in S with $x_\delta \to x$, and the continuity of f implies $f(x_\delta) \to f(x) = y$ by Part (b), hence $y \in \overline{f(S)}$ by Part (a).

(ii) Conversely, we assume that the condition in (1.6) is satisfied. Let $x \in X$ be arbitrary. To show that f is continuous at x, let $G \ni f(x)$ be open in Y. Put $S = X \setminus f^{-1}[G]$. Clearly $f[S] \subset Y \setminus G$ so also $\overline{f[S]} \subset Y \setminus G$, since $Y \setminus G$ is closed. But then $f(x) \notin \overline{f[S]}$ whence, by our assumption, $x \notin \overline{S}$. Thus there is a neighborhood $U \subset X$ of x such that $U \cap S = \emptyset$, that is, $x \in U \subset f^{-1}[G]$. So $f^{-1}[G]$ is a neighborhood of x for every open neighborhood of $f(x)$ and consequently for every neighborhood of $f(x)$ as well.

(d) This part is obvious.

(e) (i) We assume $x_\delta \to x$ in $(X, \mathcal{T}_S)$. If G is a $\mathcal{T}$–neighborhood of x in X, then $G \cap S$ is a $\mathcal{T}_S$–neighborhood. Thus we have $x_\delta \in G \cap S$ eventually, so $x_\delta \in G$ eventually, that is, $x_\delta \to s$ in $\mathcal{T}$.

(ii) Conversely we assume $x_\delta \to x$ in $(X, \mathcal{T})$. If G is a $\mathcal{T}_S$–neighborhood of x in X then there is a $\mathcal{T}$–neighborhood N of x with $G = N \cap S$ by the definition of the relative topology $\mathcal{T}_S$ of X on S. Since $x_\delta \to x$ in $(X, \mathcal{T})$, we have $x_\delta \in N$ eventually, and since $x_\delta \in S$, we also have $x_\delta \in N \cap S = G$ eventually. Thus $x_\delta \to x$ in $\mathcal{T}_S$.

(f) (i) We assume $x_\delta \to x$ and $x_\delta \to y$ in a Hausdorff space X. Let G_x and G_y be neighborhoods of x and y, respectively. Then it follows that $x_\delta \in G_x$ eventually and $x_\delta \in G_y$ eventually, and so $x_\delta \in G_x \cap G_y$ eventually by Lemma 1.10.2. In particular, we have $G_x \cap G_y \neq \emptyset$. Since X is a Hausdorff space, it follows that $x = y$.

(ii) We assume that X is not a Hausdorff space. Then there exist points x and y such that, for all neighborhoods G_x of x and for all neighborhoods G_y of y, we have $G_x \cap G_y \neq \emptyset$. Let

$$D = \{(\alpha, \beta) : \alpha \text{ is a neighborhood of } x, \beta \text{ is a neighborhood of } y\},$$

and $\geq$ be defined by

$$(\alpha, \beta) \geq (\alpha', \beta') \text{ if and only if } \alpha \subset \alpha' \text{ and } \beta \subset \beta'.$$

Then $(D, \geq)$ obviously is a directed set. For every $\delta = (\alpha, \beta) \in D$, we choose a point $x_\delta \in \alpha \cap \beta$, and we have $x_\delta \to a$. For if G is a neighborhood of x, we put $\delta_0 = (G, N)$ where N is an arbitrary neighborhood of y. Then $\delta = (\alpha, \beta) \geq \delta_0$ implies $\alpha \subset G$, hence $x_\delta \in \alpha \subset G$, and so $x_\delta \to x$.
Similarly it can be shown that $x_\delta \to y$.

$\square$

We close this section with one more important example.

Example 1.10.6 *We consider on $\mathbb{R}$ the class $\mathcal{T}$ consisting of $\emptyset$ and the complements of all countable sets in $\mathbb{R}$. Then $\mathcal{T}$ obviously is a topology on $\mathbb{R}$. A sequence (x_n) converges to x in $\mathcal{T}$ if and only if there exists $n_0 \in \mathbb{N}$ such that $x_n = x$ for all $n \geq n_0$. If $(X, \mathcal{T}')$ is a* *topological space, then every map* $f : \mathbb{R} \to X$ *is* sequentially continuous *at every point of* $\mathbb{R}$ *with respect to the topology just defined. This means that if* $x \in \mathbb{R}$ *is an arbitrary point and* (x_n) *a sequence with* $x_n \to x$ *then* $f(x_n) \to f(x)$. *The map* $f(x) = x$ *is sequentially continuous, but not necessarily continuous.*

1.11   Subnets

Subnets play a role which is equally important as that of subsequences.

Definition 1.11.1 Let D' and D be directed sets. A map $u : D' \to D$ is said to be *finalizing*, if

$$u(\delta') \geq \delta \text{ eventually, for each } \delta \in D.$$

This means that for each $\delta \in D$, there exists $\delta_0' \in D'$ such that $u(\delta') \geq \delta$ for all $\delta' \geq \delta_0'$. In the special case of $D = D' = \mathbb{N}$, the map u is finalizing if and only if $u(n) \to \infty$ for $n \to \infty$.

Example 1.11.2 *(a) The map* $u : D \to \mathbb{R}$ *is finalizing if and only if* $u(\delta) \to \infty$, *that is,* $u(\delta) > M$ *eventually for every* $M \in \mathbb{R}$.
(b) The map $u : D \to (0,1)$ *is finalizing if and only if* $u(\delta) \to 1$.
(c) Let $u : \mathbb{N} \to \mathbb{R}$ *be defined by* $u(n) = n$. *Then* u *is finalizing.*
(d) Let $u : \mathbb{R}^+ \to \mathbb{N} \cup \{0\}$ *be defined by* $u(x) = [x] = \max\{n \in \mathbb{N} \cup \{0\} : n \leq x\}$. *Then* u *is finalizing.*
(e) No function $u : (0,1] \to (0,1)$ *can be finalizing. This is because, with* $\delta = (1 + u(1))/2$, *we cannot have* $u(\delta') \geq \delta$ *eventually, since* $u(1) < \delta$.
(f) Let D *and* D' *be directed sets. We define*

$$(\delta, \delta') \geq (\tilde{\delta}, \tilde{\delta}') \text{ by } \delta \geq \tilde{\delta} \text{ and } \delta' \geq \tilde{\delta}'.$$

Then $D \times D'$ *is a directed set and the projections* P_1 *and* P_2 *are finalizing.*
To see that the projections are finalizing, let $\delta_0 \in D$ *be given. Then* $P_1(\delta, \delta') = \delta \geq \delta_0$ *whenever* $(\delta, \delta') \geq (\delta_0, \delta_0')$ *where* δ_0' *is an arbitrary element of* D'.
Similarly it can be shown that $P_2 : D \times D' \to D'$ *is finalizing.*

Now we introduce the notion of *subnets*; their role resembles that of subsequences.

Definition 1.11.3 (a) A *subnet* of a net (x_δ) is a net $(x_{u(\delta')})_{\delta' \in D'}$, where $u : D' \to D$ is finalizing. This means that if $(x_\delta)_{\delta \in D}$ and $(y_{\delta'})_{\delta' \in D'}$ are given nets with directed sets D and D', then $(y_{\delta'})$ is said to be a subnet of (x_δ) if there exists a finalizing map $u : D' \to D$ such that $y = x \circ u$, that is, $y_{\delta'} = x_{u(\delta')}$ for all $\delta' \in D'$.
(b) A subset D' of a poset D is said to be *cofinal*, if, for each $\delta \in D$, there exists $\delta' \in D'$ with $\delta' \geq \delta$.

Example 1.11.4 *(a) The set* $2\mathbb{N}$ *of even natural numbers is cofinal in* $\mathbb{N}$.
(b) Let D *be a directed set and* D' *a directed subset, then the inclusion map* $\iota : D' \to D$ *is finalizing if and only if* D' *is cofinal in* D.
(c) Let $(x_n)_{n \in \mathbb{N}}$ *be a sequence and* $\mathbb{N}$ *have the usual order. Then* $(x_n)_{n \in 2\mathbb{N}}$ *is a subnet of* $(x_n)_{n \in \mathbb{N}}$.
(d) Let $(x_n)_{n \in \mathbb{N}}$. *We consider* $(x_{[t]})_{t \in \mathbb{R}^+}$ *where* $[t] = \max\{m \in \mathbb{Z} : m \leq t\}$. *Then* $(x_{[t]})_{t \in \mathbb{R}^+}$ *is a subnet of* (x_n) *but not a subsequence.*

(e) Let $(x_t)_{t \in \mathbb{R}^+}$ be a net. We consider $(x_n)_{n \in \mathbb{N}}$ where $u : \mathbb{N} \to \mathbb{R}^+$ is the inclusion map and $(x_{u(n)})_{n \in \mathbb{N}}$ is the subnet.

(f) Let $D = [1,3] \setminus \{2\}$ as in Example 1.10.4 (c). We consider the net $(x_{2+1/n})_{n \in \mathbb{N}}$ as a subnet of the net $(x_\delta)_{\delta \in D}$. Here the map $u : \mathbb{N} \to D$ is given by $u(n) = 2 + 1/n$. The map u is finalizing, since $u(n) \to 2$ as $n \to \infty$.

(g) Natural subnet
Let $(x_\delta)_{\delta \in D}$ be a net and D' be a cofinal subset of D. Then $(x_\delta)_{\delta \in D'}$ is a subnet of $(x_\delta)_{\delta \in D}$. To see this, we define the map $u : D' \to D$ by $u(\delta') = \delta'$, the inclusion map. (Parts(a) and (c) are of that kind.) We refer to this kind of a subnet as the natural subnet, *the subnet belonging to a cofinal set.*

(h) If a net (x_δ) in a topological space X does not converge to a point x, then, as in the case of sequences, there exists a neighborhood G of x and a subnet $(x_{\delta'})_{\delta' \in D'}$ such that $x_{\delta'} \notin G$ for all $\delta' \in D'$.

The next result is analogous to the corresponding familiar one for sequences.

Lemma 1.11.5 *Let (x_δ) be a net in a set. If $x_\delta \in S$ eventually, then $x_{u(\delta')} \in S$ eventually for every subnet.*

Proof. Let $x_\delta \in S$ eventually, and $(x_{u(\delta')})$ be any subnet of the net (x_δ). Since $x_\delta \in S$ eventually, there exists $\delta_0 \in D$ such that $x_\delta \in S$ for all $\delta \geq \delta_0$. Since u is finalizing, for δ_0, there exists $\delta_0' \in D'$ such that $u(\delta') \geq \delta_0$ for all $\delta' \geq \delta_0'$. Thus it follows for all $\delta' \geq \delta_0'$ that $x_{u(\delta')} \in S$, that is, $x_{u(\delta')} \in S$ eventually. $\square$

We obtain the next result, which is a generalization of a well–known fact for sequences, as an immediate consequence of Lemma 1.11.5.

Theorem 1.11.6 *Any subnet of a convergent net in a topological space converges to the same limit as the net itself.*

Proof. Let x be a limit of a net and G be a neighborhood of x. Then the net is in G eventually, and so is every subnet by Lemma 1.11.5. $\square$

Now we define *cluster points* of nets.

Definition 1.11.7 *A point x of a topological space X is called a* cluster point of a net *(x_δ) in X, if $x_\delta \in G$ frequently for every neighborhood G of x.*

Lemma 1.11.8 *If $x_\delta \in S$ eventually for a net (x_δ) in a topological space, then each cluster point x of the net is an element of the closure of the set S.*

Proof. We assume that $x \notin \overline{S}$. Then the complement $\overline{S}^c$ of $\overline{S}$ is a neighborhood of x with $S \cap \overline{S}^c = \emptyset$. Since $x_\delta \in S$ eventually, it does not hold that $x_\delta \in \overline{S}^c$ frequently. Hence x is not a cluster point of the net (x_δ). $\square$

The following result is a characterization of cluster points in terms of the convergence of subnets.

Theorem 1.11.9 *A point x is a cluster point of a net (x_δ) in a topological space if and only if there exists a subnet of net (x_δ), which converges to x.*

Proof.

(i) We show that if the net (x_δ) has a subnet that converges to x, then x is a cluster point of the net (x_δ).

We assume that x is not a cluster point of the net (x_δ). Then there exists a neighborhood G of x such that it does not hold that $x_\delta \in G$ frequently, which implies that $x_\delta \in G^c$ eventually. By Lemma 1.11.5, each subnet has the same property and consequently cannot converge to x.

(ii) Now we show the converse implication.

Let x be a cluster point of the net (x_δ). We put

$$D' = \{(\delta, G) : \delta \in D, \ G \text{ a neighborhood of } x, \ x_\delta \in G\}.$$

Then we obviously have $D' \neq \emptyset$. We define

$$\delta' \geq \tilde{\delta}' \text{ for } \delta' = (\delta, G) \text{ and } \tilde{\delta}' = (\tilde{\delta}, \tilde{G}) \text{ by } \delta \geq \tilde{\delta} \text{ and } G \subset \tilde{G}.$$

Furthermore, we define the map $u : D' \to D$ by $u(\delta') = u(\delta, G) = \delta$. Then u is finalizing.

To see this, let $\delta \in D$ be given. We choose $\delta'_0 = (\delta, X) \in D'$ where X is the whole space. Let $\tilde{\delta}' = (\tilde{\delta}, G)$ be given with G a neighborhood of x and $\tilde{\delta}' \geq \delta'_0$, that is, $\tilde{\delta} \geq \delta$ and $G \subset X$. It follows that $u(\tilde{\delta}') = \delta' \geq \delta$.

Also the subnet $(x_{u(\delta')})$ converges to x. To see this, let N be a neighborhood of x. Since x is a cluster point of the net (x_δ), there exists $\delta_0 \in D$ with $x_{\delta_0} \in N$. We put $\delta'_0 = (\delta_0, N)$. Then we have

$$G \subset N \text{ and } x_\delta \in G \text{ for all } \delta' = (\delta, G) \geq \delta'_0 = (\delta_0, N),$$

hence $x_\delta = x_{u(\delta')} \in N$. So we have shown that $x_{u(\delta')} \in N$ eventually, that is, the subnet $(x_{u(\delta')})$ converges to x. □

Example 1.11.10 *Let $D = (0,1)$ have the natural order of $\mathbb{R}$. We put*

$$z_\delta = e^{i\pi\delta} \text{ and } w_\delta = e^{2i\pi\delta} \ (\delta \in D)$$

thus defining two nets (z_δ) and (w_δ) in $\mathbb{C}$. Then 1 is a cluster point of the net (w_δ); we have $w_\delta \to 1$. Each neighborhood of 1 contains points of the net (z_δ), however, 1 is not a cluster point of (z_δ); we have $z_\delta \to -1$.

Let $t_n = e^{i\pi/n} \to 1$ as $n \to \infty$. The terms of the sequence (t_n) are terms of the net (z_δ), however, the sequence (t_n) is no subnet of the net (z_δ), since the map

$$n \mapsto \frac{\pi}{n} \text{ from } \mathbb{N} \text{ into } D$$

is not finalizing.

Definition 1.11.11 Let $(x_\delta)_{\delta \in D}$ be a net in a topological space. The sets

$$T(\delta_0) = \overline{\{x_\delta : \delta \geq \delta_0\}} \text{ for } \delta_0 \in D$$

are called *tails of the net (x_δ).*

Example 1.11.12 *We obtain $T(4) = \{-1, 1\}$ for the net $((-1)^n)_{n \in \mathbb{N}}$.*
We obtain $T(4) = \{0, 1/4, 1/5, \dots\}$ for the net $(1/n)_{n \in \mathbb{N}}$.

Lemma 1.11.13 *The set of the tails of a net $(x_\delta)_{\delta \in D}$ in a topological space has the* finite intersection property, *that is, any finite intersection of tails is non–empty.*

Proof. Let $T(\delta_1), T(\delta_2), \dots, T(\delta_n)$ be the tails of a net $(x_\delta)_{\delta \in D}$. Since D is directed set, for $\delta_1, \delta_2, \dots, \delta_n \in D$, there exists $\delta \in D$ with $\delta \geq \delta_1, \delta_2, \dots, \delta_n$, and so $x_\delta \in T(\delta_k)$ for all $k = 1, 2, \dots, n$, that is, $x_\delta \in \bigcap_{k=1}^n T(\delta_k)$.

Lemma 1.11.14 *The set of all cluster points of a net $(x_\delta)_{\delta \in D}$ is the intersection of all its tails.*

Proof. Every cluster point of a net belongs to each one of its tails by Lemma 1.11.8. Conversely, we assume that $x \in T(\delta)$ for every $\delta \in D$. Let G be a neighborhood of x. Then we have $x \in \overline{\{x_\delta : \delta \geq \delta_0\}}$ for every $\delta_0 \in D$, hence $G \cap \{x_\delta : \delta \geq \delta_0\} \neq \emptyset$. Therefore we have $x_\delta \in G$ frequently, and so x is a cluster point of the net $(x_\delta)_{\delta \in D}$. $\qquad \square$

1.12   Compact sets

The concept of *compactness* is fundamental in topology and functional analysis. We recall the following definition.

Definition 1.12.1 (a) A subset K of a topological space $(X, \mathcal{T})$ is said to be *compact*, if every open cover of K has a finite subcover; this means that, for every subclass Γ of $\mathcal{T}$ with $\bigcup \Gamma \supset K$, there exist sets $G_1, G_2, \dots, G_n \in \Gamma$ with $\bigcup_{k=1}^n G_k \supset K$.
(b) Let A be a set and S be a subset of A. A family Φ of subsets of A is said to have the *finite intersection property relative to S*, if S and any finite subfamily of Φ have non–empty intersection; Φ is said to have the *full intersection property relative to S*, if $S \cap (\bigcap \Phi) \neq \emptyset$.
(c) A subset A of a topological space is said to be *relatively compact* or *precompact*, if its closure is compact.

The following results are well known from elementary topology.

Theorem 1.12.2 *(a) A set K in a topological space is compact if and only if every family Φ of closed sets which has the finite intersection property relative to K also has the full intersection property relative to K.*
(b) Any closed subset of a compact set is compact.
(c) Any continuous map maps compact sets onto compact sets.

Example 1.12.3 *In c_0, the closed unit ball $\overline{B} = \{x \in c_0 : \|x\|_\infty \leq 1\}$ is not compact. We consider the set $S_n = \{e^{(k)} : k \geq n\}$ for each $n \in \mathbb{N}$. It can easily be checked that each set S_n is cloesd. Then we have $\bigcap_{n=1}^\infty S_n = \emptyset$, but $\bigcap_{k=1}^n S_k \cap \overline{B} \neq \emptyset$, and so $\overline{B}$ is not compact by Theorem 1.12.2 (a).*

Theorem 1.12.4 *Let K be a subset of a topological space. Then the following statements are equivalent.*
(i) K is compact.
(ii) Each net in K has a cluster point in K.
(iii) Each net in K has a subnet which converges to a point in K.

Proof. The properties in (ii) and (iii) are equivalent by Theorem 1.11.9.

(a) We show that (i) implies (ii).

Let K be compact and $(x_\delta)_{\delta \in D}$ be a net in K. The class $\{T(\delta) : \delta \in D\}$ has the finite intersection property relative to K. Then we have

$$K \cap \left(\bigcap_{\delta \in D} T(\delta) \right) \neq \emptyset \text{ by Theorem 1.9.10 (a).}$$

Hence the net (x_δ) has a cluster point in K by Lemma 1.11.14.

(b) Finally, we show that (ii) implies (i).

We assume that the set K has the property (ii). Let Φ be a class of closed sets with the finite intersection property relative to K, and D be the class of all finite intersections of sets in Φ. Then D also is a class of closed sets with the finite intersection property relative to K. In addition, every finite intersection of sets in D is a member of D. The set D is a directed set with respect to inclusion. We choose an element $x_\delta \in \delta \cap K$ for each $\delta \in D$. Then $(x_\delta)_{\delta \in D}$ is a net in K which has a cluster point $x \in K$ by assumption. The proof is complete when we have shown that $x \in \bigcap \Phi$. We are going to show that $x \in \bigcap \{S : S \in D\}$. Since x is a cluster point, it follows from Lemma 1.11.14 that $x \in T(\delta)$ for every $\delta \in D$. But we have $T(\delta) \subset \delta$, since D is ordered by inclusion, and since $x_\delta' \in \delta'$ and δ is a closed set. Therefore we have $x \in \delta$ for each $\delta \in D$. $\qquad \square$

Corollary 1.12.5 *If K is a compact set in a topological space and (x_δ) is a net in K with exactly one cluster point x, then (x_δ) converges to x.*

Proof. If $x_\delta \not\to x$ then there exist a neighborhood G of x and a subnet $(x_{\delta'})$ outside of G. This subnet has a cluster point different from x by Theorem 1.12.4, which is also a cluster point of the original net. $\qquad \square$

Corollary 1.12.6 *Compact sets in a Hausdorff space are closed.*

Proof. Let K be a compact set and (x_δ) be a net in K with $x_\delta \to a$. Then (x_δ) has a subnet converging to $x \in K$ by Theorem 1.12.4. This subnet also converges to x by Theorem 1.11.6. Since the space is a Hausdorff space, we have $x = a$ by Theorem 1.10.5 (f), that is, $a \in K$, and so K is closed by Theorem 1.10.5 (a). $\qquad \square$

Remark 1.12.7 *A closed set remains closed when the topology is made finer. A compact set remains compact when the topology is made coarser, for an open cover in a weaker topology is an open cover in the finer topology which can be reduced to a finite subcover.*
In a situation in which closed and compact are the same, the topology cannot be changed without losing this property. For let $\mathcal{T}$ be a topology with the property that a set is compact if and only if it is closed. If $\mathcal{T}'$ is a strictly coarser topology than $\mathcal{T}$, then there exists at least one set A which is $\mathcal{T}$-closed but not $\mathcal{T}'$-closed. However, A is still $\mathcal{T}'$-compact. If $\mathcal{T}'$ is a strictly finer topology than $\mathcal{T}$, then there is at least one set A which is $\mathcal{T}'$-closed but not $\mathcal{T}$-closed. If A were $\mathcal{T}'$-compact, then A would be $\mathcal{T}$-compact, hence $\mathcal{T}$-closed, since $\mathcal{T}$-compact and $\mathcal{T}$-closed are the same. This is a contradiction.

Theorem 1.12.8 *(a) Two comparable Hausdorff topologies in which the whole space is compact are the same.*
(b) A compact Hausdorff space can neither be given a strictly weaker Hausdorff nor a strictly stronger compact topology.
(c) Let K be a compact set in a topological space, H be a Hausdorff space and $f : K \to H$ be continuous. If $S \subset K$ is closed, then $f(S)$ is also closed. In particular, it holds that if f is continuous and one to one, then f is a homeomorphism in H.

Proof.

(a) Let $\mathcal{T}$ and $\mathcal{T}'$ be compact Hausdorff topologies with $\mathcal{T}' \supset \mathcal{T}$. By Theorem 1.12.2 (b), a $\mathcal{T}'$–closed set is $\mathcal{T}'$–compact, hence $\mathcal{T}$–compact by Remark 1.12.7 and closed by Corollary 1.12.6. This implies $\mathcal{T}' \subset \mathcal{T}$.

(b) This holds by Part (a), since a topology which is stronger than a Hausdorff topology obviously is a Hausdorff topology. Also, if the whole space is compact in some topology it remains compact in any weaker topology by Remark 1.12.7.

(c) If $S \subset K$ is closed then S is compact by Theorem 1.12.2 (b). Since f is continuous, $f(S)$ is also compact by Theorem 1.12.2 (c), hence $f(S)$ is closed by Corollary 1.12.6, since H is a Hausdorff space. □

Chapter 2

Linear Topological Spaces

Chapter 2 is dedicated to the study of linear topological spaces. As far as more advanced purely algebraic concepts are concerned, thorough treatments are presented of convex and affine sets, balloons and cones, and quotient spaces. The studies of purely topological concepts mainly concentrate on supremum, weak and product topologies. The concept of linear topological spaces or topological vector spaces is very general and of great importance in functional analysis. It combines the purely topological concepts of topological spaces with the purely algebraic concepts of linear or vector spaces in a natural way such that the vector space operations of addition and multiplication by scalars are continuous. Various important properties of linear topological spaces are established and a characterization of linear topologies is given in terms of their open sets. Furthermore, the chapter deals with properties of closed maps, and presents the closed graph lemma in linear topological spaces, and Baire's category theorem in complete metric spaces. Finally, the last two sections study locally convex spaces and quotient topologies.

2.1 Introduction

The notion of *linear topological spaces* or *topological vector spaces* is very general and of great importance in functional analysis. It combines the purely topological concepts of topological spaces with the purely algebraic concepts of linear or vector spaces in a natural way such that the vector space operations of addition and multiplication by scalars are continuous.

In order to provide the necessary tools for the functional analytic results of this section, we have to develop some more advanced studies of linear and topological spaces.

Concerning purely algebraic concepts, thorough treatments are presented of convex and affine sets, balloons and cones, and quotient spaces in Sections 2.2–2.5.

The studies of purely topological concepts concentrate on supremum, weak, and product topologies in Sections 2.6 and 2.7.

After the introduction of the notion of linear topological spaces in Definition 2.8.1, various important properties of linear topological spaces are established and a characterization of linear topologies is given in terms of their open sets. Furthermore, the chapter deals with properties of closed maps, and presents the closed graph lemma, Lemma 2.9.8, in linear topological spaces, and Baire's category theorem, Theorem 2.10.9, in complete metric spaces. Finally, the last two sections study locally convex spaces and quotient topologies.

2.2 Set arithmetic and convexity

Convex sets play a vital role in many results in functional analysis and their applications, in particular, in various concepts and theorems in the theory of measures of noncompactness and fixed point theory, as we will see in Chapters 7 and 10.

It is useful to start with the definition of an arithmetic of subsets A and B of a linear space X. We denote scalars by λ and μ, write Λ for any set of scalars, and put

$$A + B = \{a + b : a \in A, b \in B\}, \ \lambda A = \{\lambda a : a \in A\},$$
$$\Lambda A = \{\lambda a : \lambda \in \Lambda, a \in A\},$$
$$-A = \{-a : a \in A\}, \ A - B = \{a - b : a \in A, b \in B\}$$

and use the conventions

$$\emptyset + A = \emptyset A = \Lambda \emptyset = \emptyset \text{ and } a + B = \{a\} + B.$$

The following simple rules are immediate consequences of the above definitions.

Remark 2.2.1 *We have for all subsets A, B and C of a linear space X and for all scalars λ and μ*

$$A + B = B + A, \ 0 + A = A, \ (A + B) + C = A + (B + C),$$

$$\lambda(A + B) = \lambda A + \lambda B, \tag{2.1}$$

$$(\lambda + \mu)A \subset \lambda A + \mu A, \ \text{where the inclusion may be strict}, \tag{2.2}$$

and

$$\mu(\lambda A) = (\mu \lambda)A.$$

Definition 2.2.2 A subset A of a linear space X is said to be
(a) *symmetric*, if $A = -A$;
(b) *balanced*, if $\lambda A \subset A$ for all scalars λ with $|\lambda| \leq 1$;
(c) *absorbing*, if, for every $x \in X$, there exists an $\varepsilon > 0$ such that $\lambda x \in A$ for all scalars λ with $|\lambda| \leq \varepsilon$;
(d) *absorbing at $x \in X$*, if $A - x$ is absorbing; *somewhere absorbing*, if A is absorbing at some point; and *nowhere absorbing*, if A is not absorbing at any point.

Remark 2.2.3 *(a) Every balanced set is symmetric.*
(b) Any absorbing subset of $\mathbb{C}$ has to contain an open disc centred at the origin.
(c) The set $A \cap (-A)$ is symmetric; the set $A \cap (-A)$ is absorbing if A is absorbing.
(d) The set A is absorbing if and only if, for each vector x, there exists a scalar $\lambda \in \mathbb{R}$ such that $x \in \mu A$ for all scalars μ with $|\mu| \geq \lambda$.
(e) If A is symmetric, then we have

(i) $x \in y + A$ if and only if $y \in x + A$;

(ii) $x \in A - B$ if and only if $(x + B) \cap A \neq \emptyset$;

(iii) $C \cap (B + A) \neq \emptyset$ if and only if $(C + A) \cap B \neq \emptyset$.

(f) If the set B is absorbing at x, then $B + A$ is absorbing at each point of $x + A$.

(g) If A is balanced and $|\mu| \geq |\lambda|$, then we have $\mu A \supset \lambda A$.

(h) If A is absorbing then we have

$$X = \bigcup_{n=1}^{\infty} nA.$$

(i) A set A is absorbing at $a \in X$ if and only if, for every $x \in X$, there exists $\varepsilon > 0$ such that

$$a + \lambda x \in A \text{ for all scalars } \lambda \text{ with } |\lambda| \leq \varepsilon.$$

Proof.

(a) Let A be balanced. Then it follows with $\lambda = -1$ that $-A \subset A$, and then $A = -(-A) \subset -A$.

(b) We assume that A does not contain any disc centred at 0, that is, for every $\delta > 0$ there exists an $x_\delta \in B_\delta(0) \setminus A$, where $B_\delta(0)$ is the open disc in $\mathbb{C}$ of radius δ and centred at 0. We choose $z = 1 \in \mathbb{C}$. Given $\varepsilon > 0$, we choose $\delta = \varepsilon/2$ and $\lambda = x_\delta$. Then we have

$$|\lambda| = |x_\delta| \leq \delta < \varepsilon \text{ and } \lambda z - x_\delta \notin A,$$

hence A is not absorbing.
This shows Part (b).

(c) (i) We show that $A \cap (-A)$ is symmetric, if A is symmetric.
We have $x \in A \cap (-A)$ if and only if $x \in A$ and $x \in -A$, if and only if $-x \in -A$ and $-x \in (-(-A)) = A$, if and only if $-x \in A \cap (-A)$, if and only if $x \in -(A \cap (-A))$.

(ii) We show that if A is absorbing then $A \cap (-A)$ is absorbing.
Let A be absorbing, and $x \in X$ be given. Then there exists $\varepsilon > 0$ such that $\lambda x \in A$ for all scalars λ with $|\lambda| \leq \varepsilon$. Putting $\mu = -\lambda$, we also obtain $\mu x \in A$, hence $-\mu x = \lambda x \in -A$, and so $\lambda x \in A \cap (-A)$.

Thus we have shown the statements in Part (c).

(d) (i) Let A be absorbing and x be a given vector. Then there exists $\delta > 0$ such that $\nu x \in A$ for all scalars ν with $|\nu| \leq \delta$. We choose $\lambda = \delta^{-1} \in \mathbb{R}$. Then it follows for all scalars μ with $|\mu| \geq \lambda$ that

$$\frac{1}{|\mu|} \leq \frac{1}{\lambda} = \delta \text{ and } x = \mu \cdot \frac{1}{\mu} \cdot x \in \mu A, \text{ since } \frac{1}{\mu} \cdot x \in A \text{ as } \frac{1}{|\mu|} \leq \delta.$$

(ii) We assume that A is not absorbing. Then there exists a vector $x \in X$ such that, for all $\delta > 0$, there exists a scalar ν with $|\nu| \leq \delta$ such that $\nu x \notin A$. Let t be an arbitrary real number. We choose $\delta > 0$ such that $1/\delta > t$ and then a scalar ν with $|\nu| < \delta$ and $\nu x \notin A$. We may assume $\nu \neq 0$, since, for $\nu = 0$, we have $0 \notin A$ and so $0 \notin tA$ for all $t \neq 0$. We put $\lambda = 1/\nu$. It follows that

$$|\lambda| = \frac{1}{|\nu|} > \frac{1}{\delta} > t \text{ and } x = \lambda \cdot \frac{1}{\lambda} \cdot x = \lambda \nu x \notin \lambda A,$$

since $\nu x \notin A$.
This completes the proof of Part (d).

(e) We assume that A is symmetric.

(i)   We have $x \in y + A$, if and only if $x = y + a$ for some $a \in A$, if and only if $y = x - a$ for some $-a \in (-A) = A$ (since A is symmetric), if and only if $y \in x + A$.

(ii) We have $(x + B) \cap A \neq \emptyset$, if and only if there exists $y \in (x + B) \cap A$, if and only if there exists y with $y = x + b$ for some $b \in B$ and $y \in A$, if and only if $x = y - b$ for some $b \in B$ and $-y \in (-A) = A$, if and only if $x \in A - B$.

(iii) We have $x \in C \cap (B + A) \neq \emptyset$, if and only if $x \in C$ and $x = b + a$ for some $b \in B$ and some $a \in A$, if and only if $b = x - a$ for some $b \in B$ and some $-a \in (-A) = A$ and some $x \in C$, if and only if $b \in (C + A) \cap B$.

Thus we have shown the statements in Part (e).

(f)   We assume that B is absorbing at x, $y \in x + A$, that is, $y = x + a$ for some $a \in A$, and $z \in X$ is an arbitrary vector. Since $B - x$ is absorbing, there exists $\varepsilon > 0$ such that, for all scalars λ with $|\lambda| \leq \varepsilon$, we have $\lambda z \in B - x$. Thus we have $\lambda z = b - x = b + a - y$ for some $b \in B$ and some $a \in A$, hence $\lambda z \in B + A - y$. Since $y \in x + A$ was arbitrary, $B + A$ is absorbing at every point of $x + A$.

(g)   Let A be balanced. Then $0 \in A$, since, for $\lambda = 0$, we have $\lambda A = \{0\} \subset A$. Thus we have $\lambda A \supset \{0\}$ for all λ with $|\lambda| \geq 0$.
If $\mu = 0$ then $|\lambda| \leq |\mu|$ implies $\lambda = 0$; but then $\lambda A = \{0\} = \mu A$.
If $|\mu| > 0$, then we put $\nu = \lambda / \mu$. Then we have $|\nu| \leq 1$ and, since A is balanced, it follows that $\nu A \subset A$, hence $\lambda A \subset \mu A$.

(h)   Obviously we have $\bigcup_{n=1}^{\infty} nA \subset X$.
Conversely, let $x \in X$ be given. Since A is absorbing, there exists $\varepsilon > 0$ such that, for some $n \in \mathbb{N}$ with $1/n \leq \varepsilon$, we have $(1/n) \cdot x \in A$, hence $x \in nA$. Thus we also have $X \subset \bigcup_{n=1}^{\infty} nA$.

(i)   Since $a + \lambda x \in A$ if and only if $\lambda x \in A - a$, Part (i) is an immediate consequence of Definition 2.2.2 (d).   □

The next example shows that sets that are balanced or absorbing in real linear spaces may not have those properties in complex linear spaces.

Example 2.2.4 *(a) Let A be a rectangle in the plane, which is not a square, and the centre of A be in the origin. Then A considered as a subset of $\mathbb{R}^2$ is balanced, but A considered as a subset of $\mathbb{C}$ is not balanced, since $iA \not\subset A$.*
(b) Schatz's apple. Let

$$A = \bar{B}_1((1,0)) \cup \bar{B}_1((-1,0)) \cup (\{0\} \times [-1,1]),$$

that is, A is the union of the closed discs of radius 1 with their centres in the points $(1,0)$ and $(-1,0)$ and the straight line segment that joins the points $(0,-1)$ and $(0,1)$.
Then A considered as a subset of $\mathbb{R}^2$ is absorbing, since every straight line through the origin meets A in an interval of positive length. However, A considered as a subset of $\mathbb{C}$ is not absorbing by Remark 2.2.3 (b), since A does not contain any disc centred at the origin.

Now we give the definition of the concept of *convexity* of a set.

Definition 2.2.5 A set A in a linear space X is said to be *convex*, if

$$\lambda A + \mu A \subset A \text{ for all scalars } \lambda, \mu \geq 0 \text{ with } \lambda + \mu = 1.$$

It is easy to show that the following statements hold.

Remark 2.2.6 *(a) A set A in a linear space is convex if and only if*

$$\lambda A + \mu A \subset (\lambda + \mu)A \text{ for all scalars } \lambda, \mu > 0.$$

(b) The intersection of any class of convex sets is a convex set.

The next theorem shows that the properties symmetric, balanced, convex and absorbing of sets are preserved under linear maps, and their preimages.

Theorem 2.2.7 *Let X and Y be linear spaces and $f : X \to Y$ be a linear map. Then we have:*
(a) If a subset A of X has any one of the properties symmetric, balanced, or convex, then the set $f(A)$ also has that same property.
(b) If f is onto and absorbing at a, then $f(A)$ is absorbing at $f(a)$.
(c) If a subset B of Y has any one of the properties symmetric, balanced, or convex, then the set $f^{-1}(B)$ also has that same property.
(d) If a subset B of Y is absorbing at $b \in f(X)$ then the set $f^{-1}(B)$ is absorbing at every point of the set $f^{-1}(\{b\})$.

Proof.

(a)   (i)   If A is symmetric, that is, if $A = -A$, then we have $-f(A) = f(-A) = f(A)$.

   (ii)   If A is balanced, that is, $\mathbb{D}A \subset A$ for $\mathbb{D} = \{\lambda : |\lambda| \leq 1\}$, then we have

$$\mathbb{D}f(A) = f(\mathbb{D}A) \subset f(A).$$

   (iii)   If A is convex and $0 \leq \lambda \leq 1$, then it follows from $\lambda A + (1 - \lambda)A \subset A$ that

$$\lambda f(A) + (1 - \lambda)f(A) = f(\lambda A + (1 - \lambda)A) \subset f(A).$$

This completes the proof of Part (a).

(b)   We assume that f is onto, the set A is absorbing at a, and $y \in Y$. Then there is $x \in X$ with $f(x) = y$ and, by Remark 2.2.3 (i), there exists $\delta > 0$ such that

$$a + \lambda x \in A \text{ for all scalars } \lambda \text{ with } |\lambda| \leq \delta.$$

This implies
$$f(a) + \lambda y = f(a) + \lambda f(x) = f(a + \lambda x) \in f(A)$$

for all $|\lambda| \leq \delta$, hence $f(A)$ is absorbing at $f(a)$.

(c)   In view of our convention concerning the empty set, we may assume $f^{-1}(A) \neq \emptyset$. The statement follows as in the proof of Part (a).

(d)   Let B be absorbing at $b \in f(X)$, $x \in f^{-1}(\{b\})$, and $\tilde{x} \in X$ be given. Since $f(X)$ is absorbing at b, by Remark 2.2.3 (i), there exists $\delta > 0$ such that $b + \lambda f(\tilde{x}) \in B$ for all scalars λ with $|\lambda| \leq \delta$, but

$$b + \lambda f(\tilde{x}) = f(x) + \lambda f(\tilde{x}) = f(x + \lambda \tilde{x}) \in B, \text{ that is, } x + \lambda \tilde{x} \in f^{-1}(B),$$

and so $f^{-1}(B)$ is absorbing at $x \in f^{-1}(\{b\})$. $\square$

We obtain the following corollary.

Corollary 2.2.8 *Let X and Y be linear spaces, A be a subset of X, and B be a subset of Y.*

(a) If A is absorbing at a and $f \in X^{\#} \setminus \{0\}$, then $f(a)$ is in the interior of $f(A)$, that is, there is $\delta > 0$ such that $B_\delta(f(a)) = \{y \in \mathbb{C} : |y - f(a)| < \delta\} \subset f(A)$.

(b) If A is absorbing at each of its points and $f \in X^{\#} \setminus \{0\}$, then $f(A)$ is an open subset of $\mathbb{C}$.

(c) If $f \in \mathcal{L}(\mathbb{F}, Y) \setminus \{0\}$ and B is absorbing at all of its points, then $f^{-1}(B) \subset \mathbb{F}$ is an open set.

(d) Let a and b be vectors and A be absorbing at all of its points. Then the set

$$M = \{\lambda \in \mathbb{R} : \lambda a + (1 - \lambda)b \in A\}$$

is an open set in $\mathbb{R}$.

   Proof.

(a) Let A be absorbing at a and $f \in X^{\#} \setminus \{0\}$.

   (i) We show that f is onto.
        Since $f \in X^{\#} \setminus \{0\}$, there exists $x \in X$ with $f(x) \neq 0$. Let $\mu \in \mathbb{C}$ be given. Putting

        $$\tilde{x} = \frac{\mu}{f(x)} \cdot x \in X,$$

   we obtain

   $$f(\tilde{x}) = f\left(\frac{\mu}{f(x)} \cdot x\right) = \frac{\mu}{f(x)} \cdot f(x) = \mu,$$

   hence f is onto.

   By Theorem 2.2.7 (b), $f(A)$ is absorbing at $f(a)$, hence $f(A)$ contains an open disc centred at $f(a)$ by Remark 2.2.3 (b).

(b) This is an immediate consequence of Part (a).

(c) Let $f \in \mathcal{L}(\mathbb{F}, Y) \setminus \{0\}$ and B be absorbing at every point. Then $f^{-1}(B)$ is absorbing at each point $f^{-1}(\{b\})$ $(b \in B)$ by Theorem 2.2.7 (d). Thus, for each $x \in f^{-1}(B) = \bigcup_{b \in B} f^{-1}(\{b\})$, there exists an open disc centred at x, which is contained in $f^{-1}(B)$, by Remark 2.2.3 (b).

(d) Let X be $\mathbb{R}$ linear. We put

   $$f(\lambda) = \lambda(a - b) \text{ for all } \lambda \in \mathbb{R}.$$

   Then $f \in \mathbb{R}^{\#}$ and $M = f^{-1}(A - b)$, since if $\lambda \in f^{-1}(A - b)$ then $f(\lambda) = \lambda(a - b) = \lambda a - \lambda b \in A - \{b\}$ which is the case if and only if $\lambda a + (1 - \lambda)b \in A$ which is equivalent to $\lambda \in M$. The statement now follows from Part (c). $\square$

2.3   Convex and affine sets

   Here we study important properties of convex and *affine* sets.

   Throughout this section, let A and B always denote subsets of a linear space X.

   We start with the definition of *convex combinations* or *weighted means* of a set.

Definition 2.3.1 A *convex combination* or *weighted mean of a set A* is a linear combination of A of the form

$$\sum_{k=1}^{n} \lambda_k a_k \text{ where } a_k \in A, \ \lambda_k \geq 0 \ (k = 1, \ldots, n) \text{ and } \sum_{k=1}^{n} \lambda_k = 1.$$

The set of all convex combinations of A is denoted by $\mathrm{cvx}(A)$.

Remark 2.3.2 *The convex combination of convex combinations of a set A again is a convex combination of A, that is, $\mathrm{cvx}(\mathrm{cvx}(A)) \subset \mathrm{cvx}(A)$.*

Proof. Let $\sum_{k=1}^{m} \lambda_k w_k(A)$ be a convex combination of convex combinations $w_k(A) = \sum_{j=1}^{n} \mu_{kj} a_j$ of A $(k = 1, 2, \ldots, m)$. Then we have $\lambda_k \geq 0$ for $k = 1, 2, \ldots, m$, $\sum_{k=1}^{m} \lambda_k = 1$, $a_j \in A$, and, for each $k = 1, 2 \ldots, m$, $\mu_{kj} \geq 0$ $(j = 1, 2, \ldots, n)$ and $\sum_{j=1}^{n} \mu_{kj} = 1$. We put $\nu_j = \sum_{k=1}^{m} \lambda_k \mu_{kj}$ for $j = 1, 2, \ldots, n$ and obtain

$$\sum_{k=1}^{m} \lambda_k w_k(A) = \sum_{k=1}^{m} \lambda_k \left(\sum_{j=1}^{n} \mu_{kj} a_j \right) = \sum_{j=1}^{n} a_j \left(\sum_{k=1}^{m} \lambda_k \mu_{kj} \right) = \sum_{j=1}^{n} \nu_j a_j,$$

where

$$\sum_{j=1}^{n} \nu_j = \sum_{j=1}^{n} \left(\sum_{k=1}^{m} \lambda_k \mu_{kj} \right) = \sum_{k=1}^{m} \lambda_k \left(\sum_{j=1}^{n} \mu_{kj} \right) = \sum_{k=1}^{m} \lambda_k = 1.$$

$\square$

Lemma 2.3.3 *A convex set contains all its convex combinations, that is, $\mathrm{cvx}(C) \subset C$ for every convex set.*

Proof. Let C be a convex set and $w_n = \sum_{k=1}^{n} \lambda_k a_k \in \mathrm{cvx}(C)$ be a convex combination of C. We prove the statement by mathematical induction with respect to $n \in \mathbb{N}$. Obviously we have $w_1, w_2 \in C$.
Now we assume that $w_n \in C$ for some $n \geq 2$. Let

$$w_{n+1} = \sum_{k=1}^{n+1} \lambda_k c_k, \text{ where } c_k \in C, \ \lambda_k \geq 0 \ (k = 1, \ldots, n+1) \text{ and } \sum_{k=1}^{n+1} \lambda_k = 1,$$

that is, $w_{n+1} \in \mathrm{cvx}(C)$. We may assume $\lambda_{n+1} \neq 1$, for otherwise $\lambda_1 = \lambda_2 = \cdots = \lambda_n = 0$. Then we have

$$w_{n+1} = \lambda_{n+1} c_{n+1} + (1 - \lambda_{n+1}) \sum_{k=1}^{n} \frac{\lambda_k}{1 - \lambda_{n+1}} \cdot c_k,$$

and $w_{n+1} \in \mathrm{cvx}(\{c_{n+1}, b\})$, where $b = \sum_{k=1}^{n} \lambda_k (1 - \lambda_{n+1})^{-1} \cdot c_k$. It follows from $\mu_k = \lambda_k (1 - \lambda_{n+1})^{-1} \geq 0$ for $k = 1, 2, \ldots, n$ and

$$\sum_{k=1}^{n} \mu_k = \sum_{k=1}^{n} \frac{\lambda_k}{1 - \lambda_{n+1}} = \frac{1}{1 - \lambda_{n+1}} \cdot \sum_{k=1}^{n} \lambda_k = \frac{1}{1 - \lambda_{n+1}} \cdot (1 - \lambda_{n+1}) = 1$$

that $b \in \mathrm{cvx}(\{c_1, c_2, \ldots, c_n\})$, hence $b \in C$ by assumption.
Since C is convex, it follows that $w_{n+1} = \lambda_{n+1} c_{n+1} + (1 - \lambda_{n+1}) b \in C$. $\square$
Next we define the convex hull of a set.

Definition 2.3.4 The intersection of all convex sets that contain a set A is called the *convex hull of A*, denoted by co(A), that is,

$$\text{co}(A) = \bigcap \{C \supset A : A \text{ convex}\}.$$

It is clear from Definition 2.3.4 and Remark 2.2.6 that the convex hull co(A) of a set A is the smallest convex set that contains A.

Theorem 2.3.5 *We have* co(A) $=$cvx(A).

Proof. It follows from Lemma 2.3.3 that cvx(A) $\subset$cvx(co(A)) $\subset$co(A). Since co(A) is a convex set, it suffices to show that cvx(A) is convex.
Let $0 < \lambda < 1$ and $a, b \in$cvx(A). Then there exist $n, m \in \mathbb{N}$, $a_k \in A$, $\lambda \alpha_k \geq 0$ for $k = 1, 2, \ldots, n$, $b_j \in A$ and $\beta_j \geq 0$ for $j = 1, 2, \ldots, m$ with $\sum_{k=1}^{n} \alpha_k = \sum_{j=1}^{m} \beta_j = 1$, $a = \sum_{k=1}^{n} \alpha_k a_k$ and $b = \sum_{j=1}^{m} \beta_j b_j$. Now

$$\sum_{k=1}^{n} \lambda \alpha_k + \sum_{j=1}^{m} \beta_j = \lambda + (1 - \lambda) = 1$$

implies $\lambda a + (1 - \lambda) b \in$cvx($A$), hence cvx($A$) is convex. $\qquad\square$

The next result asserts, among other things, that the set arithmetic operations preserve the convexity of sets.

Remark 2.3.6 *Let X be a linear space and A and B be subsets of X. Then we have:*
(a) If A and B are convex sets, so are the sets $\lambda A + \mu B$ for all scalars λ and μ.
(b) Translations and multiplications by scalars preserve the convexity of sets.
(c) Let $f \in X^{\#}$ satisfy $f(x) \leq 1$ for all $x \in A$. Then it follows that $f(x) \leq 1$ for all $x \in$co(A).
(d) Let X be a real linear space, A be a convex subset of X, $f \in X^{\#}$ and $\lambda \in \mathbb{R}$ be such that $f(a) \neq \lambda$ for all $a \in A$. Then we have

either $f(a) < \lambda$ for all $a \in A$ or $f(a) > \lambda$ for all $a \in A$.

Proof.

(a) Let $0 < \nu < 1$. Then it follows from (2.1) in Remark 2.3.2 that

$$\nu(\lambda A + \mu B) + (1 - \nu)(\lambda A + \mu B) = \nu \lambda A + \nu \mu B + (1 - \nu)\lambda A + (1 - \nu)\mu B =$$
$$\lambda(\nu A + (1 - \nu)A) + \mu(\nu B + (1 - \nu)B) \subset \lambda A + \mu B,$$

since the convexity of the sets A and B implies $\nu A + (1-\nu)A \subset A$ and $\nu B + (1-\nu)B \subset B$. Thus the set $\lambda A + \mu B$ is convex.

(b) Let A be a convex set.

  (i) First we show that convexity is preserved by translation.
       Let $A + b$ be a translation of A. Since the singleton $\{b\}$ trivially is convex, the statement follows from Part (a) with $B = \{b\}$ and $\lambda = \mu = 1$.

  (ii) Now we show that convexity is preserved by multiplication by scalars.
       This follows from Part (a) with $\mu = 0$.

Thus we have shown Part (b).

(c)   The set
$$M = f^{-1}((-\infty, 1]) = \{x \in X : f(x) \leq 1\}$$
is convex being the pre–image of the convex set $(-\infty, 1]$ under $f \in X^{\#}$ by Theorem 2.2.7 (c), and we have $M \supset A$ by assumption. It follows that $\mathrm{co}(A) \subset M$, since $\mathrm{co}(A)$ is the smallest convex set that contains A.

(d)   The set $f(A) \subset \mathbb{R}$ is convex by Theorem 2.2.7 (d), hence an interval. $\qquad\square$

Definition 2.3.7 A non–empty subset A of a linear space is called *affine*, if

$$\lambda A + (1 - \lambda)A \subset A \text{ for all scalars } \lambda.$$

The following statements are obvious.

Remark 2.3.8 *An affine set is convex and a subspace is affine.*

The next result is a characterization of affine sets.

Theorem 2.3.9 *A subset A of a linear space is affine if and only if it is a translation of a subspace S.*
We have $S = A - A$ and $A = A + (A - A)$.

Proof.

(i)   We show that if A is affine then it is the translation of a subspace.
Let A be affine. We put $S = A - A$.

  (α)   We show that S is a subspace.
Let λ be a scalar. Since A is affine, we have $\lambda A + (1 - \lambda)A \subset A$ by Definition 2.3.7 and we obtain from (2.1) and (2.2) in Remark 2.2.1 that

$$\lambda S = \lambda(A - A) = \lambda A - \lambda A = \lambda A + (1 - \lambda - 1)A$$
$$\subset \lambda A + (1 - \lambda)A - A \subset A - A = S.$$

Furthermore, since A is convex by Remark 2.3.8 it follows that $A + A \subset 2A$ by Remark 2.2.6 (a), and since A is affine, we obtain $2A + (1 - 2)A \subset A$, hence

$$S + S = (A - A) + (A - A) = (A + A) - A - A \subset 2A + (1 - 2)A - A \subset A - A = S.$$

Thus we have shown that S is a subspace.

Furthermore, since $A + A \subset 2A$ and A is affine, we have $2A - A \subset A$, and so

$$A + S = A + (A - A) = (A + A) - A \subset 2A - A \subset A. \tag{2.3}$$

We also obtain by (2.2) in Remark 2.2.1 that

$$A + S = A + (A - A) = (A + A) - A \supset 2A - A \supset (2 - 1)A = A,$$

hence with (2.3)
$$A + S = A. \tag{2.4}$$

  (β)   We show that $A = a + S$ for all $a \in A$.
Let $a \in A$ be given. Then it follows from (2.4) that $a + S \subset A + S = A$.
On the other hand, we have $A - a \subset A - A = S$, hence $A \subset a + S$.

(ii) Now we show that any translation of a subspace is affine.
Let S be a subspace, a be a vector and λ be a scalar. Then we have by (2.1) in Remark 2.2.1 and since S is a subspace

$$\lambda(a + S) + (1 - \lambda)(a + S) = \lambda a + \lambda S + (1 - \lambda)a + (1 - \lambda)S$$
$$= (\lambda + (1 - \lambda))a + \lambda S + (1 - \lambda)S \subset a + S.$$

$\square$

Remark 2.3.10 *An affine set is a translation of a unique subspace.*

Proof. Let S_1 and S_2 be subspaces with $a + S_1 = b + S_2$. Then we have $S_1 = b - a + S_2 = u + S_2$, say, and so, since S_1 and S_2 are subspaces, $S_1 = S_1 - S_1 = u + S_2 - (u + S_2) = S_2 - S_2 = S_2$. $\square$

Theorem 2.3.11 *Let A be an affine set. Then the following statements are equivalent:*

$\quad$ *(a) $0 \in A$;* $\qquad$ *(b) A is a subspace;* $\qquad$ *(c) $A = A - A$.*

Proof. By Theorem 2.3.9, (c) implies (b).
Obviously (b) implies (a).
We show that (a) implies (c). If $0 \in A$ then we have by what is shown in Part (β) of the proof of Theorem 2.3.9

$$A = 0 + A - A = A - A.$$

$\square$

2.4   Balloons and cones

Here we introduce the concepts of *balloons* and *cones*, and study their most important properties.

Definition 2.4.1 (a) A *balloon* is a convex, balanced and absorbing set.
(b) A *cone* K is a non–empty set for which $\lambda K \subset K$ for all $\lambda \geq 0$.

We start with some simple examples of balloons and cones.

Example 2.4.2 *(a) Every ball in $\mathbb{R}^n$ or $\mathbb{C}^n$ centred at the origin is a balloon.*
(b) Every region between two rays in $\mathbb{C}$ that originate in the origin is a cone.
(c) For all $a \neq 0$, the set

$$K = \{(x, y, z) \in \mathbb{R}^3 : a^2(x^2 + y^2) \leq z^2\}$$

is a cone.

Remark 2.4.3 *We have:*
(a) A cone K is convex if and only if $K + K \subset K$.
(b) If K is a convex cone then $K \cap (-K)$ is a real subspace.
(c) If K is a convex cone then the real span of K, $\mathrm{span}_{\mathbb{R}}(K)$, satisfies

$$\mathrm{span}_{\mathbb{R}}(K) = K - K.$$

(d) If K is a convex cone then the set $\bigcup_{\lambda \geq 0} \lambda K$ is a convex cone.
(e) Every convex absorbing set contains a balloon.

Proof.

(a) By Remark 2.2.6 (a), the set K is convex if and only if $\lambda K + \mu K \subset (\lambda + \mu)K$ for all scalars $\lambda, \mu > 0$. Since K is a cone, it follows that $\lambda K = \mu K = (\lambda + \mu)K = K$ and the assertion follows.

(b) (i) First we show that $K \cap (-K)$ is a cone.
Let $\lambda \geq 0$ and $\lambda x \in K \cap (-K)$. Then $\lambda x \in K$ and $\lambda x \in -K$, that is, $\lambda x \in K$ and $\lambda(-x) \in K$, and it follows that $x \in K$ and $-x \in K$, since K is a cone. Hence we have $x \in K \cap (-K)$. Thus we have shown that $\lambda(K \cap (-K)) \subset K \cap (-K)$ for all $\lambda \geq 0$, and so $K \cap (-K)$ is a cone.

Since K is convex, so obviously is $-K$, and then the intersection $K \cap (-K)$ is convex by Remark 2.2.6 (b), and $K \cap (-K)$ is symmetric by Remark 2.2.3 (c). Since $K \cap (-K)$ is a symmetric cone, it follows by Part (a) that

$$\mu(K \cap (-K)) + (K \cap (-K)) = |\mu|(K \cap (-K)) + (K \cap (-K)) \subset$$
$$(K \cap (-K)) + (K \cap (-K)) \subset K \cap (-K),$$

which shows that $K \cap (-K)$ is a real subspace.

(c) Since every element of the set $K - K$ is a real linear combination of K, we have

$$K - K \subset \mathrm{span}_{\mathbb{R}}(K).$$

Conversely, if $u \in \mathrm{span}_{\mathbb{R}}(K)$, then we have

$$u = \sum \lambda_k a_k - \sum \mu_k b_k \text{ with } \lambda_k, \mu_k > 0 \text{ and } a_k, b_k \in K.$$

Since K is a convex cone, it follows that

$$u \in \sum \lambda_k K - \sum \mu_k K \subset K - K.$$

(d) We write $V = \bigcup_{\lambda \geq 0} \lambda K$. Obviously V is a cone.
Let $x, y \in V$. Then there are scalars $\lambda, \mu \geq 0$ such that $x \in \lambda K$ and $y \in \mu K$. Since K is a convex set, it follows that

$$x + y \in \lambda K + \mu K = (\lambda + \mu)K \subset V.$$

(e) Let K be a convex absorbing set. We put

$$W = \bigcap_{|\lambda|=1} \lambda K.$$

(i) First, we show that B is convex.
Since K is convex, each set λK is convex by Remark 2.3.6 (b), and so the intersection W of convex sets is convex by Remark 2.2.6 (b).

(ii) Now we show that W is absorbing.
Let a be an arbitrary vector and λ be a scalar with $|\lambda| = 1$. Since the set K is absorbing, there exists $\delta > 0$ such that

$$\mu a \in K \text{ for all scalars } \mu \text{ with } |\mu| \leq \delta.$$

It follows from $|\mu/\lambda| = |\mu| \leq \delta$ that

$$\frac{\mu}{\lambda} \cdot a \in K, \text{ hence } \mu \cdot a \in \lambda K.$$

Thus $\mu a \in W$ for all scalars μ with $|\mu| \leq \delta$ and we have shown that W is absorbing.

(iii)   Finally we show that W is balanced.

Let $b \in W$ and μ be a scalar with $|\mu| \leq 1$. If $\mu = 0$, then $\mu b \in W$.

If $\mu \neq 0$, then we have to show that $\mu b \in \lambda K$ for all scalars λ with $|\lambda| = 1$. Since

$$\left| \frac{|\mu|}{\mu} \cdot \lambda \right| = 1, \text{ we have } b \in \frac{|\mu|}{\mu} \lambda K \text{ by the definition of } W,$$

so

$$\frac{\mu}{|\mu|\lambda} \cdot b \in K.$$

Since K is convex and $0 \in K$, it follows that

$$\frac{\mu}{\lambda} \cdot b \in K, \text{ hence } \mu b \in \lambda K.$$

$\square$

2.5   Quotient spaces

In this section, we introduce the concept of quotient spaces of linear spaces and study some of their properties which will frequently be needed throughout the book.

Definition 2.5.1 (a) Two subspaces S_1 and S_2 of a linear space X are called *complementary* if $S_1 \oplus S_2 = X$.

(b) Let S be a subspace of a linear space X. Then the set

$$X/S = \{x + S : x \in X\}$$

is called *quotient space X modulo S*.

We remark that some authors refer to X/S as the *factor space X modulo S*.

Example **2.5.2** *(a) The x–axis and the y–axis are complementary subspaces of $\mathbb{R}^2$.*

(b) In every linear space X, $\{0\}$ and X are complementary subspaces.

Remark 2.5.3 *Every subspace has a complementary subspace.*

Proof. We may assume $S \neq \{0\}$ by Example 2.5.2 (b). So let $S \neq \{0\}$ be a subspace of a linear space X, and H_S be an algebraic basis of S. We extend H_S to a basis H of X, and put $\tilde{S} =$ span$(H \setminus H_S)$.

If $a \in S \cap \tilde{S}$ then a is a linear combination of H_S, and a linear combination of $H \setminus H_S$, hence $a = 0$.

If $x \in X$ then x is a linear combination of H, hence of H_S and of $H \setminus H_S$. $\square$

Remark 2.5.4 *The elements of X/S are translations of the subspace S, hence affine sets by Theorem 2.3.9. Furthermore*

$$x + S = y + S \text{ whenever } x - y \in S.$$

We define addition and multiplication by scalars in X/S by

$$(x + S) + (y + S) = x + y + S \text{ and } \lambda(x + S) = \lambda x + S.$$

Then X/S becomes a linear space. The zero element of X/S is $0 + S = S$.
We define the natural homomorphism

$$\varphi : X \to X/S \text{ by } \varphi(x) = x + S \text{ for all } x \in X.$$

This is obviously a linear map, and $\varphi^{-1}(\{0\}) = S$. To see this let first $\varphi(x) = 0$, that is, $x + S = S$. Then $x \in S - S = S$. Conversely, if $x \in S$ then we have $\varphi(x) = x + S = S$. The natural homomorphism φ is one to one on every subspace $\tilde{S}$ which is complementary to S. For if $x \in \tilde{S}$ and $\varphi(x) = 0$, then $x \in S$, and so $x = 0$, since $\tilde{S} \cap S = \{0\}$.
Also the restriction $\varphi|_{\tilde{S}}$ of the natural homomorphism φ to $\tilde{S}$ is onto X/S. To see this, let $x + S \in X/S$ be given. Since $X = S + \tilde{S}$, we have $x = s + \tilde{s}$ for some $s \in S$ and some $\tilde{s} \in \tilde{S}$. It follows that

$$\varphi(\tilde{s}) = \varphi(x - s) = \varphi(x) - \varphi(s) = \varphi(x) = x + S.$$

The arguments in Remark 2.5.4 yield the following result.

Corollary 2.5.5 *Let X be a linear space and S and $\tilde{S}$ be complementary subspaces of X. Then X/S and $\tilde{S}$ are isomorphic.*

In view of Corollary 2.5.5, we define the *codimension* of a subspace.

Definition 2.5.6 The *codimension* of a subspace S of a linear space X is the dimension of any complementary subspace; thus the codimension of S is the dimension of X/S. We write

$$\operatorname{codim}_X(S) = \dim(X/S).$$

Example 2.5.7 *The codimension of c_0 in c is equal to 1, that is,*

$$\operatorname{codim}_c(c_0) = 1.$$

Theorem 2.5.8 *Let X be a linear space and S be a subspace of X. Then there exist a linear space X_1 and a linear map $f : X \to X_1$ such that $S = f^{-1}(\{0\})$.*

Proof. We put $X_1 = X/S$ and $f = \varphi$, where $\varphi : X \to X/S$ is the natural homomorphism. Then it follows that $S = f^{-1}(\{0\})$. □

We close this section with a very useful result which is well known from linear algebra.

Theorem 2.5.9 *Let X and Y be linear spaces over the same field $\mathbb{F}$ and $f \in \mathcal{L}(X, Y)$. Then the quotient space $X/N(f)$ is isomorphic to $R(f)$, the range of f.*

2.6 Supremum and weak topologies

Modern functional analysis deals with the study and application of various so–called *weak topologies* which we investigate in this section. Since we can compare topologies using nets by Theorem 1.10.5 (d), most of the results follow by means of characterizations of net convergence.

The first result states that net convergence in the supremum topology of a collection of topologies is equivalent to net convergence in each of the topologies of the collection.

Theorem 2.6.1 *Let Φ be a class of topologies on a set and $\mathcal{T}' = \bigvee \Phi$ denote the supremum of Φ (Definition 1.9.11). Then we have $x_\delta \to x$ in $\mathcal{T}'$ for a net (x_δ) if and only if $x_\delta \to x$ in each topology $\mathcal{T} \in \Phi$.*

Proof.

(i) We show that $x_\delta \to x$ in $\mathcal{T}'$ implies $x_\delta \to x$ in each $\mathcal{T} \in \Phi$.
Let $x_\delta \to x$ in $\mathcal{T}'$. Since $\mathcal{T}' \supset \mathcal{T}$ for each $\mathcal{T} \in \Phi$ by Definition 1.9.11, it follows that $x_\delta \to x$ in each $\mathcal{T} \in \Phi$.

(ii) Now we show that if $x_\delta \to x$ in each $\mathcal{T} \in \Phi$, then $x_\delta \to x$ in $\mathcal{T}'$.
Let $x_\delta \to x$ in each topology $\mathcal{T} \in \Phi$ and G be a $\mathcal{T}'$–neighborhood of x. By definition of $\mathcal{T}'$ and of a subbasis (Definition 1.9.8 (c)), there exist finitely many sets $G_1, G_2, \ldots, G_n$ with

$$x \in \bigcap_{k=1}^{n} G_k \subset G,$$

each of which is a $\mathcal{T}$–neighborhood for at least one $\mathcal{T} \in \Phi$. Therefore we have $x_\delta \in G_k$ eventually for all k, hence

$$x_\delta \in \bigcap_{k=1}^{n} G_k \subset G \text{ eventually,}$$

that is, $x_\delta \to x$ in $\mathcal{T}'$. □

Theorem 2.6.2 *Let $\{d_n : n \in \mathbb{N}\}$ be a countable class of semimetrics d_n. Then $\bigvee \{d_n\}$ (Definition 1.9.11) is semimetrizable and given by*

$$d = \sum_{n=1}^{\infty} \frac{1}{2^n} \cdot \frac{d_n}{1 + d_n}.$$

If the class is finite, we may use $d = \sum d_n$.

Proof. Obviously d is a semimetric, and $x_\delta \to x$ if and only if $x_\delta \to x$ with respect to each semimetric d_n. It follows from Theorem 2.6.1 that d generates the topology $\bigvee \{d_n\}$.□

If f is a map from a set A into a topological space Y then A can be given a topology by means of the preimages of f of the open sets in Y; the resulting topology is referred to as the *weak topology by f*.

Definition 2.6.3 Let A be a set, Y be a topological space and $f : A \to Y$ be a map. The topology generated by the class

$$\{f^{-1}(G) : G \text{ open in } Y\}$$

is called the *weak topology (generated) by f*, denoted by $w(A, f)$.

Remark 2.6.4 *Obviously $f : A \to Y$ is continuous when A has the topology $w(A, f)$. Furthermore, $w(A, f)$ is the weakest topology on A for which f is continuous. Finally, $w(A, f) = \{f^{-1}(G) : G \text{ open in } Y\}$.*

Proof. Let $\mathcal{T}$ be a topology on A for which f is continuous, and G be an open set in Y. Then $f^{-1}(G)$ is $\mathcal{T}$–open, and so every open set in a basis for the weak topology $w(A, f)$ is

$\mathcal{T}$-open. Consequently $w(A, f) \subset \mathcal{T}$.
Finally, since

$$\bigcap_{k=1}^{n} f^{-1}(G_k) = f^{-1}\left(\bigcap_{k=1}^{n} G_k\right) \text{ and } \bigcap_{k=1}^{n} G_k \text{ is open in } Y,$$

it follows that $w(A, f) = \{f^{-1}(G) : G \text{ open in } Y\}$. □

The next result is a simple characterization of net convergence in the weak topology by a function f.

Lemma 2.6.5 *A net (a_δ) in A satisfies $a_\delta \to a$ in the weak topology $w(A, f)$ if and only if $f(a_\delta) \to f(a)$ in Y.*

Proof.

(i) Let $a_\delta \to a$ in $w(A, f)$.
Since f is continuous by Remark 2.6.4, it follows that $f(a_\delta) \to f(a)$ by Theorem 1.10.5 (b).

(ii) We assume $f(a_\delta) \to f(a)$ in Y.
Let N be a $w(A, f)$–neighborhood of a in A. Then there exists a neighborhood G of $f(a)$ such that $f^{-1}(G) \subset N$. Hence we have $f(a_\delta) \subset G$ eventually and so $a_\delta \in N$ eventually, that is, $a_\delta \to a$ in $w(A, f)$. □

Now we generalize the concept of a weak topology by a function to that of a weak topology by a class of functions.

Definition 2.6.6 Let A be a non–empty set and Ψ be a class of topological spaces. We assume that, for every $Y \in \Psi$, there exists one or more functions $f : A \to Y$. Let Φ denote the class of all these functions. The topology

$$w(A, \Phi) = \bigvee\{w(A, f) : f \in \Phi\}$$

is called the *weak topology (generated) by Φ*.

The next result is an immediate consequence of Theorem 2.6.1 and Lemma 2.6.5.

Theorem 2.6.7 *If a set A has the weak topology $w(A, \Phi)$ by a class Φ of functions from A into a class of topological spaces, then we have $a_\delta \to a$ in $w(A, \Phi)$ if and only if $f(a_\delta) \to f(a)$ for every $f \in \Phi$.*

The next results concerns the semimetrizability of a weak topology by a sequence of functions.

Theorem 2.6.8 *The weak topology by a sequence (f_n) of functions from a set A into a class of semimetric spaces is semimetrizable.*

Proof. Since the topology is $\bigvee\{w(A, f_n) : n \in \mathbb{N}\}$, it is sufficient by Theorem 2.6.2 to show that $w(A, f)$ is semimetrizable if $f : A \to Y$ and Y is a semimetrizable space. To show this, we put $d_1(a, b) = d(f(a), f(b))$ for $a, b \in A$ where d is the semimetric on Y. We have $a_\delta \to a$ in $w(A, f)$ if and only if $f(a_\delta) \to f(a)$ in Y by Lemma 2.6.5, and this is the case by Example 1.10.4 (d) if and only if $d(f(a_\delta), f(a)) = d_1(a_\delta, a) \to 0$ which is equivalent to $a_\delta \to a$ in d_1, again by Example 1.10.4 (d). Thus the weak topology $w(A, f)$ by f is given by the semimetric d_1. □

Example **2.6.9** *Let n and k be positive integers with $1 \leq k \leq n$. We assume that $\mathbb{C}^n$ has the weak topology $w = w(\mathbb{C}^n, \{P_1, P_2, \ldots, P_k\})$ where P_j is the j–th coordinate. Then w is given by the seminorm*

$$p(a) = \sum_{j=1}^{k} |a_j|.$$

This is the same topology as the weak topology $w_f = w(\mathbb{C}^n, f)$ by the function $f : \mathbb{C}^n \to \mathbb{C}^k$ with $f(a) = (a_1, a_2, \ldots, a_k)$. All topologies are semimetrizable by Theorem 2.6.8, so it is sufficient to compare convergent sequences. Obviously we have in all topologies $a^{(m)} \to a$ if and only if $a_j^{(m)} \to a_j$ $(m \to \infty)$ for $1 \leq j \leq k$.

The next result generalizes Remark 2.6.4

Theorem 2.6.10 *Let A have the weak topology $w(A, \Phi)$. Then every $f \in \Phi$ is continuous and $w(A, \Phi)$ is the weakest topology for which this holds true.*

Proof.

(i) First we show that every function $f \in \Phi$ is continuous.
Let $f \in \Phi$, hence $f : A \to Y$ for some tolological space Y, and let G be an open set in B. Then $f^{-1}(G)$ is $w(A, \Phi)$–open by definition of $w(A, \Phi)$ (Definition 2.6.6), hence f is continuous.

(ii) Now we show that $w(A, \Phi)$ is the weakest topology for which all functions $f \in \Phi$ are continuous.
Let $\mathcal{T}$ be a topology on A such that all functions $f \in \Phi$ are continuous, and let (a_δ) be a net in A with $a_\delta \to a$. Since each f is continuous, it follows by Theorem 1.10.5 (b) that $f(a_\delta) \to f(a)$ for each $f \in \Phi$. This implies $a_\delta \to a$ in $w(A, \Phi)$ by Theorem 2.6.7. Hence we have shown that $\mathcal{T} \supset w(A, \Phi)$ by Theorem 1.10.5 (d). $\square$

Lemma 2.6.11 *Let the class $\Sigma(Y)$ be a subbasis of the topology of Y for each $Y \in \Psi$. We put*

$$\Sigma = \{f^{-1}(G) : f \in \Phi, \ f : A \to Y, \ G \in \Sigma(Y)\}.$$

Then Σ generates the topology $w(A, \Phi)$.

Proof. The inclusion $\mathcal{T} \subset w(A, \Phi)$ follows directly from Definitions 2.6.6, 1.9.11, 1.9.8 (c) and Theorem 1.9.10.
Now we show that $w(A, \Phi) \subset \mathcal{T}$.
Let $a_\delta \to a$ in $\mathcal{T}$, and let $f \in \Phi$, $f : A \to Y$, say. Then we have

$$f(a_\delta) \to f(a) \text{ in } Y. \tag{*}$$

To show that (*), let $f(a) \in G \in \Sigma(Y)$. Then we have $a \in f^{-1}(G) \in \Sigma$, and so $a_\delta \in f^{-1}(G)$ eventually, hence $f(a_\delta) \in G$ eventually. This shows (*) along with the fact that $\Sigma(Y)$ is a subbasis for Y.
Since $f \in \Phi$ was arbitrary, it follows from (*) and Theorem 2.6.7 that $a_\delta \to a$ in $w(A, \Phi)$.
Finally, $w(A, \Phi) \subset \mathcal{T}$ follows by Theorem 1.10.5 (d). $\square$

We need some more lemmas.

Lemma 2.6.12 *Let X be a linear space and $\{f_1, f_2, \ldots, f_n\} \subset X^{\#}$. If S is a subset of X containing more than n vectors, then there exists a non–trivial linear combination of S at which each function f_k vanishes.*

Proof. Let $s_1, s_2, \ldots, s_{n+1} \in S$ be distinct vectors. Then there are scalars $\lambda_1, \lambda_2, \ldots,$ λ_{n+1} not all of which are equal to 0 such that

$$\sum_{k=1}^{n+1} \lambda_k f_m(s_k) = 0 \text{ for } (m = 1, 2, \ldots, n), \tag{2.5}$$

since any homogeneous system of n linear equations in $n + 1$ unknowns has a non–trivial solution. The equations in (2.5) are equivalent to

$$f_m\left(\sum_{k=1}^{n+1} \lambda_k s_k\right) = 0 \text{ for } m = 1, 2, \ldots, n.$$

$\square$

We recall that $N(f)$ denotes the kernel or null space of a linear function f (Definition 1.6.1 (f)).

Lemma 2.6.13 *Let Φ be a family of linear functionals on a linear space X and $S = \bigcap\{N(f) : f \in \Phi\}$. Then the set*

$$\tilde{\Phi} = \{F : f = F \circ \varphi, \ f \in \Phi, \ \varphi : X \to X/S \text{ the natural homomorphism}\} \tag{2.6}$$

is total over X/S.

Proof. Let $F(x + S) = 0$ for each $F \in \tilde{\Phi}$, that is, $f(x) = F(x + S) = 0$ for each $f \in \Phi$, by the definition of the functions F. This means $x \in N(f)$ for each $f \in \Phi$, hence $x \in S$ by definition of S. Thus $x + S = S$, the zero element of X/S, and $\tilde{\Phi}$ is total over X/S by Remark 1.6.11 (a). $\square$

It is easy to see that the following result holds.

Lemma 2.6.14 *Let Φ be a finite family of linear functionals that are total over a linear space X. Then X has finite dimension, and the dimension of X does not exceed the number of elements of Φ. Furthermore, we have $\text{span}(\Phi) = X^{\#}$.*

Definition 2.6.15 A proper subspace S of a linear space X is said to be *maximal*, if $X = \text{span}(x, S)$ for some $x \in X \setminus S$.

The next result states that the kernel of every non-identically vanishing linear functional on a linear space is a maximal subspace.

Remark 2.6.16 *Let X be a linear space and $f \in X^{\#} \setminus \{0\}$. Then the kernel $N(f)$ of f is a maximal subspace of X.*

Proof. Obviously, $N(f)$ is a subspace of X.
Since $f \in X^{\#} \setminus \{0\}$, there exists $x \in X$ such that $f(x) \neq 0$, hence $N(f)$ is a proper subspace of X.

(i) We show that $X = \text{span}(x, N(f))$.
Let $y \in X$ be given. We put

$$z = y - \frac{f(y)}{f(x)} \cdot x,$$

and obtain

$$f(z) = f\left(y - \frac{f(y)}{f(x)} \cdot x\right) = f(y) - \frac{f(y)}{f(x)} \cdot f(x) = 0,$$

hence $z \in N(f)$, and so

$$y = \frac{f(y)}{f(x)} \cdot x + z \in \text{span}(x \oplus N(f)).$$

$\square$

The next theorem shows that the codimension of the intersection of a finite number of maximal subspaces cannot exceed that number.

Theorem 2.6.17 *Let $S_1, S_2, \ldots, S_n$ be maximal subspaces of X and $S = \bigcap_{k=1}^n S_k$. Then we have*

$$\text{codim}_X(S) \leq n.$$

Proof. For each $k = 1, 2, \ldots, n$, we choose a functional $f_k \in X^\# \setminus \{0\}$ with $N(f_k) = S_k$ by Theorem 1.6.5. Then $\{F_1, F_2, \ldots, F_n\}$ is total over X/S by Lemma 2.6.13, where each F_k is defined by $f_k = F_k \circ \varphi$ with $\varphi : X \to X/S$, the natural homomorphism. Now it follows from Lemma 2.6.14 that

$$\text{codim}_X(S) = \dim(X/S) \leq n.$$

$\square$

Theorem 2.6.18 *Let X be a linear space, $g, f_1, f_2, \ldots, f_n \in X^\#$, and $N(g) \supset \bigcap_{k=1}^n N(f_k)$. Then g is a linear combination of the set $\Phi = \{f_1, f_2, \ldots, f_n\}$.*

Proof. We may assume $f_k \in X^\# \setminus \{0\}$ for all k. Let $S = \bigcap_{k=1}^n N(f_k)$. Since $f_k \in X^\# \setminus \{0\}$ for $k = 1, 2, \ldots, n$, every set $N(f_k)$ is a maximal subspace by Remark 2.6.16, hence X/S is finite dimensional. By Lemma 2.6.13, the family $\tilde{\Phi}$ of (2.6) is total over X/S, and $\text{span}(\tilde{\Phi}) = (X/S)^\#$ by Lemma 2.6.14. We have $g|_S \equiv 0$ by assumption, and define the map G on X/S by $g = G \circ \varphi$ where $\varphi : X \to X/S$ is the natural homomorphism. It follows that $G \in (X/S)^\#$, and since $\text{span}(\tilde{\Phi}) = (X/S)^\#$, G is a linear combination of $\tilde{\Phi}$ on X/S,

$$G(x + S) = \sum_{k=1}^n \lambda_k F_k(x + S) \text{ for all } x \in X, \text{ where } f_k = F_k \circ \varphi,$$

hence g is a linear combination of Φ.

$\square$

Theorem 2.6.19 *Let X be a linear space and $g, f_1, f_2, \ldots, f_n \in X^\#$ with*

$$\bigcap_{k=1}^n N(f_k) \subset \{|g| < 1\} = \{x \in X : |g(x)| < 1\}.$$

Then g is a linear combination of $\{f_1, f_2, \ldots, f_n\}$.

Proof. Since g is bounded on the subspace $S = \bigcap_{k=1}^n N(f_k)$, we have $g|_S \equiv 0$ and the statement follows from Theorem 2.6.18. $\square$

Theorem 2.6.20 *Let X be a linear space and Φ be a class of linear functionals on X. Then the weak topology $w(X, \Phi)$ by Φ is semimetrizable if and only if Φ has countable dimension in $X^\#$; $w(X, \Phi)$ is seminormable if and only if Φ is of finite dimension.*

Proof. Let $S = \text{span}(\Phi) \subset X^\#$.

(i) First we show that if Φ has countable dimension then $w(X, \Phi)$ is semimetrizable. Let Φ have countable dimension. Then there exists a sequence (f_n) in S with $\mathrm{span}(\{f_n : n \in \mathbb{N}\}) = \Phi$. It follows that

$$w(X, \Phi) = w\left(X, \{f_n : n \in \mathbb{N}\}\right),$$

since each $f \in \Phi$ is a linear combination of $\{f_n : n \in \mathbb{N}\}$ and each f_n is a linear combination of Φ and thus every $f \in \Phi$ is $w(X, \{f_n : n \in \mathbb{N}\})$ continuous and each f_n is $w(X, \Phi)$ continuous. Theorem 2.6.7 implies that the net convergence in both topologies is the same, hence $w(X, \Phi) = w(A, \{f_n : n \in \mathbb{N}\})$. So $w(X, \Phi)$ is semimetrizable by Theorem 2.6.8.

(ii) Now we assume that Φ has finite dimension. Then it follows as in Part (i) of the proof that

$$w(X, \Phi) = w(X, \{f_1, f_2, \ldots, f_n\}) \text{ for finitely many } f_k \in S.$$

Then $w(X, \Phi)$ is given by the seminorm p with

$$p(x) = \sum_{k=1}^{n} |f_k(x)|.$$

(iii) Now we show that if $w(X, \Phi)$ is semimetrizable then Φ has countable dimension. Let $w(X, \Phi)$ be semimetrizable, hence first countable by Example 1.9.9 (a). Let (G_n) be a sequence of neighborhoods of 0 that is a local basis at 0. For each $n \in \mathbb{N}$, by Lemma 2.6.11, there exist functions $f_1^{(n)}, f_2^{(n)}, \ldots, f_{r(n)}^{(n)} \in \Phi$ and real numbers $\varepsilon_n > 0$ such that

$$G_n \supset \bigcap_{k=1}^{r(n)} \left\{ \left| f_k^{(n)} \right| < \varepsilon_n \right\}.$$

Let $f \in \Phi$ be given. Then we have

$$\{|f| < 1\} \supset G_n \text{ for some } n \in \mathbb{N},$$

since f is continuous and $\{G_n : n \in \mathbb{N}\}$ is a basis. Thus we have

$$\bigcap_{k=1}^{r(n)} N\left(f_k^{(n)}\right) \subset \{|f| < 1\}$$

and f is a linear combination of the set $\{f_k^{(n)} : k = 1, 2, \ldots, r(n)\}$ by Theorem 2.6.19. Consequently we obtain

$$S = \mathrm{span}\left(\left\{ f_k^{(n)} : k = 1, 2, \ldots, r(n), n \in \mathbb{N}\right\}\right).$$

(iv) Finally, we show that if $w(X, \Phi)$ is seminormable then Φ has finite dimension. The proof is similar to that of Part (iii). Now we choose $G_n = (1/n)B_1(0)$ and we can use the same set of functions for each n. In particular, if we choose $f_1, f_2, \ldots, f_r$ and $\varepsilon > 0$ such that $\bigcap_{k=1}^{r}(|f_k| < \varepsilon) \subset B_1(0)$, then $\bigcap_{k=1}^{r} N(f_k)$ is contained in every neighborhood of 0. It follows as in Part (iii) that

$$S = \mathrm{span}(\{f_1, f_2, \ldots, f_n\}). \qquad \square$$

Theorem 2.6.21 *Let X be a linear space and Φ be a class of linear functionals on X. Then every $f \in X^{\#}$ which is $w(X, \Phi)$ continuous is a linear combination of Φ.*

Proof. Now we have $\Psi = \mathbb{C}$ in Definition 2.6.6.

If g is continuous then the set $G = \{|g| < 1\}$ is open. There exist a finite class of functions $f_1, f_2, \ldots, f_n \in \Phi$ and neighborhoods $G_1, G_2, \ldots, G_n$ of 0 in $\mathbb{C}$ such that

$$N = \bigcap_{k=1}^{n} f_k^{-1}(G_k) \subset G.$$

It follows from $a \in N(f_k)$ that $f_k(a) = 0 \in G_k$, hence $N(f_k) \subset f_k^{-1}(G_k)$, and so

$$\bigcap_{k=1}^{n} N(f_k) \subset N \subset G.$$

Therefore the function g is bounded on $\bigcap_{k=1}^{n} N(f_k)$, and so g is a linear combination of $\{f_1, f_2, \ldots, f_n\}$ by Theorem 2.6.19. $\square$

2.7 Product topology

Now we are going to study *product topologies* which are a special case of the weak topologies $w(A, \Phi)$ by a class of functions of Definition 2.6.6.

Definition 2.7.1 The *product topology* for the Cartesian product $\prod X_\beta$ (Section 1.3) of a class of topological spaces $(X_\beta, \mathcal{T}_\beta)$ is the weak topology $w(\prod X_\beta, \{P_\beta\})$ by the class of all projections P_β.

Net convergence in the product topology and net convergence for each projection are equivalent.

Theorem 2.7.2 *Let (x_δ) be a net in the product of topological spaces. Then we have $x_\delta \to x$ in the product topology if and only if $P(x_\delta) \to P(x)$ for each projection P.*

Proof. The statement of the theorem is an immediate consequence of Definition 2.7.1 and Theorem 2.6.7. $\square$

Example 2.7.3 *(a) Let Y be a topological space and X be a set. If (f_δ) is a net in $Y^X = \{f : X \to Y\}$ then we have by Theorem 2.7.2 that $f_\delta \to f$ in the product topology if and only if*

$$P_x(f_\delta) = f_\delta(x) \to P_x(f) = f(x) \text{ for all } x \in X.$$

Thus convergence in the product topology is equivalent with pointwise convergence $f_\delta \to f$. In particular, if $Y = \mathbb{C}$ and $X = \mathbb{N}$, then we have $x_\delta \to x$ in $\omega = \mathbb{C}^{\mathbb{N}}$ in the product topology if and only $x_{\delta,k} \to x_k$ in $\mathbb{C}$ for each $k = 1, 2, \ldots$. Thus the product topology on ω is generated by the metric

$$d(x, y) = \sum_{k=1}^{\infty} \frac{1}{2^k} \cdot \frac{|x_k - y_k|}{1 + |x_k - y_k|}.$$

Since the coordinates in ℓ_p, c_0 and c are continuous, the topologies of these spaces are finer *than the topology of ω on them.*

(b) If we choose $k = n$ as in Example 2.6.9, then we see that the Euclidean topology on $\mathbb{C}^n$ is the product topology. To see this we consider $\mathbb{C}^n$ as the product $\mathbb{C}^B$ with $B = \{1, 2, \ldots, n\}$.

(c) Joint Continuity

Let X and Y be topological spaces and $f : X \times X \to Y$ be a map. Then f is said to be jointly continuous, if $x_\delta \to x$ and $\tilde{x}_\delta \to \tilde{x}$ implies $f(x_\delta, \tilde{x}_\delta) \to f(x, \tilde{x})$. The function f is jointly continuous if and only if $f : X \times X \to Y$ is continuous.

To see this, we first assume that $f : X \times X \to Y$ is continuous. Let $x_\delta \to x$ and $\tilde{x}_\delta \to \tilde{x}$ in X, that is, $P_k(x_\delta, \tilde{x}_\delta) \to P_k(x, \tilde{x})$ $(k = 1, 2)$ where P_k are the projections. Then Theorem 2.7.2 implies $(x_\delta, \tilde{x}_\delta) \to (x, \tilde{x})$ in $X \times X$. Since the function f is continuous, it follows from Theorem 1.10.5 (b) that $f(x_\delta, \tilde{x}_\delta) \to f(x, \tilde{x})$. Thus f is jointly continuous.

Conversely, we assume that the function $f : X \times X \to Y$ is jointly continuous. Let (z_δ) be a net in $X \times X$ with $z_\delta \to z$. We put $x_\delta = P_1(z_\delta)$, $\tilde{y}_\delta = P_2(z_\delta)$, $x = P_1(z)$ and $\tilde{x} = P_2(z)$. It follows from Theorem 2.7.2 that $x_\delta \to x$ and $\tilde{x}_\delta \to \tilde{x}$, and since f is jointly continuous, we obtain $f(z_\delta) = f(x_\delta, \tilde{x}_\delta) \to f(x, \tilde{x}) = f(z)$. Now $f : X \times X \to Y$ is continuous by Theorem 1.10.5 (b).

Theorem 2.7.4 *A non–empty open set in a product $\prod\{X_\beta : \beta \in B\}$ of topological spaces is projected onto almost all factors. This means that if $G \neq \emptyset$ is open in the product topology of $\prod X_\beta$, then we have $P_\beta(G) = X_\beta$ for all but finitely many $\beta \in B$.*

Proof. Let G be a non–empty open set in the product space $\prod X_\beta$. By the definition of a product topology (Definition 2.7.1), there are $\beta(1), \beta(2), \ldots, \beta(n) \in B$ and non–empty open sets $G_k \subset X_{\beta(k)}$ for $k = 1, 2, \ldots, n$ such that

$$\bigcap_{k=1}^{n} P_{\beta(k)}^{-1}(G_k) \subset G.$$

We put $\tilde{B} = B \setminus \bigcup_{k=1}^{n} \beta(k)$ and show that

$$P_t(G) = X_t \text{ for all } t \in \tilde{B}. \tag{*}$$

Let $t \in \tilde{B}$ and $y \in X_t$ be given. We define $x \in \prod X_\beta$ by

$$x_\beta = \begin{cases} y & (\beta = t) \\ x_{\beta(k)} \in G_k \text{ arbitrary} & (\beta = \beta(k) \text{ for } k = 1, 2, \ldots, n) \\ x_\beta \in X_\beta \text{ arbitrary} & (\beta \neq t, \beta(k)). \end{cases}$$

Then it follows that $P_t(x) = x_t = y$, $P_{\beta(k)}(x) = x_{\beta(k)} \in G_k$ $(k = 1, 2, \ldots, n)$, hence

$$x \in P_{\beta(k)}^{-1}(G_k) \text{ for } k = 1, 2, \ldots, n,$$

and so $x \in \bigcap_{k=1}^{n} P_{\beta(k)}^{-1}(G_k) \subset G$. We also have $P_t(x) = x_t = y \in P_t(G)$. Thus we have shown that $X_t \subset P_t(G)$.

Obviously we also have $P_t(G) \subset X_t$.

Hence we have shown (*). $\qquad\square$

Remark 2.7.5 *If each X_β is a linear space, so is the product $X = \prod X_\beta$ with the convention*

$$(x + \lambda y)_\beta = x_\beta + \lambda y_\beta \text{ for all } x, y \in X \text{ and all scalars } \lambda.$$

For instance, we have for $\omega = \mathbb{C}^{\mathbb{N}}$

$$(x + \lambda y)_n = x_n + \lambda y_n \text{ for all } x, y \in \omega \text{ and all scalars } \lambda.$$

The projections are linear, since

$$P_\beta(x + \lambda y) = (x + \lambda y)_\beta = x_\beta + \lambda y_\beta = P_\beta(x) + \lambda P_\beta(y).$$

Theorem 2.7.6 *A countable product of semimetric spaces is a semimetric space.*

Proof. The statement of the theoren follows from Theorem 2.6.8 if we put $f_n = P_n$ for all n. $\qquad\square$

Lemma 2.7.7 *Let $\prod X_\beta$ be an uncountable product of topological spaces which is first countable at one of its points x. Then there exists $\beta \in B$ for which $P_\beta(x)$ has no neighborhood other than X_β.*

Proof. Let $\{G_n : n \in \mathbb{N}\}$ be a countable basis of neighborhoods of x. We put

$$B_n = \{\beta \in B : P_\beta(G_n) \neq X_\beta\} \text{ for } n = 1, 2, \ldots.$$

Then each set B_n is finite by Theorem 2.7.4, and so the set $\tilde{B} = \bigcup_{n \in \mathbb{N}} B_n$ is countable. Therefore there exists $\beta \in B \setminus \tilde{B}$. Let G be a neighborhood of $P_\beta(x)$. Then $P_\beta^{-1}(G)$ is a neighborhood of x, and so there exists an $n \in \mathbb{N}$ such that $G_n \subset P_\beta^{-1}(G)$, since $\{G_n : n \in \mathbb{N}\}$ is a basis at x. So we have $A_\beta = P_\beta(G_n) \subset G$ by the choice of β. $\qquad\square$

Theorem 2.7.8 *An uncountable product of non–indiscrete topological spaces cannot be first countable.*

Proof. We can choose an element $x \in \prod_{\beta \in B} X_\beta$ such that, for each $\beta \in B$, there exists a neighborhood G_β of $P_\beta(x)$ which is a proper subset of X_β, since none of the spaces X_β has the indiscrete topology. Then the product $\prod_{\beta \in B} X_\beta$ cannot be first countable by Lemma 2.7.7. $\qquad\square$

2.8   Properties of linear topological spaces

We now connect the concepts of linearity and topology in a natural way by demanding the algebraic operations of the linear space to be continuous in the topology. Furthermore, we establish various important properties of linear topological spaces, and give a characterization of linear topologies of their open sets.

Definition 2.8.1 Let X be a linear space and $\mathcal{T}$ be a topology with the property that the algebraic operations of the linear space are continuous. Then the pair $(X, \mathcal{T})$, or X, for short, is called a *linear topological space*, or *topological vector space (TVS)*. The topology of a linear topological space is referred to as a *linear topology*.

Remark 2.8.2 *Continuity of addition means that the map $+ : X \times X \to X$ with $+(x, y) = x + y$ for all $(x, y) \in X \times X$ is continuous. This means by Example 2.7.3 (c) that if (x_δ) and (y_δ) are nets in X with $x_\delta \to x$ and $y_\delta \to y$, then it follows that $x_\delta + y_\delta \to x + y$. Continuity of multiplication by scalars means that the map $\cdot : \mathbb{C} \times X \to X$ (or $\cdot : \mathbb{R} \times X \to X$) $\cdot(\lambda, x) = \lambda x$ for all $(\lambda, x) \in \mathbb{C} \times X$ (or $(\lambda, x) \in \mathbb{R} \times X$) is continuous. This means that if (x_δ) is a net in X and (λ_δ) is a net in $\mathbb{C}$ (or $\mathbb{R}$) with $x_\delta \to a$ and $\lambda_\delta \to \lambda$ then it follows that $\lambda_\delta x_\delta \to \lambda x$.*

Example **2.8.3** *(a) If* Φ *is a class of linear topologies on a linear space* X *then* $\bigvee \Phi$ *is a linear topology.*
(b) The weak topology by a class Φ *of linear maps from a linear space into a class of linear topological spaces is a linear topology.*
(c) The product of linear topological spaces is a linear topological space.

Proof.

(a) Let (λ_δ) be a net of scalars with $\lambda_\delta \to \lambda$, and (x_δ) and (y_δ) be nets in X such that

$$x_\delta \to x \text{ and } y_\delta \to y \text{ in } \bigvee \Phi.$$

This implies by Theorem 2.6.1 that $x_\delta \to x$ and $y_\delta \to y$ in each topology $\mathcal{T} \in \Phi$, hence

$$\lambda_\delta x_\delta + y_\delta \to \lambda x + y \text{ in each } \mathcal{T} \in \Phi,$$

and so, by Theorem 2.6.1,

$$\lambda_\delta x_\delta + y_\delta \to \lambda x + y \text{ in } \bigvee \Phi.$$

(b) It suffices by Part (a) to show that if $f : X \to Y$ is a linear map from a linear space X into a linear topological space Y then the weak topology $w(X, f)$ by f is a linear topology. The proof is similar to that of Part (a), if we apply Lemma 2.6.5 instead of Theorem 2.6.1.

(c) This is a special case of Part (b). □

Now we list a few simple facts in linear topological spaces.

Remark 2.8.4 *Let* $(X, \mathcal{T})$ *be a linear topological space.*
(a) If $x_\delta \to x$, *then* $\lambda x_\delta \to \lambda x$ *for any scalar* λ, *in particular,* $-x_\delta \to -x$.
(b) If $\lambda_\delta \to \lambda$, *then* $\lambda_\delta x \to \lambda x$ *for any vector* x.
(c) If $x_\delta \to x$ *then* $x_\delta + y \to x + y$ *for every vector* y.
(d) We have $x_\delta \to x$ *if and only if* $x_\delta - x \to 0$.
(e) Translations are homeomorphisms onto, that is, for each fixed $x_0 \in X$, *the function* $f : X \to X$ *with* $f(x) = x_0 + x$ $(x \in X)$ *is a homeomorphism of* X *onto* X.
(f) Multiplication with a scalar $\lambda \neq 0$ *is a homeomorphism onto, that is, for each fixed scalar* $\lambda \neq 0$, *the function* $g : X \to X$ *with* $g(x) = \lambda x$ $(x \in X)$ *is a homeomorphism from* X *onto* X.
(g) Every translation of an open set is an open set.
(h) If G *is a neighborhood of* $x_0 \in X$ *and* $x \in X$, *then* $x + G$ *is a neighborhood of* $x + x_0$. *In particular,* $x + N$ *is a neighborhood of* x *if and only if* N *is a neighborhood of* 0.
(i) If a set S *in* X *has an interior point, then* $S - S$ *is a neighborhood of* 0.

Proof.

(a) Let $\lambda_\delta = \lambda$ for all δ and $x_\delta \to x$. Then it follows that $\lambda_\delta x = \lambda_\delta x_\delta \to \lambda x$. In particular, $\lambda = -1$ implies $-x_\delta \to -x$.

(b) Let $x_\delta = x$ for all δ and $\lambda_\delta \to \lambda$. Then it follows that $\lambda_\delta \lambda x = \lambda_\delta x_\delta \to \lambda x$.

(c) Let $x_\delta \to x$. We put $y_\delta = y$ for all δ. Then it follows that

$$x_\delta + y = x_\delta + y_\delta \to x + y.$$

(d) Applying Parts (c) and (a), we obtain that $x_\delta \to x$ implies $x_\delta - x \to x - x = 0$. Similarly, $x_\delta - x \to 0$ implies $x_\delta = (x_\delta - x) + x \to 0 + x = x$.

(e) Any translation is continuous by (c) and the inverse of a translation is also a translation.

(f) Multiplication by a scalar is continuous by Part (a). In the case of a non–zero scalar, the inverse of the multiplication is also a multiplication.

(g) This follows with Part (e).

(h) This follows with Part (g).

(i) If x is an interior point of S then S is a neighborhood of x, hence $S - x$ is a neighborhood of 0 by Part (h). Thus there exists an open set G with

$$0 \in G \subset S - x \subset S - S,$$

and so $S - S$ is a neighborhood of 0.   □

Now we establish a convenient characterization of a linear topology in terms of open sets.

Theorem 2.8.5 *A linear space X (over the field $\mathbb{F}$) with a topology is a linear topological space if and only if the following conditions are satisfied:*
(a) Every translation of an open set is open.
(b) Every neighborhood of 0 is absorbing.
(c) Every neighborhood of 0 includes a balanced neighborhood of 0.
(d) For every neighborhood G of 0, there exists a neighborhood N of 0 such that $N + N \subset G$.

Proof. We frequently apply Theorem 1.10.5 (b) which states that a function f between topological spaces is continuous at a point x, if and only if for every net (x_δ) with $x_\delta \to x$ it follows that $f(x_\delta) \to f(x)$.

(i) First we show the necessity of the conditions (a)–(d). We assume that X is a linear topological space.

   (a) This is Remark 2.8.4 (g).

   (b) Let G be a neighborhood of 0 and $x \in X$ be given. We define the function $f_x : \mathbb{F} \to X$ by $f_x(\lambda) = \lambda x$ for all $\lambda \in \mathbb{F}$. Then f_x is continuous, hence there exists $\varepsilon > 0$ such that $|\lambda| \leq \varepsilon$ implies $\lambda \in f_x^{-1}(G)$, that is, $\lambda a = f_a(\lambda) \in G$.

   (c) Let G be a neighborhood of 0. We define the map $f : \mathbb{F} \times X \to X$ by $f(\lambda, x) = \lambda x$ for all $\lambda \in \mathbb{F}$ and all $x \in X$. Then f is continuous and so, for every neighborhood G of 0 in X, there exists a neighborhood M of $(0,0) \in \mathbb{F} \times X$ such that $f(M) \subset G$. Then there exist neighborhoods N_1 of the scalar 0 and N_2 of the vector 0 such that $P_1^{-1}(N_1) \cap P_2^{-1}(N_2) \subset M$. We note that there exists $\varepsilon > 0$ such that $B_\varepsilon(0) \in N_1$ (in $\mathbb{F}$). Let $|\lambda| \leq \varepsilon$ be given and $x \in N_2$. Then we have $\lambda x = f(\lambda, x) \in f(M) \subset G$. We put $N = \varepsilon N_2$. Then it follows that $\lambda N \subset G$ for all $|\lambda| \leq 1$, since $|\lambda| \leq \varepsilon$ implies $\lambda N_2 \subset G$.
   We put $N_0 = \bigcup_{|\lambda| \leq 1} \lambda N$. Then N_0 is obviously balanced, included in G and a neighborhood of 0, since it includes N.

(d) The map $f : X \times X \to X$ with $f(x,y) = x+y$ for all $x, y \in X$ is continuous. So, given a neighborhood G of 0 in X, there exists a neighborhood M of 0 in $X \times X$ such that $f(M) \subset G$. Furthermore there exist neighborhoods N_1 and N_2 of 0 in X such that $P_1^{-1}(N_1) \cap P_2^{-1}(N_2) \subset M$. We put $N = N_1 \cap N_2$. Then we obtain $(x,y) \in M$ for $x, y \in N$, since

$$P_1(x,y) = x \in N_1 \text{ and } P_2(x,y) = y \in N_2$$

and so $(x,y) \in P_1^{-1}(N_1) \cap P_2^{-1}(N_2)$, hence $x+y = f(x,y) \in f(M) \subset G$.

This concludes Part (i) of the proof.

(ii) Now we prove the sufficiency of the conditions in (a)–(d).
We assume that (a)–(d) hold.

(α) First we show that Remark 2.8.4 (d) holds.
Let $x_\delta \to x$ and G be a neighborhood of 0. Then $x + G$ is a neighborhood of x by (a), so $x_\delta \in x + G$ eventually, hence $x_\delta - x \in G$ eventually, and consequently $x_\delta - x \to 0$.
Conversely, let $y_\delta = x_\delta - x \to 0$ and G be a neighborhood of x. Then $G - x$ is a neighborhood of 0 by (a), so $y_\delta \in G - x$ eventually, hence $x_\delta = y_\delta + x \in G$ eventually, and consequently $x_\delta \to x$.
Thus we have established Remark 2.8.4 (d) holds.

(β) Now we prove the continuity of addition.
It is sufficient to show that $x_\delta \to 0$ and $y_\delta \to 0$ imply $x_\delta + y_\delta \to 0$. This is true, since, if $x_\delta \to x$ and $y_\delta \to y$, then $x_\delta - x \to 0$ and $y_\delta - y \to 0$ by Remark 2.8.4 (d) (Part (α)), and if we prove $x_\delta - x + y_\delta - y \to 0$, then we obtain $x_\delta + y_\delta \to x+y$ by Remark 2.8.4 (d).
So let $x_\delta \to 0$ and $y_\delta \to 0$, and G be a neighborhood of 0. Then, by (d), there exists a neighborhood N of 0 such that $N + N \subset G$. Since $x_\delta, y_\delta \to 0$, it follows that $x_\delta, y_\delta \in N$ eventually, hence $x_\delta + y_\delta \in N + N \subset G$ eventually, that is, $x_\delta + y_\delta \to 0$.
Thus we have shown that addition is continuous.

(γ) Now we show that:

> *(I) For every neighborhood G of 0 and for every $k \in \mathbb{N}$, there exists a neighborhood N of 0 such that $2^k N \subset G$.*

Let G be a neighborhood of 0. We choose a neighborhood N of 0 such that $N + N \subset G$. (This is possible by (d).) It follows that $2N \subset N + N \subset G$. Now we choose a neighborhood N_2 of 0 such that $N_2 + N_2 \subset N$, and obtain $4N_2 \subset 2(N_2 + N_2) \subset 2N \subset G$. The statement in (I) follows by mathematical induction.

(δ) Now we show that:

> *(II) If G is a neighborhood of $x \in X$ and $\lambda \neq 0$, then λG is a neighborhood of λx.*

Let G be a neighborhood of 0 and $\lambda \neq 0$ be a scalar.
Case 1. $x = 0$ and $|\lambda| \geq 1$
By (c), G includes a balanced neighborhood N of 0, and so $(1/\lambda)N \subset N \subset G$, hence $N \subset \lambda G$, and consequently λG is a neighborhood of 0.
Case 2. $x = 0$ and $|\lambda| < 1$
We choose $k \in \mathbb{N}$ such that $2^k > |1/\lambda|$. By (I) there exists a neighborhood N

of 0 such that $2^k N \subset G$. Then $2^{-k}G$ is a neighborhood of 0 and it follows by **Case 1** that $\lambda G = 2^k \lambda (2^{-k}G)$ is a neighborhood of 0, since $|2^k \lambda| > 1$, and by **Case 3** that $x \neq 0$.

Then $G - x$ is a neighborhood of 0 by (a), and so $\lambda(G - x) = \lambda G - \lambda x$ is a neighborhood of 0, that is, λG is a neighborhood of λx.

Thus we have shown (II).

(ε) Now we show that

$$\lambda_\delta \to 0 \text{ implies } \lambda_\delta x \to 0 \text{ for each } x \in X. \tag{2.7}$$

Let G be a neighborhood of 0 and $x \in X$. Since G is absorbing by (b), there exists $\varepsilon > 0$ such that $\lambda x \in G$ for $|\lambda| \leq \varepsilon$. But $\lambda_\delta \to 0$ implies $|\lambda_\delta| < \varepsilon$ eventually, hence $\lambda_\delta x \in G$ eventually, that is, $\lambda_\delta x \to 0$.

Thus we have shown (2.7).

(ϕ) Finally we show that multiplication by scalars is continuous, that is, $x_\delta \to x$ and $\lambda_\delta \to \lambda$ imply $\lambda_\delta x_\delta \to \lambda x$.

Case 1. $x = 0$.

We have to show that $\lambda_\delta x_\delta \to 0$. Let G be a neighborhood of 0. We may assume by (c) that G is balanced. We put $N = (|\lambda| + 1)^{-1}G$. Then N is a neighborhood of 0 by (II). Since $x_\delta \to 0$, we have $x_\delta \in N$ eventually. Furthermore, $\lambda_\delta \to \lambda$ implies $|\lambda_\delta| < |\lambda| + 1$ eventually, and so

$$\lambda_\delta x_\delta \in \lambda_\delta N = \frac{\lambda_\delta}{|\lambda| + 1}G \subset G \text{ eventually,}$$

since G is balanced and $|\lambda_\delta|/(|\lambda| + 1) < 1$ eventually.

Case 2. $x \neq 0$

Then we have $\lambda_\delta x_\delta - \lambda x = \lambda_\delta(x_\delta - x) + (\lambda_\delta - \lambda)x$. It follows from **Case 1** and Remark 2.8.4 (d) (Part (α) of the proof) that $\lambda_\delta(x_\delta - x) \to 0$. We also have $(\lambda_\delta - \lambda)x \to 0$ by (2.7). Finally, since addition is continuous by Part (β) of the proof, we obtain $\lambda_\delta x_\delta - \lambda x \to 0$, and this implies $\lambda_\delta x_\delta \to \lambda x$.

Thus we have shown the continuity of multiplication by scalars.

This concludes the proof of the sufficiency of the conditions in (a)–(d)

This concludes the proof of the theorem. $\qquad\square$

Remark 2.8.6 *We note that $x + G$ is a neighborhood of x if and only if G is a neighborhood of 0 (Remark (2.8.4 (h)). Equivalently N is a neighborhood of x if and only if $N - x$ is a neighborhood of 0.*
A set is open if and only if it is a neighborhood of each of its points.

In Theorems 2.8.7 and 2.8.8, only the neighborhoods of 0 will be given in advance. Then the open sets will be defined by Remark 2.8.6.

Theorem 2.8.7 *Let X be a linear space and Γ be a nonempty collection of balanced absorbing sets such that*
(i) given $G_1, G_2 \in \Gamma$, there exists $G_3 \in \Gamma$ such that $G_3 \subset G_1 \cap G_2$;
(ii) given $G_1 \in \Gamma$, there exists $G_2 \in \Gamma$ such that $G_2 + G_2 \subset G_1$.
Let any set which includes a member of Γ be called a neighborhood of 0 and let the terms neighborhood and open set be defined as in Remark 2.8.6.
Then X becomes a linear topological space whose neighborhoods are precisely those described above.

Proof.

(a) We prove that the collection $\mathcal{T}$ of open sets defined in the theorem is a topology. Obviously we have $\emptyset \in \mathcal{T}$. Also $X \in \mathcal{T}$, since X is a neighborhood of 0, and so $x + X$ is a neighborhood of x for each $x \in X$, but $x + X = X$. Consequently X is a neighborhood of each of its points, and hence open.

Now let $\mathcal{O}$ be a subclass of $\mathcal{T}$. We put

$$G_0 = \bigcup \{G : G \in \mathcal{O}\},$$

and let $x \in G_0$. Then $x \in G$ for some $G \in \mathcal{O}$. Hence we have $x \in G \subset G_0$, and so G_0 is a neighborhood of x. Consequently G_0 is open.

Finally, let $G_1, G_2, \ldots, G_n \in \mathcal{T}$. We put

$$G = \bigcap_{k=1}^{n} G_k,$$

and let $x \in G$. For each $k = 1, 2, \ldots, n$, let $G_k = x + N_k$, where N_k is a neighborhood of 0. By assumption (i), there exists $N \in \Gamma$ with $N \subset \bigcap_{k=1}^{n} N_k$, and it follows that $x + N \subset G$, hence G is a neighborhood of x, and consequently G is open.

Thus we have shown that $\mathcal{T}$ is a topology. This completes Part (a) of the proof.

(b) We refer to the neighborhoods in the sense of the topology $\mathcal{T}$ of Part (a) as $\mathcal{T}$–neighborhoods and to the neighborhoods in the sense of Remark 2.8.6 as $\mathcal{R}$–neighborhoods, and have to show that $\mathcal{T}$– and $\mathcal{R}$–neighborhoods are the same.

(α) First we show that every $\mathcal{T}$–neighborhood is an $\mathcal{R}$–neighborhood.

Let G be a $\mathcal{T}$–neighborhood of x. Then, by definition of a $\mathcal{T}$–neighborhood, there exists a $\mathcal{T}$–open set G_0 such that $x \subset G_0 \subset G$. By the definition of open (in the sense of Remark 2.8.6), G is an $\mathcal{R}$–neighborhood of x.

(β) Conversely, we show that every $\mathcal{R}$–neighborhood is a $\mathcal{T}$–neighborhood of x.

Let G be an $\mathcal{R}$–neighborhood of x. We have to find a $\mathcal{T}$–open set G_0 such that $x \in G_0 \subset G$. For this purpose, we put

$$G_0 = \{z \in X : G \text{ is an } \mathcal{R}\text{–neighborhood of } z\}.$$

Since $x \in G_0$, we have $G_0 \neq \emptyset$. Let $y \in G_0$. We choose $N \in \Gamma$ such that $N + N \subset G - y$. This is possible by Theorem 2.8.7 (ii), since G is an $\mathcal{R}$–neighborhood of x, and so $G - y$ is an $\mathcal{R}$–neighborhood of 0. Then we have

$$y + N + N \subset G, \tag{*}$$

and so

$$y + N \subset G_0. \tag{**}$$

To prove (**), let $u \in y + N$. Then it follows by (*) that $u + N \subset G$, and so G is an $\mathcal{R}$–neighborhood of u, that is, $u \in G_0$ by the definition of G_0. Thus we have shown (**).

By (**), G_0 is an $\mathcal{R}$–neighborhood of y. Since G_0 is an $\mathcal{R}$–neighborhood of all of its points, G is a $\mathcal{T}$–neighborhood of x.

Thus we have shown Part (β).

This completes the Part (b) of the proof.

(c) It remains to show that our topology is linear. This can be achieved by checking the conditions in (a)–(d) of Theorem 2.8.5.   □

The following result based on Theorems 2.8.5 and 2.8.7 is useful for defining examples of linear topologies.

Theorem 2.8.8 *Let X be a linear space and Γ be a nonempty collection of balanced absorbing sets such that the condition in (ii) of Theorem 2.8.7 holds. Furthermore, let every subset of X which includes a finite intersection of members of Γ be called a neighborhood of 0, and let neighborhood and open set be defined as in Remark 2.8.6. Then X becomes a linear topological space whose neighborhoods are precisely those just described.*

Proof. Let Γ' be the set of all finite intersections of members of Γ. Then Γ' satisfies the conditions in (i) and (ii) of Theorem 2.8.7.   □

Definition 2.8.9 The topology obtained from the collection Γ of neighborhoods of 0 as in Theorem 2.8.8 is called the *linear topology generated by* Γ and Γ is called the *generating subbase of neighborhoods of* 0.

Now we are going to define the concept of *weak linear topology by a family of maps* of a certain kind, similar to that of the weak topology $w(A, \Phi)$ in Definition 2.6.6.

Definition 2.8.10 Let X be a linear space and Φ be a family of seminorms on X, or of linear maps from X to a collection of linear topological spaces. Then the *weak topology by* Φ, denoted by $\sigma(X, \Phi)$, or $\sigma(\Phi)$, for short, is the linear topology generated by

$$\{f^{-1}(G) : f \in \Phi, \ G \text{ a balanced, absorbing neighborhood of } 0 \text{ in the range space of } f\}.$$

Example 2.8.11 *(a) If Φ is a family of seminorms, then $\sigma(\Phi)$ is the linear topology generated by*

$$\left\{ \{p < \varepsilon\} : p \in \Phi, \ \varepsilon > 0 \right\}, \ \text{where } \{p < \varepsilon\} = p^{-1}([0, \varepsilon))) = \{x : p(x) < \varepsilon\}.$$

In particular, $\sigma(X, p) = \sigma(X, \{p\})$ is a seminormed space.
(b) The weak topology by a family $\mathcal{F}$ of linear functionals can be written as the weak linear topology by a family of seminorms. If $\Phi = \{|f| : f \in \mathcal{F}\}$, then we have

$$w(X, \mathcal{F}) = \sigma(X, \mathcal{F}) = \sigma(X, \Phi).$$

We list a few properties of linear topological spaces, the proofs of which are left to the reader.

Remark 2.8.12 *(a) In a linear topological space generated by Γ as in Definition 2.8.9, net convergence is described as $x_\delta \to 0$ if and only if $x_\delta \in G$ eventually for every $G \in \Gamma$, together with Remark 2.8.4 (d).*
(b) Net convergence in $\sigma(X, \Phi)$ (Definition 2.8.10) is described by $x_\delta \to x$ if and only if $f(x_\delta) \to f(x)$ for all $f \in \Phi$.
(c) The weak topology $\sigma(X, \Phi)$ is the weakest linear topology such that all members of Φ are continuous.
(d) If X is a linear space and Φ is a collection of linear maps from X to linear topological spaces, then $\sigma(X, \Phi) = w(X, \Phi)$.

We need more properties of linear topological spaces.

Lemma 2.8.13 *Let X be a linear topological space. Then there exists a directed set D with the following property.*
(i) Given a set S and a point $x \in \overline{S}$, there exists a net $(x_\delta)_{\delta \in D}$ in S with $x_\delta \to x$.

Proof. We know from Theorem 1.10.5 (a) that for every $x \in \overline{S}$, there exists a net $(x_\delta)_{\delta \in D}$ in S with $x_\delta \to x$ where D may depend on S and x. Here D depends only on X, but not on S and x. We choose D as the class of all neighborhoods of 0, directed by inclusion. For every $\delta \in D$, we choose $x_\delta \in (x + \delta) \cap S$ and it follows as in Example 1.10.4 (e) that $x_\delta \to x$. □

Remark 2.8.14 *We have*

$$\overline{S_1} + \overline{S_2} \subset \overline{S_1 + S_2} \text{ for all subsets } S_1 \text{ and } S_2 \text{ of a linear topological space.} \tag{2.8}$$

Equality in (2.8) need not even hold in $\mathbb{R}$. If, however, one of the sets is compact, then equality holds in (2.8).

Proof.

(i) First we show the set inclusion in (2.8).
Let $x \in \overline{S_1} + \overline{S_2}$. Then there are $s \in \overline{S_1}$ and $t \in \overline{S_2}$ such that $x = s + t$. By Lemma 2.8.13, there exist nets (s_δ) in S_1 and (t_δ) in S_2 such that $s_\delta \to s$ and $t_\delta \to t$. This implies $s_\delta + t_\delta \to s + t$, and so we have $s + t \in \overline{S_1 + S_2}$ by Theorem 1.10.5 (a).

(ii) Now we give an example that equality in (2.8) does not hold in $\mathbb{R}$.
We consider the sets $S_1 = \mathbb{N}$ and $S_2 = \{1/n - n : n \in \mathbb{N}\}$ in $\mathbb{R}$. Then $1/n \in S_1 + S_2$ for all $n \in \mathbb{N}$, but $0 \notin S_1 + S_2$, hence $S_1 + S_2$ is not closed, however, S_1 and S_2 are closed.

(iii) Finally we show that equality holds in (2.8) if one of the sets is compact.
Let S_2 be compact and $x \in \overline{S_1 + S_2}$. Then there exists a net in $S_1 + S_2$ by Lemma 2.8.13, $(s_\delta + t_\delta)$, say, with $s_\delta \in S_1$, $t_\delta \in S_2$ and $s_\delta + t_\delta \to x$. Since S_2 is compact there exists a subnet $(t_{u(\delta')})$ by Theorem 1.12.4 (iii) such that $t_{u(\delta')} \to t \in S_2$. It also follows that $s_{u(\delta')} + t_{u(\delta')} \to x$, and so

$$s_{u(\delta')} = \left(s_{u(\delta')} + t_{u(\delta')}\right) - t_{u(\delta')} \to x - t.$$

Therefore we have $x - t \in \overline{S_1}$ and $x \in \overline{S_1} + S_2 \subset \overline{S_1} + \overline{S_2}$. Thus we have shown that $\overline{S_1 + S_2} \subset \overline{S_1} + \overline{S_2}$. □

Theorem 2.8.15 *The closure of a subspace of a linear topological space is a subspace.*

Proof.

(i) We show that if S is a subset of a linear topological space, and λ and μ are scalars, then we have
$$\lambda \overline{S} + \mu \overline{S} \subset \overline{\lambda S} + \overline{\mu S}. \tag{2.9}$$

Let $x \in \lambda \overline{S}$. Then there exists $s \in \overline{S}$ with $x = \lambda s$. Also, by Lemma 2.8.13, there exists a directed set D and a net $(s_\delta)_{\delta \in D}$ in S with $s_\delta \to s$. Let $(x_\delta)_{\delta \in D}$ be the net with $x_\delta = \lambda s_\delta$ for all $\delta \in D$. This is a net in λS with $x_\delta \to \lambda s = x$, that is, $x \in \overline{\lambda S}$ by Theorem 1.10.5 (a). Thus we have $\lambda \overline{S} \subset \overline{\lambda S}$. Similarly we obtain $\mu \overline{S} \subset \overline{\mu S}$ and then (2.9) follows.

Furthermore it follows from the inclusion in (2.8) of Remark 2.8.14 that

$$\overline{\lambda S + \mu S} \subset \overline{\lambda S + \mu S},$$

hence by (2.9)

$$\lambda \overline{S} + \mu \overline{S} \subset \overline{\lambda S + \mu S}. \tag{2.10}$$

Finally, if S is a subspace, then we have $\lambda S + \mu S \subset S$ for all scalars λ and μ, and hence $\lambda \overline{S} + \mu \overline{S} \subset \overline{S}$ by (2.10). □

The proof of the following result is similar to that of Theorem 2.8.15.

Theorem 2.8.16 *The closure of and interior (with 0 adjoined) of a balanced set are balanced.*

Definition 2.8.17 (a) A set S in a linear topological space is said to be *bounded*, if it is absorbed by any neighborhood of 0, that is, if, for every neighborhood N of 0 there exists $\varepsilon > 0$ such that $\delta S \subset N$ for all δ with $0 \leq \delta \leq \varepsilon$.
(b) A set S in a metric space (X, d) is said to be *d–bounded*, or *metrically bounded* if there exists an $M \in \mathbb{R}$ such that $d(x, y) \leq M$ for all $x, y \in S$.

Example **2.8.18** *A set in a seminormed space is bounded if and only if it is contained in an open disc. Hence the concepts of boundedness and d–boundedness are the same in seminormed spaces.*

Remark 2.8.19 *(a) Every subset of a bounded set S in a linear topological space is bounded.*
(b) If a set is bounded in a stronger topology then it is bounded.

Proof.

(a) If a neighborhood of 0 absorbs a set S, then obviously it also absorbs every subset of S.

(b) If S is absorbed by every $\mathcal{T}'$–neighborhood of 0 and if $\mathcal{T}' \supset \mathcal{T}$, then obviously S is also absorbed by every $\mathcal{T}$–neighborhood of 0. □

Theorem 2.8.20 *Let $(X, \mathcal{T})$ and $(Y, \mathcal{T}')$ be linear topological spaces. Then we have:*
(a) Every continuous linear map preserves bounded sets, that is, if $f : X \to Y$ is continuous, then $f(S)$ is a bounded set in Y for every bounded set S in X.
(b) If a linear map $f : X \to Y$ is continuous at a point, then it is continuous (everywhere).
(c) If a linear map $f : X \to Y$ is bounded on a neighborhood N of 0, that is, if $f(N)$ is bounded, then f is continuous. The converse does not hold, in general; it does, however, hold if B contains a bounded neighborhood of 0.

Proof.

(a) Let $f : X \to Y$ be a continuous linear map and $S \subset X$ be bounded. If N is a neighborhood of 0 in Y then $f^{-1}(N)$ is a neighborhood of 0 in X by Definition 1.9.6 (b), since f is continuous, and since S is bounded, there exists $\varepsilon > 0$ such that

$$\delta S \subset f^{-1}(N) \text{ for all } \delta \text{ with } 0 \leq \delta \leq \varepsilon.$$

This implies $\delta f(S) = f(\delta S) \subset N$ or all δ with $0 \leq \delta \leq \varepsilon$, since f is linear, hence $f(S)$ is bounded.

(b) Let f be continuous at $x_0 \in X$, $x \in X$ be given and $(x_\delta)_{\delta \in D}$ be a net in X with $x_\delta \to x$. Then it follows that $x_0 + x_\delta - x \to x_0$, hence $f(x_\delta) - f(x) = f(x_0 + x_\delta - x) - f(x_0) \to 0$, hence f is continuous at x (here we apply Theorem 1.10.5 (b)).

(c) (i) First we show that if f is linear and bounded on a neighborhood of 0, then it is continuous.

Let $f : X \to Y$ be linear, G be a neighborhood of 0 in X, and $f(G)$ be bounded in Y. If N is a neighborhood of 0 in Y, then we have

$$\lambda N \supset f(G) \text{ for some } \lambda > 0, \text{ hence } f^{-1}(N) \supset \frac{1}{\lambda} \cdot G.$$

Thus $f^{-1}(N)$ is a neighborhood of 0 in X which implies by Part (b) that f is continuous everywhere.

(ii) Now we show that the converse implication of Part (i) does not hold, in general. We consider the map id: $\omega \to \omega$. This map is continuous, but does not map any open set onto a bounded set, since ω has no bounded neighborhoods (Example 2.8.21 below).

(iii) Finally, we assume that Y has a bounded neighborhood N of 0 and $f : X \to Y$ is continuous. Then f is bounded on $f^{-1}(N)$ which is a neighborhood of 0 in X, since f is continuous. $\square$

Example 2.8.21 *The set ω of all complex sequences does not contain any bounded sets.*

Proof. We assume that ω contains a bounded neighborhood N of some point $x \in \omega$. Then all projections $P_n(N)$ are bounded by Theorem 2.8.20 (a), since the projections P_n are continuous by Theorem 2.7.2 and Theorem 1.10.5 (b). Since N is a neighborhood of x in ω, there exists an open set G such that $x \in G \subset N$. By Theorem 2.7.4, there exists at least one $n \in \mathbb{N}$ such that $P_n(G) = \mathbb{C}$, but $\mathbb{C}$ is not bounded. $\square$

The next result is an immediate consequence of Theorem 2.8.20 (a) and (b).

Corollary 2.8.22 *A linear functional on a linear topological space is continuous if and only if it is bounded on a neighborhood of 0.*

In view of Definitions 2.5.6 and 2.6.15, we obtain that a maximal subspace is a subspace of codimension 1.

Theorem 2.8.23 *A maximal subspace of a linear topological space is either closed or dense.*

Proof. Let S be a maximal subspace of a linear topological space X. Then $\overline{S}$ is a subspace by Theorem 2.8.15, hence we have $\overline{S} = S$, or $\overline{S} = X$, since S is a maximal subspace. $\square$

Theorem 2.8.24 *A linear functional f on a linear topological space is continuous if and only if $N(f)$ is closed.*

Proof.

(i) First we show that if f is continuous then $N(f)$ is closed.

Let f be continuous, then $N(f) = f^{-1}(\{0\})$ is closed, since $\{0\}$ is closed.

(ii)   Now we show that if $N(f)$ is closed then f is continuous.
We assume that f is not continuous. We show that $N(f)$ is dense. Then $N(f)$ is not closed by Theorem 2.8.23 and Remark 2.6.16.
Let x be an arbitrary vector and N be a balanced neighborhood of 0. Since f is not bounded on N by 2.8.20 (c) and $f(N)$ is a balanced subset of $\mathbb{C}$ by Theorem 2.2.7 (a), it follows that $f(N) = \mathbb{C}$. In particular, there is $y \in N$ such that $f(y) = -f(x)$. Then it follows that $x + y \in N(f) \cap (x + N)$. Thus $N(f)$ meets every open set by Theorem 2.8.5 (c), that is, $N(f)$ is dense. $\square$

2.9   Closed maps

The *closed graph theorem*, Theorem 3.6.1 of Section 3.6, is one of the four most important theorems in functional analysis; it concerns *closed maps*. It is useful to define the concept of a closed map in this section, and study some related properties. Linearity is not relevant for a start.

Definition 2.9.1 (a) Let X and Y be sets and $f : X \to Y$ be a map. The set

$$graph(f) = \{(x, f(x)) : x \in X\} \subset X \times Y$$

is called the *graph of* f.
(b) Let X and Y be topological spaces and $f : X \to Y$ be a map. Then f is said to be *closed*, if graph(f) is a closed subset of $X \times Y$.

Remark 2.9.2 *(a) Sometimes a map which preserves closed sets is said to be closed. Our definition comes from the fact that a function and its graph formally are the same.*
(b) We have $(x, y) \in$ graph(f) if and only if $y = f(x)$.

The first theorem states, among other things, that every continuous map from a topological space in a Hausdorff space is closed.

Theorem 2.9.3 *If X and Y are topological spaces then $f : X \to Y$ is closed if and only if for every net $(x_\delta)_{\delta \in D}$ in X with $x_\delta \to x$ and $f(x_\delta) \to y$ it follows that $y = f(x)$. In particular, if Y is a Hausdorff space and f is continuous then f is closed.*

   Proof.

(i)   We assume that the map $f : X \to Y$ is closed. Let $(x_\delta)_{\delta \in D}$ be a net in X such that $x_\delta \to x$ and $f(x_\delta) \to y$. It follows by Theorem 2.7.2 that

$$(x_\delta, f(x_\delta)) \to (x, y) \text{ in } X \times Y.$$

Since graph(f) is a closed set in $X \times Y$, it follows that $(x, y) \in$ graph(f) by Theorem 1.10.5 (a), and so $y = f(x)$ by Remark 2.9.2 (b).

(ii)   Conversely, we assume that for any net $(x_\delta)_{\delta \in D}$ in X with $x_\delta \to x$ and $f(x_\delta) \to y$ it follows that $y = f(x)$.
Let (u_δ) be a net in $G = $ graph(f) such that $u_\delta \to v$. Then we have $u_\delta = (x_\delta, y_\delta)$ and $v = (x, y)$ with $y_\delta = f(x_\delta)$, $x_\delta \to x$ and $y_\delta \to y$. Since $y = f(x)$ by assumption, it follows that $v = (x, y) \in$ graph$(f) = G$, and so G is closed by Theorem 1.10.5 (a).

(iii)   Finally, we assume that Y is a Hausdorff space and $f : X \to Y$ is continuous. If $(x_\delta)_{\delta \in D}$ is a net in X with $x_\delta \to x$, then it follows that $f(x_\delta) \to f(x)$ by Theorem 1.10.5 (b). Since we also have $f(x_\delta) \to y$ and Y is a Hausdorff space, it follows that $y = f(x)$ by Theorem 1.10.5 (f), hence f is closed by Part (ii) of the proof. □

Example 2.9.4 *(a) If X and Y are semimetrizable then Theorem 2.9.3 holds with nets replaced by sequences, since $X \times Y$ is semimetrizable by Theorem 2.7.6.*
(b) We define $f : \mathbb{R} \to \mathbb{R}$ by

$$f(x) = \begin{cases} \dfrac{1}{x} & (x \neq 0) \\ 0 & (x = 0) \end{cases}.$$

Then f is closed, since $\operatorname{graph}(f) = \mathbb{R} \times \mathbb{R}$ is closed, but f is not continuous.

Remark 2.9.5 *Let X and Y be topological spaces and $f : X \to Y$ be a closed map.*
(a) If the inverse map $f^{-1} : Y \to A$ exists then it is closed.
(b) If $K \subset X$ is compact, then $f(K)$ is closed, but not compact, in general. Thus any closed map on a compact space preserves closed sets.
(c) If $K \subset Y$ is compact, then $f^{-1}(K)$ is closed.
(d) The little closed graph theorem
If Y is compact, then f is continuous.
(e) If X is compact and the inverse map f^{-1} exists, then f^{-1} is continuous.
(f) The little open mapping theorem
If X is compact and f is one to one and onto, then f is an open mapping. Neither one to one nor onto can be omitted.

   Proof.

(a)   The map $g : X \times Y \to Y \times X$ with $g(x, y) = (y, x)$ for all $(x, y) \in X \times Y$ obviously is a homeomorphism from $X \times Y$ onto $Y \times X$. We also have $g(\operatorname{graph}(f)) = \operatorname{graph}(f^{-1})$, since $(x, y) \in \operatorname{graph}(f^{-1})$, if and only if $y = f^{-1}(x)$, if and only if $x = f(y)$, if and only if $(y, x) \in \operatorname{graph}(f)$, if and only if $(x, y) = g(x, y) \in g(\operatorname{graph}(f))$. If one of the sets $\operatorname{graph}(f)$ or $\operatorname{graph}(f^{-1})$ is closed, so is the other.

(b)   (i)   First we show that if $K \subset X$ is compact then $f(K)$ is closed. Let (y_δ) be a net in $f(K)$ with $y_\delta \to y$. Then, for every y_δ, there exists $x_\delta \in K$ such that $f(x_\delta) = y_\delta$. Since K is compact, by Theorem 1.12.4, there exists a subnet (x_α) of the net (x_δ) with $x_\alpha \to x \in K$. It follows that $f(x_\alpha) \to y$ by Theorem 1.11.6, since $(f(x_\alpha))$ is a subnet of $f(x_\delta)$ and $f(x_\delta) \to y$. Now we have $y = f(x)$ by Theorem 2.9.3, hence $y \in f(K)$, and so $f(K)$ is closed by Theorem 1.10.5 (a).

   (ii)   We consider the map $f : \mathbb{R} \to \mathbb{R}$ of Example 2.9.4. Then $K = [0, 1]$ is a compact set, but $f(K) = \{0\} \cup [1, \infty)$ is not compact.

   (iii)   Let K be a compact space and F be a closed subset of K. Then F is compact by Theorem 1.12.2 (b), hence $f(F)$ is closed by Part (i).

(c)   Let $(x_\delta)_{\delta \in D}$ be a net in $f^{-1}(K)$ with $x_\delta \to x$. Since K is compact there is a subnet $(f(x_{u(\delta')}))$ of the net $(f(x_\delta))$ with $f(x_{u(\delta')}) \to y \in K$ by Theorem 1.12.4, and also $x_{u(\delta')} \to x$ by Theorem 1.11.6. It follows by Theorem 2.9.3 that $y = f(x)$, hence $x \in f^{-1}(K)$, and so $f^{-1}(K)$ is closed by Theorem 1.10.5 (a).

(d)   If $K \subset Y$ is closed, then K is compact by Theorem 1.12.2 (b), and so $f^{-1}(K)$ is closed by Part(c). Thus f is continuous.

(e) By Part (a), f^{-1} is closed, and hence continuous by Part (d), since X is compact by hypothesis.

(f) (i) By Part (e), f^{-1} is continuous, and hence f is open.

    (ii) We consider the map $f : [0,1] \to [0,1]$ with

$$f(x) = \begin{cases} 0 & (x \in [0,1/2)) \\ 2x - 1 & (x \in [1/2,1]). \end{cases}$$

Then f is continuous, so closed by Theorem 2.9.3, and onto, but not open, since $f((1/4,1/3)) = \{0\}$. Thus one to one cannot be omitted.

    (iii) We consider $Y = \mathbb{R}$ with the usual topology $\mathcal{T}$, $X = [0,1]$ with the topology $\mathcal{T}_X$ of $\mathbb{R}$ on X and $\iota : (X, \mathcal{T}_X) \to (Y, \mathcal{T})$, the inclusion map. Then ι is one to one and continuous, so closed by Theorem 2.9.3, but $\iota(X) = [0,1]$ and ι is not open. Thus onto cannot be omitted.

$\square$

Corollary 2.9.6 *A closed linear functional on a linear topological space must be continuous.*

Proof. Since $N(f) = f^{-1}(\{0\})$, $N(f)$ is closed by Remark 2.9.5 (c), and so f is continuous by Theorem 2.8.24. $\square$

Example 2.9.7 *If X has two Hausdorff topologies $\mathcal{T}$ and $\mathcal{T}'$ with $\mathcal{T}'$ strictly stronger than $\mathcal{T}$, then the identity map* id: $(X, \mathcal{T}) \to (X, \mathcal{T}')$ *is not continuous. However* id *is closed, since the inverse map* id^{-1} $: (X, \mathcal{T}') \to (X, \mathcal{T})$ *is continuous, for* id^{-1} *continuous implies that* id^{-1} *is closed by Theorem 2.9.3, and so* id *is closed by Remark 2.9.5 (a).*

In particular, if $X = \ell_1$, $\mathcal{T}'$ is the natural topology on ℓ_1 (Example 1.9.2 (b)) and $\mathcal{T}$ is the natural topology of c_0 on ℓ_1, then id: $(X, \mathcal{T}) \to (X, \mathcal{T}')$ *is an example of a closed map between normed spaces that is not continuous.*

Lemma 2.9.8 (The closed graph lemma) *Let X and Y be topological spaces and $f : X \to Y$ be a closed map. If both X and Y are given stronger topologies then f remains closed.*

Proof. The topology on $X \times Y$ becomes stronger, since the convergence of a net (x_δ) in $X \times Y$ is equivalent to the convergence of the nets $(P_1(x_\delta))$ in X and $(P_2(x_\delta))$ in Y. Thus graph(f) remains closed. $\square$

Remark 2.9.9 *Let X and Y be linear topological spaces and $f : X \to Y$ be a linear map with the property that every $f(N)$ is a neighborhood of 0 in Y for every neighborhood N of 0 in X. Then f is an open map.*

Proof. Let $G \neq \emptyset$ be an open set in X and $y \in f(G)$. Then there exists $x \in G$ with $f(x) = y$. Since G is open and $x \in G$, G is a neighborhood of x in X, and so $N = G - x$ is a neighborhood of 0 in X by Remark 2.8.4 (h). Now $f(N)$ is a neighborhood of 0 in Y by hypothesis, and we have

$$f(G) = f(N + x) = f(N) + f(x) = f(N) + y;$$

therefore $f(G)$ is a neighborhood of y in Y, again by Remark 2.8.4 (h). Since $y \in f(G)$ was arbitrary, $f(G)$ is open. Thus the image $f(G)$ of every open set G in X is an open set in Y. But this means that f is an open map. $\square$

2.10 Baire's category theorem

Baire's category theorem is needed in the proof of the *closed graph theorem* which is one of the major results in functional analysis. We also need some results from the theory of linear metric spaces.

Completeness plays an important role. Therefore we recall the concept of completeness of a semimetric space introduced at the end of Section 1.7.

Definition 2.10.1 A semimetric space (X, d) is said to be *complete* if every Cauchy sequence in X converges.

We list a few familiar examples.

Example **2.10.2** *(a) The metric space $\mathbb{Q}$ with the natural metric of $\mathbb{R}$ on $\mathbb{Q}$ is not complete. For instance, the sequence (x_n) with $x_n = (1 + 1/n)^n$ for all $n \in \mathbb{N}$ is a Cauchy sequence which does not converge in $\mathbb{Q}$.*
(b) The space $C[0, 1]$ is complete with

$$d(f, g) = \max\{|f(t) - g(t)| : t \in [0, 1]\} \text{ for all } f, g \in C[0, 1].$$

(c) The space $C[0, 1]$ is not complete with

$$\rho(f, g) = \int_0^1 |f(t) - g(t)| \, dt \text{ for all } f, g \in C[0, 1].$$

(d) Let $p \geq 1$. Then the space ℓ_p is complete with the metric given by its natural norm $\|\cdot\|_p$.

The following simple results are needed.

Lemma 2.10.3 *Let (X, d) be a semimetric space.*
(a) Then every Cauchy sequence is d–bounded.
(b) Then every convergent sequence is a Cauchy sequence.
(c) Then every Cauchy sequence is contained in a disc, and the disc can be chosen such that its centre is an arbitrary point.
(d) If a Cauchy sequence (x_n) has a convergent subsequence, then the sequence itself converges.

   Proof.

(a) Let (x_n) be a Cauchy sequence. Then there exists $n_0 \in \mathbb{N}$ such that $d(x_p, x_q) < 1$ for all $p, q > n_0$. We put

$$r_1 = \max\{d(x_n, x_{n_0+1}) : 1 \leq n \leq n_0\} \text{ and } r = 1 + 2r_1.$$

Then we have $d(x_m, x_n) \leq r$ for all $m, n \in \mathbb{N}$, and so the Cauchy sequence (x_n) is d–bounded.

(b) Let $x_n \to a$ and $\varepsilon > 0$ be given. Then there exists $n_0 \in \mathbb{N}$ such that $d(x_n, a) < \varepsilon/2$ for all $n > n_0$. Hence we have for all $p, q > n_0$

$$d(x_p, x_q) \leq d(x_p, a) + d(x_q, a) < \frac{\varepsilon}{2} + \frac{\varepsilon}{2} = \varepsilon,$$

and so the sequence (x_n) is a Cauchy sequence.

(c) Let $x \in X$ be given. Using the notations of Part (a) of the proof, we put $t = d(x, x_{n_0+1}) + r$. Then we have $x_n \in \overline{B}_t(x)$ for all $n \in \mathbb{N}$.

(d) Let (x_n) be a Cauchy sequence and $(x_{n(k)})$ a convergent subsequence, $x_{n(k)} \to x$ as $k \to \infty$, say. Furthermore, let $\varepsilon > 0$ be given. Since (x_n) is a Cauchy sequence, there exists $m_0 \in \mathbb{N}$ such that $d(x_q, x_p) < \varepsilon/2$ for all $p, q \geq m_0$. Also, since $x = \lim_{k \to \infty} x_{n(k)}$, there exists $k_0 \in \mathbb{N}$ such that $d(x_{n(k)}, x) < \varepsilon/2$ for all $k \geq k_0$. Since n_k tends to ∞ as k increases, we can take $k_1 > k_0$ such that $n_{k-1} \geq m_0$. Then we have for all $n \geq m_0$

$$d(x_n, x) \leq d(x_n, x_{n(k_1)}) + d(x_{n(k_1)}, x) < \frac{\varepsilon}{2} + \frac{\varepsilon}{2} = \varepsilon,$$

that is, $x_n \to x$ as $n \to \infty$. □

Theorem 2.10.4 (Cantor's intersection theorem) *Let (B_n) be a decreasing sequence of nonempty, closed and bounded subsets B_n of a complete metric space (X, d) and $\lim_{n \to \infty} \mathrm{diam}(B_n) = 0$, then the intersection B_∞ of all B_n is nonempty and consists of exactly one point.*

Proof. First, $\lim_{n \to \infty} \mathrm{diam}(B_n) = 0$ implies $\mathrm{diam}(B_\infty) = 0$, hence $B_\infty = \emptyset$ or B_∞ consists of a single point. We choose $x_n \in B_n$ for each n. Since $B_{n+1} \subset B_n$ for all n and $\lim_{n \to \infty} \mathrm{diam}(B_n) = 0$, the sequence (x_n) is a Cauchy sequence which is convergent by the completeness of X, $x = \lim_n x_n$ say. Also, since each set B_n is closed, and x is the limit of a sequence in B_n, we must have $x \in B_n$. This is true for all n, hence we get $x \in B_\infty$ □

Remark 2.10.5 *It can be shown that the converse implication of Cantor's intersection theorem also holds true, that is, if X is a metric space with the property that every decreasing sequence of closed bounded subsets with $\lim_{n \to \infty} B_n = 0$ has nonempty intersection, then X is complete.*

Definition 2.10.6 Let S be a subset of a topological space.
(a) If S is dense in an open non–empty set, then S is said to be *somewhere dense*, otherwise it is said to be *nowhere dense*.
(b) If S is the union of a countable family of nowhere dense sets, then S is said to be of *first category*; otherwise it is said to be of *second category*.

Remark 2.10.7 *(a) A set S in a semimetric space is somewhere dense if and only if $\overline{S}$ contains an open disc.*
(b) The set $\mathbb{Q}$ is dense in $\mathbb{R}$; $\mathbb{R}$ is not dense in $\mathbb{R}^2$; $\mathbb{R}$ is nowhere dense in $\mathbb{R}^2$.
(c) The set of all polynomials is dense in $C[0,1]$ with the metric of Example 2.10.2 (b).
(d) Every subset of a set of first category is of first category.
(e) Every countable union of first category sets is of first category.
(f) Let S be a subset of a topological space X. Then the following statements are equivalent:

 (i) S is nowhere dense

 (ii) $\overline{S}$ has no interior points

 (iii) The complement $\overline{S}^c = X \setminus \overline{S}$ of the closure of S is dense in X.

Proof.

(a) (i) First we show that if S is somewhere dense in a metric space, then $\overline{S}$ contains an open disc.
     Let S be somewhere dense in a metric space. Then there exists an open set $G \neq \emptyset$ such that $G \subset \overline{S}$. Since G is open, there exists an open disc $B_r(a) \subset G$.

(ii) Now we show that if $\overline{S}$ contains an open disc then S is somewhere dense. We assume that S is not somewhere dense. Then $\overline{S}$ does not contain any nonempty open set, that is, $G \not\subset \overline{S}$ for all open sets $G \neq \emptyset$, in particular, $B_r(a) \not\subset \overline{S}$ for all open discs.

(b) (i) It is known from elementary analysis that $\mathbb{Q}$ is dense in $\mathbb{R}$.

    (ii) The set $\mathbb{R} = \overline{\mathbb{R}}$ does not contain any open disc in $\mathbb{R}^2$, and hence is nowhere dense in $\mathbb{R}^2$.

(c) This is the statement of the Weierstrass approximation theorem.

(d) Let $S_0 \subset S$ and S be of first category. Then

$$S = \bigcup_{n \in \mathbb{N}} M_n,$$

where all the sets M_n are nowhere dense. Thus every set $\overline{M_n}$ does not contain an open disc, and the same holds true for the closure of the sets $M'_n = M_n \cap S_0$, and we have

$$S_0 = \bigcup_{n \in \mathbb{N}} M'_n.$$

(e) Let $V = \bigcup_{n \in \mathbb{N}} S_n$ be a countable union of first category sets S_n. Then we have

$$S_n = \bigcup_{m \in \mathbb{N}} M_m^{(n)} \text{ for each } n \in \mathbb{N},$$

where each set $M_m^{(n)}$ is nowhere dense, and

$$V = \bigcup_{n \in \mathbb{N}} \bigcup_{m \in \mathbb{N}} M_m^{(n)}$$

is a countable union of countable unions of nowhere dense sets, hence a countable union of nowhere dense sets, and so of first category.

(f) (α) We show that (ii) implies (i).
We assume that $\overline{S}$ has no interior points, but S is not nowhere dense. Then S is somewhere dense in a nonempty open set G by Definition 2.10.6 (a), and so $G \subset \overline{S}$ by Definition 1.9.6 (d). This contradicts the assumption that $\overline{S}$ has no interior points.

    (β) We show that (i) implies (iii).
We assume that S is nowhere dense, that is, $\overline{S}$ contains no nonempty open set, but $\overline{S^c}$ is not dense in X. Since $\overline{S^c}$ is not dense in X, $\overline{(\overline{S^c})} \neq X$, hence there exists $x \in \overline{(\overline{S^c})}^c$, and since $\overline{(\overline{S^c})}^c$ is open there exists a nonempty open set G with $G \subset \overline{(\overline{S^c})}^c \subset (\overline{S^c})^c = \overline{S}$, a contradiction to the assumption that $\overline{S}$ contains no nonempty open set.

    (γ) Finally, we show that (iii) implies (ii).
We assume that $\overline{S^c}$ is dense in X, that is, $\overline{(\overline{S^c})} = X$, but $\overline{S}$ has an interior point x. Thus there exists an open set G with $x \in G \subset \overline{S}$, hence $\overline{S^c} \subset G^c$. Since G^c is closed, it follows that $\overline{(\overline{S^c})} \subset G^c$, and so $x \in \overline{(\overline{S^c})}^c$, a contradiction to $\overline{(\overline{S^c})} = X$.

$\square$

The following lemma is needed in the proof of Baire's category theorem.

Lemma 2.10.8 *Any sequence of dense open sets in a complete metric space has non–empty intersection.*

Proof. Let (G_n) be a sequence of open dense sets in a complete metric space (X, d). Let $x_1 \in G_1$. Since G_1 is open, there exists an open disc $B_r(x_1)$ with $B_r(x_1) \subset G_1$. Now we choose a closed disc $\bar{B}_{r(1)}(x_1) \subset B_r(x_1)$. Since G_2 is dense in X, there exists $x_2 \in G_2 \cap \bar{B}_{r(1)/2}(x_1)$, and since G_2 is open, there exists an open ball $B_{\tilde{r}(1)}(x_2) \subset G_2$. We choose a closed ball $\bar{B}_{r(2)}(x_2)$ with $r(2) < \min\{\tilde{r}(1), r(1)/2\}$. Then we have

$$r(2) < \frac{r(1)}{2} \text{ by definition,}$$

and also

$$\bar{B}_{r(2)}(x_2) \subset \bar{B}_{r(1)}(x_1),$$

since if $x \in \bar{B}_{r(2)}(x_2)$, then $d(x, x_2) \leq r(2) < r(1)/2$, and so $d(x, x_1) \leq d(x, x_2) + d(x_2, x_1) < r(1)/2 + r(1)/2 = r(1)$, hence $x \in \bar{B}_{r(1)}(x_1)$; we also have

$$\bar{B}_{r(2)}(x_2) \subset G_2.$$

Continuing in this way, we can find closed balls $\bar{B}_{r(n)}(x_n) \subset G_n$ with

$$\bar{B}_{r(n)}(x_n) \subset \bar{B}_{r(n-1)}(x_{n-1}) \text{ for } n = 2, 3, \ldots \text{ and } \lim_{n \to \infty} r(n) = 0.$$

Since $x_m \in \bar{B}_{r(n)}(x_n)$ for $m > n$, we have $d(x_m, x_n) \leq r(n)$, hence (x_n) is a Cauchy sequence, which converges by the completeness of X, $x_n \to x$, say. It follows that $x \in \bar{B}_{r(n)}(x_n)$ for each $n \subset \mathbb{N}$, since $x_m \in \bar{B}_{r(n)}(x_n)$ for each $m > n$, $x_m \to x$ and $\bar{B}_{r(n)}(x_n)$ is closed. This implies

$$x \in \bigcap_{n \in \mathbb{N}} \bar{B}_{r(n)}(x_n) \subset \bigcap_{n \in \mathbb{N}} G_n.$$

$\square$

Now we can prove the famous *Baire category theorem.*

Theorem 2.10.9 (Baire's category theorem) *Any complete metric space is of second category in itself.*

Proof. Let (S_n) be a sequence of nowhere dense sets in a complete metric space X. Then $(X \setminus \bar{S}_n)$ is a sequence of open dense sets, since $\bar{S}_n$ contains no interior points. It follows by Lemma 2.10.8 that

$$\bigcap_{n \in \mathbb{N}} (X \setminus \bar{S}_n) \neq \emptyset, \text{ hence } \bigcup_{n \in \mathbb{N}} S_n \neq X.$$

$\square$

2.11   Locally convex spaces

Locally convex spaces play an important role in the theory of linear topological spaces. It will turn out that the topology of a locally convex space can be described by a family of functionals, the seminorms.

Definition 2.11.1 A linear topology is said to be *locally convex* if every neighborhood of 0 includes a convex neighborhood of 0.

We give a few examples of locally convex spaces.

Example **2.11.2** *(a) Any seminormed space is locally convex.*
(b) Let Φ be a family of locally convex topologies on a linear space. Then $\bigvee \Phi$ is locally convex.
(c) If in Example 2.8.3 (b) the range spaces of the functions in Φ all are locally convex, then the weak linear topology $\sigma(\Phi)$ by Φ (Definition 2.8.10) is locally convex; in particular, a product of locally convex space is locally convex.

Proof.

(a)   This holds, since the open balls in a seminormed space are convex.

(b)   Any neighborhood of 0 includes a finite intersection G of convex neighborhoods of 0 in the topologies of Φ. Since the intersection of convex sets is convex, G is convex.

(c)   By Part (b), it suffices to check this for a single function f. If C is a convex set, then so is its preimage $f^{-1}(C)$ by Theorem 2.2.7 (c). $\square$

We have just seen in Example 2.11.2 (c) that the weak linear topology $\sigma(X, \Phi)$, where X is a linear space and Φ is a family of seminorms defined on X, is locally convex. At a later stage, we will see that this is the only kind there is (Theorem 3.10.1).

2.12   Quotient topologies

In Definition 2.6.3, we used a function f from a set A into a topological space Y to introduce a topology on A, namely the weak topology $w(A, f)$ by f. Now we use a linear function from a linear topological space X onto a linear space Y, to introduce a topology on Y, the so-called *quotient topology*.

Definition 2.12.1 Let X be a linear topological space, Y be a linear space and $f \in \mathcal{L}(X, Y)$ be onto. The *quotient topology $\mathcal{Q}_f$ (on Y by f)* is the topology generated by

$$\mathcal{B} = \{f(N) : N \text{ is a balanced neighborhood of 0 in } X\}$$

(as in Theorem 1.9.10).
We also write *quotient topology*, for short, when the meaning is clear from the context.

Theorem 2.12.2 *Let X be a linear topological space, Y be a linear space and $f \in \mathcal{L}(X, Y)$ be onto. Then*
(a) $f : X \to (Y, \mathcal{Q}_f)$ is continuous and open;
(b) $\mathcal{Q}_f$ is the only topology for Y for which this is true.

Proof.

(a)   First we prove Part (a).

(i)   We show that f is continuous.
      Let V be a $\mathcal{Q}_f$–neighborhood of 0 in Y. Then, by Definition 2.12.1, there exists
      a neighborhood N of 0 in X such that $f(N) \subset V$. Hence, f is continuous at 0
      by Definition 1.9.6 (b), and since f is linear, f is continuous by Theorem 2.8.20
      (b).

(ii) Now we show that f is open.
      Let N be a balanced neighborhood of 0 in X. Then $f(N)$ is a $\mathcal{Q}_f$–neighborhood
      of 0 in Y by Remark 2.9.9.

(b)   Now we prove Part (b).
      First let $\mathcal{T}$ be a topology of Y such that f is continuous. Then $\mathcal{T} \subset \mathcal{Q}_f$.
      On the other hand, if $f : X \to (Y, \mathcal{T})$ is open, then $\mathcal{T} \supset \mathcal{Q}_f$. $\square$

Remark 2.12.3 *Part (b) of the proof of Theorem 2.12.2 shows that $\mathcal{Q}_f$ is the largest topology that makes f continuous and the smallest that makes f open.*

Definition 2.12.4 A linear map q from a linear topological space X onto a linear space Y is said to be a *quotient map*, if Y has the quotient topology $\mathcal{Q}_q$. If Y has the quotient topology by some map, then Y is called the *quotient of X (by q)*; we also write *quotient of X*, for short.

Remark 2.12.5 *(a) By Theorem 2.12.2, the term quotient map is synonymous with continuous open linear onto map.*
(b) We note that a one-to-one quotient map is a homeomorphism onto (Definition 1.9.6 (c) and subsequent Warning 2). Thus the concept of a quotient map is a generalization of that of a homeomorphism onto which is not one to one.

Theorem 2.12.6 *Let Y be the quotient of a linear topological space X by the quotient map $q : X \to Y$, and f be a linear map from Y into a linear topological space Z. Then $f : Y \to Z$ is continuous if and only if $f \circ q : X \to Z$ is continuous.*

Proof.

(i)   First we show that the continuity of f implies that of $f \circ q$.
      Let $f : Y \to Z$ be continuous. Since $q : X \to Y$ is continuous by Theorem 2.12.2 (a),
      the map $f \circ q : X \to Z$ is continuous.

(ii) Now we show that the continuity of $f \circ q$ implies that of f.
      Let $f \circ q : X \to Z$ be continuous, and V be a neighborhood of 0 in Z. Since $f \circ q$ is
      continuous, there exists a neighborhood u of 0 in X such that $f(q(U)) \subset V$. Since q
      is open by Theorem 2.12.2 (a), $q(U)$ is a neighborhood of 0 in Y, and so $f : Y \to Z$
      is continuous. $\square$

We need the following lemma.

Lemma 2.12.7 *Let A be any subset of a topological space.*
(a) If G is any neighborhood of 0, then $\bar{A} \subset A + G$.
(b) We have

$$\bar{A} = \bigcap \{A + G : G \text{ is a neighborhood of } 0\}.$$

Proof.

(a) First we show the statement in (a).

Let $a \in \bar{A}$, G be a neighborhood of 0 and N be a symmetric neighborhood of 0 with $N \subset G$; such a neighborhood exists by Theorem 2.8.5 (c) and Remark 2.2.3 (a). Since $a + N$ is a neighborhood of a by Remark 2.8.4 (h), it follows that $A \cap (a + N) \neq \emptyset$, and so $a \in A + N \subset A + G$.

(b) Finally we show the statement in (b).

We put

$$B = \bigcap \{A + G : G \text{ is a neighborhood of } 0\}.$$

It follows from Part (a) that $\bar{A} \subset B$.

To prove the converse implication, let $x \notin \bar{A}$ and G_1 be a neighborhood of 0 such that $A \cap (x + G) = \emptyset$. Also let N be a symmetric neighborhood of 0 with $N \subset G$ (such a neighborhood exists by the same argument as in Part (a) of the proof). Then $A \cap (x + N) = \emptyset$, hence $x \notin A + N$, and so $x \notin B$. $\qquad \square$

Theorem 2.12.8 *Every linear topological space X is regular (Definition 1.9.13 (c)).*

Proof. Let S be a closed set in X, and $x \in X \setminus S$. By Lemma 2.12.7 (b), there exists a neighborhood G of 0 such that $x \notin S + G$. Let N be a symmetric neighborhood of 0 with $N + N \subset G$; such a neighborhood exists by Theorem 2.8.5 (c) and (d). We put $G_1 = S + N$ and $G_2 = x + N$. Then $G_1 = \bigcup \{s + N : s \in S\}$ is the union of the sets $s + N$, which are open by Remark 2.8.4 (g), hence G_1 is open. Also $G_1 \cap G_2 = \emptyset$, since $x \notin S + G \supset S + N + N$, hence $(x + N) \cap (S + N) = \emptyset$. $\qquad \square$

Theorem 2.12.9 *The following conditions on a linear topological space X are equivalent:*
(1) X is a T_0 space;
(2) X is a T_3 space;
(3) $\overline{\{0\}} = \{0\}$;
(4) for each $x \in X \setminus \{0\}$, there exists a neighborhood N of 0 such that $x \notin N$.

Proof.

(i) First we show that the statement in (1) implies that in (2).

Let $a, b \in X$ and $a \neq b$. Since X is a T_0 space, by Definition 1.9.13 (a), there exists an open set G which is a neighborhood of one of the points, a say, but does not contain b. Then $b - G + a$ is a neighborhood of b by Remark 2.8.4 (h) and $a \notin b - G + a$, hence X is a T_1 space by Definition 1.9.13 (b). By Theorem 2.12.8, X is also regular, and so a T_3 space by Definition 1.9.13 (d).

(ii) Now we show that the statement in (2) implies that in (3).

Clearly a T_3 space is a T_1 space, and obviously every singleton in a T_1 space is closed.

(iii) The statements in (3) and (4) are equivalent by Lemma 2.12.7 (b) and the fact that $a \notin \overline{\{0\}}$.

(iv) Finally, we show that the statement in (3) implies that in (1).

Since $\overline{\{0\}}$ is a closed set by the statement in (3), every singleton in X is a closed set, because translation in a linear topological space is a homeomorphism onto by Remark 2.8.4 (f). Thus X is a T_1 space. $\qquad \square$

Definition 2.12.10 A linear topological space is said to be *separated* if one of the equivalent conditions of Theorem 2.12.9 holds. The topology of a separated linear topological space is said to be *separated*.

Theorem 2.12.11 *Let X be a linear topological space, Y be a linear space and q be a quotient map. Then $\{0\}$ is closed in the quotient Y of X if and only if $N(q)$ is closed in X.*

Proof.

(i) First we show that if $\{0\}$ is closed in Y then $N(q)$ is closed in X.
Let $\{0\}$ be closed in Y. Since q is continuous by Theorem 2.12.2 (a), the set $N(f) = q^{-1}(\{0\})$ is closed in X.

(ii) Now we show that if $N(q)$ is closed in X then $\{0\}$ is closed in Y.
We assume that $N(q)$ is closed in X. Since q is onto by Definition 2.12.1, there exists $x \in X \backslash N(q)$ with $y = q(x)$. Also, since $N(q)$ is closed in X, there exists a neighborhood U of 0 in X such that $(x-U) \cap N(q) = \emptyset$. This implies $y \notin q(U)$. For if there were some $u \in U$ with $y = q(u)$, then it would follow that $q(x - u) = q(x) - q(u) = y - y = 0$, that is $x - u \in N(q)$, which is a contradiction to the choice of U.
Since q is open by Theorem 2.12.2 (a), $q(U)$ is a neighborhood of 0 in Y, hence $\{0\}$ is closed in Y by Theorem 2.12.9. □

Chapter 3

Linear Metric Spaces

Chapter 3 deals with linear metric and linear semimetric spaces, the concepts of paranormed spaces and Schauder bases, and their most important properties. Among other things, it is shown that the quotient of a paranormed space is a paranormed space, and that quotient paranorms preserve completeness. The highlights of the chapter are the open mapping theorem, the closed graph theorem, the uniform boundedness principle and the Banach–Steinhaus theorem, which are generally considered as being the main results in functional analysis, apart from the Hahn–Banach extension theorem which is presented in Chapter 4. Furthermore, the chapter contains studies of useful properties of seminorms, further results related to local convexity, and the Minkowski functional and its role in defining a seminorm or norm on a linear space. Finally, a sufficient condition is established for the metrizability of a linear topology, and also a criterion is given for a linear topology to be generated by a seminorm.

3.1 Introduction

In this chapter, we study linear semimetric and metric spaces. We introduce the related concepts of paranormed spaces in Definition 3.2.1, and the notion of Schauder bases in Definition 3.4.1, and study their most important properties. The highlights of the chapter are the open mapping theorem, Theorem 3.5.3, the closed graph theorem, Theorem 3.6.1, the uniform boundedness principle, Theorem 3.7.4, and the Banach–Steinhaus theorem, Theorem 3.7.5. Furthermore, the chapter contains further studies of useful properties of seminorms and the concept of local convexity in Sections 3.3 and 3.10, and of the Minkowski functional and its role in defining a seminorm or norm on a linear space in Theorem 3.9.3. Also a sufficient condition is established for the metrizability of a linear topology in Theorem 3.11.2. It was mentioned in Example 2.11.2 (c) of Section 2.11 that the weak linear topology $\sigma(X, \Phi)$, where X is a linear space and Φ is a family of seminorms defined on X, is locally convex. In Theorem 3.10.1, we are going to prove that this is the only kind there is.

From now on, we are going to use the same symbol $\| \cdot \|$ for both seminorms and norms, instead of p for seminorms, as previously. This is done to avoid conflict with the use of p for paranorms which will be introduced in Definition 3.2.

3.2 Paranormed spaces

Before we define the concept of a *linear semimetric space* we deal with the notion of a *paranorm* which is a generalization of the absolute value, and can be used in the definition of linear semimetrics.

Definition 3.2.1 (a) A *paranorm* is a real–valued function p defined on a linear space X, which satisfies the following conditions for all $x, y \in X$

(P.1) $p(0) = 0$,
(P.2) $p(x) \geq 0$,
(P.3) $p(-x) = p(x)$,
(P.4) $p(x + y) \leq p(x) + p(y)$ *(triangle inequality)*.

(P.5)   If (λ_n) is a sequence of scalars with $\lambda_n \to \lambda$ and (x_n) is a sequence of vectors with $p(x_n - x) \to 0$, then it follows that $p(\lambda_n x_n - \lambda x) \to 0$ *(continuity of multiplication by scalars)*.

A linear space X with a paranorm p is said to be a *paranormed space*, denoted by (X, p), or simply by X.

(b) A paranorm p is said to be *total*, if $p(x) = 0$ implies $x = 0$.

(c) If p is a paranorm, we write $\{p < \lambda\} = p^{-1}([0, \lambda))$ and $\{p \leq \lambda\} = p^{-1}([0, \lambda])$ for $\lambda \geq 0$.

Remark 3.2.2 *(a) By Remark 1.6.11 (a), the definition of the term total in Definition 3.2.1 (b) is equivalent to the fact that the set $\{p\}$ is total in the sense of Definition 1.6.9 (b).*

(b) Every paranorm p defines a semimetric d by

$$d(x, y) = p(x - y) \text{ for all } x \text{ and } y.$$

The semimetric defined in this way is translation invariant, *that is,*

$$d(a + c, b + c) = d(a, b) \text{ for all } a, b \text{ and } c.$$

The conditions (P.4) and (P.5) in Definition 3.2.1 mean that addition and multiplication by scalars are continuous.

Now we define the notions of *linear semimetric, linear metric, seminormed, normed, Fréchet and Banach spaces.*

Definition 3.2.3 (a) A semimetric space that is also a linear space is said to be a

$$linear \left\{ \begin{array}{c} semimetric \\ metric \\ seminormed \\ normed \end{array} \right\} space$$

if the semimetric is generated by a

$$\left\{ \begin{array}{c} paranorm \\ total\ paranorm \\ seminorm \\ norm \end{array} \right\}$$

(b) A complete normed space is said to be a *Banach space*; a complete linear metric space is said to be a *Fréchet space*.

Example 3.2.4 *(a) The space c of all convergent complex sequences is a Banach space with its natural norm of Example 1.9.2 (b).*
(b) The space ω is a Fréchet space with its natural metric of Example 1.9.2 (b).

Now we are going to define the concepts of *convergence* and *absolute convergence of series*, which are meaningful in paranormed spaces.

Definition 3.2.5 Let (a_k) be a sequence in a linear semimetric space and

$$s_n = \sum_{k=1}^{n} a_k \ (n = 1, 2, \dots).$$

(a) The sequence (s_n) is called a *series*, denoted by

$$\sum_{n=1}^{\infty} a_n.$$

If $s_n \to a \ (n \to \infty)$, then we say that the *series converges to a*. In a linear metric space, a sequence can at most have one limit in which case we write

$$\sum_{n=1}^{\infty} a_n = a;$$

a is referred to as the *limit* or *sum of the series*.
(b) If the semimetric is given by a paranorm p and the series $\sum_{n=1}^{\infty} a_n$ satisfies

$$\sum_{n=1}^{\infty} p(a_n) < \infty,$$

then the series is said to be *absolutely convergent*.

The following remark shows that absolute convergence does not imply convergence, in general.

Remark 3.2.6 *A convergent series need not be absolutely convergent. For instance, $\sum_{n=1}^{\infty}(1/n)e^{(n)}$ is convergent in c_0, but not absolutely convergent.*
An absolutely convergent series need not converge. For instance, if ϕ has the natural norm of c_0, that is, $\|a\| = \sup_k |a_k|$ for all $a \in \phi$, then the series $\sum_{n=1}^{\infty}(1/n^2)e^{(n)}$ is absolutely convergent, but not convergent.

However, the property that absolute convergence implies convergence is a characterization of the completeness of a linear metric space, as the following result will show.

Theorem 3.2.7 *A linear metric space (X, d) is complete if and only if every absolutely convergent series converges.*

Proof. Let the metric d be given by the total paranorm p.

(i) If the series $\sum_{n=1}^{\infty} a_n$ is absolutely convergent, then we have

$$p(s_m - s_n) = p\left(\sum_{k=n+1}^{m} a_k\right) \leq \sum_{k=n+1}^{m} p(a_k) \text{ for all } m > n,$$

and so $\left(\sum_{k=1}^{n} a_k\right) = (s_n)$ is a Cauchy sequence which converges by the completeness of the space X.

(ii) Conversely, we assume that every absolutely convergent series in X converges. Let (a_n) be a Cauchy sequence in X. Then there exists a strictly increasing sequence $((n(k))_{k=1}^{\infty}$ of natural numbers such that

$$p\left(a_{n(k+1)} - a_{n(k)}\right) < \frac{1}{2^k} \text{ for } k = 1, 2, \dots,$$

and so

$$\sum_{k=1}^{\infty} p\left(a_{n(k+1)} - a_{n(k)}\right) < \infty.$$

It follows from our assumption that the series $\sum_{k=1}^{\infty}(a_{n(k+1)} - a_{n(k)})$ converges, that is, the sequence $(s_m)_{m=1}^{\infty}$ with

$$s_m = \sum_{k=1}^{m}(a_{n(k+1)} - a_{n(k)}) = a_{n(m+1)} - a_{n(1)} \text{ for } m = 1, 2, \dots$$

converges, and so $(a_{n(k)})_{k=1}^{\infty}$ converges. Hence the Cauchy sequence (a_n) has a convergent subsequence $(a_{n(k)})_{k=1}^{\infty}$, and so the whole sequence (a_n) converges by Lemma 2.10.3. Consequently, the space X is complete. □

Remark 3.2.8 *The statement of Theorem 3.2.7 extends to linear semimetric spaces, although in this case the limits may not be unique.*

3.3   Properties of paranormed and seminormed spaces

Next we are going to prove a few more very important results in functional analysis in linear metric and seminormed spaces. In view of Definition 3.2.1, it makes sense, first to study some fundamental properties of paranormed spaces.

As we mentioned at the beginning of this chapter, the symbol $\|\cdot\|$ is also going to be used to denote seminorms; the symbol $p(a)$ will denote the paranorm of a.

We need the following definition.

Definition 3.3.1 Let p and q be paranorms.
(a) Then p is said to be *stronger* than q, if $p(x_n) \to 0$ for any sequence (x_n) implies $q(x_n) \to 0$; p is said to be *weaker* than q, if q is stronger than p.
(b) If p is stronger than q and q is stronger than p, then p and q are said to be *equivalent*.
(c) If p is stronger than q and p and q are not equivalent, then p is said to be *strictly stronger* than q, and q is said to be *strictly weaker* than p.

Remark 3.3.2 *If the paranorm p is stronger than the paranorm q, then we have $N(p) \subset N(q)$.*

Proof. Let $x \in N(p)$ be given. Then we have $p(x) = 0$. It follows for the sequence (x_n) with $x_n = x$ for all n that $p(x_n) \to 0$ $(n \to \infty)$ by the condition in (P.5) of Definition 3.2.1. This implies $q(x) = q(x_n) \to 0$, since p is stronger than q. Therefore we have $x \in N(q)$. □

Example **3.3.3** *We define p, q, r and h on $\mathbb{R}^2$ by*

$$p(x, y) = |x| + |y|, \ q(x, y) = \frac{|x|}{1 + |x|} + |y|, \ r(x, y) = |x| \text{ and } h(x, y) = |y|.$$

Then p is strictly stronger than r, q is strictly stronger than r, p and q are equivalent, h is not stronger than r, and r is not stronger than h.

Proof. We have $p(x, y) \geq r(x, y)$ for all $(x, y) \in \mathbb{R}^2$ and $r(0, 1) = 0 \neq 1 = p(0, 1)$; $q(x_n, y_n) \to 0$ implies $|x_n|/(1 + |x_n|) \to 0$ and $|y_n| \to 0$, hence $r(x_n, y_n) = |x_n| \to 0$, but $r(0, 1) = 0 \neq 1 = q(0, 1)$.

The equivalence of p and q is obvious.

We have $N(h) = \mathbb{R} \times \{0\} \not\subset N(r) = \{0\} \times \mathbb{R}$ and $N(r) \not\subset N(h)$. $\qquad\square$

Remark 3.3.4 *Let X be a linear space, x, y and x_n be vectors, λ and λ_n be scalars, and p and q be paranorms on X. Then we have*

(a) $\qquad p(x - y) \geq |p(x) - p(y)|$;

(b) $\qquad p(nx) \leq np(x)$ *and* $p(x/n) \geq (1/n)p(x)$ *for all* $n \in \mathbb{N}$.

(c) If the sequence (λ_n) is bounded and $p(x_n) \to 0$ then we have $p(\lambda_n x_n) \to 0$.

(d) The set $N(p)$ is a subspace of X; the set $\{p = \lambda\} = \{x \in X : p(x) = \lambda\}$, however, need not be an affine set.

(e) For each x, p is constant on the affine set $x + N(p)$.

(f) If there exists a constant M such that $q(x) \leq M \cdot p(x)$ for all x, then p is stronger than q. The converse is not true, in general.

(g) Let $p \equiv 0$ on a subspace S of X. We define $\tilde{p}$ on X/S by $\tilde{p}(x + S) = p(x)$ for all $x \in X$. Then $\tilde{p}$ is a paranorm on X/S, and $\tilde{p}$ is total over $X/N(p)$.

(h) Let $p \equiv 0$ and $q \equiv 0$ on a subspace S of X and $\tilde{p}$ be stronger than $\tilde{q}$ on X/S. Then p is stronger than q.

(j) The set $\{p < \varepsilon\} = \{x \in X : p(x) < \varepsilon\}$ is absorbing for each $\varepsilon > 0$.

Proof.

(a) We have $p(x) \leq p(x - y) + p(x)$, hence $p(x) - p(y) \leq p(x - y)$. Interchanging x and y, we also obtain $p(y) - p(x) \leq p(y - x) = p(x - y)$, hence

$$-p(x - y) \leq p(x) - p(y) \leq p(x - y), \text{ that is, } |p(x) - p(y)| \leq p(x - y).$$

(b) We have

$$p(nx) = p(x + x + \cdots + x) \leq p(x) + p(x) + \cdots + p(x) = np(x).$$

This implies

$$p\left(\frac{x}{n}\right) = \frac{n}{n}p\left(\frac{x}{n}\right) \geq \frac{1}{n}p\left(n \cdot \frac{x}{n}\right) = \frac{1}{n}p(x).$$

(c) We assume $p(\lambda_n x_n) \not\to 0$. Then there exists $c > 0$ such that $p(\lambda_n x_n) \geq c$ for infinitely many n. Since the sequence (λ_n) is bounded, there exists a subsequence $(\lambda_{n(k)})$ with $\lambda_{n(k)} \to \lambda$ $(k \to \infty)$. Then we have

$$p(x_{n(k)}) \to 0, \ p(\lambda_{n(k)} x_{n(k)}) \to 0, \text{ but } p(\lambda_{n(k)} x_{n(k)}) \geq c,$$

a contradiction to the condition in (P.5) of Definition 3.2.1.

(d) (i) We show that $N(p)$ is a subspace.

Let $x \in N(p)$ and λ be a scalar. We define the sequences (λ_n) and (x_n) by $\lambda_n = \lambda$ and $x_n = x$ for all $n \in \mathbb{N}$. It follows that $p(x_n) \to 0$ and $\lambda_n \to \lambda$ $(n \to \infty)$, and so, by the condition in (P.5) of Definition 3.2.1,

$$p(\lambda x) = p(\lambda_n x_n - \lambda \cdot 0) \to 0,$$

that is, $\lambda x \in N(p)$. Furthermore, if $x, y \in N(p)$ then we have

$$0 \leq p(x + y) \leq p(x) + p(y) = 0, \text{ that is, } x + y \in N(p).$$

(ii) We give an example in which $\{p=1\}$ is not an affine set.
We consider $p(x) = |x|$ on $\mathbb{C}$. We choose $x = 1$, $\tilde{x} = i$ and $\lambda = i$. Then we have $x, \tilde{x} \in \{p = 1\}$, but

$$y = \lambda x + (1 - \lambda)\tilde{x} = i + (1 - i)i = 2i + 1 \text{ and } y \notin \{p = 1\}.$$

(e) The set $x + N(p)$ is affine by Part (d) and Theorem 2.3.9.

(f) Let $q(x) \le M \cdot p(x)$ for all $x \in A$ and $p(x_n) \to 0$. Then it follows that $0 \le q(x_n) \le M \cdot p(x_n) \to 0$, that is, $q(x_n) \to 0$, and so p is stronger than q.
In Example 3.3.3, the paranorm q is stronger than r, but if M is any positive real, then we have for $|x| > M$

$$r(x, 0) = |x| > M > M \cdot \frac{|x|}{1 + |x|} = M \cdot q(x, 0).$$

(g) It follows from the definition of $\tilde{p}$ and the conditions in (P.1) to (P.5) of Definition 3.2.1 for the paranorm p that

$$\tilde{p}(0 + S) = p(0) = 0, \ \tilde{p}(-x + S) = p(-x) = p(x) = \tilde{p}(x + S),$$
$$\tilde{p}((x + S) + (y + S)) = \tilde{p}(x + y + S) = p(x + y)$$
$$\le p(x) + p(y) = \tilde{p}(x + S) + \tilde{p}(y + S),$$

and if $\lambda_n \to \lambda$ and $\tilde{p}(x_n + S - (x + S)) = \tilde{p}(x_n - x + S) = p(x_n - x) \to 0$, then

$$\tilde{p}(\lambda_n x_n + S - (\lambda x + S)) = \tilde{p}(\lambda_n x_n - \lambda x + S) = p(\lambda_n x_n - \lambda x) \to 0.$$

Thus $\tilde{p}$ is a paranorm.
Putting $S = N(p)$, we conclude from $\tilde{p}(x + S) = p(x) = 0$, that $x \in S$, hence $x + S = S$, and so $\tilde{p}$ is total.

(h) Let $p(x_n) \to 0$. Then we have $\tilde{p}(x_n + S) \to 0$. This implies that $\tilde{q}(x_n + S) = q(x_n) \to 0$, since $\tilde{p}$ is stronger on X/S than $\tilde{q}$. Thus we have shown that p is stronger than q on X.

(j) If x is a vector which is not absorbed by $\{p < \varepsilon\}$, then there exists a sequence (λ_n) of scalars with $\lambda_n \to 0$ $(n \to \infty)$ and $\lambda_n x \notin \{p < \varepsilon\}$ for all n, that is $p(\lambda_n x) \ge \varepsilon$ for all n, which is a contradiction to the condition in (P.5) of Definition 3.2.1. □

Theorem 3.3.5 *Let (p_k) be a sequence of paranorms on a linear space and*

$$p(x) = \sum_{k=1}^{\infty} \frac{1}{2^k} \cdot \frac{p_k(x)}{1 + p_k(x)} \quad \text{(Fréchet combination of (p_k))}.$$

(a) Then p is a paranorm and

$$\left\{ \begin{array}{c} p(x_n) \to 0 \ (n \to \infty) \\ \text{if and only if} \\ p_k(x_n) \to 0 \ (n \to \infty) \text{ for each } k \in \mathbb{N}; \end{array} \right\} \qquad (3.1)$$

(b) p is the weakest paranorm which is stronger than each p_k;
(c) p is total if and only if $\{p_k : k \in \mathbb{N}\}$ is a total set.

Proof.

(i)   First we show (3.1).

(α)   We show that $p(x_n) \to 0$ $(n \to \infty)$ implies $p_k(x_n) \to 0$ $(n \to \infty)$ for each $k \in \mathbb{N}$. Let $p(x_n) \to 0$ and $k \in \mathbb{N}$ be given. Then

$$p(x_n) \geq \frac{1}{2^k} \cdot \frac{p_k(x_n)}{1 + p_k(x_n)}$$

implies

$$2^k \cdot p(x_n) + 2^k \cdot p_k(x_n) p(x_n) \geq p_k(x_n),$$

and

$$2^k \cdot p(x_n) \geq p_k(x_n) \left(1 - 2^k \cdot p(x_n) \right).$$

Since $p(x_n) \to 0$, there exists $n_0 \in \mathbb{N}$ such that $p(x_n) < 2^{-k}$ for all $n \geq n_0$, and so we have for all $n \geq n_0$

$$p_k(x_n) \leq \frac{2^k \cdot p(x_n)}{1 - 2^k \cdot p(x_n)} \to 0 \ (n \to \infty).$$

(β)   Now we show that $p_k(x_n) \to 0$ $(n \to \infty)$ for each $k \in \mathbb{N}$ implies $p(x_n) \to 0$ $(n \to \infty)$.
Let $p_k(x_n) \to 0$ $(n \to \infty)$ for each $k \in \mathbb{N}$. Let $m \in \mathbb{N}$ be given. Then we have

$$0 \leq p(x_n) = \sum_{k=0}^{\infty} \frac{1}{2^k} \cdot \frac{p_k(x_n)}{1 + p_k(x_n)}$$

$$\leq \sum_{k=1}^{m} \frac{1}{2^k} \cdot \frac{p_k(x_n)}{1 + p_k(x_n)} + \sum_{k=m+1}^{\infty} \frac{1}{2^k} \to \frac{1}{2^m} \ (n \to \infty).$$

Since $m \in \mathbb{N}$ was arbitrary, it follows that $p(x_n) \to 0$ $(n \to \infty)$.

Thus we have shown (3.1).

Because of (3.1), p generates the weak topology $w(X, (p_k))$ on X by Theorem 2.6.7 which is semimetrizable by Theorem 2.6.8 and which is the weakest topology stronger than every p_k.
Thus we have shown Part (b).
The condition in (P.5) of Definition 3.2.1 for a paranorm is an immediate consequence of (3.1). Thus we have also shown Part (a).
Part (c) is obvious. □

In Section 2.12, we introduced the concept of quotient topologies (Definition 2.12.1) and studied some of their basic properties. We will see in the next result that the quotient of a paranormed space is paranormed.

Theorem 3.3.6 *A quotient Y of a paranormed space X is a paranormed space.*

Proof. Let p_1 denote the paranorm on X and $q : X \to Y$ be the quotient map. We define

$$p_2(y) = \inf\{p_1(x) : y = q(x)\} \text{ for all } y \in Y$$

and observe that p_2 is defined on all of Y, since $q : X \to Y$ is onto by Definition 2.12. Obviously $p_2(0) = 0$, $p_2(y) \geq 0$ and $p_2 = p_2(-y)$ for all $y \in Y$.

(i) Now we show that p_2 satisfies the triangle inequality.
Let $y, \tilde{y} \in Y$ and $\varepsilon > 0$ be given. By the definition of p_2 there exist $x, \tilde{x} \in X$ such that

$$y = q(x), \ \tilde{y} = q(\tilde{x}), \ p_1(x) < p_2(y) + \frac{\varepsilon}{2} \text{ and } p_1(\tilde{x}) < p_2(\tilde{y}) + \frac{\varepsilon}{2}.$$

Since q is linear by Definition 2.12, it follows that $y + \tilde{y} = p_1(x) + p_1(\tilde{x}) = p_1(x + \tilde{x})$, hence we obtain by the definition of p_2 and the triangle inequality for the paranorm p_1

$$p_2(y + \tilde{y}) \le p_1(x + \tilde{x}) \le p_1(x) + p_1(\tilde{x}) < p_2(y) + p_2(\tilde{y}) + \varepsilon.$$

Since $\varepsilon > 0$ was arbitrary we conclude $p_2(y) + p_2(\tilde{y})$.

(ii) We put $d(y, \tilde{y}) = p_2(y - \tilde{y})$ for all $y, \tilde{y} \in Y$. Then d is a translation invariant semimetric on Y by what we have shown so far. We show that d generates the quotient topology. Since this is a linear topology, it follows that p_2 is a paranorm.

(α) First we show that $q : X \to (Y, d)$ is continuous.
Let $x_n \to x \in X$ as $n \to \infty$. Then it follows by the linearity of q and the definition of p_2 that

$$d(q(x_n), q(x)) = p_2(q(x_n) - q(x)) = p_2(q(x_n - x)) \le p_1(x_n - x) \to 0$$
$$\text{as } n \to \infty,$$

and so q is continuous.

(β) Now we show that q is open.
Let G be an open set in X, $b \in q(G)$ and $b = q(a)$ for some $a \in G$. Since G is open, there exists $\varepsilon > 0$ such that $p_1(x - a) < \varepsilon$ implies $x \in G$. Now let $y \in Y$ be such that $p_2(y - b) < \varepsilon/2$. Also let $w \in X$ with $y - b = q(w)$ and

$$p_1(w) < p_2(y - b) + \frac{\varepsilon}{2}.$$

Since

$$p_1(w + a - a) = p_1(w) + p_1(0) = p_1(w) < p_2(y - b) + \frac{\varepsilon}{2} < \frac{\varepsilon}{2} + \frac{\varepsilon}{2} = \varepsilon,$$

it follows that $w + a \in G$, and

$$y = q(w) + b = q(w) + q(a) = q(w + a) \in q(G).$$

Thus $p_2(y - b) < \varepsilon/2$ implies $y \in q(G)$, so b is an interior point of $q(G)$, and consequently $q(G)$ is open.

By Parts (α) and (β) of the proof, the map q continuous and open, hence a quotient map by Remark 2.12.5 (a).
Thus we have shown that d generates the quotient topology.
This concludes the proof of the theorem.　　　　　　　　　　□

The next result is the analogue of Theorem 3.3.6 for seminormed spaces.

Theorem 3.3.7 *A quotient Y of a seminormed space X is a seminormed space.*

Proof. Now let p_1 denote the seminorm on X. We define p_2 as in the proof of Theorem 3.3.6. It remains to show that p_2 is homogeneous.
We obtain for any scalar $\lambda \neq 0$ and all $y \in Y$

$$
\begin{aligned}
p_2(\lambda y) &= \inf\{p_1(x) : \lambda y = q(x)\} \\
&= \inf\{p_1(x) : y = (1/\lambda)q(x) = q(x/\lambda) \\
&= |\lambda| \inf\{p_1(x/\lambda) : y = q(x/\lambda)\} = |\lambda| p_2(y).
\end{aligned}
$$
□

The next result states that quotient paranorms preserve completeness.

Theorem 3.3.8 *A quotient Y of a complete paranormed space X is complete.*

Proof. Let p_1 denote the paranorm on X, q be the quotient map and the paranorm on Y be defined as in the proof of Theorem 3.3.6. Furthermore, let $\sum_{n=1}^{\infty} y_n$ be an absolutely convergent series in Y. For each n, we choose $x_n \in X$ with $y_n = q(x_n)$ such that $p_1(x_n) < p_2(y_n) + 2^{-n}$; such a choice is possible by the definition of p_2. Then it follows that

$$
\sum_{n=1}^{\infty} p_1(x) < \sum_{n=1}^{\infty} \left(p_2(y) + \frac{1}{2^n} \right),
$$

that is, the series $\sum_{n=1}^{\infty} x_n$ is absolutely convergent in X. Since X is complete, it follows by Theorem 3.2.7 that the series $\sum_{n=1}^{\infty} x_n$ converges. Since the quotient map q is continuous by Theorem 2.12.2 (a), we obtain

$$
q\left(\sum_{n=1}^{\infty} y_n \right) = \sum_{n=1}^{\infty} q(y_n) = \sum_{n=1}^{\infty} x_n,
$$

that is, the series $\sum_{n=1}^{\infty} y_n$ converges in Y. Thus we have shown that absolute convergence in Y implies convergence in Y, and consequently Y is complete by Theorem 3.2.7. □

We obtain the following corollary.

Corollary 3.3.9 *The separated quotient of every Fréchet space (Banach space) is a Fréchet space (Banach space).*

Proof. Let $(Y, \mathcal{Q}_q)$ be the separated quotient space of a Fréchet space. Then Y is a T_0 space by Definition 2.12.10 (and Theorem 2.12.9 (1)), hence, by Example 1.9.15, the quotient topology $\mathcal{Q}_q$ is generated by a metric d, and (Y, d) is complete by Theorem 3.3.8. The statement on Banach spaces is proved in exactly the same way. □

3.4 Schauder basis

The notion of an algebraic basis was exclusively based on a linear concept which only involved finite sums. The concept of a *Schauder basis*, however, involves series, hence also convergence, which is a topological concept.

Definition 3.4.1 A *Schauder basis* for a linear metric space X is a sequence (b_n) of vectors $b_n \in X$ such that, for each vector $x \in X$, there exists a unique sequence (λ_n) of scalars such that

$$
x = \sum \lambda_n b_n.
$$

The series $\sum \lambda_n b_n$ that converges to x is said to be the *expansion of x* (with respect to the Schauder basis (b_n)).

Remark 3.4.2 *(a) If (b_n) is a Schauder basis in a linear metric space then $b_n \neq 0$ for all $n \in \mathbb{N}$.*
(b) A space with a Schauder basis is separable, that is, it contains a countable dense subset.

Proof.

(a) If $b_n = 0$ for some $n \in \mathbb{N}$, then 0 has the representations $0 = 1 \cdot b_n = 0 \cdot b_n$, and so (b_n) cannot be a Schauder basis.

(b) Let (X, d) be a linear metric space with a Schauder basis (b_n) and

$$S = \mathrm{span}_{\mathbb{Q}}\{b_n : n \in \mathbb{N}\} = \left\{ \sum_{n=1}^{N} (\rho_n + i\sigma_n)b_n : \rho_n, \sigma_n \in \mathbb{Q},\ N \in \mathbb{N} \right\}$$

denote the set of all finite linear combinations of $\{b_n : n \in \mathbb{N}\}$ with rational coefficients. Obviously S is countable.
Now we show that S is dense in X.
Let p denote the total paranorm given by the metric d, that is, $p(x - y) = d(x, y)$ for all $x, y \in X$. Let $x \in X$ and $\varepsilon > 0$ be given. Since (b_n) is a Schauder basis for X, there exist $N \in \mathbb{N}$ and unique scalars $\lambda_k \in \mathbb{C}$ such that

$$p\left(x - \sum_{n=1}^{N} \lambda_n b_n \right) < \frac{\varepsilon}{2}.$$

Furthermore, for each $n = 1, 2, \ldots, N$, there exists a sequence $(\nu_n^{(m)}) = (\rho_n^{(m)} + i\sigma_n^{(m)})$ with $\rho_n^{(m)}, \sigma_n^{(m)} \in \mathbb{Q}$ for $m = 1, 2, \ldots$ such that $\nu_n^{(m)} \to \lambda_n$ $(m \to \infty)$. Since X is a linear metric space, for each $n = 1, 2, \ldots, N$, there exists $m_0 = m_0(n; \varepsilon)$ such that

$$p\left(\left(\nu_n^{(m)} - \lambda_n \right) b_n \right) < \frac{\varepsilon}{2N},$$

by the continuity of multiplication by scalars. We put $M = \max\{m_0(n; \varepsilon) : 1 \leq n \leq N\}$ and $\nu_n = \nu_n^{(M)}$ for $1 \leq n \leq N$. Since f is translation invariant, it follows that

$$p\left(x - \sum_{n=1}^{N} \nu_n b_n \right) \leq p\left(x - \sum_{n=1}^{N} \lambda_n b_n \right) + p\left(\sum_{n=1}^{N} (\lambda_n - \nu_n) b_n \right)$$

$$< \frac{\varepsilon}{2} + \sum_{n=1}^{N} p\left((\lambda_n - \nu_n) b_n \right) < \frac{\varepsilon}{2} + \sum_{n=1}^{N} \frac{\varepsilon}{2N} = \varepsilon.$$

$\square$

Example 3.4.3 *(a) The concepts of a Schauder basis and an algebraic basis coincide in finite dimensional spaces. This also holds true for the set ϕ of all finite sequences with the metric of c_0. Then $(e^{(n)})$ is a Schauder basis and $\{e^{(n)} : n \in \mathbb{N}\}$ is an algebraic basis.*
(b) In general, the two concepts of a basis differ; $(e^{(n)})$ is a Schauder basis for c_0, but $\{e^{(n)} : n \in \mathbb{N}\}$ is not an algebraic basis for c_0, since $\mathrm{span}(\{e^{(n)} : n \in \mathbb{N}\}) = \phi$ and ϕ is a proper subspace of c_0. On the other hand, any algebraic basis of c_0 is uncountable, and therefore cannot be a Schauder basis.
(c) The spaces ℓ_p for $0 < p < \infty$, c_0 and ω all have $(e^{(n)})$ as a Schauder basis; more precisely, every sequence $x = (x_k)$ in those spaces has a unique representation $x = \sum_{k=1}^{\infty} x_k e^{(k)}$. The

space c has $(e, e^{(1)}, e^{(2)}, \dots)$ as a *Schauder basis*; more precisely, every sequence $x = (x_k) \in c$ has a unique representation

$$x = \xi e + \sum_{k=1}^{\infty} (x_k - \xi) e^{(k)}, \quad \text{where } \xi = \lim_{k \to \infty} x_k.$$

(d) We know from Theorem 1.5.9 that every linear space has an algebraic basis; ℓ_∞, however, has no Schauder basis.

Proof.

(a) (i) We show the statement for ℓ_p and $1 \leq p < \infty$.
Let $x \in \ell_p$. Then given $\varepsilon > 0$, there is $M \in \mathbb{N}$ such that

$$\left(\sum_{n=m}^{\infty} |x_n|^p \right)^{1/p} < \varepsilon \quad \text{for all } m > M.$$

Thus we have for all $m > N$

$$\left\| x - \sum_{n=1}^{m} x_n e^{(n)} \right\| = \left(\sum_{n=m+1}^{\infty} |x_n|^p \right)^{1/p} < \varepsilon,$$

that is, $x = \sum_{n=1}^{\infty} x_n e^{(n)}$.
Now we show that the representation is unique.
Let $x = \sum_{n=1}^{\infty} \lambda_n e^{(n)}$ be one more representation of the sequence x. Given $\varepsilon > 0$, there exists $N \in \mathbb{N}$ such that

$$\left\| x - \sum_{n=1}^{\infty} x_n e^{(n)} \right\| < \frac{\varepsilon}{2} \quad \text{and} \quad \left\| x - \sum_{n=1}^{\infty} \lambda_n e^{(n)} \right\| < \frac{\varepsilon}{2}.$$

Let $k \in \mathbb{N}$ be arbitrary. Then we have for all $m > \max\{M, N\}$

$$0 \leq |\lambda_k - x_k| \leq \left(\sum_{n=1}^{m} |\lambda_n - x_n|^p \right)^{1/p} = \left\| \sum_{n=1}^{m} \lambda_n e^{(n)} - \sum_{n=1}^{m} x_n e^{(n)} \right\|$$

$$\leq \left\| x - \sum_{n=1}^{m} \lambda_n e^{(n)} \right\| + \left\| x - \sum_{n=1}^{m} x_n e^{(n)} \right\| < \frac{\varepsilon}{2} + \frac{\varepsilon}{2} < \varepsilon.$$

Since $\varepsilon > 0$ and $k \in \mathbb{N}$ were arbitrary, it follows that $\lambda_k = x_k$ for all $k \in \mathbb{N}$.

(ii) Now we show the statement for c_0.
Let $x \in c_0$ and $\varepsilon > 0$ be given. Then there is $n_0 \in \mathbb{N}$ such that $|x_n| < \varepsilon/2$ for all $n > n_0$. Thus we have for all $m > n_0$

$$\left\| x - \sum_{n=1}^{m} x_n e^{(n)} \right\| = \sup_{n>m} |x_n| \leq \sup_{n>n_0} |x_n| \leq \frac{\varepsilon}{2} < \varepsilon.$$

The uniqueness of the representation is shown in the same way as in Part (i).

(iii) Let $x \in c$. Then $x = \xi \cdot e + x^{(0)}$ where $\xi = \lim_{n \to \infty} x_n$ and $x^{(0)} \in c_0$. Since $x^{(0)} = (x_n^{(0)}) \in c_0$ with $x_n^{(0)} = x_n - \xi$ for all $n \in \mathbb{N}$, it follows from Part (ii) that

$$x^{(0)} = \sum_{n=1}^{\infty} x_n^{(0)} e^{(n)} = \sum_{n=1}^{\infty} (x_n - \xi) e^{(n)}, \quad \text{hence } x = \xi \cdot e + \sum_{n=1}^{\infty} (x_n - \xi) e^{(n)}.$$

(iv)   Finally, we show the statement for ω.
    Let $x \in \omega$ and $\varepsilon > 0$ be given. We choose $n_0 \in \mathbb{N}$ such that $\sum_{n=n_0}^{\infty} 2^{-n} < \varepsilon$.
    Then we have for all $m > n_0$

$$d\left(x, \sum_{n=1}^{m} x_n e^{(n)}\right) = \sum_{n=m+1}^{\infty} \frac{1}{2^n} \cdot \frac{|x_n|}{1 + |x_n|} \leq \sum_{n=n_0}^{\infty} \frac{1}{2^k} < \varepsilon.$$

The uniqueness of the representation is shown in the same way as in Part (i).

(d)   We show that the space ℓ_∞ is not separable; then it does not have a Schauder basis by Remark 3.4.2 (b).
    Let $A = \{a \in \ell_\infty : a_n \in \{0,1\}, \ n = 1,2,\dots\}$. Then A is uncountable. Also we have $d(a^{(1)}, a^{(2)}) = 1$ for any two distinct elements $a^{(1)}$ and $a^{(2)}$ of A.
    If ℓ_∞ were separable, then ℓ_∞ would contain a countable dense set S. We would have

$$\ell_\infty \subset \bigcup_{s \in S} B_{1/3}(s),$$

and $\bigcup_{s \in S} B_{1/3}(s)$ is a countable set. Since A is uncountable, there must be $\tilde{s} \in S$ such that $B_{1/3}(\tilde{s})$ contains two distinct elements $a, \tilde{a}$ of A. This implies

$$1 = d(a, \tilde{a}) \leq d(a, \tilde{s}) + d(\tilde{s}, \tilde{a}) = \frac{1}{3} + \frac{1}{3},$$

which is a contradiction. $\square$

Remark 3.4.4 *If X is a linear metric space with a Schauder basis (b_n), then, for each $x \in X$, there exists a unique sequence (λ_n) of scalars with $x = \sum \lambda_n b_n$. We define the map $\ell^{(n)} : X \to \mathbb{C}$ for each $n \in \mathbb{N}$ by*

$$\ell^{(n)}(x) = \lambda_n \ (x \in X).$$

Then we have $\ell^{(n)} \in X^{\#}$ for each $n \in \mathbb{N}$. For if $x, y \in X$ and λ is a scalar, then we have $x = \sum \lambda_n b_n = \sum \ell^{(n)}(x) b_n$, $y = \sum \mu_n b_n = \sum \ell^{(n)}(y) b_n$ and $\lambda x + y = \sum \ell^{(n)}(\lambda x + y) b_n$, and so

$$\lambda x + y = \sum \ell^{(n)}(\lambda x + y) b_n = \lambda \sum \ell^{(n)}(x) + \sum \ell^{(n)}(y) b_n$$
$$= \sum \left(\lambda \ell^{(n)}(x) + \ell^{(n)}(y)\right) b_n,$$

and the uniqueness of the representation implies

$$\ell^{(n)}(\lambda x + y) = \lambda \ell^{(n)}(x) + \ell^{(n)}(y).$$

The functionals $\ell^{(n)}$ must not be confused with the coordinates P_n, where $P_n(y) = y_n \ (n \in \mathbb{N})$ for each sequence $y = (y_k)$.
The functionals $\ell^{(n)}$ are referred to as coordinate functionals *with respect to the Schauder basis (b_n).*

Example 3.4.5 *We define in c*

$$b_n = \begin{cases} e & (n = 1) \\ e^{(n-1)} & (n \geq 2). \end{cases}$$

Then we have for $x \in c$ with $\xi = \lim_{n\to\infty} x_n$ by Part (iii) of the proof of Example 3.4.3 (c)

$$x = \xi \cdot e - \sum_{n=1}^{\infty}(x_n - \xi)e^{(n)} = \xi \cdot b_1 + \sum_{n=2}^{\infty}(x_{n-1} - \xi)b_n$$

It follows that

$$\ell^{(1)}(x) = \xi \ \text{and} \ \ell^{(n)}(x) = x_{n-1} - \xi \ \text{for} \ n = 2, 3, \dots .$$

Lemma 3.4.6 *Let X be a linear semimetric space, $x_0 \in X$, $p(x_0) \neq 0$, where p denotes the paranorm given by the semimetric, $f \in X^\#$ and $g : X \to X$ be defined by $g(x) = f(x)x_0$ for all $x \in X$. If g is continuous, then f is also continuous.*

Proof. If f is not continuous, then, in particular, f is not continuous at 0. Hence there exists a sequence (x_n) with $x_n \to 0$ and $f(x_n) \not\to f(0)$. Since f is linear, it follows that $f(0) = 0$. Thus we can choose the sequence (x_n) with $x_n \to 0$ such that $(1/(f(x_n)))$ is bounded. Since g is continuous by assumption and linear by definition, we have $g(x_n) \to 0$, and so $g(x_n)/f(x_n) \to 0$. But we have

$$\frac{g(x_n)}{f(x_n)} = x_0 \ \text{for all} \ n, \ \text{hence} \ p(x_0) - 0,$$

contradicting $p(x_0) \neq 0$. □

We note that if X is a linear metric space, then the condition $p(x) \neq 0$ can be replaced by $x \neq 0$.

Definition 3.4.7 A Schauder basis (b_n) is said to be *monotone*, if the sequence

$$\left(p\left(\sum_{k=1}^{n}\lambda_k b_k\right)\right)_{n=1}^{\infty} \ \text{is monotone increasing.}$$

Theorem 3.4.8 *Let (b_n) be a monotone Schauder basis for a linear metric space X. Then the coordinate functionals of Definition 3.4.4 with respect to this basis are continuous.*

Proof. We define the map $g_n : X \to X$ by $g_n(x) = \ell^{(n)}(x)b_n$ $(x \in X)$ for each $n \in \mathbb{N}$, where $\ell^{(n)}$ is the n–th coordinate functional. Then we obtain from the monotony of (b_n)

$$p(g_n(x)) = p\left(\sum_{k=1}^{n}\ell^{(k)}(x)b_k - \sum_{k=1}^{n-1}\ell^{(k)}(x)b_k\right)$$

$$\leq p\left(\sum_{k=1}^{n}\ell^{(k)}(x)b_k\right) + p\left(\sum_{k=1}^{n-1}\ell^{(k)}(x)b_k\right) \leq 2 \cdot p(x),$$

and so g_n is continuous at 0. Since g_n is linear, g_n is continuous by Theorem 2.8.20 (b). By Remark 3.4.2 (a), $b_n \neq 0$, and so $\ell^{(n)}$ is continuous by Lemma 3.4.6. □

3.5 Open mapping theorem

In this and the following section, we will prove the *open mapping*, and *closed graph theorems* which are among the most important results in functional analysis.

We need a few lemmas for their proofs.

Lemma 3.5.1 *Let X and Y be linear metric spaces, X be complete, and $f : X \to Y$ be a closed linear map with the following property.*

(I) For every neighborhood N of 0 in X, the set $\overline{f(N)}$ is a neighborhood of 0 in Y.

Then f is open; in particular, f is onto.

Proof.

(i) To prove that f is open we show that for every $r > 0$ the set $f(B_r(0))$ is a neighborhood of 0 in Y, where, as usual, $B_r(0) = \{x \in X : d(x, 0) < r\}$.

   We put $r(n) = r/2^n$ and $G_n = B_{r(n)}(0) \subset X$ for $n \in \mathbb{N} \cup \{0\}$. Then G_n is a neighborhood of 0 in X for each n, hence $\overline{f(G_n)}$ is a neighborhood of 0 in Y for each n by Property (I). So, for each n, there exists $\varepsilon_n > 0$ such that

$$B_{\varepsilon_n}(0) \subset \overline{f(G_n)} \text{ in } Y. \tag{3.2}$$

By choosing $\varepsilon_n < 1/n$, we achieve $\varepsilon_n \to 0$ as $n \to \infty$. The proof of Part (i) is completed when we have shown that

$$\overline{f(G_1)} \subset f(G_0).$$

For each $n \in \mathbb{N} \cup \{0\}$, let $H_n = B_{\varepsilon_n}(0)$ in Y. Let $y \in H_1$, hence $y \in \overline{f(G_1)}$ by (3.2). Since $\overline{f(G_1)}$ is closed, there is a sequence (y_n) in $f(G_1)$ with $y_n \to y$ $(n \to \infty)$. In particular, for ε_2, there is $x_1 \in G_1$ such that $p(y - f(x_1)) < \varepsilon_2$, where p is the paranorm that generates the metric on Y. This implies $y - f(x_1) \in H_2$. As above there exists $x_2 \in G_2$ such that $p(y - f(x_1) - f(x_2)) < \varepsilon_3$. Continuing in this way, we obtain $x_n \in G_n$ for each n such that

$$p\left(y - \sum_{k=1}^{n} f(x_k)\right) < \varepsilon_{n+1}.$$

Therefore we have

$$\sum_{k=1}^{\infty} f(x_k) = y. \tag{3.3}$$

Since $x_n \in G_n$ for all n, we have $p_X(x_n) < r/2^n$ for all n, where p_X is the paranorm on X. Thus the series $\sum_{n=1}^{\infty} x_n$ is absolutely convergent, hence convergent by Theorem 3.2.7, since X is complete. We put

$$x = \sum_{n=1}^{\infty} x_n. \tag{3.4}$$

Since f is closed, we obtain from (3.3) and (3.4) by Theorem 2.9.3 that $y = f(x)$. Also, (3.4) implies

$$p_X(x) \leq \sum_{n=1}^{\infty} p_X(x_n) < \sum_{n=1}^{\infty} \frac{r}{2^n} = r,$$

that is, $x \in G_0$, hence $y = f(x) \in f(G_0)$. Since $y \in H_1$ was arbitrary, we have $H_1 \subset f(G_0)$. Thus $f(B_r(0)) = f(G_0)$ is a neighborhood of 0 in Y. This completes proof of Part (i).

Since every neighborhood of 0 contains a set $B_r(0)$, we have shown that f maps neighborhoods of 0 in X onto neighborhoods of 0 in Y. Since f is linear, it follows that f is open by Remark 2.9.9. Thus we have shown the first statement of the lemma, namely that f is open.

(ii) Now we show that f is onto.

Let $y \in Y$ be given. Since the whole space X is open and f is open, as we have just shown, $f(X)$ is open. Also $f(X)$ is a linear space and contains an interior point y_0, so there exists $r > 0$ such that $B_r(y_0) \subset f(X)$. Furthermore, there is $\lambda > 0$ such that $\lambda y \in B_r(b) - y_0$, for if $\lambda y \notin B_r(b) - y_0$ for all $\lambda > 0$, then it would follow that

$$p(\lambda y + y_0 - y_0) = p(\lambda y) \geq r,$$

in particular, for $\lambda_n = 1/n$ ($n \in \mathbb{N}$), we would have $\lambda_n \to 0$ as $n \to \infty$, but $p(\lambda_n y) \geq r$, a contradiction to Condition (P.5) in Definition 3.2.1. Therefore it follows that

$$y \in \frac{1}{\lambda} \cdot B_r(0) - \frac{1}{\lambda} \cdot y_0 \in f\left(\frac{1}{\lambda} \cdot X\right) - f\left(\frac{1}{\lambda} \cdot X\right) \subset f(X).$$

So we have $Y \subset f(X)$, hence f is onto. □

Lemma 3.5.2 *Let X and Y be linear metric spaces, X be complete, and $f : X \to Y$ be a closed linear map with the following property.*

(II) For every neighborhood N of 0 in X, the set $f(N)$ is somewhere dense in Y.

Then f is open; in particular, f is onto.

Proof. Let N be a neighborhood of 0 in X and G be a symmetric neighborhood of 0 with $G + G \subset N$. Such a symmetric neighborhood exists as the following argument shows. Since $0 + 0 = 0$ and addition is continuous, there are neighborhoods V_1 and V_2 of 0 such that $V_1 + V_2 \subset N$. We put $G = V_1 \cap V_2 \cap (-V_1) \cap (-V_1)$. Since $f(G)$ is somewhere dense by Property (II), $\overline{f(G)}$ has an interior point. By the symmetry of G, we have $G + G = G - G$, and so by Remark 2.8.14

$$\overline{f(N)} \supset \overline{f(G+G)} = \overline{f(G-G)} = \overline{f(G) - f(G)} \supset \overline{f(G)} - \overline{f(G)}.$$

Since $\overline{f(G)}$ has an interior point, $\overline{f(G)} - \overline{f(G)}$ is a neighborhood of 0 by Remark 2.8.4 (i), and so $\overline{f(N)}$ is a neighborhood of 0. The statement now follows from Lemma 3.5.1. □

Now we are able to prove the open mapping theorem.

Theorem 3.5.3 (The open mapping theorem)
Let X and Y be Fréchet spaces and $f : X \to Y$ be a closed linear surjective map. Then f is open.

Proof. Let N be a neighborhood of 0 in X. Since N is absorbing, we have by Theorem 2.2.3 (b)

$$\bigcup_{k=1}^{\infty} k \cdot N = X,$$

and so, since f is surjective and linear,

$$\bigcup_{k=1}^{\infty} k \cdot f(N) = f\left(\bigcup_{k=1}^{\infty} k \cdot N\right) = f(X) = Y.$$

The complete metric space B is of second category by Baire's category theorem, Theorem 2.10.9. Therefore not every set $k \cdot f(N)$ is nowhere dense. Therefore $f(N)$ is somewhere dense, and the statement follows by Lemma 3.5.2. □

Remark 3.5.4 *In the little open mapping theorem, Remark 2.9.5 (f), we still had to assume that the map was one to one.*

Many textbooks present the following version of the open mapping theorem, where the assumption of f being a closed map is replaced by f being continuous (for instance, [176, 213, 208]).

Theorem 3.5.5 (Open mapping theorem II)
Let X and Y be Fréchet spaces and $f : X \to Y$ be a continuous linear surjective map. Then f is open.

Proof. Since the Fréchet spaces X and Y are Hausdorff spaces, the continuity of the map $f : X \to Y$ implies by Theorem 2.9.3 that f is closed, and consequently open by Theorem 3.5.3. $\qquad\square$

We obtain the following useful corollaries. The first one states that, continuous, linear, bijective maps between Fréchet spaces are homeomorphisms.

Corollary 3.5.6 *Let X and Y be Fréchet spaces, and $f : X \to Y$ be linear, continuous, one to one and onto. Then f is a homeomorphism.*

Proof Since f is continuous and Y, being a metric space, is a Hausdorff space, f is closed by Theorem 2.9.3. By the open mapping theorem, Theorem 3.5.3, f is open, and so f^{-1} is continuous. $\qquad\square$

The next corollary states that two comparable complete invariant metrics are equivalent.

Corollary 3.5.7 *If a linear space has two complete invariant metrics one of which is stronger than the other one, then both metrics are equivalent; that is, two comparable complete invariant metrics are equivalent.*

Proof. Let d_1 and d_2 be two metrics for X, both of which are complete and invariant. If d_1 is stronger than d_2, then the identity map id: $(X, d_1) \to (X, d_2)$ is linear, continuous, one to one and onto, hence a homeomorphism by Corollary 3.5.6. $\qquad\square$

3.6 The closed graph theorem

Now we are able to prove the *closed graph theorem*. Later we are also going to establish the *uniform boundedness principle*. Both results are of great importance in functional analysis.

First we prove the closed graph theorem; a special case of this is Corollary 2.9.6 where the final space is $\mathbb{C}$.

Theorem 3.6.1 (The closed graph theorem)
Any closed linear map between Fréchet spaces is continuous.

Proof. Let q and $\tilde{q}$ be the paranorms that define the metrics of the Fréchet spaces X and Y; also let $f : X \to Y$ be a closed linear map. We put

$$p(x) = q(x) + \tilde{q}(f(x)) \text{ for all } x \in X.$$

(i) We show that p is a total paranorm on X.

It is clear that the conditions in (P.1) to (P.4) for a paranorm in Definition 3.2.1 are satisfied. To show that the condition in (P.5) of Definition 3.2.1 is also satisfied, we assume that $\lambda_n \to \lambda$ and $p(x_n - x) \to 0$. Since p is stronger than q, this implies $q(x_n - x) \to 0$, and then also

$$\tilde{q}(f(x_n) - f(x)) = \tilde{q}(f(x_n - x)) = p(x_n - x) - q(x_n - x) \to 0.$$

Since q and $\tilde{q}$ are paranorms, it follows that

$$q(\lambda_n x_n - \lambda x) \to 0 \text{ and } \tilde{q}(\lambda_n f(x_n) - \lambda f(x)) \to 0,$$

hence

$$p(\lambda_n x_n - \lambda x) = q(\lambda_n x_n - \lambda x) + \tilde{q}\left(f(\lambda_n x_n - \lambda x)\right)$$
$$= q(\lambda_n x_n - \lambda x) + \tilde{q}\left(\lambda_n f(x_n) - \lambda f(x)\right) \to 0.$$

Finally, since p is total, p defines a metric on A.

(ii) Now we show that X is complete with respect to p.

Let (x_n) be a Cauchy sequence in (X, p). Then (x_n) is a Cauchy sequence in (X, q) by the definition of p. Also $(f(x_n))$ is a Cauchy sequence in X. Since (X, q) and $(Y, \tilde{q})$ both are complete, both sequences converge, $q(x_n - x) \to 0$ and $\tilde{q}(f(x_n) - y) \to 0$, say. Since f is closed, it follows that $y = f(x)$ by Theorem 2.9.3. This implies $p(x_n - x) \to 0$.

Since p defines an invariant complete metric on X, stronger than q, the metrics defined by p and q are equivalent by Corollary 3.5.7. Finally $p(x_n) \to 0$ implies $\tilde{q}(f(x_n)) \to 0$, and so f is p–continuous by Theorem 1.10.5 (b). $\qquad\square$

3.7 Uniform boundedness principle

Now we are going to prove the *uniform boundedness principle*. We need the following definition.

Definition 3.7.1 A set Φ of linear maps of a linear topological space X into a linear topological space Y is said to be *equicontinuous* if, for every neighborhood G of 0 in Y, the set

$$\bigcap_{f \in \Phi} f^{-1}(G)$$

is a neighborhood of 0 in X; this means that, for every neighborhood G of 0 in Y, there exists a single neighborhood N of 0 in X such that $f(N) \subset G$ for all $f \in \Phi$.

The set Φ is said to be *pointwise bounded*, if the set $\Phi_x = \{f(x) : f \in \Phi\}$ is a bounded set for every $x \in X$.

Example **3.7.2** *Let Y be a seminormed space. We use the notations of Definition 3.7.1. Then Φ is equicontinuous if and only if Φ is uniformly bounded on a neighborhood G of 0, that is, if there exists $M \in \mathbb{R}$ such that $\|f(x)\| \le M$ for all $x \in G$ and for all $f \in \Phi$.*

Proof.

(i) First we assume that Φ is equicontinuous. Since $B_1(0)$ is a neighborhood of 0 in Y, the set $G = \bigcap_{f \in \Phi} f^{-1}(B_1(0))$ is a neighborhood of 0 in X, and $f(x) \in B_1(0)$ for all $x \in G$ and all $f \in \Phi$, that is, $\|f(x)\| \leq 1$ for all $x \in G$ and all $f \in \Phi$. Hence Φ is uniformly bounded.

(ii) Now we assume that Φ is uniformly bounded. Then there are a neighborhood G of 0 in X and a constant $M \in \mathbb{R}$ such that $\|f(a)\| < M$ for all $x \in G$ and all $f \in \Phi$. Hence we have

$$f\left(\frac{\varepsilon}{M} \cdot G\right) \subset B_\varepsilon(0) \text{ for each } f \in \Phi \text{ and for all } \varepsilon > 0.$$

$\square$

We obtain as a special case:

Remark 3.7.3 *If X and Y are seminormed spaces, then Φ is equicontinuous if and only if there exists a constant M such that*

$$\|f\| = \sup\{\|f(x)\| : \|x\| \leq 1\} \leq M \text{ for all } f \in \Phi.$$

Now we prove the uniform boundedness principle. The proof uses the closed graph theorem, Theorem 3.6.1.

Theorem 3.7.4 (Uniform boundedness pinciple)
Any pointwise bounded sequence of continuous linear functionals on a Fréchet space is equicontinuous.

Proof. Let X be a Fréchet space, and (f_n) be a pointwise bounded sequence of continuous linear functionals on X. We define the map $g : X \to \ell_\infty$ by $g(x) = (f_n(x))$ for all $x \in X$ and show that g is continuous. Then (f_n) is equicontinuous, since if $G = g^{-1}(B_1(0))$, then G is a neighborhood of 0 in X, by the continuity of g. Furthermore $x \in G$ implies that $\|g(x)\|_\infty = \sup_n |f_n(x)| < 1$, hence $x \in f_n^{-1}(B_1(0))$ for all $n \in \mathbb{N}$, that is, $a \in \bigcap_{n=1}^\infty f_n^{-1}(B_1(0))$. Therefore we have $G \subset \bigcap_{n=1}^\infty f_n^{-1}(B_1(0))$, and so the sequence (f_n) is equicontinuous.

(i) First, we show that the function $g : X \to \omega$ is continuous. Let $x_k \to x$ $(k \to \infty)$ in X. Since each functional f_n is continuous, this implies $f_n(x_k) \to f(x)$ $(k \to \infty)$. Also we obtain from the equivalence of convergence in ω and coordinatewise convergence (Theorem 3.3.5 (a)) that $g(x_k) \to g(x)$ in ω, and consequently $g : X \to \omega$ is continuous.

(ii) Now we show that $g : X \to \ell_\infty$ is continuous.
Since ω is a Hausdorff space and $g : X \to \omega$ is continuous, $g : X \to \omega$ is closed by Theorem 2.9.3. Furthermore the map $g : X \to \ell_\infty$ is closed, since, by Definition 3.3.1 (a) and Theorem 3.3.5 (a), the natural topology on ℓ_∞ is stronger than the topology of ω on ℓ_∞. Finally, since X and ℓ_∞ are Fréchet spaces, the map $g : X \to \ell_\infty$ is continuous by the closed graph theorem, Theorem 3.6.1. $\square$

The famous *Banach-Steinhaus theorem* is closely related to the uniform boundedness principle, Theorem 3.7.4. It states that the limit function of a sequence of pointwise convergent linear functionals on a Frechet space is continuous.

Theorem 3.7.5 (Banach–Steinhaus theorem)
Let (f_n) be a pointwise convergent sequence of continuous linear functionals on a Fréchet space X. Then the limit function f with $f(x) = \lim_{n \to \infty} f_n(x)$ for all $x \in X$ is continuous.

Proof. The function $g : X \to c$ of the proof of Theorem 3.7.4 is continuous by the same argument as in the proof of Theorem 3.7.4 with ℓ_∞ replaced by c. The composite function $f = \lim \circ g$ of continuous functions is continuous. $\square$

We close this section with the so–called *convergence lemma*.

Theorem 3.7.6 (Convergence lemma) ([208, 9-3-104])
Let X and Y be linear topological spaces and $(f_\delta(x))_\delta$ be an equicontinuous map of linear functions $f_\delta : X \to Y$. Then the set $\{x \in X : f_\delta \to 0\}$ is a closed subspace of X.

Proof. The proof is left to the reader. $\square$

3.8   Properties of seminorms

Now we are going to consider a specialization of the concept of a paranorm, namely that of a *seminorm*; the notion of a seminorm was defined in Definition 1.8.1. We recall that a norm is a total seminorm.

We start with two simple observations.

Remark 3.8.1 *(a) A norm satisfies $\|x\| > 0$ whenever $x \neq 0$.*
(b) A seminorm is a paranorm.

Proof.

(a)   This is obvious (Remark 1.8.2).

(b)   Putting $\lambda = 0$ in the condition in (S.3) of Definition 1.8.1, we obtain for all a

$$p(0) = p(\lambda \cdot x) = |\lambda| p(x) = 0,$$

that is, the condition in (P.1) of Definition 3.2.1 is satisfied. The condition in (S.1) of Definition 1.8.1 is the condition in (P.2) of Definition 3.2.1. Putting $\lambda = -1$ in (S.3) of Definition 1.8.1, we obtain for all x

$$p(-x) = p(\lambda x) = |\lambda| p(x) = p(x),$$

that is, (P.3) Definition 3.2.1 is satisfied. The condition in (S.2) of Definition 1.8.1 is the condition in (P.4) of Definition 3.2.1. Finally let $\lambda_n \to \lambda$ and $p(x_n - x) \to 0$ $(n \to \infty)$. Then the conditions in (S.1), (S.2) and (S.3) of Definition 1.8.1 yield

$$0 \le p(\lambda_n x_n - \lambda x) \le p((\lambda_n - \lambda)x) + p(\lambda_n(x_n - x))$$
$$= |\lambda_n - \lambda| p(x) + |\lambda_n| p(x_n - x).$$

Since $\lambda_n \to \lambda$ $(n \to \infty)$, there exists a constant M such that $|\lambda_n| \le M$ for all $n \in \mathbb{N}$, and so $|\lambda_n| p(x_n - x) \to 0$ $(n \to \infty)$. Thus it follows that $p(\lambda_n x_n - \lambda x) \to 0$ $(n \to \infty)$, and the condition in (P.5) of Definition 3.2.1 is also satisfied. $\square$

*Example **3.8.2** (a) Let f be a linear functional on a linear space X. We put*

$$p(x) = |f(x)| \text{ for all } x \in X.$$

Then p is a seminorm, which is never a norm unless X is 0– or 1–dimensional.
(b) Let $\mathcal{B}[X]$ be the space of all bounded complex–valued functions on a set X. We put

$$\|f\| = \sup\{|f(x)| : x \in X\} \text{ for all } f \in \mathcal{B}[X].$$

Then $\|\cdot\|$ obviously is a norm on $\mathcal{B}[X]$.
(c) Obviously, ℓ_p with $\|\cdot\|_p$ for $1 \le p < \infty$, and c_0, c and ℓ_∞ with $\|\cdot\|_\infty$ are normed spaces (Example 1.8.2 (b)).

Proof. We only have to show Part (a).
We have $p(x) = |f(x)| \ge 0$ for all $x \in X$, that is, (SN.1) of Definition 1.8.1,

$$p(x+y) = |f(x+y)| = |f(x) + f(y)| \le |f(x)| + |f(y)| = p(x) + p(y)$$

for all $x, y \in X$, that is, (SN.2) of Definition 1.8.1, and

$$p(\lambda x) = |f(\lambda x)| = |\lambda f(x)| = |\lambda| \cdot |f(x)| = |\lambda| p(x)$$

for all $x \in X$ and all scalars λ, that is, (SN.3) of Definition 1.8.1.
If $f \not\equiv 0$ then $N(f)$ is a maximal subspace by Remark 2.6.16. $\qquad\square$

Remark 3.8.3 *Let p and q be seminorms and f be a linear functional on a linear space X.*
(a) If p is bounded on an affine set S, then p is constant on S and equal to zero on the subspace $S - S$.
In particular, if p is bounded on a subspace, then p vanishes there.
A norm cannot be bounded on any affine set with more than one point.
(b) If $\mathrm{Re} f(x) \le p(x)$ for all $x \in X$, then $|f(x)| \le p(x)$ for all $x \in X$.
(c) If $\{p < 1\} \cap \{f = 1\} = \emptyset$, then $|f(x)| \le p(x)$ for all $xx \in X$. The converse implication also holds true.
(d) If $\{p < 1\} \subset \{q \le 1\}$ then $p \ge q$, that is, $p(x) \ge q(x)$ for all $x \in X$.
(e) The seminorm p is stronger than the seminorm q if and only if there exists a constant M such that $q(x) \le M \cdot p(x)$ for all $x \in X$.
(f) If q is bounded on the set $\{p < 1\}$, then p is stronger than q.

Proof.

(a) Let p be bounded on S.

   (i) First we assume that S is a subspace.
        If there exists $x \in S$ with $p(x) \ne 0$, then p is unbounded on S, since we have $\lambda x \in S$ for any scalar λ and also $p(\lambda x) = |\lambda| \cdot p(x)$ by (SN.3) of Definition 1.8.1.

   (ii) Now we assume that S is an affine set.
        Since

        $$0 \le p(x - y) \le p(x) + p(y) \text{ for all } x, y \in S,$$

        the boundedness of p on S implies that p is bounded on $S - S$. By Theorem 2.3.9, $S - S$ is a subspace, and so p vanishes identically on $S - S$ by Part (i). Since S is a translation of $S - S \subset N(p)$, p is constant on S by Remark 3.3.4 (e).

(b) Let x be a vector. Then there exist $r \ge 0$ and $\phi \in \mathbb{R}$ such that $f(x) = re^{i\phi}$. We have $f(x \cdot e^{(-i\phi)}) = r \in \mathbb{R}$, hence $f(x \cdot e^{(-i\phi)}) \le p(x \cdot e^{(-i\phi)})) = p(x)$, and so $r = |f(x)| \le p(x)$.

(c) (i) First we show that $\{p < 1\} \cap \{f = 1\} = \emptyset$ implies $|f(x)| \le p(x)$ for all x. We assume $|f(x)| > p(x)$ for some vector x. Then we have

$$f\left((f(x))^{-1} \cdot x\right) = |f(x)|^{-1} p(x) < 1,$$

hence $y = (f(x))^{-1} \cdot x \in \{p < 1\}$, and

$$f(y) = f\left((f(x))^{-1} \cdot x\right) = (f(x))^{-1} \cdot f(x) = 1,$$

hence $y \in \{f = 1\}$. Thus we have $y \in \{p < 1\} \cap \{f = 1\}$.

(ii) Now we show the converse part, that is, $|f(x)| \le p(x)$ for all x implies $\{p < 1\} \cap \{f = 1\} = \emptyset$.
We assume that there exists $y \in \{p < 1\} \cap \{f = 1\}$. Then we have $p(y) < 1$ and $f(y) = 1$, that is, $p(y) < |f(y)|$.

(d) We assume that $p \ge q$ does not hold. Then there exist vector x and a scalar λ such that $q(x) > \lambda > p(x)$. Since $\lambda > 0$, we may put $y = \lambda^{-1} \cdot x$ and obtain

$$p(y) = p\left(\lambda^{-1} \cdot x\right) = \lambda^{-1} \cdot p(x) < 1 \text{ and } q(y) = q\left(\lambda^{-1} \cdot x\right) = \lambda^{-1} \cdot q(x) > 1,$$

and so $y \in \{p < 1\} \setminus \{q \le 1\}$.

(e) The sufficiency follows by Remark 3.3.4 (f).
Conversely, we assume that p is stronger than q. If such a constant M does not exists then, for every $n \in \mathbb{N}$, there exists a vector x_n such that $q(x_n) \ge n^2 \cdot p(x_n)$. Now $p(x_n) \ne 0$ for each n, for otherwise, if $p(x_n) = 0$ for some n, then it would follow by Remark 3.3.2 that $q(x_n) = 0$, since p is stronger than q. Putting $y_n = (np(x_n))^{-1} \cdot x_n$ for $n = 1, 2, \ldots$, we obtain

$$p(y_n) = p\left((np(x_n))^{-1} \cdot x_n\right) = \frac{1}{np(x_n)} \cdot p(x_n) = \frac{1}{n} \to 0 \; (n \to \infty)$$

and

$$q(y_n) = q\left((np(x_n))^{-1} \cdot x_n\right) = \frac{1}{np(x_n)} \cdot q(x_n) \ge n \to \infty \; (n \to \infty),$$

and consequently p is not stronger than q.

(f) We assume that q is bounded on the set $\{p < 1\}$. Then there is a constant M such that $q(x) \le M$ for all $x \in X$ with $p(a) < 1$. If $M \ne 0$ then we obtain by Part (d) applied to q/M that $q(x) \le M \cdot p(x)$. Now p is stronger than q by Part (e).
If $M = 0$, then $N(q) \supset \{p < 1\}$. The set $\{p < 1\}$, however, is absorbing, and $N(q)$ is a subspace. Thus $N(q)$ is the whole space, that is, $q \equiv 0$. $\square$

Seminorms are slightly better behaved than paranorms. The importance of seminorms p in functional analysis arises from the fact that the set $\{p < 1\}$ is convex.

Definition 3.8.4 Let p be a seminorm on a seminormed space (X, q). Then

$$\|p\| = \sup\{p(x) : q(x) \le 1\}$$

is called the *norm of p*.
If f is a linear map from a seminormed space (X, p) into a seminormed space (Y, q) then

$$\|f\| = \sup\{q(f(x)) : p(x) \le 1\}$$

is called the *norm of f*.

Remark 3.8.5 *(a) The norm of a linear functional f is*

$$\|f\| = \sup\{|f(x)| : p(x) \le 1\}.$$

(b) If f is a linear functional and $p(x) = |f(x)|$ for all x, then we have that $\|p\| = \|f\|$; $\|p\|$ measures the value of f on the unit disc.
(c) Obviously $\|p\| = 0$ implies $p = 0$; so $\|\cdot\|$ indeed is a norm.
(d) If p is defined on a seminormed space (X, q) and if S is a subset of X then we write

$$\|p\|_S = \{p(s) : s \in S \text{ and } q(s) \le 1\}.$$

Definition 3.8.6 (a) A seminorm p is said to be *bounded*, if $\|p\| < \infty$.
(b) A linear map f between two seminormed spaces is said to be *bounded linear map*, if $\|f\| < \infty$.

Lemma 3.8.7 *Let p be a bounded seminorm on a seminormed space (X, q) and f be a bounded linear map from (X, q) into a seminormed space $(Y, \tilde{q})$. Then we have*

$$p(x) \le \|p\| \cdot q(x) \text{ and } \tilde{q}(f(x)) \le \|f\| \cdot q(x) \text{ for all } x \in X.$$

  Proof.

(i)   We show that $p(x) \le \|p\| \cdot q(x)$ for all $x \in X$.
       It follows from Definition 3.8.4 that

$$q(x) \le 1 \text{ implies } p(x) \le \|p\|. \tag{3.5}$$

       From this we obtain $p(x) \le \|p\| \cdot q(x)$ for all $x \in X$, for otherwise, there exist a vector x and a scalar λ with

$$p(x) > \lambda \ge \|p\| \cdot q(x).$$

       We put $y = (\|p\|/\lambda)x$ and obtain

$$q(y) = q\left(\frac{\|p\|}{\lambda} \cdot x\right) = \frac{\|p\|}{\lambda} \cdot q(x) < 1$$

       and

$$p(y) = p\left(\frac{\|p\|}{\lambda} \cdot x\right) = \frac{\|p\|}{\lambda} \cdot p(x) > \|p\|,$$

       which contradicts (3.5).

(ii) The inequality $\tilde{q}(f(x)) \le \|f\| \cdot q(x)$ for all $x \in X$ is shown exactly as the statement for q in Part (i) of the proof. $\square$

Example 3.8.8 *(a) We define $h(x, y) = 3x + 4y$ for all $(x, y) \in \mathbb{R}^2$. Then we have $h \in (\mathbb{R}^2)^\#$ and $\|h\| = 5$.*
(b) We define on c_0 (with the natural norm)

$$h(x) = \sum_{n=1}^{\infty} \frac{x_n}{n!} \text{ for all } x = (x_n) \in c_0.$$

Then we have $h \in c_0^\#$ and $\|h\| = e - 1$. (Of course, here e denotes Euler's number.)
(c) We define on ϕ

$$\|x\| = \sup_k |x_k| \text{ for all } x = (x_k) \in \phi.$$

Then ϕ is a normed subspace of c_0. We put

$$h(x) = \sum_{k=1}^{\infty} |x_k| \text{ for all } x = (x_k) \in \phi.$$

Then h is a norm on ϕ with $\|h\| = \infty$.

Proof.

(a) Obviously we have $h \in (\mathbb{R}^2)^{\#}$.
Furthermore, it follows from

$$h\left(\frac{3}{5}, \frac{4}{5}\right) = 5 \text{ and } \left\|\left(\frac{3}{5}, \frac{4}{5}\right)\right\| = \sqrt{\frac{9+16}{25}} = 1$$

that $\|h\| \geq 5$. Let $a = (x, y) \in \mathbb{R}^2$ be given with $\|a\| \leq 1$. Applying Hölder's inequality (1.1) with $p = q = 2$, we obtain

$$|h(x,y)| = |3x + 4y| \leq \sqrt{3^2 + 4^2} \cdot \sqrt{x^2 + y^2} = 5 \cdot \|a\| \leq 5,$$

and so $\|h\| \leq 5$.

(b) Obviously we have $h \in c_0^{\#}$.
It follows for all $x \in c_0$ with $\|x\| = \sup_k |x_k| \leq 1$ that

$$|h(x)| \leq \sum_{k=1}^{\infty} \frac{|x_k|}{k!} \leq \sum_{k=1}^{\infty} \frac{1}{k!} = e - 1, \text{ hence } \|h\| \leq e - 1.$$

To show the converse inequality, let $m \in \mathbb{N}$ be given. We put $x = \sum_{k=1}^{m} e^{(k)}$. Then we have $x \in c_0$, $\|x\| = 1$ and

$$h(x) = \sum_{k=1}^{m} \frac{1}{k!}, \text{ that is, } \|h\| \geq \sum_{k=1}^{m} \frac{1}{k!}.$$

Since $m \in \mathbb{N}$ was arbitrary, we conclude

$$\|h\| \geq \sum_{k=1}^{\infty} \frac{1}{k!} = e - 1.$$

(c) Obviously ϕ is a linear subspace of c_0. Let $M > 0$ be given. We choose $n \in \mathbb{N}$ with $n > M$ and $x = \sum_{k=1}^{n} e^{(k)}$. Then we have $x \in \phi$, $\|x\| = 1$ and

$$h(x) = \sum_{k=1}^{\infty} |x_k| = n > M, \text{ hence } \|h\| > M.$$

Since $M > 0$ was arbitrary, it follows that $\|h\| = \infty$. $\qquad\square$

Definition 3.8.9 Let X and Y be seminormed spaces and Φ be a family of linear functions $f : X \to Y$. Then Φ is said to be a *uniformly bounded*, if there exists a number $M \in \mathbb{R}$ such that

$$\|f\| \leq M \text{ for all } f \in \Phi.$$

Remark 3.8.10 *(a) By Remark 3.7.3, a family Φ of linear functions between seminormed spaces is uniformly bounded if and only if Φ is equicontinuous.*
(b) Obviously a family Φ of linear functionals on a seminormed space X is pointwise bounded if and only if, for every $x \in X$, there exists $M_x \in \mathbb{R}$ such that

$$|f(x)| \leq M_x \text{ for all } f \in \Phi.$$

We obtain as an immediate consequence of Theorem 3.7.4.

Theorem 3.8.11 (Uniform boundedness principle for Banach spaces)
Every pointwise bounded sequence of continuous linear functionals on a Banach space is uniformly bounded.

3.9   The Minkowski functional

The *Minkowski functional* plays an important role in the connection with the metrizability of topological spaces. It can be used to introduce a norm on a linear space by means of a balloon. We recall that a balloon is a convex, balanced and absorbing set (Definition 2.4.1 (a)).

Definition 3.9.1 Let B be a balloon in a linear space X. Then the *Minkowski functional* or *gauge μ_B of B* is defined by

$$\mu_B(x) = \inf\{t > 0 : x \in tB\} \text{ for every } x \in X.$$

We need the following properties of the Minkowski functional.

Remark 3.9.2 *Let B be a balloon in a linear space X, μ_B be the Minkowski functional of B, and x be a vector. Then we have*
(a) $\lambda > \mu_B(x)$ *implies* $x \in \lambda B$,
(b) $\lambda < \mu_B(x)$ *implies* $x \notin \lambda B$,
(c) $x \in \lambda B$ *implies* $\mu_B(x) \leq \lambda$,
(d) $\{\mu_B < 1\} \subset B \subset \{\mu_B \leq 1\}$.

  Proof.

(a) Let $\lambda > \mu_B(x)$. Then, by the definition of the infimum, there exists a scalar μ such that $x \in \mu B$ and $\lambda > \mu > \mu_B(x)$. Since B is balanced, we obtain for $\nu = \mu/\lambda$, that is, $|\nu| \leq 1$

$$\mu B = (\lambda \cdot \nu)B = \lambda(\nu B) \subset \lambda B.$$

(b) Let $\lambda < \mu_B(x)$ then, by the definition of the Minkowski functional, $\lambda < \mu$ for all $\mu > 0$ for which $x \in \mu B$, thus $x \notin \lambda B$.

(c) This is equivalent to Part (b).

(d) The first inclusion is Part (a) with $\lambda = 1$, and the second inclusion is Part (c) with $\lambda = 1$.   □

The next result states that the Minkowski functional of a balloon in a linear space is a seminorm which is a norm if and only if B does not contain a maximal subspace of positive dimension.

Theorem 3.9.3 *Let B be a balloon in a linear space X. Then we have:*
(a) The Minkowski functional μ_B of B is a seminorm for X.
(b) The Minkowski functional μ_B of B is a norm if and only if B does not contain any subspace of positive dimension.

Proof.

(a) We show that μ_B is a seminorm.

Since B is absorbing, we have $\{\lambda > 0 : x \in \lambda B\} \neq \emptyset$ for each $x \in X$. Thus μ_B is defined for each $x \in X$, and $\mu_B(x) \geq 0$.

(i) We show that $\mu_B(\lambda x) = |\lambda|\mu_B(x)$ for all scalars λ and all $x \in X$.
We put
$$s = \mu_B(\lambda x) \text{ and } t = |\lambda|\mu_B(x).$$

It follows by Remark 3.9.2 (a) that $\lambda x \in s'B$ for all $s' \geq s$. Since B is balanced, so is $s'B$, and thus it follows that $|\lambda|x \in s'B$ for all $s' \geq s$.
We also have $t = |\lambda|\mu_B(x) \leq s'$ for all $s' \geq s$. This obviously is true for $\lambda = 0$. If $\lambda \neq 0$ then $|\lambda|x \in s'B$ implies $x \in |\lambda|^{-1} \cdot s'B$, hence $\mu_B(a) \leq |\lambda|^{-1} \cdot s'$ by Remark 3.9.2 (c), therefore $t = |\lambda|\mu_B(x) \leq s'$. Now this implies

$$t \leq s. \tag{3.6}$$

We also have $|\lambda|x \in t'B$ for all $t' \geq t$. This is obvious for $\lambda = 0$. If $\lambda \neq 0$, then $t' \geq t = |\lambda|\mu_B(x)$ implies $t'|\lambda|^{-1} \geq \mu_B(x)$, so $x \in t'|\lambda|^{-1} \cdot B$ by Remark 3.9.2 (a), hence $|\lambda|x \in t'B$. This implies $\lambda x \in t'B$ for all $t \geq t'$, since B is balanced, and so $t'B$ is balanced, too. From this we obtain $s = \mu_B(\lambda x) \leq t'$ for all $t' \geq t$ by Remark 3.9.2 (c), hence

$$s \leq t. \tag{3.7}$$

Finally (3.6) and (3.7) imply

$$\mu_B(\lambda x) = s = t = |\lambda|\mu_B(x).$$

(ii) Now we show that $\mu_B(x + y) \leq \mu_B(x) + \mu_B(y)$ for all $x, y \in X$.
Let $x, y \in X$ and $\varepsilon > 0$ be given. Since $\mu_B(x) + \varepsilon > \mu_B(x)$ and $\mu_B(y) + \varepsilon > \mu_B(y)$, it follows by Remark 3.9.2 (a) that

$$x \in (\mu_B(x) + \varepsilon)B \text{ and } y \in (\mu_B(y) + \varepsilon)B.$$

The convexity of the set B implies by Part Remark 2.2.6 (a)

$$x + y \in (\mu_B(x) + \varepsilon)B + (\mu_B(y) + \varepsilon)B \subset (\mu_B(x) + \mu_B(y) + 2\varepsilon)B,$$

hence by Remark 3.9.2

$$\mu_B(x + y) \leq \mu_B(x) + \mu_B(y) + 2\varepsilon.$$

Since $\varepsilon > 0$ was arbitrary, we have $\mu_B(x + y) \leq \mu_B(x) + \mu_B(y)$.

This completes the proof of Part (a).

(b) If μ_B is not a norm, then $N(\mu_B)$ is a subspace with a positive dimension. If follows by Remark 3.9.2 (d) that $N(\mu_B) \subset \{\mu_B < 1\} \subset B$.
Conversely, if B contains a subspace S with a positive dimension, then we obtain by Remark 3.9.2 (d) that $S \subset B \subset \{\mu_B \leq 1\}$. Thus μ_B is bounded on S and so vanishes identically on S by Remark 3.8.3 (a). Consequently μ_B is not a norm. $\square$

3.10   Local convexity

We mentioned in Section 2.11 that the weak linear topology $\sigma(X, \Phi)$, where X is a linear space and Φ is a family of seminorms defined on X, is locally convex. Now we are going to prove that this is the only kind there is.

Theorem 3.10.1 *Every locally convex linear topology for a linear space is $\sigma(X, \Phi)$ for some family Φ of seminorms defined on a linear space X.*

Proof. Let Γ be the set of convex balanced neighborhoods of 0 and Φ be the set of Minkowski functionals of the members of Γ. If G is a neighborhood of 0, then it includes a member of Γ, hence $G \supset \{p < 1\}$ where p is the gauge of that member.
Conversely, given $\varepsilon > 0$, $p \in \Phi$, let $G = \{p < 1\}$. Then G is a neighborhood of 0, in fact, $G \in \Gamma$, and $\varepsilon G \subset \{p < \varepsilon\}$. So we have shown that the linear topology generated by Φ is stronger than, and weaker than the original topology. $\qquad\square$

Now we are going to describe a locally convex space by means of a family Φ of seminorms which generates its topology.
We need the following concept.

Definition 3.10.2 *A family Φ of seminorms on a linear space X is said to be* determining, *or to* determine the topology of X, *if the topology of X is $\sigma(\Phi)$.*

The following result is useful.

Theorem 3.10.3 *Let $q, p_1, p_2, \ldots, p_n$ be seminorms on a linear space X, and M and $\varepsilon > 0$ be such that $p_k(x) < \varepsilon$ for all $k = 1, 2, \ldots, n$ implies $q(x) \leq M$. Then we have*

$$q(x) \leq \frac{M}{\varepsilon} \sum_{k=1}^{n} p_k(x) \text{ for all } x \in X. \tag{3.8}$$

Proof. We assume that (3.8) does not hold. Then there exists a vector x and a scalar λ such that

$$q(x) > \lambda > \frac{M}{\varepsilon} \sum_{k=1}^{n} p_k(x).$$

We put $y = (M/\lambda)a$. Then we have for each r with $1 \leq r \leq n$

$$p_r(y) = \frac{M}{\lambda} p_r(x) \leq \frac{M}{\lambda} \sum_{k=1}^{n} p_k(x) < \varepsilon, \text{ while } q(y) > M.$$

$\qquad\square$

We need the following results.

Remark 3.10.4 *Let Φ and Ψ be families of seminorms on a linear space X. Then we have:*
(a) $x_\delta \to x$ in $\sigma(\Phi)$ if and only if $p(x_\delta - x) \to 0$ for every $p \in \Phi$.
(b) If $\Phi = \bigcup_{\beta \in I} \Phi_\beta$, where I is an arbitrary indexing set, then $\sigma(\Phi) = \bigvee \sigma(\{\Phi_\beta : \beta \in I\})$; in particular, $\sigma(\Phi) = \bigvee\{\sigma(p) : p \in \Phi\}$, where we write $\sigma(p) = \sigma(\{p\})$, for short.
(c) If $\Phi \supset \Psi$, then $\sigma(\Phi)$ is stronger than $\sigma(\Psi)$.
(d) $\sigma(\Phi)$ is stronger than $\sigma(\Psi)$ if and only if every $p \in \Psi$ is continuous in $\sigma(\Phi)$.
(e) $\sigma(\{p_1, p_2, \ldots, p_n\}) = \sigma(\sum_{k=1}^{n} p_k)$.
(f) Let q be a seminorm on X. Then q is $\sigma(\Phi)$ continuous if and only if there exists a finite subset $\{p_1, p_2, \ldots, p_n\}$ of Φ and there exists a scalar M such that $q \leq M \sum_{k=1}^{n} p_k$ pointwise.

Proof.

(a) Let $x_\delta \to x$, $p \in \Phi$ and $\varepsilon > 0$ be given. Then we have $x_\delta - x \in \{p < \varepsilon\}$ eventually, since $\{p < \varepsilon\}$ is a neighborhood of 0. This means $p(x_\delta - x) < \varepsilon$ eventually.

Conversely, if $p(x_\delta - x) \to 0$ for every $p \in \Phi$, let G be a neighborhood of 0. Then there exist $\varepsilon > 0$ and $p_1, p_2, \ldots, p_n \in \Phi$ such that $G \supset \bigcap_{k=1}^n \{p_k < \varepsilon\}$. Since $p_k(x_\delta - x) \to 0$ for each k, we have $p_k(x_\delta - xx) < \varepsilon$ for all k eventually, and so $x_\delta - x \in G$ eventually.

(b) Let (x_δ) be a net in X and $x_\delta \to 0$ in $\sigma(\Phi)$. Then $p(x_\delta) \to 0$ for all $p \in \Phi$ by Part (a), and so $p(x_\delta) \to 0$ for each $p \in \Phi_\beta$ and each $\beta \in I$. Consequently we have $x_\delta \to 0$ in $\sigma(\Phi_\beta)$ for each $\beta \in I$ by Part (a), and so $x_\delta \to 0$ in $\bigvee\{\sigma(\Phi_\beta) : \beta \in I\}$ by Theorem 2.6.1.

Conversely, it is proved similarly that $x_\delta \to 0$ in $\bigvee\{\sigma(\Phi_\beta) : \beta \in I\}$ implies $x_\delta \to 0$ in $\sigma(\Phi)$.

(c) We have $x_\delta \to 0$ in $\sigma(\Phi)$ implies $p(x_\delta) \to 0$ for every $p \in \Phi$, hence, for each $p \in \Phi$.

(d) If $\sigma(\Phi)$ is stronger than $\sigma(\Psi)$, then every $p \in \Phi$ is continuous in $\sigma(\Phi)$, since it is continuous in $\sigma(\Psi)$.

To prove the converse implication, let $x_\delta \to 0$ in $\sigma(\Phi)$. Then we have $p(x_\delta) \to 0$ for every $p \in \Psi$.

(e) This is clear from Part (a).

(f) If the inequality holds, then q is continuous at 0, hence everywhere. To see this, let $x_\delta \to x$, hence $y_\delta = x_\delta - x \to 0$ by Remark 2.8.4 (d). Since q is a seminorm, it is a paranorm by Remark 3.8.1 (b), and so we obtain by Remark 3.3.4 (a)

$$0 \le |q(x_\delta) - q(x)| \le q(x_\delta - x) = q(y_\delta),$$

and $q(y_\delta) \to 0$, since q is continuous at 0. This implies $q(x_\delta) \to q(x)$, and so q is continuous at x.

Conversely, if q is continuous, the set $\{q < 1\}$ is a neighborhood of 0, and hence includes $\bigcap_{k=1}^n \{p_k < \varepsilon\}$ for some finite selection $p_1, p_2, \ldots, p_n \in \Phi$ and $\varepsilon > 0$. Part (f) now follows from Theorem 3.10.3. $\square$

Theorem 3.10.5 *Let S be a set in a locally convex space X with topology $\sigma(\Phi)$. Then S is bounded if and only if $p(S)$ is bounded for every $p \in \Phi$.*

Proof. Let S be bounded. Sine X has the topology $\sigma(\Phi)$, every $p \in \Phi$ is continuous by Remark 3.10.4 (a), hence $p(S)$ is bounded for each $p \in \Phi$ by Theorem 2.8.20 (a).

Conversely, let $p(S)$ be bounded for every $p \in \Phi$, and G be a neighborhood of 0. Then

$$G \supset \bigcap_{k=1}^n \{p_k < \varepsilon\} \text{ for some } p_1, p_2, \ldots, p_n \in \Phi \text{ and } \varepsilon > 0.$$

Then there exists M such that $p_k(x) < M$ for $x \in S$ and $k = 1, 2, \ldots$, and so

$$\frac{\varepsilon}{M} \cdot S \subset \frac{\varepsilon}{M} \bigcap_{k=1}^n \{p_k < M\} = \bigcap_{k=1}^n \{p_k < \varepsilon\} \subset G.$$

Hence S is bounded. $\square$

Theorem 3.10.6 *Let X and Y be linear topological spaces, $\sigma(\Phi)$ be the topology of Y and $f : X \to Y$ be a linear map. Then f is continuous if and only if $p \circ f$ is continuous for each $p \in \Phi$.*

Proof. If f is continuous, then $p \circ f$ is continuous for each $p \in \Phi$, since each $p \in \Phi$ is continuous.

Conversely, we assume that $p \circ f$ is continuous for each $p \in \Phi$. Let $x_\delta \to 0$. Then we have $(p \circ f)(x_\delta) \to p(f(0)) = p(0) = 0$ for each $p \in \Phi$, since $p \circ f$ is continuous, f is linear and p is a seminorm. It follows by Remark 3.10.4 (a), that $f(x_\delta) \to 0$. Consequently f is continuous at 0, hence continuous everywhere by Theorem 2.8.20 (b), since f is linear.   □

We now study conditions for the semimetrizability and seminormability in terms of determining families of seminorms. The following concept is useful for this task.

Definition 3.10.7 Let Φ and Ψ be families of seminorms on a linear space. If $\Psi \subset \Phi$ and $\sigma(\Psi) = \sigma(\Phi)$, then Φ is said to be *reducible* to Ψ. If Φ is not reducible to a proper subset of itself, then Φ is said to be *irreducible*.

Theorem 3.10.8 *Let Φ be a family of seminorms on a linear space X.*
(a) If Φ is countable, then $\sigma(\Phi)$ is semimetrizable.
(b) If $\sigma(\Phi)$ is semimetrizable, then Φ is reducible to a countable set.
(c) If ϕ is finite, then $\sigma(\Phi)$ is seminormable.
(d) If $\sigma(\Phi)$ is seminormable, then Φ is reducible to a finite set.

Proof.

(a) If $\Phi = \{p_n : n \in \mathbb{N}\}$, then $\sigma(\Phi) = \bigvee_n \sigma(\{p_n\})$ by Remark 3.10.4 (b), and $\bigvee_n \sigma(\{p_n\})$ is semimetrizable by Theorem 2.6.2. The paranorm q which yields the invariant semi-metric for $\bigvee_n \sigma(\{p_n\})$ is given by

$$q(x) = \sum_{n=1}^{\infty} \frac{1}{2^n} \cdot \frac{p_n(x)}{1 + p_n(x)} \text{ for all } x \in X. \tag{3.9}$$

(b) Let $\sigma(\Phi)$ be semimetrizable and (G_n) be a basic sequence of neighborhoods of 0. For each n, there exist $p_1^{(n)}, p_2^{(n)}, \ldots, p_{r(n)}^{(n)} \in \Phi$, and $\varepsilon > 0$ such that

$$G_n \supset \bigcap_{k=1}^{r(n)} \left\{ p_k^{(n)} < \varepsilon \right\}.$$

We put $\Psi = \{p_k^{(n)} : 1 \leq k \leq r(n); \ n = 1, 2, \ldots\}$. If (x_δ) is a net in X such that $p(x_\delta) \to 0$ for all $p \in \Psi$, then $x_\delta \in G_n$ eventually for each n, hence $x_\delta \to 0$. Thus $\sigma(\Psi)$ is stronger than $\sigma(\Phi)$. It is also weaker by Remark 3.8.1 (c), since $\Psi \subset \Phi$.

(c) If $\Phi = \{p_1, p_2, \ldots, p_n\}$, then $\sigma(\Phi)$ is seminormable by 3.8.1 (e).

(d) Let $\sigma(\Phi)$ be seminormable, $\sigma(\Phi) = \sigma(\{q\})$ say. Then it follows by Remark 3.8.1 (f) that there exist $p_1, p_2, \ldots, p_n \in \Phi$ and a constant M such that $q \leq M \sum_{k=1}^n p_k$. We put $\Psi = \{p_1, p_2, \ldots, p_n\}$. Then $\sigma(\Psi)$ is stronger than $\sigma(\{q\})$ by Remark 3.8.1 (d), and weaker by Remark 3.8.1 (c).   □

The following result is an immediate consequence of Theorem 3.10.8.

Corollary 3.10.9 *A locally convex semimetric space has defined on it a sequence (p_n) of seminorms such that its topology is $\sigma(\{p_n\})$ and is given by the paranorm defined in (3.9). The semimetric is a metric if and only if a total sequence (p_n) can be defined, in which all determining sequences are total.*

Corollary 3.10.10 *Let $\{p_n\}$ be an increasing family of seminorms on a linear space, that is, p_{n+1} is stronger than p_n for each n, such that no p_n is stronger than all the others. Then $\sigma(\{p_n\})$ is not seminormable.*

Proof. If $\sigma(\{p_n\})$ were seminormable, then it would be reducible to a finite set by Theorem 3.10.8 (d). From this finite set we would be able to select a strongest seminorm. □

We close this section with an example which shows that ω is not normable.

Example 3.10.11 *We define $p_n(x) = |x_n|$ and $q_n(x) = \sum_{k=1}^{n} |x_k|$ ($n = 1, 2, \ldots$) for all $x = (x_n)_{n=1}^{\infty} \in \omega$. By Remark 3.8.1 (f), each q_n is $\sigma(\{p_1, p_2, \ldots, p_n\})$ continuous, and each p_n is $\sigma(\{q_k\})$ continuous for $k \geq n$. Thus $\sigma(\{p_n\}) = \sigma(\{q_n\})$. The seminorms q_n are increasing and there is no strongest seminorm, since q_{n+1} is strictly stronger than q_n, which is true, since $q_{n+1} \geq q_n$, $q_{n+1}(e^{(n+1)}) = 1$ and $q_n(e^{(n+1)}) = 0$. It follows from Corollary 3.10.10 that ω is not a normed space.*

3.11   Metrizability

We recall that any metric defines a topology on a set, and that the converse implication is not true, in general. The question naturally arises: When is a linear topology metrizable? In this section, we give a sufficient condition for the metrizability of a linear topology.

We need the following results from Section 2.8, which we state here again for the reader's convenience.

Lemma 3.11.1 *Let X be a linear topological space.*
(a) Each neighborhood of 0 is absorbing (Theorem 2.8.5 (b)).
(b) Each neighborhood G of 0 contains a neighborhood N of 0 such that $\lambda N \subset G$ for all λ with $|\lambda| \leq 1$ ((I) in Part (i)(c) of the proof of Theorem 2.8.5).
(c) Each neighborhood G of 0 contains a balanced neighborhood of 0 (Theorem 2.8.5 (c)).
(d) For every neighborhood G of 0 there exists a neighborhood N of 0 such that $N + N \subset G$ (Theorem 2.8.5 (d)).

Theorem 3.11.2 *Let $(X, \mathcal{T})$ be a first countable linear topological space. Then there exists a metric d on X such that*
(a) d generates the topology $\mathcal{T}$ of X,
(b) the open discs centred at the origin are balanced,
(c) d is invariant, that is, $d(x + z, y + z) = d(x, y)$ for all $x, y, z \in X$.

Proof.

(i)   Construction of d.
    Since X is first countable, we can construct, by induction and using Lemma 3.11.1 (d) and (c), a local basis (V_n) of balanced sets V_n with

$$V_{n+1} + V_{n+1} \subset V_n. \qquad (3.10)$$

If we take $G_0 = X$, we can achieve that (3.10) also holds true for $n = 0$.
Now let D be the set of all rational numbers r with

$$r = \sum_{n=1}^{\infty} c_n(r)2^{-n} \text{ where } c_n(r) = 0 \text{ except for finitely many } c_n(r) = 1.$$

Then every $r \in D$ satisfies the inequality $0 \leq r < 1$. We put

$$X(r) = \begin{cases} X & (r \geq 1) \\ \sum_{n=1}^{\infty} c_n(r)V_n & (r \in D). \end{cases}$$

(We note that the sums $\sum_{n=1}^{\infty} c_n(r)V_n$ for $r \in D$ are finite.) We also put

$$f(x) = \inf\{r : x \in X(r)\} \text{ for all } x \in X \tag{3.11}$$

and

$$d(x,y) = f(x - y) \text{ for all } x, y \in X.$$

(ii) We show that if

$$X(r) + X(t) \subset X(r + t) \text{ for all } r, t \in D, \tag{3.12}$$

then d is a translation invariant metric for X.
First (3.12) implies

$$X(r) + X(t - r) \subset X(r + t - r) = X(t) \text{ for } t > r,$$

and since $0 \in X(s)$ for all s, it follows that

$$X(r) \subset X(r) + 0 \subset X(r) + X(t - r) \subset X(t) \text{ for } t > r,$$

that is,

$$X(r) \subset X(t) \text{ for } t > r.$$

Thus the set $\{X(r)\}$ is totally ordered by set inclusion.

(α) We show that

$$f(x + y) \leq f(x) + f(y) \text{ for all } x, y \in X. \tag{3.13}$$

Obviously we may assume $f(x + y) < 1$. Let $\varepsilon > 0$ be given. Then there are $r, s \in D$ such that $f(x) < r$ and $f(y) < s$ and $r + s < f(x) + f(y) + \varepsilon$. It follows from (3.11) that $x \in X(r)$ and $y \in X(s)$, and we obtain by (3.12)

$$x + y \in X(r + s).$$

This implies

$$f(x + y) \leq r + s < f(x) + f(y) + \varepsilon.$$

Since $\varepsilon > 0$ was arbitrary, we obtain (3.13).

(iii) Now we show Parts (a) and (b).
The open discs centred at 0 are the open sets

$$B_\delta(0) = \{x \in X : f(a) < \delta\} = \bigcup_{r < \delta} X(r).$$

If $\delta < 2^{-n}$, then we have $B_\delta(0) \subset V_n$. Hence $\{B_\delta(0)\}$ is a local basis at 0. This shows Part (a).
Since each set $X(r)$ is balanced, $B_\delta(0)$ is also balanced.

(iv) Now we prove (3.12) by mathematical induction.

Let P_N be the statement: *If $r + s < 1$ and $c_n(r) = c_n(s) = 0$ for all $n > N$, then we have*

$$X(r) + X(s) \subset X(r + s). \tag{3.14}$$

It is easy to see that P_1 holds true.

We assume that P_{N-1} holds true for some $N > 1$. We choose $r, s \in D$ such that $r + s < 1$ and $c_n(r) = c_n(s) = 0$ for all $n > N$. Now we define r' and s' by

$$r = r' + c_N(r) \cdot 2^{-N} \text{ and } s = s' + c_N(s) \cdot 2^{-N}.$$

Then we have

$$X(r) = X(r') + c_N(r)V_N \text{ and } X(s) = X(s') = c_N(s)V_N.$$

Since we assume that P_{N-1} holds true, we have $X(r') + X(s') \subset X(r' + s')$, hence

$$X(r) + X(s) \subset X(r' + s') + c_N(r)V_N + c_N(s)V_N. \tag{3.15}$$

If $c_N(r) = c_N(s) = 0$, then $r = r'$ and $s = s'$, and (3.15) yields (3.14).
If $c_N(r) = 0$ and $c_N(s) = 1$, then

$$X(r) + X(s) \subset X(r' + s') + V_N = X(r' + s' + 2^{-N}) = X(r + s),$$

and (3.14) again holds.

The case $c_N(r) = 1$ and $c_N(s) = 0$ is treated in the same way. Finally, if $c_N(r) = c_N(s) = 1$, then, by P_{N-1}

$$X(r' + s') + V_N + V_N \subset X(r' + s') + V_{N-1} = X(r' + s') + X(2^{-N+1}) \subset$$
$$X(r' + s' + 2^{-N+1}) = X(r + s).$$

Thus P_{N-1} implies P_N in each case, and consequently (3.14) holds. □

Theorem 3.11.3 *If a linear topological space contains a bounded convex neighborhood of 0 then the topology is generated by a seminorm.*

Proof. Let X be a linear topological space and G be a bounded convex neighborhood of 0.

(i) First we show that X contains a bounded neighborhood N of 0 which is a balloon. There exists a balanced neighborhood N_0 of 0 with $N_0 \subset G$, by Lemma 3.11.1. Let $B = \text{conv}N_0$ be the convex hull of N_0. Then B is convex, and $B \subset G$, since $G \supset N_0$ is convex, and B is the smallest convex set that contains N_0. Also B is absorbing, since it is a neighborhood of 0 and every neighborhood of 0 is absorbing by Lemma 3.11.1 (a). The convex, absorbing set B contains a balloon by Remark 2.4.3 (e), namely

$$N = \bigcap_{|\lambda|=1} \lambda B, \text{ by Part (e) of the proof of Remark 2.4.3.}$$

It also follows that $N_0 \subset N$. For if $N_0 \not\subset N$, then there exist $x \in X$ and a scalar λ with $|\lambda| = 1$, such that $x \in N_0$ and $x \notin \lambda B$. Since N_0 is balanced, it follows that $\lambda^{-1}x \in N_0$ and $\lambda^{-1}x \notin B$, contradicting $N_0 \subset B$.

So N is a neighborhood of 0 which is a balloon, and is bounded, since $N \subset B \subset G$. This completes the proof of (i).

Since N is a balloon, we may define the Minkowski functional μ_N of N which is a seminorm by Theorem 3.9.3 (a).

Let $\mathcal{T}_N$ denote the seminorm–topology. It remains to show that $\mathcal{T} = \mathcal{T}_N$.

(ii) First we show that $\mathcal{T} \subset \mathcal{T}_N$.
      Let M be a $\mathcal{T}$–neighborhood of 0. Since N is bounded, there exists $\varepsilon > 0$ such that $\varepsilon N \subset M$. Thus M is a $\mathcal{T}_N$–neighborhood of 0.

(iii) Now we show that $\mathcal{T}_N \subset \mathcal{T}$.
      Let M be a $\mathcal{T}_N$–neighborhood of 0. Then there exists $\varepsilon > 0$ such that $B_\varepsilon(0) \subset M$. We have, however, $N = B_1(0)$, so $M \supset B_\varepsilon(0) = \varepsilon N$, and consequently M is a $\mathcal{T}$–neighborhood of 0.   □

Chapter 4

Banach Spaces

Chapter 4 concentrates on the studies of Banach spaces. It contains the most important examples of Banach spaces and results on the bounded linear operators between normed and Banach spaces. Highlights are the fundamental Hahn–Banach extension theorem and several of its corollary theorems which are vitally important in the proof of many results in functional analysis. Furthermore, the versions in Banach spaces are given of the open mapping and closed graph theorems, Banach's theorem of the bounded inverse, and the Banach–Steinhaus theorem. Two important applications of the closed graph theorem are the criterion for a closed subspace in a Banach space to have a topological complement, and the Eni–Karauš theorem. Further applications of the Hahn–Banach and Banach–Steinhaus theorems are the classical representation theorems for the continuous linear functionals on the classical sequence spaces, including the representation theorem for the space of bounded real sequences, and the Riesz representation theorem for the continuous linear functionals on the space of continuous real functions on the unit interval. The other topics focus on the reflexivity of spaces, the studies of adjoint operators, quotient spaces, Cauchy nets, the equivalence of norms, compactness and the Riesz lemma, compact operators and operators with closed range.

4.1 Introduction

A normed space is both a vector and metric space. In this space, we use the relation between two structures. A normed space which is complete as a metric space is called a *Banach space* (Definition 3.2.3). The axioms defining a Banach space were introduced by A. A. Bennett in 1916 [19]. Using these axioms, F. Riesz in 1918 [174] gave a very important extension of the theory of Fredholm integral equations. Independently of one another, S. Banach in 1922 [9], N. Wiener in 1922 [206], and H. Hahn in 1922 [81] used similar axioms. S. Banach gave fundamental results for the spaces defined by those axioms. Today these spaces are the well–known Banach spaces.

This chapter deals mainly with the studies of Banach spaces. It contains the most important examples of Banach spaces and results on the bounded linear operators between normed and Banach spaces. The first highlight is the fundamental Hahn–Banach extension theorem, Theorem 4.5.2, which is of vital importance in the proof of many results in functional analysis; also many of its corollaries are given; it can be found in [10, Théorème 1, Théorème 2]. Furthermore, the versions of Banach spaces are given in Section 4.6 of the open mapping and closed graph theorems, Banach's theorem of the bounded inverse, and the Banach–Steinhaus theorem, an alternative proof of which is added that does not use Baire's theorem. Two important applications of the closed graph theorem are the criterion for a closed subspace in a Banach space to have a topological complement, and the Eni–Karauš theorem, Theorem 4.6.13. Further applications of the Hahn–Banach and Banach–Steinhaus theorems are the classical representation theorems for the continuous linear functionals on the classical sequence spaces, including the representation theorem in the case of the space of bounded real sequences, and the Riesz representation theorem, Theorem 4.7.7, for the continuous linear functionals on the space of continuous real functions on the unit interval. The other topics focus on the reflexivity of spaces, the studies of adjoint operators, quotient spaces, Cauchy nets, the equivalence of norms, compactness and the Riesz lemma, Lemma 4.13.2, compact operators and operators with closed range.

4.2   Some basic properties

We recall a few facts concerning normed spaces.

A norm $\| \cdot \|$ on a normed space $(X, \| \cdot \|)$ defines a metric d on X by (see (1.5))

$$d(x, x') = \|x - x'\| \text{ for all } x, x' \in X. \tag{4.1}$$

The following statements are immediate consequences of (4.1).

Remark 4.2.1 *Let $(X, \| \cdot \|)$ be a normed spaces. Then the metric d defined in (4.1) is translation invariant, that is,*

$$d(x + y, x' + y) = d(x, x') \text{ for all } x, x', y \in X;$$

we also have

$$d(\lambda x, \lambda x') = |\lambda| d(x, x') \text{ for all scalars } \lambda \text{ and all } x, x' \in X.$$

Also, by Theorem 2.8.15, the closure of a subspace of a normed space is a subspace.
A sequence (x_n) in a normed space $(X, \| \cdot \|)$ converges to $x \in X$ if and only if $\|x_n - x\| \to 0$ as $n \to \infty$. Furthermore, since

$$| \|x\| - \|x'\| | \leq \|x - x'\| \text{ for all } x, x' \in X,$$

the function $\| \cdot \| : X \to \mathbb{R}$ is continuous everywhere.

The following result is fundamental.

Lemma 4.2.2 (Lemma on the linear combination)
Let $\{x_1, x_2, \ldots, x_n\}$ be a linearly independent set of vectors in a normed space X. Then there exists a positive number c such that we have for all scalars $\lambda_1, \lambda_2, \ldots, \lambda_n$

$$\left\| \sum_{k=1}^{n} \lambda_k x_k \right\| \geq c \left(\sum_{k=1}^{n} |\lambda_k| \right). \tag{4.2}$$

Proof. To simplify the proof we put $t = |\lambda_1| + \cdots + |\lambda_n|$.
If $t = 0$, then the inequality in (4.2) is obviously satisfied for each c.

So we assume $t > 0$. In this case the inequality in (4.2) is equivalent with the inequality obtained from (4.2) when it is divided by t and with $\mu_k = \lambda_k/t$ for all k, that is,

$$\left\| \sum_{k=1}^{n} \mu_k x_k \right\| \geq c, \text{ where } \sum_{k=1}^{n} |\mu_k| = 1. \tag{4.3}$$

So it suffices to prove (4.3) for each n tuple of scalars $\mu_1, \mu_2 \ldots, \mu_n$ with $\sum_{k=1}^{n} |\mu_k| = 1$. If the inequality in (4.3) is not satisfied, then there exists a sequence (y_m) in X such that

$$y_m = \sum_{k=1}^{n} \alpha_k^{(m)} x_k, \ \sum_{k=1}^{n} \left| \alpha_k^{(m)} \right| = 1 \text{ for } m = 1, 2, \ldots,$$
$$\text{and } \|y_m\| \to 0, \ (m \to \infty). \tag{4.4}$$

It follows from $\sum_{k=1}^{n} |\alpha_k^{(m)}| = 1$ that $|\alpha_k^{(m)}| \leq 1$ for $k = 1, \ldots, n$. Hence the sequences $(\alpha_k^{(m)})_{m=1}^{\infty}$ are bounded for all $k = 1, \ldots, n$. By the Bolzano–Weierstrass theorem, the sequence $(\alpha_1^{(m)})_{m=1}^{\infty}$ has a convergent subsequence. We denote by α_1 the limit of this subsequence and by $(y_{1,m})$ the corresponding subsequence of the sequence (y_m). Furthermore, analogously we conclude that the sequence $(y_{1,m})$ has a subsequence $(y_{2,m})$ such that the corresponding sequence $(\alpha_2^{(m)})_{m=1}^{\infty}$ converges to the limit α_2, say. Continuing in this way, after n steps, we obtain a subsequence $(y_{n,m})_{m=1}^{\infty}$ of the sequence (y_m) with the terms $y_{n,m} = \sum_{k=1}^{n} \beta_k^{(m)} x_k$ with $\sum_{k=1}^{n} |\beta_k^{(m)}| = 1$, and $\beta_k^{(m)} \to \alpha_k$ as $m \to \infty$ for $k = 1, \ldots, n$. It follows that $y_{n,m} \to y = \sum_{k=1}^{n} \alpha_k x_k$ as $m \to \infty$, that is, $\sum_{k=1}^{n} |\alpha_k| = 1$. Now, since the set $\{x_1, \ldots, x_n\}$ is linearly independent, it follows that $y \neq 0$, but (4.4) implies $y = 0$. This is a contradiction. $\square$

If $(X, \|\cdot\|)$ is a normed space and Y is a subspace of the vector space X, then the restriction of the norm $\|\cdot\|$ on Y obviously is a norm on Y and the normed space $(Y, \|\cdot\|)$ is referred to as *subspace of the normed space* X. Usually we say Y *is a subspace of* X, or Y *is a subspace of the normed space* X, and we suppose that we consider the normed subspace $(Y, \|\cdot\|)$. We remark that the closure of a subspace is also a subspace.

Theorem 4.2.3 *Every finite dimensional subspace Y of a normed space X is complete, that is, a Banach space. In particular, every finite dimensional normed space is complete, that is, a Banach space.*

Proof. Let (y_n) be a Cauchy sequence in Y and $H = \{h_1, \ldots, h_m\}$ be an algebraic basis of Y. Every vector y_n has a unique representation as a linear combination of the basis H, that is, $y_n = \sum_{k=1}^{m} \lambda_k^{(n)} h_k$. Since (y_n) is a Cauchy sequence, for every $\varepsilon > 0$, there exists an n_0, such that $n, j \geq n_0$ implies $\|y_n - y_j\| < \varepsilon$. By Lemma 4.2.2, there exists $c > 0$ such that for all $n, j \geq n_0$,

$$\|y_n - y_j\| = \left\| \sum_{k=1}^{m} \left(\lambda_k^{(n)} - \lambda_k^{(j)} \right) h_k \right\| \geq c \sum_{k=1}^{m} \left| \lambda_k^{(n)} - \lambda_k^{(j)} \right|, \tag{4.5}$$

hence

$$\left| \lambda_k^{(n)} - \lambda_k^{(j)} \right| \leq \sum_{l=1}^{m} \left| \lambda_l^{(n)} - \lambda_l^{(j)} \right| < \frac{\varepsilon}{c}. \tag{4.6}$$

It follows from (4.6) that, for every k, the sequence $(\lambda_k^{(n)})_{n=1}^{\infty}$ is a Cauchy sequence in $\mathbb{R}$ or in $\mathbb{C}$, hence convergent, $\lambda_k = \lim_{n \to \infty} \lambda_k^{(n)}$ say for $k = 1, \ldots, m$. Now $y = \sum_{k=1}^{m} \lambda_k h_k \in Y$, and it is easy to show that $y_n \to y$ as $n \to \infty$. $\square$

The following corollary is an immediate consequence of the previous theorem.

Corollary 4.2.4 *Every finite dimensional subspace Y of a normed space X is closed in X.*

Lemma 4.2.5 *Let Y be a subspace of a normed space Y and $Y \neq X$. Then Y has no interior points.*

Proof. We assume that x_0 is an interior point of Y and $x_1 \in X \setminus Y$. But $x_0 + n^{-1}x_1 \in Y$ for n large enough. This is a contradiction to the assumption $x_1 \notin Y$.   □

We continue with a result concerning the algebraic dimensions of infinite dimensional Banach spaces.

Theorem 4.2.6 *Let X be an infinite dimensional Banach space. Then $\dim X > \aleph_0$.*

Proof. We assume that an infinite dimensional Banach space X has a countable algebraic basis $\{h_1, h_2, \ldots, h_n, \ldots\}$. Let X_n denote the linear hull of $\{h_1, h_2, \ldots, h_n\}$ for $n = 1, 2, \ldots$. By Theorem 4.2.4, X_n is a closed subspace of X for $n = 1, 2, \ldots$. Since $X = \cup_{n=1}^{\infty} X_n$, Baire's category theorem (Theorem 2.10.9) implies that at least one of the subspaces X_n has an interior point. This is a contradiction to Lemma 4.2.5.   □

Definition 4.2.7 The normed spaces $(X, \|\cdot\|_X)$ and $(Y, \|\cdot\|_Y)$ over the same field $\mathbb{F}$ are said to be *congruent* or *isometrically isomorphic* (we also say that they are *isomorphic*), if there exists a linear bijection $T : X \to Y$ such that

$$\|Tx\|_Y = \|x\|_X \text{ for all } x \in X;$$

this is denoted by $X \cong Y$. The map T is referred to as *congruence* or an *isometric isomorphism*.

Lemma 4.2.8 *If X and Y are congruent normed spaces and one of them is a Banach space, so is the other one.*

Proof. The proof is left to the readers.   □

4.3   Examples

In this section, we consider some of the most important examples of normed and Banach spaces. The proofs are left to the readers. The norms defined in the examples below are usually referred to as the *standard* or *natural* norms.

Example 4.3.1 *The spaces $\mathbb{R}$ or $\mathbb{C}$ with $\|\cdot\| = |\cdot|$, where $|\cdot|$ is the absolute value, are Banach spaces.*

Example 4.3.2 (Example 2.10.2 (d)) *The spaces ℓ_p for $1 \leq p < \infty$ are Banach spaces with respect to*

$$\|x\|_p = \left(\sum_{k=1}^{\infty} |x_k|^p \right)^{1/p} \quad \text{for } 1 \leq p < \infty.$$

The triangle inequality for $\|\cdot\|_p$ follows by Minkowski's inequality (1.2). Using the convention $\mathbb{C}^n, \mathbb{R}^n \subset \phi$, we also obtain that, for all $n \in \mathbb{N}$, the spaces $\mathbb{C}^n$ or $\mathbb{R}^n$ of all n tuples of complex or real numbers are Banach spaces with respect to $\|\cdot\|_p$.

Example 4.3.3 *The spaces* c_0, c *(Example 3.2.4) and* ℓ_∞ *are Banach spaces with respect to*

$$\|x\|_\infty = \sup_k |x_k|;$$

$\|\cdot\|_\infty$ *is referred to as the* supremum norm.

Example 4.3.4 *For every measure space* $(X, \mathcal{R}, \mu)$, *the spaces* $L_p(\mu)$ *or* L_p $(1 \le p < \infty)$ *of all measurable functions* f *for which* $\int |f|^p \, d\mu < \infty$, *are Banach spaces (see Theorem 4.10.5) with respect to*

$$\|f\| = \left(\int |f|^p \, d\mu \right)^{1/p}. \tag{4.7}$$

We remark that in L_p *we identify the functions which are equal a.e. with respect to* μ *on* X. *By* Minkowski's inequality for integrals *for* $1 \le p < \infty$,

$$\left(\int |f + g|^p \, d\mu \right)^{1/p} \le \left(\int |f|^p \, d\mu \right)^{1/p} + \left(\int |g|^p \, d\mu \right)^{1/p} \quad \text{for all } f, g \in L_p, \tag{4.8}$$

it follows that $\|\cdot\|$ *defined in (4.7) satisfies the triangle inequality. We remark that Example 4.3.2 is a special case of Example 4.3.4 if we choose* $X = \{1, 2, \ldots, n, \ldots\}$ *or* $X = \{1, 2, \ldots, n\}$ *and the measure* μ *as the* counting *or* discrete measure.

Example 4.3.5 *For each measure space* $(X, \mathcal{R}, \mu)$ *let* $L_\infty(E, \mu)$ *for* $E \subset \mathcal{R}$ *or* L_∞ *be the space of all measurable functions* f *on* E *for which there exists a real number* M *such that*

$$|f(x)| \le M \text{ for almost all } x \in E. \tag{4.9}$$

It is easy to show that $L_\infty(E, \mu)$ *is a linear space with the natural algebraic operations. We remark that we identify in* $L_\infty(E, \mu)$ *the functions that are equal a.e. on* E *with respect to* μ. *It can be shown that* $L_\infty(E, \mu)$ *is a Banach space with respect to*

$$\|f\| = \inf_{0 \le M < \infty} \{M : |f(x)| \le M \text{ for almost all } x \in E\},$$

$\|\cdot\|$ *is called the* essential supremum *of* f. *A function* f, *which satisfies (4.9) is called* essentially bounded *on the set* E, *and* L_∞ *is referred to as the* space of essentially bounded functions.

Example 4.3.6 *The space* $\mathrm{C}[a, b]$ *of all real or complex-valued functions on the interval* $[a, b] \subset \mathbb{R}$ *is a Banach space with respect to (Example 2.10.2 (b))*

$$\|f\|_\infty = \sup_{x \in [a, b]} |f(x)|;$$

and $\mathrm{C}[a, b]$ *is a normed space, but not a Banach space with respect to (Example 2.10.2 (c))*

$$\|f\|_1 = \int_a^b |f(t)| \, dt.$$

Example 4.3.7 *Let* $\mathrm{BV}[a, b]$ *be the linear space of all real– or complex–valued functions* f *of bounded variation on the interval* $[a, b] \subset \mathbb{R}$; $\mathrm{BV}[a, b]$ *is a Banach space with respect to*

$$\|f\| = |f(a)| + V_a^b(f),$$

where $V_a^b(f)$ *denotes the total variation of the function* f *on the interval* $[a, b]$.

Example 4.3.8 *The space* $\mathrm{B}[a, b]$ *of all bounded real– or complex–valued functions on the interval* $[a, b] \subset \mathbb{R}$ *is a Banach space with respect to*

$$\|f\|_\infty = \sup_{x \in [a, b]} |f(x)|.$$

4.4   Bounded linear operators

In this section, we study one more important example of a Banach space, namely the *Banach space of bounded linear operators.*

Definition 4.4.1 Let X and Y be normed linear spaces over the same field $\mathbb{F}$ of scalars. An operator $A \in \mathcal{L}(X,Y)$ is said to be *bounded* if there exists a real number $M \geq 0$ such that

$$\|Ax\| \leq M\|x\| \text{ for all } x \in X. \tag{4.10}$$

If $x \neq 0$, then (4.10) implies

$$\sup_{x \neq 0} \frac{\|Ax\|}{\|x\|} \leq M, \tag{4.11}$$

and so

$$\|Ax\| \leq \left(\sup_{x \neq 0} \frac{\|Ax\|}{\|x\|} \right) \|x\| \text{ for all } x \in X. \tag{4.12}$$

Hence we have

$$\begin{aligned}
&\inf\{M \geq 0 : \|Ax\| \leq M\|x\| \text{ for all } x \in X\} \\
&= \min\{M \geq 0 : \|Ax\| \leq M\|x\| \text{ for all } x \in X\} \\
&= \sup_{x \neq 0} \frac{\|Ax\|}{\|x\|}.
\end{aligned} \tag{4.13}$$

It follows from (4.10), (4.11), (4.12) and (4.13) that:

$$\text{an operator } A \text{ is bounded if and only if } \sup_{x \neq 0} \frac{\|Ax\|}{\|x\|} < \infty. \tag{4.14}$$

Definition 4.4.2 Let X and Y be normed spaces and $A \in \mathcal{L}(X,Y)$ be a bounded operator. The *norm of the operator* A is denoted by $\|A\|$ defined as

$$\|A\| = \sup_{x \neq 0} \frac{\|Ax\|}{\|x\|}. \tag{4.15}$$

It follows from (4.13) and (4.15) that a bounded operator $A \in \mathcal{L}(X,Y)$ satisfies

$$\|Ax\| \leq \|A\|\,\|x\| \text{ for all } x \in X, \tag{4.16}$$

and for every $\epsilon > 0$ there exist $x_\epsilon \in X$ such that

$$\|Ax_\epsilon\| > (\|A\| - \epsilon)\|x_\epsilon\|.$$

The set of all bounded linear operators from X into Y is denoted by $\mathcal{B}(X,Y)$. If $X = Y$, then we write $\mathcal{B}(X) = \mathcal{B}(X,X)$, for short. We write $X^* = \mathcal{B}(X,\mathbb{F})$ and refer to X^* as the *space of bounded linear functionals on* X, or the *dual space of* X.

We note that if $X \neq \{0\}$ is a normed space, then we obviously have for the identity $I : X \to X$

$$I \in \mathcal{B}(X) \text{ and } \|I\| = 1.$$

Let Z be a normed space over the field $\mathbb{F}$, $A \in \mathcal{B}(X,Y)$ and $B \in \mathcal{B}(Y,Z)$. Then $BA = B \circ A \in \mathcal{L}(X,Z)$, and since A and B are bounded operators, (4.16) implies

$$\|BA(x)\| = \|B(Ax)\| \leq \|B\| \, \|Ax\|$$
$$\leq \|B\| \, \|A\| \, \|x\| \text{ for all } x \in X. \tag{4.17}$$

Now (4.17) implies

$$BA \in \mathcal{B}(X,Z) \text{ and } \|BA\| \leq \|B\| \, \|A\|.$$

Theorem 4.4.3 *Let X and Y be normed spaces over the same field $\mathbb{F}$. Then $\mathcal{B}(X,Y)$ is a linear subspace of $\mathcal{L}(X,Y)$ and the operator norm is a norm on $\mathcal{B}(X,Y)$.*

Proof. It follows for $A, B \in \mathcal{B}(X,Y)$ and $\lambda \in \mathbb{F}$ that

$$\|(A+B)x\| \leq \|Ax\| + \|Bx\| \leq \|A\| \, \|x\| + \|B\| \, \|x\|$$
$$\leq (\|A\| + \|B\|)\|x\| \text{ for all } x \in X, \tag{4.18}$$

and

$$\|(\lambda A)x\| = |\lambda| \, \|Ax\| \leq |\lambda| \, \|A\| \, \|x\| \text{ for all } x \in X. \tag{4.19}$$

It follows from (4.18), (4.15) and (4.13) that $A + B \in \mathcal{B}(X,Y)$ and

$$\|A + B\| \leq \|A\| + \|B\|. \tag{4.20}$$

Analogously (4.19), (4.15) and (4.13) imply $\lambda A \in \mathcal{B}(X,Y)$ and

$$\|\lambda A\| = |\lambda| \, \|A\|. \tag{4.21}$$

So we have proved that $\mathcal{B}(X,Y)$ is a linear subspace of $\mathcal{L}(X,Y)$. Clearly, we have

$$\|A\| = 0 \text{ if and only if } A = 0. \tag{4.22}$$

It follows from (4.20), (4.21) and (4.22) that the operator norm (4.15) is a norm on the space $\mathcal{B}(X,Y)$. $\qquad\square$

If not stated otherwise, we always assume that $\mathcal{B}(X,Y)$ is a normed space with the operator norm (4.15).

Theorem 4.4.4 *Let X be a normed space and Y be a Banach space. Then $\mathcal{B}(X,Y)$ is a Banach space.*

Proof. Let (A_n) be a Cauchy sequence in $\mathcal{B}(X,Y)$ and $x \in X$. It follows from

$$\|A_n x - A_m x\| = \|(A_n - A_m)x\| \leq \|A_n - A_m\| \, \|x\|$$

that $(A_n x)$ is a Cauchy sequence in Y, and hence convergent since Y is complete. We put

$$Ax = \lim_{n \to \infty} A_n x.$$

Obviously $A \in \mathcal{L}(X,Y)$. Since the Cauchy sequence (A_n) is bounded, there exists L with $0 \leq L < \infty$ such that $\|A_n\| \leq L$ for each $n \in \mathbb{N}$. Thus we have

$$\|Ax\| = \lim_{n \to \infty} \|A_n x\| \leq \left(\limsup_{n \to \infty} \|A_n\| \right) \|x\|$$
$$\leq L \, \|x\| \text{ for all } x \in X. \tag{4.23}$$

It follows from (4.23) and (4.10) that $A \in \mathcal{B}(X, Y)$.

Now we show that $\lim A_n = A$. Let $\varepsilon > 0$ and n_0 be a natural number such that

$$\|A_n - A_m\| < \varepsilon \text{ for all } n, m \geq n_0. \tag{4.24}$$

It follows from (4.24) and (4.15) that

$$\frac{\|(A_n - A_m)x\|}{\|x\|} < \varepsilon \text{ for each } x \neq 0 \text{ and for all } n, m \geq n_0,$$

hence, letting $m \to \infty$, we obtain

$$\frac{\|(A_n - A)x\|}{\|x\|} \leq \varepsilon \text{ for each } x \neq 0 \text{ and all } n \geq n_0. \tag{4.25}$$

Finally, (4.25) and (4.15) imply

$$\|A_n - A\| = \sup_{x \neq 0} \frac{\|(A_n - A)x\|}{\|x\|} \leq \epsilon \text{ for all } n \geq n_0. \qquad \square$$

Corollary 4.4.5 *The dual X^* of each normed space X is a Banach space.*

Proof. This is an immediate consequence of Theorem 4.4.4. $\qquad \square$

Theorem 4.4.6 *Let X and Y be normed spaces and $A \in \mathcal{L}(X, Y)$. Then we have*

$$\sup_{\|x\| \leq 1} \|Ax\| = \sup_{\|x\| = 1} \|Ax\| = \sup_{x \neq 0} \frac{\|Ax\|}{\|x\|}. \tag{4.26}$$

Proof. Since

$$\{\|Ax\| : \|x\| \leq 1\} \supset \{\|Ax\| : \|x\| = 1\} = \left\{ \frac{\|Ax\|}{\|x\|} : x \neq 0 \right\},$$

we obtain

$$\sup_{\|x\| \leq 1} \|Ax\| \geq \sup_{\|x\| = 1} \|Ax\| = \sup_{x \neq 0} \frac{\|Ax\|}{\|x\|}. \tag{4.27}$$

If $\sup_{x \neq 0} \|Ax\|/\|x\| = \infty$, then (4.27) implies (4.26). From

$$\sup_{x \neq 0} \frac{\|Ax\|}{\|x\|} < \infty \text{ and (4.14), it follows that } A \in \mathcal{B}(X, Y).$$

So we have

$$\sup_{\|x\| \leq 1} \|Ax\| \leq \|A\|. \qquad \square$$

The next theorem shows, among other things, that every bounded operator is continuous.

Theorem 4.4.7 *Let X and Y be normed spaces and $A \in \mathcal{L}(X, Y)$. Then the following statements are equivalent:*

$$A \text{ is uniformly continuous on } X. \tag{4.28}$$

$$A \text{ is continuous at } 0. \tag{4.29}$$

$$A \in \mathcal{B}(X, Y). \tag{4.30}$$

Proof.

(i) Obviously (4.28) implies (4.29).

(ii) Now we show that (4.30) implies (4.28).

We assume $\varepsilon > 0$ and $M > 0$ satisfy the condition in (4.10). Then it follows for $\delta = \varepsilon/M$ and $x, y \in X$ that

$$\|x - y\| < \delta \text{ implies } \|Ax - Ay\| \leq M\,\|x - y\| < \varepsilon.$$

(iii) Finally we prove that (4.29) implies (4.30).

Since the map A is continuous at 0, for $\varepsilon = 1$, there exists $\delta > 0$ such that $\|Ax\| < 1$ whenever $\|x\| < \delta$. Now, if $0 \neq x \in X$ and $0 < \alpha < \delta$, then $\|\,\alpha x/\|x\|\,\| < \delta$ and

$$\|Ax\| = \left\| \frac{\|x\|}{\alpha} A\left(\frac{\alpha x}{\|x\|} \right) \right\| < \alpha^{-1}\|x\|.$$

So we have $A \in \mathcal{B}(X, Y)$. $\qquad\square$

Theorem 4.4.8 *Let X and Y be normed spaces and $A \in \mathcal{L}(X, Y)$. Then the operator A be bounded if and only if $A(Q)$ is a bounded subset of Y for every bounded subset Q of X.*

Proof. This is left as an exercise for the readers. $\qquad\square$

Theorem 4.4.9 *Let X be a finite dimensional normed space and Y be a normed space. Then we have $\mathcal{B}(X, Y) = \mathcal{L}(X, Y)$.*

Proof. Let $\dim X = n$, $\{h_1, h_2 \ldots, h_n\}$ be an algebraic basis of X and $x \in X$. Then there exist scalars λ_k for $k = 1, 2, \ldots, n$ such that $x = \sum_{k=1}^{n} \lambda_k h_i$, and so

$$\|Ax\| \leq \sum_{k=1}^{n} |\lambda_k| \|Ah_k\| \leq \left(\max_k \|Ah_k\| \right) \sum_{k=1}^{n} |\lambda_k|. \tag{4.31}$$

It follows from (4.2) and (4.31) that there exists $c > 0$ such that

$$\|Ax\| \leq \left(c^{-1} \max_k \|Ah_k\| \right) \|x\| \text{ for all } x \in X. \qquad\square$$

The following theorem shows that there exist unbounded linear operators.

Theorem 4.4.10 *Let X be an infinite dimensional normed space and $Y \neq \{0\}$ be a normed space. Then there exists an unbounded linear operator $A : X \to Y$.*

Proof. Let $\{h_1, h_2, \ldots\}$ be a subset of an algebraic basis $\{h_i\}_{i \in I}$ of the space X and $0 \neq y \in Y$. We define a function B on the basis $\{h_i\}_{i \in I}$ by

$$B(h_i) = \begin{cases} i\|h_i\|y & i \in \mathbb{N} \\ 0 & i \in I \setminus \mathbb{N}. \end{cases}$$

Since any linear map is uniquely defined by its images on the basis of the space, let $A : X \to Y$ be the linear map which is the extension of B. It follows from $Ah_n = Bh_n$ for $n = 1, 2, \ldots$ and

$$\sup_{x \neq 0} \frac{\|Ax\|}{\|x\|} \geq \sup_n \frac{\|Ah_n\|}{\|h_n\|} = \infty$$

that A is not a bounded operator. $\qquad\square$

We close this section with one more important result.

Theorem 4.4.11 *A linear functional f on a normed space X is continuous if and only if its kernel $N(f)$ is a closed subspace of X.*

Proof. We note that $N(f)$ is a subspace of X by Remark 1.6.2.

(i) If f is continuous then $N(f) = f^{-1}(\{0\})$ is closed in X (being the pre–image of the closed set $\{0\}$ in $\mathbb{C}$) by Remark 1.9.7 (b).

(ii) Now we prove the converse implication.

We assume that $N(f)$ is a closed subspace of X. If $f \equiv 0$, then f is trivially continuous. So we assume $f \not\equiv 0$. Then the complement $N(f)^c = X \setminus N(f)$ of $N(f)$ is open and nonempty, hence there exists $x \in N(f)^c$ with $f(x) \neq 0$. We put $y = x/f(x)$. Then $f(y) = f(x/f(x)) = f(x)/f(x) = 1$, hence $y \in N(f)^c$. Since $N(f)^c$ is open, there exists a positive real number r such that the open ball $B_r(y)$ is contained in $N(f)^c$.

(α) We put $S = \{x \in X : |f(x)| < 1\}$, and show that

$$B_r(0) \subset S. \tag{*}$$

Let $z \in B_r(0)$ be given. If $z \notin S$, then $|f(z)| \geq 1$ and so we obtain for $w = -z/f(z)$,

$$\|w\| = \left\| -\frac{z}{f(z)} \right\| = \frac{1}{|f(z)|} \cdot \|z\| < r, \tag{**}$$

that is, $w \in B_r(0)$. Furthermore we have

$$f(y + w) = f(y) + f(w) = 1 - 1 = 0, \text{ that is, } y + w \in N(f). \tag{***}$$

Also $y + w \in B_r(y)$, since $\|w\| < r$ by (**). Together with (***), we have $y + w \in N(f) \cap B_r(y)$, that is, $N(f) \cap B_r(y) \neq \emptyset$. This is a contradiction to $B_r(y) \subset N(f)^c$. Thus we have established (*).

Now it follows from (*) that $\|x\| < r$ implies $|f(x)| < 1$. Let $y \in X \setminus \{0\}$ be given. We put $z = ry/(2\|y\|)$ and obtain $\|z\| \leq r/2 < r$, hence $|f(z)| < 1$ and finally, putting $M = 2/r$

$$|f(y)| = \left| f\left(\frac{2\|y\|}{r} \cdot z \right) \right| = \frac{2\|y\|}{r} \cdot |f(z)| \leq M \cdot \|y\|.$$

Since the last inequality trivially also holds for $x = 0$, we have shown that the linear functional f is bounded, and hence continuous by Theorem 4.4.7. $\qquad \square$

4.5 The Hahn–Banach theorem

In mathematics (as in everyday life) some results (events) are more important than others. Sometimes it takes a number of years or even decades before the importance of some results is really recognized and appreciated. Today, three theorems in functional analysis are generally rated as the most important ones. We study them in this and the next section. The second and third theorems were already considered in a more general spaces than Banach spaces in Sections 3.5, 3.6 and 3.7.

The first fundamental theorem is the *Hahn-Banach theorem*, which asserts the existence of a nontrivial extension of a continuous linear functional from a subspace to the whole space.

This theorem is of vital importance in the proofs of many important results in functional analysis. It was first proved by Hahn in 1927 for a real normed space, and later, in 1929, by Banach for real linear spaces (without topology) in [10, Théorème 1, Théorème 2]. The complex version of this theorem was proved by Bohnenblust and Sobczkyk in 1938, and independently by Soukhomlinoff in the same year.

The second fundamental theorem is the *open mapping theorem* (Theorem 3.5.3 for closed linear surjective maps between Fréchet spaces). It asserts that a continuous surjective linear map A between Banach spaces X and Y is open, that is, the image $A(G)$ of any open set in X is open in Y. The *closed graph theorem* (Theorem 3.6.1 for closed linear maps between Fréchet spaces) is a corollary of the open mapping theorem, and asserts that a linear map between two Banach spaces with a closed graph is continuous. One version of the open mapping theorem and the closed graph theorem was proved by Banach in 1929. A more general version of these theorems was proved by Schauder in 1930.

The third fundamental theorem is the *uniform boundedness principle* (Theorem 3.7.4 for pointwise bounded sequences of continuous linear functionals on a Fréchet space). It asserts that any set of pointwise bounded continuous linear maps from one Banach space to another is uniformly bounded on the unit ball. The uniform boundedness principle was proved by Hahn in 1922 for continuous linear functionals on a Banach space. Later it was proved by Hildebrandt for continuous linear maps between Banach spaces in 1923 and also proved in a more general version by Banach and Steinhaus in 1927. The uniform boundedness principle is also referred to as the *Banach-Steinhaus theorem* (Theorem 3.7.5 for pointwise convergent sequences of continuous linear functionals on a Fréchet).

In this section, we prove the first fundamental theorem, namely the *Hahn-Banach theorem* and some important corollaries.

We recall that if S is a subspace of a linear space X then $S^{\#}$ denotes the set of all linear functionals on S.

We need the following result (Remark 3.8.3 (b)), which we state again for the reader's convenience.

Lemma 4.5.1 *Let X be a linear space, and p be a seminorm on X and $f \in X^{\#}$. Then $\mathrm{Re}\,f(x) \leq p(x)$ for all $x \in X$ implies $|f(x)| \leq p(x)$ for all $x \in X$.*

Theorem 4.5.2 (Hahn–Banach extension theorem) *Let S be a subspace of a linear space X, p be a seminorm defined on X and $f \in S^{\#}$ with*

$$|f(s)| \leq p(s) \text{ for all } s \in S.$$

Then there exists an extension $F \in X^{\#}$ of f with

$$|F(x)| \leq p(x) \text{ for all } x \in X.$$

Proof. We may assume $S \neq X$, for otherwise there is nothing to show.

(a) First we consider the case in which X is a real linear space and f is a real linear functional.

 (i) Let $x \in X \setminus S$. First we extend f to $\mathrm{span}(x, S)$. Let $s, t \in S$. Then we have

$$f(s) - f(t) = f(s - t) \leq p(s - t) \leq p(s + x) + p(-t - x),$$

  hence

$$-p(-t - x) - f(t) \leq p(s + x) - f(s) \text{ for all } s, t \in S.$$

This implies

$$\tau = \sup_{t \in S} \left(-p(-t-x) - f(t)\right) \leq p(s+x) - f(s) \text{ for all } s \in S,$$

hence

$$\tau \leq \sigma = \inf_{s \in S} \left(p(s+x) - f(s)\right).$$

Consequently there exists $\mu \in \mathbb{R}$ such that

$$-p(-t-x) - f(t) \leq \mu \leq p(t+x) - f(t) \text{ for all } t \in S. \qquad (4.32)$$

Now let $y \in \operatorname{span}(x, S)$, that is, $y = \lambda x + s$ with uniquely determined $s \in S$ and $\lambda \in \mathbb{R}$. We put

$$h(y) = \lambda \mu + f(s) \text{ with } \mu \text{ from } (4.32).$$

Obviously we have $h \in (\operatorname{span}(a, S))^{\#}$ and $h|_S = f$. Furthermore, we obtain

$$|h(y)| \leq p(y). \qquad (4.33)$$

If $\lambda = 0$, then $y = s$ and there is nothing to show.
If $\lambda > 0$, then (4.32) applied to $t = \lambda^{-1} \cdot s$ yields

$$\begin{aligned}
h(y) = h(\lambda x + s) &= \lambda \cdot h(x + \lambda^{-1} \cdot s) = \lambda \left(\mu + f(\lambda^{-1} \cdot s)\right) \\
&\leq \lambda \left(p(\lambda^{-1} \cdot s + x) - f(\lambda^{-1} \cdot s) + f(\lambda^{-1} \cdot s)\right) \\
&= p(s + \lambda x) = p(y).
\end{aligned}$$

If $\lambda < 0$, then we similarly obtain

$$\begin{aligned}
h(y) = h(\lambda x + s) &= \lambda \cdot h(x + \lambda^{-1} \cdot s) = \lambda \left(\mu + f(\lambda^{-1} \cdot s)\right) \\
&\leq -\lambda \left(p(-\lambda^{-1} \cdot s - x) - f(\lambda^{-1} \cdot s) + f(\lambda^{-1} \cdot s)\right) \\
&= |\lambda| \cdot p(\lambda^{-1} \cdot s + x = p(s + \lambda x) = p(y).
\end{aligned}$$

Therefore we have $h(y) \leq p(y)$. Finally, it follows that

$$-p(y) = -p(-y) \leq -h(-y) = h(y),$$

hence altogether $-p(y) \leq h(y) \leq p(y)$, that is, (4.33) holds true.

If $X = \operatorname{span}(x, S)$, then we are done. One might consider to continue the above process. It may, however, not terminate. So we are going to apply Zorn's lemma (or the axiom of transfinite induction, Axiom 1) instead.

(ii) We put

$$\mathcal{P} = \{g \in T^{\#} : |g(x)| \leq p(x) \text{ on } T, \ T \supset S \text{ linear space, } g|_S = f\},$$

and define the relation $\geq$ on $\mathcal{P}$ by

$$g_2 \geq g_1 \text{ if and only if } g_2 \text{ is an extension of } g_1.$$

Then $(\mathcal{P}, \geq)$ is a partially ordered set with $\mathcal{P} \neq \emptyset$, since $f \in \mathcal{P}$. By Zorn's lemma (or the axiom of transfinite induction, Axiom 1), $\mathcal{P}$ contains a maximal chain, that is, a maximal totally ordered subset $\mathcal{C}$. We put

$$W = \{x \in X : g(x) \text{ is defined for some } g \in \mathcal{C}\}.$$

(α) We show that W is a subspace of X.
Let $x, \tilde{x} \in W$ and $\lambda \in \mathbb{R}$. Then there exist $g, \tilde{g} \in \mathcal{C}$, such that $g(x)$ and $\tilde{g}(\tilde{x})$ are defined, and $g \in T^{\#}$, $\tilde{g} \in \tilde{T}^{\#}$ for some subspaces $T, \tilde{T} \supset S$. Then $g(\lambda x)$ is also defined. Since $\mathcal{C}$ is a chain, we have $\tilde{g} \geq g$, say, and so $\tilde{g}(\lambda x + \tilde{x})$ is defined, hence $\lambda x + \tilde{x} \in W$.

We define F on W by

$$F(x) = g(x) \text{ for some } g \in \mathcal{C} \text{ for which } g(x) \text{ is defined.}$$

Since $\mathcal{C}$ is totally ordered, $F(x)$ is uniquely defined.
It can be seen that $F \in W^{\#}$. We also have

$$|F(x)| = |g(x)| \leq p(x) \text{ for all } x \in W.$$

(β) It remains to show that $W = X$.
If $W \neq X$, then there would exist $x \in X \setminus W$, and we would be able to extend F, as in Part (i), to some $\tilde{F} \in (\text{span}(a, W))^{\#}$ with $|\tilde{F}(y)| \leq p(y)$ for all $y \in \text{span}(a, W)$. This is a contradiction to the maximality of $\mathcal{C}$.

This completes the proof of Part (ii).

Thus we have proved the theorem for real linear functionals in real linear spaces.

(b) Now we prove the complex case.
We write
$$f = f_1 + i f_2 \text{ with } f_1 = \text{Re}(f) \text{ and } f_2 = \text{Im}(f).$$

Then f_1 and f_2 are real linear functionals on S, and we have

$$f_1(ix) = \frac{1}{2}\left(f(ix) + \overline{f(ix)}\right) = \frac{1}{2}\left(if(x) - i\overline{f(x)}\right) = -f_2(x)$$

and

$$|f_1(x)| \leq |f(x)| \leq p(x) \text{ on } S.$$

By Part (a), we can extend f_1 to F_1 with $|F_1(x)| \leq p(x)$ on A. Now we define $F(x) = F_1(x) - iF_1(ix)$. Then we have $F|_S = f$ and

$$F(ix) = F_1(ix) - iF_1(-x) = F_1(ix) + iF_1(x)$$
$$= i\left(F_1(x) - iF_1(ix)\right) = iF(x).$$

We finally obtain $|F(x)| \leq p(x)$ for all $x \in A$ by Lemma 4.5.1. $\square$

If X is a seminormed space, then we denote by X^* the space X' of continuous linear functionals on X with the norm

$$\|f\| = \sup\{|f(x)| : \|x\| \leq 1\} \text{ for all } f \in X'.$$

Theorem 4.5.3 *Every linear functional on a subspace of a seminormed space X can be extended to a linear functional on X such that its norm is preserved.*

Proof. If $\|f\|_S = \infty$, then every extension has an infinite norm. If $\|f\|_S < \infty$, then the statement follows from Theorem 4.5.2 with

$$p(x) = \|f\|_S \cdot \|x\| \ (x \in X). \square$$

The next corollary is due to Banach [10, Théorème 4].

Corollary 4.5.4 *Let X be a seminormed space, S be a proper subspace of X, $x \in X \setminus S$ and*

$$\text{dist}(x, S) = \inf\{\|x - s\| : s \in S\}.$$

Then there exists $f \in X^{\#}$ with

$$f(x) = 1, \ f|_S \equiv 0 \ and \ \|f\| = \frac{1}{\text{dist}(x, S)}.$$

(Here $\|f\| = \infty$ if $\text{dist}(x, S) = 0$.)

Proof. We define f on $\text{span}(x, S)$ by

$$f(y) = \lambda \ \text{for} \ y = \lambda x + s \in \text{span}(x, S).$$

Then we have $f \in \text{span}(x, S)^{\#}$, $f(x) = 1$ and $f|_S = 0$. By Theorem 4.5.3, f can be extended to all of X such that its norm is preserved. If $\|f\| < \infty$, then it follows for each $s \in S$

$$\|f\| \cdot \|x - s\| \geq |f(x - s)| = |f(x) - f(s)| = |f(x)| = 1,$$

hence

$$\|x - s\| \geq \frac{1}{\|f\|}, \ \text{and so} \ \text{dist}(x, S) \geq \frac{1}{\|f\|}.$$

This shows in particular that $\text{dist}(x, S) = 0$ implies $\|f\| = \infty$.

On the other hand, if $y \in \text{span}(x, S)$, then $f(y) = 0$ for $y \in S$, and for $y \notin S$, we have $y = \lambda x + s$ for some $s \in S$ and some scalar $\lambda \neq 0$. It follows that $f(y) = \lambda$, that is, $y = f(y)x + s$ for some $s \in S$. Since $(-f(y))^{-1} \cdot s \in S$, this implies

$$\frac{\|y\|}{|f(y)|} = \left\|(f(y))^{-1} \cdot y\right\| = \left\|x - (f(y))^{-1} \cdot (-s)\right\| \geq \text{dist}(x, S),$$

hence

$$|f(y)| \leq \frac{\|y\|}{\text{dist}(x, S)} \ \text{and so} \ \|f\| \leq \frac{1}{\text{dist}(x, S)}. \qquad \square$$

Corollary 4.5.5 *For every vector $x \neq 0$ in a seminormed space X, there exists a linear functional f on X with $f(x) = \|x\|$ and $\|f\| = 1$. In particular, if X is normed, then $X^{\#}$ is total over X.*

Proof. We choose d as in Corollary 4.5.4, $S = \{0\}$ and multiply afterwards with $\|x\|$. $\square$

Corollary 4.5.6 *Let x_0 be an element of a seminormed space X. Then we have*

$$\|x_0\| = \sup_{\substack{\|f\| = 1 \\ f \in X^*}} |f(x_0)|.$$

Proof. If $x_0 = 0$, then the proof is obvious. If $x_0 \neq 0$, then we apply Corollary 4.5.5. $\square$

The next corollary is due to Banach [10, Théorème 3].

Corollary 4.5.7 *Let S be a set in a seminormed space X and f be a functional on S. If there exists a number M such that for any finite set $\{\lambda_1, \lambda_2, \ldots, \lambda_n\}$ of scalars and for any finite subset $\{s_1, s_2, \ldots, s_n\}$ of S*

$$\left|\sum_{k=1}^{n} \lambda_k f(s_k)\right| \leq M \cdot \left\|\sum_{k=1}^{n} \lambda_k s_k\right\|, \tag{4.34}$$

then f can be extended to a linear functional F on all of X with $\|F\| \leq M$. Conversely, if such a linear functional F exists, then (4.34) holds with $M = \|F\|$.

Proof.

(i) The last part is trivial because of

$$\sum_{k=1}^{n} \lambda_k f(s_k) = F\left(\sum_{k=1}^{n} \lambda_k s_k\right).$$

(ii) Now we assume that (4.34) is satisfied. We extend f to the span of S, by putting

$$F(t) = \sum_{k=1}^{n} \lambda_k s_k \text{ for all } t = \sum_{k=1}^{n} \lambda_k s_k \in \text{span}(S).$$

The value of $F(t)$ is independent of the chosen representation of t, for if $t = \sum_{k=1}^{n} \lambda'_k s'_k$ then it follows from (4.34) that

$$\left|\sum_{k=1}^{n} \lambda_k f(s_k) - \sum_{k=1}^{n} \lambda'_k f(s'_k)\right| \leq M \cdot \left\|\sum_{k=1}^{n} \lambda_k s_k - \sum_{k=1}^{n} \lambda'_k s'_k\right\| = M \cdot \|t - t\| = 0.$$

By Theorem 4.5.3, we can extend F on the whole space without increasing its norm. $\square$

The Hahn–Banach theorem can be used to prove the so–called *separation theorem*. We need the following lemma.

Lemma 4.5.8 *Let X be a normed space and $f \neq 0$ a linear (bounded or unbounded) functional on X. Then there exists a number $c > 0$ such that*

$$\lambda \in \{f(x) : \|x\| \leq c|\lambda|\} \text{ for each scalar } \lambda.$$

Proof. If $x_0 \in X$ such that $f(x_0) = 1$, then we obviously have $f(\alpha x_0) = \alpha f(x_0) = \alpha$ for every scalar α. If λ is any scalar, then $f(\lambda x_0) = \lambda$, and since $\|\lambda x_0\| = \|x_0\| |\lambda|$, the conclusion of the lemma holds true for $c = \|x_0\|$. $\square$

Theorem 4.5.9 (Separation theorem) *Let X be a normed space and $E \neq \emptyset$ be a closed and absolutely convex (that is, convex and balanced) subset of X. If $x_0 \in X \setminus E$, then there exists an $f \in X^*$ such that*

$$|f(x_0)| > \sup_{x \in E} |f(x)|. \tag{4.35}$$

Proof. If $0 < r < \inf_{x \in E} \|x_0 - x\|$, then we have

$$x_0 \notin E_1 \equiv \{x + z : x \in E, \|z\| \leq 1\}.$$

It is easy to see that E_1 is a balloon, that is (Definition 2.4.1), an absolutely convex and absorbing subset of X, and that $0 \in E_1$. Let $X_0 = \mathbb{F}x_0$ and the functional $f_0 : X_0 \to \mathbb{F}$ be defined by $f_0(\lambda x_0) = \lambda$ for all $\lambda \in \mathbb{F}$ and μ_{E_1} the Minkowski functional of the set E_1. Obviously f_0 is a linear functional, and it follows from Remark 3.9.2 (d) that

$$|f_0(\lambda x_0)| \leq \mu_{E_1}(\lambda x_0) \text{ for all } \lambda \in \mathbb{F}.$$

We note that $\mu_{E_1}(x) \leq \|x\|/r$ for all $x \in X$. By the Hahn-Banach theorem there exists $f \in X^*$ such that

$$f(x) \leq \mu_{E_1}(x) \leq \frac{\|x\|}{r} \text{ for all } x \in X$$

and
$$f(x) = f_0(x) \text{ for all } x \in X_0.$$

It follows from Corollary 3.9.2 and the homogeneity of the Minkowski functional (Theorem 3.9.3 (a)) that there exists $c > 0$ such that

$$f(x_0) = 1 \geq \sup_{x \in E_1} \mu_{E_1}(x) \geq \sup_{x \in E_1} |f(x)| = \sup_{x \in E, \|z\| \leq 1} |f(x+z)| \geq c + \sup_{x \in E} |f(x)|.$$

Thus we have proved (4.35).   □

The Hahn–Banach theorem and its corollaries can be used to prove some results on the topological complements in normed spaces.

Lemma 4.5.10 *Every finite dimensional subspace F of a normed space X has a* topological complement, *that is, there exists a closed subspace E in X such that*

$$X = F \oplus E. \tag{4.36}$$

Proof. Let $\dim F = m$ and $\{h_1, h_2, \ldots, h_m\}$ be an algebraic basis of the subspace F. If $F_1 = \text{span}\{h_2, \ldots, h_m\}$, then F_1 is a closed subspace of X and $h_1 \notin F_1$. By Corollary 4.5.4, there exists a functional $f_1 \in X^*$ such $f_1(h_1) = 1$ and $f_1(x) = 0$ on F_1. Analogously we prove that there exist functionals $f_k \in X^*$ for $1 \leq k \leq m$ and subspaces $F_k = \text{span}(\{h_1, h_2 \ldots, h_m\} \setminus \{h_k\})$ for $1 \leq k \leq m$ such that

$$f_k(h_k) = 1 \text{ and } f_k(x) = 0 \text{ for all } x \in F_k \text{ and } k = 1, 2, \ldots, m. \tag{4.37}$$

We put $E = \bigcap_{k=1}^{m} N(f_k)$. Then E is a closed subspace of X and it is easy to show that $F \cap E = \{0\}$. To prove (4.36), we assume $x \in X$. We note that

$$x = \sum_{k=1}^{m} f_k(x)h_k + \left(x - \sum_{k=1}^{m} f_k(x)h_k \right),$$

and

$$\sum_{k=1}^{m} f_k(x)h_i \in F \text{ and } \left(x - \sum_{k=1}^{m} f_k(x)h_k \right) \in E. \qquad \square$$

The following result is related to the previous lemma and its proof, which does not use the Hahn–Banach theorem, is simple.

Lemma 4.5.11 *If E is a closed subspace of the normed space X and $\text{codim} E = \dim X/E < \infty$, then there exists a closed subspace F in X such that*

$$X = F \oplus E.$$

Proof. The proof is left to the readers.   □

Corollary 4.5.12 *Let X be a normed space. If the vectors $x_1, x_2, \ldots, x_n \in X$ are linearly independent, then there exists n functionals $f_1, f_2 \ldots, f_n \in X^*$ such that*

$$f_k(x_j) = 0 \text{ for } k \neq j \text{ and } f_k(x_k) = 1. \tag{4.38}$$

Proof. This holds by (4.37).   □

We need Theorem 2.6.18 in the proof of the next result.

Lemma 4.5.13 *Let X be a normed space. If the vectors $f_1, f_2 \ldots, f_n \in X^*$ are linearly independent, then there exist n vectors $x_1, x_2 \ldots, x_n \in X$ such that the conditions in (4.38) hold.*

Proof. Let $N_1 = \cap_{k=2}^n N(f_k)$. It follows from Theorem 2.6.18 that there exists a vector $x_1 \in N_1$ such that $f_1(x_1) \neq 0$, (we may choose x_1 such that $f_1(x_1) = 1$). It follows from $x_1 \in N_1$, that $f_2(x_1) = \ldots = f_n(x_1) = 0$. Analogously we can define the vectors $x_2, \ldots, x_n$. $\square$

Definition 4.5.14 *Let X be a normed space. If the vectors $x_1, x_2 \ldots, x_n \in X$ and the functionals $f_1, f_2 \ldots, f_n \in X^*$ satisfy the conditions in (4.38). Then the vectors $x_1, x_2 \ldots, x_n$ and functionals $f_1, f_2 \ldots, f_n$ form the so–called biorthogonal sequence.*

It is easy to prove that if the vectors $x_1, x_2 \ldots, x_n$ and the functionals $f_1, \ldots, f_n$ form a biorthogonal sequence, then both the vectors $x_1, \ldots, x_n$ and the functionals $f_1, \ldots, f_n$ are linearly independent.

We close this section with an example that will show that the Hahn–Banach extension theorem cannot be generalized to linear maps. We need the following results.

Lemma 4.5.15 *Let S be a subspace of a linear space X. A* projection *or* projector *is a* linear *idempotent map $P : X \to X$, that is, a linear map with $P^2 = P \circ P = P$.*
(a) Let $P : X \to X$ be a projection, and $S = \{x \in X : P(x) = x\}$ and $T = N(P)$. Then S and T are complementary subspaces of X, and we have

$$P(X) = P(S) = S \text{ and } P(T) = \{0\}.$$

(b) Let $P : X \to X$ be a projection and $Q = \text{id} - P$ where id: $X \to X$ is the identity on X. Then Q is a projection, we have $P \circ Q = Q \circ P = 0$, and Q yields the same pair of complementary subspaces as P.
(c) Every projection P of c onto c_0 satisfies $\|P\| > 1$.
(d) Let $f : c_0 \to c_0$ be defined by $f(x) = x$ for all $x \in c_0$. Then every linear extension $F : c \to c_0$ of f is a projection onto c_0.

Proof.

(a) The set T is a subspace by Remark 1.6.2 (a). It is easy to see that S is a subspace and $S \cap T = \{0\}$.
Now we show that
$$P(X) = P(S) = S. \tag{*}$$

 (i) We trivially have $P(S) \subset P(X)$.
 (ii) Let $y \in P(X)$. Then there exists $x \in X$ with $y = P(x)$, and so $P(y) = P(P(x)) = P(x) = y$, that is, $y \in S$. Hence we have shown that $P(X) \subset S$.
 (iii) Now let $s \in S$. Then it follows that $s = P(s) \in P(S)$, so $S \in P(S)$. We also have by Parts (i) and (ii) that
$$P(S) \subset P(X) \subset S \subset P(S), \text{that is, (*) holds.}$$

 (iv) Finally we show that $X = S + T$.
 Since obviously $S + T \subset X$, it suffices to show that $X \subset S + T$.
 So let $x \in X$ be given. We put $y = x - t$ for some $t \in T$. Then we have $P(P(y)) = P(y)$, hence $P(y) \in S = P(S)$, by (*). It follows that $y \in S$, and so $x = y + t \in S + T$.

This completes the proof of Part (a).

(b) (i) First we show that Q is a projection.
Obviously $Q : X \to X$ is a linear map, being the difference of the linear maps id and Q. Also it follows from the linearity of Q that

$$Q^2 = (\mathrm{id} - P) \circ (\mathrm{id} - P) = \mathrm{id} - P - P \circ (\mathrm{id} - P) = \mathrm{id} - P = Q.$$

Thus Q is a projection.

(ii) Now we show that $P \circ Q = Q \circ P = 0$.
We have

$$P \circ Q = P \circ (\mathrm{id} - P) = P - P \circ P = P - P = 0$$

and

$$Q \circ P = (\mathrm{id} - P) \circ P = P - P \circ P = P - P = 0.$$

(iii) Now we show that P and Q have the same pairs of complementary subspaces. We put $\tilde{S} = \{x \in X : Q(x) = x\}$ and $\tilde{T} = N(Q)$. Then we have $x \in \tilde{S}$ if and only if $Q(x) = x - P(x) = x$, that is, if and only if $P(x) = 0$. Hence we have $\tilde{S} = T$. We also have $x \in \tilde{T}$ if and only if $Q(x) = x - P(x) = 0$, that is, if and only if $P(x) = x$. Hence we have $\tilde{T} = S$.

This completes the proof of Part (b).

(c) Let P be a projection from c onto c_0. We put

$$S = \{x \in c : P(x) = x\} \text{ and } T = N(P).$$

It follows from Part (a) and since P is onto that

$$c_0 = P(c) = P(S) = S.$$

We also have $c = S + T$, hence $T = \mathrm{span}(\{e\})$.
Now let $x \in c$ be given. Then there exist $\lambda \in \mathbb{C}$ and $x_0 \in c_0$ such that $x = x_0 + \lambda e$, and so $P(x) = P(x_0) + \lambda P(e) = x_0$. We choose $x_0 = 2e^{(1)} \in c_0$ and $x = x_0 - e$. It follows that $\|x\| = \sup_k |x_k| = 1$ and $\|P(x)\| = \|x_0\| = 2$, hence $\|P\| \geq 2 > 1$.

(d) Let F be a linear extension of f to all of c. Obviously $F(c) = c_0$ and $F(e) \in c_0$, hence

$$F(F(e)) = f(F(e)) = F(e).$$

It follows from the linearity of F on c that $F(F(x)) = F(x)$ for all $x \in c$ and consequently F is a projection. $\qquad\square$

Example 4.5.16 *Let f be defined on c_0 by $f(x) = x$. Then we have $\|f\| = 1$. By Lemma 4.5.15 (d), every linear extension $F : c \to c_0$ of f is a projection from c onto c_0 with norm greater than 1. Thus the Hahn–Banach extension theorem does not hold for linear maps, in general.*

4.6 Important theorems

In this section, we study the versions for Banach of the open mapping, closed graph and Banach–Steinhaus theorems and Banach's theorem for the bounded inverse.

First we list the special cases for Banach spaces of Theorems 3.5.3, 3.6.1 and 3.7.5.

Theorem 4.6.1 (Open mapping theorem) (Theorem 3.5.5)
Every surjective bounded linear map between Banach spaces is an open mapping.

Theorem 4.6.2 (Closed graph theorem) (Theorem 3.6.1)
Every closed linear operator between Banach spaces is bounded.

Theorem 4.6.3 (Banach's theorem of the bounded inverse) (Corollary 3.5.6) *Let X and Y be Banach spaces and $A \in \mathcal{B}(X,Y)$ be bijective. Then $A^{-1} \in \mathcal{B}(Y,X)$.*

Theorem 4.6.4 (Banach–Steinhaus theorem) *Let X be a Banach space, Y be a normed space and (A_n) be a sequence of operators in $\mathcal{B}(X,Y)$. If $\sup_n \|A_n x\| < \infty$ for each $x \in X$, then we have $\sup_n \|A_n\| < \infty$, that is, if the sequence (A_n) is pointwise bounded, then it is norm bounded.*

Proof. For each $m, n \in \mathbb{N}$, we put

$$E_{m,n} = \{x \in X : \|A_n x\| \le m\} \text{ and } E_m - \bigcap_{n=1}^{\infty} E_{m,n}.$$

Since $\|\cdot\| : Y \to \mathbb{R}$ is continuous by Remark 4.2.1 and $A_n \in \mathcal{B}(X,Y)$ for each n, the map $f_n : \|A_n(\cdot)\| = \|\cdot\| \circ A_n : X \to \mathbb{R}$ is continuous for each n, and so $E_{m,n} = f_n^{-1}([0,m])$ is a closed set in X for each n, hence E_m is closed being the intersection of the closed sets $E_{m.n}$. Obviously we have $X = \bigcup_{m=1}^{\infty} E_m$. Since X is complete, it follows by Baire's category theorem (Theorem 2.10.9) that there exists a natural number m_0 such that E_{m_0} is not nowhere dense. Since E_{m_0} is a closed set, there exists an open ball $B_0 = B_r(x_0) \subset E_{m_0}$. Let $x \in X \setminus \{0\}$. We put $y = x_0 + \lambda x$ where $\lambda = r/(2\|x\|)$. Then we have $y \in B_0$, since $\|x - x_0\| = \|\lambda x\| = \lambda \|x\| = r/2 < r$. Now it follows from $x_0 \in B_0 \subset E_{m_0}$ that

$$\|A_n y\| \le m_0 \text{ and } \|A_n x_0\| \le m_0 \text{ for each } n \in \mathbb{N}. \tag{4.39}$$

Hence (4.39) implies

$$\|A_n x\| = \frac{1}{\lambda} \|A_n(y - x_0)\| \le \frac{1}{\lambda} (\|A_n y\| + \|A_n x_0\|) \tag{4.40}$$

$$\le \frac{2m_0}{\lambda} = \frac{4m_0}{r} \|x\| \text{ for all } n \in \mathbb{N},$$

and so

$$\sup_n \|A_n\| \le \frac{4m_0}{r}. \qquad \square \tag{4.41}$$

Remark 4.6.5 *Theorem 4.6.1 extends to collections of $A_\alpha \in \mathcal{B}(X,Y)$ where J is an arbitrary indexing set. For if card $J > \aleph_0$ and $\sup_{\alpha \in J} = \infty$, then there exists a sequence (A_n) in $\mathcal{B}(X,Y)$ such that $\lim_{n \to \infty} \|A_n\| = \infty$ which contradicts (4.41).*

Remark 4.6.6 *The proof we gave of the Banach–Steinhaus theorem, Theorem 4.6.4, is the standard one used in most textbooks and goes back to S. Banach, H. Steinhaus and S. Saks in 1927 [11]; it uses the Baire category theorem, Theorem 2.10.9. The original proofs given by H. Hahn [81] and S. Banach [9] do not use Baire's theorem, but are elementary and apply the method of the* gliding hump *in the contraposition, that is, under the assumption that the conclusion does not hold, a sequence (A_n) in $\mathcal{B}(X,Y)$ and an element $x \in X$ are constructed such that $\lim_{n\to\infty} \|A_n x\| = \infty$. We present a similar proof given in [192]. We need the next lemma.*

Lemma 4.6.7 *Let X be a Banach space, Y be a normed space and $A \in \mathcal{B}(X,Y)$. Then we have*

$$\sup_{x' \in B_r(x)} \|Ax'\| \geq \|A\|r \text{ for any } x \in X \text{ and } r > 0. \tag{4.42}$$

Proof. We have for all $\xi \in X$

$$\max\{\|A(x+\xi)\|, \max\|A(x-\xi)\|\} \geq \frac{1}{2} \cdot (\|A(x+\xi)\| + \|A(x-\xi)\|) \geq \|A\xi\|. \tag{4.43}$$

Taking the supremum over $\xi \in B_r(0)$ in (4.43) we obtain (4.42).   □

Alternative proof of Theorem 4.6.4. We assume that $\sup_n \|A_n\| = \infty$. Then we can choose a subsequence $A_{n_k} \in \mathcal{B}(X,Y)$ such that $\|A_{n_k}\| \geq 4^k$. We choose $x_0 = 0$ and use Lemma 4.6.7 to choose $x_k \in X$ inductively for $k \geq 1$ such that $\|x_k - x_{k-1}\| \leq 3^{-k}$ and $\|A_{n_k} x_k\| \geq (2/3) \cdot 3^{-k}\|A_{n_k}\|$. The sequence (x_k) is Cauchy, and hence convergent to some $x \in X$. It is easy to see that $\|x - x_k\| \leq 3^{-k}/2$ and hence

$$\|A_{n_k} x\| \geq \frac{1}{6} \cdot 3^{-k}\|A_{n_k}\| \geq \frac{1}{6} \cdot \left(\frac{4}{3}\right)^k \to \infty \text{ as } k \to \infty.   □$$

The next result states that the limit function of a pointwise convergent sequence of bounded linear operators from a Banach space into a normed space is a bounded linear operator. This result extends the special case of Theorem 3.7.5 for continuous linear functionals on a Fréchet space to bounded linear operators on a Banach space.

Corollary 4.6.8 *Let X be a Banach space, Y be a normed space and (A_n) a sequence in $\mathcal{B}(X,Y)$. If the limit $Ax = \lim_n Ax$ exists for each $x \in X$ then $A \in \mathcal{B}(X,Y)$.*

Proof. Since the sequence $(A_n x)$ is bounded for each $x \in X$, it follows from Theorem 4.6.4 that $\sup_n \|A_n\| < \infty$. It can be shown as in the proof of Theorem 4.6.4 that A is a bounded linear operator.   □

The next example shows that Theorem 4.6.4 need not hold if X is not a Banach space.

Example 4.6.9 *We consider ϕ, the space of all finite sequences, with the norm $\|\cdot\|_2$ of ℓ_2. We define the sequence (A_n) of linear operators $A_n : \phi \to \ell_2$ by*

$$A_n(e^{(k)}) = \begin{cases} 0 & \text{for } k \neq n \\ ne^{(n)} & \text{for } k = n. \end{cases}$$

Then we have $\sup_n \|A_n x\| < \infty$ for each $x \in \phi$, but $\sup_n \|A_n\| = \infty$.

Proof. The proof is left to the readers.   □

An application of Corollary 4.6.8 yields the following result.

Example **4.6.10** *Let* $a = (a_k)_{k=1}^\infty$ *be a complex sequence,* $1 \le p \le \infty$ *and* q *be the cojugate number of* p. *If the series* $\sum_{k=1}^\infty a_k x_k$ *converges for all sequences* $x = (x_k)_{k=1}^\infty \in \ell_p$, *then we have* $a \in \ell_q$. *Furthermore if we define the functional* $f : \ell_p \to \mathbb{C}$ *by* $f(x) = \sum_{k=1}^\infty a_k x_k$ *for all* $x \in \ell_p$, *then we have* $f \in \ell_p^*$ *and*

$$\|f\| = \|a\|_q. \tag{4.44}$$

Proof. We define the linear functionals $f_n : \ell_1 \to \mathbb{C}$ for each $n \in \mathbb{N}$ by

$$f_n(x) = \sum_{k=1}^n a_k x_k \text{ for all } x \in \ell_p.$$

(i) First let $p = 1$, and so $q = \infty$.
   It follows from

$$|f_n(x)| \le \sum_{k=1}^n |a_k x_k| \le \max_{1 \le k \le n} |a_k| \sum_{k=1}^\infty |x_k| = \max_{1 \le k \le n} |a_k| \cdot \|x\|_1 \tag{4.45}$$

that $f_n \in \ell_1^*$ and $\|f_n\| \le \max_{1 \le k \le n} |a_k|$ for all n. By assumption the limit $f(x) = \lim_{n \to \infty} f_n(x) = \sum_{k=1}^\infty a_k x_k$ exists for all $x \in \ell_1$, hence $f \in \ell_1^*$ by Corollary 4.6.8. Since $|a_n| = |f(e^{(n)})| \le \|f\| \|e^{(n)}\|_1 = \|f\| < \infty$ for all n, we obtain $a \in \ell_\infty$ and so $\|a\|_\infty \le \|f\|$. We also obtain the converse inequality from (4.45). Thus we have shown (4.44) for $p = 1$.

(ii) Now let $1 < p < \infty$, and so $q = p/(p-1)$.
   Applying Hölder's inequality, we obtain

$$|f_n(x)| \le \sum_{k=1}^n |a_k x_k| \le \left(\sum_{k=1}^n |a_k|^q \right)^{1/q} \cdot \left(\sum_{k=1}^\infty |x_k|^p \right)^{1/p}$$
$$\left(\sum_{k=1}^n |a_k|^q \right)^{1/q} \cdot \|x\|_p, \tag{4.46}$$

which implies $f_n \in \ell_p^*$ for all n. As in Part (ii), this yields $f \in \ell_p^*$. Let $n \in \mathbb{N}$ be given. We put $x^{(n)} = \sum_{k=1}^n \text{sgn}(a_k) |a_k|^{q-1}$, where $\text{sgn}(z) = z/|z|$ for $z \in \mathbb{C} \setminus \{0\}$ and $\text{sgn}(0) = 0$ denotes the *sign of the complex number* z. It follows that

$$|f(x^{(n)})| = \sum_{k=1}^n |a_k|^q \le \|f\| \|x^{(n)}\|_p = \|f\| \left(\sum_{k=1}^n |a_k|^q \right)^{1/p},$$

hence

$$\left(\sum_{k=1}^n |a_k|^q \right)^{1/q} \le \|f\|.$$

Since $n \in \mathbb{N}$ was arbitrary, it follows that $a \in \ell_q$ and $\|a\|_q \le \|f\|$. We also obtain the converse inequality from (4.46). Thus we have shown (4.44) for $1 < p < \infty$.

(iii) Finally let $p = \infty$, hence $q = 1$. Interchanging x and a in (4.45), we obtain

$$|f_n(x)| \le \sum_{k=1}^n |a_k x_k| \le \left(\sum_{k=1}^n |a_k| \right) \|x\|_\infty. \tag{4.47}$$

Again, this implies $f \in \ell_\infty^*$. Let $n \in \mathbb{N}$ be given. We put $x^{(n)} = \sum_{k=1}^n \mathrm{sgn}(a_k)$. It follows that

$$|f(x^{(n)})| = \sum_{k=1}^n |a_k| \leq \|f\| \, \|x^{(n)}\|_\infty \leq \|f\|.$$

Since $n \in \mathbb{N}$ was arbitrary, it follows that $a \in \ell_1$ and $\|a\|_1 \leq \|f\|$. We also obtain the converse inequality from (4.47). Thus we have shown (4.44) for $p = \infty$. $\qquad\square$

We close this section with two important applications of the closed graph theorem, Theorem 4.6.2.

We need the following definition.

Definition 4.6.11 If X is a normed space, M and N are closed subspaces of X and $X = M \oplus N$, then N is said to be the *topological complement of M*, and X is referred to as the *topological direct sum of M and N*.

In the next theorem, we give a necessary and sufficient condition for a closed subspace in a Banach space to have a topological complement.

Theorem 4.6.12 *Let M be a closed subspace of a Banach space X. Then there exists a closed subspace N of X such that*

$$X = M \oplus N \tag{4.48}$$

if and only if there exists a projector $P \in \mathcal{B}(X)$ from X onto M.

Proof. We assume that there exists a projector $P \in \mathcal{B}(X)$ from X onto M. Then the null space of P is a topological complement of M.

Conversely, we assume that N is a closed subspace in X and $X = M \oplus N$. Every $x \in X$ has a unique representation $x = y + z$, $y \in M$ and $z \in N$. Let $Q : X \to X$ be the map defined by $Q(x) = y$. Obviously we have $Q^2 = Q$, $R(Q) = M$ and Q is a linear operator. We show that $Q \in \mathcal{B}(X)$. By the closed graph theorem, it suffices to prove that Q is a closed operator. Let (x_n) be a sequence in X, such that $x_n \to x \in X$ and $Q(x_n) \to y \in X$. It follows from $Q(x_n) \in M$ and $M = \overline{M}$ that $y \in M$. Furthermore $x_n - Q(x_n) \in N$ and $x_n - Q(x_n) \to x - y$. From $N = \overline{N}$ we conclude $x - y \in N$. Hence $x = y + (x - y)$, $y \in M$ and $x - y \in N$. Thus $Q(x) = y$, that is, Q is a closed operator. $\qquad\square$

Theorem 4.6.13 (V. M. Eni - G. T. Karauš) ([E–K]) *A normed X is finite dimensional if and only if it is a Banach space for any norm on X.*

Proof. It is well known that any finite dimensional normed space is a Banach space (Theorem 4.2.3). Hence to prove the theorem it suffices to show that on each infinite dimensional Banach space $(X, \|\cdot\|)$ we can define a new norm $\|\cdot\|_1$ such that $(X, \|\cdot\|_1)$ is not a Banach space. Let $A \in L(X)$ and $A \notin \mathcal{B}(X)$ (here X is normed by $\|\cdot\|$). We define the norm $\|\cdot\|_1$ on X by

$$\|x\|_1 = \|x\| + \|Ax\|, \ x \in X. \tag{4.49}$$

(It is easy to see that $\|\cdot\|_1$ is a norm on X.) We are going to show that $(X, \|\cdot\|_1)$ is not a Banach space. If $(X, \|\cdot\|_1)$ were a Banach space, then it would be easy to prove that A is a closed operator. Suppose that (x_n) is a sequence in X, and $x_n \to x \in X$ and $A(x_n) \to y \in X$. It follows from (4.49) that (x_n) is a Cauchy sequence in $(X, \|\cdot\|_1)$, and if $(X, \|\cdot\|_1)$ is a Banach space, again (4.49) implies $y = Ax$. Hence A is a closed operator, and so $A \in \mathcal{B}(X)$ by Theorem 4.6.2, which is a contradiction. $\qquad\square$

We will use the following result.

Remark 4.6.14 *In lemma 4.12-3, it is stated that an operator $A \in \mathcal{B}(X, Y)$, where X and Y are Banach spaces, has the property that the image under A of the unit ball $(B_1(0))$ in X contains an open ball about 0 in Y.*

4.7 Representation theorems

In this section we prove some representation theorems for continuous linear functionals on certain Banach spaces.

Theorem 4.7.1 *We have $\ell_p^* \cong \ell_q$ for $1 \leq p < \infty$ where $q = \infty$ for $p = 1$ and $q = p/(p-1)$ for $1 < p < \infty$; this means, $f \in \ell_p^*$ if and only if there exists a sequence $b \in \ell_q$ such that*

$$f(a) = \sum_{n=1}^{\infty} a_n b_n \ \text{for all } a \in \ell_p;$$

furthermore,

$$\|f\| = \|b\|_q = \begin{cases} \sup_n |b_n| & \text{for } p = 1 \\ \left(\sum_{n=1}^{\infty} |b_n|^q \right)^{1/q} & \text{for } 1 < p < \infty. \end{cases}$$

Proof. For each $f \in \ell_p^*$, we define $Tf = b$ where $b = b^{(f)} = (f(e^{(k)}))_{k=1}^{\infty}$.

(i) First we show that $T : \ell_p^* \to \ell_q$.

Let $f \in \ell_p^*$ and $x = (x_k)_{k=1}^{\infty} \in \ell_p$ be given. Then we have $x = \sum_{k=1}^{\infty} x_k e^{(k)}$ uniquely by Example 3.4.3 (c), and since $f \in \ell_p^*$, we obtain

$$f(x) = f\left(\sum_{k=1}^{\infty} x_k e^{(k)} \right) = \sum_{k=1}^{\infty} x_k f(e^{(k)}) = \sum_{k=1}^{\infty} b_k x_k,$$

and the convergence of the series $\sum_{k=1}^{\infty} b_k x_k$ for all $x \in \ell_p$ implies $b \in \ell_q$ by Example 4.6.10.

It is easy to see that T is linear, so $T \in \mathcal{L}(\ell_p^*, \ell_q)$.

(ii) Now we show that T is one to one.

If $Tf = b = 0$, then $f(e^{(k)}) = b_k = 0$ for all k, and so $f(x) = 0$ for all $x \in \ell_p$, hence $f \equiv 0$. Since T is linear, it follows that T is one to one.

(iii) Now we show that T is onto.

Let $b \in \ell_q$ be given. We put $f(x) = \sum_{k=1}^{\infty} b_k x_k$ for all $x \in \ell_p$. Then $f \in \ell_p^*$ by Example 4.6.10 and $b_k = f(e^{(k)})$ for all k, that is $Tf = b$.

(iv) It remains to show that $T \in \mathcal{B}(\ell_p^*, \ell_q)$ and $\|Tf\| = \|f\|$.

Let $f \in \ell_p^*$ be given. Then we have, by Part (i), $f(x) = \sum_{k=1}^{\infty} b_k x_k$ for all $x \in \ell_p^*$, and it follows from Example 4.6.10 that $\|f\| = \|b\|_q = \|Tf\|$, whence $\|T\| \leq 1$ and $T \in \mathcal{B}(\ell_p^*, \ell_q)$. $\qquad \square$

Theorem 4.7.2 *We have $c^* \cong \ell_1$, that is, $f \in c^*$ if and only if there exists a complex number a and there exists a sequence $b \in \ell_1$ such that*

$$f(x) = a\xi + \sum_{k=1}^{\infty} b_k x_k \ \text{for all } x \in c, \text{ where } \xi = \lim_{n \to \infty} x_n; \tag{4.50}$$

moreover,

$$\|f\| = |a| + \|b\|_1. \tag{4.51}$$

Proof.

(i) First we show the sufficiency, that is, we show that if there exist $a \in \mathbb{C}$ and $b \in \ell_1$ then (4.50) defines a functional $f \in c^*$.

Let $x \in c$ be given and $\xi = \lim_{k \to \infty} x_k$. Then $x \in \ell_\infty$, $|\xi| = |\lim_{k \to \infty} x_k| \le \sup_k |x_k| = \|x\|_\infty$ and

$$\sum_{k=0}^{\infty} |b_k x_k| \le \|b\|_1 \|x\|_\infty,$$

that is, $\sum_{k=1}^{\infty} b_k x_k$ converges for all $x \in c$. Thus f is well defined and trivially $f \in c^\#$. Furthermore we obtain

$$|f(x)| \le |a\xi| + \sum_{k=0}^{\infty} |b_k x_k| \le (|a| + \|b\|_1) \|x\|_\infty,$$

hence $c \in c^*$ and

$$\|f\| \le |a| + \|b\|_1. \tag{4.52}$$

(ii) Now we show the necessity.

Let $f \in c^*$ and $x \in c$ be given. Then we have by Part (iii) in the proof of Example 3.4.3 (c) that x has a unique representation $x = \xi e + \sum_{k=1}^{\infty}(x_k - \xi)e^k$ where $\xi = \lim_{k \to \infty} x_k$. Now $f \in c^*$ implies

$$f(x) = \xi f(e) + \sum_{k=1}^{\infty}(x_k - \xi)f(e^{(k)}). \tag{4.53}$$

Let $m \in \mathbb{N}$ be given and $x^{(m)} = \sum_{k=1}^{m} \operatorname{sgn}(f^{(k)}) \in c_0$. Then we have $\|x^{(m)}\|_\infty \le 1$, and

$$|f(x^{(n)})| = \sum_{k=1}^{m} |f(e^{(k)})| \le \|f\| \cdot |x^{(n)}|_\infty \le \|f\|. \tag{4.54}$$

Since $m \in \mathbb{N}$ was arbitrary, this implies $(f(e^{(k)}))_{k=1}^{\infty} \in \ell_1$ and so we can write (4.53) as

$$f(x) = a\xi + \sum_{k=1}^{\infty} b_k x_k$$

$$\text{with } a = f(e) - \sum_{k=1}^{\infty} f(e^{(k)}) \text{ and } b = (f(e^{(k)}))_{k=1}^{\infty} \in \ell_1. \tag{4.55}$$

(iii) Now we show (4.51).

In view of (4.52) it suffices to show that

$$|a| + \|b\|_1 \le \|f\|. \tag{4.56}$$

Let $m \in \mathbb{N}$ be given. We define the sequence $x^{(m)}$ by $x_k^{(m)} = \operatorname{sgn}(b_k)$ for $1 \le k \le$ and $\operatorname{sgn}(a)$ for $k > m$. Then we have $x^{(m)} \in c$, $\xi = \lim_{k \to \infty} x_k^{(m)} = \operatorname{sgn}(a)$, $\|x^{(m)}\|_\infty \le 1$ and

$$|f(x^{(m)})| = \left| |a| + \sum_{k=1}^{m} |b_k| + \sum_{k=m+1}^{\infty} b_k \operatorname{sgn}(a) \right| \le |f|.$$

Now $b \in \ell_1$ implies $\sum_{k=m+1}^{\infty} b_k \operatorname{sgn}(a) \to 0$ as $m \to \infty$ and so we obtain $|a| + \|b\|_1 \le \|f\|$, that is, (4.56).

(iv)   We define $T : c^* \to \ell_1$ by $T(f) = (a, b_1, b_2, \dots)$ with $a \in \mathbb{C}$ and $b \in \ell_1$ as in (4.55). Then we have $\|Tf\|_1 = |a| + \|b\|_1 = \|f\|$, hence T is norm preserving. The map T is onto by Part (i) of the proof, and it is easy to see that T is linear and one to one. $\square$

Similarly, as in the proof of the previous theorem, the following result can be shown.

Theorem 4.7.3 *We have $c_0^* \cong \ell_1$, that is, $f \in c_0^*$ if and only if there exists a sequence $b \in \ell_1$ such that*

$$f(x) = b_k x_k \text{ for all } x \in c_0; \tag{4.57}$$

moreover,

$$\|f\| = \|b\|_1. \tag{4.58}$$

Now we determine the continuous dual ℓ_∞^* of the space ℓ_∞. This case is essentially different from all the previous cases in that ℓ_∞^* cannot be identified with any sequence spaces. The crucial fact in the previous representation theorems was that each of the spaces had a Schauder basis. The space ℓ_∞, however, does not have a Schauder basis by Example 3.4.3 (d). Here we consider the special case of the space $\ell_{\infty,\mathbb{R}}$ of bounded real sequences.

First we have to introduce a few notions.

Definition 4.7.4 Let H be a set and $\mathcal{P}(H) = \{S : S \subset H\}$ be its power set.

(a)   A *partition* Π *of* H is a finite class

$$\Pi = \{E_k : E_k \subset H \text{ for } k = 1, 2, \dots, n\} \subset \mathcal{P}(H)$$

such that

$$E_k \cap E_j = \emptyset \text{ and } \bigcup \Pi = \bigcup_{k=1}^{n} E_k = H.$$

(b)   By $\mathcal{B}[H]$, we denote the set of all bounded real–valued functions defined on H. We put

$$\|f\| = \sup\{|f(h)| : h \in H\} \text{ for all } f \in \mathcal{B}[H].$$

Then $(\mathcal{B}[H], \|\cdot\|)$ is a normed space. In particular, we have

$$\ell_{\infty,\mathbb{R}} = \mathcal{B}(\mathbb{N}).$$

(c)   Let $f \in \mathcal{B}[H]$, $\Pi = \{E(1), E(2), \dots, E(n)\}$ be a partition of H, and $h_k \in E(k) \neq \emptyset$ be arbitrary points for $k = 1, 2, \dots, n$. Then the *Riemann sums* $\sigma(f; \Pi, (h_k))$ *of* f *with respect to the partition* Π *and the points* h_k are defined by

$$\sigma(f; \Pi, (h_k)) = \sum_{k=1}^{n} f(h_k) \chi_{E(k)},$$

where $\chi_{E(k)}$ denotes the *characteristic function of* $E(k)$ defined by

$$\chi_{E(k)}(h) = \begin{cases} 1 & (h \in E(k)) \\ 0 & (h \notin E(k)). \end{cases}$$

If Π_1 and Π_2 are partitions of H, the Π_2 is said to be *finer* than Π_1 if each set in Π_2 is contained in a set of Π_1; in such a case, we write

$$\Pi_2 \geq \Pi_1.$$

(d)   A map $\mu : \mathcal{P}(H) \to \mathbb{R}$ is said to be an *additive set function*, if

$$\mu(\emptyset) = 0 \text{ and } \mu(E_1 \cup E_2) = \mu(E_1) + \mu(E_2) - \mu(E_1 \cap E_2) \text{ for all } E_1, E_2 \in \mathcal{P}(H).$$

We define for every additive set function μ on H

$$\|\mu\| = \int_H |d\mu| = \sup\left\{ \sum_{E \in \Pi} |\mu(E)| : \Pi \text{ is a partition of } H \right\}.$$

By $\mathcal{M}[H]$ we denote the set of all additive set functions μ on H for which $\|\mu\| < \infty$; $(\mathcal{M}[H], \|\cdot\|)$ is a normed space.

First we characterize the continuous dual $(\mathcal{B}[H])^*$ of the space of all bounded real–valued functions of a non–empty set H.

Theorem 4.7.5 *Let H be a non–empty set. Then we have*

$$(\mathcal{B}[H])^* \cong \mathcal{M}[H].$$

Proof. We define the map T on $(\mathcal{B}[H])^*$ as follows. If $f \in (\mathcal{B}[H])^*$ then we put

(i)   First we show that $T : (\mathcal{B}[H])^* \to \mathcal{M}[H]$, that is, for each $f \in (\mathcal{B}[H])^*$, we have $\mu_f \in \mathcal{M}[H]$.
It follows from $\chi_\emptyset \equiv 0$ that $\mu_f(\emptyset) = f(\chi_\emptyset) = 0$, by the linearity of f. Let $E_1, E_2 \in \mathcal{P}(H)$ be given. Then we have $\chi_{E_1 \cup E_2} = \chi_{E_1} + \chi_{E_2} - \chi_{E_1 \cap E_2}$, and, by the linearity of f,

$$\mu_f(E_1 \cup E_2) = f(\chi_{E_1}) + f(\chi_{E_2}) - f(\chi_{E_1 \cap E_2})$$
$$= \mu(E_1) + \mu(E_2) - \mu(E_1 \cap E_2).$$

Thus μ_f is an additive set function on H.

(α)   We have to show that $\|\mu_f\| < \infty$.
Let $\Pi = \{E(1), E(2), \ldots, E(n)\}$ be an arbitrary partition of the set H. We define the function φ on H by

$$\varphi(h) = \operatorname{sgn}(\mu_f(E(k))) \text{ for } h \in E(k); \ k = 1, 2, \ldots, n.$$

Then we have $\varphi \in \mathcal{B}[H]$, $\|g\| \leq 1$ and

$$\varphi = \sum_{k=1}^n \varphi(h)\chi_{E(k)}, \tag{4.59}$$

since for each $h \in H$ there exists one and only one $j \in \{1, 2, \ldots, n\}$ for which $h \in E(j)$ and it follows that

$$\varphi(h) = \sum_{k=1}^n \varphi(h)\chi_{E(k)}(h) = \varphi(h)\chi_{E(j)}(h).$$

Now, since $f \in (\mathcal{B}[H])^*$, we obtain by (4.59)

$$\sum_{k=1}^n |\mu_f(E(k))| = \sum_{k=1}^n \operatorname{sgn}(\mu_f(E(k))) \cdot \mu_f(E(k)) = \sum_{k=1}^n \varphi(h)f(\chi_{E(k)})$$

$$= f\left(\sum_{k=1}^{n} \varphi(h)\chi_{E(k)}\right) = f(\varphi) \leq \|f\| < \infty,$$

hence

$$\|\mu_f\| = \sup\left\{\sum_{E\in\Pi} |\mu_f(E)| : \Pi \text{ is a partition of } H\right\} \leq \|f\| < \infty. \qquad (4.60)$$

Thus we have shown that $Tf = \mu_f \in \mathcal{M}[H]$ for each $f \in \mathcal{B}[H]$.

(ii) Now we show that T is linear.

Let $f, g \in (\mathcal{B}[H])^*$ and $\lambda \in \mathbb{R}$ be given. Then we have for each set $E \in \mathcal{P}(H)$

$$\begin{aligned}
(T(\lambda f + g))(E) &= \mu_{\lambda f+g}(E) = (\lambda f + g)(\chi_E) = \lambda f(\chi_E) + g(\chi_E) \\
&= \lambda \mu_f(E) + \mu_g(E) = \lambda(T_f)(E) + (T_g)(E) \\
&= (\lambda(Tf) + Tg)(E),
\end{aligned}$$

hence $T(\lambda f + g) = \lambda Tf + Tg$.

(iii) Now we show that T is one to one.

Let $Tf = Tg$. Then we have for all $E \in \mathcal{P}(E)$

$$(Tf)(E) = \mu_f(E) = \mu_g(E) = (Tg)(E),$$

hence

$$f(\chi_E) = g(\chi_E) \text{ for all } E \in \mathcal{P}(E). \qquad (4.61)$$

Let $\varphi \in \mathcal{B}[H]$ be given and $\Pi = \{E(1), E(2), \ldots, E(n)\}$ be a partition of H. Then it follows by (4.59) and the linearity of f and g from (4.61)

$$\begin{aligned}
f(\varphi) = f\left(\sum_{k=1}^{n} \varphi(h)\chi_{E(k)}\right) &= \sum_{k=1}^{n} \varphi(h)f(\chi_{E(k)}) \\
&= \sum_{k=1}^{n} \varphi(h)g(\chi_{E(k)}) = g\left(\sum_{k=1}^{n} \varphi(h)\chi_{E(k)}\right) = g(\varphi),
\end{aligned}$$

hence $f(\varphi) = g(\varphi)$ for all $\varphi \in \mathbf{B}[H]$, and so $f = g$. Thus we have shown that T is one to one.

(iv) Now we show that T is onto $\mathcal{M}[H]$.

Let $\mu \in \mathcal{M}[H]$ be given, and let $\varphi \in \mathcal{B}[H]$ be an arbitrary step function, that is, there exists a partition $\Pi = \{E(1), E(2), \ldots, E(n)\}$ of H and $\lambda_1, \lambda_2, \ldots, \lambda_n \in \mathbb{R}$ such that

$$\phi = \sum_{k=1}^{n} \lambda_k \chi_{E(k)}.$$

We define the map $\tilde{f}$ by

$$\tilde{f}(\phi) = \sum_{k=1}^{n} \lambda_k \mu(E(k)).$$

Then $\tilde{f}$ is linear on the subspace of step functions, and furthermore

$$|\tilde{f}(\phi)| \leq \sum_{k=1}^{n} |\lambda_k| \cdot |\mu(E(k))|.$$

Let $h \in H$ be given. Then there exists one and only one $j \in \{1, 2, \ldots, n\}$ such that $h \in E(j)$, and we obtain

$$\varphi(h) = |\lambda_j| \le \|\varphi\| = \sup_{h \in H} |\varphi(h)|,$$

that is, $|\lambda_k| \le \|\varphi\|$ for all $k = 1, 2, \ldots$. Therefore we have

$$|\tilde{f}(\phi)| \le \sum_{k=1}^{n} |\mu(E(k))| \cdot \|\varphi\| \le \|\mu\| \cdot \|\varphi\|.$$

This means $\|\tilde{f}\| \le \|\mu\|$. Now we extend $\tilde{f}$ to $f \in (\mathcal{B}[H])^*$ with $\|f\| \le \|\mu\|$ by Theorem 4.5.3. Finally we have for any set $E \in \mathcal{P}(H)$

$$f(\chi_E) = \tilde{f}(\chi_E) = \mu(E), \text{ hence } Tf = \mu.$$

Thus we have shown that T is onto.

(v)   Finally, we show that $\|Tf\| = \|\mu_f\| = \|f\|$. In view of (4.60), it remains to show that

$$\|f\| \le \|\mu_f\|. \tag{4.62}$$

(β)   To prove (4.62), we first show that, for each $\varphi \in \mathcal{B}[H]$ and for each $\delta > 0$, there exists a partition $\Pi = \Pi(\varphi, \delta) = \{E(1), E(2), \ldots, E(n)\}$ of H and there exist points $\tilde{h}_k \in E(k) \neq \emptyset$ for $k = 1, 2, \ldots, n$ such that

$$\left\|\varphi - \sigma(\varphi; \Pi, (\tilde{h}_k))\right\| < \delta. \tag{4.63}$$

Let $\varphi \in \mathcal{B}[H]$ and $\delta > 0$ be given. We put

$$\lambda = -\|\varphi\| - 1 \text{ and } \nu = \|\varphi\| + 1,$$

choose $n_0 \in \mathbb{N}$ with $n_0 > 2(\nu - \lambda)/\delta$ and put

$$y_k = \lambda + k \frac{\nu - \lambda}{n_0}, \ E_k = \varphi^{-1}((y_{k-1}, y_k]) = \{h \in H : y_{k-1} < h \le y_k\}$$

for $k = 1, 2, \ldots, n_0$, and $\Pi = \{E(1), E(2), \ldots, E(n_0)\}$. Then Π is a partition of H. For arbitrary $h \in H$ there exists one and only one $j \in \{1, 2, \ldots, n_0\}$ such that $h \in E(j)$. Then we have for arbitrary points $\tilde{h}_k \in E(k) \neq \emptyset$

$$\left(\sigma\left(\varphi; \Pi, (\tilde{h}_k)\right)\right)(h) = \sum_{k=1}^{n_0} \varphi(\tilde{h}_k) \chi_{E(k)}(h) = \varphi(\tilde{h}_j).$$

It follows from $h, h_j \in E(j)$ that

$$|\varphi(h) - \varphi(\tilde{h})| \le y_j - y_{j-1} = \frac{\nu - \lambda}{n_0} < \frac{\delta}{2},$$

hence

$$\left\|\varphi - \sigma\left(\varphi; \Pi, (\tilde{h}_k)\right)\right\| = \sup_{h \in H} \left|\varphi(h) - \left(\sigma\left(\varphi; \Pi, (\tilde{h}_k)\right)\right)(h)\right| \le \frac{\delta}{2} < \delta.$$

This completes the proof of Part (β).

Now let $f \in (\mathcal{B}[H])^*$, $\varphi \in \mathcal{B}[H]$ and $\varepsilon > 0$ be given. Since f is continuous there exists $\delta > 0$ such that for all

$$\|f(\psi) - f(\varphi)\| < \varepsilon \text{ for all } \psi \in B[H] \text{ with } \|\psi - \varphi\| < \delta.$$

For $\delta > 0$, we choose a partition $\Pi = \{E(1), E(2), \ldots, E(n)\}$ of H and points $\tilde{h}_k \in E(k) \neq \emptyset$ for $k = 1, 2, \ldots, n$ by Part (β) such that (4.63) is satisfied. Then we have

$$|f(\varphi)| \leq \left| f(\varphi) - f\left(\sigma\left(\varphi; \Pi, (\tilde{h}_k)\right)\right)\right| + \left| f\left(\sigma\left(\varphi; \Pi, (\tilde{h}_k)\right)\right)\right|$$

$$< \varepsilon + \left|\sum_{k=1}^{n} \varphi(\tilde{h}_k) f\left(\chi_{E(k)}\right)\right| = \varepsilon + \left|\sum_{k=1}^{n} \varphi(\tilde{h}_k) \mu_f(E(k))\right|$$

$$\leq \varepsilon + \|\varphi\| \cdot \sum_{k=1}^{n} |\mu_f(E(k))| \leq \varepsilon + \|\varphi\| \cdot \|\mu_f\|.$$

Since $\varepsilon > 0$ was arbitrary, it follows that $|f(\varphi)| \leq \|\mu_f\| \cdot \|\varphi\|$, that is, (4.62). This completes the proof of Part (v). $\qquad\square$

Remark 4.7.6 *(a) We obtain as a special case of Theorem 4.7.5 that*

$$(\ell_{\infty,\mathbb{R}})^* \cong \mathcal{M}[\mathbb{N}].$$

(b) For each $f \in (\mathcal{B}[H])^*$ there exists $\mu \in \mathcal{M}[H]$ such that

$$f(a) = \int_H \varphi(h) \, d\mu \text{ for all } \varphi \in \mathcal{B}[H] \text{ and } \|f\| = \int_H |d\mu|.$$

Now we prove the famous *Riesz representation theorem*.

Theorem 4.7.7 (Riesz representation theorem) *For each $f \in (C[0,1])^*$ there exists a function $g \in BV[0,1]$ such that*

$$f(\varphi) = \int_0^1 \varphi(t) \, dg(t) \text{ for all } \varphi \in C[0,1];$$

moreover, we have

$$\|f\| = \bigvee_0^1 g = \int_0^1 |dg(t)|.$$

Proof. Let $f \in (C[0,1])^*$. We extend f by Theorem 4.5.3 to $F \in (\mathcal{B}[0,1])^*$ with $\|F\| = \|f\|$ and $F|_{C[0,1]} = f$. By Remark 4.7.6 (b), there exists $\mu \in \mathcal{M}[0,1]$ such that

$$F(\psi) = \int_0^1 \psi(t) \, d\mu \text{ for all } \psi \in \mathcal{B}[0,1] \text{ and } \|F\| = \int_0^1 |d\mu|.$$

Now let $\phi \in C[0,1]$. We put

$$g(t) = \begin{cases} F\left(\chi_{[0,t]}\right) = \mu([0,t]) & (t \in (0,1]) \\ 0 & (t = 0). \end{cases}$$

Then we have $g \in BV[0,1]$,

$$\sigma(\varphi; \Pi, (\tilde{t}_k)) \to \int_0^1 \varphi(t)\, dg(t) = f(\phi) \text{ for } \|\Pi\| \to 0,$$

and

$$\|f\| = \int_0^1 |dg(t)|. \qquad \square$$

4.8   Reflexivity

Using the Hahn–Banach theorem, Theorem 4.5.2, we shall prove that any normed space X can be identified with some subspace $\tilde{X}$ of its second dual space $X^{**} = (X^*)^*$, the so–called *bidual of X*.

This means that X and $\tilde{X}$ are isometrically isomorphic.

Theorem 4.8.1 *Let X be a normed space. Then X is isometrically isomorphic to a subspace $\tilde{X}$ of the bidual X^{**}.*

Proof. For $x \in X$, we define the function $\tilde{x} : X^* \to \mathbb{C}$ by

$$\tilde{x}(f) = f(x) \text{ for all } f \in X^*.$$

It is easy to see that $\tilde{x}$ is a linear functional. We obtain from this

$$|\tilde{x}(f)| = |f(x)| \leq \|f\|\,\|x\| \text{ for all } f \in X^*,$$

which implies $\tilde{x} \in X^{**}$ and $\|\tilde{x}\| \leq \|x\|$. It follows from Corollary 4.5.6 that $\|\tilde{x}\| = \|x\|$. The correspondence $x \to \tilde{x}$ for all $x \in X$ defines a map $J : X \to X^{**}$. We leave it to the reader to prove that J is linear. We remark that $\|Jx - Jy\| = \|J(x-y)\| = \|x-y\|$. Hence J is an isometric isomorphism from X onto $R(J)$, the range of J. We put $\tilde{X} = R(J)$. $\qquad \square$

Definition 4.8.2 A normed space X is said to be *reflexive* if the map J in the proof of Theorem 4.8.1 is surjective, that is, $R(J) = X^{**}$. The map J is called a *canonical map* of the space X into the space X^{**}. Sometimes the canonical map $J : X \to X^{**}$ is denoted by J_X, and when X is a subspace of its bidual X^{**}, then $X \subset X^{**}$ means that $x \in X$ is a functional in X^{**} defined by $x(f) = f(x)$ for each $f \in X^*$.

If a normed space X is reflexive then X is isomorphic (and also isometric) to the bidual X^{**}. It is interesting that R. C. James [100] proved that if X is isometric with X^{**}, then X need not be reflexive.

Corollary 4.8.3 *If a normed space X is reflexive then it is complete, and hence a Banach space.*

Proof. We know from Corollary 4.4.5 that X^{**} is a Banach space. Since J is an isomorphism and isometric and $R(J) = X^{**}$, it follows that X is a Banach space. $\qquad \square$

To prove that the spaces ℓ_p are reflexive for $1 < p < \infty$, but ℓ_1 is not reflexive, we need the following theorem.

Theorem 4.8.4 *If the dual space X^* of a normed space X is separable, then the space X is separable.*

Proof. Let $\{x_1^*, x_2^*, \ldots,\}$ be a countable dense set in X^*. Then there exists a sequence $(x_k)_{k=1}^\infty$ in X such that $\|x_k\| = 1$ and $|x_k^*(x_k)| \geq \|x_k^*\|/2$ for each k. Let Y be a closed subspace generated by the elements x_k for $k = 1, 2, \ldots$. Then Y is a separable space. If $Y = X$, then the theorem is proved. If $Y \neq X$, then, by Corollary 4.5.4, there exists a functional $x^* \in X^*$ such $\|x^*\| = 1$ and $x^*(y) = 0$ for all $y \in Y$. Since $\{x_1^*, x_2^*, \ldots\}$ is dense in X^*, there exists a subsequence $(x_{k(m)}^*)$ of the sequence (x_k^*) such that $x_{k(m)}^* \to x^*$ as $m \to \infty$. From

$$\|x_{k(m)}^* - x^*\| \geq |x_{k(m)}^*(x_{k(m)}) - x^*(x_{k(m)})| = |x_{k(m)}^*(x_{k(m)})| \geq \frac{1}{2}\|x_{k(m)}^*\|,$$

we first have $\|x_{k(m)}^*\| \to 0$ as $m \to \infty$, and then $x^* = 0$. This is a contradiction. $\qquad\square$

Example 4.8.5 *The spaces ℓ_p are reflexive for $1 < p < \infty$, but ℓ_1 is not reflexive.*

Proof. It follows from Theorem 4.7.1 that the spaces ℓ_p are reflexive for $1 < p < \infty$. We show that ℓ_1 is not reflexive. We know that ℓ_1 is separable (Remark 3.4.2 (b) and Example 3.4.3 (c)), but ℓ_∞ is not separable (Example 3.4.3 (c)). Theorem 4.7.1 implies that ℓ_1^* is not separable. If ℓ_1 were reflexive, then ℓ_1^{**} would also be separable and, by Theorem 4.8.4, ℓ_1^* would be separable. This is a contradiction. $\qquad\square$

Now we give a characterization of reflexive Banach spaces.

Theorem 4.8.6 *A normed space X is reflexive if and only if the dual space X^* is reflexive.*

Proof. We assume that X is reflexive and $x^{***} \in X^{***}$. Then $x^* = x^{***} \circ J_X \in X^*$, and we have for all $x \in X$

$$J_{X^*}(x^*)(J_X(x)) = J_X(x)(x^*) = x^*(x) = x^{***}(J_X(x)). \tag{4.64}$$

Since J_X is onto, (4.64) implies $J_{X^*}(x^*) = x^{***}$, hence X^* is reflexive. Conversely we assume that the space X^* is reflexive, and X is not reflexive. Then $J_X(X)$ is a closed and proper subspace in X^{**}. Let $x^{**} \in X^{**} \setminus J_X(X)$. By Corollary 4.5.4, there exists a functional $x^{***} \in X^{***}$ such that $x^{***}(x^{**}) \neq 0$ and $x^{***}(J_X(x)) = 0$ for each $x \in X$. Since X^* is reflexive, there exists $x^* \in X^*$ such that $J_{X^*}(x^*) = x^{***}$. Hence we have

$$0 = J_{X^*}(x^*)(J_X(z)) = J_X(z)(x^*) = x^*(z) \text{ for all } z \in X,$$

and so $x^* = 0$, hence $x^{***} = 0$. This is a contradiction. $\qquad\square$

Theorem 4.8.7 *Every closed subspace Y of a reflexive space X is reflexive.*

Proof. Let $J_Y : Y \to Y^{**}$ and $J_X : X \to X^{**}$ be canonical maps. If $y^{**} \in Y^{**}$, then there exists a functional $x^{**} \in X^{**}$ defined by

$$x^{**}(z^*) = y^{**}(z^*|_Y) \text{ for each } z^* \in X^*, \tag{4.65}$$

where $z^*|_Y$ is the restriction of the functional z^* on the subset Y. Since J_X is a map onto, there exists $x \in X$ such that $J_X(x) = z^*$. We are going to prove that $x \in Y$ and $J_Y(x) = y^{**}$. If $x \notin Y$, then by Corollary 4.5.4, there exists a functional $u^* \in X^*$ such that

$$u^*(x) \neq 0 \text{ and } u^*(y) = 0 \text{ for all } y \in Y. \tag{4.66}$$

It follows from (4.65) and (4.66) that $x^{**}(u^*) = y^{**}(u^*|_Y) = 0$, and so $J_X(x)(u^*) = u^*(x) = 0$, which is a contradiction. Thus we have shown that $x \in Y$. It is easy to prove $J_Y(x) = y^{**}$. $\qquad\square$

4.9 Adjoint operators

Let X, Y and Z be normed spaces, $A \in \mathcal{B}(X,Y)$ and $B \in \mathcal{B}(Y,Z)$. The composition of A and B is the operator $BA = B \circ A : X \to Z$ defined by $BAx = B(Ax)$ for all $x \in X$. It is clear that $BA \in \mathcal{B}(X,Z)$ and easy to show that $\|BA\| \le \|B\| \, \|A\|$. Hence if $Z = \mathbb{C}$, $\mathcal{B}(Y,\mathbb{C}) = Y^*$ and $g \in Y^*$, then we have

$$gA \in X^* \text{ and } \|gA\| \le \|g\| \, \|A\|. \tag{4.67}$$

Let $A' : Y^* \to X^*$ be the map defined by

$$A'(g) = gA \text{ for all } g \in Y^*, \tag{4.68}$$

that is,

$$(A'g)(x) = g(Ax) \text{ for all } x \in X \text{ and } g \in Y^*.$$

It is easy to prove that A' is a linear operator, and (4.67) implies

$$A' \in \mathcal{B}(Y^*, X^*) \text{ and } \|A'\| \le \|A\|. \tag{4.69}$$

Definition 4.9.1 Let X and Y be normed spaces and $A \in \mathcal{B}(X,Y)$. The operator $A' \in \mathcal{B}(Y^*, X^*)$ defined by (4.68) is referred to as an *adjoint (dual, conjugate) operator of the operator A*.

If $A \in \mathcal{B}(X,Y)$ and $y \in R(A)$, then there exists $x \in X$ such that

$$Ax = y. \tag{4.70}$$

Now (4.70) implies

$$g(Ax) = g(y) \text{ for all } g \in Y^*. \tag{4.71}$$

We have from (4.68) and (4.71)

$$A'g(x) = g(y) \text{ for all } g \in Y^*. \tag{4.72}$$

If $g \in N(A')$, then (4.72) implies $g(y) = 0$. Hence

$$y \in R(A) \text{ implies } g(y) = 0 \text{ for all } g \in N(A'). \tag{4.73}$$

The implication in (4.73) shows that when we solve the equation $Ax = y$ we meet the adjoint operator. This is one of the reasons for studying adjoint operators.

The first result states that the converse inequality of (4.69) also holds.

Theorem 4.9.2 *Let X and Y be normed spaces and $A \in \mathcal{B}(X,Y)$. Then we have*

$$A' \in \mathcal{B}(Y^*, X^*) \text{ and } \|A'\| = \|A\|.$$

Proof. By (4.69) it suffices to prove $\|A\| \le \|A'\|$. Let $x \in X$ and $Ax \ne 0$. By Corollary 4.5.5, there exists a functional $g_x \in Y^*$ such that $\|g_x\| = 1$ and $g_x(Ax) = \|Ax\|$. Now we have

$$\|Ax\| = |g_x(Ax)| = |(A'g_x)(x)| \le \|A'g_x\| \, \|x\| \le \|A'\| \, \|g_x\| \, \|x\|. \tag{4.74}$$

Since $\|g_x\| = 1$, (4.74) implies $\|A\| \le \|A'\|$. $\qquad\square$

Theorem 4.9.3 *Let X, Y and Z be normed spaces, $A, B \in \mathcal{B}(X,Y)$, $C \in \mathcal{B}(Y,Z)$ and $\lambda \in \mathbb{C}$. Then we have*

$$O' = O \text{ and } I' = I,$$
$$(A + B)' = A' + B',$$
$$(\lambda A)' = \lambda A',$$
$$(CB)' = B'C'.$$
$$\text{If } A^{-1} \in \mathcal{B}(Y, X) \text{ exists then } (A')^{-1} \text{ exists,}$$
$$(A')^{-1} \in \mathcal{B}(X^*, Y^*) \text{ and } (A')^{-1} = (A^{-1})'.$$

Proof. We leave the proof to the reader. □

Before we continue studying properties of adjoint operators we need the next definition and several auxiliary results.

Definition 4.9.4 *Let X be a normed space, $U \subset X$ and $W \subset X^*$. An element $x^* \in X^*$ is called an* annihilator of the set U *if $x^*(u) = 0$ for all $u \in U$. The set of all annihilators of U is denoted by U^0. An element $x \in X$ is called an* annihilator of the set W *if $w^*(x) = 0$ for all $w^* \in W$. The set of all annihilators of W is denoted by ^0W.*

Lemma 4.9.5 *Let X be a normed space, $U \subset X$ and $W \subset X^*$. Then U^0 is a closed subspace of X^*, and ^0W is a closed subspace of X.*

Proof. The proof is left to the reader. □

Lemma 4.9.6 *Let M be a subspace of a normed space X. Then we have*

$$\overline{M} = {}^0(M^0).$$

Proof. Obviously $M \subset {}^0(M^0)$, and so $\overline{M} \subset {}^0(M^0)$.
To prove the converse inclusion we assume that $x_o \in X \setminus \overline{M}$. Then $\inf\{\|x_0 - y\| : y \in \overline{M}\} = d > 0$, and by Corollary 4.5.4 there exists a functional $x^* \in X^*$ such that

$$\|x^*\| = 1, \ x^*(x_0) = d \text{ and } x^*(y) = 0 \text{ for all } y \in \overline{M}. \tag{4.75}$$

Now (4.75) implies $x^*(x_0) \neq 0$ and $x^* \in M^0$. Hence $x_0 \notin {}^0(M^0)$. Thus we have shown that $\overline{M} \supset {}^0(M^0)$. □

Lemma 4.9.7 *Let S be a subset of a normed space X, and M be the closed subspace of X generated by S, that is, $M = \overline{\text{span}(S)}$. Then we have*

$$M^0 = S^0 \text{ and } M = {}^0(S^0). \tag{4.76}$$

Proof. Obviously $S \subset M$, and so $M^0 \subset S^0$. To prove the converse inclusion, we assume $x^* \in S^0$ and $y \in M$. Then there exists a sequence (y_n) in $\text{span}(S)$ such that $\lim_n y_n = y$. Since $x^*(y_n) = 0$, it follows that $x^*(y) = 0$, and so $x^* \in M^0$. Hence $S^0 \subset M^0$, and we have proved $M^0 = S^0$. The second result in (4.76) follows from the first result, which we have just proved, and from Lemma 4.9.6. □

Lemma 4.9.8 *Let X and Y be normed spaces $A \in \mathcal{B}(X,Y)$, then we have*

$$R(A)^0 = N(A') \text{ and } \overline{R(A)} = {}^0[R(A)^0] = {}^0[N(A')]. \tag{4.77}$$

Proof. Since $y^* \in R(A)^0$ is equivalent to $y^*(Ax) = 0$ for each $x \in X$, which is equivalent to $A'y^*(x) = 0$ for each $x \in X$, which is equivalent to $A'y^* = 0$, the first result in (4.77) follows. Now the remainder of the proof follows from Lemma 4.9.7. □

Corollary 4.9.9 *Let X and Y be normed spaces and $A \in \mathcal{B}(X, Y)$, then we have*

$$R(A) = {}^0N(A') \text{ if and only if } R(A) = \overline{R(A)}.$$

Proof. The proof is by Lemma 4.9.8. □

If $A \in \mathcal{B}(X, Y)$, then $A' \in \mathcal{B}(Y^*, X^*)$ and $A'' \in \mathcal{B}(X^{**}, Y^{**})$. Since $X \subset X^{**}$ and $Y \subset Y^{**}$, we can consider the restriction of the operator A'' on X, that is, the operator $A''|_X$. The following lemma is related to this.

Lemma 4.9.10 *Let X and Y be normed spaces and $A \in \mathcal{B}(X, Y)$. Then we have*

$$A''|_X = A.$$

Proof. As we already mentioned in Definition 4.8.2, when $x \in X$ considered as a functional in X^{**}, then $x(f) = f(x)$ for all $f \in X^*$; when we consider Ax as a functional in Y^{**}, then $(Ax)(g) = g(Ax)$ for all $g \in Y^{**}$. With the comments we just made, we have for all $x \in X$ and all $g \in X^*$

$$(A''x)(g) = x(A'g) = (A'g)x = g(Ax) = (Ax)(g),$$

that is,

$$A''x = Ax \text{ for all } x \in X. \qquad \square$$

4.10   Quotient spaces

We recall the definition of a quotient space (Definition 2.5.1 (b)).

If Y is a subspace of a linear space X, then the *quotient or factor space*

$$X/Y = \{x + Y : x \in X\}$$

is a linear space with the *addition of vectors*

$$(x + Y) + (x' + Y) = (x + x') + Y \text{ for all } x, x' \in X,$$

and the *multiplication of vectors by scalars*

$$\lambda(x + Y) = \lambda x + Y \text{ for all } \lambda \in \mathbb{F} \text{ and all } x \in X.$$

We remark that Y is the *zero element in the vetor space X/Y*. When the spaces are normed then we have.

Theorem 4.10.1 *Let Y be a closed subspace of the normed space X and the function $\|\cdot\|_Y : X/Y \to \mathbb{R}$ be defined by*

$$\|x + Y\|_Y = \inf\{\|x + y\| : y \in Y\} \text{ for all } x \in X. \tag{4.78}$$

Then $\|\cdot\|_Y$ is a norm on the linear space X/Y.

Proof. Let $x \in X$.

(i) First we show that $x + Y$ is a closed subspace of X.
If (x_n) is a sequence $x + Y$ which converges to $z \in X$, then there exists a sequence (y_n) in Y such that $x_n = x + y_n$ for each n. Hence we have $y_n = x_n - x \to z - x \in Y$. So we have proved that $z \in x + Y$, that is, $x + Y$ is a closed subspace of X.

(ii) Now we show that $\| \cdot \|_Y$ is the norm on the quotient space X/Y.
If $\|x + Y\|_Y = 0$, then (4.78) implies that there exists a sequence (x_n) in $x + Y$ such that $\lim \|x_n\| = 0$. Since $x + Y$ is a closed subspace of X, it follows that $0 \in x + Y$. So we have shown that $x + Y$ is the zero element in the quotient space X/Y. Clearly we have $\|Y\|_Y = 0$.

(iii) We leave it to the reader to prove the homogeneity condition.

(iv) To prove the triangle inequality, we assume that $L, M \in X/Y$ and $\varepsilon > 0$. It follows from (4.78) that there exist $x \in L$ and $y \in M$ such $\|x\| < \|L\|_Y + \varepsilon$ and $\|y\| < \|M\|_Y + \varepsilon$. Now $x + y \in L + M$ implies

$$\|L + M\|_Y \le \|x + y\| \le \|x\| + \|y\| < \|L\|_Y + \|M\|_Y + 2\varepsilon. \qquad \square$$

Definition 4.10.2 Let Y be a closed subspace of the normed space X. The norm $\| \cdot \|_Y$ defined by (4.78) is called the *quotient norm on the linear space X/Y*.

Sometimes the quotient norm $\| \cdot \|_Y$ is simply denoted by $\| \cdot \|$, that is, the index Y is omitted.

The concepts of a series and its convergence in normed spaces are defined analogously as those in Definition 3.2.5 in linear metric spaces.

Remark 4.10.3 *It is clear from Theorem 3.2.7 that a normed space is a Banach space if and only if every absolutely convergent series converges.*

A closed subspace Y of a Banach space X is a Banach space. It follows from the next theorem that the quotient space X/Y is a Banach space.

Theorem 4.10.4 *Let Y be a closed subspace of a Banach space X and $\| \cdot \|_Y$ be the quotient norm of the quotient space X/Y defined in (4.78). Then $(X/Y, \| \cdot \|_Y)$ is a Banach space.*

Proof. Let $\varepsilon > 0$ and (L_n) be a sequence in X/Y such that the series $\sum_{n=1}^{\infty} \|L_n\|_Y$ converges. For each n, there exists $x_n \in L_n$, such that

$$\|x_n\| < \|L_n\|_Y + \varepsilon 2^{-n} \text{ for } n = 1, 2, \ldots. \tag{4.79}$$

It follows from (4.79) and Theorem 4.10.2 that the series $\sum_{n=1}^{\infty} x_n$ converges; let $x \in X$ denote its sum. Now we have for $L = x + Y$

$$\left\| \sum_{n=1}^{k} L_n - L \right\|_Y = \inf_{y \in Y} \left\| \sum_{n=1}^{k} x_n - x + y \right\|$$

$$\le \left\| \sum_{n=1}^{k} x_n - x \right\| \text{ for all } k. \tag{4.80}$$

It follows from (4.80) and Theorem 4.10.2 that X/Y is a Banach space. $\qquad \square$

In Example 4.3.4, we considered the normed space $(L_p, \|.\|_p)$. Now, using Theorem 4.10.4, we are going to show that L_p is a Banach space.

Theorem 4.10.5 (Riesz–Fischer) *Let $(X, \mathcal{R}, \mu)$ be a measure space with measure μ. The normed space $(L_p, \|\cdot\|_p)$ for $1 \leq p < \infty$ is a Banach space.*

Proof. Let (f_n) be a sequence in L_p such that $\sum_{n=1}^{\infty} \|f_n\| = \alpha < \infty$. By Theorem 4.10.2, it is enough to show that the series $\sum_{n=1}^{\infty} f_n$ converges. We consider the sequence (g_n) with $g_n(x) = \sum_{k=1}^{n} |f_k(x)|$ for $x \in X$ and $n = 1, 2, \ldots$. Minkowski's inequality for integrals (4.8) implies

$$\|g_n\|_p \leq \sum_{k=1}^{n} \|f_k\|_p \leq \alpha,$$

that is,

$$\int g_n^p \, d\mu \leq \alpha^p. \tag{4.81}$$

Let $g(x) = \lim_{n \to \infty} g_n(x)$. It follows from (4.81) and the monotone convergence theorem that

$$\int g^p \, d\mu \leq \alpha^p. \tag{4.82}$$

Consequently, (4.82) implies $g \in L_p$, and the function g is finite almost everywhere, g is finite a.e., for short. So $\sum_{k=1}^{\infty} |f_k(x)| < \infty$ a.e. and there exists a function f, which is finite a.e., such that $\sum_{k=1}^{\infty} f_k(x) = f(x)$ a.e. (clearly we may choose f to be a measurable function). Hence we have $f \in L_p$ and

$$\left| \sum_{k=1}^{n} f_k(x) - f(x) \right|^p \leq 2^p [g(x)]^p, \ a.e. \text{ for } n = 1, 2, \ldots. \tag{4.83}$$

It follows from (4.82), (4.83) and Lebesgue's dominated convergence theorem that

$$\lim_{n \to \infty} \int \left| \sum_{k=1}^{n} f_k(x) - f(x) \right|^p d\mu = \int \lim_{n \to \infty} \left| \sum_{k=1}^{n} f_k(x) - f(x) \right|^p d\mu = 0. \qquad \square$$

4.11 Cauchy nets, summable families

We start with the definition of the concept of a *Cauchy net in a topological space*.

Definition 4.11.1 *Let $(D, \geq)$ be a directed set. A net $(x_\delta)_{d \in D}$ in a normed space X is a Cauchy net if, for every $\varepsilon > 0$, there exists $\delta_0 = \delta_0(\varepsilon) \in D$ such that $\|x_\delta - x_{\delta'}\| < \varepsilon$ for all $\delta, \delta' \geq \delta_0$.*

The following theorem shows that Banach spaces can be defined by the use of Cauchy nets.

Theorem 4.11.2 *In a Banach space X, every Cauchy net converges.*

Proof. Let $(x_\delta)_{\delta \in D}$ be a Cauchy net in X. Analogously, as in the proof of Theorem 3.2.7, we choose $\delta(1) \in D$ such that $\|x_\delta - x_{\delta(1)}\| < 1/2$ for all $\delta \geq \delta(1)$. Using the method of mathematical induction, we define a sequence $(\delta(k))_{k=1}^{\infty}$ in D as follows: for $k > 1$ and already chosen $\delta(1), \ldots, \delta(k) \in D$, we choose $\delta(k+1) \in D$ such that $\delta(k+1) \geq \delta(k)$ and

$$\|x_\delta - x_{\delta(k+1)}\| < \frac{1}{2^{k+1}} \text{ for all } \delta \geq \delta(k+1). \tag{4.84}$$

It is easy to show that the sequence $(x_{\delta(n)})_{n=1}^{\infty}$ is a Cauchy sequence. Hence it is convergent, $x \in X$ say. We are going to show that the net $(x_\delta)_{\delta \in D}$ converges to x. Let $\varepsilon > 0$ be given and n be a natural number such that

$$\frac{1}{2^n} < \frac{\varepsilon}{2} \text{ and } \|x_{\delta(n)} - x\| < \frac{\epsilon}{2}. \tag{4.85}$$

Now (4.84) and (4.85) imply

$$\|x_\delta - x\| \le \|x_\delta - x_{\delta(n)}\| + \|x_{\delta(n)} - x\| < \frac{1}{2^n} + \frac{\varepsilon}{2} < \varepsilon \text{ for all } \delta \ge \delta(n). \qquad \square$$

The notations we are going to introduce in the next definition we will use with the same meaning until the end of this section.

Definition 4.11.3 Let $D \neq \emptyset$ a directed set, X be a normed space, $(x_\delta)_{\delta \in D}$ be a family of vectors in X and $\mathcal{F} = \{F \subset D : F \text{ has finitely many elements}\}$. The set $\mathcal{F}$ is a directed set with respect to the ordering by inclusion (Example 1.2.2). For every $F \in \mathcal{F}$, we put $z_F = \sum_{\delta \in F} x_\delta$. If the $(z_F)_{F \in \mathcal{F}}$ converges to $x \in X$ say, then we say that the family $\{x_\delta : \delta \in D\}$ of vectors is *summable* and has the *sum* x. Symbolically we denote this by

$$\sum_{\delta \in D} x_\delta = x.$$

Hence the family $\{x_\delta : \delta \in D\}$ of vectors is summable and has the sum $x \in X$, if, for every $\varepsilon > 0$, there exists $F(\varepsilon) \in \mathcal{F}$ such that

$$\left\| \sum_{\delta \in F} x_\delta - x \right\| < \varepsilon \text{ for all } F \in \mathcal{F} \text{ with } F \supset F(\varepsilon).$$

Theorem 4.11.4 *If the family $\{x_\delta : \delta \in D\}$ of vectors in a normed space X is summable and has the sum $x \in X$, then the set $D_0 = \{\delta \in D : x_\delta \neq 0\}$ is at most countable, and the family $\{x_\delta : \delta \in D_0\}$ is summable and has the sum x.*

Proof. Let $\varepsilon > 0$. If the family $\{x_\delta : \delta \in D\}$ is summable with the sum x, then there exists $F(\varepsilon/2) \in \mathcal{F}$ such that

$$\left\| \sum_{\delta \in F} x_\delta - x \right\| < \frac{\varepsilon}{2} \text{ for all } F \in \mathcal{F} \text{ with } F \supset F(\varepsilon/2). \tag{4.86}$$

Let $E \in \mathcal{F}$ and $E \cap F(\varepsilon/2) = \emptyset$. Then (4.86) implies

$$\left\| \sum_{\delta \in E} x_\delta \right\| = \left\| \sum_{\delta \in E \cup F(\varepsilon/2)} x_\delta - \sum_{\delta \in F(\varepsilon/2)} \delta \right\|$$

$$\le \left\| x - \sum_{\delta \in E \cup F(\varepsilon/2)} x_\delta \right\| + \left\| x - \sum_{\delta \in F(\varepsilon/2)} x_\delta \right\| < \varepsilon. \tag{4.87}$$

It follows from (4.87) that, for each $\delta \in D \setminus F(\varepsilon/2)$, we have $\|x_\delta\| < \varepsilon$, and $\{\delta \in D : \|x_\delta\| \ge \varepsilon\} \subset F(\varepsilon/2)$. Hence, for each natural number n, the set $D_n = \{\delta \in D : \|x_\delta\| \ge 1/n\} \in \mathcal{F}$. Since $D_0 = \bigcup_{n=1}^{\infty} D_n$, it follows that D_0 is a countable set.

For every $\varepsilon > 0$, we put $F_0(\varepsilon) = F(\varepsilon) \cap D_0$. If $F \in \mathcal{F}$, $F \subset D_0$ and $F \supset F_0(\varepsilon)$, then we have

$$F \cup F(\varepsilon) \in \mathcal{F} \text{ and } \left\| x - \sum_{\delta \in F \cup F(\varepsilon)} x_\delta \right\| < \varepsilon. \tag{4.88}$$

Since $\sum_{\delta \in F} x_\delta = \sum_{\delta \in F \cup F(\varepsilon)} x_\delta$, it follows from (4.88) that the family $\{x_\delta : \delta \in D\}$ of vectors is summable and has the sum x. $\square$

Lemma 4.11.5 *Let $\{x_\delta : \delta \in D\}$ and $\{y_\delta : \delta \in D\}$ be summable families in the normed space X. Then the family $\{x_\delta + y_\delta : \delta \in D\}$ is summable and*

$$\sum_{\delta \in D} (x_\delta + y_\delta) = \sum_{\delta \in D} x_\delta + \sum_{\delta \in D} y_\delta.$$

For every $\lambda \in \mathbb{F}$ the family $\{\lambda x_\delta : \delta \in D\}$ is summable and

$$\sum_{\delta \in D} \lambda x_\delta = \lambda \sum_{\delta \in D} x_\delta.$$

Proof. The proof is left to the reader. $\square$

Theorem 4.11.6 *Let $\{x_\delta : \delta \in D\}$ be a family of vectors in the Banach space X. If the family $\{\|x_\delta\| : \delta \in D\}$ of scalars is summable, then the family $\{x_\delta : \delta \in D\}$ of vectors is summable.*

Proof. We use the notations of Definition 4.11.3. It follows from Theorem 4.11.2 that it suffices to show that $\{z_F\}_{F \in \mathcal{F}}$ is a Cauchy net. Let $\varepsilon > 0$ be given. Since the family of scalars is summable, there exists $F_0 \in \mathcal{F}$ such that

$$\sum_{\delta \in F} \|x_\delta\| - \sum_{\delta \in F_0} \|x_\delta\| < \varepsilon \text{ for all } F \in \mathcal{F} \text{ with } F \geq F_0. \tag{4.89}$$

From (4.89), for each $F_1, F_2 \geq F_0$ with $F_1, F_2 \in \mathcal{F}$ it follows that

$$\|z_{F_1} - z_{F_2}\| = \left\| \sum_{\delta \in F_1} x_\delta - \sum_{\delta \in F_2} x_\delta \right\| = \left\| \sum_{\delta \in F_1 \setminus F_2} x_\delta - \sum_{\delta \in F_2 \setminus F_1} x_\delta \right\|$$

$$\leq \sum_{\delta \in F_1 \setminus F_2} \|x_\delta\| + \sum_{\delta \in F_2 \setminus F_1} \|x_\delta\| \leq \sum_{\delta \in F_1 \cup F_2} \|x_\delta\| - \sum_{\delta \in F_0} \|x_\delta\| < \varepsilon. \quad \square$$

Theorem 4.11.7 *A family $\{x_\delta : \delta \in D\}$ of vectors in a Banach space X is summable if and only if, for every $\varepsilon > 0$, there exists $F(\varepsilon) \in \mathcal{F}$ such that*

$$E \in \mathcal{F} \text{ and } E \cap F(\varepsilon) = \emptyset \text{ implies } \left\| \sum_{\delta \in E} x_\delta \right\| < \varepsilon. \tag{4.90}$$

Proof. If the family $\{x_\delta : \delta \in D\}$ of vectors is summable, then (4.90) follows from the proof of Theorem 4.11.4 (see (4.87)). On the other hand, from (4.90), we have, for each $F_1, F_2 \geq F(\epsilon)$ with $F_1, F_2 \in \mathcal{F}$,

$$\|z_{F_1} - z_{F_2}\| = \left\| \sum_{\delta \in F_1} x_\delta - \sum_{\delta \in F_2} x_\delta \right\| = \left\| \sum_{\delta \in F_1 \setminus F_2} x_\delta - \sum_{\delta \in F_2 \setminus F_1} x_\delta \right\|$$

$$\leq \left\| \sum_{\delta \in F_1 \setminus F_2} x_\delta \right\| + \left\| \sum_{\delta \in F_2 \setminus F_1} x_\delta \right\| < 2\varepsilon. \tag{4.91}$$

Clearly (4.91) implies $(z_F)_{F \in \mathcal{F}}$ is a Cauchy net; and so, by Theorem 4.11.2, the net $(z_F)_{F \in \mathcal{F}}$ converges. $\square$

4.12 Equivalent norms

We recall that if (X, d_1) and (X, d_2) are metric spaces and τ_1 and τ_2 are the topologies defined by the metrics d_1 and d_2, respectively, then we say that the metrics d_1 and d_2 are *equivalent* if the topologies τ_1 and τ_2 are equivalent, that is, if $\tau_1 = \tau_2$.

Definition 4.12.1 Let $(X, \|\cdot\|_1)$ and $(X, \|\cdot\|_2)$ be normed spaces and d_1 and d_2 be the metrics defined by the norms $\|\cdot\|_1$ and $\|\cdot\|_2$, respectively. Then the norms $\|\cdot\|_1$ and $\|\cdot\|_2$ are said to be *equivalent* if the metrics d_1 and d_2 are equivalent.

Equivalent norms, with corresponding equivalent metrics, determine the same family of open sets, or, equivalently, the same family of closed sets. In metric spaces with the same topology, not only the families of open and closed sets are the same, but also the closures, interiors, exteriors and boundaries of sets are the same. The next theorem gives conditions for the equivalence of norms.

Theorem 4.12.2 *The norms $\|\cdot\|_1$ and $\|\cdot\|_2$ on a linear space X are equivalent if and only if there exist positive real numbers a and b such that*

$$a\,\|x\|_1 \le \|x\|_2 \le b\,\|x\|_1 \text{ for all } x \in X. \tag{4.92}$$

Proof. First we assume that the norms are equivalent. We have to prove that (4.92) is satisfied.
It is enough to prove the second inequality in (4.92), that is, there exists $b > 0$ such that

$$\|x\|_2 \le b\,\|x\|_1 \text{ for all } x \in X. \tag{4.93}$$

If there does not exist such a number b for which (4.93) is satisfied, then there exists a sequence (x_n) in X such that

$$\|x_n\|_2 > n\,\|x_n\|_1 \text{ for } n = 1, 2, \dots. \tag{4.94}$$

Let

$$y_n = \frac{x_n}{\|x_n\|_2} \text{ for } n = 1, 2, \dots, \tag{4.95}$$

and $E = \{y_n : n = 1, 2, \dots\}$. Let $\overline{E}_{(1)}$ and $\overline{E}_{(2)}$ denote the closures of the set E with respect to the topologies defined by the norms $\|\cdot\|_1$ and $\|\cdot\|_2$. It follows from (4.94) that $\|y_n\|_1 < 1/n$, for all $n = 1, 2, \dots$. So we have $\|y_n\|_1 \to 0$ as $n \to \infty$, that is, $0 \in \overline{E}_{(1)}$. Clearly, (4.95) implies $\|y_n\|_2 = 1$ for all $n = 1, 2, \dots$ and so $0 \notin \overline{E}_{(2)}$. Hence the set $\overline{E}_{(2)}$ is closed in the topology defined by the norm $\|\cdot\|_2$, and not closed in the topology defined by the norm $\|\cdot\|_1$. This contradicts the assumption of the equivalence of the two norms. Consequently we have shown the inequality in (4.93).
Now we assume that the inequality (4.92) is satisfied, and show that the norms are equivalent. It is enough to prove that the norms define the same families of closed sets. Let $F \subset X$ and F be closed with respect to the topology defined by the norm $\|\cdot\|_1$. If $\overline{F}_{(2)}$ is the closure of the set F in the topology defined by the norm $\|\cdot\|_2$ and $x \in \overline{F}_{(2)}$, then there exists a sequence (x_n) in F such that $\|x_n - x\|_2 \to 0$ as $n \to \infty$. The inequality (4.92) implies $\|x_n - x\|_1 \to 0$ as $n \to \infty$. Since F is a closed set in the topology defined by the norm $\|\cdot\|_1$ it follows that $x \in F$, hence $\overline{F}_{(2)} \subset F$. Therefore $\overline{F}_{(2)} = F$ is a closed set in the topology defined by the norm $\|\cdot\|_2$. $\square$

The following result is an immediate consequence of Theorem 4.12.2 and Corollary 3.5.7

Corollary 4.12.3 *Let X be a Banach space with respect to the norms $\|\cdot\|_1$ and $\|\cdot\|_2$. If there exists a constant $a > 0$ such that*

$$\|x\|_2 \leq a\|x\|_1 \text{ for all } x \in X,$$

then the norms are equivalent.

All norms on finite dimensional spaces are equivalent.

Theorem 4.12.4 *Any two norms $\|\cdot\|_1$ and $\|\cdot\|_2$ on a finite dimensional space are equivalent.*

Proof. Let $\dim X = n$ and $\{e_1, \ldots, e_n\}$ be a basis of X. Every $x \in X$ can be represented as a linear combination $x = \lambda_1 e_1 + \cdots + \lambda_n e_n$. By Lemma 4.2.2, there exists $c > 0$ such that

$$\|x\|_2 \geq c(|\lambda_1| + \cdots + |\lambda_n|). \tag{4.96}$$

We put $M = \max\{\|e_k\|_1 : k = 1, 2, \ldots, n\}$. Then we have by the triangle inequality

$$\|x\|_1 \leq \sum_{k=1}^{n} |\lambda_k| \, \|e_k\|_k \leq M \sum_{k=1}^{n} |\lambda_k|. \tag{4.97}$$

It follows from (4.96) and (4.97) that, for $a = c/M$, we have $a\|x\|_1 \leq \|x\|_2$. So we have proved the first inequality in (4.92), and the second inequality can be established similarly. $\square$

4.13 Compactness and the Riesz lemma

In this section, we study certain properties of sets in normed spaces.

We recall the following well–known characterization of compact sets in finite dimensional normed spaces.

Theorem 4.13.1 *A subset of a finite dimensional normed space is compact if and only if it is closed and bounded.*

Here we are interested in the compactness in infinite dimensional normed spaces. The following *Riesz lemma* of 1918 [174] is important for these studies.

Lemma 4.13.2 (Riesz lemma) *Let Y be a closed proper subspace of a normed space X. Then, for each $\varepsilon \in (0, 1)$, there exists $x_\varepsilon \in X$ such that*

$$\|x_\varepsilon\| = 1 \text{ and } d(x_\varepsilon, Y) = \inf_{y \in Y} \|x_\varepsilon - y\| > \varepsilon.$$

Proof. It follows from the assumption of the lemma that there exist $x_0 \in X \setminus Y$ such that

$$d(x_0, Y) = \inf_{y \in Y} \|x_0 - y\| = d > 0.$$

Since $d < d/\varepsilon$, there exists $y_1 \in Y$ such that $d \leq \|x_0 - y_1\| < d/\varepsilon$. We put $x_\varepsilon = (x_0 - y_1)/\|x_0 - y_1\|$ and choose $y \in Y$. Obviously we have $\|x_\varepsilon\| = 1$ and

$$\|x_\varepsilon - y\| = \frac{\|x_0 - (y_1 + y\|x_0 - y_1\|)\|}{\|x_0 - y_1\|} \geq \frac{d}{\|x_0 - y_1\|} > \varepsilon. \qquad \square$$

The following theorem is proved by applying the Riesz lemma. It shows that the compactness of the closed unit ball (or the unit sphere) in a normed space is rather a strong condition, and implies that the normed space is finite dimensional.

Theorem 4.13.3 *If the closed unit ball $\overline{B}_X$ in a normed space X is compact, then X is finite dimensional.*

Proof. We assume that $\overline{B}_X$ is a compact subset of X and that X is an infinite dimensional space. Let $x_1 \in X$, $\|x_1\| = 1$ and $X_1 = \mathrm{span}(\{x_1\})$. Then X_1 is a proper closed subspace of X. By the Riesz lemma, there exists $x_2 \in X$ such that $\|x_2\| = 1$ and $d(x_2, X_1) > 1/2$. Then we have $\|x_2 - x_1\| > 1/2$. We write $X_2 = \mathrm{span}\{x_1, x_2\}$ for the span of the set $\{x_1, x_2\}$. Then X_2 is a proper closed subspace X. Again, by the Riesz lemma, there exists $x_3 \in X$ such that $\|x_3\| = 1$ and $d(x_3, X_2) > 1/2$. We note that $\|x_3 - x_1\| > 1/2$ and $\|x_3 - x_2\| > 1/2$. Applying the method of mathematical induction and the Riesz lemma, we conclude that there exists a sequence $(X_n)_{n=1}^{\infty}$ of finite dimensional subspaces X_n of X and a sequence $(x_n)_{n=1}^{\infty}$ in X such that

$$X_1 \subsetneq X_2 \subsetneq \cdots, \ x_n \in X_n, \ \|x_n\| = 1 \text{ and } d(x_{n+1}, X_n) > \frac{1}{2} \text{ for } n = 1, 2, \ldots . \quad (4.98)$$

Now (4.98) implies

$$\|x_n - x_m\| > \frac{1}{2} \text{ for } m \neq n,$$

hence, the sequence $(x_n)_{n=1}^{\infty}$ has no convergent subsequence. This contradicts the assumption that $\overline{B}_X$ is a compact subset of X (Theorem 1.12.4). $\qquad\square$

Corollary 4.13.4 *If the unit sphere $S_0(1)$ in a normed space X is compact, then X is finite dimensional.*

Proof. This follows by the proof of Theorem 4.13.3. $\qquad\square$

4.14 Compact operators

In Section 4.4, we studied the space $\mathcal{B}(X,Y)$ of bounded linear operators from the normed space X into the normed space Y. Now we are interested in operators in $\mathcal{B}(X,Y)$ with some special properties. We start with the study of two subsets of $\mathcal{B}(X,Y)$, namely the *set of compact operators* and the *set of operators with finite dimensional range*.

Definition 4.14.1 Let X and Y be normed spaces and $A \in \mathcal{L}(X,Y)$. Then the operator A is said to be *compact* if $A(Q)$ is a relatively compact subset (Definition 1.12.1 (c)) of Y for every bounded subset Q of X.

Let $\overline{B}_X$ denote the closed unit ball in X. Then clearly

the operator A is compact if and only if $\overline{A(\overline{B}_X)}$ is compact in Y.

The *set of all compact operators from X into Y* is denoted by $\mathcal{K}(X,Y)$. If $X = Y$, we write $\mathcal{K}(X) = \mathcal{K}(X,X)$, for short. It follows from Theorem 4.4.8 that each compact operator is bounded. The next example shows that the converse implication is not true, in general.

Example **4.14.2** *Let X be an infinite dimensional normed space and $I \in \mathcal{L}(X)$ be the identity operator. Then I is bounded, but not compact.*

Proof. Obviously we have $I \in \mathcal{B}(X)$. The closed unit ball $\overline{B}_X$ is a bounded subset of X, but $I(\overline{B}_X) = \overline{B}_X$ is not a relatively compact subset of X by Theorem 4.13.3. □

Theorem 4.14.3 *Let X and Y be Banach spaces and $A \in \mathcal{K}(X,Y)$. Then the range of the operator A does not contain any infinite dimensional closed subspace of Y.*

Proof. We assume that $Z \subset R(A)$ is a closed infinite dimensional subspace of Y. Let $U = A^{-1}(Z)$ denote the preimage of Z and $A_0 = A|_U$ denote the restriction of A on U. Obviously U and Z are Banach spaces, $A_0 \in \mathcal{K}(U,Z)$ and the map A_0 is onto. It follows from Remark 4.6.16 that the closed unit ball $\overline{B}_Z$ in Z is compact, hence Z is finite dimensional by Theorem 4.13.3. This is a contradiction. □

The next theorem states that every bounded linear operator with finite dimensional range is compact, and that every linear operator on a finite dimensional normed space is compact.

Theorem 4.14.4 *Let X and Y be normed spaces and $A \in \mathcal{L}(X,Y)$. Then we have:*

$$A \in \mathcal{B}(X,Y) \text{ and } \dim A(X) < \infty \text{ implies } A \in \mathcal{K}(X,Y); \qquad (4.99)$$

$$\dim X < \infty \text{ implies } A \in \mathcal{K}(X,Y). \qquad (4.100)$$

Proof.

(i) First wc show the inclusion in (4.99).
    Let $A \in \mathcal{B}(X,Y)$, $\dim A(X) < \infty$ and Q be a bounded subset of X. It follows from Theorem 4.4.8 that $A(Q)$ is a bounded subset of Y. So $\overline{A(Q)}$ is a bounded and closed subset of Y. Since $\dim A(X) < \infty$ it follows that $A(X)$ is a closed subset of Y by Theorem 4.2.4, hence $\overline{A(Q)}$ is a compact subset of $A(X)$ by Theorem 4.13.1. Thus we have proved the implication in (4.99).

(ii) The implication in (4.100) follows from Theorem 4.4.9 and (4.99). □

Definition 4.14.5 *An operator $A \in\in \mathcal{B}(X,Y)$ is said to be a finite rank operator, or a finite dimensional operator if $\dim A(X) < \infty$.*
The set of all finite rank operators in $\mathcal{B}(X,Y)$ is denoted by $\mathcal{F}(X,Y)$. If $X = Y$, we write $\mathcal{F}(X) = \mathcal{F}(X,X)$, for short.

We note that follows from (4.99) that

$$\mathcal{F}(X,Y) \subset \mathcal{K}(X,Y). \qquad (4.101)$$

Corollary 4.14.6 *Let X and Y be Banach spaces and $A \in \mathcal{K}(X,Y)$. Then $R(A) = \overline{R(A)}$ implies $A \in \mathcal{F}(X,Y)$.*

Proof. This follows from Theorem 4.14.3. □

Theorem 4.14.7 *Let X and Y be normed spaces and $A \in \mathcal{L}(X,Y)$. Then A is compact if and only if the sequence (Ax_n) has a convergent subsequence for each bounded sequence (x_n) in X.*

Proof. Let A be a compact operator and (x_n) be a bounded sequence in X. Since $Q = \{x_n : n = 1, 2, \ldots\}$ is a bounded subset of X, the set $A(Q)$ is a relatively compact subset of Y; so the sequence (Ax_n) has a convergent subsequence.

Conversely, we assume that the sequence (Ax_n) has a convergent subsequence for each bounded sequence (x_n) in X. If Q is a bounded subset of X and (y_n) is a sequence in $A(Q)$, then there exists a sequence (x_n) in Q such that $Ax_n = y_n$ for all $n = 1, 2, \ldots$. Hence the sequence (y_n) has a convergent subsequence. $\square$

Corollary 4.14.8 *If* $A, B \in \mathcal{K}(X, Y)$ *and* $\lambda \in \mathbb{F}$, *then*

$$A + B \in \mathcal{K}(X, Y) \text{ and } \lambda A \in \mathcal{K}(X, Y). \tag{4.102}$$

Proof. If (x_n) is a bounded sequence in X, then Theorem 4.14.7 implies that the sequence (x_n) contains a subsequence $(x_{1,k})$ such that the sequence $(Ax_{1,k})$ converges. Analogously, the sequence $(x_{1,k})$ contains a subsequence $(x_{2,k})$ such that the sequence $(Bx_{2,k})$ converges. So the sequence $((A+B)x_{2,k})$ converges. By Theorem 4.14.7, the operator $A+B$ is compact. The second part in (4.102) is obvious. $\square$

We know that the composition of bounded linear operators again is a bounded linear operator. If one of the operators is compact, then the next result shows that their composition is a compact operator. We use the following notation. If $\Theta(X, Y)$ and $\Psi(Y, Z)$ are collections of functions $\phi : X \to Y$ and $\psi : Y \to Z$, then we write

$$\Psi(Y, Z) \circ \Theta(X, Y) = \{f = \psi \circ \theta : X \to Z : \theta \in \Theta(X, Y) \text{ and } \psi \in \Psi(Y, Z)\}.$$

Corollary 4.14.9 *If* X, Y *and* Z *are normed spaces over the same field* $\mathbb{F}$, *then we have*

$$\mathcal{K}(Y, Z) \circ \mathcal{B}(X, Y) \subset \mathcal{K}(X, Z) \text{ and } \mathcal{B}(Y, Z) \circ \mathcal{K}(X, Y) \subset \mathcal{K}(X, Z). \tag{4.103}$$

Proof. Let $A \in \mathcal{K}(Y, Z)$ and $B \in \mathcal{B}(X, Y)$. If (x_n) is a bounded sequence in X, then the sequence (Bx_n) is bounded and, by Theorem 4.14.7, the sequence (x_n) contains a subsequence $(x_{n(k)})$ such that the sequence $(A(Bx_{1,k}))$ converges. Hence we have $A \circ B \in \mathcal{K}(X, Z)$.

Analogously, if $A \in \mathcal{B}(Y, Z)$, $B \in \mathcal{K}(X, Y)$ and (x_n) is a bounded sequence in X, then the sequence (x_n) has a subsequence $(x_{n(k)})$ such that the sequence $(Bx_{n(k)})$ converges by Theorem 4.14.7. Hence the sequence $(A(Bx_{n(k)}))$ converges, and again Theorem 4.14.7 implies $A \circ B \in \mathcal{K}(X, Z)$. $\square$

The next theorem is of practical importance. It shows how it can be shown in many concrete situations that an operator is compact.

Theorem 4.14.10 *Let* X *be a normed space and* Y *be a Banach space. If the sequence* (A_n) *in* $\mathcal{K}(X, Y)$ *converges to the limit* A, *then* $A \in \mathcal{K}(X, Y)$. *Hence the set* $\mathcal{K}(X, Y)$ *is a closed subset of* $\mathcal{B}(X, Y)$.

Proof. Let (x_m) be a bounded sequence in X. By Theorem 4.14.7 the sequence (x_m) contains a subsequence $(x_{1,m})$ such that the sequence $(A_1 x_{1,m})$ converges. Analogously, the sequence $(x_{1,m})$ contains a subsequence $(x_{2,m})$ such that the sequence $(A_2 x_{2,m})$ converges. Using the method of mathematical induction and Theorem 4.14.7, we see that, for every $n \geq 2$ there exists a subsequence $(x_{n,m})_{m=1}^{\infty}$ of the sequence $(x_{n-1,m})_{m=1}^{\infty}$ such that the sequence $(A_n x_{n,m})_{m=1}^{\infty}$ converges. So the *diagonal sequence* $(y_m) = (x_{m,m})$ clearly is a subsequence of the sequence (x_m) and has the property that the sequence $(A_n y_m)_{m=1}^{\infty}$ converges for each $n \in \mathbb{N}$. We are going to show that the sequence (Ay_m) converges. Then $A \in \mathcal{K}(X, Y))$ by Theorem 4.14.7. For this it suffices to show that (Ay_m) is a Cauchy sequence. (Here we use that Y is a Banach space for the first time.) Let $\varepsilon > 0$ be given. Since the sequence (y_m) is bounded, there exists a positive real number c such that

$$\|y_m\| \le c \text{ for each } m = 1, 2, \ldots. \tag{4.104}$$

It follows from $A_n \to A$ that there exists a natural number n_0 such that

$$\|A_p - A\| < \frac{\varepsilon}{3c} \text{ and all } p \ge n_0. \tag{4.105}$$

Let $p \ge n_0$ and m_o be a natural number such that

$$\|A_p y_j - A_p y_k\| < \frac{\varepsilon}{3} \text{ for all } j, k \ge m_o. \tag{4.106}$$

It follows from (4.104), (4.105) and (4.106) that we have for all $j, k \ge m_o$

$$\|Ay_j - Ay_k\| \le \|Ay_j - A_p y_j\| + \|A_p y_j - A_p y_k\| + \|A_p y_k - Ay_k\|$$
$$< \|A - A_p\| \|y_j\| + \frac{\epsilon}{3} + \|A_p - A_k\| \|y_k\| < \varepsilon. \qquad \square$$

Remark 4.14.11 *It follows from Corollary 4.14.8 that $\mathcal{K}(X,Y)$ is a subspace of the space $\mathcal{B}(X,Y)$. Furthermore the set $\mathcal{F}(X,Y)$ of all finite dimensional operators clearly is a subspace of $\mathcal{B}(X,Y)$. It follows from (4.101) that $\mathcal{F}(X,Y) \subset \mathcal{K}(X,Y)$. If Y is a Banach space, then we know from Theorem 4.14.10 that $\mathcal{K}(X,Y)$ is a closed subset of $\mathcal{B}(X,Y)$ and it follows that*

$$\overline{\mathcal{F}(X,Y)} \subset \mathcal{K}(X,Y). \tag{4.107}$$

One of the most interesting problems in the theory of Banach spaces was, for a very long time, the following approximation problem: Does equality hold in (4.107) when Y is a Banach space? P. Enflo [65] showed that the answer is negative.

4.15 Operators with closed range

If X and Y are Banach spaces and $A \in \mathcal{K}(X,Y) \setminus \mathcal{F}(X,Y)$, then the range $R(A)$ of the operator A is not a closed subspace of Y (Theorem 4.14.3). It is important to know the operators with closed range. As usual we denote by $\overline{B}_Z$ the close unit ball in the normed space Z.

Definition 4.15.1 Let X and Y be normed spaces and $A \in \mathcal{B}(X,Y)$. The *minimum modulus (modulus of injectivity)* of the operator A is defined as

$$j(A) = \inf_{\|x\|=1} \|Ax\|.$$

An operator $A \in \mathcal{B}(X,Y)$ is said to be bounded-below, if $j(A) > 0$. The set of all bounded-below operators $A \in \mathcal{B}(X,Y)$ is denoted by $\mathcal{J}(B,X)$.

It follows from the definition that

$$\|Ax\| \ge j(A)\|x\| \text{ for all } x \in X, \tag{4.108}$$

$$j(A) = \sup\{c \ge 0 : \|Ax\| \ge c\|x\| \text{ for all } x \in X\},$$

$$0 \le j(A) \le \|A\|.$$

Theorem 4.15.2 *Let X and Y be Banach spaces and $A \in \mathcal{B}(X, Y)$. Then*

$$j(A) > 0 \text{ if and only if } R(A) = \overline{R(A)} \text{ and } N(A) = \{0\}.$$

Proof. We assume $j(A) > 0$. Obviously $N(A) = \{0\}$. It follows from $y \in \overline{R(A)}$ that there exists a sequence (x_n) in X, such that $\lim_n Ax_n = y$. We obtain from (4.108)

$$\|Ax_n - Ax_k\| \geq j(A)\|x_n - x_k\|,$$

and (x_n) is a Cauchy sequence. Hence there exists $x \in X$ with $x = \lim_n x_n$, and so $y = \lim_n Ax_n = Ax$. Thus we have proved the inclusion $\overline{R(A)} \subset R(A)$, and so $\overline{R(A)} = R(A)$. On the other hand, if $R(A) = \overline{R(A)}$ and $N(A) = \{0\}$, then the map $A_1 : X \to R(A)$ defined by $A_1 x = Ax$ for each $x \in X$ is one to one and onto. By the theorem of the bounded inverse, Theorem 4.6.3, we have $A_1^{-1} \in \mathcal{B}(R(A), X)$,

$$A_1^{-1} A_1 x = x \text{ and } \|x\| \leq \|A_1^{-1}\| \|A_1 x\| \text{ for each } x \in X.$$

Obviously

$$\|A_1 x\| \geq \frac{1}{\|A_1^{-1}\|} \|x\| \text{ for each } x \in X,$$

and so $j(A) > \|A_1^{-1}\|^{-1} > 0$. $\qquad\square$

If the operator is not one to one, then, by considering the quotient space, we may study when such an operator has closed range. This is connected with the following definition.

Definition 4.15.3 (Kato) ([111, p. 271–272]) Let X and Y be normed spaces and $A \in \mathcal{B}(X, Y)$. The *reduced minimum modulus* of the operator A, is defined as

$$\gamma(A) = \sup\{c \geq 0 : \|Ax\| \geq c \cdot d(x, N(A)) \text{ for all } x \in X\}, \qquad (4.109)$$

where $d(x, N(A)) = \inf\{\|x - u\| : u \in N(A)\} = \|x\|_{X/N(A)}$ is the quotient norm on $X/N(A)$. We use the convention that $\gamma(A) = \infty$ if and only if $N(A) = X$, and we have

$$\|Ax\| \geq \gamma(A) d(x, N(A)) \text{ for all } x \in X.$$

Theorem 4.15.4 (Kato) ([111, Lemma 322.]) *Let X and Y be Banach spaces and $A \in \mathcal{B}(X, Y)$. Then*

$$\gamma(A) > 0 \text{ if and only if } R(A) = \overline{R(A)}.$$

Proof. If $N(A) = \{0\}$, then obviously $j(A) = \gamma(A)$, and the proof follows by Theorem 4.15.2. We assume that $\{0\} \neq N(A) \neq X$. The quotient space $\tilde{X} = X/N(A)$ is a Banach space with the quotient norm by Theorem 4.10.4, and

$$\|\tilde{u}\| = \inf_{u \in \tilde{u}} \|u\| = \inf_{z \in N(A)} \|u - z\| = d(u, N(A)) \text{ for all } u \in \tilde{u} \in \tilde{X}.$$

Let $\tilde{A} : \tilde{X} \to Y$ be the operator defined by

$$\tilde{A}\tilde{u} = Au \text{ for all } u \in \tilde{u} \in \tilde{X}.$$

It is easy to show that $\tilde{A}$ is well defined and $R(\tilde{A}) = R(A)$. Since $j(\tilde{A}) = \gamma(A)$, we only have to apply Theorem 4.15.2 again. $\qquad\square$

Theorem 4.15.5 (Kato) ([111, Lemma 332]) *Let X and Y be Banach spaces and $A \in \mathcal{B}(X, Y)$. We assume that Z is a closed subspace in Y such that $R(A) \oplus Z$ is a closed subspace in Y. Then $R(A)$ is a closed subspace in Y.*

Proof. We shall use the notations with the same meanings as in the proof of Theorem 4.15.4. Let $A_0 : \tilde{X} \times Z \to R(A) \oplus Z$ be the operator defined by

$$A_0(\tilde{u}, z) = Au + z \text{ for some } u \in \tilde{u} \in \tilde{X} \text{ and all } z \in Z.$$

Here we mention that $\tilde{X} \times Z$ is a Banach space, for example with the norm defined by

$$\|(\tilde{x}, z)\|_{\tilde{X} \times Z} = \sqrt{\|\tilde{x}\|_{\tilde{X}}^2 + \|z\|_Z^2},$$

and the operator A_0 is well defined. It is easy to prove that A_0 is continuous, one to one and onto. By the theorem of the bounded inverse, Theorem 4.6.3, it follows analogously as in the proof of Theorem 4.15.2 that

$$\|Au + z\| \geq \frac{1}{\|A_0^{-1}\|} \|(\tilde{u}, z)\| = \frac{1}{\|A_0^{-1}\|} (\|\tilde{u}\|^2 + \|z\|^2)^{1/2},$$

so that for $z = 0$

$$\|Au\| \geq \frac{1}{\|A_0^{-1}\|} \|\tilde{u}\|, \text{ for all } u \in X.$$

Hence $\gamma(A) \geq \|A_0^{-1}\|^{-1} > 0$, and to complete the proof we only have to apply Theorem 4.15.4. $\qquad\square$.

Corollary 4.15.6 *Let X and Y be Banach spaces and $A \in \mathcal{B}(X, Y)$. If $Y/R(A)$ is a finite dimensional space, then $R(A)$ is a closed subspace in Y.*

Proof. This follows from Theorem 4.15.5. $\qquad\square$

There exists a function which *measures* how close an operator is to surjectivity. The next definition is related to this.

Definition 4.15.7 ([154, B.3.4.]) *Let X and Y be normed spaces and $A \in \mathcal{B}(X, Y)$. The modulus (coefficient) of surjectivity of the operator A, is defined as*

$$q(A) = \sup\{a \geq 0 : a\overline{B}_Y \subset A\left(\overline{B}_X\right)\}.$$

We use the convention $q(O) = 0$. The set of all operators $A \in \mathcal{B}(X, Y)$ for which $q(A) > 0$ is denoted by $\mathcal{Q}(B, X)$.

The next lemma gives us another formula for $q(A)$.

Lemma 4.15.8 ([154, B.3.5.]) *Let X and Y be Banach spaces and $A \in \mathcal{B}(X, Y)$. Then we have*

$$q(A) = \sup\{a \geq 0 : a\overline{B}_Y \subset \overline{A\left(\overline{B}_X\right)}\}. \tag{4.110}$$

Proof. Let the right-hand side of (4.110) be denoted by $\bar{q}(A)$. Obviously $q(A) \leq \bar{q}(A)$. Now we prove the converse inequality. We assume $\bar{q}(A) > 0$. Let $y_1 = y \in \overline{B}_Y$, $0 < \varepsilon < 1$ and $a = (1 - \varepsilon)\bar{q}(A)$. Then we have

$$a\overline{B}_Y \subset \overline{A\left(\overline{B}_X\right)},$$

and there exists $x_1 \in \overline{B}_X$ such that

$$\|y_1 - a^{-1}Ax_1\| \leq \epsilon \text{ and } \|x_1\| \leq \|y_1\| \leq 1.$$

Analogously, for $y_2 = y_1 - a^{-1}Ax_1$ there exists $x_2 \in X$ such that

$$\|y_2 - a^{-1}Ax_2\| \leq \epsilon^2 \text{ and } \|x_2\| \leq \|y_2\|.$$

Continuing this process and using mathematical induction we can easily see that there exists a sequence $x_n \in X$ such that

$$\|x_1\| \leq 1, \ \|x_n\| \leq \|y_n\| \text{ and } \|y_n - a^{-1}Ax_n\| \leq \epsilon^n \text{ for all } n = 1, 2, \ldots,$$

where $y_n = y_1 - \sum_{i=1}^{n-1} a^{-1}Ax_i$ for $n = 2, 3, \ldots$. Then we have $\|x_n\| \leq \|y_n\| = \|y_{n-1} - a^{-1}Ax_{n-1}\| \leq \varepsilon^n$ for $n = 2, 3, \ldots$. It follows that $\sum_{n=1}^{\infty} \|x_n\| \leq (1 - \varepsilon)^{-1}$. We put $x = \sum_{n=1}^{\infty} x_n$. Now

$$y = \sum_{n=1}^{\infty} a^{-1}Ax_n = a^{-1}Ax \text{ and } \|x\| \leq (1 - \varepsilon)^{-1},$$

implies

$$(1 - \varepsilon)a\overline{B}_Y \subset A\left(\overline{B}_X\right),$$

that is, $q(A) \geq (1 - \epsilon)^2 \bar{q}(A)$. $\qquad\square$

Corollary 4.15.9 *Let X and Y be Banach spaces and $A \in \mathcal{B}(X,Y)$. Then we have*

$$q(A) > 0 \text{ if and only of } R(A) = Y,$$

that is, $A \in \mathcal{Q}(X,Y)$ if and only if A is surjective.

Proof. This follows by Remark 4.6.14 and the previous lemma $\qquad\square$

The next theorem shows that there exists some duality between the functions j and q.

Theorem 4.15.10 ([154, B.3.8.]) *Let X and Y be Banach spaces, $A \in \mathcal{B}(X,Y)$ and $A' \in \mathcal{B}(Y^*, X^*)$. Then we have*

$$q(A') = j(A) \text{ and } j(A') = q(A).$$

Proof. We assume $j(A) > c > 0$. Then we have

$$\|Ax\| \geq c\|x\| \text{ for each } x \in X.$$

Let $A_1 : X \to R(A)$ be the operator defined as in the proof of Theorem 4.15.2. If $f \in \overline{B}_{X^*}$, we define the function $g \in R(A)^*$ by

$$g(y) = f(A_1^{-1}y) \text{ for each } y \in R(A).$$

It is easy to show that $\|g\| \leq c^{-1}$. By the Hahn-Banach theorem there exists an extension of g to Y, say $g_1 \in Y^*$, such that $\|g_1\| \leq c^{-1}$. We note that $A'g_1 = f$. It follows that $c\overline{B}_{X^*} \subset A'(\overline{B}_{Y^*})$, that is, $q(A') \geq j(A)$.

To prove the converse inequality, we assume $q(A') > a > 0$. For $x \in X$, let $f_0 \in \overline{B}_{X^*}$ be such that $|f_0x| = \|x\|$. From $a\overline{B}_{X^*} \subset A'(\overline{B}_{Y^*})$, we obtain the existence of $g_0 \in \overline{B}_{Y^*}$ such that $A'g_0 = af_0$. Hence we have

$$\|Ax\| \geq |g_0(Ax)| = |A'g_0(x)| = af_0(x) = a\|x\|,$$

and so $q(A') \leq j(A)$.

Analogously we can prove $q(A) \leq j(A')$ (we leave this to the reader). We only have to prove

$q(A) \geq j(A')$. Let $j(A') > c > 0$ and assume that $y \in \overline{B}_Y$ such that $cy \notin \overline{A(\overline{B}_X)}$. By Theorem 4.2.4 there exists $g \in Y^*$ such that

$$|g(cy)| > 1 \text{ and } |g(Ax)| \leq 1 \text{ for all } x \in \overline{B}_X,$$

and so

$$\|A'g\| = \sup\{|g(Ax)| : x \in \overline{B}_X\} \leq 1 < |g(cy)| \leq c\|g\|.$$

This is a contradiction and it follows that $\overline{A(\overline{B}_X)} \supset c\overline{B}_Y$. Now Lemma 4.15.8 implies $j(A') \leq q(A)$. $\qquad\square$

The next lemma shows that the functions j and q are continuous. The lemma is rather elementary.

Lemma 4.15.11 ([154, B.3.11.]) *Let X and Y be Banach spaces and $A, B \in \mathcal{B}(X,Y)$. Then we have*

$$|j(A+B) - j(A)| \leq \|B\| \text{ and } |q(A+B) - q(A)| \leq \|B\|. \tag{4.111}$$

Proof. If $x \in X$ and $\|x\| = 1$, then

$$\|(A+B)x\| - \|Ax\| \leq \|Bx\|,$$

implies

$$\|(A+B)x\| - j(A) \leq \|B\|,$$

and so

$$j(A+B) - j(A) \leq \|B\|.$$

Obviously now $j(A) - j(A+B) \leq \|B\|$ and we have proved the first inequality in (4.111). The second inequality follows from the first, and Theorems 4.15.10 and 4.9.2. $\qquad\square$

We obtain the following result from the previous lemma.

Corollary 4.15.12 *Let X and Y be Banach spaces. Then the sets $\mathcal{J}(X,Y)$ and $\mathcal{Q}(X,Y)$ are open subsets of $\mathcal{B}(X,Y)$.*

We proved in a simple way that the functions j and q are continuous. The proof of the continuity of γ is much more complicated. We refer the readers who are interested in the results related to the continuity of the function γ to the results in [7, 77, 111, 112, 127, 137, 163, 219, 220].

We close this section with the following theorem.

Theorem 4.15.13 (Banach) *Let X and Y be Banach spaces, $A \in \mathcal{B}(X,Y)$ and $A' \in \mathcal{B}(Y^*, X^*)$. Then the following statements are equivalent:*

$$R(A) = \overline{R(A)}, \tag{4.112}$$

$$R(A') = N(A)^0, \tag{4.113}$$

and

$$R(A') = \overline{R(A')}. \tag{4.114}$$

Proof.

(i) First we prove that (4.112) implies (4.113).

Let $f \in R(A')$ and $x \in N(A)$. Then there exists $g \in Y^*$ such that $A'g = f$, and so $f(x) = (A'g)x = g(Ax) = 0$. So we have shown that $f \in N(A)^0$, that is, $R(A') \subset N(A)^0$.

To prove the converse inclusion, we assume $f \in N(A)^0$. Let $g_0 : R(A) \to \mathbb{C}$ be the functional defined by

$$g_0(y) = f(x) \text{ for all } y \in R(A), \text{ where } x \in X \text{ is chosen such that } Ax = y. \quad (4.115)$$

We note that g_0 is well defined: namely, $Ax_1 = y$ implies $x - x_1 \in N(A)$, and so $f(x) = f(x_1)$. It is easy to show that g_0 is linear. We prove that g_0 is a continuous functional. It follows from (4.115) that

$$|g_0(y)| \leq \|f\|\|x - z\| \text{ for all } z \in N(A),$$

hence

$$|g_0(y)| \leq \|f\|d(x, N(A)). \quad (4.116)$$

Now we have by Theorem 4.15.4, (4.109) and (4.116)

$$|g_0(y)| \leq \|f\|\frac{\|Ax\|}{\gamma(A)} = \frac{\|f\|}{\gamma(A)}\|y\|,$$

and so g_0 is a continuous functional. Let $g_1 \in Y^*$ be an extension of the linear functional g_0 (by the Hahn–Banach theorem). Since we have for each $x \in X$

$$f(x) = g_0(Ax) = g_1(Ax) = (A'g_1)x,$$

it follows that $A'g_1 = f$, that is, $f \in R(A')$.

(ii) Obviously (4.113) implies (4.114).

(iii) Now we prove that (4.114) implies (4.112).

Let $A_1 : X \to \overline{R(A)}$ be the operator defined by $A_1x = Ax$ for all $x \in X$. Obviously $A_1 \in \mathcal{B}(X, \overline{R(A)})$, and by Lemmas 4.9.7 and 4.9.8 it follows that $A'_1 \in \mathcal{B}(R(A)^*, X^*)$ is one to one. Let us prove $R(A'_1) = \overline{R(A'_1)}$. If $g \in \overline{R(A'_1)}$, then there exists a sequence (f_n) in $R(A)^*$ such that $\lim_{n\to\infty} A'_1(f_n) = g$. Let h_n be extensions of the functionals f_n on Y by the Hahn–Banach theorem. Then $\lim_{n\to\infty} A'(h_n) = g$, and since $R(A')$ is a closed subspace of X^*, it follows that there exists $h \in Y^*$ such that $A'h = g$. If h_0 is the restriction of the functional h on the subspace $\overline{R(A_1)}$, then we have $A'h_0 = g$. So we showed $R(A'_1) = \overline{R(A'_1)}$. Now by the following well–known result: if $B \in \mathcal{B}(X, Y)$, B' is one to one, and $R(B') = \overline{R(B')}$ then $\overline{R(B)} = Y$; (see [65, Lemma 3, p. 488] for the proof), it follows that $R(A_1) = R(A) = \overline{R(A)}$. $\qquad\square$

Chapter 5

Hilbert Spaces

Chapter 5 contains the study of inner product and Hilbert spaces, and properties of operators between Hilbert spaces. The theory of Hilbert spaces provides essential tools in the theories of partial differential equations, quantum mechanics, Fourier analysis including applications to signal processing, and thermodynamics. First, many standard examples are given and the more familiar results are established, such as the Cauchy–Schwarz inequality and the parallelogram law and the P. Jordan–J. von Neumann theorem. Also the concepts of orthogonality and orthonormality are introduced, and the theorem of the elements of minimal norm, the theorem on the orthogonal decomposition, the Riesz representation theorem, Bessel's inequality, Fourier coefficients, Parseval's equality and the Gram–Schmidt orthogonalization process are proved. Additional topics are Hilbert adjoint and Hermitian, normal, positive and unitary operators. Finally, the chapter contains a detailed study of projectors and orthogonal projectors, and formulas and results on the norm of idempotent operators.

5.1   Introduction

Hilbert spaces are the most important examples of Banach spaces. Their study was initiated by Hilbert's results at the beginning of the twentieth century; his results are related to the spaces ℓ_2 and L_2. Banach spaces have some properties of the Euclidean space $\mathbb{C}^n$ (Theorem 3.2.7), while some geometric properties of Euclidean spaces such as, for instance, orthogonality, also exist in Hilbert spaces. Furthermore, the study of Hilbert spaces is connected with the operator theory on Hilbert spaces. Hence we also study operators on Hilbert spaces in this chapter.

Hilbert spaces naturally arise in mathematics and physics, typically as infinite-dimensional function spaces. They played an important role in the development of quantum mechanics. Hilbert spaces were initially studied in the first decade of the 20th century by David Hilbert, Erhard Schmidt, and Frigyes Riesz. In 1929, John von Neumann introduced the notion of the Hilbert spaces and gave their abstract axiomatic definition.

The theory of Hilbert spaces provides essential tools in the theories of partial differential equations, quantum mechanics, Fourier analysis including applications to signal processing, and thermodynamics. Apart from the classical Euclidean spaces, the spaces ℓ_2 and L_2, examples of Hilbert spaces include Sobolev spaces consisting of generalized functions, and Hardy spaces of holomorphic functions.

In addition to the concept of orthogonality in Euclidean space $\mathbb{R}^n$, geometric properties such as the validity of the parallelogram law and the Pythagorean theorem can be extended to general Hilbert spaces.

This chapter contains the standard examples and familiar results such as the Cauchy–Schwarz inequality, Theorem 5.3.1, the parallelogram law, Theorem 5.5.2, the P. Jordan–J. von Neumann theorem, Theorem 5.5.3 the concepts of orthogonality, the theorem of the elements of minimal norm, Theorem 5.7.1, the theorem on the orthogonal decomposition, Theorem 5.8.1, the Riesz representation theorem, Theorem 5.9.1, Bessel's inequality, Theorem 5.10.1, Fourier coefficients, Parseval's equality (5.47), and the Gram–Schmidt orthogonalization process, Theorem 5.12.1. The topics are Hilbert adjoint and Hermitian, normal, positive and unitary operators, and the characterization of positive bounded linear operators. Finally, the chapter contains a detailed study of projectors and orthogonal projectors, formulas and results on the norm of idempotent operators.

5.2 Inner product spaces

One important property of Hilbert spaces is that they are linear spaces with an additional binary operation referred to as the *inner* or *scalar product*; inner products are generalizations of the well–known dot product in Euclidean $\mathbb{C}^n$.

Definition 5.2.1 *An inner or scalar product on a (complex) linear space X is a function $s : X \times X \to \mathbb{C}$ which satisfies the following conditions*

$$s(\lambda_1 x_1 + \lambda_2 x_2, y) = \lambda_1 s(x_1, y) + \lambda_2 s(x_2, y) \tag{5.1}$$
$$\text{for all } \lambda_1, \lambda_2 \in \mathbb{C} \text{ and all } x_1, x_2, \, y \in X,$$

$$s(x, y) = \overline{s(y, x)} \text{ for all } x, y \in X, \tag{5.2}$$
$$s(x, x) \geq 0 \text{ for all } x \in X, \tag{5.3}$$
$$s(x, x) = 0 \text{ if and only if } x = 0. \tag{5.4}$$

A linear space X with an inner product s, that is, the ordered pair (X, s), is referred to as *inner product (pre-Hilbert) space, or space with a scalar product*.

To be more precise, by Definition 5.2.1 we introduce a *complex inner product space* or *complex unitary space*.

If the linear space X is real, the function s is real and has the mentioned properties, then X is called a *real unitary* or *real inner product space*.

The inner product in an inner product space is usually denoted by $\langle \cdot, \cdot \rangle$, that is, $\langle x, y \rangle = s(x, y)$ for all $x, y \in X$, and we shall use this notation. Analogously, as for normed spaces, we normally say that X *is an inner product space*, in which case we suppose that X is a linear space with an inner product $\langle x, y \rangle$ for all $x, y \in X$.

Remark 5.2.2 *We note that the conditions in (5.1) and (5.2) imply*

$$s\langle x, \lambda_1 y_1 + \lambda_2 y_2 \rangle = \overline{\lambda}_1 s\langle x, y_1 \rangle + \overline{\lambda}_2 s\langle x, y_2 \rangle \tag{5.5}$$
$$\text{for all } \lambda_1, \lambda_2 \in \mathbb{C} \text{ and all } x, y_1, y_2 \in X.$$

In our further work, we always suppose that our inner space is complex, unless explicitly stated otherwise.

Lemma 5.2.3 *Let X be an inner product space and $x_1, x_2 \in X$. If $\langle x_1, y \rangle = \langle x_2, y \rangle$ for all $y \in X$, then $x_1 = x_2$. In particular, if $\langle x_1, y \rangle = 0$ for all $y \in X$, then $x_1 = 0$.*

Proof. By the assumption, we have for each $y \in X$

$$0 = \langle x_1, y \rangle - \langle x_2, y \rangle = \langle x_1 - x_2, y \rangle. \tag{5.6}$$

For $y = x_1 - x_2$ it follows from (5.6) that $\langle x_1 - x_2, x_1 - x_2 \rangle = 0$, that is, $x_1 = x_2$. From $\langle x_1, y \rangle = 0$ for all $y \in Y$, taking $y = x_1$, we obtain $\langle x_1, x_1 \rangle = 0$, that is, $x_1 = 0$. $\square$

Lemma 5.2.4 *Let X and Y be inner product spaces. If $A \in \mathcal{L}(X, Y)$, then*

$$A = O \text{ if and only if } \langle Ax, y \rangle = 0 \text{ for all } x \in X \text{ and all } y \in Y. \tag{5.7}$$

If $A \in \mathcal{L}(X)$, then

$$\langle Ax, x \rangle = 0 \text{ for all } x \in X \text{ implies } A = O. \tag{5.8}$$

Proof.

(i) First we prove the statement in (5.7).
We note that $A = O$ implies $Ax = 0$ for all $x \in X$. Hence $\langle Ax, y \rangle = \langle 0, y \rangle = 0$ for all $x \in X$ and all $y \in Y$.
Conversely, if $\langle Ax, y \rangle = 0$ for all $x \in X$ and all $y \in Y$, then Lemma 5.2.3 implies $Ax = 0$ for all $x \in X$, that is, $A = O$. Thus we have shown (5.7).

(ii) Now we prove (5.8).
We suppose that $\langle Az, z \rangle = 0$ for all $z = \lambda x + y \in X$ and $\lambda \in \mathbb{C}$. Hence we obtain

$$\begin{aligned} 0 &= \langle A(\lambda x + y), \lambda x + y \rangle \\ &= |\lambda|^2 \langle Ax, x \rangle + \langle Ay, y \rangle + \lambda \langle Ax, y \rangle + \overline{\lambda} \langle Ay, x \rangle. \end{aligned} \tag{5.9}$$

It follows from (5.9) that for all $x, y \in X$ and $\lambda \in \mathbb{C}$, we have

$$0 = \lambda \langle Ax, y \rangle + \overline{\lambda} \langle Ay, x \rangle. \tag{5.10}$$

If we first take $\lambda = 1$ and then $\lambda = i$, then we have by (5.10)

$$\langle Ax, y \rangle + \langle Ay, x \rangle = 0, \tag{5.11}$$

and

$$\langle Ax, y \rangle - \langle Ay, x \rangle = 0. \tag{5.12}$$

Summing we obtain from (5.11) and (5.12),

$$2 \langle Ax, y \rangle = 0 \text{ for all } x, y \in X. \tag{5.13}$$

Then (5.13) and (5.7) imply $A = 0$. $\square$

Remark 5.2.5 *We already mentioned that we always work in complex inner product spaces unless stated otherwise. Of course, many results hold true for real inner product spaces. For example, from the proof of (5.7) we can see that this is true for real inner product spaces. We remark that (5.8) does not hold true for real inner product spaces. For instance, if $X = \mathbb{R}^2$, a well–known inner product space, and $A \in \mathcal{L}(X)$ is the rotation around 0 by $\pi/2$, then the vector Ax is orthogonal to the vector x, that is, $\langle Ax, x \rangle = 0$ for all $x \in X$. But clearly $A \neq 0$.*

Corollary 5.2.6 *Let X be an inner product space, $A, B \in \mathcal{L}(X)$ and $\langle Ax, x \rangle = \langle Bx, x \rangle$ for all $x \in X$. Then we have $A = B$.*

Proof. This follows by (5.8). $\square$

5.3   Elementary properties and Hilbert spaces

In this section, we are going to prove some elementary properties of inner product spaces and define Hilbert spaces.

Theorem 5.3.1 (Cauchy–Schwarz inequality) *Let X be an inner product space. Then we have*

$$|\langle x,y\rangle| \leq \langle x,x\rangle^{1/2}\langle y,y\rangle^{1/2} \text{ for all and } x,y \in X. \tag{5.14}$$

Proof. If $y = 0$, then $\langle y,y\rangle = 0$ and $\langle x,y\rangle = 0$, that is, in this case the theorem is true. So we assume $y \neq 0$, and have for each $\lambda \in \mathbb{C}$

$$0 \leq \langle x - \lambda y, x - \lambda y\rangle = \langle x,x\rangle - \overline{\lambda}\langle x,y\rangle - \lambda\langle y,x\rangle + |\lambda|^2\langle y,y\rangle. \tag{5.15}$$

From (5.15), it follows for $\lambda = \langle x,y\rangle/\langle y,y\rangle$, that

$$0 \leq \langle x,x\rangle - \frac{|\langle x,y\rangle|^2}{\langle y,y\rangle}. \qquad \square$$

We remark that equality holds in (5.14) if and only if x and y are linearly independent.

Theorem 5.3.2 (Minkowski inequality) *Let X be an inner product space. Then we have*

$$\langle x+y, x+y\rangle^{1/2} \leq \langle x,x\rangle^{1/2} + \langle y,y\rangle^{1/2} \text{ for all } x,y \in X. \tag{5.16}$$

Proof. Applying the Cauchy–Schwarz inequality we obtain

$$\begin{aligned}
0 \leq \langle x+y, x+y\rangle &= |\langle x,x\rangle + \langle x,y\rangle + \langle y,x\rangle + \langle y,y\rangle| \\
&\leq \langle x,x\rangle + |\langle x,y\rangle| + |\langle y,x\rangle| + \langle y,y\rangle \\
&\leq \langle x,x\rangle + 2\langle x,x\rangle^{1/2}\langle y,y\rangle^{1/2} + \langle y,y\rangle = \left(\langle x,x\rangle^{1/2} + \langle y,y\rangle^{1/2}\right)^2.
\end{aligned} \qquad \square$$

Any inner product defines a norm. Namely, the function $x \to \langle x,x\rangle^{1/2}$ from an inner product space X into $\mathbb{R}$ is nonnegative and satisfies the triangle inequality (5.16). Also, we have $\langle \lambda x, \lambda x\rangle^{1/2} = |\lambda|\langle x,x\rangle^{1/2}$ for all $\lambda \in \mathbb{C}$ and all $x \in X$. Therefore the function with $\langle x,x\rangle^{1/2}$ for all $x \in X$ is a norm on X.

Definition 5.3.3 Let X be an inner product space. The norm $\|\cdot\|$ with

$$\|x\| = \langle x,x\rangle^{1/2} \text{ for all } x \in X \tag{5.17}$$

is said to be *a norm defined by an inner (scalar) product*. If we do not explicitly mention otherwise, then we always assume that an inner product space X is a normed space with its norm defined by (5.17).
If an inner product space X is a Banach space then it is called a *Hilbert space*.

Theorem 5.3.4 *Let X be an inner product space. Then the inner product is a continuous function, that is, if $x,y \in X$, and (x_n) and (y_n) are sequences in X, then*

$$x_n \to x \text{ and } y_n \to y \text{ imply } \langle x_n, y_n\rangle \to \langle x,y\rangle. \tag{5.18}$$

Proof. It follows from the Cauchy–Schwarz inequality that

$$|\langle x_n, y_n \rangle - \langle x, y \rangle| \leq |\langle x_n, y_n - y \rangle| + |\langle x_n - x, y \rangle|$$
$$\leq \|x_n\| \, \|y_n - y\| + \|x_n - x\| \, \|y\|$$

and the conclusion follows, since the sequence (x_n) is bounded by Lemma 2.10.3. $\square$

Corollary 5.3.5 *Let X be an inner product space and $y \in X$. Then the functions $f, g :$ $X \to \mathbb{C}$ with $f(x) = \langle y, x \rangle$ and $g(x) = \langle x, y \rangle$ for all $x \in X$ are continuous on X.*

Proof. This follows from Theorem 5.3.4. $\square$

5.4 Examples

In this section, we give some examples of inner product and Hilbert spaces. We leave the proofs to the reader.

Example 5.4.1 *For every natural number n, the space $\mathbb{C}^n$ $(\mathbb{R}^n)$ of ordered n–tuples $x = (x_1, \dots, x_n)$ of complex (real) numbers is a Hilbert space with the inner product*

$$\langle x, y \rangle = \sum_{k=1}^{n} x_k \overline{y_k},$$

which in the real case is the well-known dot product *of calculus with $x \bullet y = \langle x, y \rangle$.*

Example 5.4.2 *The space ℓ_2 of all complex (real) sequences $x = (x_k)$ such that $\sum_{k=1}^{\infty} |x_k|^2 < \infty$ is a Hilbert space with the inner product*

$$\langle x, y \rangle = \sum_{k=1}^{\infty} x_k \overline{y_k}.$$

Example 5.4.3 *For any measurable space $(X, \mathcal{R}, \mu)$, the space $L_2(\mu)$ of all measurable functions f such that $\int |f|^2 \, d\mu < \infty$ is a Hilbert space with the scalar product*

$$\langle f, g \rangle = \int f \overline{g} \, d\mu.$$

(We note that Examples 5.4.1 and 5.4.2 are special cases of Example 5.4.3 when $X = \{1, 2, \dots, n\}$ and $X = \{1, 2, \dots, n, \dots\}$, respectively, and the measure μ is the counting or discrete measure.)

Example 5.4.4 *The linear space of all continuous functions f on the interval $[a, b]$ is an inner product space (but not a Hilbert space) with the inner product*

$$\langle f, g \rangle = \int_a^b f(t) \overline{g(t)} \, dt.$$

Example 5.4.5 *Let S be a nonempty set and with cardinality card $S = \alpha$. For each $s \in S$, let $C_s = \mathbb{C}$, and*

$$\prod_{s \in S} C_s$$

be the Cartesian product of the sets C_s (Section 1.3). We define the subset $\ell_2(S)$ of $\prod_{s\in S} C_s$, which consists of all functions f for which the family $\{|f(s)|^2\}_{s\in S}$ is summable. Then $\ell_2(S)$ is a linear space with respect to the natural linear operations

$$(f + g)(s) = f(s) + g(s) \text{ for all } s \in S$$

and

$$(\lambda f)(s) = \lambda(f(s)) \text{ for all } \lambda \in \mathbb{C} \text{ and all } s \in S.$$

For $f, g \in \ell_2(S)$ the family $\{|f(s)\overline{g(s)}|\}_{s\in S}$ is summable (this follows from the inequality $|f(s)g(s)| \leq (1/2)(|f(s)|^2 + |g(s)|^2)$). It is easy to show that $\ell_2(S)$ is a unitary space with the inner product

$$\langle f, g \rangle = \sum_{s\in S} f(s)\overline{g(s)} \text{ for all } f, g \in \ell_2(s).$$

It can be shown that $\ell_2(S)$ is a Hilbert space (see the proof of Theorem 5.11.8 in Section 5.11).

Example 5.4.6 *Let X be the set of all absolutely continuous functions $f : [0, 1] \to \mathbb{C}$ such that $f(0) = 0$ and $f' \in L_2(0, 1)$. Then X is a linear space with the usual linear operations and X with the inner product*

$$\langle f, g \rangle = \int_0^1 f'(t)\overline{g'(t)}\, dt$$

is a Hilbert space.

Example 5.4.7 *Let G be an open subset of the complex plane $\mathbb{C}$, and let $L_2^a(G)$ be the set of all analytic functions $f : G \to \mathbb{C}$ such that*

$$\iint\limits_G |f(x + iy)|^2\, dx\, dy < \infty;$$

$L_2^a(G)$ is the well–known Bergmann space of the set G. We remark that $L_2^a(G) \subset L_2(\mu)$, where μ is the area measure of the set G and $L_2^a(G)$ has the inner product and norm from $L_2(\mu)$. It can be shown that $L_2^a(G)$ is a Hilbert space. (We refer to [47, Proposition 1.13]).

5.5   Theorem P. Jordan-J. von Neumann

We saw in Section 5.3 that every inner product space is a normed space. The question naturally arises, under which conditions does the converse implication also hold true, if at all? We consider this question in this section.

We need the following two theorems as preliminary results.

Theorem 5.5.1 (Polarization equation) *If X is a complex inner product space then we have for all $x, y \in X$*

$$\langle x, y \rangle = \frac{1}{4}\left(\|x + y\|^2 - \|x - y\|^2\right) + \frac{i}{4}\left(\|x + iy\|^2 - \|x - iy\|^2\right). \qquad (5.19)$$

If X is a real inner product space then we have for all $x, y \in X$

$$\langle x, y \rangle = \frac{1}{4}\left(\|x + y\|^2 - \|x - y\|^2\right). \qquad (5.20)$$

Proof. If we replace the norm on the right-hand sides in (5.19) and (5.20) by the inner product, and use the elementary properties of an inner product, then we easily obtain the left-hand sides. □

Theorem 5.5.2 (Parallelogram law) *Let X be an inner product space. Then we have*

$$\|x + y\|^2 + \|x - y\|^2 = 2\|x\|^2 + 2\|y\|^2 \text{ for all } x, y \in X. \tag{5.21}$$

Proof. The proof is similar to that of the previous theorem. □

If X is a normed space and the vectors $x, y \in X$ satisfy the identity in (5.21), then we say that x and y satisfy the *parallelogram law*. This name is connected to a result in elementary geometry on the edges and diagonals of a parallelogram. Of course, in this case the norm of a vector is its length. We say the *norm satisfies the parallelogram law* if any two vectors satisfy the parallelogram law.

The following theorem is the converse of Theorem 5.5.2. It is due to P. Jordan and J. von Neumann (1935) [102].

Theorem 5.5.3 (P. Jordan-J. von Neumann) *Let $(X, \|\cdot\|)$ be a normed space and let the norm $\|\cdot\|$ satisfy the parallelogram law.*
(a) If X is a real normed space, then

$$\langle x, y \rangle_1 = \frac{1}{4}\left(\|x + y\|^2 - \|x - y\|^2\right) \text{ for all } x, y \in X \tag{5.22}$$

defines an inner product on X, and

$$\|x\| = \langle x, x \rangle_1^{1/2} \text{ for all } x \in X.$$

(b) If X is a complex normed space, then

$$\langle x, y \rangle_2 = \frac{1}{4}\left(\|x + y\|^2 - \|x - y\|^2\right) + \frac{i}{4}\left(\|x + iy\|^2 - \|x - iy\|^2\right) \tag{5.23}$$

for all $x, y \in X$

defines an inner product on X, and

$$\|x\| = \langle x, x \rangle_2^{1/2} \text{ for all } x \in X.$$

Proof.

(a) First we assume that X is a real normed space, and let $x \in X$. It follows from (5.22) that $\langle x, x \rangle_1 = \|(x + x)/2\|^2$. Hence $\langle x, x \rangle_1 \geq 0$ and $\langle x, x \rangle_1 = 0$ if and only if $x = 0$. Furthermore (5.22) implies that for $x, y \in X$

$$\langle x, y \rangle_1 = \frac{1}{4}(\|y + x\|^2 - \|y - x\|^2) = \langle y, x \rangle_1.$$

Thus we have shown the conditions in (5.3), (5.4) and (5.2) in Definition 5.2.1.

(i) Now we show that

$$\langle x + y, z \rangle_1 = \langle x, z \rangle_1 + \langle y, z \rangle_1 \text{ for all } x, y, z \in X. \tag{5.24}$$

The parallelogram law yields for all $x, y, z \in X$

$$\langle x, y \rangle_1 + \langle z, y \rangle_1 =$$

$$= \frac{1}{4} \left(\|x+y\|^2 - \|x-y\|^2 + \|z+y\|^2 - \|z-y\|^2 \right)$$

$$= \frac{1}{4} \left(\left(\|x+y\|^2 + \|z+y\|^2 \right) - \left(\|x-y\|^2 + \|z-y\|^2 \right) \right)$$

$$= \frac{1}{4} \left(\frac{1}{2} \|x+y+z+y\|^2 + \|x-z\|^2 - \frac{1}{2} \|x-y+z-y\|^2 - \|x-z\|^2 \right)$$

$$= \frac{1}{8} \left(\|x+z+2y\|^2 - \|x+z-2y\|^2 \right)$$

$$= \frac{1}{2} \left(\left\| \frac{x+z}{2} + y \right\|^2 - \left\| \frac{x+z}{2} - y \right\|^2 \right) = 2 \left\langle \frac{x+z}{2}, y \right\rangle_1.$$

Thus we have proved

$$\langle x, y \rangle_1 + \langle z, y \rangle_1 = 2 \left\langle \frac{x+z}{2}, y \right\rangle_1 \quad \text{for all } x, y, z \in X. \tag{5.25}$$

We have for $z = 0$

$$\langle x, y \rangle_1 = 2 \left\langle \frac{x}{2}, y \right\rangle_1.$$

If we replace x by $2x$ we obtain

$$\langle 2x, y \rangle_1 = 2 \langle x, y \rangle_1. \tag{5.26}$$

Finally, if we replace x and z by $2x$ and $2z$, respectively, then we have by (5.25) and (5.26)

$$\langle x + z, y \rangle_1 = \langle x, y \rangle_1 + \langle z, y \rangle_1. \tag{5.27}$$

Hence we have shown (5.24).

(ii)   Now we show that

$$\langle \lambda x, y \rangle_1 = \lambda \langle x, y \rangle_1 \text{ for all } x, y \in X \text{ and all } \lambda \in \mathbb{R}. \tag{5.28}$$

The function $f_y = \langle \cdot, y \rangle_1 : X \to \mathbb{R}$ is additive for each fixed $y \in X$ by (5.24), and hence rational homogeneous by Theorem 1.6.4, so real homogeneous by the continuity of the norm in (5.22).
Thus we have shown (5.28).

Now (5.24) and (5.28) yield the condition in (5.1).
This completes the proof of Part (a).

(b)   Now we assume that X is a complex normed space. We note that

$$\langle x, y \rangle_2 = \langle x, y \rangle_1 + i \langle x, iy \rangle_1 \text{ for all } x, y \in X. \tag{5.29}$$

Since we have $(x, ix)_1 = 0$ for all $x \in X$, it follows that

$$\langle x, x \rangle_2 = \langle x, x \rangle_1 + i \langle x, ix \rangle_1 = \langle x, x \rangle_1 = \|x\|^2.$$

Hence $\langle x, x \rangle_2 = 0$ if and only if $x = 0$.
Since $\langle ix, iy \rangle_1 = \langle x, y \rangle_1$ for all $x, y \in X$, it follows from (5.29) that for all $x, y \in X$,

$$\langle y, x \rangle_2 = \langle y, x \rangle_1 + i \langle y, ix \rangle_1 = \langle y, x \rangle_1 + i \langle iy, -x \rangle_1$$

$$= \langle x, y \rangle_1 - i \langle x, iy \rangle_1 = \overline{\langle x, y \rangle_2}.$$

Also we have for all $x, y \in X$,

$$\langle ix, y \rangle_2 = \langle ix, y \rangle_1 + i\langle ix, iy \rangle_1 = \langle ix, y \rangle_1 + i\langle x, y \rangle_1$$
$$= i[\langle x, y \rangle_1 - i\langle ix, y \rangle_1] = i[\langle x, y \rangle_1 - i\langle -x, iy \rangle_1]$$
$$= i[\langle x, y \rangle_1 + i\langle x, iy \rangle_1] = i\langle x, y \rangle_2.$$

Hence, if $\lambda = \alpha + i\beta \in \mathbb{C}$, $\alpha, \beta \in \mathbb{R}$ and $x, y \in X$, then we obtain

$$\langle \lambda x, y \rangle_2 = \langle (\alpha + i\beta)x, y \rangle_2 = \langle \alpha x + i\beta x, y \rangle_2$$
$$= \langle \alpha x, y \rangle_2 + \langle i\beta x, y \rangle_2 = \langle \alpha x, y \rangle_2 + i\langle \beta x, y \rangle_2$$
$$= \alpha \langle x, y \rangle_2 + i\beta \langle x, y \rangle_2 = \lambda \langle x, y \rangle_2.$$

The function $\langle \cdot, \cdot \rangle_2$ is additive in both arguments (that is, it is additive in one argument when the other one is fixed, and vice versa), because the function $\langle \cdot, \cdot \rangle_1$ is additive in both arguments. $\qquad\square$

Further results related to Theorem 5.5.3 and their generalizations can be found in the works of S. Kurepa, [124] and [125].

5.6   Orthogonality

Any inner product can be used to define orthogonality in inner product spaces, and to study the geometry of inner product spaces.

We consider the following motivation from elementary geometry in Euclidean $\mathbb{R}^n$ with the dot product $\bullet$ of Example 5.4.1. Let Δ be the triangle spanned by the vectors v, w and $v - w$ along its edges. Then we have

$$\|w - v\|^2 = (w - v) \bullet (w - v) = w \bullet w - 2v \bullet w + v \bullet v = \|w\|^2 - 2v \bullet w + \|v\|^2.$$

On the other hand, by the elementary *law of cosines*, the angle ϕ between the edges along w and v satisfies

$$\|w - v\|^2 = (w - v) \bullet (w - v) = \|w\|^2 + \|v\|^2 - 2\|v\| \cdot \|w\| \cdot \cos \phi.$$

It follows that

$$v \bullet w = 2\|v\| \cdot \|w\| \cdot \cos \phi,$$

and the vectors v and w are orthogonal if $\phi = \pi/2$, that is, $v \bullet w = 0$.

This leads to the following generalization.

Definition 5.6.1 Let X be an inner product space and $x, y \in X$. If $\langle x, y \rangle = 0$, then x and y are said to be *orthogonal*, denoted by $x \perp y$. If E and F are subsets of X and if each vector in E is orthogonal to each vector in F, then E is said to be orthogonal to F, denoted by $E \perp F$; in this case, obviously, we also have $F \perp E$.

If $E \subset X$, then $E^\perp$ is the set of all $y \in X$ such $y \perp x$ for each $x \in E$. The set $E^\perp$ is called the *orthogonal complement of E*.

Clearly, we have $E \cap E^\perp \subset \{0\}$. Usually we write $E^{\perp\perp} = (E^\perp)^\perp$, and $E^{\perp\perp\perp} = ((E^\perp)^\perp)^\perp$.

Lemma 5.6.2 *If E is a subset of an inner product space X, then $E^\perp$ is a closed subspace of X.*

Proof. Let $x, y, z \in X$, $\lambda \in \mathbb{C}$, $x \perp y$ and $x \perp z$. Obviously, we have $x \perp (y + z)$ and $x \perp \lambda y$. Hence, $x^{\perp}$ is a subspace of X. Since the function $f : X \to \mathbb{C}$ with $f(u) = \langle x, u \rangle$ for all $u \in X$ is continuous by Corollary 5.3.5, and $\{x\}^{\perp} = f^{-1}\{0\}$, it follows that $\{x\}^{\perp}$ is a closed subspace of X. So we proved that $\{x\}^{\perp}$ is a closed subspace of X for each $x \in X$. Since $E^{\perp} = \cap \{\{x\}^{\perp} : x \in E\}$, it follows that $E^{\perp}$ is a closed subspace of X. $\square$

Definition 5.6.3 A subset E of an inner product space X is said to be *orthogonal* if $x, y \in E$ and $x \neq y$ imply $x \perp y$. An orthogonal set E is said to be *orthonormal* if the norm of each element in E is equal to 1. An orthonormal set E is said to be *complete (maximal)* if it is not a proper subset of any orthonormal set in X.

Theorem 5.6.4 *If X is an inner product space and $X \neq \{0\}$, then X contains a complete orthonormal set.*

Proof. Let $\mathcal{O}$ be the family of all orthonormal subsets of X. The family $\mathcal{O}$ is not empty, since if $x \in X$ and $\|x\| = 1$, then $\{x\} \in \mathcal{O}$. Let $\preceq$ be the partial order of inclusion on $\mathcal{O}$ (Example 1.2.2 (i)), that is,

$$E_1 \preceq E_2 \text{ if and only if } E_1 \subset E_2. \tag{5.30}$$

If $U = \{E_i : i \in I\}$ is a chain in $\mathcal{O}$, then it is easy to prove that $E_0 = \cup_{i \in I} E_i$ is an upper bound of U. Hence, by Zorn's lemma, there exists a maximal element $E \in \mathcal{O}$ with respect to the partial order defined in (5.30). It follows by Definition 5.6.3 that E is a complete orthonormal set. $\square$

The next result is the generalization of Pythagoras's law in Euclidean $\mathbb{R}^2$ to any finite orthogonal set in a general inner product space.

Theorem 5.6.5 (Pythagoras's law) *Let $\{x_1, \ldots, x_n\}$ be an orthogonal subset of an inner product space X. Then we have*

$$\left\| \sum_{k=1}^{n} x_k \right\|^2 = \sum_{k=1}^{n} \|x_k\|^2. \tag{5.31}$$

Proof. We obtain

$$\left\| \sum_{k=1}^{n} x_k \right\|^2 = \left\langle \sum_{k=1}^{n} x_k, \sum_{j=1}^{n} x_j \right\rangle = \sum_{k=1}^{n} \langle x_k, x_k \rangle + \sum_{k \neq j} \langle x_k, x_j \rangle$$

$$= \sum_{k=1}^{n} \langle x_k, x_k \rangle = \sum_{k=1}^{n} \|x_k\|^2. \qquad \square$$

The next lemma is easy to prove and we are going to apply it several times.

Lemma 5.6.6 *Let $\{x_i\}_{i \in I}$ be a summable family in an inner product space X. If $z \in X$, then the families $\{(x_i, z)\}_{i \in I}$ and $\{(z, x_i)\}_{i \in I}$ of scalars are summable, and*

$$\sum_{i \in I} \langle x_i, z \rangle = \left\langle \sum_{i \in I} x_i, z \right\rangle \text{ and } \sum_{i \in I} \langle z, x_i \rangle = \left\langle \sum_{i \in I} z, x_i \right\rangle.$$

Proof. We leave the proof to the reader. $\square$

In Hilbert space, the identity in (5.31) of Pythagoras's law, Theorem 5.6.5, extends to any summable orthogonal family $\{x_i\}_{i \in I}$ of vectors.

Theorem 5.6.7 (Generalized Pythagoras's theorem) *Let X be a Hilbert space. An orthogonal family $\{x_i\}_{i \in I}$ of vectors in X is summable if and only if the family of $\{\|x_i\|^2\}_{i \in I}$ of scalars is summable; in this case, we have*

$$\left\| \sum_{i \in I} x_i \right\|^2 = \sum_{i \in I} \|x_i\|^2.$$

Proof. The *if and only if* part of the theorem follows from Theorems 4.11.7 and 5.6.5. We assume $x = \sum_{i \in I} x_i$, and have by Lemma 5.6.6

$$\|x\|^2 = \left\langle \sum_{i \in I} x_i, x \right\rangle = \sum_{i \in I} \left\langle x_i, \sum_{j \in I} x_j \right\rangle$$

$$= \sum_{i \in I} \left(\sum_{j \in I} \langle x_i, x_j \rangle \right) = \sum_{i \in I} \|x_i\|^2. \qquad \square$$

The next lemma gives us a necessary and sufficient condition for the orthogonality of two vectors in an inner product space.

Lemma 5.6.8 *Let X be an inner product space and $x, y \in X$. Then we have*

$$x \perp y \text{ if and only if } \|y\| \leq \|\lambda x + y\| \text{ for all } \lambda \in \mathbb{C}. \tag{5.32}$$

Proof. It follows from $x \perp y$ that $\lambda x \perp y$ for all $\lambda \in \mathbb{C}$. Then we have

$$\|\lambda x + y\|^2 = \|\lambda x\|^2 + \|y\|^2 \geq \|y\|^2.$$

Conversely, we obtain for all $x, y \in X$ and all $\lambda \in \mathbb{C}$

$$0 \leq \|\lambda x + y\|^2 = \langle \lambda x + y, \lambda x + y \rangle$$
$$= |\lambda|^2 \langle x, x \rangle + \lambda \langle x, y \rangle + \overline{\lambda} \langle y, x \rangle + \|y\|^2$$
$$= |\lambda|^2 \|x\|^2 + 2 \operatorname{Re} \lambda \langle x, y \rangle + \|y\|^2.$$

Obviously, $x = 0$ implies $x \perp y$. If $x \neq 0$, we obtain from the inequality above, putting $\lambda = -\overline{\langle x, y \rangle} / \|x\|^2$,

$$\|y\|^2 \leq \|\lambda x + y\|^2 = \|y\|^2 - \frac{|\langle x, y \rangle|^2}{\|x\|^2}. \qquad \square$$

Remark 5.6.9 *There are many ways to introduce orthogonality in normed spaces to preserve some properties of orthogonality in inner product spaces (for instance, see [18, 24, 99, 113, 139, 140]). In connection with (5.32), it is said that, in a real normed space X, a vector $x \in X \setminus \{0\}$ is orthogonal to $y \in X$ in the sense of Birkhoff [30], denoted by $x \perp y$ (B), if*

$$\|x\| \leq \|x + \lambda y\| \text{ for all } \lambda \in \mathbb{R}$$

in a complex normed space X, a vector $x \in X \setminus \{0\}$ is said to be orthogonal to $y \in X$ in the sense of James [128], denoted by $x \perp y$ (J), if

$$\|x\| \leq \|x + \lambda y\| \text{ for all } \lambda \in \mathbb{C}.$$

5.7   Theorem of the element of minimal norm

Convex sets in Hilbert spaces have the important property described in the next theorem.

Theorem 5.7.1 (Theorem on the element of minimal norm) *Let E be a nonempty, closed and convex subset of a Hilbert space X. Then there is a unique vector in E with minimal norm, that is, there is one and only one element $x \in E$ such that*

$$\|x\| = \inf\{\|z\| : z \in E\}.$$

Proof. We put $d = \inf\{\|z\| : z \in E\}$.

(i)   First we prove the uniqueness of the element of minimal norm.
We assume that there exist vectors $x, y \in E$ such that $\|x\| = \|y\| = d$. Then we have by the parallelogram law (Theorem 5.5.2) for the vectors $x/2$ and $y/2$

$$\left\|\frac{x+y}{2}\right\|^2 + \left\|\frac{x-y}{2}\right\|^2 = 2\left\|\frac{x}{2}\right\|^2 + 2\left\|\frac{y}{2}\right\|^2.$$

Hence we obtain after an elementary computation

$$\|x - y\|^2 = 2\|x\|^2 + 2\|y\|^2 - 4\left\|\frac{x+y}{2}\right\|^2. \tag{5.33}$$

Since E is a convex set, we have $(x+y)/2 \in E$, and so $\|(x+y)/2\| \geq d$, and so

$$\|x - y\|^2 \leq 2\|x\|^2 + 2\|y\|^2 - 4d^2 = 0.$$

Consequently $x = y$, that is, the vector in E with minimal norm is unique, if it exists.

(ii)   Now we show the existence.
It follows from the definition of the number d that there exists a sequence (x_n) in E with $\lim_n \|x_n\| = d$. Because of the parallelogram law (Theorem 5.5.2) for the vectors $x_n/2$ and $x_m/2$ (similarly as we showed in (5.33) for the vectors $x/2$ and $y/2$, respectively) we obtain the inequality

$$\|x_n - x_m\|^2 \leq 2\|x_n\|^2 + 2\|x_m\|^2 - 4d^2. \tag{5.34}$$

It follows from (5.34) that (x_n) is a Cauchy sequence. Therefore there exists $x \in E$ with $\lim_n x_n = x$. By the continuity of the norm, we have $\|x\| = \lim_n \|x_n\| = d$.   □

Corollary 5.7.2 *Let E be a nonempty, closed and convex subset of a Hilbert space X. Then, for each $z \in X$, there exists only one $x \in E$ such that*

$$\|z - x\| = \inf\{\|z - y\| : y \in E\}.$$

Proof. We note that $z - E \equiv \{z - y : y \in E\}$ is a nonempty, closed and convex subset of the Hilbert space X. The proof now follows from Theorem 5.7.1.   □

5.8   Theorem on the orthogonal decomposition

In Euclidean space $\mathbb{R}^3$, let π be a plane and p be a line, both through the origin, and p be orthogonal to π. It is well known that each vector in $\mathbb{R}^3$ can uniquely be represented as a sum of two vectors, one from the plane π and one in the direction of the line p, that is, we have the well–known orthogonal decomposition.

The next theorem shows that all Hilbert spaces have that property.

Theorem 5.8.1 (Theorem on the orthogonal decomposition)
Let M be a closed subspace of a Hilbert space X. If $z \in X$, then there exist unique vectors $x \in M$ and $y \in M^\perp$ such that $z = x + y$. Hence

$$X = M \oplus M^\perp, \tag{5.35}$$

and this direct sum is called the orthogonal sum.

Proof.

(i)   First we show the existence of an orthogonal decomposition.

Since M is a nonempty, closed and convex subset of the Hilbert space X, it follows by Corollary 5.7.2 that there exists a unique $x \in M$ such that

$$\|z - x\| = \inf\{\|z - u\| : u \in M\}. \tag{5.36}$$

Let $y = z - x$, hence $z = x + y$. We are going to show that $y \in M^\perp$. Obviously $y \perp 0$ for $0 \in M$. Let $0 \neq s \in M$, and we show that $y \perp s$. We may assume $\|s\| = 1$. Since $y - \langle y, s \rangle s \in z - M$, (5.36) implies

$$\|y\|^2 \leq \|y - \langle y, s \rangle s\|^2 = \|y\|^2 - 2\overline{\langle y, s \rangle}\langle y, s \rangle + \langle y, s \rangle \overline{\langle y, s \rangle}\|s\|^2 = \|y\|^2 - |\langle y, s \rangle|^2. \tag{5.37}$$

It follows from (5.37) that $|\langle y, s \rangle|^2 \leq 0$, hence $y \perp s$. Thus we have $y \in M^\perp$.

(ii)   Now we prove the uniqueness of the decomposition. If $z = x_1 + y_1$, $x_1 \in M$ and $y_1 \in M^\perp$ is one more decomposition then $x - x_1 = y_1 - y \in M \cap M^\perp$. Therefore $x - x_1 = y_1 - y = 0$, that is, $x = x_1$ and $y = y_1$. $\qquad\qquad\square$

Corollary 5.8.2 *Let M be a subspace of a Hilbert space X. Then we have $M^{\perp\perp} = \overline{M}$.*

Proof. Since $M \subset M^{\perp\perp}$ and $M^{\perp\perp}$ is a closed subspace of X, it follows that $\overline{M} \subset M^{\perp\perp}$. If $z \in M^{\perp\perp}$, by Theorem 5.8.1, there exist $x \in \overline{M}$ and $y \in \overline{M}^\perp$ such that $z = x + y$. Since $\overline{M}^\perp \subset M^\perp$, it follows that $0 = \langle z, y \rangle = \langle x, y \rangle + \langle y, y \rangle = \langle y, y \rangle$. Hence $y = 0$ and $z = x \in \overline{M}$. So we proved $M^{\perp\perp} \subset \overline{M}$. $\qquad\qquad\square$

Corollary 5.8.3 *A subspace M of a Hilbert space X is dense if and only if $M^\perp = \{0\}$.*

Proof. From $M^\perp = \{0\}$, we obtain $M^{\perp\perp} = X$, and so the subspace M is dense by Corollary 5.8.2. Conversely, if M is dense, again we obtain $\overline{M} = M^{\perp\perp} = X$ by Corollary 5.8.2 . It follows from $X = M^\perp \oplus M^{\perp\perp}$ (Theorem 5.8.1) that $M^\perp = \{0\}$. $\qquad\qquad\square$

Corollary 5.8.4 *Let $\{e_i\}_{i \in I}$ be an orthonormal set in the Hilbert space X, and M be the smallest closed subspace in X containing all the vectors e_i $(i \in I)$. Then either $M = X$ or there exists $e \in X$ with $\|e\| = 1$ and $e \perp e_i$ for all $i \in I$.*

Proof. If $M \neq X$, then Theorem 5.8.1 implies $M^\perp \neq \{0\}$. Hence, there exists $e \in M^\perp$ such that $\|e\| = 1$. $\qquad\qquad\square$

We note that if $M \neq X$ in Corollary 5.8.4, then the set $\{e_i\}_{i \in I}$ is not complete, and if we add the element e then the new set is orthonormal.

5.9   Riesz representation theorem

Let y be an element of the Hilbert space X. Then the functional f_y, defined by $f_y(x) = \langle x, y \rangle$ for all $x \in X$, is linear and continuous. The next theorem asserts that each bounded linear functional on X is of that form.

Theorem 5.9.1 (Riesz representation theorem) *If f is a continuous linear functional on a Hilbert space X then there exists a unique vector $y \in X$ such that*

$$f(x) = \langle x, y \rangle \text{ for all } x \in X. \tag{5.38}$$

In this case, we have $\|f\| = \|y\|$.

    Proof.

(i)   First we show the existence of an element $y \in X$ which satisfies (5.38).
Since $f \in X^*$, the null space $N(f)$ of f, is a closed subspace of X.
If $N(f) = X$, then $f \equiv 0$ and we can choose $y = 0$ (obviously the identity in (5.38) is satisfied).
If $N(f) \neq X$, then Corollary 5.8.3 implies $N(f)^\perp \neq \{0\}$. Now we choose $z \in N(f)^\perp$, $z \neq 0$ and $x \in X$. For $v = f(x)z - f(z)x$, we see that $f(v) = f(x)f(z) - f(z)f(x) = 0$, that is, $v \in N(f)$. Since $z \in N(f)^\perp$, it follows

$$0 = \langle v, z \rangle = \langle f(x)z - f(z)x, z \rangle = f(x)\langle z, z \rangle - f(z)\langle x, z \rangle,$$

hence

$$f(x) = \frac{f(z)}{\langle z, z \rangle}\langle x, y \rangle = \left\langle x, \frac{\overline{f(z)}}{\|z\|^2}z \right\rangle = \langle x, y \rangle,$$

where $y = (\overline{f(z)}/\|z\|^2)z$.
Thus we have shown that y satisfies (5.38).

(ii)   Now we show the uniqueness of the element $y \in X$ which satisfies (5.38).
We assume that there exists $w \in X$ such that $f(x) = \langle x, w \rangle$ for all $x \in X$. Then we have $f(w - y) = \langle w - y, w \rangle = \langle w - y, y \rangle$, and $w = y$.
The Cauchy-Schwarz inequality implies

$$|f(x)| = |\langle x, y \rangle| \leq \|x\|\|y\| \text{ for all } x \in X,$$

so we have $\|f\| \leq \|y\|$. Now $\|y\|^2 = f(y) \leq \|f\|\|y\|$ implies $\|y\| \leq \|f\|$, and we have shown that $\|f\| = \|y\|$. $\qquad\square$

Remark 5.9.2 *Let X be a Hilbert space and $A : X^* \to X$ be the map defined by $A(f) = y$, where $f(x) = \langle x, y \rangle$ for all $f \in X^*$. The map A is one to one and onto. Furthermore, if $f_1(x) = \langle x, x_1 \rangle$ and $f_2(x) = \langle x, x_2 \rangle$, $f_1, f_2 \in X^*$ for all $x \in X$, then we have for all $\alpha, \beta \in \mathbb{C}$*

$$(\alpha f_1 + \beta f_2)(x) = \alpha f_1(x) + \beta f_2(x) = \alpha\langle x, x_1 \rangle + \beta\langle x, x_2 \rangle$$
$$= \langle x, \overline{\alpha}x_1 + \overline{\beta}x_2 \rangle. \tag{5.39}$$

Hence, (5.39) implies

$$A(\alpha f_1 + \beta f_2) = \overline{\alpha}Af_1 + \overline{\beta}Af_2,$$

and A is a conjugate–linear operator. Theorem 5.9.1 implies that A is an isometry. We put

$$\langle f_1, f_2 \rangle_1 = \langle Af_2, Af_1 \rangle. \tag{5.40}$$

It is is easy to show that $\langle \cdot, \cdot \rangle_1$ is an inner product on X^*. For instance,

$$\langle \alpha f_1, f_2 \rangle_1 = \langle Af_2, A\alpha f_1 \rangle = \langle Af_2, \overline{\alpha} Af_1 \rangle = \alpha \langle Af_2, Af_1 \rangle = \alpha \langle f_1, f_2 \rangle.$$

The norm defined by this inner product (5.40) is equal to the norm on the dual space X^* by the Riesz representation theorem. Hence $(X^*, \langle \cdot, \cdot \rangle_1)$ is a Hilbert space.

Theorem 5.9.3 *Every Hilbert space X is reflexive.*

Proof. We are going to show that the canonical map $J : X \to X^{**}$ is onto, that is, for every $g \in X^{**}$ there exists $x \in X$ such that $Jx = g$.
Let $A : X^* \to X$ be the conjugate linear operator defined in Remark 5.9.2. For $f_1, f_2 \in X^*$, we define

$$\langle f_1, f_2 \rangle_1 = \langle Af_2, Af_1 \rangle.$$

It follows from Remark 5.9.2 that X^* is a Hilbert space with the scalar product $\langle \cdot, \cdot \rangle_1$. We know by Theorem 5.9.1 that there exists a functional $f_0 \in X^*$ such that

$$g(f) = \langle f, f_0 \rangle_1 \text{ for each } f \in X^*.$$

Let $Af_0 = x_0$. We are going to show that $Jx_0 = g$. We have for $f \in X^*$

$$(Jx_0)f = f(x_0) = \langle x_0, Af \rangle = \langle Af_0, Af \rangle = \langle f, f_0 \rangle_1 = g(f). \qquad \square$$

5.10 Bessel's inequality, Fourier coefficients

The first important result in this section is Bessel's inequality in the next theorem, which implies that any vector in an inner product space can have at most a countable number of nonzero Fourier coefficients with respect to a given orthonormal set basis.

Theorem 5.10.1 (Bessel's inequality) *Let $\{e_n : n \in \mathbb{N}\}$ be an orthonormal set in an inner product space X. Then we have*

$$\sum_{k=1}^{\infty} |\langle x, e_k \rangle|^2 \leq \|x\|^2 \text{ for all } x \in X. \tag{5.41}$$

Proof. For each $n \in \mathbb{N}$, we put $\alpha_n = \langle x, e_n \rangle x$. Then we have for all $x \in X$

$$0 \leq \left\| x - \sum_{k=1}^{n} \alpha_k e_k \right\|^2 = \|x\|^2 - \sum_{k=1}^{n} \langle x, \alpha_k e_k \rangle - \sum_{k=1}^{n} \langle \alpha_k e_k, x \rangle + \sum_{k,j=1}^{n} \langle \alpha_k e_k, \alpha_j e_j \rangle$$

$$= \|x\|^2 - \sum_{k=1}^{n} |\alpha_k|^2. \qquad \square$$

Theorem 5.10.2 *Let $\mathcal{E}$ be an orthonormal set in an inner product space X and $x \in X$. Then $\langle x, e \rangle \neq 0$ for at most countably many vectors $e \in \mathcal{E}$.*

Proof. For each $n \in \mathbb{N}$, let $\mathcal{E}_n = \{e \in \mathcal{E} : |\langle x, e \rangle| > \|x\|/n\}$. By Theorem 5.10.1, the set $\mathcal{E}_n$ is finite. We note $\cup_{n=1}^{\infty} \mathcal{E}_n = \{e \in \mathcal{E} : \langle x, e \rangle \neq 0\}$. $\square$

Definition 5.10.3 Let $\{e_i\}_{i \in I}$ be an orthonormal set in an inner product space X and $x \in X$. The scalars $\alpha_i = \langle x, e_i \rangle$ for $i \in I$ are called *Fourier coefficients* of the element x with respect to the orthonormal set $\{e_i\}_{i \in I}$.

The inequality in (5.42) of the next theorem generalizes Bessel's inequality (5.41) in Theorem 5.10.1.

Theorem 5.10.4 *Let $\{e_i\}_{i \in I}$ be an orthonormal set in an inner product space X. If $\alpha_i = \langle x, e_i \rangle$ and $\beta_i = \langle y, e_i \rangle$ are the Fourier coefficients of the elements $x, y \in X$ with respect to the orthonormal set $\{e_i\}_{i \in I}$, then the families $\{|\alpha_i|^2\}_{i \in I}$, $\{|\alpha_i \overline{\beta}_i|\}_{i \in I}$ and $\{\alpha_i \overline{\beta}_i\}_{i \in I}$ are summable and the following inequalities hold*

$$\sum_{i \in I} |\langle x, e_i \rangle|^2 \leq \|x\|^2 \text{ (Bessel's inequality)}, \tag{5.42}$$

$$\left| \sum_{i \in I} \alpha_i \overline{\beta}_i \right| \leq \sum_{i \in I} |\alpha_i \overline{\beta}_i| \leq \|x\|^2 \|y\|^2. \tag{5.43}$$

Proof. The summability of the families follow by Theorems 4.11.7, 5.10.1 and 5.10.2. The inequalities (5.42) and (5.43) follow by Theorems 4.11.4, 5.10.1 and 5.10.2. $\square$

Theorem 5.10.5 *Let $\{e_i\}_{i \in I}$ be an orthonormal set in the inner product space X. If $\alpha_i = \langle x, e_i \rangle$ are the Fourier coefficients of the element $x \in X$ with respect to the orthonormal set $\{e_i\}_{i \in I}$, then we have*

$$\inf_{\lambda_i \in \mathbb{C}, i \in I} \left\| x - \sum_{i \in I} \lambda_i e_i \right\| = \left\| x - \sum_{i \in I} \alpha_i e_i \right\|,$$

where the infimum is taken over all summable families $\{\lambda_i e_i\}_{i \in I}$.

Proof. Let $\{\lambda_i e_i\}_{i \in I}$ be a summable family. Then we have

$$\left\| x - \sum_{i \in I} \lambda_i e_i \right\|^2 = \left\langle x - \sum_{i \in I} \lambda_i e_i, x - \sum_{j \in I} \lambda_j e_j \right\rangle$$

$$= \langle x, x \rangle - \sum_{i \in I} \lambda_i \langle e_i, x \rangle - \sum_{j \in I} \overline{\lambda}_j \langle x, e_j \rangle + \sum_{i,j \in I} \lambda_i \overline{\lambda}_j \langle e_i, e_j \rangle$$

$$= \|x\|^2 + \sum_{i \in I} |\lambda_i - \langle x, e_i \rangle|^2 - \sum_{i \in I} |\langle x, e_i \rangle|^2.   \square$$

Theorem 5.10.6 *Let $\{e_i\}_{i \in I}$ be an orthonormal set in the Hilbert space X. If $\alpha_i = (x, e_i)$ are the Fourier coefficients of the element $x \in X$ with respect to $\{e_i\}_{i \in I}$, then the family of vectors $\{\alpha_i e_i\}_{i \in I}$ is summable. If z denotes their sum then $(x - z) \perp e_i$ for all $i \in I$.*

Proof. It follows from Theorems 5.6.7 and 5.10.4 that the families of vectors $\{\alpha_i e_i\}_{i \in I}$ are summable. Lemma 5.6.6 implies $\langle z, e_i \rangle = \alpha_i$ for all $i \in I$. $\square$

5.11 Parseval's equality, Hilbert basis

The following theorem shows the importance of the completeness of orthonormal sets and Fourier coefficients in Hilbert spaces.

Theorem 5.11.1 *Let $\{e_i\}_{i \in I}$ be an orthonormal set in the Hilbert space X. The following conditions are equivalent:*

> *The set $\{e_i\}_{i \in I}$ is complete.* (5.44)
>
> *If $x \in X$ and $\langle x, e_i \rangle = 0$ for all $i \in I$, then $x = 0$.* (5.45)
>
> *X is the smallest closed subspace of X that contains the set $\{e_i\}_{i \in I}$.* (5.46)
>
> *Let $x \in X$ and $\alpha_i = \langle x, e_i \rangle$ for all $i \in I$. Then we have*
>
> $$x = \sum_{i \in I} \alpha_i e_i. \text{ (Fourier representation)}.$$ (5.47)
>
> *If $x, y \in X$, $\alpha_i = \langle x, e_i \rangle$ and $\beta_i = \langle y, e_i \rangle$ for all $i \subset I$, then we have*
>
> $$\langle x, y \rangle = \sum_{i \in I} \alpha_i \overline{\beta}_i. \text{ (Parseval's equality)}.$$ (5.48)
>
> *If $x \in X$ and $\alpha_i = \langle x, e_i \rangle$ for all $i \in I$, then we have*
>
> $$\|x\|^2 = \sum_{i \in I} |\alpha_i|^2. \text{ (Parseval's equality)}$$ (5.49)

Proof. We are going to prove the following implications:

$$(5.44) \text{ implies } (5.45) \text{ implies } (5.46) \text{ implies } (5.47)$$
$$\text{implies } (5.48) \text{ implies } (5.49) \text{ implies } (5.44).$$

(i) First we show that (5.44) implies (5.45).
   If $x \in X$ and, for each $i \in I$, we have $\langle x, e_i \rangle = 0$, then $x \neq 0$ implies that $\{e_i\}_{i \in I} \cup \{x/\|x\|\}$ is an orthonormal set. This in a contradiction to (5.44), and so $x = 0$.

(ii) Now we show that (5.45) implies (5.46).
   This implication was proved in Corollary 5.8.4.

(iii) Now we show that (5.46) implies (5.47).
   It follows from Theorem 5.10.6 that, for $z = \sum_{i \in I} \alpha_i e_i$, we have $(x - z) \perp e_i$ for all $i \in I$. Hence $x - z$ is orthogonal to the smallest closed subspace of X which contains the set $\{e_i\}_{i \in I}$. By assumption this subspace is X, and so $\langle x - z, x - z \rangle = 0$, that is, $x = z$.

(iv) Now we show that (5.47) implies (5.48).
   This follows from Theorem 5.6.7, its proof and Lemma 5.6.6.

(v) Now we show that (5.48) implies (5.49).
   This follows from (5.48), if we replace y by x.

(vi) Finally, we show that (5.49) implies (5.44).
   If the set $\{e_i\}_{i \in I}$ were not complete, then there would exist a vector $e \in X$ such that $\{e_i\}_{i \in I} \cup \{e\}$ is an orthonormal set. Now it follows from (5.49) that $\|e\|^2 = \sum_{i \in I} |\langle e, e_i \rangle|^2 = 0$. This is a contradiction to $\|e\| = 1$. $\qquad \square$

In the following definition, we introduce one more concept of a basis, namely that of an *orthonormal Hilbert basis.*

Definition 5.11.2 (Hilbert basis) Let $\{e_i\}_{i \in I}$ be an orthonormal set in the Hilbert space X. If X is a minimal closed subspace of X which contains the set $\{e_i\}_{i \in I}$, then the set $\{e_i\}_{i \in I}$ is called the *Hilbert (orthonormal) basis of the space* X.

It follows from Theorems 5.6.4 and 5.11.1 that every Hilbert space has an orthonormal basis. It is well known that all algebraic (Hamel) bases of a linear space have the same cardinality. An analogous result holds true for Hilbert bases; this is stated in the next theorem.

Theorem 5.11.3 *If* $\{e_i\}_{i \in I}$ *and* $\{g_j\}_{j \in J}$ *are orthogonal bases of the Hilbert space* X, *then the cardinality of* I *is equal to that of* J, *that is,*

$$\mathrm{card}I = \mathrm{card}J.$$

Proof. If at least one of the sets I and J is finite, then the result follows from the theory of linear algebra. We assume that $\mathrm{card}I \geq \aleph_0$ and $\mathrm{card}J \geq \aleph_0$. We put $J_i = \{j \in J : \langle e_i, g_j \rangle \neq 0\}$ for all $i \in I$. We know from Theorem 5.10.2 that $\mathrm{card}J_i \leq \aleph_0$. From (5.47) it follows $g_j = \sum_{i \in I} \langle g_j, e_i \rangle e_i$, and so $\langle e_i, g_j \rangle \neq 0$ for some $i \in I$. Hence we have proved $J = \cup_{i \in I} J_i$, and so $\mathrm{card}J \leq \sum_{i \in I} \mathrm{card}J_i \leq \sum_{i \in I} \aleph_0 = \mathrm{card}I$. Analogously it can be shown that $\mathrm{card}I \leq \mathrm{card}J$. $\qquad\square$

In view of Theorem 5.11.3, we can define the *orthogonal Hilbert dimension* of a Hilbert space, similarly as the algebraic dimension of a linear space.

Definition 5.11.4 (Orthogonal (Hilbert) dimension)
The *orthogonal (Hilbert) dimension* of a Hilbert space X is the cardinality of any orthonormal basis of X.

In Hilbert space, if not specifically mentioned, we always mean *Hilbert dimension* by *dimension.* Since every Hilbert space is a linear space, it has an algebraic dimension. If at least one of these dimensions is finite, then so is the other dimension and the two are equal. Since each orthonormal set is linearly independent, the orthogonal dimension is not greater than the algebraic dimension. The next theorem shows that we can have $<$ in the mentioned inequality.

Theorem 5.11.5 *If the orthogonal dimension of the Hilbert space* X *is infinite then the algebraic dimension of* X *is not less than* $2^{\aleph_0}$.

Proof. First we prove that there exists a family $\{N_t\}_{t \in \mathbb{R}}$ of infinite subsets of $\mathbb{N}$ such that $N_t \cap N_s$ is a finite set for all $t \neq s$. Since there exists a bijection between the set $\mathbb{N}$ and the set of rational numbers, it is sufficient to prove that there exists a set of rational numbers with the mentioned property. In this case, we take N_t for each real number t as a set of rational numbers which has only one accumulation point and this point is t.
We assume that $\{e_n\}_{n \in \mathbb{N}}$ is an orthonormal set in X, and $x \in X$ with $x = \sum_{n \in \mathbb{N}} \alpha_n e_n$, where $\alpha_n = \langle x, e_n \rangle \neq 0$ for each $n \in \mathbb{N}$. Let $x_t = \sum_{n \in N_t} \alpha_n e_n$, where N_t is the set defined above. We prove that the set $\{x_t\}_{t \in \mathbb{R}}$ is linearly independent. If $\lambda_1 x_{t_1} + \cdots + \lambda_k x_{t_k} = 0$, where $\lambda_i \in \mathbb{C}$ for $i = 1, 2 \ldots, k$, then it follows from the definition of the sets N_t that there exists at least one element, $n \in N_{t_1}$ say, such that $n \notin N_{t_i}$ for $i = 2, \ldots, k$. Hence $\langle \lambda_1 x_{t_1} + \cdots + \lambda_k x_{t_k}, e_n \rangle = \lambda_1 \alpha_n = 0$. Since $\alpha_n \neq 0$, it follows that $\lambda_1 = 0$. Analogously we can prove $\lambda_i = 0$ for all $i = 2, \ldots k$. Hence the set $\{x_t\}_{t \in \mathbb{R}}$ is linearly independent. $\qquad\square$

It is well known that the linear spaces X and Y over the same field of scalars are isomorphic if there exists a linear bijection between them. We recall that, by Theorem 1.6.8, linear spaces over the same field of scalars are isomorphic if they have the same algebraic dimension. We are going to show that a similar result holds true for Hilbert spaces.

Definition 5.11.6 The inner product spaces $(X, \langle \cdot, \cdot \rangle_1)$ and $(Y, \langle \cdot, \cdot \rangle_2)$ are said to be *isomorphic* if there exists an isomorphism A between the linear spaces X and Y such that

$$\langle Ax, Ay \rangle_2 = \langle x, y \rangle_1, \text{ for all } x, y \in X.$$

Theorem 5.11.7 *Hilbert spaces are isomorphic if and only if they have the same orthogonal dimension.*

Proof. First we assume that the Hilbert spaces X and Y have the same dimension. Let $\{e_i\}_{i \in I}$ and $\{g_i\}_{i \in I}$ be orthonormal bases for X and Y, respectively. We put $Ax = \sum_{i \in I} \langle x, e_i \rangle g_i$ for $x \in X$. It is easy to show that A is a linear operator from X into Y. Since $x = \sum_{i \in I} \langle x, e_i \rangle e_i$ by Theorem 5.11.1, it follows from Theorems 5.6.7 and 5.11.1 that

$$\|Ax\|^2 = \sum_{i \in I} |\langle x, e_i \rangle|^2 = \|x\|^2. \tag{5.50}$$

Now (5.50) implies that A is an isometry, and so $A(X)$ is a closed subspace of Y. It follows from $\{g_i\}_{i \in I} \subset A(X)$, that $A(X) = Y$. From (5.19) and the fact that $\|Ax\|^2 = \|x\|^2$ for each $x \in X$, it follows that $\langle Ax, Ay \rangle = \langle x, y \rangle$ for all $x, y \in X$. Hence X and Y are isomorphic Hilbert spaces.

It is easy to prove the converse implication and we leave the proof to the reader. $\square$

The next theorem is similar to the well–known result for linear spaces.

Theorem 5.11.8 *For every cardinal number k, there exists a Hilbert space X such that its orthogonal dimension is equal to k.*

Proof. Let S be a nonempty set, $\text{card} S = k$ and $\ell_2(S)$ be the inner product space of Example 5.4.5. We show that $\ell_2(S)$ is a Hilbert space. Let (f_n) be a Cauchy sequence in $\ell_2(S)$. We have for each $s \in S$

$$|f_n(s) - f_m(s)|^2 \leq \sum_{s \in S} |f_n(s) - f_m(s)|^2 = \|f_n - f_m\|^2,$$

hence $(f_n(s))_{n=1}^\infty$ is a Cauchy sequence of scalars. So, for each $s \in S$, there exists a scalar $f(s)$ such that $\lim_{n \to \infty} f_n(s) = f(s)$. In this case, we define $f \in \prod_{s \in S} C_s$, and show that $f \in \ell_2(S)$ and $f_n \to f$ in $\ell_2(S)$.

Let I be a finite subset of S and $\epsilon > 0$. There exists a natural number n_0 such that

$$n, m > n_0 \text{ implies } \sum_{s \in I} |f_n(s) - f_m(s)|^2 \leq \sum_{s \in S} |f_n(s) - f_m(s)|^2 < \epsilon^2. \tag{5.51}$$

If we let $n \to \infty$ in (5.51) then we obtain

$$\sum_{s \in I} |f(s) - f_m(s)|^2 \leq \epsilon^2 \text{ for all } m > n_0. \tag{5.52}$$

Since I is an arbitrary finite subset of S, (5.52) implies

$$\sum_{s \in S} |f(s) - f_m(s)|^2 \leq \epsilon^2 \text{ for all } m > n_0, \tag{5.53}$$

that is, we have $f - f_m \in \ell_2(S)$ for each $m > n_0$. Now $f = (f - f_m) + f_m$ implies $f \in \ell_2(S)$. Again, it follows from (5.53) that $\lim_{n \to \infty} f_n = f$. Thus we have proved that $\ell_2(S)$ is a Hilbert space.

It is easy to prove that the family of functions $\{e_s\}_{s \in S}$ defined by $e_s(t) = 1$ for $t = s$ and $e_s(t) = 0$ for $t \neq s$, is a Hilbert basis for $\ell_2(S)$. $\square$

5.12   Gram–Schmidt orthogonalization method

Let X be an inner product space, and (e_n) be an orthonormal sequence in X and $x \in \mathrm{span}(\{e_1, \dots, e_n\})$. Then we have

$$x = \sum_{k=1}^{n} \lambda_k e_k, \tag{5.54}$$

where λ_k are scalars for $k = 1, \dots, n$. It follows from (5.54) that

$$\langle x, e_j \rangle = \left\langle \sum_{k=1}^{n} \lambda_k e_k, e_j \right\rangle = \sum_{k=1}^{n} \lambda_k \langle e_k, e_j \rangle = \lambda_j, \tag{5.55}$$

and so

$$x = \sum_{k=1}^{n} \langle x, e_k \rangle e_k. \tag{5.56}$$

We have for $x + \lambda_{n+1} e_{n+1} \in \mathrm{span}(\{e_1, \dots, e_{n+1}\})$

$$x + \lambda_{n+1} e_{n+1} = \sum_{k=1}^{n+1} \langle x, e_k \rangle e_k. \tag{5.57}$$

The formulas (5.56) and (5.57) show that the orthonormal sequence (e_n) has advantages over other linear independent sequences in X. These advantages are that we can define the scalars λ_i in (5.56) and for this, in (5.57), it is necessary to compute only one scalar $\langle x, e_{n+1} \rangle$ more and the other scalars are the same as in (5.56). The problem is how to get an orthonormal sequence from a linearly independent sequence. The next theorem will show us how this can be achieved.

Theorem 5.12.1 (Gram-Schmidt) *If (v_n) is a linearly independent sequence of vectors in an inner product space X, then there exists an orthonormal sequence (e_n) in X such that*

$$\mathrm{span}(\{e_1, \dots, e_n\}) = \mathrm{span}(\{v_1, \dots, v_n\}) \text{ for each } n \in \mathbb{N}.$$

Proof. The first term of the sequence (e_n) is

$$e_1 = \frac{1}{\|v_1\|} v_1.$$

The vector

$$x_2 = v_2 - \langle v_2, e_1 \rangle e_1$$

is unequal to 0 and $x_2 \perp e_1$. We put

$$e_2 = \frac{1}{\|x_2\|} x_2$$

and

$$x_3 = v_3 - \langle v_3, e_1 \rangle e_1 - \langle v_3, e_2 \rangle e_2.$$

Then $x_3 \neq 0$, $x_3 \perp e_1$ and $x_3 \perp e_2$. We put

$$e_3 = \frac{1}{\|x_3\|} x_3.$$

In the n–th step we consider the vector

$$x_n = v_n - \sum_{k=1}^{n-1} \langle v_n, e_k \rangle e_k.$$

Since $x_n \neq 0$ (why?) and x_n are orthogonal to $e_1, \ldots, e_{n-1}$, we define

$$e_n = \frac{1}{\|x_n\|} x_n.$$

It is easy to prove

$$\operatorname{span}(\{e_1, \ldots, e_n\}) = \operatorname{span}(\{v_1, \ldots, v_n\}) \qquad \square$$

The method to construct an orthogonal sequence as in Theorem 5.12.1 is known as the *Gram–Schmidt orthogonalization method*. We mention that we obtain the orthonormal sequence (e_n).

The statements of following examples are obtained by applying the Gram–Schmidt orthogonalization process. For the proofs, we refer the interested reader to any textbook on orthogonal polynomials listed at the end of this section.

Example 5.12.2 *If the sequence $1, x, x^2, \ldots$ is orthogonalized by the Gram–Schmidt process in $L_2(-1, 1)$, we obtain the orthonormal sequence of polynomials*

$$e_n(x) = \sqrt{\frac{(2n+1)}{2}} P_n(x)$$

with

$$P_n(x) = \frac{1}{2^n n!} \left(\frac{d}{dx} \right)^n [(x^2 - 1)^n].$$

The functions $P_n(x)$ are called Legendre polynomials.

Example 5.12.3 *If the sequence $e^{-x^2/2}$, $xe^{-x^2/2}$, $x^2 e^{-x^2/2}, \ldots$ is orthogonalized by the Gram–Schmidt process in $L_2(-\infty, \infty)$, we obtain the sequence*

$$e_n(x) = [2^n n! \sqrt{\pi}]^{-1/2} H_n(x) e^{-x^2/2},$$

with

$$H_n(x) = (-1)^n e^{x^2} \left(\frac{d}{dx} \right)^n (e^{-x^2}).$$

The functions $e_n(x)$ are called Hermite functions, *and the polynomials $H_n(x)$ are called* Hermite polynomials.

Example 5.12.4 *If the sequence $e^{-x/2}$, $xe^{-x/2}$, $x^2e^{-x/2}, \ldots$ is orthogonalized by the Gram–Schmidt process in $L_2(0, \infty)$, we obtain the sequence $e_n(x) = e^{-x/2}L_n(x)/n!$, where*

$$L_n(x) = e^x \left(\frac{d}{dx} \right)^n (x^n e^{-x}).$$

The functions $e_n(x)$ are called Laguerre functions, *and the polynomials $L_n(x)$ are called* Laguerre polynomials.

The theories of orthogonal polynomials, orthonormal and Fourier series are important fields in mathematics. The interested readers are referred to the monographs [5, 70, 85, 98, 107, 202, 221].

5.13 Hilbert adjoint operators

Each bounded linear operator on a complex Hilbert space has a corresponding *adjoint operator*. Hilbert adjoint operators generalize conjugate transposes of square matrices, and play the roles of the complex conjugates of complex numbers. In this section, we study some of the basic properties of Hilbert adjoint operators.

Theorem 5.13.1 *Let X and Y be Hilbert spaces and $A \in \mathcal{B}(X, Y)$. Then there exists exactly one operator $B \in \mathcal{B}(Y, X)$ such that*

$$\langle Ax, y \rangle = \langle x, By \rangle \text{ for each } x \in X \text{ and each } y \in Y. \tag{5.58}$$

Proof.

(i) First we show the existence of an operator $B \in \mathcal{B}(X, Y)$ which satisfies (5.58). Let $y \in Y$ and $f : X \to \mathbb{C}$ be the map defined by

$$f(x) = \langle Ax, y \rangle \text{ for all } x \in X.$$

Then $f \in X^*$ and, by the Riesz representation theorem, there exists a unique vector $z \in X$ such that

$$f(x) = \langle x, z \rangle \text{ for all } x \in X.$$

Let $B : Y \to X$ be the map defined by $B(y) = z$. Then B is a linear operator that satisfies the condition in (5.58). It follows from

$$\|By\|^2 = \langle By, By \rangle = \langle ABy, y \rangle \le \|A\| \, \|By\| \, \|y\| \text{ for all } y \in Y$$

that $\|B\| \le \|A\|$ and $B \in \mathcal{B}(Y, X)$.

(ii) Finally, we show that the operator which satisfies the condition in (5.58) is unique. If $B_1 \in \mathcal{B}(Y, X)$ satisfies the condition in (5.58), then

$$\langle x, By - B_1y \rangle = 0 \text{ for each } x \in X \text{ and each } y \in Y. \tag{5.59}$$

It follows from (5.59) for $x = By - B_1y$ that

$$\langle By - B_1y, By - B_1y \rangle = 0 \text{ for all } y \in Y,$$

and so

$$By = B_1y \text{ for all } y \in Y. \qquad \square$$

Definition 5.13.2 Let X and Y be Hilbert spaces and $A \in \mathcal{B}(X,Y)$. The operator $B \in \mathcal{B}(Y,X)$, defined in (5.58) and denoted by A^*, is called the *Hilbert adjoint operator of the operator A.*

The next theorem gives some basic properties of Hilbert adjoint operators.

Theorem 5.13.3 *Let X, Y and Z be Hilbert spaces, $A, B \in \mathcal{B}(X,Y)$, $C \in \mathcal{B}(Y,Z)$ and $\lambda \in \mathbb{C}$. Then we have*

$$\langle A^* y, x \rangle = \langle y, Ax \rangle \text{ for each } x \in X \text{ and each } y \in Y, \tag{5.60}$$

$$(A + B)^* = A^* + B^*, \tag{5.61}$$

$$(\lambda A)^* = \bar{\lambda} A^*, \tag{5.62}$$

$$(A^*)^* = A, \tag{5.63}$$

$$\|A\| = \|A^*\|, \tag{5.64}$$

$$\|A^* A\| = \|AA^*\| = \|A\|^2, \tag{5.65}$$

$$A^* A = O \text{ if and only if } A = O, \tag{5.66}$$

$$(CB)^* = B^* C^*, \tag{5.67}$$

$$O^* = O \text{ and } I^* = I. \tag{5.68}$$

Proof.

(i) The statement in (5.60) follows from $\langle A^* y, x \rangle = \overline{\langle x, A^* y \rangle} = \overline{\langle Ax, y \rangle} = \langle y, Ax \rangle$.

(ii) To prove (5.61), we remark that we have for each $x \in X$ and each $y \in Y$

$$\langle x, (A+B)^* y \rangle = \langle (A+B)x, y \rangle = \langle Ax, y \rangle + \langle Bx, y \rangle$$
$$= \langle x, A^* y \rangle + \langle x, B^* y \rangle = \langle x, (A^* + B^*)y \rangle. \tag{5.69}$$

It follows from (5.69) and Theorem 5.13.1 that (5.61) holds.
Analogously, it follows from

$$\langle (\lambda A)x, y \rangle = \langle \lambda Ax, y \rangle = \lambda \langle Ax, y \rangle = \lambda \langle x, A^* y \rangle = \langle x, \bar{\lambda} A^* y \rangle$$

and Theorem 5.13.1 that (5.62) holds.

(iii) To prove (5.63), we observe that we obtain for each $x \in X$ and each $y \in Y$ using (5.60)
$$\langle (A^*)^* x, y \rangle = \langle x, A^* y \rangle = \langle Ax, y \rangle. \tag{5.70}$$

It follows from (5.70) and Lemma 5.2.3 that (5.63) holds and we obtain $A^{**} = (A^*)^*$.

(iv) Now we show (5.64).
We know $\|A^{**}\| \le \|A^*\| \le \|A\|$ from the proof of Theorem 5.13.1. Now, obviously (5.63) implies (5.64).

(v)   Now we show (5.65).
Using the Cauchy–Schwarz inequality and (5.60) we obtain

$$\|Ax\|^2 = \langle Ax, Ax \rangle = \langle A^*Ax, x \rangle \leq \|A^*Ax\| \, \|x\| \leq \|A^*A\| \, \|x\|^2,$$

that is, $\|A\|^2 \leq \|A^*A\|$. It follows from (5.64) that

$$\|A\|^2 \leq \|A^*A\| \leq \|A^*\| \, \|A\| = \|A\|^2,$$

and so

$$\|A^*A\| = \|A\|^2. \tag{5.71}$$

We have from (5.71) and (5.63)

$$\|AA^*\| = \|A^{**}A^*\| = \|A^*\|^2. \tag{5.72}$$

Hence, (5.71), (5.72) and (5.64) imply (5.65).

(vi)   Obviously (5.65) implies (5.66).

(vii)   Now we prove (5.67).
We observe that we obtain from the definition of the Hilbert adjoint operator for all $x \in X$ and $z \in Z$,

$$\langle x, (CB)^*z \rangle = \langle (CB)x, z \rangle = \langle Bx, C^*z \rangle = \langle x, B^*C^*z \rangle. \tag{5.73}$$

Hence, it follows from Lemma 5.2.3 and (5.73) that $(CB)^*z = B^*C^*z$ (why?).
Thus we have proved (5.67).

(viii)   We leave the proof of (5.68) to the reader.   □

Theorem 5.13.4 *Let X and Y be Hilbert spaces and $A \in \mathcal{B}(X,Y)$. If $A^{-1} \in \mathcal{B}(Y,X)$, then $(A^*)^{-1} \in \mathcal{B}(X,Y)$ exists and*

$$(A^*)^{-1} = (A^{-1})^*.$$

Proof. We obtain from (5.67) and (5.68)

$$A^*(A^{-1})^* = (A^{-1}A)^* = I = (AA^{-1})^* = (A^{-1})^*A^*. \square$$

Theorem 5.13.5 *Let X and Y be Hilbert spaces and $A \in \mathcal{B}(X,Y)$. Then we have*

$$N(A) = R(A^*)^{\perp}, \tag{5.74}$$

$$N(A^*) = R(A)^{\perp}, \tag{5.75}$$

$$X = N(A) \oplus \overline{R(A^*)}, \tag{5.76}$$

$$Y = N(A^*) \oplus \overline{R(A)}, \tag{5.77}$$

$$N(A) = N(A^*A), \tag{5.78}$$

$$N(A^*) = N(AA^*), \tag{5.79}$$

$$\overline{R(A)} = \overline{R(AA^*)}, \tag{5.80}$$

$$\overline{R(A^*)} = \overline{R(A^*A)}. \tag{5.81}$$

Proof.

(i) First we show the identity in (5.74).
If $x \in N(A)$, then $0 = \langle Ax, y \rangle = \langle x, A^*y \rangle$ for each $y \in Y$, and so $x \in R(A^*)^\perp$. Hence we have shown that $N(A) \subset R(A^*)^\perp$. On the other hand, if $x \in R(A^*)^\perp$, then $\langle Ax, y \rangle = \langle x, A^*y \rangle = 0$ for all $y \in Y$ and $Ax = 0$. Hence we have also shown that $R(A^*)^\perp \subset N(A)$, and so (5.74) holds.

(ii) The identity in (5.75) follows from that in (5.74) and (5.63).

(iii) The proofs of the identities in (5.76) and (5.77) are left to the reader.

(iv) Now we prove the identity in (5.78). We note that if $x \in N(A^*A)$, then $\langle Ax, Ax \rangle = \langle x, A^*Ax \rangle = 0$, and so $x \in N(A)$. Thus we have shown that $N(A) \supset N(A^*A)$. Since the converse inclusion is obvious, we have proved (5.78).

(v) The identity in (5.79) follows from that in (5.78) and (5.63).

(vii) Now we show the identity in (5.80).
Obviously we have $\overline{R(A)} \supset \overline{R(AA^*)}$.
To prove the converse inclusion, we assume $x \in R(A)$. By Theorem 5.8.1, there exist $y \in \overline{R(AA^*)}$ and $z \in \overline{R(AA^*)}^\perp$ such that $x = y + z$. It follows from (5.75) and (5.79) that $0 = \langle x, z \rangle = \langle y, z \rangle + \langle z, z \rangle = \langle z, z \rangle$. Hence $z = 0$ and it follows that $x = y \in \overline{R(AA^*)}$. So $R(A) \subset \overline{R(AA^*)}$ and $\overline{R(A)} \subset \overline{R(AA^*)}$.
This completes the proof of (5.80).

(viii) Finally, the identity in (5.81) follows from that in (5.80) and from (5.63). $\square$

5.14 Hermitian, normal, positive and unitary operators

In this section, we study some fundamental properties of important operators on Hilbert spaces, namely of Hermitian, normal, positive and unitary operators.

Definition 5.14.1 Let X be a Hilbert space and $A \in \mathcal{B}(X)$. Then A is

$$\textit{normal, if } AA^* = A^*A;$$

$$\textit{self–adjoint (Hermitian), if } A = A^*;$$

$$\textit{unitary, if } A^*A = AA^* = I.$$

Theorem 5.14.2 *The operator* $A \in \mathcal{B}(X)$ *is self–adjoint if and only if* $\langle Ax, x \rangle$ *is a real number for each* $x \in X$.

Proof. Let A be self–adjoint and $x \in X$. Then we have

$$\langle Ax, x \rangle = \langle x, A^*x \rangle = \langle x, Ax \rangle = \overline{\langle Ax, x \rangle}.$$

Hence $\langle Ax, x \rangle$ is a real number.
Conversely, if $\langle Ax, x \rangle$ is a real number for each $x \in X$, then we have

$$\langle Ax, x \rangle = \overline{\langle Ax, x \rangle} = \overline{\langle x, A^*x \rangle} = \langle A^*x, x \rangle,$$

and so, by (5.8), we have $A = A^*$. $\square$

Definition 5.14.3 Let X be a Hilbert space and $A \in \mathcal{B}(X)$. Then A is said to be a *positive operator* if

$$\langle Ax, x \rangle \geq 0 \text{ for all } x \in X, \tag{5.82}$$

and in this case we write $A \geq 0$ or $0 \leq A$. (Perhaps it would be more suitable to use the term *nonnegative operator*, but the standard name is *positive operator*.)

Corollary 5.14.4 *Every positive operator is self–adjoint.*

Proof. This follows from Theorem 5.14.2 and Definition 5.14.3. $\square$

Theorem 5.14.5 *If X is a Hilbert space and $A \in \mathcal{B}(X)$, then A^*A is a positive operator.*

Proof. We have $\langle A^*Ax, x \rangle = \langle Ax, Ax \rangle \geq 0$ for every $x \in X$. $\square$

Definition 5.14.6 Let X be a Hilbert space, and A and B be self–adjoint operators in $\mathcal{B}(X)$. If $A - B \geq 0$, then we write $A \geq B$ or $B \leq A$. It follows from (5.82) that

$$B \leq A \text{ if anf only if } \langle Bx, x \rangle \leq \langle Ax, x \rangle \text{ for all } x \in X. \tag{5.83}$$

It is easy to show that (5.83) defines a partial order which is referred to as the *standard partial order on the set of all self–adjoint operators*.

In view of Theorem 5.14.2, the following definition makes sense.

Definition 5.14.7 Let X be a Hilbert space, $A \in \mathcal{B}(X)$ be a self–adjoint operator and

$$m(A) = \inf_{\|x\|=1} \langle Ax, x \rangle \text{ and } M(A) = \sup_{\|x\|=1} \langle Ax, x \rangle.$$

The numbers $m(A)$ and $M(A)$ are called, respectively, the *lower* and *upper bound of the self–adjoint operator* A. We note

$$-\|A\| \leq m(A) \leq M(A) \leq \|A\|.$$

Theorem 5.14.8 *Let X be a Hilbert space and $A \in \mathcal{B}(X)$ be self–adjoint. Then we have*

$$\|A\| = \max\{|m(A)|, |M(A)|\} = \sup_{\|x\|=1} |\langle Ax, x \rangle|. \tag{5.84}$$

Proof. It follows from the Cauchy-Schwarz inequality that

$$L \equiv \sup_{\|x\|=1} |\langle Ax, x \rangle| \leq \sup_{\|x\|=1} \|Ax\| \, \|x\| = \|A\|.$$

Thus to prove (5.84) it suffices to show that $\|A\| \leq L$. If $Ax = 0$ for each x with $\|x\| = 1$, then $A = O$ and we are done.

We note that we have for all $y, z \in X$

$$\langle A(y+z), y+z \rangle - \langle A(y-z), y-z \rangle = 2\langle Ay, z \rangle + 2\langle Az, y \rangle. \tag{5.85}$$

Since A is self–adjoint, (5.85) yields

$$2\langle Ay, z \rangle + 2\langle z, Ay \rangle \leq L\|y+z\|^2 + L\|y-z\|^2 = 2L(\|y\|^2 + \|z\|^2). \tag{5.86}$$

If $x \in X$, $Ax \neq 0$ and $0 \neq \lambda \in \mathbb{R}$, then it follows from (5.86) for $y = \lambda x$ and $z = Ax/\lambda$ that

$$4\|Ax\|^2 = 4\langle Ax, Ax \rangle = 2\left\langle A\lambda x, \frac{Ax}{\lambda} \right\rangle + 2\left\langle \frac{Ax}{\lambda}, A\lambda x \right\rangle$$

$$\leq 2L\left(\lambda^2\|x\|^2 + \frac{\|Ax\|^2}{\lambda^2}\right). \tag{5.87}$$

We obtain for $\lambda^2 = \|Ax\|/\|x\|$ from (5.87)

$$4\|Ax\|^2 \leq 2L(\|Ax\|\,\|x\| + \|Ax\|\,\|x\|) = 4L\|Ax\|\,\|x\|,$$

that is, $\|A\| \leq L$. $\qquad\square$

Theorem 5.14.9 *Let X be a Hilbert space, and $A, B \in \mathcal{B}(X)$ be self-adjoint. Then AB is a self-adjoint operator if and only if A and B commute, that is, if and only if $AB = BA$.*

Proof. From (5.67) and the assumptions of the theorem, we obtain $(AB)^* = B^*A^* = BA$. Hence $AB = (AB)^*$ if and only if $AB = BA$. $\qquad\square$

Theorem 5.14.10 *Let X be a Hilbert space, (A_n) be a sequence of self-adjoint operators in $\mathcal{B}(X)$, $A \in \mathcal{B}(X)$ and $A_n \to A$ for $(n \to \infty)$. Then A is self-adjoint.*

Proof. It follows from (5.61) and (5.64) that

$$\begin{aligned}
\|A - A^*\| &\leq \|A - A_n\| + \|A_n - A_n^*\| + \|A_n^* - A^*\| \\
&= \|A - A_n\| + 0 + \|(A_n - A)^*\| \\
&= 2\|A_n - A\| \to 0 \ (n \to \infty),
\end{aligned}$$

hence $A = A^*$. $\qquad\square$

Next we are going to show some basic properties of unitary operators.

Theorem 5.14.11 *Let X be a Hilbert space, $U, V \in \mathcal{B}(X)$ be unitary operators and $A \in \mathcal{B}(X)$. Then*

$$U \text{ is an isometry;} \tag{5.88}$$
$$X \neq \{0\} \text{ implies } \|U\| = 1; \tag{5.89}$$
$$U^{-1}(= U^*) \text{ is an isometry;} \tag{5.90}$$
$$UV \text{ is unitary;} \tag{5.91}$$
$$A \text{ is unitary if and only if } A \text{ is an isometry and onto.} \tag{5.92}$$

Proof.

(i) The statement in (5.88) follows from

$$\|Ux\|^2 = \langle Ux, Ux\rangle = \langle x, U^*Ux\rangle = \langle x, Ix\rangle = \|x\|^2.$$

(ii) Obviously (5.88) implies (5.89) and (5.88) implies (5.90).

(iii) Now we prove the statement in (5.91). We note that we have from (5.67)

$$(UV)^*UV = (V^*U^*)UV = V^*(U^*U)V = V^*V = I.$$

Analogously we obtain $UV(UV)^* = I$.

(iv) Finally, we show the statement in (5.92).

(α) First we show that if A is unitary then it is an isometry and onto.
If A is unitary, it is onto by definition, and (5.88) implies that A is an isometry.

(β) Now we show that if A is an isometry and onto, then it is unitary.

We assume that A is an isometry and onto. Obviously there exists $A^{-1} \in \mathcal{B}(X)$. We are going to show that $A^* = A^{-1}$. Since A is an isometry, we have for all $x \in X$

$$\langle A^*Ax, x \rangle = \langle Ax, Ax \rangle = \langle x, x \rangle = \langle Ix, x \rangle,$$

hence

$$\langle (A^*A - I)x, x \rangle = 0. \tag{5.93}$$

It follows from (5.8) and (5.93) that $A^*A - I = O$, that is, $A^*A = I$. We observe $A^* = A^*(AA^{-1}) = (A^*A)A^{-1} = IA^{-1} = A^{-1}$. □

Theorem 5.14.12 *Let X be a Hilbert space and $C \in \mathcal{B}(X)$. Then there exist self–adjoint operators $A, B \in \mathcal{B}(X)$ such $C = A + iB$. The operators A and B are uniquely defined by the operator C.*

Proof. Let $A = (1/2)(C + C^*)$ and $B = (1/2i)(C - C^*)$. Then A and B are self–adjoint and $C = A + iB$. To show the uniqueness, we assume that there exist self–adjoint operators $A_1, B_1 \in \mathcal{B}(X)$ such that $C = A_1 + iB_1$. Then $A - A_1 = i(B_1 - B)$ is self–adjoint, and so $A - A_1 = -i(B_1 - B)$, hence $A = A_1$ and $B = B_1$. □

Definition 5.14.13 *Let X be a Hilbert space and $A \in \mathcal{B}(X)$. The real and imaginary parts of the operator A, denoted, respectively, by ReA and ImT, are defined as*

$$\text{Re}A = \frac{A + A^*}{2} \text{ and } \text{Im}A = \frac{A - A^*}{2i}.$$

Theorem 5.14.14 *Let X be a Hilbert space and $A \in \mathcal{B}(X)$. The following conditions are equivalent:*

$$A \text{ is normal}; \tag{5.94}$$

$$\text{Re}A \text{ and } \text{Im}A \text{ commute}; \tag{5.95}$$

$$\|Ax\| = \|A^*x\| \text{ for each } x \in X. \tag{5.96}$$

Proof. It is easy to check

$$\text{Re}A\,\text{Im}A = \frac{A^2 - AA^* + A^*A - A^{*2}}{4i} \tag{5.97}$$

and

$$\text{Im}A\,\text{Re}A = \frac{A^2 + AA^* - A^*A - A^{*2}}{4i}. \tag{5.98}$$

(i) The equivalence of (5.94) and (5.95) obviously follows from (5.97) and (5.98).

(ii) To prove the equivalence of (5.94) and (5.96) we obtain for $x \in X$

$$\|Ax\|^2 - \|A^*x\|^2 = \langle Ax, Ax \rangle - \langle A^*x, A^*x \rangle = \langle (A^*A - AA^*)x, x \rangle, \tag{5.99}$$

and now the proof follows from (5.8) and (5.99). □

The sum of positive operators clearly is positive. The next theorem concerns the product of positive operators.

Theorem 5.14.15 *Let X be a Hilbert space and $A, B \in \mathcal{B}(X)$ be positive operators. If A and B commute, that is, if $AB = BA$, then AB is a positive operator.*

Proof. We may assume $A \neq 0$. Let (A_n) be a sequence of operators in $\mathcal{B}(X)$ defined as follows:

$$A_1 = \|A\|^{-1}A \text{ and } A_{n+1} = A_n - A_n^2 \text{ for } n = 1, 2, \ldots.$$

(i) First we show that

$$0 \leq A_n \leq I \text{ for each } n. \tag{5.100}$$

We apply the method of mathematical induction. For $n = 1$, the inequality in (5.100) is true, since $0 \leq A$ implies $0 \leq A_1$. The inequality $A_1 \leq I$, follows by the Cauchy–Schwarz inequality from

$$\langle A_1 x, x \rangle = \|A\|^{-1}\langle Ax, x \rangle \leq \|A\|^{-1}\|Ax\|\,\|x\| \leq \|x\|^2 = \langle Ix, x \rangle.$$

Now we assume that (5.100) is true $n = k$. Then $0 \leq I - A_k \leq I$, and since A_k is self–adjoint, we have for each $x \in X$

$$\langle A_k^2(I - A_k)x, x \rangle = \langle A_k(I - A_k)x, A_k x \rangle = \langle (I - A_k)A_k x, A_k x \rangle \geq 0.$$

Hence

$$A_k^2(I - A_k) \geq 0; \tag{5.101}$$

and analogously we can prove

$$A_k(I - A_k)^2 \geq 0. \tag{5.102}$$

It follows from (5.101) and (5.102) that

$$0 \leq A_k^2(I - A_k) + A_k(I - A_k)^2 = A_k - A_k^2 = A_{k+1}. \tag{5.103}$$

Thus we have proved $0 \leq A_{k+1}$. The inequality $A_{k+1} \leq I$ follows from $A_k^2 \geq 0$ and $I - A_k \geq 0$; when we sum the last two inequalities we obtain

$$0 \leq I - A_k + A_k^2 = I - A_{k+1}.$$

So we have proved (5.100).

We note that

$$A_1 = A_1^2 + A_2 = A_1^2 + A_2^2 + A_3 = \cdots = A_1^2 + A_2^2 + \cdots + A_n^2 + A_{n+1}. \tag{5.104}$$

It follows from $A_{n+1} \geq 0$ and (5.104) that

$$A_1^2 + A_2^2 + \cdots + A_n^2 = A_1 - A_{n+1} \leq A_1. \tag{5.105}$$

Since the operators A_k are self–adjoint, (5.105) and (5.83) imply

$$\sum_{k=1}^{n} \|A_k x\|^2 = \sum_{k=1}^{n} \langle A_k x, A_k x \rangle = \sum_{k=1}^{n} \langle A_k^2 x, x \rangle \leq \langle A_1 x, x \rangle. \tag{5.106}$$

It follows from (5.106) that $\sum_{k=1}^{\infty} \|A_k x\|^2 < \infty$, and so $A_n x \to 0$ as $n \to \infty$. Now, we obtain from (5.104)

$$\left(\sum_{k=1}^{n} A_k^2 \right) x = (A_1 - A_{n+1})x \to A_1 x, \ (n \to \infty). \tag{5.107}$$

Since the operators A_k commute with B, it follows from $B \geq 0$, (5.107) and the continuity of the inner product that

$$\langle ABx, x \rangle = \|A\| \langle BA_1 x, x \rangle = \|A\| \lim_{n \to \infty} \sum_{k=1}^{n} \langle BA_k^2 x, x \rangle$$

$$= \|A\| \lim_{n \to \infty} \sum_{k=1}^{n} \langle BA_k x, A_k x \rangle \geq 0. \qquad \square$$

5.15   Projectors and orthogonal projectors

Let M and N be subspaces of a linear space X. Then

$$Z = M + N = \{z : z = x + y, \ x \in M, \ y \in N\}$$

denotes the *sum of the subspaces* M and N. If $M \cap N = \{0\}$, then Z is called the *direct sum of M and N*, and in this case we write

$$Z = M \oplus N.$$

If $X = M \oplus N$, then N is said to be the *algebraic complement of M*. It is well known that in every linear space each subspace has an algebraic complement.

We recall from Lemma 4.5.15 that a map $P : X \to X$ is said to be *idempotent* if $P^2 = P$, and linear idempotent operators usually are called *(algebraic) projectors*. The operators O and I are projectors, the so–called *trivial projectors*. For each projector, P, the subspaces $R(P)$ and $N(P)$ are algebraically complementary, that is,

$$X = R(P) \oplus N(P).$$

Obviously P is a projector if and only if $I - P$ is a projector; in this, case we have

$$R(P) = N(I - P) \text{ and } N(P) = R(I - P).$$

On the other hand, each direct sum of X defines a projector. Namely, if $X = M \oplus N$, then each $x \in X$ can be written in the form $x = x_1 + x_2$, where $x_1 \in M$ and $x_2 \in N$. The map $P : X \to X$, defined $Px = x_1$ is a projector, $R(P) = M$ and $N(P) = N$. We say that P is a *projector on M parallel to N*, and sometimes we use the notation $P_{M,N}$.

It is well known that in a Hilbert space X, each closed subspace M has a topological complement; by Theorem 5.8.1, such a complement is $M^\perp$. The situation in Banach spaces has already been considered in Theorem 4.6.12.

Definition 5.15.1 Let X be a Hilbert space and M be a closed subspace of X. Since $X = M \oplus M^\perp$, each $x \in X$ can uniquely be written as $x = x_1 + x_2$ with $x_1 \in M$ and $x_2 \in M^\perp$. The map $P : X \to X$, defined by $Px = x_1$, is called an *orthogonal projector*, or, more precisely, an *orthogonal projector on M*, and is usually denoted by P_M.

Definition 5.15.2 Let X be a normed space. An operator $P \in \mathcal{B}(X)$ such that $P^2 = P$ is called a *projector* or *idempotent*.

If P is a projector, then $I - P$ is a projector and

$$X = R(P) \oplus N(P) \tag{5.108}$$

is a topological sum (this follows from the fact that $R(P) = N(I - P)$).

From $\|P\| = \|P^2\| \le \|P\|\,\|P\|$, if $0 \ne P$, it follows that $1 \le \|P\|$. All projectors of norm 1 in Hilbert spaces are characterized in the following theorem.

Theorem 5.15.3 *Let X be a Hilbert space, and $P \in \mathcal{B}(X)$ be a projector with $P \ne 0$. Then the following statements are equivalent:*

$$P \text{ is an orthogonal projector}; \tag{5.109}$$

$$\|P\| = 1; \tag{5.110}$$

$$P \text{ is a self-adjoint operator}; \tag{5.111}$$

$$P \text{ is a normal operator}; \tag{5.112}$$

$$P \text{ is a positive operator}. \tag{5.113}$$

Proof.

(i) First we show that (5.109) implies (5.110).

Since $X = R(P) \oplus R(P)^\perp$ and $R(P)^\perp = N(P)$, each $x \in X$ has a unique representation $x = x_1 + x_2$ with $x_1 \in R(P)$ and $x_2 \in R(P)^\perp$. Hence $\|Px\| = \|x_1\| \le \|x\|$ and so $\|P\| \le 1$, thus $\|P\| = 1$.

(ii) Now we show that (5.109) implies (5.113).

Using the notations from the proof of (5.109) implies (5.110), we have $\langle Px, x \rangle = \langle x_1, x_1 + x_2 \rangle = \langle x_1, x_1 \rangle \ge 0$, and P is a positive operator.

(iii) Now we show that (5.113) implies (5.111).

This follows by Corollary 5.14.4.

(iv) It is obvious that (5.111) implies (5.112).

(v) Now we show that (5.112) implies (5.109).

It follows from (5.96) that $N(P) = N(P^*)$, and we know from (5.75) that $N(P^*) = R(P)^\perp$. Hence $N(P) = R(P)^\perp$, and P is an orthogonal projector.

(vi) Finally, we show that (5.110) implies (5.109).

Let $x \in N(P)^\perp$. Since $x - Px \in N(P)$, it follows that $0 = \langle x - Px, x \rangle = \|x\|^2 - (Px, x)$. So $\|x\|^2 = \langle Px, x \rangle \le \|Px\| \|x\| \le \|P\| \|x\| \|x\| = \|x\|^2$, that is, $\|Px\| = \|x\| = \langle Px, x \rangle^{1/2}$. Hence we have

$$\|x - Px\|^2 = \|x\|^2 - 2\mathrm{Re}\langle Px, x \rangle + \|Px\|^2 = 0. \tag{5.114}$$

It follows from (5.114) that $x = Px$, and we have proved $N(P)^\perp \subset R(P)$. Now we prove the converse inclusion. We assume that $z \in R(P)$. Let $z = u + v$ with $u \in N(P)$ and $v \in N(P)^\perp$. Since $z = Pz = Pv = v$ (why?), we have $R(P) \subset N(P)^\perp$. Thus we have proved $R(P) = N(P)^\perp$, and P is an orthogonal projector. $\qquad\square$

Orthogonal projectors are very useful to work with and are used in many proofs.

Theorem 5.15.4 *Let X be a Hilbert space, M be a closed subspace of X and P_M be the orthogonal projector on M. Then we have*

$$\|x - P_M x\| = \inf\{\|x - y\| : y \in M\} \text{ for all } x \in X.$$

Proof. This follows by (5.35), Definition 5.15.1 and (5.108). $\qquad\square$

If X is a normed space and Y is a Banach space, then we know (see (4.107)) that $\overline{\mathcal{F}(X,Y)} \subset \mathcal{K}(X,Y)$. As one application of orthogonal projectors, we shall now prove that if Y is a Hilbert space, then $\overline{\mathcal{F}(X,Y)} = \mathcal{K}(X,Y)$.

Theorem 5.15.5 *Let X be a Banach space and Y be a Hilbert space. Then we have*

$$\overline{\mathcal{F}(X,Y)} = \mathcal{K}(X,Y).$$

Proof. It is enough to prove

$$\mathcal{K}(X,Y) \subset \overline{\mathcal{F}(X,Y)}. \tag{5.115}$$

We assume that $A \in \mathcal{K}(X,Y)$, $S = \{x \in X : \|x\| = 1\}$ and $\epsilon > 0$. Since AS is a relatively compact subset of Y, the set AS can be covered by a finite number of open balls of radius less than ϵ. Let $P \in \mathcal{B}(Y)$ be the orthogonal projector on the subspace generated by the centres of the covering balls. Hence it follows from Lemma 5.15.4 that

$$\|Ax - PAx\| < \epsilon \text{ for each } x \in S,$$

that is,

$$\|A - PA\| \leq \epsilon. \tag{5.116}$$

Since $PA \in \mathcal{F}(X,Y)$, (5.116) implies (5.115). $\qquad\square$

The next theorem gives a necessary and sufficient condition for the composition of orthogonal projectors to be an orthogonal projector.

Theorem 5.15.6 *Let X be a Hilbert space and let P_1 and P_2 be orthogonal projectors in $\mathcal{B}(X)$. Then $P_1 P_2$ is an orthogonal projector if and only if $P_1 P_2 = P_2 P_1$. In this case, we have*

$$P_1 P_2 = P_{R(P_1) \cap R(P_2)}. \tag{5.117}$$

Proof. If P_1 and P_2 commute, then obviously $P_1 P_2$ is idempotent. Now (5.67) and Theorem 5.15.3 imply

$$(P_1 P_2)^* = P_2^* P_1^* = P_2 P_1 = P_1 P_2,$$

and by Theorem 5.15.3, $P_1 P_2$ is an orthogonal projector. Conversely, if $P_1 P_2$ is an orthogonal projector, then Theorem 5.15.3 and (5.67) imply $P_1 P_2 = P_2 P_1$.

To prove (5.117), we note that $P_1 P_2 = P_2 P_1$ implies $R(P_1 P_2) \subset R(P_1) \cap R(P_2)$. For the same reason, $(P_1 P_2)x = P_1(P_2 x) = P_1 x = x$ for each $x \in R(P_1) \cap R(P_2)$, and so $R(P_1) \cap R(P_2) \subset R(P_1 P_2)$. $\qquad\square$

The next theorem gives a necessary and sufficient condition for the sum of orthogonal projectors to be an orthogonal projector.

Theorem 5.15.7 *Let X be a Hilbert space and P_1 and P_2 be orthogonal projectors in $\mathcal{B}(X)$. Then $P_1 + P_2$ is an orthogonal projector if and only if $P_1 P_2 = 0$. In this case, $R(P_1) + R(P_2)$ is a closed subspace of X and*

$$P_1 + P_2 = P_{R(P_1) + R(P_2)}. \tag{5.118}$$

Proof. Obviously, $P_1 + P_2$ is a self–adjoint operator. It follows from $P_1 P_2 = 0$ that $P_2 P_1 = 0$ (since $(P_1 P_2)^* = P_2 P_1 = 0$). Now it is easy to check that $P_1 + P_2$ is idempotent. It follows by Theorem 5.15.3 that $P_1 + P_2$ is an orthogonal projector.

Conversely, if $P_1 + P_2$ is an orthogonal projector, then $(P_1 + P_2)^2 = P_1 + P_2$ implies

$$P_1 P_2 + P_2 P_1 = 0. \tag{5.119}$$

Multiplying (5.119) from the left by P_1 we obtain

$$P_1 P_2 + P_1 P_2 P_1 = 0, \tag{5.120}$$

and multiplying (5.120) from the right by P_1 we obtain

$$2 P_1 P_2 P_1 = 0. \tag{5.121}$$

Obviously (5.120) and (5.121) imply $P_1 P_2 = 0$.

To prove (5.118), we note that $P_1 P_2 = 0$ implies $R(P_1) \perp R(P_2)$; clearly, we have for every $x_1 \in R(P_1)$ and every $x_2 \in R(P_2)$

$$\langle x_1, x_2 \rangle = \langle P_1 x_1, P_2 x_2 \rangle = \langle P_2 P_1 x_1, x_2 \rangle = 0. \tag{5.122}$$

It follows from the orthogonality of $R(P_1)$ and $R(P_2)$ that $R(P_1) + R(P_2)$ is a closed subspace of X. It is easy to show that $R(P_1 + P_2) = R(P_1) + R(P_2)$. $\qquad\square$

Lemma 5.15.8 *If M_1 and M_2 are closed subspaces of the Hilbert space X and P_{M_1} and P_{M_2} are the corresponding projectors, then*

$$M_1 \perp M_2 \text{ if and only if } P_{M_1} P_{M_2} = 0.$$

Proof. We know from (5.122) that $P_{M_1} P_{M_2} = 0$ implies $M_1 \perp M_2$. Conversely, $M_1 \perp M_2$ implies $P_{M_1}(P_{M_2} x) = 0$ for all $x \in X$. Hence we have $P_{M_1} P_{M_2} = 0$. $\square$

Definition 5.15.9 *Let X be a Hilbert space and P and Q be orthogonal projectors in $\mathcal{B}(X)$. The projector P is said to be* orthogonal *to the projector Q, denoted by $P \perp Q$, if $PQ = 0$.*

The difference of orthogonal projectors is an orthogonal projector under certain conditions. These conditions are related to partial ordering (5.83).

Theorem 5.15.10 *Let X be a Hilbert space and P_1 and P_2 be orthogonal projectors in $\mathcal{B}(X)$. Then the following conditions are equivalent:*

$$R(P_1) \subset R(P_2), \tag{5.123}$$
$$P_2 P_1 = P_1, \tag{5.124}$$
$$P_1 P_2 = P_1, \tag{5.125}$$
$$\|P_1 x\| \le \|P_2 x\| \text{ for each } x \in X, \tag{5.126}$$
$$P_1 \le P_2, \tag{5.127}$$
$$N(P_1) \supset N(P_2), \tag{5.128}$$
$$P_2 - P_1 \text{ is an orthogonal projector.} \tag{5.129}$$

If one of the conditions above is satisfied (and consequently each of them) then we have

$$P_2 - P_1 = P_{R(P_2) \cap [R(P_1)]^\perp}. \tag{5.130}$$

Proof.

(i) First we note that obviously (5.123) implies (5.124).

(ii) Now we show that (5.124) implies (5.125).
It follows from (5.124), (5.67) and (5.111) that $P_1 = P_1^* = (P_2 P_1)^* = P_1^* P_2^* = P_1 P_2$.
So we have shown that (5.124) implies (5.125).

(iii) Now we show that (5.125) implies (5.126).
It follows from (5.110) and (5.125) that

$$\|P_1 x\| = \|P_1 P_2 x\| \le \|P_1\| \, \|P_2 x\| \le \|P_2 x\| \text{ for all } x \in X.$$

(iv) Now we show that (5.126) implies (5.127).
From the already proved implication that (5.110) implies (5.109), and from (5.126) it follows that

$$\langle P_1 x, x \rangle = \|P_1 x\|^2 \le \|P_2 x\|^2 = \langle P_2 x, x \rangle \text{ for all } x \in X. \tag{5.131}$$

(v)   Now we show that (5.127) implies (5.128).
      We have for all $x \in X$

$$\|P_1 x\|^2 = \langle P_1 x, x \rangle \leq \langle P_2 x, x \rangle = \|P_2 x\|^2,$$

hence $x \in N(P_2)$ implies $x \in N(P_1)$, and we have shown that $N(P_1) \supset N(P_2)$.

(vi) Obviously (5.128) implies (5.123).

(vii) Now we show that (5.123) is equivalent to (5.129).
      If $P_2 - P_1$ is an orthogonal projector, then we have

$$P_2 - P_1 = (P_2 - P_1)^2 = P_2^2 - P_2 P_1 - P_1 P_2 + P_1^2,$$

and it follows that
$$P_1 P_2 + P_2 P_1 = 2 P_1. \qquad (5.132)$$

Multiplying (5.132) from the left by P_2 we obtain

$$P_2 P_1 P_2 + P_2 P_1 = 2 P_2 P_1, \text{ that is, } P_2 P_1 P_2 = P_2 P_1, \qquad (5.133)$$

and multiplying (5.132) from the right by P_2 we get

$$P_1 P_2 + P_2 P_1 P_2 = 2 P_1 P_2, \text{ that is, } P_2 P_1 P_2 = P_1 P_2. \qquad (5.134)$$

It follows from (5.132), (5.133) and (5.134) that

$$P_2 P_1 = P_1 P_2 = P_1, \qquad (5.135)$$

and so $R(P_1) \subset R(P_2)$.
Conversely, if $R(P_1) \subset R(P_2)$, we conclude from the already proved part of the theorem that (5.135) holds, and so (5.134). Obviously, (5.134) implies that $P_2 - P_1$ is an orthogonal projector (why?). So we have shown the equivalence of the conditions in $(5.123) - (5.129)$.

(viii) To prove (5.130), we note that (5.124) and (5.125) imply

$$P_2 - P_1 = P_2(I - P_1) = (I - P_1)P_2. \qquad (5.136)$$

Finally, (5.117) and (5.136) imply

$$R(P_2 - P_1) = R(P_2(I - P_1)) = R(P_2) \cap R(P_1)^\perp. \qquad \square$$

The next theorem is established by an application of Theorems 5.15.6 and 5.15.7. In this theorem we consider orthogonal commuting projectors which are not necessarily mutually orthogonal.

Theorem 5.15.11 *Let X be a Hilbert space, P_1 and P_2 be orthogonal projectors in $\mathcal{B}(X)$, and $P_1 P_2 = P_2 P_1$. Then $P = P_1 + P_2 - P_1 P_2$ is an orthogonal projector and*

$$R(P) = \overline{R(P_1) + R(P_2)}.$$

Proof. We note that
$$P = P_1 + (I - P_1)P_2.$$

Since $(I - P_1)P_2 = P_2(I - P_1)$, it follows by Theorem 5.15.6 that $(I - P_1)P_2$ is an orthogonal projector on $R(P_1)^\perp \cap R(P_2)$. It follows from

$$P_1(I - P_1)P_2 = \left(P_1 - P_1^2\right)P_2 = 0,$$

that $P_1 \perp (I - P_1)P_2$. By Theorems 5.15.7 and 5.15.6, P is an orthogonal projector and

$$R(P) = R(P_1) + (R(P_1)^\perp \cap R(P_2))$$
$$= \overline{R(P_1) + (R(P_1)^\perp \cap R(P_2))} \subset \overline{R(P_1) + R(P_2)}. \tag{5.137}$$

Analogously, it can be shown that

$$R(P) = R(P_2) + (R(P_2)^\perp \cap R(P_1)). \tag{5.138}$$

Obviously, it follows from (5.137), (5.138), and the fact that P is an orthogonal projector that

$$\overline{R(P_1) + R(P_2)} \subset R(P) \subset \overline{R(P_1) + R(P_2)}. \qquad \square$$

5.16 On the norm of idempotent operators

In this section, we assume that $A \neq I, 0$ is a bounded linear idempotent operator on a Hilbert space X with range R and kernel K, and let P_R and P_K denote the orthogonal projections of X onto these subspaces. In the next theorem we give a formula for the norm of $\|A\|$ using P_R and P_K.

Theorem 5.16.1 *Let A be a bounded linear idempotent operator on a Hilbert space X with range R and kernel K. Then we have*

$$\|A\| = \frac{1}{\sqrt{1 - \|P_K P_R\|^2}}. \tag{5.139}$$

Proof (from [166]). We have for each $x \in X$

$$\|P_R x\|^2 = \|AP_R x - AP_K P_R x\|^2$$
$$\leq \|A\|^2 \|P_R x - P_K P_R x\|^2$$
$$\leq \|A\|^2 \left(\|P_R x\|^2 - \|P_K P_R x\|^2\right),$$

hence,

$$\|P_K P_R x\|^2 \leq \left(1 - \|A\|^{-2}\right)\|P_R x\|^2 \leq \left(1 - \|A\|^{-2}\right)\|x\|^2,$$

and

$$\|P_K P_R\| < 1.$$

It follows that

$$\|A\| \geq \frac{1}{\sqrt{1 - \|P_K P_R\|^2}}. \tag{5.140}$$

Let $x \in X$. Then we have

$$\|x\|^2 = \|Ax + P_K(I - A)x\|^2 \geq \|Ax - P_K(Ax)\|^2$$
$$= \|A\|^2 - \|P_K(A)\|^2 = \|Ax\|^2 - \|P_K(P_R Ax)\|^2$$
$$\geq (1 - \|P_K P_R\|^2)\|Ax\|^2,$$

hence

$$\|A\| \leq \frac{1}{\sqrt{1 - \|P_K P_R\|^2}}. \tag{5.141}$$

From (5.141) and (5.140) we get (5.139). □

Now we obtain the following corollaries.

Corollary 5.16.2 *If P_R and P_K are projections onto the range and the null space of a bounded idempotent operator A, then*

$$\|P_R P_K\| = \frac{\sqrt{\|A\|^2 - 1}}{\|A\|}. \tag{5.142}$$

Proof. Since $\|P_R P_K\| = \|P_K P_R\|$, (5.139) implies (5.142). □

Vidav [203, Proof of Theorem 1] (see also [117, (2.8)]) proved the following estimate:

$$\|P_R P_K\| \leq \frac{\|A\|}{\sqrt{1 + \|A\|^2}}. \tag{5.143}$$

Clearly (5.142) is sharper than (5.143).

Corollary 5.16.3 *Let A be a bounded linear idempotent operator on a Hilbert space X. Then we have*

$$\|A\| = \|I - A\|. \tag{5.144}$$

Proof. It is well known that $I - A \in \mathcal{B}(X)$ is an idempotent operator with range K and kernel R. Now a proof is a direct consequence of Theorem 5.16.1 and the fact that $\|P_R P_K\| = \|(P_R P_K)^*\| = \|P_K P_R\|$. □

Corollary 5.16.4 *Let A be a bounded linear idempotent operator on a Hilbert space X. Then we have*

$$\|P_R P_K\| = \|(I - P_R)(I - P_K)\|. \tag{5.145}$$

Proof. It is well known that $\|A\| = \|A^*\|$ and that $A^* \in \mathcal{B}(X)$ is an idempotent operator with range $K^\perp$ and kernel $R^\perp$. Now the proof is a direct consequence of Theorem 5.16.1. □

The following result is the well-known *Akhiezer-Glazman equality*.

Theorem 5.16.5 *If P and Q are orthogonal projections in $\mathcal{B}(X)$, where X is a Hilbert space, then*

$$\|P - Q\| = \max\{\|P(I - Q)\|, \|Q(I - P)\|\}. \tag{5.146}$$

Proof. We note that $P(I - Q) = P(P - Q)$, and so $\|P(I - Q)\| = \|P(P - Q)\| \leq \|P\|\|P - Q\| \leq \|P - Q\|$. To prove the converse inequality in (5.146) we note that

$$P - Q = P(P - Q) + (I - P)(P - Q) = P(I - Q) - (I - P)Q.$$

Thus we have for each $x \in X$

$$\begin{aligned}
\|(P - Q)x\|^2 &= \|P(I - Q)x\|^2 + \|(I - P)Qx\|^2 \\
&\leq \|P(I - Q)\|^2 \cdot \|(I - Q)x\|^2 + \|(I - P)Q)\|^2 \cdot \|Qx\|^2 \\
&\leq \max\{\|P(I - Q)\|^2, \|Q(I - P)\|^2\} \left(\|(I - Q)x\|^2 + \|Qx\|^2\right) \\
&= \max\{\|P(I - Q)\|^2, \|Q(I - P)\|^2\}\|x\|^2.
\end{aligned}$$

Now, we get (5.146). □

Following Kato [111, p. 197], for any closed subspaces R and K of H, we define the *gap* (or *opening*) *between R and K* by

$$\text{gap}(R, K) = \|P_R - P_K\|.$$

Now we obtain a relation between the norm of $P_R P_K$ and the gap between R and $K^\perp$.

Corollary 5.16.6 *If P_R and P_K are projections onto the range and the null space of a bounded idempotent operator A, then*

$$\|P_R P_K\| = \|P_R - P_K^\perp\|.$$

Proof. We apply (5.139), (5.145) and (5.146). □

Remark 5.16.7 *Identity (5.139) is due to Ljance [131]; see also [156, 103]. Here we offer a new proof of an old theorem. Labrousse [126] was unaware of the work of Ljance [131], and proved (5.139) with the squared norm of $P_K P_R$ replaced by the squared norm of $P_R - P_K^\perp$, which is not a straightforward exercise. We remark that V. Pták [156] mentioned that T. Ando and B. Sz. Nagy called his attention to the fact that this result had been contained in a paper of Ljance [131]. Pták admitted the journal had not been accessible to him, and gave a proof of this result. Finally, (5.144) is due to Del Pasqua [29]; see also [111]. Ljance [131] was unawere of the work of Del Pasqua [29], and he also proved (5.144).*

Chapter 6

Banach Algebras

Chapter 6 is dedicated to the study of Banach algebras. The introduction of the concept of a Banach algebra is followed by a great number of examples and a study of the invertibility of elements in a complex Banach algebra with identity, and of the spectrum, resolvent and spectral radius. Also the concept of the topological divisor of zero is introduced and several important results related to this concept are established. Additional highlights include the study of subalgebras, one theorem by Hochwald–Morell as well as two theorems by Harte for regular elements in Banach algebras, conditions for the invertibility of operators in the Banach algebra of bounded linear operators from a Banach space into itself. Also detailed studies of the spectra of adjoint, normal and compact operators are included. Finally, the concept of C^*–algebras is introduced, their most important properties are established and conditions for the invertibility of the difference of projections are given.

6.1 Introduction

A great interest in Banach algebras started after the publication of Gelfand's paper in 1941 [71], which now is a classic paper. Until 1935, nothing was known about Banach algebras. C. Rickart [173] suggested that *Banach algebras are between analysis and algebra (or more precisely, with legs in analysis and head in algebra)*. Here we present some fundamental results. We refer for further reading to the books [20] by Berberian, [26] by Bonsall and J. Duncan, and [47] by Conway.

This chapter is dedicated to the study of Banach algebras. The introduction of the concept of a Banach algebra is followed by a great number of examples and a study of the invertibility of elements in a complex Banach algebra with identity, and of the spectrum, resolvent and spectral radius. Also, the concept of the topological divisor of zero is introduced and several results related to this concept are established. Further highlights include the study of subalgebras, one theorem by Hochwald–Morell, Theorem 6.9.1, and two theorems by Harte for regular elements in Banach algebras, Theorems 6.9.4 and 6.9.6, conditions for the invertibility of operators in the Banach algebra of bounded linear operators from a Banach space into itself in Section 6.10, and detailed studies of the spectra of adjoint, normal and compact operators in Sections 6.11 and 6.12. Finally, the concept of C^*–algebras is introduced, their most important properties are established and conditions for the invertibility of the difference of projections are given in Sections 6.13 and 6.14.

6.2   The concept of a Banach algebra

A vector space $\mathcal{A}$ over the field $\mathbb{F}$ is an *algebra* over $\mathbb{F}$ if the binary operations with

$$(a,b) \mapsto a \cdot b = ab : \mathcal{A} \times \mathcal{A} \to \mathcal{A} \text{ and } (\lambda, a) \mapsto \lambda a : \mathbb{F} \times \mathcal{A} \to \mathcal{A}$$

satisfy the following conditions for all $a, b, c \in \mathcal{A}$ and $\lambda \in \mathbb{F}$

$$a \cdot (b \cdot c) = (a \cdot b) \cdot c,$$

$$a \cdot (b + c) = (a \cdot b) + (a \cdot c), \qquad (a + b) \cdot c = (a \cdot c) + (b \cdot c),$$

$$(\lambda a) \cdot b = \lambda(a \cdot b) = a \cdot (\lambda b).$$

Obviously, the algebra $\mathcal{A}$ is a ring $(\mathcal{A}, +, \cdot)$, and the corresponding results for rings are transferred to algebras.

We say that an algebra $\mathcal{A}$ has a *unit (identity) element*, if the ring $\mathcal{A}$ has an identity, that is, if there exists an element $1 \in \mathcal{A}$ with $1 \neq 0$ such that

$$1 \cdot a = a = a \cdot 1 \text{ for all } a \in \mathcal{A}.$$

It should be noted that 1 denotes an element in $\mathbb{F}$ and at the same time an element in $\mathcal{A}$, and usually, when $\lambda \in \mathbb{F}$ and $1 \in \mathcal{A}$, we simply write $\lambda \cdot 1 = \lambda \in \mathcal{A}$. If the algebra $\mathcal{A}$ has an identity, then $a \in \mathcal{A}$ is said to be *left invertible in $\mathcal{A}$* if there exists $y \in \mathcal{A}$ such that $ya = 1$, and in this case, y is a *left inverse of the element* a. The set of all left invertible elements in $\mathcal{A}$ is denoted by $\mathcal{A}_l^{-1}$. Analogously an element $a \in \mathcal{A}$ is said to be *right invertible in $\mathcal{A}$* if there exists $z \in \mathcal{A}$ such that $az = 1$, and in this case, z is a *right inverse element of* a. The set of all right inverse elements in $\mathcal{A}$ is denoted by $\mathcal{A}_r^{-1}$. If an element $a \in \mathcal{A}$ is left and right invertible in $\mathcal{A}$, that is, if there exist $y, z \in \mathcal{A}$ such that $ya = 1 = az$, then a is said to be an *invertible element in $\mathcal{A}$*. In this case, we have

$$y = y \cdot 1 = y(az) = (ya)z = 1 \cdot z = z,$$

and y is denoted by a^{-1} and called an *inverse of the element* a. The set of all invertible elements in $\mathcal{A}$ is denoted by $\mathcal{A}^{-1}$. An algebra is *commutative* if the corresponding ring is commutative. If $\mathbb{F} = \mathbb{R}$, then the algebra is said to be *real*, and if $\mathbb{F} = \mathbb{C}$, the algebra is said to be *complex*. If $\mathcal{B}$ is a subset of the algebra $\mathcal{A}$, such that $\mathcal{B}$ is an algebra with the same algebraic operations as the algebra $\mathcal{A}$, then $\mathcal{B}$ is said to be a *subalgebra* in $\mathcal{A}$. A subspace $\mathcal{I}$ of an algebra $\mathcal{A}$ is a *left (right) ideal in $\mathcal{A}$* if $ax \in \mathcal{I}$ ($xa \in \mathcal{I}$) for all $a \in \mathcal{A}$ and all $x \in \mathcal{I}$. Also $\mathcal{I}$ is a *two-sided ideal in $\mathcal{A}$* if $\mathcal{I}$ is a left and right ideal in $\mathcal{A}$.

The algebras $\mathcal{A}_1$ and $\mathcal{A}_2$ over the same scalar field $\mathbb{F}$ are *isomorphic* if there exists an isomorphism $f : \mathcal{A}_1 \to \mathcal{A}_2$ of the linear spaces $\mathcal{A}_1$ and $\mathcal{A}_2$ such that

$$f(ab) = f(a)f(b) \text{ for all } a, b \in \mathcal{A}_1.$$

Definition 6.2.1 An algebra $\mathcal{A}$ is called a *normed algebra* if there exists a norm on $\mathcal{A}$, that is, if $(\mathcal{A}, \| \cdot \|)$ is a normed space such that

$$\|a \cdot b\| \leq \|a\| \, \|b\| \text{ for all } a, b \in \mathcal{A}. \tag{6.1}$$

If $\mathbb{F} = \mathbb{R}$ ($\mathbb{F} = \mathbb{C}$), then $\mathcal{A}$ is said to be a *real (complex)* normed algebra.
A normed algebra $(\mathcal{A}, \| \cdot \|)$ is a *Banach algebra* if $(\mathcal{A}, \| \cdot \|)$ is a Banach space.

Remark 6.2.2 *It is of special interest to study Banach algebras with identity. But if $\mathcal{A}$ is an algebra over $\mathbb{F}$ (without identity), then on the linear space $\tilde{\mathcal{A}} = \mathbb{F} \times \mathcal{A}$, we can define a structure of an algebra by putting for all $\lambda, \mu \in \mathbb{F}$ and $a, b \in \mathcal{A}$*

$$(\lambda, a)(\mu, b) = (\lambda\mu, \lambda b + \mu a + ab).$$

The element $(1, 0) \in \tilde{\mathcal{A}}$ is the identity in the algebra $\tilde{\mathcal{A}}$, and the algebra $\tilde{\mathcal{A}}$ is said to be obtained from the algebra $\mathcal{A}$ by joining the identity. If $\mathcal{A}$ is a normed algebra, then $\tilde{\mathcal{A}}$ is a normed space with the norm

$$\|(\lambda, x)\| = |\lambda| + \|x\|.$$

Since we have for all $\lambda, \mu \in \mathbb{F}$ and all $x, y \in \mathcal{A}$

$$\begin{aligned}
\|(\lambda, x)(\mu, y)\| &= |\lambda\mu| + \|xy + \mu x + \lambda y\| \\
&\leq |\lambda|\,|\mu| + \|x\|\,\|y\| + |\mu|\|x\| + |\lambda|\,\|y\| \\
&= \|(\lambda, x)\|\,\|(\mu, y)\|,
\end{aligned}$$

$\tilde{\mathcal{A}}$ is a normed algebra. A normed algebra $\tilde{\mathcal{A}}$ is said to be a normed algebra with identity obtained from $\mathcal{A}$ by joining an identity. It is easy to prove that the map $x \to (0, x)$ is an isomorphism and an isometry of the algebra $\mathcal{A}$ into the subalgebra of the algebra $\tilde{\mathcal{A}}$.

Remark 6.2.3 *If the normed algebra $\mathcal{A}$ has an identity 1, then $1^2 = 1$ implies $\|1\| \leq \|1\|^2$, that is, $\|1\| \geq 1$. It is convenient to have $\|1\| = 1$. If $\|1\| > 1$, then we define a new norm $\|\cdot\|_0 : \mathcal{A} \to \mathbb{R}$*

$$\|x\|_0 = \sup_{\|z\|=1} \|xz\| \text{ for all } x \subset \mathcal{A},$$

and $\mathcal{A}$ is a normed algebra. It follows from $\|x\|_0 = \sup_{z \neq 0} \|xz\| / \|z\|$ that

$$\frac{\|x\|}{\|1\|} \leq \|x\|_0 \leq \sup_{z \neq 0} \frac{\|x\|\,\|z\|}{\|z\|} = \|x\|,$$

and $\|\cdot\|$ and $\|\cdot\|_0$ are equivalent norms (Theorem 4.12.2). Obviously, we have $\|1\|_0 = 1$.

The inequality in (6.1) shows that the product in a normed algebra is a continuous function.

Lemma 6.2.4 *If (a_n) and (b_n) are sequences in the normed algebra $\mathcal{A}$ and $a, b \in \mathcal{A}$, then*

$$a_n \to a \text{ and } b_n \to b \text{ implies } a_n b_n \to ab.$$

Proof. We leave the proof to the readers. □

Finally, keeping in mind Remarks 6.2.2 and 6.2.3, usually when we study general properties of a Banach algebra $\mathcal{A}$ we always assume that $\mathcal{A}$ is a *complex algebra with identity* 1 and $\|1\| = 1$.

6.3 Examples

Here we give a few examples of normed and Banach algebras and we leave it to the readers to check the statements.

Example **6.3.1** *Let* $\mathcal{A}(D)$ *be the set of all functions* $f : \{z \in \mathbb{C} : |z| \le 1\} \to \mathbb{C}$ *such that* f *is continuous on* $\{z \in \mathbb{C} : |z| \le 1\}$ *and analytic on* $\{z : |z| < 1\}$. *If we consider* $\mathcal{A}(D)$ *as an algebra with the usual algebraic operations and the sup–norm*

$$\|f\| = \sup\{|f(z)| : |z| \le 1, z \in \mathbb{C}\},$$

then $\mathcal{A}(D)$ *is a commutative Banach algebra with identity.*

Example **6.3.2** *Let* X *be a Banach space over the complex field* $\mathbb{C}$ *and* $\mathcal{B}(X)$ *be the Banach space of all bounded linear operators on* X. *We remark that we have for all* $A, B \in \mathcal{B}(X)$ *and* $\lambda \in \mathbb{C}$,

$$(A + B)(x) = A(x) + B(x) \text{ for all } x \in X;$$
$$(\lambda A)(x) = \lambda A(x) \text{ for all } x \in X;$$
$$(AB)(x) = A(B(x)) \text{ for all } x \in X;$$

and

$$\|A\| = \sup\{\|Ax\| : \|x\| \le 1\}.$$

Since

$$\|AB\| \le \|A\| \, \|B\|,$$

it follows that $\mathcal{B}(X)$ *is a Banach algebra with identity* I.

Example **6.3.3** *Let* Ω *be a compact Hausdorff space and* $C(\Omega)$ *be the set of all continuous functions* $f : \Omega \to \mathbb{C}$. *It is easy to see that* $C(\Omega)$ *is a commutative Banach algebra with identity with respect to the usual algebraic operations and the sup–norm*

$$\|f\| = \sup\{|f(s)| : s \in \Omega\}.$$

Example **6.3.4** *Let* G *be a group and*

$$L(G) = \{x : G \to \mathbb{C} : \sum_{g \in G} |x(g)| < \infty\}.$$

Then $L(G)$ *is a linear space with the usual sum and multiplication by scalars, and a Banach space with the norm*

$$\|x\| = \sum_{g \in G} |x(g)|.$$

We define the product in $L(G)$ *by*

$$xy(h) = \sum_{g \in G} x(g)y(g^{-1}h) = \sum_{g \in G} x(hg^{-1})y(g) \text{ for all } x, y \in L(G) \text{ and all } h \in G.$$

This product is associative and called the convolution product. *The space* $L(G)$ *is a Banach algebra, referred to as the* group algebra on G; *we have*

$$\|xy\| = \sum_{g \in G} |xy(g)| = \sum_{g \in G} \left| \sum_{h \in G} x(h)y(h^{-1}g) \right|$$
$$\le \sum_{g \in G} \sum_{h \in G} |x(h)| \, |y(h^{-1}g)| = \sum_{h \in G} |x(h)| \sum_{g \in G} |y(h^{-1}g)|$$
$$= \sum_{h \in G} |x(h)| \|y\| = \|x\| \|y\|.$$

The identity in $L(G)$ *is the function which maps the identity in* G *to 1 and all other elements to zero.*

Example 6.3.5 *Let $L_1(\mathbb{R})$ be the algebra of all complex Lebesgue integrable functions on the real line. We define the multiplication by the convolution*

$$(f * g)(x) = \int_{-\infty}^{+\infty} f(x-t)g(t)\,dt,$$

and the norm as the L_1-norm

$$\|f\| = \int_{-\infty}^{+\infty} |f(t)|\,dt.$$

It can be shown that $L_1(\mathbb{R})$ is a commutative Banach algebra without identity.

Example 6.3.6 *Let $\mathcal{W}$ be the algebra of all absolutely convergent Fourier series*

$$f(x) = \sum_{n=-\infty}^{+\infty} a_n \exp(inx),$$

with the product $(fg)(x) = f(x)g(x)$ and norm

$$\|f\| = \sum_{n=-\infty}^{+\infty} |a_n|.$$

Then $\mathcal{W}$ is a commutative Banach algebra with identity, and usually $\mathcal{W}$ is called a Wiener *algebra.*

Example 6.3.7 *Let $\mathbf{F}$ be a bounded subset of the complex plane $\mathbb{C}$, and $\mathbf{P}(\mathbf{F})$ be the algebra of all complex polynomials p of the complex variable z and norm*

$$\|p\|_{\mathbf{F}} = \sup_{z \in \mathbf{F}} |p(z)|.$$

Then $\mathbf{P}(\mathbf{F})$ is a commutative normed algebra with identity.

6.4   Invertibility

In this section and later, $\mathcal{A}$ is always a complex Banach algebra with identity 1 and we assume $\|1\| = 1$. We are going to prove the basic properties of the sets $\mathcal{A}_l^{-1}$, $\mathcal{A}_r^{-1}$ and $\mathcal{A}^{-1}$.

Lemma 6.4.1 *If $z \in \mathcal{A}$ and $\|z\| < 1$, then*

the sequence $y_n = \sum_{k=0}^{n-1} z^k$, $(z^0 = 1)$ converges, that is, there exists $y \in \mathcal{A}$ such that

$$\lim_{n \to \infty} y_n = y = \sum_{k=0}^{\infty} z^k, \tag{6.2}$$

$$1 - z \in \mathcal{A}^{-1} \text{ and } y = (1-z)^{-1} = \sum_{k=0}^{\infty} z^k, \tag{6.3}$$

$$\|(1-z)^{-1}\| \le \frac{1}{1 - \|z\|}. \tag{6.4}$$

Proof.

(i) First we show (6.2).
Since $\|z\| < 1$ and $\|z^n\| \leq \|z\|^n$, it follows that the series $\sum_{k=0}^{\infty} \|z^k\|$ converges, and by Theorem 3.2.7, the series $\sum_{k=0}^{\infty} z^k$ converges.
Thus we have shown (6.2).

(ii) Now we show (6.3).
It follows from $\|z\| < 1$ and $\|z^k\| \leq \|z\|^k$ that $z^k \to 0$ as $k \to \infty$. Since $(1-z)y_n = 1 - z^n$, we obtain $\lim_{n\to\infty}(1-z)y_n = (1-z)y = 1$.
Analogously, it can be shown that $y(1-z) = 1$. Hence $1 - z \in \mathcal{A}^{-1}$ and $(1-z)^{-1} = \sum_{k=0}^{\infty} z^k$, and we have shown (6.3).

(iii) Finally, we show (6.4).
We remark that

$$\|y_n\| = \left\|\sum_{k=0}^{n-1} z^k\right\| \leq \sum_{k=0}^{n-1} \|z^k\| \leq \sum_{k=0}^{\infty} \|z\|^k = \frac{1}{1 - \|z\|},$$

(6.3) and $y_n \to y$ imply (6.4). $\qquad\square$

Corollary 6.4.2 *If $x \in \mathcal{A}$ and $\|1 - x\| < 1$, then*

$$x \in \mathcal{A}^{-1}, \ x^{-1} = \sum_{n=0}^{\infty} (1-x)^n \ and \ \|x^{-1}\| \leq \frac{1}{1 - \|1-x\|}.$$

Proof. This follows from Lemma 6.4.1, if we take $z = 1 - x$. $\qquad\square$

Corollary 6.4.3 *If $x \in \mathcal{A}$, $\lambda \in \mathbb{C}$ and $|\lambda| > \|x\|$, then we have*

$$\lambda - x \in \mathcal{A}^{-1}, \ (\lambda - x)^{-1} = \sum_{n=0}^{\infty} \lambda^{-n-1} x^n \ and \ \|(\lambda - x)^{-1}\| \leq \frac{1}{|\lambda| - \|x\|}.$$

Proof. It follows from $\lambda - x = \lambda(1 - x/\lambda)$ and $\|x/\lambda\| < 1$ by Lemma 6.4.1 that $\lambda - x \in \mathcal{A}^{-1}$. Since $(\lambda - x)^{-1} = (1/\lambda) \cdot (1 - x/\lambda)^{-1}$, applying Lemma 6.4.1 we obtain

$$(\lambda - x)^{-1} = \frac{1}{\lambda} \sum_{n=0}^{\infty} \lambda^{-n} x^{-n} = \sum_{n=0}^{\infty} \lambda^{-n-1} x^n$$

and

$$\|(\lambda - x)^{-1}\| \leq \frac{1}{|\lambda|} \frac{1}{1 - \|x/\lambda\|} = \frac{1}{|\lambda| - \|x\|}. \qquad\square$$

Theorem 6.4.4 *The sets $\mathcal{A}^{-1}$, $\mathcal{A}_l^{-1}$ and $\mathcal{A}_r^{-1}$ are open subsets of $\mathcal{A}$.*

Proof. Let $a \in \mathcal{A}^{-1}$ and $B_{\|a^{-1}\|^{-1}}(a)$ be the open ball in $\mathcal{A}$ with centre in a and radius $1/\|a^{-1}\|$. We have to prove $B_{\|a^{-1}\|^{-1}}(a) \subset \mathcal{A}^{-1}$.
Let $b \in B_{\|a^{-1}\|^{-1}}(a)$. Since

$$\|1 - ba^{-1}\| = \|(a-b)a^{-1}\| \leq \|a - b\| \, \|a^{-1}\| < 1,$$

it follows by Corollary 6.4.2 that $ba^{-1} \in \mathcal{A}^{-1}$. Hence $b = (ba^{-1})a \in \mathcal{A}^{-1}$, and we have shown that $\mathcal{A}^{-1}$ is an open subset of $\mathcal{A}$.

We prove that $\mathcal{A}_l^{-1}$ is an open subset of $\mathcal{A}$. If $a \in \mathcal{A}_l^{-1}$, then there exists $c \in \mathcal{A}$ such that $ca = 1$. We are going to show that $B_{\|c\|^{-1}}(a) \subset \mathcal{A}_l^{-1}$. Let $b \in B_{\|c\|^{-1}}(a)$. Since

$$\|1 - cb\| = \|c(a - b)\| \le \|c\| \, \|a - b\| < 1,$$

it follows by Corollary 6.4.2 that $cb \in \mathcal{A}^{-1}$. It follows from $1 = (cb)^{-1}cb = ((cb)^{-1}c)b$ that $b \in \mathcal{A}_l^{-1}$. So we have shown that $\mathcal{A}_l^{-1}$ is an open subset in $\mathcal{A}$. Analogously we can prove that $\mathcal{A}_r^{-1}$ is also an open subset of $\mathcal{A}$. $\qquad\square$

Lemma 6.4.5 *Let $\mathcal{A}$ be a Banach algebra, $a \in \mathcal{A}$ and (a_n) be a sequence in $\mathcal{A}^{-1}$ which converges to a. If $\{\|a_n^{-1}\| : n = 1, 2, \ldots\}$ is a bounded set, then $a \in \mathcal{A}^{-1}$.*

Proof. Let $\sup_n \{\|a_n^{-1}\| : n = 1, 2, \ldots\} = L$ and n be a natural number with $\|a_n - a\| L < 1$. Since $a = a_n(1 + a_n^{-1}(a - a_n))$, Lemma 6.4.1 implies $1 + a_n^{-1}(a - a_n) \in \mathcal{A}^{-1}$. Hence we have $a \in \mathcal{A}^{-1}$. $\qquad\square$

Corollary 6.4.6 *If $a \in \partial(\mathcal{A}^{-1})$, $a_n \in \mathcal{A}^{-1}$ for $n = 1, 2, \ldots$ and $\lim_{n \to \infty} a_n = a$, then $\lim_{n \to \infty} \|a_n^{-1}\| = \infty$.*

Proof. If the sequence $\|a_n^{-1}\|$ does not converge to ∞, then there exists a real number $L > 0$, such that $\|a_n^{-1}\| < L$ for infinitely many n. Hence there exists a natural number n such that $\|a_n - a\| < 1/L$. Then we have

$$\|1 - a_n^{-1}a\| = \|a_n^{-1}(a_n - a)\| \le \|a_n^{-1}\| \, \|a_n - a\| < 1. \tag{6.5}$$

It follows from Corollary 6.4.2 and the inequality in (6.5) that $a_n^{-1}a \in \mathcal{A}^{-1}$. Since $a = a_n(a_n^{-1}a)$ and $\mathcal{A}^{-1}$ is a group, it follows that $a \in \mathcal{A}^{-1}$. But since $\mathcal{A}^{-1}$ is an open subset $\mathcal{A}$ by Theorem 5.4.4) and since a is in the boundary of the set $\mathcal{A}^{-1}$ by assumption, we have $a \notin \mathcal{A}^{-1}$. This is a contradiction. $\qquad\square$

Theorem 6.4.7 *Let $a_n \in \mathcal{A}^{-1}$ for $n = 1, 2, \ldots$ and $a \in \mathcal{A}^{-1}$. Then*

$$a_n \to a \text{ implies } a_n^{-1} \to a^{-1}.$$

Proof. Let $b \in \mathcal{A}$ and $\|a - b\| < 1/(2\|a^{-1}\|)$. Then $b \in \mathcal{A}^{-1}$, and

$$\|b^{-1}\| - \|a^{-1}\| \le \|b^{-1} - a^{-1}\| = \|b^{-1}(a - b)a^{-1}\| \le \|b^{-1}\| \, \|a - b\| \, \|a^{-1}\| < \frac{1}{2}\|b^{-1}\|,$$

implies

$$\|b^{-1}\| < 2\|a^{-1}\|.$$

Hence we have

$$\|a^{-1} - b^{-1}\| = \|a^{-1}(a - b)b^{-1}\| \le \|a^{-1}\| \, \|a - b\| \, \|b^{-1}\| < 2\|a^{-1}\|^2 \|a - b\|.$$

Since $a_n \to a$, there exists a natural number n_0 such that $n \ge n_0$ implies $\|a - a_n\| < 1/(2\|a^{-1}\|)$. For $n \ge n_0$, it follows from the already proved part of the theorem that

$$\|a_n^{-1} - a^{-1}\| < 2\|a^{-1}\|^2 \|a_n - a\|. \qquad\square$$

The next theorem shows that the map $a \to a^{-1}$ $(a \in \mathcal{A}^{-1})$ is differentiable in some sense.

Theorem 6.4.8 *The map $a \to a^{-1}$ for $a \in \mathcal{A}^{-1}$ is differentiable on $\mathcal{A}^{-1}$ in the following sense: If $a \in \mathcal{A}^{-1}$ and $0 \neq h \in \mathbb{C}$ such that $a + h \in \mathcal{A}^{-1}$, then*

$$\lim_{h \to 0} \frac{(a+h)^{-1} - a^{-1}}{h} = -(a^{-1})^2.$$

Proof. We assume $0 < |h| < \|a^{-1}\|^{-1}$. Since $\|a + h - a\| = |h| < \|a^{-1}\|^{-1}$, it follows from the proof of Theorem 6.4.4 that $a + h \in \mathcal{A}^{-1}$. Now

$$(a+h)^{-1} - a^{-1} = (a+h)^{-1}[a - (a+h)]a^{-1}$$
$$= -h(a+h)^{-1}a^{-1},$$

and Theorem 6.4.7 imply

$$\frac{(a+h)^{-1} - a^{-1}}{h} = -(a+h)^{-1}a^{-1} \to -(a^{-1})^2 \text{ as } h \to 0. \qquad \square$$

6.5 Spectrum and resolvent

The concept of a *spectrum* is algebraic in nature, but many of its properties are studied by analytical methods.

Definition 6.5.1 Let $a \in \mathcal{A}$. The *spectrum of a*, denoted by $\sigma(a)$, is the set of all complex numbers λ such that $\lambda - a$ is not invertible in $\mathcal{A}$, that is,

$$\sigma(a) = \{\lambda \in \mathbb{C} : \lambda - a \notin \mathcal{A}^{-1}\}.$$

The *left spectrum of the element a* is denoted by $\sigma_l(a)$ and defined by

$$\sigma_l(a) = \{\lambda \in \mathbb{C} : \lambda - a \notin \mathcal{A}_l^{-1}\}.$$

The *right spectrum of an element a* is denoted by $\sigma_r(a)$ and defined by

$$\sigma_r(a) = \{\lambda \in \mathbb{C} : \lambda - a \notin \mathcal{A}_r^{-1}\}.$$

The complement of the set $\sigma(a)$, denoted by $\rho(a)$, is called the *resolvent set of the element a*. Hence we have

$$\rho(a) = \{\lambda \in \mathbb{C} : \lambda - a \in \mathcal{A}^{-1}\} = \mathbb{C} \setminus \sigma(a).$$

Since $\lambda - a = -(a - \lambda)$, the condition $\lambda - a \notin \mathcal{A}^{-1}$ in the definition of the spectrum of the element a could be replaced (and we sometimes do this) by $a - \lambda \notin \mathcal{A}^{-1}$. An analogous remark holds for the right and left spectrum of an element a, the resolvent set of a, etc.

Theorem 6.5.2 *Let $a \in \mathcal{A}$. Then $\sigma(a)$ is a closed and bounded subset of $\mathbb{C}$.*

Proof. Let $f : \mathbb{C} \to \mathcal{A}$ be the map defined by $f(\lambda) = \lambda - a$ for all $(\lambda \in \mathbb{C})$. Then f is continuous on $\mathbb{C}$. Since $\rho(a) = f^{-1}(\mathcal{A}^{-1})$ and $\mathcal{A}^{-1}$ is an open subset of $\mathcal{A}$ by Theorem 6.4.4, it follows that $\rho(a)$ is an open subset of $\mathbb{C}$. Hence $\sigma(a)$ is a closed subset of $\mathbb{C}$. It follows from Corollary 6.4.3 that $\sigma(a) \subset \{\lambda \in \mathbb{C} : |\lambda| \leq \|a\|\}$, so $\sigma(a)$ is a bounded subset of $\mathbb{C}$. $\square$

Definition 6.5.3 Let $a \in \mathcal{A}$. The *resolvent function (resolvent)* of the element a is the function $R_a(\lambda) : \rho(a) \to \mathcal{A}$ defined by

$$R_a(\lambda) = (\lambda - a)^{-1} \text{ for } \lambda \in \rho(a).$$

Obviously, we have $\sigma(0) = \{0\}$, $\sigma(1) = \{1\}$, and $R_0(\lambda) = \lambda^{-1}$ for all $\lambda \in \rho(0)$, that is, $R_1(\lambda) = (\lambda - 1)^{-1}$ for all $\lambda \in \rho(1)$. We recall that $e \in \mathcal{A}$ is an *idempotent element (idempotent)* if $e^2 = e$. An idempotent e is *nontrivial* if $e \neq 0, 1$.

Lemma 6.5.4 *Let $e \in \mathcal{A}$ be a nontrivial idempotent. Then we have*

$$\sigma(e) = \{0, 1\} \text{ and } R_e(\lambda) = \frac{1 - e}{\lambda} + \frac{e}{\lambda - 1} \text{ for all } \lambda \in \rho(e).$$

Proof. The proof is left to the readers. □

Lemma 6.5.5 (Resolvent equality) *Let $a, b \in \mathcal{A}$, $\lambda, \mu \in \rho(a)$ and $\eta \in \rho(a) \cap \rho(b)$. Then*

$$R_a(\lambda) - R_a(\mu) = -(\lambda - \mu)R_a(\lambda)R_a(\mu), \tag{6.6}$$
$$R_a(\eta) - R_b(\eta) = R_a(\eta)(a - b)R_b(\eta). \tag{6.7}$$

The equalities in (6.6) and (6.7) are called, respectively, the *first and second resolvent equalities*.

Proof. To prove (6.6), we remark that we have for $\lambda, \mu \in \rho(a)$

$$(\lambda - a)^{-1} - (\mu - a)^{-1} = (\lambda - a)^{-1}[(\mu - a) - (\lambda - a)](\mu - a)^{-1}$$
$$- -(\lambda - \mu)R_a(\lambda)R_a(\mu).$$

Concerning (6.7), it follows from $\eta \in \rho(a) \cap \rho(b)$ that

$$(\eta - a)^{-1} - (\eta - b)^{-1} = (\eta - a)^{-1}[(\eta - b) - (\eta - a)](\eta - b)^{-1}$$
$$= R_a(\eta)(a - b)R_b(\eta). \qquad □$$

Theorem 6.5.6 *If $a \in \mathcal{A}$, $\lambda \in \mathbb{C}$ and $|\lambda| > \|a\|$, then we have*

$$\lim_{\lambda \to \infty} R_a(\lambda) = 0. \tag{6.8}$$

If $a \in \mathcal{A}$ and $\lambda \in \rho(a)$, then we have

$$\lim_{h \to 0} \frac{R_a(\lambda + h) - R_a(\lambda)}{h} = -(\lambda - a)^{-2}. \tag{6.9}$$

Proof. It follows from $|\lambda| > \|a\|$ by Corollary 6.4.3 that $\lambda \in \rho(a)$. Since $R_a(\lambda) = (\lambda - a)^{-1} = \frac{1}{\lambda}(1 - \frac{a}{\lambda})^{-1}$, Theorem 6.4.7 implies that $R_a(\lambda) \to 0$, $(\lambda \to \infty)$, that is, we have shown (6.8).

The identity in (6.9) follows from Theorems 6.4.7, 6.4.8 and (6.6). □

Corollary 6.5.7 *If f is a continuous linear functional on $\mathcal{A}$, that is, $f \in \mathcal{A}^*$, $a \in \mathcal{A}$ and $|\lambda| > \|a\|$, $\lambda \in \mathbb{C}$, then we have*

$$\lim_{\lambda \to \infty} f(R_a(\lambda)) = 0.$$

The function

$$F(\lambda) = f(R_a(\lambda)) \ (\lambda \in \rho(a))$$

is differentiable for each $\lambda \in \rho(a)$, and the derivative satisfies

$$\frac{dF(\lambda)}{d\lambda} = -f((\lambda - a)^{-2}) \ (\lambda \in \rho(a)).$$

Proof. This follows from Theorem 6.5.6. □

The following theorem gives us an important property of the spectrum.

Theorem 6.5.8 *For each element $a \in \mathcal{A}$ the spectrum of a is a nonempty and compact subset of $\mathbb{C}$.*

Proof. We assume that $\sigma(a)$ is an empty set. Then $\rho(a) = \mathbb{C}$, and for $f \in \mathcal{A}^*$, by Corollary 6.5.7, the function $F(\lambda) = f(R_a(\lambda))$ for $(\lambda \in \mathbb{C})$ is differentiable for each $\lambda \in \mathbb{C}$ and $\lim_{\lambda \to \infty} F(\lambda) = 0$. Hence it follows by Liouville's theorem (which asserts that every bounded entire function, that is, every bounded function that is differentiable on all of $\mathbb{C}$, is constant) that $F(\lambda)$ is constant for all $\lambda \in \mathbb{C}$. From $a \in \mathcal{A}^{-1}$ and Corollary 4.5.4, it follows that there exists a functional $g \in \mathcal{A}^*$ such that $\|g\| = 1$ and $g(a^{-1}) = \|a^{-1}\|$. So we have $g(R_a(0)) = \|a^{-1}\| \neq 0$, and by the already proved part ot the theorem we obtain $g(R_a(\lambda)) = 0$ for $(\lambda \in \mathbb{C})$. This is a contradiction. Hence $\sigma(a)$ is not an empty set. That $\sigma(a)$ is a compact subset of $\mathbb{C}$ follows from Theorem 6.5.2. □

Theorem 6.5.9 (Gelfand-Mazur) *Let $\mathcal{A}$ be a complex Banach algebra with identity 1, in which every element $a \neq 0$ is invertibile. Then $\mathcal{A}$ is a one-dimensional complex algebra and the map $\lambda \to \lambda \cdot 1$ is an isometric isomorphism from $\mathbb{C}$ onto $\mathcal{A}$.*

Proof. Let $a \in \mathcal{A}$ and $\lambda \in \sigma(a)$ ($\sigma(a) \neq \emptyset$ by Theorem 6.5.8). Hence $a - \lambda \cdot 1$ is not invertibile, and we have $a - \lambda \cdot 1 = 0$ by assumption, that is, $a = \lambda \cdot 1$. □

If $a \in \mathcal{A}$ and $p_n(t) = \alpha_0 t^n + \alpha_1 t^{n-1} + \cdots + \alpha_n$ is a polynomial then $p_n(a)$ is an element in $\mathcal{A}$ defined by $p_n(a) = \alpha_0 a^n + \alpha_1 a^{n-1} + \cdots + \alpha_n \cdot 1 \in A$. The next theorem shows us a relation between $\sigma(p_n(a))$ and $p_n(\sigma(a))$.

Theorem 6.5.10 (Spectral mapping theorem for polynomials) *If $a \in A$ and p is a polynomial, then we have*

$$\sigma(p(a)) = p(\sigma(a)),$$

that is, $\mu \in \sigma(p(a))$ if and only if there exists $\lambda \in \sigma(a)$ such that $p(\lambda) = \mu$.

Proof. Let $\lambda \in \sigma(a)$ and $q(t) = p(t) - p(\lambda)$. Then $q(\lambda) = 0$ and there exist complex numbers $\alpha, \lambda_1, \ldots, \lambda_n$ such that

$$q(t) = \alpha(t - \lambda)(t - \lambda_1) \cdots (t - \lambda_n).$$

Since $q(a) = \alpha(a - \lambda)(a - \lambda_1) \cdots (a - \lambda_n)$ and $a - \lambda$ is not invertible, it follows that $q(a)$ is not invertible. From $q(a) = p(a) - p(\lambda)$, we have $p(\lambda) \in \sigma(p(a))$. So we have proved $p(\sigma(a)) \subset \sigma(p(a))$. Let us prove the converse inclusion. If $\mu \in \sigma(p(a))$ and $g(t) = p(t) - \mu$, then there exist complex numbers $\mu_1, \ldots, \mu_m, \alpha$ such that

$$g(t) = \alpha(t - \mu_1) \cdots (t - \mu_m).$$

Since $g(a) = \alpha(a - \mu_1) \cdots (a - \mu_m)$ and $g(a) = p(a) - \mu \notin \mathcal{A}^{-1}$, there exist $i \in \{1, \ldots, m\}$ such that $a - \mu_i \notin \mathcal{A}^{-1}$. Hence $\mu_i \in \sigma(a)$ and $g(\mu_i) = 0$, and so $p(\mu_i) = \mu$. Thus we have proved $\sigma(p(a)) \subset p(\sigma(a))$. □

Theorem 6.5.11 *Let $a, b \in \mathcal{A}$. Then we have*

$$\sigma(ab) \cup \{0\} = \sigma(ba) \cup \{0\}.$$

Proof. Let $\lambda \neq 0$ and $ab - \lambda \in \mathcal{A}^{-1}$. For $u = (ab - \lambda)^{-1}$, we see that $(ab - \lambda)u = 1$ implies $abu = \lambda u + 1$. Hence, $(ba - \lambda)(bua - 1) = b(abu)a - ba - \lambda bua + \lambda = b(\lambda u + 1)a - ba - \lambda bua + \lambda = \lambda$. Analogously we can show that $(bua - 1)(ba - \lambda) = \lambda$. So $ba - \lambda$ is invertibile. □

For $a \in \mathcal{A}^{-1}$, the next theorem determines a relation between $\sigma(a)$ and $\sigma(a^{-1})$.

Theorem 6.5.12 *Let $a \in \mathcal{A}$. If $a \in \mathcal{A}^{-1}$ then we have*

$$\sigma(a^{-1}) = \{\lambda^{-1} : \lambda \in \sigma(a)\}.$$

Proof. We have for $\lambda \neq 0$

$$\lambda^{-1} - a^{-1} = -\lambda^{-1}a^{-1}(\lambda - a).$$
□

6.6   Spectral radius

We know from Theorem 6.5.2 that the spectrum of the element $a \in \mathcal{A}$, that is, $\sigma(a)$ is a nonempty and compact subset of the set of complex numbers $\mathbb{C}$. It is of interest to find the closed ball of the smallest radius with its centre in the origin that contains the set $\sigma(a)$. The following definition is related to this.

Definition 6.6.1 *Let $a \in \mathcal{A}$. The spectral radius of the element a, denoted by $r(a)$, is defined by*

$$r(a) = \sup\{|\lambda| : \lambda \in \sigma(a)\}.$$

We note that $0 \leq r(a) \leq \|a\|$ by Corollary 6.4.3.

Corollary 6.6.2 *If $a, b \in \mathcal{A}$, then $r(ab) = r(ba)$.*

Proof. This follows from Theorem 6.5.11. □

Corollary 6.6.3 *If $x \in \mathcal{A}$ and $y \in \mathcal{A}^{-1}$, then $r(y^{-1}xy) = r(x)$.*

Proof. It follows from Corollary 6.6.2 that

$$r((y^{-1}x)y) = r(y(y^{-1}x)) = r((yy^{-1})x) = r(x).$$
□

The next theorem gives us a method for computing the spectral radius.

Theorem 6.6.4 *For each $a \in \mathcal{A}$, there exists the limit $v(a) = \lim_{n \to \infty} \|a^n\|^{1/n}$. The number $v(a)$ has the following properties:*

$$v(a) = \inf\{\|a^n\|^{1/n} : n = 1, 2, \ldots\}, \tag{6.10}$$

$$0 \leq v(a) \leq \|a\|, \tag{6.11}$$

$$v(\lambda a) = |\lambda|v(a) \text{ for each } \lambda \in \mathbb{C}, \tag{6.12}$$

$$v(ab) = v(ba) \text{ for each } a, b \in \mathcal{A}, \tag{6.13}$$

$$v(a^k) = (v(a))^k \text{ for each natural number } k, \tag{6.14}$$

$$\text{if } a, b \in \mathcal{A} \text{ and } ab = ba, \text{ then}$$

$$v(ab) \leq v(a)v(b) \text{ and } v(a+b) \leq v(a) + v(b). \tag{6.15}$$

Proof.

(i) First we show (6.10).
We put $v = \inf\{\|a^n\|^{1/n} : n = 1, 2, \ldots\}$, and prove $v = \lim \|a^n\|^{1/n}$. If $\epsilon > 0$, then there exists m such that $\|a^m\|^{1/m} < v + \epsilon$. For each natural number n, there exist integers p and q with $0 \leq q \leq m - 1$ such that $n = pm + q$. Hence we have

$$\|a^n\|^{1/n} \leq \|a^m\|^{p/n}\|a\|^{q/n} \leq (v + \epsilon)^{pm/n}\|a\|^{q/n}. \tag{6.16}$$

Since $pm/n \to 1$ and $q/n \to 0$ as $n \to \infty$, (6.16) implies

$$\limsup \|a^n\|^{1/n} \leq v + \epsilon,$$

hence

$$\limsup \|a^n\|^{1/n} \leq v. \tag{6.17}$$

It follows from $v \equiv \inf\{\|a^n\|^{1/n} : n = 1, 2, \ldots\}$ that $v \leq \|a^n\|^{1/n}$ for each natural number n. Hence we have

$$v \leq \liminf \|a^n\|^{1/n}. \tag{6.18}$$

It follows from (6.17) and (6.18) that the limit $\lim_{n\to\infty} \|a^n\|^{1/n}$ exists and is equal to $\inf\{\|a^n\|^{1/n} : n = 1, 2, \ldots\}$.
So we have shown (6.10).

(ii) The proofs of (6.11) and (6.12) are easy and left to the readers.

(iii) Now we prove (6.13).
We know that we have for each natural number n

$$\|(ab)^n\| \leq \|a\|\, \|(ba)^{n-1}\|\, \|b\|,$$

that is,

$$\|(ab)^n\|^{1/n} \leq \|a\|^{1/n}\|(ba)^{n-1}\|^{1/n}\|b\|^{1/n}. \tag{6.19}$$

If we let $n \to \infty$ in (6.19), we obtain $v(ab) \leq v(ba)$. Interchanging a and b yields the second inequality $v(ba) \leq v(ab)$, and so $v(ab) = v(ba)$.
So we have shown (6.13).

(iv) Now we prove (6.14).
It follows from the fact that $\lim_n \|a^n\|^{1/n} = v(a)$ that we have for each natural number k

$$v(a^k) = \lim_{n\to\infty} \|(a^k)^n\|^{1/n} = \lim_{n\to\infty} \left(\|a^{kn}\|^{1/kn}\right)^k = (v(a))^k.$$

So we have proved (6.14).

(v) Now we prove (6.15).
It follows from $ab = ba$ that $\|(ab)^n\|^{1/n} \leq \|a^n\|^{1/n}\|b^n\|^{1/n}$ for each natural number n.
Now (6.10) implies $v(ab) \leq v(a)v(b)$.
To prove the second inequality in (6.15), we assume that $t > v(a)$, $s > v(b)$, $a \neq 0$, $b \neq 0$, $x = a/t$ and $y = b/s$. We have for each natural number n

$$\|(a+b)^n\| = \left\|\sum_{k=0}^{n} \binom{n}{k} a^k b^{n-k}\right\| \leq \sum_{k=0}^{n} \binom{n}{k} \|a^k\|\, \|b^{n-k}\|$$

$$= \sum_{k=0}^{n} \binom{n}{k} t^k s^{n-k} \|x^k\|\, \|y^{n-k}\|,$$

that is,

$$\|(a+b)^n\|^{1/n} \leq \left(\sum_{k=0}^{n} \binom{n}{k} t^k s^{n-k} \|x^k\| \|y^{n-k}\| \right)^{1/n}. \tag{6.20}$$

It is clear that for each natural number n, there exist natural numbers n' and n'' such that $n = n' + n''$ and $\|x^{n'}\| \|y^{n''}\| = \max\{\|x^k\| \|y^{n-k}\| : 0 \leq k \leq n\}$. So (6.20) implies

$$\|(a+b)^n\|^{1/n} \leq \left(\sum_{k=0}^{n} \binom{n}{k} t^k s^{n-k} \right)^{1/n} \left(\|x^{n'}\| \|y^{n''}\| \right)^{1/n},$$

and so we have by (6.10)

$$v(a+b) \leq ((t+s)^n)^{1/n} \left(\|x^{n'}\| \|y^{n''}\| \right)^{1/n}. \tag{6.21}$$

If (n'/n) is a bounded sequence, there exists a subsequence (n'_m/n_m) which converges to p where $0 \leq p \leq 1$. Then we obviously have $\lim_{m \to \infty} n''_m/n_m = 1 - p$.
First we assume $p \neq 0$. It follows from $v(x) = v(a/t) = v(a)/t < 1$ (see (6.12)) that

$$\lim_{m \to \infty} \|x^{n'_m}\|^{1/n_m} = \lim_{m \to \infty} \left(\|x^{n'_m}\|^{1/n'_m} \right)^{n'_m/n_m} = (v(x))^p < 1. \tag{6.22}$$

If $p = 0$, we have

$$\limsup_{m \to \infty} \|x^{n'_m}\|^{1/n_m} \leq \lim_{m \to \infty} \|x\|^{n'_m/n_m} = 1. \tag{6.23}$$

So, for $p \neq 0$ and for $p = 0$, it follows from (6.22) and (6.23) that

$$\limsup_{m \to \infty} \|x^{n'_m}\|^{1/n_m} \leq 1.$$

Analogously we can show that $\limsup_{m \to \infty} \|y^{n''_m}\|^{1/n_m} \leq 1$. So it follows from (6.21) that $v(a+b) \leq t + s$, and so $v(a+b) \leq v(a) + v(b)$. $\qquad \square$

The next theorem gives a method to calculate the spectral radius.

Theorem 6.6.5 (Theorem of the spectral radius) *We have for each $a \in \mathcal{A}$*

$$r(a) = \lim_{n \to \infty} \|a^n\|^{1/n}.$$

Proof. If $\lambda \in \sigma(a)$, then we have $\lambda^n \in \sigma(a^n)$ for each natural number n by Theorem 6.5.9. Hence Corollary 6.4.3 implies $|\lambda|^n = |\lambda^n| \leq \|a^n\|$, that is, $|\lambda| \leq \|a^n\|^{1/n}$. So we have

$$|\lambda| \leq \liminf \|a^n\|^{1/n},$$

and we have shown that

$$r(a) \leq \liminf \|a^n\|^{1/n}. \tag{6.24}$$

If $\lambda \in \mathbb{C}$ and $|\lambda| > \|a\|$, then $\lambda - a \in \mathcal{A}^{-1}$ and $(\lambda - a)^{-1} = \sum_{n=0}^{\infty} \lambda^{-n-1} a^n$ by Corollary 6.4.3. Let $R_a : \rho(a) \to \mathcal{A}$ be the resolvent function of the element a, $f \in \mathcal{A}^*$ and $F : \rho(a) \to \mathbb{C}$ be the function defined by $F(\lambda) = f(R_a(\lambda)) = f((\lambda - a)^{-1})$ for $\lambda \in \rho(a)$. Then we have

$$F(\lambda) = \sum_{n=0}^{\infty} \lambda^{-n-1} f(a^n) \text{ for all } |\lambda| > \|a\|.$$

Since the function F is differentiable for each $|\lambda| > r(a)$, and since its *Laurent series* is unique in the set $\{\lambda \in \mathbb{C} : |\lambda| > r(a)\}$, it follows that

$$F(\lambda) = \sum_{n=0}^{\infty} \lambda^{-n-1} f(a^n) \text{ for all } |\lambda| > r(a). \qquad (6.25)$$

The series (6.25) converges for $|\lambda| > r(a)$, and so

$$\sup_n |f(\lambda^{-n-1} a^n)| < \infty. \qquad (6.26)$$

Since $f \in \mathcal{A}^*$ is an arbitrary element, by the uniform boundedness principle, (6.26) implies

$$\sup_n \|\lambda^{-n-1} a^n\| < \infty. \qquad (6.27)$$

For each natural number n, (6.27) implies

$$\|a^n\| \leq |\lambda|^{n+1} \sup_n \|\lambda^{-n-1} a^n\|,$$

and so

$$\limsup \|a^n\|^{1/n} \leq |\lambda|,$$

and

$$\limsup \|a^n\|^{1/n} \leq r(a). \qquad (6.28)$$

It follows from (6.24) and (6.28) that

$$r(a) = \liminf \|a^n\|^{1/n} = \limsup \|a^n\|^{1/n} = \lim_{n \to \infty} \|a^n\|^{1/n}. \qquad \square$$

Corollary 6.6.6 *Let $a, b \in \mathcal{A}$. Then we have*

$$r(a) = \inf\{\|a^n\|^{1/n} : n = 1, 2, \ldots\},$$

$$r(\lambda a) = |\lambda| r(a) \text{ for each scalar } \lambda,$$

$$r(a^k) = (r(a))^k \text{ for each natural number } k,$$

$$r(ab) = r(ba),$$

$$ab = ba \text{ implies } r(ab) \leq r(a) r(b) \text{ and } r(a + b) \leq r(a) + r(b). \qquad (6.29)$$

Proof. This follows from Theorems 6.6.4 and 6.6.5. $\qquad \square$

Remark 6.6.7 *We know from Theorem 6.6.4 that $\lim_{n \to \infty} \|a^n\|^{1/n}$ exists; but we did not use this result in the proof of Theorem 6.6.5, where we showed the existence of the mentioned limit in a different way. Furthermore, Theorem 6.6.5 shows that the spectral radius $r(a)$ of the element a, which is defined algebraically by $\sigma(a)$, was described analytically using this limit $\lim_{n \to \infty} \|a^n\|^{1/n}$. This is a nice example of the interaction of analysis and algebra.*

Definition 6.6.8 *If $a \in \mathcal{A}$ and $r(a) = 0$, then a is called a quasi–nilpotent element.*

We note that $a \in \mathcal{A}$ is *nilpotent* if there exists a natural number n such that $a^n = 0$. Obviously any nilpotent element is also quasi-nilpotent, but the converse implication is not true, in general. The next result describes properties of the commuting quasi–nilpotent elements.

Corollary 6.6.9 *If $a, b \in \mathcal{A}$ are quasi–nilpotent elements and $ab = ba$, then $a + b$ and ab are quasi–nilpotent.*

Proof. This follows from (6.29). $\qquad \square$

6.7   Topological divisor of zero

Let $\mathcal{R}$ be a ring or an algebra. An element $a \in \mathcal{R}$ is a *left (right) divisor of zero in $\mathcal{R}$* if there exists $b \in \mathcal{R}$ such that $b \neq 0$ and $ab = 0$ $(ba = 0)$. If $a \in \mathcal{R}$ is a left or right divisor of zero in $\mathcal{R}$, then we say that a is a *divisor of zero in $\mathcal{R}$*. From now on, as usual, $\mathcal{A}$ is a Banach algebra with identity 1.

In this section, we study the topological generalizations of left and right divisors of zero and divisors of zero in rings.

Definition 6.7.1 An element $a \in \mathcal{A}$ is a *left (right) topological divisor of zero in $\mathcal{A}$* if there exists a sequence (b_n) in $\mathcal{A}$ such that $\|b_n\| = 1$ for all n and $\lim ab_n = 0$ $(\lim b_n a = 0)$. If $a \in \mathcal{A}$ is a left or right topological divisor of zero in $\mathcal{A}$, then it is called a *topological divisor of zero in $\mathcal{A}$*. If there exists a sequence (b_n) in $\mathcal{A}$, such that $\|b_n\| = 1$ for all n with $\lim ab_n = 0$ and $\lim b_n a = 0$, then a is called a *two–sided topological divisor of zero in $\mathcal{A}$*. We denote by $\mathcal{Z}^l$, $\mathcal{Z}^r$ and $\mathcal{Z}$, respectively, the sets of all left, right and topological divisors of zero in $\mathcal{A}$. It is clear that $\mathcal{Z} = \mathcal{Z}^l \cup \mathcal{Z}^r$. We write $\mathcal{H}^l = \mathcal{A} \setminus \mathcal{Z}^l$, $\mathcal{H}^r = \mathcal{A} \setminus \mathcal{Z}^r$ and $\mathcal{H} = \mathcal{A} \setminus \mathcal{Z}$.

Obviously, each left (right) divisor of zero in $\mathcal{A}$ is a left (right) topological divisor of zero in $\mathcal{A}$. The converse implication is not true, in general, as is shown in the next example.

Example 6.7.2 *Let $C[0, \pi/2]$ be the Banach algebra of real continuous functions on the interval $[0, \pi/2]$ with the sup–norm. Then $u \in C[0, \pi/2]$, defined by $a(t) = \sin t$ for $t \in [0, \pi/2]$, is a left and right topological divisor of zero in $C[0, \pi/2]$, but it is not a divisor of zero in $C[0, \pi/2]$.*

Proof. Obviously a is not a divisor of zero in $C[0, \pi/2]$. Let b_n be a sequence in A defined by $b_n(t) = 1 - nt$, $t \in [0, 1/n]$ and $b_n(t) = 0$ for all $t \in [1/n, 1]$. Then $\|b_n\| = 1$ for all $n = 1, 2, \ldots$ and $\lim ab_n = 0$. $\square$

Lemma 6.7.3 *Let $\mathcal{A}$ be a Banach algebra. Then*

$$a \in \mathcal{Z}^l \text{ implies } a \notin \mathcal{A}_l^{-1}, \tag{6.30}$$

$$a \in \mathcal{Z}^r \text{ implies } a \notin \mathcal{A}_r^{-1}. \tag{6.31}$$

Proof.

(i) First we show the implication in (6.30).
We assume $a \in \mathcal{Z}^l$ and $a \in \mathcal{A}_l^{-1}$. Then there exists $b \in \mathcal{A}$ such that $ba = 1$, and there exists a sequence (b_n) in $\mathcal{A}$ such that $\|b_n\| = 1$ for each n, and $\lim ab_n = 0$. Since

$$1 = \|b_n\| = \|(ba)b_n\| \leq \|b\| \|ab_n\|,$$

which is a contradiction, $a \notin \mathcal{A}_l^{-1}$ and we have shown (6.30).

(ii) Analogously, we can show the implication in (6.31). $\square$

Theorem 6.7.4 *Let $\mathcal{A}$ be a Banach algebra and $\partial(\mathcal{A}^{-1})$ be the boundary of $\mathcal{A}^{-1}$. Now $a \in \partial(\mathcal{A}^{-1})$ implies that a is a two–sided topological divisor of zero in $\mathcal{A}$, and lies in $\mathcal{Z}^l \cap \mathcal{Z}^r$.*

Proof. If $a \in \partial(\mathcal{A}^{-1})$, then, by Corollary 6.4.6, there exists a sequence $a_n \in \mathcal{A}^{-1}$ such that $\lim a_n = a$ and $\lim \|a_n^{-1}\| = \infty$. We put $b_n = a_n^{-1} \|a_n^{-1}\|^{-1}$ for $n = 1, 2, \ldots$. It follows from

$$\|ab_n\| = \|(a - a_n)b_n + a_n b_n\|$$
$$\leq \|(a - a_n)b_n\| + \|a_n b_n\| \leq \|a - a_n\| + \|a_n^{-1}\|^{-1}$$

that $\lim ab_n = 0$. Analogously it can be proved that $\lim b_n a = 0$. $\qquad \square$

Corollary 6.7.5 *Let $a \in \mathcal{A}$ and $\lambda \in \partial\sigma(a)$. Then $a - \lambda$ is a two–sided topological divisor of zero in $\mathcal{A}$.*

Proof. It follows from $\lambda \in \partial\sigma(a)$ that $\lambda \in \partial\rho(a)$. Hence we have $a - \lambda \in \partial(\mathcal{A}^{-1})$, and the proof follows from Theorem 6.7.4. $\qquad \square$

We are going to show that $\mathcal{H}^l$, $\mathcal{H}^r$ and $\mathcal{H}$ are open subsets of $\mathcal{A}$. First let us prove the next lemma.

Lemma 6.7.6 *Let λ and μ be functions from $\mathcal{A}$ into $[0, +\infty)$ defined by*

$$\lambda(x) = \inf_{y \neq 0} \frac{\|xy\|}{\|y\|} \quad and \quad \mu(x) = \inf_{y \neq 0} \frac{\|yx\|}{\|y\|}.$$

The functions λ and μ have the following properties for all $x, y \in \mathcal{A}$:

$$|\lambda(x) - \lambda(y)| \leq \|x - y\|; \tag{6.32}$$

$$|\mu(x) - \mu(y)| \leq \|x - y\|; \tag{6.33}$$

$$\lambda(x)\lambda(y) \leq \lambda(xy) \leq \|x\|\lambda(y); \tag{6.34}$$

$$\mu(x)\mu(y) \leq \mu(xy) \leq \mu(x)\|y\|.$$

Proof. It is sufficient to prove the results for the function λ only, that is, the inequalities in (6.32) and (6.34), since the corresponding results for the function μ can be proved analogously.

(i) First we prove the inequality in (6.32).
     Let $x, y, z \in \mathcal{A}$ and $z \neq 0$. Then we have

$$\lambda(x) \leq \frac{\|xz\|}{\|z\|} = \frac{\|(x - y)z + yz\|}{\|z\|} \leq \|x - y\| + \frac{\|yz\|}{\|z\|},$$

hence $\lambda(x) \leq \|x - y\| + \lambda(y)$. Obviously $\lambda(x) - \lambda(y) \leq \|x - y\|$, and we obtain the symmetry $\lambda(y) - \lambda(x) \leq \|x - y\|$. Thus we have proved (6.32).

(ii) Finally, we prove (6.34).
      We remark

$$\inf_{z \neq 0} \frac{\|xz\|}{\|z\|} \cdot \inf_{z \neq 0} \frac{\|yz\|}{\|z\|} \leq \inf_{\substack{z \neq 0 \\ yz \neq 0}} \frac{\|xyz\|}{\|yz\|} \cdot \inf_{z \neq 0} \frac{\|yz\|}{\|z\|}$$

$$\leq \inf_{z \neq 0} \frac{\|xyz\|}{\|z\|} \leq \|x\| \cdot \inf_{z \neq 0} \frac{\|yz\|}{\|z\|},$$

and so (6.34) follows. $\qquad \square$

Corollary 6.7.7 *The sets $\mathcal{H}^l$, $\mathcal{H}^r$ and $\mathcal{H}$ are open subsets of $\mathcal{A}$.*

Proof. We remark that $\lambda(x) = 0$ if and only if $x \in \mathcal{Z}^l$, and $\mu(x) = 0$ if and only if $x \in \mathcal{Z}^r$. It follows from (6.32) and (6.33) that λ and μ are continuous functions. Hence the sets $\mathcal{Z}^l$, $\mathcal{Z}^r$ and $\mathcal{Z}$ are closed subsets of $\mathcal{A}$. $\qquad \square$

6.8   Spectrum and subalgebras

Let $\mathcal{A}$ be a Banach algebra with identity 1, and $\mathcal{B}$ be a Banach subalgebra in $\mathcal{A}$, also with identity 1. If $a \in \mathcal{B}$, then we may consider the *spectrum of a with respect to* $\mathcal{B}$, denoted by $\sigma_{\mathcal{B}}(a)$, as a conventional spectrum of a with respect to $\mathcal{A}$, $\sigma(a) = \sigma_{\mathcal{A}}(a)$. Let $r_{\mathcal{B}}(a)$ be the *spectral radius of a with respect to* $\mathcal{B}$, that is, $r_{\mathcal{B}}(a) = \sup\{|\lambda| : \lambda \in \sigma_{\mathcal{B}}(a)\}$; obviously, we have $r(a) = r_{\mathcal{A}}(a)$.

Lemma 6.8.1 *Let $\mathcal{A}$ be a Banach algebra with identity 1, and $\mathcal{B}$ be Banach subalgebra of $\mathcal{A}$, also with identity 1. If $a \in \mathcal{B}$, then we have*

$$\sigma_{\mathcal{A}}(a) \subset \sigma_{\mathcal{B}}(a), \tag{6.35}$$

$$r_{\mathcal{A}}(a) = r_{\mathcal{B}}(a). \tag{6.36}$$

Proof. The inclusion in (6.35) follows from the elementary fact $\mathcal{B}^{-1} \subset \mathcal{A}^{-1}$. The equality in (6.36) follows from Theorem 6.6.5.   □

We shall prove later that we know some more information for the inclusion (6.35); but for this we need the following definitions and lemma.

Definition 6.8.2 Let $\mathbf{K}$ be a compact subset of the complex plane $\mathbb{C}$. The complement of $\mathbf{K}$, denoted by $\mathbf{K}^c$, is an open subset of $\mathbb{C}$, and its connected components are open subsets of $\mathbb{C}$ (since $\mathbb{C}$ is a locally connected space). Since $\mathbb{C}$ is separable, the set $\mathbf{K}^c$ can at most have countably many components. We put $\rho(\mathbf{K}) = \sup\{|\lambda| : \lambda \in \mathbf{K}\}$. Obviously, there exists only one component of $\mathbf{K}^c$ which contains $\{\lambda \in \mathbb{C} : |\lambda| > \rho(\mathbf{K})\}$. This component is an unbounded component of the set $\mathbf{K}^c$. The other components of $\mathbf{K}^c$ bounded, and each of them is called a *hole in* $\mathbf{K}$. If $\mathbf{K}^c$ has only one component, (obviously, this is an unbounded component of $\mathbf{K}^c$), then we say that the *set $\mathbf{K}$ has no hole*.

Definition 6.8.3 Let $\mathbf{K}$ be a compact subset of the complex plane $\mathbb{C}$, $f : \mathbf{K} \to \mathbb{C}$ and

$$\|f\|_{\mathbf{K}} = \sup\{|f(z)| : z \in \mathbf{K}\}.$$

The *polynomially convex cover of* $\mathbf{K}$, denoted by $\widehat{\mathbf{K}}$, is the set

$$\widehat{\mathbf{K}} = \{z \in \mathbb{C} : |p(z)| \leq \|p\|_{\mathbf{K}} \text{ for each polynomial }\}.$$

The set $\mathbf{K}$ is *polynomially convex* if $\mathbf{K} = \widehat{\mathbf{K}}$.

Obviously $\mathbf{K} \subset \widehat{\mathbf{K}}$; if $\mathbf{K}_1$ and $\mathbf{K}_2$ are compact subsets of $\mathbb{C}$, then $\mathbf{K}_1 \subset \mathbf{K}_2$ implies $\widehat{\mathbf{K}_1} \subset \widehat{\mathbf{K}_2}$.

Lemma 6.8.4 *If $\mathbf{K}$ is a compact subset of $\mathbb{C}$, then the complement of $\widehat{\mathbf{K}}$ is an unbounded component of the set $\mathbf{K}^c$. Hence we have*

$$\widehat{\mathbf{K}} = \mathbf{K} \cup W, \text{ where } W \text{ is the union of all holes in } \mathbf{K}.$$

Proof. The set $\mathbf{K}^c$ has at most countably many components. Let these components be the sets $C_0, C_1, C_2, \ldots$, where C_0 is the unbounded component. It is sufficient to prove that $C_0^c = \widehat{\mathbf{K}}$. For each natural number n, the set C_n is open and bounded, and $\partial C_n \subset \mathbf{K}$, and by the maximum principle, we have $C_n \subset \widehat{\mathbf{K}}$ for $n = 1, 2, \ldots$. Hence, it follows from

$C_0^c = \mathbf{K} \cup (\cup_{n=1}^{\infty} C_n)$ that $C_0^c \subset \widehat{\mathbf{K}}$, so $\widehat{\mathbf{K}}^c \subset C_0$.

Now we prove the converse inclusion. We assume that $\lambda \in C_0$. The function $(z - \lambda)^{-1}$ is analytic in some neighborhood of the set $C_0^c \equiv L$, and by Runge's corollary (see [47, Corollary 8.5, p. 88]), there exists a sequence of polynomials (p_n) such that

$$\|p_n - (z - \lambda)^{-1}\|_L \to 0, \text{ as } n \to \infty.$$

Hence it follows for $q_n = (z - \lambda)p_n$ that $\|q_n - 1\|_L \to 0$, and there exists a natural number m such that $\|q_m - 1\|_L < 1/2$. It follows from $|(q_m - 1)(\lambda)| = 1$ that $\lambda \notin \widehat{L}$. Also $\mathbf{K} \subset L$ implies $\lambda \notin \widehat{\mathbf{K}}$. Thus we have shown that $C_0 \subset \widehat{\mathbf{K}}^c$. $\qquad\square$

Theorem 6.8.5 *Let $\mathcal{A}$ be a Banach algebra with identity 1, and $\mathcal{B}$ be a Banach subalgebra of $\mathcal{A}$, also with identity 1. If $a \in \mathcal{A}$ and $\mathcal{A}(a)$ denotes the smallest Banach subalgebra of $\mathcal{A}$ which contains $\{1, a\}$, it is easy to show that $\mathcal{A}(a) = cl\{p(a) : p$ runs through the set of all polynomials$\}$. If $a \in \mathcal{B}$, then we have*

$$\sigma_{\mathcal{A}}(a) \subset \sigma_{\mathcal{B}}(a) \text{ and } \partial\sigma_{\mathcal{B}}(a) \subset \partial\sigma_{\mathcal{A}}(a); \tag{6.37}$$

$$\widehat{\sigma_{\mathcal{A}}(a)} = \widehat{\sigma_{\mathcal{B}}(a)}; \tag{6.38}$$

$$U \text{ is a hole in } \sigma_{\mathcal{A}}(a) \text{ implies } U \subset \sigma_{\mathcal{B}}(a) \text{ or } U \cap \sigma_{\mathcal{B}}(a) = \emptyset; \tag{6.39}$$

$$\widehat{\sigma_{\mathcal{A}}(a)} = \sigma_{\mathcal{A}(a)}(a). \tag{6.40}$$

Proof.

(i)   The inclusions in (6.37) follow from (6.35) and Corollary 6.7.5.

(ii) The identity in (6.38) follows from (6.37) and the maximum principle.

(iii) Now prove the implication in (6.39). We assume that U is a hole in $\sigma_{\mathcal{A}}(a)$ and we put $U_1 = U \cap \sigma_{\mathcal{B}}(a)$. It follows from this, $\partial\sigma_{\mathcal{B}}(a) \subset \partial\sigma_{\mathcal{A}}(a) \subset \sigma_{\mathcal{A}}(a)$ and $U \cap \sigma_{\mathcal{A}}(a) = \emptyset$ that $U_1 = U \cap \text{int}(\sigma_{\mathcal{B}(a)})$, and U_1 is an open subset in $\mathbb{C}$. The set $U_2 = U \setminus \sigma_{\mathcal{B}}(a)$ is also open in $\mathbb{C}$. Since $U = U_1 \cup U_2$, $U_1 \cap U_2 = \emptyset$, and since U is connected, it follows that at least one of the sets U_1 and U_2 is empty. Thus we have shown (6.39).

(iv) Finally we prove the identity in (6.40).

First we note that the inclusion $\widehat{\sigma_{\mathcal{A}}(a)} \supset \sigma_{\mathcal{A}(a)}(a)$ is obvious.

To prove the converse inclusion, we assume that $\lambda \in \widehat{\sigma_{\mathcal{A}}(a)} \setminus \sigma_{\mathcal{A}(a)}(a)$. Now $(a - \lambda)^{-1} \in \mathcal{A}(a)$, and let (p_n) be a sequence of polynomials such that $p_n(a) \to (a - \lambda)^{-1}$. It follows from $q_n(z) = (z - \lambda)p_n(z)$ for $z \in \mathbb{C}$ that $\|q_n(a) - 1\| \to 0$, and so there exists a natural number m such that $\|q_m(a) - 1\| < 1/2$. We obtain from $\lambda \in \widehat{\sigma_{\mathcal{A}}(a)}$ and Theorem 6.5.10

$$\begin{aligned}
1 = |(q_m - 1)(\lambda)| &\leq \sup\{|(q_m - 1)(z)| : z \in \sigma_{\mathcal{A}}(a)\} \\
&= \sup\{|w| : w \in \sigma_{\mathcal{A}}((q_m - 1)(a))\} \\
&= r_{\mathcal{A}}((q_m - 1)(a)) \leq \|q_m(a) - 1\| < 1/2.
\end{aligned}$$

This is a contradiction, and so the inclusion $\widehat{\sigma_{\mathcal{A}}(a)} \subset \sigma_{\mathcal{A}(a)}(a)$ must also hold. Thus we have shown the identity in (6.40). $\qquad\square$

Definition 6.8.6 Let $(\mathcal{A}, \|\cdot\|)$ be a Banach algebra with identity 1, p be a nontrivial idempotent in $\mathcal{A}$ and

$$p\mathcal{A}p = \{pap : a \in \mathcal{A}\} = \{a \in \mathcal{A} : a = ap = pa\}.$$

Then $(p\mathcal{A}p, \|\cdot\|)$ is a Banach algebra with identity p. If $a \in \mathcal{A}$ and $ap = pa$ then p is said to *decompose* a. The element $b \in \mathcal{A}$ is said to be *invertible with respect to* p if

$$pbp \in (p\mathcal{A}p)^{-1},$$

and analogously for left and right invertibility.

The next theorem shows that if p decomposes a, then the invertibility in $\mathcal{A}$ could be obtained by the invertibility of $p\mathcal{A}p$ and $(1-p)\mathcal{A}(1-p)$.

Theorem 6.8.7 *Let $\mathcal{A}$ be a Banach algebra with identity 1 and p be a nontrivial idempotent in $\mathcal{A}$. If $a \in \mathcal{A}$ and $ap = pa$, then we have*

$$a \in \mathcal{A}^{-1} \text{ if and only if } pap \subset (p\mathcal{A}p)^{-1} \text{ and } (1-p)a(1-p) \in ((1-p)\mathcal{A}(1-p))^{-1}. \quad (6.41)$$

Proof.

(i) First we show the sufficiency of the condition $a \in \mathcal{A}^{-1}$ in (6.41).
 We assume $a \in \mathcal{A}^{-1}$. Then there exists $b \in \mathcal{A}$ such that $ba = 1$. From

$$(pbp)(pap) = pbap = p,$$

we obtain that pbp is a left inverse of pap in $p\mathcal{A}p$. Since

$$(1-p)b(1-p)(1-p)a(1-p) = (1-p)ba(1-p) = 1 - p,$$

$(1-p)b(1-p)$ is a left inverse of $(1-p)a(1-p)$ in $(1-p)\mathcal{A}(1-p)$.
 Analogously we can prove the results for the right inverses. Hence we have shown the sufficiency of the condition $a \in \mathcal{A}^{-1}$ in (6.41).

(ii) First we show the necessity of the condition $a \in \mathcal{A}^{-1}$ in (6.41).
 We assume that there exist $c \in p\mathcal{A}p$ and $d \in (1-p)\mathcal{A}(1-p)$ such that

$$cap = p \text{ and } da(1-p) = 1 - p.$$

It follows from

$$1 = p + (1-p) = (cp + d(1-p))a,$$

that $a \in \mathcal{A}_l^{-1}$.
 Analogously we can show that $a \in \mathcal{A}_r^{-1}$ follows. $\qquad\square$

Theorem 6.8.8 *Let $\mathcal{A}$ be a Banach algebra with identity 1, and p be a nontrivial idempotent in $\mathcal{A}$. Then $p\mathcal{A}p$ is a closed subalgebra of $\mathcal{A}$, p is the identity in $p\mathcal{A}p$ and, for each $x \in p\mathcal{A}p$, we have*

$$\sigma_\mathcal{A}(x) = \sigma_{p\mathcal{A}p}(x) \cup \{0\}.$$

Proof. It is easy to show that $p\mathcal{A}p$ is a subalgebra of $\mathcal{A}$ with identity p. We have to prove that $p\mathcal{A}p$ is a closed subset of $\mathcal{A}$. Let (x_n) be a sequence in $\mathcal{A}$ such that $\lim_n px_np = x \in \mathcal{A}$. Then we have $\lim_n ppx_npp = pxp \in \mathcal{A}$, and this implies $x = pxp \in p\mathcal{A}p$. Hence $p\mathcal{A}p$ is a closed subset of $\mathcal{A}$.

Let $x \in p\mathcal{A}p$. Then $x(1 - p) = 0$, and it follows that $0 \in \sigma_{\mathcal{A}}(x)$. Now $\lambda \in \sigma_{p\mathcal{A}p}(x)$ and $x - \lambda \in \mathcal{A}^{-1}$ imply $x - \lambda p \in (p\mathcal{A}p)^{-1}$ by Theorem 6.8.7. Hence we have shown that

$$\sigma_{\mathcal{A}}(x) \supset \sigma_{p\mathcal{A}p}(x) \cup \{0\}.$$

To show the converse inclusion, we assume $0 \neq \lambda \in \sigma_{\mathcal{A}}(x)$. Since $p(x - \lambda)p = x - \lambda p$ and $(1 - p)(x - \lambda)(1 - p) = -\lambda(1 - p) \in ((1 - p)\mathcal{A}(1 - p))^{-1}$, it follows by Theorem 6.8.7 that $\lambda \in \sigma_{p\mathcal{A}p}(x)$. $\qquad\square$

We remark that an ideal (left, right or two–sided) $\mathcal{I}$ in $\mathcal{A}$ is a *proper (left, right, two–sided) ideal* in $\mathcal{A}$ if $\{0\} \neq \mathcal{I} \neq \mathcal{A}$.

Theorem 6.8.9 *Let $\mathcal{A}$ be a Banach algebra with identity 1 and $\mathcal{L}$ be a proper two–sided ideal in $\mathcal{A}$ with identity f. Then $\mathcal{L}$ is a closed subalgebra of $\mathcal{A}$ and for each $a \in \mathcal{L}$ and each $\lambda \in \mathbb{C}$, the following equality holds*

$$\sigma_{\mathcal{A}}(a + \lambda) = \sigma_{\mathcal{L}}(a + \lambda f) \cup \{\lambda\}.$$

Proof. By assumption, f is a nontrivial idempotent in $\mathcal{A}$ and $f\mathcal{A}f \subset \mathcal{L}$. Since f is an identity in $\mathcal{L}$, we have $\mathcal{L} = f\mathcal{L}f \subset f\mathcal{A}f$. Hence we have shown that $\mathcal{L} = f\mathcal{A}f$. It follows from Theorem 6.8.8 that $\mathcal{L}$ is a closed subalgebra of $\mathcal{A}$. Now, for each $a \in \mathcal{L}$ and each $\lambda \in \mathbb{C}$, it follows from Theorem 6.5.10 that

$$\sigma_{\mathcal{A}}(a + \lambda) = \sigma_{\mathcal{A}}(a) + \lambda \text{ and } \sigma_{\mathcal{L}}(a + \lambda f) = \sigma_{\mathcal{L}}(a) + \lambda. \tag{6.42}$$

Again by Theorem 6.8.8, keeping in mind (6.42), we complete the proof. $\qquad\square$

6.9   One Hochwald-Morell and two Harte theorems

Let $\mathcal{A}$ be a Banach algebra with identity 1. If $a, b \in \mathcal{A}^{-1}$, then $(ab)^{-1} = b^{-1}a^{-1}$ and $a^{-1}a = 1$. These conditions are equivalent to the existence of a function $g : \mathcal{A}^{-1} \to \mathcal{A}$ with $g(ab) = g(b)g(a)$ and $g(a)a = 1$. Hochwald and Morell [94] asked, could the domain of the function g be extended, that is, does there exist $\mathcal{L} \subset \mathcal{A}$, such that $\mathcal{L}$ is a multiplicative semigroup, $\mathcal{L} \setminus \mathcal{A}^{-1} \neq \emptyset$, and the function $g : \mathcal{L} \to \mathcal{A}$ satisfies $g(ab) = g(b)g(a)$ and $g(a)a = 1$ for all $a, b \in \mathcal{L}$? The answer to this question is negative, as the following theorem shows.

Theorem 6.9.1 (S. H. Hochwald–B. B. Morell) ([94, Theorem 2.1]) *Let $\mathcal{A}$ be a Banach algebra with identity 1 and $\mathcal{A}_l^{-1}$ be a multiplicative semigroup of left invertible elements in $\mathcal{A}$ such that $\mathcal{A}_l^{-1} \setminus \mathcal{A}^{-1} \neq \emptyset$. Then there does not exist a function $g : \mathcal{A}_l^{-1} \to \mathcal{A}$ such that for all $a \in \mathcal{A}_l^{-1}$*

$$g(a)a = 1, \tag{6.43}$$

and

$$b \in \mathcal{A}_l^{-1} \text{ and } ab = ba \text{ imply } g(a)g(b) = g(b)g(a). \tag{6.44}$$

Proof. We assume that such a function g exists and $a \in \mathcal{A}_l^{-1} \setminus \mathcal{A}^{-1}$. If $|\lambda| > \|a\|$, then $\lambda - a \in \mathcal{A}^{-1}$ by Corollary 6.4.3. Since $(\lambda - a)^{-1}(\lambda - a) = 1 = (\lambda - a)(\lambda - a)^{-1}$, it follows that $(\lambda - a)^{-1}a = a(\lambda - a)^{-1}$, and from (6.44), we obtain

$$g((\lambda - a)^{-1})g(a) = g(a)g((\lambda - a)^{-1}). \tag{6.45}$$

It follows from (6.43) and (6.45) that $(\lambda - a)g(a) = g(a)(\lambda - a)$, hence $ag(a) = g(a)a$. Now, we get from (6.43) $ag(a) = 1 = g(a)a$, which is a contradiction to $a \notin \mathcal{A}^{-1}$. $\square$

Harte [89, Theorem 1] generalized the previous theorem, and before we present Harte's result, we have to introduce the following notions.

Definition 6.9.2 An element $a \in \mathcal{A}$ is *regular (in the sense of J. von Neumann) in* $\mathcal{A}$, or simply *regular*, if there exists $x \in \mathcal{A}$ such that

$$axa = a.$$

An algebra $\mathcal{A}$ is *regular* if every element is $\mathcal{A}$ is regular. The set of all regular elements in $\mathcal{A}$ is denoted by $\mathcal{A}^{\sqcap}$. We note

$$\mathcal{A}^{\sqcap} = \{a \in \mathcal{A} : a \in a\mathcal{A}a\},$$

$$\mathcal{A}_l^{-1} \subset \mathcal{A}^{\sqcap} \text{ and } \mathcal{A}_r^{-1} \subset \mathcal{A}^{\sqcap}.$$

The next lemma is useful and easy to prove.

Lemma 6.9.3 *If an element $a \in \mathcal{A}$ is regular, then there exists $x \in \mathcal{A}$ such that*

$$axa = a,$$

and

$$xax = x.$$

Proof. Since $a \in \mathcal{A}$ is regular, there exists $y \in \mathcal{A}$ such that

$$aya = a. \tag{6.46}$$

We put $x = yay$. Applying (6.46) several times, it is easy to prove that

$$axa = a(yay)a = (aya)ya = aya = a,$$

and

$$xax = (yay)a(yay) = y(aya)(yay) = y(aya)y = yay = x. \qquad \square$$

Theorem 6.9.4 (Harte) ([89, Theorem 1]) *Let $\mathcal{A}$ be a Banach algebra with identity 1 and $\mathcal{D}$ be a multiplicative semigroup in $\mathcal{A}$ with $\mathcal{A}^{-1} \subset \mathcal{D} \subset \mathcal{A}^{\sqcap}$. Then, if the function $g : \mathcal{D} \to \mathcal{A}$ has the properties for all $a \in \mathcal{D}$*

$$a = ag(a)a, \tag{6.47}$$

and

$$b \in \mathcal{A}^{-1} \text{ and } ab = ba \text{ imply } g(a)g(b) = g(b)g(a), \tag{6.48}$$

then

$$a \in \mathcal{D} \text{ implies } a \in a\mathcal{A}^{-1}a.$$

Proof. If $|\lambda| > \|a\|$, then $\lambda - a \in \mathcal{A}^{-1}$ by Corollary 6.4.3 and $(\lambda - a)^{-1}a = a(\lambda - a)^{-1}$. Thus it follows from (6.47) and (6.48) that

$$g((\lambda - a)^{-1}) = \lambda - a, \tag{6.49}$$

and

$$g((\lambda - a)^{-1})g(a) = g(a)g((\lambda - a)^{-1}). \tag{6.50}$$

Now (6.49) and (6.50) imply $(\lambda - a)g(a) = g(a)(\lambda - a)$, that is,

$$ag(a) = g(a)a. \tag{6.51}$$

We put

$$x = g(a)ag(a) + (1 - ag(a)). \tag{6.52}$$

It is easy to show that

$$axa = a[g(a)ag(a) + (1 - ag(a))]a = ag(a)ag(a)a + a(1 - ag(a))a = a.$$

It follows from (6.51), (6.52) and (6.47) that

$$x[a + (1 - ag(a))] = [g(a)ag(a) + (1 - ag(a))][a + (1 - ag(a))]$$

$$= g(a)ag(a)a + (1 - ag(a))a + g(a)ag(a)(1 - ag(a)) + (1 - ag(a))(1 - ag(a))$$

$$= g(a)a + a - ag(a)a + g(a)g(a)a(1 - g(a)a) + 1 - ag(a) = g(a)a + 1 - ag(a) = 1,$$

that is, we have shown that $x \in \mathcal{A}_r^{-1}$. Analogously, we can prove

$$[a + (1 - ag(a))]x = [a + (1 - ag(a))][g(a)ag(a) + (1 - ag(a))]$$

$$= ag(a)ag(a) + a(1 - ag(a)) + (1 - ag(a))g(a)ag(a) + (1 - ag(a))(1 - ag(a))$$

$$= ag(a) + a(1 - g(a)a) + (1 - ag(a))ag(a)g(a) + (1 - ag(a)) =$$

$$= ag(a) + 1 - ag(a) = 1,$$

that is, $x \in \mathcal{A}_l^{-1}$. $\qquad \square$

We remark that Theorem 6.9.1 follows from Theorem 6.9.4, since $a \in a\mathcal{A}^{-1}a$ and $a \in \mathcal{A}_l^{-1}$ imply $a \in \mathcal{A}^{-1}$.

Lemma 6.9.5 *Let $\mathcal{A}$ be a Banach algebra with identity 1 and*

$$\mathcal{A}^\bullet = \{a \in \mathcal{A} : a^2 = a\}$$

be the set of idempotents in $\mathcal{A}$. Then

$$\{a \in \mathcal{A} : a \in a\mathcal{A}^{-1}a\} = \mathcal{A}^\bullet \mathcal{A}^{-1} = \mathcal{A}^{-1}\mathcal{A}^\bullet.$$

Proof. The proof is left to the readers. $\qquad \square$

Now we are going to prove one more theorem by Harte (see [87], or [88, Theorem 7.3.4]).

Theorem 6.9.6 (Harte) ([87, 1.1 Theorem]) *Let $\mathcal{A}$ be a Banach algebra with identity 1. Then*

$$\mathcal{A}^\sqcap \cap \overline{\mathcal{A}^{-1}} = \mathcal{A}^{-1}\mathcal{A}^\bullet.$$

Proof. If $a \in \mathcal{A}^\sqcap \cap \overline{\mathcal{A}^{-1}}$, then there exist $x \in \mathcal{A}$ and $b \in \mathcal{A}^{-1}$ such that $\|b - a\| \, \|x\| < 1$,

$$a = axa \text{ and } 1 + (b - a)x \in \mathcal{A}^{-1}.$$

We put $p = xa$ and $c = (1 + (b - a)x)^{-1}b \in \mathcal{A}^{-1}$. It is easy to see that p is idempotent and

$$cp = (1 + (b - a)x)^{-1}bxa = (1 + (b - a)x)^{-1}(1 + (b - a)x)a = a.$$

Thus we have shown that $\mathcal{A}^\sqcap \cap \overline{\mathcal{A}^{-1}} \subset \mathcal{A}^{-1}\mathcal{A}^\bullet$. Concerning the converse inclusion, it follows from $a = cp, c \in \mathcal{A}^{-1}$ and $p \in \mathcal{A}^\bullet$ that $a_n = c(p + (1/n)(1 - p)) \in \mathcal{A}^{-1}$ for $n = 1, 2, \dots$ and $\lim_n a_n = a$. $\qquad \square$

6.10 $\mathcal{B}(X)$ as a Banach algebra

In the previous sections we have studied some properties of general Banach algebras. In this section we study the Banach algebra $\mathcal{B}(X)$ of all bounded linear operators on a Banach space X in more detail. We remark that many results which we present here are true with an adequate explanation for operators in $\mathcal{B}(X, Y)$ when X and Y are Banach spaces.

Theorem 6.10.1 *An operator $A \in \mathcal{B}(X)$ is left invertible, that is, $A \in \mathcal{B}(X)_l^{-1}$ if and only if A is one to one and there exists a projector $P \in \mathcal{B}(X)$ such that $R(P) = R(A)$.*

Proof.

(i) First we show the sufficiency of the condition $A \in \mathcal{B}(X)_l^{-1}$.
It follows from $A \in \mathcal{B}(X)_l^{-1}$ that there exists $B \in \mathcal{B}(X)$ such that $BA = I$. Then $ABA = A$ and, for $P = AB$, we have $P^2 = P \in \mathcal{B}(X)$. Obviously, $R(AB) \subset R(A)$. It follows from $PA = A$ that $R(A) \subset R(P)$. So we have shown that $R(P) = R(A)$. Obviously A is one to one.
This concludes the Part (i) of the proof.

(ii) Now we show the necessity of the condition $A \in \mathcal{B}(X)_l^{-1}$.
We assume that A is one to one. There exists a projector $P \in \mathcal{B}(X)$ such that $R(P) = R(A)$, and we show that $A \in \mathcal{B}(X)_l^{-1}$. Since $R(P) = R(A)$ is a closed subspace of the Banach space X, $R(A)$ is a Banach subspace. Let A_0 be the operator from X into $R(A)$ defined by $A_0 x = Ax$ for $x \in X$. Obviously, we have $A_0 \in \mathcal{B}(X, R(A))$. Now, by the theorem of the bounded inverse (Theorem 4.6.3), there exists $C \in \mathcal{B}(R(A), X)$ such that $CA_0 = I \in \mathcal{B}(X)$. Since $P \in \mathcal{B}(X)$ and $R(P) = R(A)$, it follows that $CP \in \mathcal{B}(X)$ and $(CP)A_0 = I$. Obviously, we have $(CP)A = I$. $\qquad\square$

Theorem 6.10.2 *An operator $A \in \mathcal{B}(X)$ is right invertible, that is, $A \in \mathcal{B}(X)_r^{-1}$ if and only if $R(A) = X$ and there exists a projector $P \in \mathcal{B}(X)$ such that $R(P) = N(A)$.*

Proof.

(i) First we show the sufficiency of the condition $A \in \mathcal{B}(X)_r^{-1}$.
It follows from $A \in \mathcal{B}(X)_r^{-1}$ that there exists $B \in \mathcal{B}(X)$ such that $AB = I$. Hence we have $R(A) = X$ and $ABA = A$. Furthermore, $BABA = BA$ and, for $Q = BA$, we have $Q^2 = Q \in \mathcal{B}(X)$. Obviously $N(BA) \supset N(A)$. If $x \in N(BA)$, then $Ax = (ABA)x = A(BAx) = 0$. So we have shown that $N(BA) \subset N(A)$, and so $N(Q) = N(A)$. We remark that $P = I - Q \in \mathcal{B}(X)$ is a projector and $R(P) = N(Q) = N(A)$.

(ii) First we show the necessity of the condition $A \in \mathcal{B}(X)_r^{-1}$.
Conversely, let $R(A) = X$ and $P \in \mathcal{B}(X)$ be a projector such that $R(P) = N(A)$. It follows from Theorem 4.6.12 that there exists a decomposition of the space X into closed subspaces $N(P)$ and $R(P)$, that is, $X = N(P) \oplus R(P)$. Let A_1 be the operator from $N(P)$ onto X defined by $A_1 x = Ax$ for all $x \in N(P)$. Obviously, A_1 is one to one and $A_1 \in \mathcal{B}(N(P), X)$. By the theorem of the bounded inverse (Theorem 4.6.3), there exists an operator $C \in \mathcal{B}(X, N(P))$ such that $A_1 S = I \in \mathcal{B}(X)$. Since $(I - P)S \in \mathcal{B}(X)$, $R(I - P) = N(P)$ and the restriction of the operator $I - P$ on $N(P)$ is the identity, it follows that $A(I - P)C = I$. So we have shown that $A \in \mathcal{B}(X)_r^{-1}$. $\qquad\square$

The invertibility of elements in a Banach algebra $\mathcal{B}(X)$ is characterized by Banach's theorem on the bounded inverse, Theorem 4.6.3, and it is convenient to mention it here.

Corollary 6.10.3 *An operator $A \in \mathcal{B}(X)$ is invertible if and only if it is one to one and $R(A) = X$.*

Proof. This follows from 6.10.1 and Theorem 6.10.2.   □

Theorems 6.10.1 and 6.10.2 imply the one–sided invertibility of elements in $\mathcal{B}(X)$, among other things, it depends on the existence of special projectors in $\mathcal{B}(X)$. Namely, if $A \in \mathcal{B}(X)$ is left (right) invertible then there exist projectors $P, Q \in \mathcal{B}(X)$, such that $R(P) = R(A)$ and $(R(Q) = N(A)$. These properties of P and Q are related to the regularity of elements in Definition 6.9.2.

Remark 6.10.4 *An operator $A \in \mathcal{B}(X)$ is regular in $\mathcal{B}(X)$ if there exists $B \in \mathcal{B}(X)$ such that*

$$ABA = A. \tag{6.53}$$

In this case, the operator B is a generalized inverse (pseudo–inverse) *of A.*

Unfortunately, the terminology is not standardized. Sometimes *regular* is referred to as *invertible* and having a generalized inverse is described as *relatively regular*. Some authors also refer to B in (6.53) as an *inner inverse of the operator of A, or as a g_1–inverse of A,* and then A is said to be a *g_1–invertible operator*.

If for $A \in \mathcal{B}(X)$, there exists an operator $B \in \mathcal{B}(X)$ such that

$$BAB = B, \tag{6.54}$$

then B is said to be an *outer inverse of A*.

If, for an operator $A \in \mathcal{B}(X)$, there exists an operator $C \in \mathcal{B}(X)$ such that

$$ACA = A, \tag{6.55}$$

and

$$CAC = C, \tag{6.56}$$

then many authors refer to C as a *g_2–inverse of A or as a* generalized inverse *of A (C as a g–inverse of A, for short), and A is referred to as a g_2–invertible operator*, or simply as a *g–invertible operator*.

We note that if an operator A is g_1–invertible, then it is also g_2-invertible by Lemma 6.9.3. The next theorem gives a characterization of g–invertible operators.

Theorem 6.10.5 *Let $A \in \mathcal{B}(X)$. The following statements are equivalent*

$$A \text{ is a } g\text{–invertible operator}; \tag{6.57}$$

$$R(A) = \overline{R(A)} \text{ and there exist projectors } P, Q \in \mathcal{B}(X) \text{ such that}$$
$$R(P) = R(A) \text{ and } R(Q) = N(A). \tag{6.58}$$

Proof.

(i) First we show that the condition in (6.57) implies those in (6.58).
     We assume that A is g–invertible. Then there exists $B \in \mathcal{B}(X)$ such $ABA = A$ and $BAB = B$. Obviously $ABAB = AB$ and $BABA = BA$. Furthermore, if $P = AB$ and $F = BA$, we obtain $P^2 = P$, $F^2 = F$ and $P, F \in \mathcal{B}(X)$. It follows from the proofs of Theorems 6.10.1 and 6.10.2) that $R(AB) = R(A)$ and $N(BA) = N(A)$, respectively. Hence, $R(A) = R(P)$ implies that $R(A)$ is a closed subspace of X. We put $Q = I - F$ and remark that $Q^2 = Q$ and $R(Q) = R(I - F) = N(A)$.
     So we have proved that (6.57) implies (6.58).

(ii) Finally, we prove that the conditions in (6.58) imply (6.57).

We have from (6.58) that $X = R(Q) \oplus N(Q) = N(A) \oplus N(Q)$, $R(P) = R(A)$ and $R(Q) = N(A)$. Let A_1 be the operator from $N(Q)$ to $R(P)$ defined by $A_1 x = Ax$ for all $x \in N(Q)$. Obviously A_1 is a bijection and $A_1 \in \mathcal{B}(N(Q), R(P))$. By the theorem of the bounded inverse (Theorem 4.6.3), there exists the inverse operator A_1^{-1} of A_1 and $A_1^{-1} \in \mathcal{B}(R(P), N(Q))$. We put $B = A_1^{-1}P$. Obviously, we have $B \in \mathcal{B}(X)$. Furthermore, $ABA = A(I - Q + Q)BA = A(I - Q)BA = A_1 A_1^{-1} PA = A$ and $BAB = BA(I - Q + Q)B = BA(I - Q)B = BA_1 A_1^{-1} P = BP = B$. □

If A is a g–invertible operator, then A can have several g–inverses. The next theorem gives more information on the number of g–inverses.

Theorem 6.10.6 (Caradus) ([37, Theorem 1 (v)]) *Let $A \in \mathcal{B}(X)$ be a g–invertible operator and $A\{1,2\}$ be the set of all g–inverses of A. Also let $\mathcal{P}$ be the set of all projectors in $\mathcal{B}(X)$ such that each range is equal to $R(A)$ and $\mathcal{Q}$ be the set of all projectors in $\mathcal{B}(X)$ with range equal to $N(A)$. Then $\mathcal{P} \times \mathcal{Q}$ and $A\{1,2\}$ have the same cardinality, that is,*

$$\operatorname{card}(\mathcal{P} \times \mathcal{Q}) = \operatorname{card}(A\{1,2\}).$$

Proof. It is sufficient to prove that there exists a bijection from $\mathcal{P} \times \mathcal{Q}$ onto $A\{1,2\}$. We define the map $f : \mathcal{P} \times \mathcal{Q} \to A\{1,2\}$ such that $f(P,Q) = A_1^{-1}P$ for $(P,Q) \in \mathcal{P} \times \mathcal{Q}$, and A_1^{-1} is the operator from the Part (ii) of the proof of Theorem 6.10.5, namely that (6.58) implies (6.57). By Theorem 6.10.5, the map f is well defined. Let $g : A\{1,2\} \to \mathcal{P} \times \mathcal{Q}$ be defined by $g(B) = (AB, I - BA)$ for all $B \in A\{1,2\}$. We remark that it follows from Part (i) of the proof of Theorem 6.10.5, namely that (6.57) implies (6.58), that $AB \in \mathcal{P}$ and $(I - BA) \in \mathcal{Q}$. We note

$$(gf)(P,Q) = g(f(P,Q)) = g(A_1^{-1}P) = (AA_1^{-1}P, I - A_1^{-1}PA) = (A(I - Q + Q)A_1^{-1}P,$$

$$I - A_1^{-1}PA(I - Q + Q)) = (A(I - Q)A_1^{-1}P,$$

$$I - A_1^{-1}PA(I - Q)) = (A_1 A_1^{-1}P, I - A_1^{-1}A(I - Q)) = (P, I - (I - Q)) = (P,Q)$$

and

$$(P,Q) \in \mathcal{P} \times \mathcal{Q}.$$

Furthermore, we have

$$(fg)(B) = f(g(B)) = f(AB, I - BA) = A_1^{-1}AB = A_1^{-1}A(I - (I - BA) + (I - BA))B$$
$$= A_1^{-1}A(I - (I - BA))B = (I - (I - BA))B = B, \ B \in A\{1,2\}.$$

Thus we have shown that f is a bijection. □

Theorem 6.10.5 gives conditions for existence of the inner inverse of $A \in \mathcal{B}(X)$. It is easy to see that each operator $A \in \mathcal{B}(X)$ has an outer inverse; obviously, we can take $B = 0$ in (6.54). It is of interest to determine when an operator $A \in \mathcal{B}(X)$ has an outer inverse $B \neq 0$. The answer is given in the next theorem.

Theorem 6.10.7 *Let $A \in \mathcal{B}(X)$. Then the operator A has an outer inverse $B \in \mathcal{B}(X)$ with $B \neq 0$ if and only if $A \neq 0$.*

Proof. If $B \neq 0$ is an outer inverse of A, then obviously $A \neq 0$. Conversely, we assume $A \neq 0$ and $z \in X$ such that $Az \neq 0$. By Corollary 4.5.4, there exists $f \in X^*$ such that $f(Az) = 1$. Let $B : X \to X$ be the operator defined by

$$Bx = f(x)z \text{ for all } x \in X.$$

It is easy to show that $B \in \mathcal{B}(X)$ and $B \neq 0$. Since

$$(BAB)x = B(Af(x)z) = f(x)(B(Az)) = f(x)f(Az)z = Bx \text{ for all } x \in X,$$

it follows that B is an outer inverse of A.   □

Although g–invertibile operators have some properties of invertible operators, we can expect that in many properties they may differ. The next example shows that the product of g–invertibile operators need not necessarily be g–invertible.

Example **6.10.8 (S. R. Caradus)** ([36, Example, p.29]) *Let* $X = \ell_2$ *and* $B, C \in \mathcal{B}(X)$ *be defined as follows*

$$Be^{(k)} = \begin{cases} e^{(2k)} & \text{if } k \text{ is an odd number} \\ 0 & \text{if } k \text{ is an even number.} \end{cases}$$

$$Ce^{(k)} = \frac{1}{k^2}e^{(k+1)} \text{ for all } k = 1, 2 \ldots.$$

(Here $e^{(k)}$, *as usual, is the vector with* 1 *in the* k–*th coordinate and* 0 *otherwise.) Let* λ *be an element of the resolvent set of* C *and* $A = B(C - \lambda I)$. *Then* A *is* g–*invertibile and* A^2 *is not* g–*invertibile.*

Proof. We remark that $B^2 = 0$ and C is a compact but not finite dimensional operator. The operator BCB is compact, and since $BCBe^{(k)} = (1/4k^2)e^{(4k+2)}$, where k is an odd number, BCB is not a finite dimensional operator. Hence $R(BCB)$ is not a closed subset of X by Theorem 4.14.3. Since $R(A) = R(B)$ is a closed subspace, A is g–invertible (ℓ_2 is a Hilbert space). It follows from

$$A^2 = B(C - \lambda I)B(C - \lambda I) = (BC - \lambda B)(BC - \lambda B)$$
$$= BCBC - \lambda BCB - \lambda B^2 C + \lambda^2 B^2 = BCBC - \lambda BCB$$
$$= BCB(C - \lambda I),$$

that $R(A^2) = R(BCB)$ is not a closed set. Hence, by Theorem 6.10.5, the operator A^2 is not g-invertible.   □

We are going to study which properties of an operator in $\mathcal{B}(X)$ describe its singularity. The next theorems describe the divisors of zero in $\mathcal{B}(X)$.

Theorem 6.10.9 *Let* $A \in \mathcal{B}(X)$. *Then*

$$A \text{ is a left divisor of zero in } \mathcal{B}(X) \text{ if and only if } A \text{ is not one to one,} \qquad (6.59)$$

$$A \text{ is a right divisor of zero in } \mathcal{B}(X) \text{ if and only if } \overline{R(A)} \neq X. \qquad (6.60)$$

Proof.

(i) First we show the equivalence in (6.59).
    If A is a left divisor of zero in $\mathcal{B}(X)$, then there exists $B \in \mathcal{B}(X)$ such that $B \neq 0$ and $AB = 0$. Let $x \in X$ and $Bx \neq 0$. It follows from $A(Bx) = 0$ that A is not one to one. Conversely, we assume that A is not one to one. Then there exists $z \in X$ such that $z \neq 0$ and $Az = 0$. By Corollary 4.5.4, there exists $f \in X^*$ such that $\|f\| = 1$ and $fz = \|z\|$. We define the operator B by $Bx = f(x)z$ for all $x \in X$. It is easy to show that $B \in \mathcal{B}(X)$ and $\|B\| = \|z\|$. It follows that $B \neq 0$. Also $(AB)x = A(Bx) = A(f(x)z) = f(x)Az = 0$ for all $x \in X$ implies that A is a left divisor of zero in $\mathcal{B}(X)$. So we have shown the equivalence in (6.59).

(ii) Finally, we show the equivalence in (6.60).
If A is a right divisor of zero $\mathcal{B}(X)$, then there exists $B \in \mathcal{B}(X)$ such that $B \neq 0$ and $BA = 0$. It follows that $R(A) \subset N(B)$, that is, $\overline{R(A)} \subset N(B)$. Hence $\overline{R(A)} \neq X$. Conversely, if $\overline{R(A)} \neq X$, then there exists $y \in X \setminus \overline{R(A)}$ and, by Corollary 4.5.4, there exists $f \in X^*$ such that $\|f\| = 1$, $f(y) = \inf\{\|y - z\| : z \in \overline{R(A)}\} > 0$ and $f(z) = 0$ for each $z \in \overline{R(A)}$. We define the operator B by $Bx = f(x)y$ for all $x \in X$. It is easy to show that $B \in \mathcal{B}(X)$, $\|B\| = \|y\|$ and $(BA)x = B(Ax) = f(Ax) = 0$, $x \in X$. $\square$

In the following theorem we describe the topological divisors of zero in $\mathcal{B}(X)$. In these results, we use the functions j from Definition 4.15.1 and q from Definition 4.15.7.

Theorem 6.10.10 *Let $A \in \mathcal{B}(X)$. The following statements are equivalent:*

$$A \text{ be a left topological divisor of zero in } \mathcal{B}(X), \tag{6.61}$$

$$j(A) = 0. \tag{6.62}$$

There exists a sequence (x_n) in X, such that $\|x_n\| = 1$ for $n = 1, 2, \ldots$, and

$$\lim_{n \to \infty} \|Ax_n\| = 0. \tag{6.63}$$

Proof. Obviously (6.62) and (6.63) are equivalent. Hence, it suffices to show the equivalence of (6.61) and (6.63).

(i) First we show that the condition in (6.61) implies the statement in (6.63).
Let A be a left topological divisor of zero in $\mathcal{B}(X)$. Then there exists a sequence B_n in $\mathcal{B}(X)$ such that $\|B_n\| = 1$ for each n, that is, $\lim_{n \to \infty} AB_n = 0$. It follows from $\|B_n\| = 1$ that there exists a sequence (y_n) in X such that $\|y_n\| = 1$ and $\|B_n(y_n)\| \geq 1/2$ for each n. We put $x_n = B_n(y_n/\|B_n y_n\|)$ for $n = 1, 2, \ldots$. Obviously we have $\|x_n\| = 1$ for each n. It follows from

$$\|Ax_n\| = \left\| \frac{1}{\|B_n y_n\|} AB_n y_n \right\| = \frac{1}{\|B_n y_n\|} \|AB_n y_n\|$$
$$\leq 2\|AB_n\| \, \|y_n\| = 2\|AB_n\| \to 0 \text{ as } n \to \infty, \tag{6.64}$$

that (6.61) implies (6.63).

(ii) Now we show that the statement in (6.63) implies (6.61).
If (x_n) is the sequence in (6.63), then, by Corollary 4.5.4, there exists $f \in X^*$ such that $\|f\| = 1$. We define the sequence B_n by $B_n(x) = f(x)x_n$ for all $x \in X$. It is easy to prove that $B_n \in \mathcal{B}(X)$ and $\|B_n\| = \|x_n\| = 1$ for all n. It follows from

$$\|AB_n\| = \sup_{\|x\|=1} \|AB_n(x)\| = \sup_{\|x\|=1} \|f(x)\| \|Ax_n\| = \|Ax_n\|, \tag{6.65}$$

that $\lim_{n \to \infty} AB_n = 0$ and we have shown that (6.63) implies (6.61). $\square$

Definition 6.10.11 Let $A \in \mathcal{B}(X)$. We put

$$\sigma_p(A) = \{\lambda \in \mathbb{C} : A - \lambda I \text{ is not one to one}\},$$

$$\sigma_a(A) = \{\lambda \in \mathbb{C} : j(A - \lambda I) = 0\}, \tag{6.66}$$

$$\sigma_l(A) = \{\lambda \in \mathbb{C} : A - \lambda I \notin \mathcal{B}(X)_l^{-1}\}.$$

The sets $\sigma_p(A)$, $\sigma_a(A)$ and $\sigma_l(A)$ are called, respectively, the *point spectrum, approximate*

point spectrum and *left spectrum of A*.
It follows from Theorems 6.10.1, 6.10.9 and 6.10.10 that

$$\sigma_p(A) \subset \sigma_a(A) \subset \sigma_l(A).$$

The complex number λ is *eigenvalue of the operator A* if there exists $x \in X$ with $x \neq 0$ and $Ax = \lambda x$. In this case, the vector x is called an *eigenvector of A corresponding to the eigenvalue λ*. Obviously λ is an eigenvalue of A if and only if $\lambda \in \sigma_p(A)$.

There exist operators whose point spectra are empty. We shall prove that the approximate spectrum is never empty.

Corollary 6.10.12 *Let $A \in \mathcal{B}(X)$. Then $\lambda \in \sigma_a(A)$ if and only if one of the following equivalent conditions is satisfied.*

$$A - \lambda I \text{ is a left topological divisor of zero in } \mathcal{B}(X),$$

$$j(A - \lambda I) = 0.$$

There exists a sequence (x_n) in X such that $\|x_n\| = 1$ for $n = 1, 2, \ldots,$ and
$$\lim_{n \to \infty} \|(A - \lambda I)x_n\| = 0.$$

Proof. This follows from Theorem 6.10.10 and (6.66).   □

Theorem 6.10.13 *Let $A \in \mathcal{B}(X)$. Then*

$$\sigma_a(A) \text{ is a compact subset of } \mathbb{C}, \tag{6.67}$$

$$\partial\sigma(A) \subset \sigma_a(A), \tag{6.68}$$

$$\sigma_a(A) \text{ is a nonempty subset off } \mathbb{C}. \tag{6.69}$$

Proof.

(i)   First we show (6.67).
      Since the minimum modulus is a continuous function by Lemma 4.15.11, it follows that $g : \mathbb{C} \to \mathbb{R}$ defined by $g(\lambda) = j(A - \lambda I)$ for $\lambda \in \mathbb{C}$ is continuous. It follows from $\sigma_a(A) = g^{-1}(0)$ that $\sigma_a(A)$ is a closed subset of $\mathbb{C}$, and $\sigma_a(A) \subset \sigma(A)$ implies that $\sigma(A)$ is a compact subset of $\mathbb{C}$. Hence we have shown (6.67).

(ii) Now we show (6.68).
      It follows from $\lambda \in \partial\sigma(A)$ that $A - \lambda I \in \partial(\mathcal{B}(X)^{-1})$, and by Theorem 6.7.4, we see that $A - \lambda I$ is a two–sided topological divisor of zero in $\mathcal{B}(X)$. So it follows from Corollary 6.10.12 that $\lambda \in \sigma_a(A)$. Hence we have shown (6.68).

(iii) Finally, we show (6.69).
      It follows from $\sigma(A) \neq \emptyset$ (Theorem 6.5.8) and (6.68) that $\sigma_a(A) \neq \emptyset$.   □

Theorem 6.10.14 *Let $A \in \mathcal{B}(X)$ and $A' \in \mathcal{B}(X^*)$ be the adjoint operator of A. Then the following statements are equivalent*

$$A \text{ is not surjective}, \tag{6.70}$$

$$A \text{ is a right topological divisor of zero in } \mathcal{B}(X), \tag{6.71}$$

$$A' \text{ is a left topological divisor of zero in } \mathcal{B}(X^*). \tag{6.72}$$

Proof. Since $j(A') = q(A)$, by Theorem 4.15.10, it follows from Corollaries 4.15.9 and 6.10.12 that the conditions in (6.70) and (6.72) are equivalent. Hence it is sufficient to prove that the conditions in (6.71) are (6.72) are equivalent.

(i) First we show that the condition in (6.71) implies that in (6.72).
   If A is a right topological divisor of zero in $\mathcal{B}(X)$, then there exists a sequence (B_n) in $\mathcal{B}(X)$ such $\|B_n\| = 1$ and $\lim_n \|B_n A\| = 0$. Since $\|B'_n\| = \|B_n\| = 1$ by Theorem 4.9.2, and $\|A'B'_n\| = \|(B_n A)'\| = \|B_n A\|$ by Theorem 4.9.3, it follows that A' is a left topological divisor of zero in $\mathcal{B}(X)$. Thus we have shown that (6.71) implies (6.72).

(ii) Finally, we show that the condition in (6.72) implies that in (6.71).
   If A' is a left topological divisor of zero in $\mathcal{B}(X^*)$, then it follows from Theorem 6.10.10 that there exists a sequence (f_n) in $\in X^*$ such that $\|f_n\| = 1$ and $\lim_{n\to\infty} \|A'f_n\| = 0$. Let $y \in X$ and $\|y\| = 1$. We define the sequence (B_n) such that $B_n(x) = f_n(x)y$ for $x \in X$. It is easy to show that $B_n \in \mathcal{B}(X)$ and $\|B_n\| = \|f_n\|\,\|y\| = 1$. Since

$$B_n Ax = f_n(Ax)y = [(f_n A)(x)]y = [(A'f_n)(x)]y,$$

it follows that $\|B_n A\| = \|A'f_n\|$. So we have proved that A is a right topological divisor of zero in $\mathcal{B}(X)$. $\qquad\square$

Definition 6.10.15 Let $A \in \mathcal{B}(X)$. We write

$$\sigma_d(A) = \{\lambda \in \mathbb{C} : \overline{\{R(A - \lambda I)}} \neq X\},$$

$$\sigma_\delta(A) = \{\lambda \in \mathbb{C} : q(A - \lambda I) = 0\},$$

$$\sigma_r(A) = \{\lambda \in \mathbb{C} : A - \lambda I \notin \mathcal{B}(X)_r^{-1}\}.$$

The sets $\sigma_d(A)$, $\sigma_\delta(A)$ and $\sigma_r(A)$ are called, respectively, the *defect spectrum*, *approximate defect spectrum* and *right spectrum of A*.

It follows from Theorem 6.10.2, (6.60), and Theorem 6.10.14 and (6.31) that

$$\sigma_d(A) \subset \sigma_\delta(A) \subset \sigma_r(A).$$

Corollary 6.10.16 *Let $A \in \mathcal{B}(X)$. Then $\lambda \in \sigma_\delta(A)$ if and only if one of the following equivalent conditions is satisfied:*

$$A - \lambda I \text{ is a right topological divisor of zero in } \mathcal{B}(X);$$

$$q(A - \lambda I) = 0;$$

$$\lambda \in \sigma_a(A').$$

Proof. This follows from Theorem 4.15.10, Corollary 6.10.12 and Theorem 6.10.14. $\quad\square$

Theorem 6.10.17 *Let X be a Banach space and $A \in \mathcal{B}(X)$. Then we have that*

$$\sigma(A') = \sigma(A). \tag{6.73}$$

If X is a Hilbert space and $A \in \mathcal{B}(X)$, then

$$\sigma(A^*) = \sigma(A)^*,$$

where, for $W \subset \mathbb{C}$, we use the notation $W^* = \{\overline{w} : w \in W\}$.

Proof. The proof is left to the readers.   □

Corollary 6.10.18 *Let $A \in \mathcal{B}(X)$. Then we have that*

$$\sigma_\delta(A) \text{ is a compact subset of } \mathbb{C};$$

$$\partial\sigma(A) \subset \sigma_\delta(A);$$

$$\sigma_\delta(A) \text{ is a nonempty subset of } \mathbb{C}.$$

Proof. This follows from Theorems 6.10.16 and (6.73).   □

The following results hold for bounded linear operators on Hilbert spaces.

Theorem 6.10.19 *If X is a Hilbert space and $A \in \mathcal{B}(X)$, then we have*

$$A \in \mathcal{B}(X)_l^{-1} \text{ if and only if } j(A) > 0;$$

$$A \in \mathcal{B}(X)_r^{-1} \text{ if and only if } q(A) > 0;$$

$$A \text{ is } g- \text{ is invertible if and only if } R(A) = \overline{R(A)};$$

$$\sigma_a(A) = \sigma_l(A);$$

$$\sigma_\delta(A) = \sigma_r(A).$$

Proof. The proof is left to the readers.   □

For an operator $A \in \mathcal{B}(X)$, we studied its spectrum and also some subsets of the spectrum. We will show that the spectrum of an operator can be represented as a union of some of its special subsets. We start with the next theorem.

Theorem 6.10.20 *Let $A \in \mathcal{B}(X)$. Then the following conditions are equivalent:*

$$A \notin \mathcal{B}(X)^{-1}; \tag{6.74}$$

$$A \text{ is either a right or a left topological divisor of zero in } \mathcal{B}(X). \tag{6.75}$$

Proof. We assume that $A \notin \mathcal{B}(X)^{-1}$ and A is neither a right nor left topological divisor of zero in $\mathcal{B}(X)$. Then (6.60) implies $\overline{R(A)} = X$, and Theorem 6.10.10 implies $j(A) > 0$. Hence, A is one to one, and onto this is a contradiction to $A \notin \mathcal{B}(X)^{-1}$. Thus we have shown that (6.74) implies (6.75).
Obviously (6.75) implies (6.74).   □

Corollary 6.10.21 *Let $A \in \mathcal{B}(X)$. Then we have*

$$\sigma(A) = \sigma_d(A) \cup \sigma_a(A).$$

Proof. This follows by Corollary 6.10.12 and Theorem 6.10.20.   □

Definition 6.10.22 Let $A \in \mathcal{B}(X)$. The *residual spectrum of the operator A,* denoted by $\sigma_{res}(A)$, is the set of all complex numbers λ such that $A - \lambda I$ is one to one and $\overline{R(A - \lambda I)} \neq X$. Hence we have

$$\sigma_{res}(A) = \sigma_d(A) \setminus \sigma_p(A).$$

The *continuous spectrum of the operator A,* denoted by $\sigma_c(A)$, is the set of all complex numbers λ such that $A - \lambda I$ is one to one, $\overline{R(A - \lambda I)} = X$ and $R(A - \lambda I) \neq X$. Hence we have

$$\sigma_c(A) = \sigma(A) \setminus \{\sigma_p(A) \cup \sigma_d(A)\}.$$

Corollary 6.10.23 *If $A \in \mathcal{B}(X)$, then the sets $\sigma_p(A)$, $\sigma_c(A)$ and $\sigma_{res}(A)$ are mutually disjoint and*

$$\sigma(A) = \sigma_p(A) \cup \sigma_c(A) \cup \sigma_{res}(A).$$

Proof. It follows from Definition 6.10.22 that the sets $\sigma_p(A)$, $\sigma_c(A)$ and $\sigma_{res}(A)$ are mutually disjoint. For the same reason, we obtain

$$\sigma(A) = \sigma_c(A) \cup \{\sigma_p(A) \cup \sigma_d(A)\} = \sigma_c(A) \cup \sigma_p(A) \cup \sigma_{res}(A). \qquad \square$$

Remark 6.10.24 *Let X be a Hilbert space and $A \in \mathcal{B}(X)$. If A is a g-invertible operator, then in the set $A\{1,2\}$ of all g-inverses of A, there exists a very important operator. This operator is $A^\dagger$, the well–known Moore-Penrose pseudo–inverse of the operator A. The operator $A^\dagger$ can be defined by several equivalent conditions.*

For instance, by Desoer and Whalen [53], $A^\dagger$ is the unique operator defined by

$$A^\dagger A x = x \text{ for all } x \in N(A)^\perp$$

and

$$A^\dagger y = 0 \text{ for all } y \in R(A)^\perp.$$

Furthermore, the operator $A^\dagger$ can be defined as the unique operator that satisfies the following Penrose equations

$$A A^\dagger A = A, \ A^\dagger A A^\dagger = A^\dagger, (A A^\dagger)^* = A A^\dagger \text{ and } (A^\dagger A)^* = A^\dagger A.$$

If $y \in R(A)$, then there exists $u \in X$ such that $Au = y$, but if $y \in X \setminus R(A)$, then the equation $Ax = y$ has no solution. Then it is of interest to estimate the value of

$$\inf_{x \in X} \|Ax - y\|.$$

It can be proved ([78, 150]) that

$$\inf_{x \in X} \|Ax - y\| = \|Ax_0 - y\|, \tag{6.76}$$

where $A^\dagger y = x_0$. Furthermore, if there exists $w \in X$ and $w \neq x_0$ then we have

$$\inf_{x \in X} \|Ax - y\| = \|Aw - y\|,$$

and

$$\|x_0\| < \|w\|. \tag{6.77}$$

The properties in (6.76) and (6.77) are the well–known approximation properties of the minimal norm *of the Moore-Penrose pseudo–inverse $A^\dagger$.*

The study of Moore–Penrose pseudo–inverses, as that of many other pseudo–inverses, is ongoing research (see, for instance, [8, 16, 36, 37, 87, 90, 91, 97, 110, 123, 128, 144, 145, 161, 162]). Pseudo–inverses are used in the case when the conventional inverse does not exist, and in many situations, they act in the same way as conventional inverses, if they exist.

6.11 The spectrum of self–adjoint and normal operators

In this section, X is always a Hilbert space, and we study some of the properties of self–adjoint and normal operators.

Theorem 6.11.1 *Let $A \in \mathcal{B}(X)$ be a normal operator and $\lambda \in \mathbb{C}$. Then*

$$\lambda \in \rho(A) \text{ if and only if } j(A - \lambda I) > 0. \tag{6.78}$$

Proof. It suffices to show that the right-hand side implies the left-hand side in (6.78). It follows from $j(A - \lambda I) > 0$ and Theorem 4.15.2 that $R(A - \lambda I) = \overline{R(A - \lambda I)}$ and $N(A - \lambda I) = \{0\}$. We assume $R(A - \lambda I) \neq X$. Then there exists $z \in X$ such that $z \neq 0$ and $z \in R(A - \lambda I)^{\perp}$. Hence

$$0 = \langle (A - \lambda I)x, z \rangle = \langle x, (A^* - \overline{\lambda}I)z \rangle \text{ for each } x \in X. \tag{6.79}$$

If we choose $x = (A^* - \overline{\lambda}I)z$, then (6.79) implies $\|(A^* - \overline{\lambda}I)z\| = 0$, and then Theorem 5.14.14 implies $\|(A - \lambda I)z\| = 0$. Hence $z = 0$, and so $R(A - \lambda I) = X$. $\square$

Corollary 6.11.2 *Let $A \in \mathcal{B}(X)$ be a normal operator. Then $\sigma(A) = \sigma_a(A)$.*

Proof. This follows by Theorem 6.11.1. $\square$

The next theorem describes a relation between the spectral radius and the norm of a normal operator.

Theorem 6.11.3 *Let $A \in \mathcal{B}(X)$ be a normal operator. Then we have $r(A) = \|A\|$.*

Proof. Let $B = A^*A$. It follows from $B^* = B$ that $\|B\|^2 = \|B^*B\| = \|B^2\|$. It is easy to prove by induction that

$$\|B\|^{2^k} = \|B^{2^k}\| \text{ for all } k = 1, 2, \ldots,$$

and then it follows from Theorem 6.6.5 that

$$\|B\| = \lim_{k \to \infty} \|B^{2^k}\|^{1/2^k} = r(B). \tag{6.80}$$

Now (5.65) and (6.80) imply

$$\|A\|^2 = \|A^*A\| = r(A^*A). \tag{6.81}$$

Analogously, it follows from $A^*A = AA^*$ that for all $n = 1, 2, \ldots$

$$(A^*A)^n = (A^n)^*A^n,$$

$$\|(A^*A)^n\| = \|(A^n)^*A^n\| = \|A^n\|^2,$$

and

$$\|(A^*A)^n\|^{1/n} = (\|A^n\|^{1/n})^2. \tag{6.82}$$

It follows from Theorem 6.6.5 and (6.81), when we let $n \to \infty$ in (6.82), that

$$\|A\|^2 = r(A^*A) = (r(A))^2, \text{ that is, } \|A\| = r(A). \quad \square$$

If $A \in \mathcal{B}(X)$ is a self–adjoint operator, then $\langle Ax, x \rangle \in \mathbb{R}$ for all $x \in X$ by Theorem 5.14.2. The numbers $m(A) = \inf_{\|x\|=1} \langle Ax, x \rangle$ and $M(A) = \sup_{\|x\|=1} \langle Ax, x \rangle$, are, respectively, the lower and upper bounds of the operator A, and the inequality $-\|A\| \leq m(A) \leq M(A) \leq \|A\|$ holds by Definition 5.14.7. The next theorem shows that the numbers $m(A)$ and $M(A)$ can be used in the study of some properties of the spectrum of A.

Theorem 6.11.4 *Let $A \in \mathcal{B}(X)$ be a self–adjoint operator. Then we have*

$$\sigma(A) \subset [m(A), M(A)] \text{ and } m(A), \ M(A) \in \sigma(A).$$

Proof. If $\lambda \notin [m(A), M(A)]$, then $d \equiv \inf\{|\lambda - t| : t \in [m(A), M(A)]\} > 0$. It follows for each $x \in X$ and $\|x\| = 1$ from

$$\|(A - \lambda I)x\| = \|(A - \lambda I)x\| \, \|x\| \geq$$
$$\geq |\langle (A - \lambda I)x, x \rangle| = |\langle Ax, x \rangle - \lambda \|x\|^2| = |\langle Ax, x \rangle - \lambda| \geq d,$$

that $j(A - \lambda I) \geq d > 0$. Now, we have by Theorem 6.11.1, $\lambda \in \rho(A)$, hence $\sigma(A) \subset [m(A), M(A)]$.

We are going to show that $m(A), M(A) \in \sigma(A)$. If $0 \leq m(A) \leq M(A)$, then $M(A) = \|A\|$ by Theorem 5.14.8. Let (x_n) be a sequence in X such that $\|x_n\| = 1$ and $\lim_n \langle Ax_n, x_n \rangle = \|A\| = M(A)$. Then we have

$$\|Ax_n - M(A)x_n\|^2 = \|Ax_n\|^2 - M(A)\langle Ax_n, x_n \rangle + M(A)^2$$
$$\leq M(A)^2 - 2M(A)\langle Ax_n, x_n \rangle + M(A)^2 \to 0, \ (n \to \infty).$$

It follows from this and Theorem 6.11.1 that $M(A) \in \sigma(A)$.

To prove the general case, that is, without the assumption $0 \leq m(A) \leq M(A)$, we note that $-m(A)I + A$ is a self–adjoint operator and $\langle [-m(A)I + A]x, x \rangle \geq 0$ for each $x \in X$. From this and the already proved part of the theorem, it follows that

$$-m(A) + M(A) = \sup_{\|x\|=1, \ x \in X} \langle [-m(A)I + A]x, x \rangle \in \sigma(-m(A)I + A)$$
$$= -m(A) + \sigma(A),$$

and so $M(A) \in \sigma(A)$. For the same reason, we have $-m(A) \in \sigma(-A) = -\sigma(A)$, that is, $m(A) \in \sigma(A)$. $\qquad \square$

Theorem 6.11.5 *If $A \in \mathcal{B}(X)$ is a self–adjoint operator, then A is a positive operator if and only if $\sigma(A) \subset [0, \infty)$.*

Proof. This follows from Theorem 6.11.4 and Definition 5.14.3. $\qquad \square$

Lemma 6.11.6 *If $A \in \mathcal{B}(X)$ is a self–adjoint operator, λ_1 and λ_2 with $\lambda_1 \neq \lambda_2$ are the eigenvalues of the operator A, and x_1 and x_2 are the corresponding eigenvectors then $x_1 \perp x_2$.*

Proof. It follows by the assumption of the theorem that $Ax_1 = \lambda_1 x_1$ and $Ax_2 = \lambda_2 x_2$, and we know from Theorem 6.11.4 that $\lambda_1, \lambda_2 \in \mathbb{R}$. Hence we have

$$\lambda_1 \langle x_1, x_2 \rangle = \langle Ax_1, x_2 \rangle = \langle x_1, Ax_2 \rangle = \langle x_1, \lambda_2 x_2 \rangle = \lambda_2 \langle x_1, x_2 \rangle,$$

that is $\langle x_1, x_2 \rangle = 0$. $\qquad \square$

6.12 Spectrum of compact operators

In this section, we present some results related to the spectrum of compact operators.

Lemma 6.12.1 *Let X be a Banach space and $A \in \mathcal{B}(X)$ be a compact operator. If (λ_n) is a sequence of distinct elements in $\sigma_p(A)$, then $\lim_{n\to\infty} \lambda_n = 0$.*

Proof. For each n and each $x_n \in N(A - \lambda_n I)$, let $x_n \neq 0$. It is easy to show that $\{x_1, x_2, \dots\}$ is a linearly independent set. If $X_n = \text{span}(\{x_1, \dots, x_n\})$, then $\dim X_n = n$ and $X_n \subsetneqq X_{n+1}$. By the Riesz lemma, there exist $y_n \in X_n$ for $n = 2, 3, \dots$ such that $\|y_n\| = 1$ and $d(y_n, X_{n-1}) > 1/2$. We put $y_n = \alpha_1 x_1 + \cdots + \alpha_n x_n$. It follows from

$$(A - \lambda_n)y_n = \alpha_1(\lambda_1 - \lambda_n)x_1 + \cdots + \alpha_{n-1}(\lambda_{n-1} - \lambda_n)x_{n-1} \in X_{n-1}, \qquad (6.83)$$

for all $n > m$, that

$$A(\lambda_n^{-1} y_n) - A(\lambda_m^{-1} y_m) = \lambda_n^{-1}(A - \lambda_n)y_n - \lambda_m^{-1}(A - \lambda_m)y_m + y_n - y_m$$
$$= y_n - [y_m + \lambda_m^{-1}(A - \lambda_m)y_m - \lambda_n^{-1}(A - \lambda_n)y_n].$$

Since $[y_m + \lambda_m^{-1}(A - \lambda_m)y_m - \lambda_n^{-1}(A - \lambda_n)y_n] \in X_{n-1}$, it follows from (6.83) that

$$\|A(\lambda_n^{-1} y_n) - A(\lambda_m^{-1} y_m)\| \geq d(y_n, X_n) > \frac{1}{2},$$

that is, the sequence $(A(\lambda_n^{-1} y_n))$ has no convergent subsequence. Hence, since A is a compact operator, it follows that the sequence $(\lambda_n^{-1} y_n)$ has no bounded subsequence. Hence, $\|y_n\| = 1$ implies

$$\lim_{n\to\infty} \|\lambda_n^{-1} y_n\| = \lim_{n\to\infty} |\lambda_n^{-1}| = \infty, \text{ that is, } \lim_{n\to\infty} |\lambda_n| = 0. \qquad \square$$

Lemma 6.12.2 *Let X be a Banach space and $A \in \mathcal{B}(X)$ be a compact operator. Then it follows from $\lambda \in \sigma(A)$ for $\lambda \neq 0$ that $\lambda \in \sigma_p(A)$, that is, λ is an eigenvalue of the operator A.*

Proof. Let $\lambda \in \partial\sigma(A)$ and $\lambda \neq 0$. It follows from Theorem 6.10.13 that $\lambda \in \sigma_a(A)$, and so there exists a sequence (x_n) in X such that $\|x_n\| = 1$ and $\lim_{n\to\infty}(A - \lambda I)x_n = 0$. Since A is a compact operator, there exists a subsequence (x_{n_k}) of the sequence (x_n) such that the sequence (Ax_{n_k}) converges to $x \in X$, say. Since

$$x_{n_k} - \lambda^{-1} Ax_{n_k} \to 0 \text{ and } \lambda^{-1} Ax_{n_k} \to \lambda^{-1} x \text{ imply } x_{n_k} \to \lambda^{-1} x,$$

it follows that

$$Ax_{n_k} \to \lambda^{-1} Ax, \text{ that is, } \lambda^{-1} Ax = x.$$

Hence $Ax = \lambda x$, and since $\|x\| = |\lambda| \neq 0$, λ is an eigenvalue of A. Now, by Lemma 6.12.1, the boundary $\partial\sigma(A)$ of $\sigma(A)$ is countable with the only one possible limit point zero. Hence the compact set $\partial\sigma(A)$ has a connected complement. Now it follows that the interior of $\sigma(A)$ is an empty set that is, $\partial\sigma(A) = \sigma(A)$. $\qquad \square$

The next theorem follows mainly from the previous two lemmas.

Theorem 6.12.3 *Let X be an infinite dimensional Banach space and $A \in \mathcal{B}(X)$ be a compact operator, then the following cases are possible:*

$$\sigma(A) = \{0\};$$

$\sigma(A) = \{0, \lambda_1, \dots, \lambda_n\},$ *where for each $1 \leq k \leq n$, $\lambda_k \neq 0$, each λ_k is an eigenvalue of the operator A and $\dim N(A - \lambda_k I) < \infty$;*

$\sigma(A) = \{0, \lambda_1, \lambda_2, \dots\},$ *where $k \geq 1$, $\lambda_k \neq 0$, each λ_k is an eigenvalue of the operator A, $\dim N(A - \lambda_k I) < \infty$ and $\lim_{k\to\infty} \lambda_k = 0$.*

Proof. The proof follows from Lemmas 6.12.1 and 6.12.2, and Corollary 4.14.9. □

Lemma 6.12.4 *Let X be a Hilbert space, $A \in \mathcal{K}(X)$, $A \neq O$ and A be a self–adjoint operator. Then $\|A\|$ or $-\|A\|$ are eigenvalues of A and there exists a corresponding eigenvector $x \in X$ such that $\|x\| = 1$ and $|\langle Ax, x \rangle| = \|A\|$.*

Proof. It follows by Theorems 5.14.8 and 6.11.4, and Lemma 6.12.2 that $\|A\|$ or $-\|A\|$ are eigenvalues of A. Hence, there exist $\lambda \in \mathbb{R}$ such that $|\lambda| = \|A\|$, and $x \in X$ with $\|x\| = 1$ and $Ax = \lambda x$. Obviously, we have $|\langle Ax, x \rangle| = |\langle \lambda x, x \rangle| = \|A\|$. □

The next theorem is usually referred to as the *spectral theorem for a self–adjoint compact operator*.

Theorem 6.12.5 (Spectral theorem for a self–adjoint compact operator)
Let X be a Hilbert space, $A \in \mathcal{K}(X)$, $A \neq O$ and A be a self–adjoint operator.
If A is a finite rank operator, then the non–zero eigenvalues of A form a finite set of real numbers $\lambda_1, \ldots, \lambda_n$, and

$$Ax = \sum_{k-1}^{n} \lambda_k \langle x, x_k \rangle \text{ for all } x \in X, \tag{6.84}$$

where $\{x_1, x_2 \ldots, x_n\}$ is an orthonormal set of the corresponding eigenvectors, that is, x_k is the eigenvector corresponding to the eigenvalue of λ_k.
If A is not a finite rank operator, then the nonzero eigenvalues of A form a sequence of real numbers $\lambda_1, \lambda_2, \ldots$ such that $\lim_{n \to \infty} |\lambda_n| = 0$ and

$$Ax = \sum_{k=1}^{\infty} \lambda_k \langle x, x_k \rangle \text{ for all } x \in X, \tag{6.85}$$

where $\{x_1, x_2, \ldots\}$ is the orthonormal set of the corresponding eigenvectors.

Proof. We assume that A is not a finite rank operator. We put $X_1 = X$ and $A_1 = A$. We prove that there exists a sequence of closed subspaces X_n of X with $X_n \neq \{0\}$, a sequence of eigenvalues λ_n of the operator A such that $|\lambda_n| = \|A_{|X_n}\|$, and a sequence of eigenvectors x_n with $\|x_n\| = 1$ corresponding to the eigenvalues λ_n such that for each $n = 1, 2, \ldots$

$$|\lambda_n| \geq |\lambda_{n+1}|, \ x_n \in X_n,$$
$$X_{n+1} = \{x \in X : \langle x, x_i \rangle = 0 \text{ for all } i = 1, \ldots, n\}. \tag{6.86}$$

The proof uses the method of mathematical induction. By Lemma 6.12.4 there exist $x_1 \in X_1$ with $\|x_1\| = 1$ and a scalar λ_1 such that $|\lambda_1| = \|A_1\|$ and $Ax_1 = \lambda_1 x_1$. We assume that, for each $k = 1, 2, \ldots, n-1$, there exist subspaces X_k, scalars λ_k and vectors x_k with the properties in (6.86). We put

$$X_n = \{x \in X : \langle x, x_k \rangle = 0 \text{ for each } k = 1, \ldots, n-1\}.$$

Obviously X_n is a closed subspace of X, $X_n \subset X_{n-1}$ and $X_n \neq \{0\}$. It follows from $x \in X_n$ that, for each $k = 1, 2, \ldots, n-1$,

$$\langle Ax, x_k \rangle = \langle x, Ax_k \rangle = \langle x, \lambda_k x_k \rangle = 0,$$

and so $Ax \in X_n$. Hence, $A : X_n \mapsto X_n$. We put $A_n = A_{|X_n}$. It is easy to show that A_n is a compact, self–adjoint operator, and we prove that $A_n \neq O$. If $x \in X$ and $A_n = O$, then we have

$$x - \sum_{k=1}^{n-1} \langle x, x_k \rangle x_k \in X_n \text{ and } Ax = \sum_{k=1}^{n-1} \lambda_k \langle x, x_k \rangle x_k,$$

which is impossible, since we assumed that A is not a finite rank operator. Hence, $A_n \neq O$. Now again, by Lemma 6.12.4, there exist a vector $x_n \in X_n$ with $\|x_n\| = 1$, and a scalar λ_n such that $|\lambda_n| = \|A_n\|$ and $Ax_n = \lambda_n x_n$. Now $X_n \subset X_{n-1}$ implies

$$|\lambda_n| = \|A_n\| \leq \|A_{n-1}\| = |\lambda_{n-1}|.$$

(i) We show that

$$\lim_{n \to \infty} \lambda_n = 0. \tag{6.87}$$

If (6.87) were not true, then the sequence (x_n/λ_n) would be bounded, hence the sequence $x_n = A(x_n/\lambda_n)$ would have a convergent subsequence, which is a contradiction, since x_n is an orthonormal sequence. Hence, we have shown (6.87).

(ii) To prove (6.85), we assume $x \in X$ and $z_n = x - \sum_{k=1}^{n} \langle x, x_k \rangle x_k$ for $n = 1, 2, \ldots$. We note that, for each n, $z_n \in X_{n+1}$,

$$\|z_n\|^2 = \|x\|^2 - \sum_{i=1}^{n} |\langle x, x_i \rangle|^2 \leq \|x\|^2,$$

and so

$$\|Az_n\| \leq \|A_{|X_{n+1}}\| \|z_n\| \leq |\lambda_{n+1}| \|x\|. \tag{6.88}$$

It follows from (6.87) and (6.88) that $\lim Az_n = 0$, that is,

$$Ax = \sum_{k=1}^{\infty} \lambda_k \langle x, x_k \rangle x_k \tag{6.89}$$

Now we assume that $\lambda \neq 0$ is an eigenvalue of A, and show that $\lambda = \lambda_n$ for some $n \in \mathbb{N}$. Now there exists $x \in X$ with $\|x\| = 1$ such that $Ax = \lambda x$. If, for each $k = 1, 2, \ldots$, $\lambda \neq \lambda_k$, then it follows from Lemma 6.11.6 that $\langle x, x_k \rangle = 0$, and then we have from (6.89) that $x = 0$. This a contradiction. Thus we have shown (6.85).

(iii) Finally, we show (6.84).
Since A is a finite rank operator, there exist only a finite number of distinct eigenvalues of A. Similarly, as in (6.89), let $\lambda_1, \lambda_2, \ldots, \lambda_n$ be a finite set of eigenvalues and $\{x_1, x_2, \ldots, x_n\}$ be the orthonormal set of the corresponding eigenvectors. It follows that $\|A_{|X_{n+1}}\| = 0$, hence, analogously as above, we obtain

$$Ax = \sum_{k=1}^{n} \lambda_k \langle x, x_k \rangle x_k \text{ for each } x \in X. \qquad \square$$

6.13   C*-algebras

C^*-algebras are Banach algebras of special interest. Their importance among Banach algebras is comparable to that of Hilbert spaces among other Banach spaces.

Definition 6.13.1 An algebra $\mathcal{A}$ over the field $\mathbb{C}$ is an algebra with *involution* if a map

$$a \mapsto a^* : \mathcal{A} \mapsto \mathcal{A}$$

is defined with the properties that for each $a, b \in \mathcal{A}$ and each $\lambda \in \mathbb{C}$

$$(a + b)^* = a^* + b^*,$$

$$(ab)^* = b^* a^*,$$

$$(\lambda a)^* = \overline{\lambda} a^*,$$

$$a^{**} = a.$$

A map $a \mapsto a^*$ is an *involution* on the algebra $\mathcal{A}$.
A subalgebra $\mathcal{B}$ of the algebra $\mathcal{A}$ is a *subalgebra with involution of the algebra* $\mathcal{A}$, if $b \in \mathcal{B}$ implies $b^* \in \mathcal{B}$.

Definition 6.13.2 A normed algebra (Banach algebra) $\mathcal{A}$ with involution is called a *normed $*$-algebra (Banach $*$-algebra)*. A Banach $*$–algebra $\mathcal{A}$ is called a C^*-*algebra* if

$$\|aa^*\| = \|a\|^2 \text{ for all } a \in \mathcal{A}. \tag{6.90}$$

Sometimes, in the definition of C^* algebras, the condition

$$\|a^* a\| = \|a\|^2 \text{ for all } a \in \mathcal{A} \tag{6.91}$$

is used instead of (6.90). The conditions in (6.90) and (6.91) are equivalent by the next Lemma 6.13.3.

Examples of C^*–algebras are the algebra of all complex numbers $\mathbb{C}$ with the involution $\lambda \to \overline{\lambda}$, the algebra of continuous functions of a compact Hausdorff space (see Example 6.3.3) with the involution $f \to f^*$ and $f^*(t) = \overline{f(t)}$, and the algebra of all bounded linear operators $\mathcal{B}(X)$ on a Hilbert space X with the involution $A \to A^*$, where A^* is the Hilbert adjoint operator of A.

Throughout this section, $\mathcal{A}$ will always be a C^*-algebra with identity 1 and we assume $\|1\| = 1$.

Lemma 6.13.3 *For each $a \in \mathcal{A}$, the following equality holds*

$$\|a^*\| = \|a\|. \tag{6.92}$$

Proof. It follows from $\|a\|^2 = \|aa^*\| \leq \|a\| \|a^*\|$ that $\|a\| \leq \|a^*\|$, hence $\|a^*\| \leq \|a^{**}\| = \|a\|$, and we have established the equality (6.92). □

Lemma 6.13.4 *Let $a \in \mathcal{A}$. Then*

$$a \in \mathcal{A}^{-1} \text{ implies } a^* \in \mathcal{A}^{-1} \text{ and } (a^*)^{-1} = (a^{-1})^*; \tag{6.93}$$

$$\sigma(a^*) = \overline{\sigma(a)} = \{\overline{\lambda} : \lambda \in \sigma(a)\}. \tag{6.94}$$

Proof.

(i)   First we prove (6.93).
      It follows from $a \in \mathcal{A}^{-1}$ that there exists $a^{-1} \in \mathcal{A}$ and we have $a^{-1} a = a a^{-1} = 1$. Now it follows from the fact $1^* = 1 \cdot 1^* = (1 \cdot 1^*)^* = 1^{**} = 1$, that $a^* (a^{-1})^* = (a^{-1})^* a^* = 1$. Hence $(a^{-1})^*$ is an inverse of a^*. So we have shown (6.93).

(ii) Finally, we prove (6.94).
      We assume $\lambda \in \rho(a)$. Hence $a - \lambda \in \mathcal{A}^{-1}$, and (6.93) implies $a^* - \overline{\lambda} \in \mathcal{A}^{-1}$. So we have $\overline{\lambda} \in \rho(a^*)$. In the same way, we can show that $\lambda = \overline{\overline{\lambda}} \in \rho(a^{**}) = \rho(a)$. So we have shown (6.94). □

Definition 6.13.5 An element $a \in \mathcal{A}$ is *self–adjoint* or *Hermitian* if $a = a^*$, *normal* if $aa^* = a^*a$, and *unitary* if $aa^* = a^*a = 1$.

Theorem 6.13.6 *If $a \in \mathcal{A}$ is a normal element, then $r(a) = \|a\|$.*

Proof. The proof is left to the reader, (see the proof of Theorem 6.11.3). □

Corollary 6.13.7 *If $a \in \mathcal{A}$ is a normal and quasi–nilpotent element, then $a = 0$.*

Proof. This follows from Theorem 6.13.6. □

Theorem 6.13.8 *If $a \in \mathcal{A}$ is a unitary element, then*

$$\sigma(a) \subset \{\lambda \in \mathbb{C} : |\lambda| = 1\}. \tag{6.95}$$

Proof. Since a unitary element a has norm 1, it follows that

$$\sigma(a) \subset \{\lambda \in \mathbb{C} : |\lambda| \leq 1\}. \tag{6.96}$$

The element $a^{-1} = a^*$ is unitary, and it follows that

$$\sigma(a^{-1}) \subset \{\lambda \in \mathbb{C} : |\lambda| = 1\}. \tag{6.97}$$

Theorem 6.5.11, and (6.96) and (6.97) imply (6.95). □

Theorem 6.13.9 *If $a \in \mathcal{A}$ is a self–adjoint element, then $\sigma(a) \subset \mathbb{R}$.*

Proof. Let $\lambda \in \mathbb{C} \setminus \mathbb{R}$. Then there exist $\alpha, \beta \in \mathbb{R}$, $\beta \neq 0$, such that $\lambda = \alpha + i\beta = \beta(\alpha\beta^{-1} + i)$. Since

$$a - \lambda = \beta((\beta^{-1}a - \alpha\beta^{-1}) - i),$$

and since $\beta^{-1}a - \alpha\beta^{-1}$ is a self–adjoint element, it suffices to show that

$$a - i \in \mathcal{A}^{-1}.$$

Let $t \in \mathbb{R}$ and p be a polynomial defined by $p(z) = t - iz$ for $z \in \mathbb{C}$. If $i \in \sigma(a)$, then it follows by Theorem 6.5.9 that

$$t + 1 = p(i) \in p(\sigma(a)) = \sigma(p(a)) = \sigma(t - ia),$$

and so

$$|t + 1| \leq \|t - ia\|,$$

hence

$$(t+1)^2 \leq \|t - ia\|^2 = \|(t - ia)(t - ia)^*\|$$
$$= \|t^2 + a^2\| \leq t^2 + \|a\|^2.$$

Thus we have

$$1 + 2t \leq \|a\|^2 \text{ for all } t \in \mathbb{R}.$$

This is a contradiction. It follows that $i \in \rho(a)$. □

Definition 6.13.10 A *C^*-subalgebra* is a closed subalgebra with involution of a C^*-algebra.

Theorem 6.13.11 *Let $\mathcal{A}$ be a C^*-algebra with identity 1, and $\mathcal{B}$ be a C^*-subalgebra of $\mathcal{A}$, also with identity 1. Then*

$$\sigma_{\mathcal{A}}(a) = \sigma_{\mathcal{B}}(a) \text{ for each } a \in \mathcal{B}.$$

Proof. It follows from (6.35) that

$$\sigma_{\mathcal{A}}(a) \subset \sigma_{\mathcal{B}}(a) \text{ for each } a \in \mathcal{B}.$$

To show the converse inclusion, we assume $a \in \mathcal{B}$, $\lambda \in \sigma_{\mathcal{B}}(a)$ and $\lambda \notin \sigma_{\mathcal{A}}(a)$. We may assume $\lambda = 0$. Then $a^* \in \mathcal{B}$ and $a^* \in \mathcal{A}^{-1}$. Hence, $aa^* \in \mathcal{B}$ and $aa^* \in \mathcal{A}^{-1}$. So, since aa^* is a self–adjoint element, Theorem 6.13.9 and (6.37) imply $(aa^*)^{-1} \in \mathcal{B}$. Hence $a^{-1} = a^*(aa^*)^{-1} \in \mathcal{B}$. This is a contradiction. $\qquad\square$

6.14 Inverting the difference of projections

Let R and K be subspaces of a Hilbert space X, and let P_R and P_K denote the orthogonal projections of X onto these subspaces. Buckholtz proved that the operator $P_R - P_K$ is invertible if and only if X is the direct sum of R and K [35]. In this case there exists a linear idempotent operator A with range R and kernel K; see [117, 126, 203].

Theorem 6.14.1 *Let R and K be closed subspaces of a Hilbert space X, and let P_R and P_K denote the orthogonal projections of X onto these subspaces. Then the following statements are equivalent:*

(1)   The operator $P_R - P_K$ is invertible.

(2) X is direct sum of R and K.

(3)   There exists an idempotent operator $A \in \mathcal{B}(X)$ with range R and kernel K.

(4) $\|P_R + P_K - I\| < 1$.

(5) $P_R + P_K$ and $I - P_R P_K$ are invertible.

(6) $I - P_R P_K$ and $I - (I - P_R)(I - P_K) = P_R + P_K - P_R P_K$ are invertible.

If $P_R - P_K$ is invertible, then

$$(P_R - P_K)^{-1} = A + A^* - I. \tag{6.98}$$

The idempotent operator A is given by

$$A = (I - P_R P_K P_R)^{-1}(P_R - P_P P_K) = (P_R - P_K)^{-1}(I - P_K). \tag{6.99}$$

Proof.

(i)   The equivalence of (2) and (3) is well known (see Theorem 4.6.12).

(ii)   Now we prove that (1) and (3) are equivalent.

(α) We show that (3) implies (1).

We suppose that there exists an idempotent operator $A \in \mathcal{B}(X)$ with range R and kernel K. It follows from $N(I - A) = R(A) = R$ and $R(I - A) = N(A) = K$ that

$$(I - A)P_R = O \text{ and } P_K(I - A) = I - A,$$

hence

$$(A + A^* - I)(P_R - P_K) = (P_R A + P_K(I - A))^* - (I - A)P_R - AP_K$$
$$= (A + I - A)^* = I. \tag{6.100}$$

Taking the adjoint in (6.100) we get

$$(P_R - P_K)(A + A^* - I) = I.$$

Thus, we obtain (1) and (6.98).

(iii) Now we prove that (1) implies (2) (which is equivalent to (3) as we know from Part (i) of the proof).

If $x \in R \cap K$ then $(P_R - P_K)x = 0$, and so $x = 0$. If $y \in X$, then there exists $x \in X$ such that $(P_R - P_K)x = P_R x - P_K x = y$. So we have $R \oplus K = X$, and we obtain (2). Thus we get (1) and (6.98).

Thus we have established the equivalence of (1) and (3).

(iv) Now we prove that (1) implies (4).

We note that we have for each $x \in X$

$$\langle (P_R - P_K)x, (P_R - P_K)x \rangle + \langle (P_R + P_K - I)x, (P_R + P_K - I)x \rangle$$
$$= \langle x, (P_R - P_K)^2 x \rangle + \langle x, (P_R + P_K - I)^2 x \rangle = \langle x, x \rangle.$$

Thus we obtain for each $x \in X$

$$\|(P_R - P_K)x\|^2 + \|(P_R + P_K - I)x\|^2 = \|x\|^2,$$

and so

$$\inf_{x \in H, \|x\|=1} \|(P_R - P_K)x\|^2 = 1 - \sup_{x \in H, \|x\|=1} \|(P_R + P_K - I)x\|^2. \tag{6.101}$$

Since $P_R - P_K$ is invertible, (4) follows from (6.101).

(v) Now we prove that (4) implies (2).

If $x \in R \cap K$, then $\|(P_R + P_K - I)x\| = \|x\|$, and so $x = 0$. It follows from (6.101) that $P_R + P_K$ is invertible. Therefore, if $x \in X$, then we obtain $x = P_R(P_R + P_K)^{-1} + P_K(P_R + P_K)^{-1}$, and (2) follows.

(vi) Now we show that (4) implies (5).

We note that (4) implies $I + (P_R + P_K - I) = P_R + P_K$ is invertible. Furthermore, by the Akhiezer-Glazman equality $\|P_R P_K\| < 1$, and so $I - P_R P_K$ is invertible.

(vii) Now we prove that (5) implies (2).

If $x \in R \cap K$, then we have $(I - P_R P_K)x = 0$, and so $x = 0$. If $x \in X$, then we obtain

$$x = (P_R + P_K)(P_R + P_K)^{-1}x = P_R(P_R + P_K)^{-1}x + P_K(P_R + P_K)^{-1}x,$$

and we get (2).

(viii) Now we show that (4) implies (6), and (2).

It follows from (4) by the Akhiezer-Glazman equality that $\|P_R P_K\| < 1$ and $\|(I - P_R)(I - P_K)0\| < 1$, and so we get (6).

As in the previous case, the invertibility of $I - P_R P_K$ implies $R \cap K = \{0\}$. If $x \in X$, then

$$x = (P_R + P_K - P_R P_K)(I - (I - P_R)(I - P_K))^{-1}x$$
$$= P_R(I - P_K)(I - (I - P_R)(I - P_K))^{-1}x + P_K(I - (I - P_R)(I - P_K))^{-1}x,$$

and so we obtain (2).

(ix) Finally, we prove (6.99).

We note that M is defined by these formulae, since in these expressions, the operators are O on K, and I on R. $\quad\square$

Chapter 7

Measures of Noncompactness

Chapter 7 deals with measures of noncompactness, their properties and some applications. Measures of noncompactness are very useful tools in functional analysis, for instance in metric fixed point theory and the theory of operator equations in Banach spaces, and in the characterizations of compact operators between Banach spaces. After a motivation, results from fixed point theory and the discussion of compact operators and the Hausdorff distance, an axiomatic introduction of measures of noncompactness in complete metric spaces is given, and their most important properties are established, such as monotonicity and the generalized Cantor intersection property. There also are detailed studies of the most important properties of the Kuratowski and Hausdorff measures of noncompactness on bounded sets of Banach spaces related to the linear structure of Banach spaces, in particular, the invariance under the passage to the convex hull, which is crucial in the proof of Darbo's fixed point theorem. Furthermore, the famous Goldenštein–Goh'berg–Markus theorem is proved which gives a very useful estimate for the Hausdorff measure of noncompactness of bounded sets in Banach spaces with a Schauder basis. Finally, the notion of measures of noncompactness of operators and some of their properties are considered, and Fredholm's alternative is proved.

7.1   Introduction

Measures of noncompactness are very useful tools in functional analysis, for instance in metric fixed point theory and the theory of operator equations in Banach spaces. They are also used in the studies of functional equations, ordinary and partial differential equations, fractional partial differential equations, integral and integro–differential equations, optimal control theory, and in the characterizations of compact operators between Banach spaces.

After a motivation, various results from fixed point theory and a discussion of compact operators and the Hausdorff distance in Sections 7.2, 7.3 and 7.4, an axiomatic introduction of measures of noncompactness in complete metric spaces is provided, and their most important properties are established in Section 7.5, such as monotonicity in (7.11) and the generalized Cantor intersection property in (7.14); also the alternative axiomatic approaches for measures in Banach spaces by Banaś and Goebel [12], and by Akhmerov et al. [4] are presented. In particular, we study the *Kuratowski, Hausdorff* and *separation measures of noncompactness* in Sections 7.6–7.8. We also deal with an application of measures of noncompactness in metric fixed point theory. Furthermore, we prove the famous

Goldenštein–Goh'berg–Markus theorem, Theorem 7.9.3, which gives an estimate for the Hausdorff measure of noncompactness of bounded sets in Banach spaces with a Schauder basis. We also use the Kuratowski measure of noncompactness to prove *Darbo's fixed point theorem*, Theorem 7.6.9, which is a generalization of *Schauder's fixed point theorem*, and the fixed point theorem by Darbo–Sadovskiĭ, Theorem 7.10.6. Finally, we study the measure of noncompactness of operators and establish Fredholm theorems for operators with a Hausdorff measure of noncompactness less than 1 in Section 7.13, in particular, we prove the famous Fredholm alternative, Theorem 7.13.8.

We will apply the results of the current section in Chapter 9 to establish estimates or identities for the Hausdorff measure of noncompactness of linear operators between the classical sequence spaces, and use them for the characterization of compact linear operators between those spaces.

The first measure of noncompactness, denoted by α, was defined and studied by Kuratowski [121] in 1930. In 1955, G. Darbo [51] used the function α to prove his fixed point theorem (Theorem 7.3.11), which is a very important generalization of Schauder's fixed point theorem (Theorem 7.3.8), and includes the existence part of Banach's fixed point theorem (Theorem 7.3.2).

Other measures of noncompactness were introduced by Goldenštein, Goh'berg and Markus [GGM1], the ball or Hausdorff measure of noncompactness, which was later studied by Goldenštein and Markus [GGM2] in 1968, Istrăţescu [95] in 1972, and others. Apparently Goldenštein, Goh'berg and Markus were not aware of Kuratowski's and Darbo's work. It is surprising that Darbo's theorem was almost never noticed and applied until in the 1970s, when mathematicians working in functional analysis, operator theory and differential equations started to apply Darbo's theorem and developed the theory connected with measures of noncompactness.

These measures of noncompactness are studied in detail and their use is discussed, for instance, in the monographs [AKP, 201, 12, 96, 122, 134, 135, 13].

7.2   Preliminary results

Now we recall and list some useful definitions and results which are well known and needed throughout this section.

Let (X, d) be a metric space and M and S be subsets of X. If $\varepsilon > 0$, then S is called an *ε–net of M* if, for every $x \in M$, there exists $s \in S$ such that $d(x, s) < \varepsilon$; if the set S is finite then the ε–net S of M is called a *finite ε–net of M*. The set M is said to be *totally bounded* if it has a finite ε–net for every $\varepsilon > 0$.

We know from Theorem 1.12.4 that *a subset M in a metric space is compact if and only if every sequence (x_n) in M has a convergent subsequence with its limit in M*. A subset M of a topological space is said to be *relatively compact* or *precompact* (Definition 1.12.1 (c)) if its closure $\overline{M}$ is compact. The following facts are known.

Every relatively compact subset of a metric space X is totally bounded; if the space X is complete then every totally bounded subset of X is relatively compact.

It is easy to show that *a subset M of a metric space is relatively compact if and only if every sequence (x_n) has a convergent subsequence; in this case the limit of the subsequence need not be in M.*

We also recall some standard notations from (1.4) in Section 1.7.

If S and S' are subsets of a metric space (X, d) and $x \in X$, then

$$\text{dist}(x, S) = \inf\{d(x, s) : s \in S\}, \ \text{dist}(S, S') = \inf\{d(s, s') : s \in S, s' \in S'\} \text{ and}$$
$$\text{diam}(S) = \sup\{d(s, \tilde{s}) : s, \tilde{s} \in S\}$$

denote the *distance of x and S*, *distance of S and S'* and *diameter of S*, respectively. It is clear that $\text{diam}(S) = \text{diam}(\bar{S})$. We are going to frequently use this result throughout the remainder of this chapter.

Convexity plays an important role in this section and its applications in fixed point theory. We recall a few notations and basic facts from Section 2.3.

Let M be a subset of a linear space. Then the set

$$\text{co}(M) = \bigcap \{C \supset M : C \text{ convex }\}$$

is called the *convex hull of M* (Definition 2.3.4 and Theorem 2.3.5, see also Remark 7.2.1). A *convex combination of elements of M* is an element of the form (Definition 2.3.1)

$$\sum_{k=1}^{n} \lambda_k x_k \text{ where } x_k \in M, \ \lambda_k \geq 0 \ (k = 1, 2, \dots, n) \text{ and } \sum_{k=1}^{n} \lambda_k = 1;$$

we write $\text{cvx}(M)$ for the set of all convex combinations of elements of M.

Remark 7.2.1 *Obviously the intersection of any family of convex sets is a convex set; so the convex hull of a set M is the smallest convex set that contains M. Also $\text{co}(M)$ is the set of all convex combinations of M (cf. (7.2)).*

The following results hold.

Theorem 7.2.2 *Let X be a linear space, $C, C_1, C_2, \dots, C_n$ be convex subsets of X, and M be a subset of X. Then we have*

$$\text{cvx}(C) \subset C, \tag{7.1}$$
$$\text{co}(M) = \text{cvx}(M) \tag{7.2}$$

and

$$\text{co} \left(\bigcup_{k=1}^{n} C_k \right) = \left\{ \sum_{k=1}^{n} \lambda_k C_k : \lambda_k \geq 0 \ (k = 1, 2, \dots, n) \text{ and } \sum_{k=1}^{n} \lambda_k = 1 \right\}. \tag{7.3}$$

Proof.

(i)   The inclusion in (7.1) is the statement in Lemma 2.3.3.

(ii) The identity in (7.2) is the statement in Theorem 2.3.5.

(iii) Finally we show the identity in (7.3).
      We put $D = \bigcup_{k=1}^{n} C_k$ and

$$S = \left\{ \sum_{k=1}^{n} \lambda_k C_k : \lambda_k \geq 0 \text{ for } k = 1, 2, \dots, n \text{ and } \sum_{k=1}^{n} \lambda_k = 1 \right\}.$$

It follows by (7.1) that $S \subset \text{co}(D)$. Since trivially $D \subset S$, to prove (7.3) it suffices to show that S is convex.

We assume that $\lambda \in (0, 1)$ and $x, y \in S$. Then there exist $\alpha_k, \beta_k \geq 0$, $x_k, y_k \in C_k$ for $k = 1, 2, \ldots, n$ with $\sum_{k=1}^{n} \alpha_k = \sum_{k=1}^{n} \beta_k = 1$ such that

$$x = \sum_{k=1}^{n} \alpha_k x_k \text{ and } y = \sum_{k=1}^{n} \beta_k y_k.$$

We put $\gamma_k = \lambda \alpha_k + (1 - \lambda) \beta_k$ for $k = 1, 2, \ldots, n$. Since the sets $C_1, C_2, \ldots, C_n$ are convex, there exist $z_k \in C_k$ for $k = 1, 2, \ldots, n$ such that

$$\lambda \alpha_k x_k + (1 - \lambda) \beta_k y_k = \gamma_k z_k \text{ for } k = 1, 2, \ldots, n. \tag{7.4}$$

To see this, let k be an integer with $1 \leq k \leq n$.
If $\gamma_k = 0$, then we may choose $z_k = x_k \in C_k$. If $\gamma_k \neq 0$, then we put

$$z_k = \gamma_k^{-1} \left(\lambda \alpha_k x_k + (1 - \lambda) \beta_k y_k \right), \ \mu_k = \gamma_k^{-1} \lambda \alpha_k \text{ and } \nu_k = \gamma_k^{-1} (1 - \lambda) \beta_k,$$

and obtain
$$\mu_k + \nu_k = \gamma_k^{-1} \left(\lambda \alpha_k + (1 - \lambda) \beta_k \right) = \lambda_k^{-1} \lambda_k = 1,$$
$$z_k = \mu_k x_k + \nu_k y_k \in C_k \text{ (since } C_k \text{ is convex), and}$$
$$\gamma_k z_k = \gamma_k \left(\mu_k x_k + \nu_k y_k \right) = \lambda \alpha_k x_k + (1 - \lambda) \beta_k y_k,$$

that is, (7.4) holds.
We also have

$$\sum_{k=1}^{n} \gamma_k = \lambda \sum_{k=1}^{n} \alpha_k + (1 - \lambda) \sum_{k=1}^{n} \beta_k = \lambda + (1 - \lambda) = 1, \tag{7.5}$$

and (7.4) and (7.5) imply

$$\lambda x + (1 - \lambda) y = \sum_{k=1}^{n} \left(\lambda \alpha_k x_k + (1 - \lambda) \beta_k y_k \right) = \sum_{k=1}^{n} \gamma_k z_k \in S.$$

This concludes the proof of (iii). $\qquad\qquad\qquad\qquad\qquad\qquad\qquad\qquad\square$

7.3 Fixed point theorems

Compactness and measures of noncompactness play an important role in fixed point theory. There are, however, cases when the operators are not compact and the results have to be extended to noncompact operators. Here we consider some fixed point theorems, which we do not prove, and their applications as a motivation for the introduction of measures of noncompactness.

We start with the definition of the concept of a fixed point.

Definition 7.3.1 Let X be a nonempty set and $f : X \to X$ be a function. Then $x_0 \in X$ is called a *fixed point of f* if $f(x_0) = x_0$. We denote the set of all fixed points of f by $\mathcal{F}(f)$.

By a *metric fixed point theorem*, we mean an existence result for a fixed point of a function f under conditions which depend on the metric d, and which are not invariant when we replace d by an equivalent metric.

The best-known metric fixed point theorem is *Banach's fixed point theorem*, also called the *contractive mapping theorem*.

We recall that a map f from a metric space (X, d) into itself is said to be *contractive*, or called a *contraction* if there exists a constant $c \in (0, 1)$ such that

$$d(f(x), f(y)) \leq c \cdot d(x, y) \text{ for all } x, y \in X.$$

Theorem 7.3.2 (Banach's fixed point theorem; contractive mapping theorem)
Every contraction from a complete metric space into itself has a unique fixed point.

The proof will be given in Section 10.2.

It is clear that a contractive map can lose this property if d is replaced by an equivalent metric. Banach's theorem is a basic tool in functional analysis, nonlinear analysis and differential equations; it can, for instance, be used to prove the Picard–Lindelöf existence and uniqueness theorem for initial value problems of certain systems of first-order ordinary differential equations with initial conditions

$$y'_k = f_k(x, y_1, y_2, \ldots, y_n) \text{ with } y_k(\xi) = \eta_k \ (k = 1, 2, \ldots, n). \tag{7.6}$$

Theorem 7.3.3 (Picard–Lindelöf) *Let the functions $f_k : \mathbb{R}^{n+1} \to \mathbb{R}$ be continuous on the rectangle*

$$R = \left\{ (x, y_1, y_2, \ldots, y_n) \subset \mathbb{R}^{n+1} : |x - \xi| \leq \alpha, \ |y_k - \eta_k| \leq \beta \ (k = 1, 2, \ldots, n) \right\}$$

and satisfy a Lipschitz condition in $y_1, y_2, \ldots, y_n$, that is, there exists a constant M such that

$$|f_k(x, y_1, y_2, \ldots, y_n) - f_k(x, \tilde{y}_1, \tilde{y}_2, \ldots, \tilde{y}_n)| \leq M \cdot \sum_{j=1}^{n} |y_j - \tilde{y}_j| \ (k = 1, 2, \ldots, n).$$

If we put

$$K = \max_{1 \leq k \leq n} \left\{ \max_R |f_k(x, y_1, y_2, \ldots, y_n)| \right\} \text{ and } \delta = \min \left\{ \alpha, \frac{\beta}{K} \right\},$$

then there exists one and only one solution on $U_\delta(\xi) = \{x : |x - \xi| < \delta\}$ of the initial value problem (7.6).

If the contractive condition of f in Theorem 7.3.2 is relaxed, that is, if we consider so–called *nonexpansive mappings* f, that is, functions $f : X \to X$ satisfying

$$d(f(x), f(y)) \leq d(x, y) \text{ for all } x, y \in X,$$

then Banach's fixed point theorem need no longer hold.

In 1965, Browder proved a fixed point theorem for nonexpansive maps. The concept of *uniform convexity* is needed there; a normed space X is said to be *uniformly convex*, if for every ε with $0 < \varepsilon \leq 2$, there exists $\delta > 0$ such that

$$\|x - y\| \geq \varepsilon \text{ implies } \left\| \frac{x + y}{2} \right\| \leq 1 - \delta \text{ for all } x, y \in X \text{ with } \|x\| = \|y\| = 1.$$

The geometric meaning of this is that the midpoint of the straight line segment of two points on the unit sphere is deep inside the unit ball, unless the segment is short.

Theorem 7.3.4 (Browder's fixed point theorem)
Let X be a Banach space, C be a convex and bounded subset of X and $f : C \to C$ be a nonexpansive map. If X is either a Hilbert space, or a uniformly convex or a reflexive Banach space, then f has a fixed point.

This result uses the convexity hypothesis which is more usual in topological fixed point theory, and the geometric properties of Banach spaces commonly used in linear functional analysis.

The following text between and including Theorems 7.3.5 and 7.3.7, and Remark 7.3.6 can be found in [201, p. 8].

Brouwer's fixed point theorem should be considered in a different setting.

Theorem 7.3.5 (Brouwer's fixed point theorem) ([201, Theorem 1.3])
Every continuous map from the closed unit ball of $\mathbb{R}^n$ into itself has a fixed point.

Remark 7.3.6 ([201, Remark 1.5]) *In the case of one variable, the Brouwer fixed point theorem is the following:*
Every continuous function of the interval $[-1, 1]$ onto itself has a fixed point,
or equivalently,
every continuous function of the interval $[-1, 1]$ onto itself intersects the main diagonal at some point.

One cannot expect uniqueness of the fixed point in Brouwer's theorem (Theorem 7.3.5), in general. So we must consider the non–empty set $\mathcal{F}(f)$ of fixed points of a function f. If f is continuous, then the set

$$\mathcal{F}(f) = \ker(f - \mathrm{id}) = (f - \mathrm{id})^{-1}(\{0\}), \text{ where id is the identity map,}$$

is closed. It is natural to study what other properties the set $\mathcal{F}(f)$ has. The following theorem shows that no other special features can be inferred, since we will see that for any given non–empty closed subset of the closed unit ball $\overline{B}_1^n(0)$ of Euclidean $\mathbb{R}^n$ there exists a continuous function $f : \overline{B}_1^n(0) \to \overline{B}_1^n(0)$ which has $\mathcal{F}(f)$ as the set of its fixed points.

Theorem 7.3.7 ([201, Theorem 1.6]) *Let $F \neq \emptyset$ be a closed subset of $\overline{B}_1^n(0)$. Then there exists a continuous function $f : \overline{B}_1^n(0) \to \overline{B}_1^n(0)$ with $F = \mathcal{F}(f)$.*

Proof. For every $x \in \overline{B}_1^n(0)$, let $\mathrm{dist}(x, F) = \inf\{\|x - y\| : y \in F\}$. Obviously this function is continuous. We define the function $f : \overline{B}_1^n(0) \to \overline{B}_1^n(0)$ by

$$f(x) = \begin{cases} x - \mathrm{dist}(x, F)\dfrac{x - x_0}{\|x - x_0\|} & (x \neq x_0) \\ x_0 & (x = x_0), \end{cases}$$

where x_0 is an arbitrary point in F.
It is easy to show that f is well defined and continuous. Moreover, $\mathcal{F}(f) = F$, and the theorem is proved. $\square$

An important generalization of Brouwer's fixed point theorem was obtained by Schauder.

Theorem 7.3.8 (Schauder's fixed point theorem) ([181, Theorem I])
Every continuous map from a nonempty, compact and convex subset C of a Banach space X into C has a fixed point.

Clearly the conditions in the hypothesis are preserved if the norm of X is replaced by an equivalent norm, so Theorem 7.3.8 cannot be viewed as a metric fixed point theorem. Schauder's fixed point theorem can be used to prove Peano's existence theorem for the solution of systems of first-order ordinary differential equations with initial conditions.

The situation is completely different when certain generalizations are considered in infinite dimensional Banach spaces, in particular, those concerning *condensing maps*, where a condensing map is one under which the image of any set is, in a certain sense, more compact than the set itself. The degree of noncompactness of a set is measured by certain functions called *measures of noncompactness*.

Important applications of measures of noncompactness arise in the theory of differential equations in infinite Banach spaces.

Remark 7.3.9 *Let* $(X, \| \cdot \|)$ *be a finite dimensional Banach space, and the function* f : $[0, t] \times X$ *be continuous and bounded, that is,*

$$\|f(t, x)\| \leq M \ \text{on} \ [0, t] \times X \ \text{for some constant} \ M.$$

Then it is well known by Peano's theorem that the initial value problem

$$x' = f(t, x) \ \text{with} \ x(0) = 0 \tag{7.7}$$

has at least one solution.
This statement, however, does not hold in infinite dimensional Banach spaces, in general, as the following Example 7.3.10 will show.
We will study at the end of Section 7.5 after Example 7.5.7 and Remark 7.5.6, which concerns Banaś and Goebel's axioms of a measure of noncompactness in Banach spaces, how Peano's theorem can be extended to the infinite dimensional case by the use of a measure of noncompactness. This generalization of Peano's theorem was proved by the use of Bielecki's method of weighted means.

Example **7.3.10 (Dieudonné)** ([54] or [12, Example, p.67]) *Let* $f : c_0 \to c_0$ *be defined by*

$$(f(x))_k = \sqrt{|x_k|} + \frac{1}{k} \ \text{for} \ x = (x_k)_{k=1}^{\infty}; \ (k = 1, 2, \dots).$$

Then the differential equation in (7.7) is equivalent to the infinite system of equations

$$x'_k = \sqrt{|x_k|} + \frac{1}{k} \ \text{for} \ k = 1, 2, \dots.$$

It is easy to see that if $x_k = 0$ *for all* k, *then* $x_k \geq t^2/4$ *for* $t \geq 0$. *Hence the initial value problem has no solution in* c_0.

Similar examples of this type have been given by Cellina in [40] for the case of arbitrary nonreflexive spaces, Yorke [212] in a Hilbert space and finally by Godunov [72] who proved that in any Banach space there exists an equation of the form (7.7) without solutions.

One more important application of a measure of noncompactness is Darbo's fixed point theorem, which uses *Kuratowski's measure of noncompactness* α mentioned in the introduction. Darbo's theorem is a generalization of Schauder's fixed point theorem.

Theorem 7.3.11 (Darbo's fixed point theorem) ([51])
Let C *be a non–empty bounded, closed and convex subset of a Banach space* X *and* α *be the Kuratowski measure of noncompactness on* X. *If* $f : C \to C$ *is a continuous map such that there exists a constant* $k \in [0, 1)$ *with*

$$\alpha(f(Q)) \leq k \cdot \alpha(Q) \ \text{for every} \ Q \subset C, \tag{7.8}$$

then f *has a fixed point in* C.

Darbo's fixed point theorem generalizes from compact sets to bounded and closed sets in infinite dimensional Banach spaces, and needs the additional hypothesis of the *condensing property* in (7.8). As is well known, when we pass from finite to infinite dimensional Banach spaces, bounded and closed subsets need not necessarily be compact. So it is natural to ask if Schauder's fixed point theorem (Theorem 7.3.8) holds in infinite dimensional Banach spaces for convex, closed and bounded subsets. The following example provides a strong negative answer to this question.

Example **7.3.12 (Kakutani)** ([201, Proposition 2.3, p. 11])   There is a fixed point free continuous map on the unit ball of

$$\ell_2(\mathbb{Z}) = \left\{ x = (x_n) : \sum_{n \in \mathbb{Z}} |x_n|^2 < \infty \right\}.$$

Proof. We consider $\ell_2(\mathbb{Z})$ with the standard Schauder basis $(e^{(n)})_{n \in \mathbb{Z}}$, where for each $n \in \mathbb{N}$, $e^{(n)}$ is the sequence with $e_n^{(n)} = 1$ and $e_k^{(n)} = 0$ for $k \neq n$, and with the natural norm given by

$$\|x\| = \|x\|_2 = \left(\sum_{n \in \mathbb{Z}} |x_n|^2 \right)^{1/2} \quad \text{for all } x \in \ell_2(\mathbb{Z}).$$

We write $\overline{B}_{\ell_2(\mathbb{Z})}$ for the closed unit ball in $\ell_2(\mathbb{Z})$. Every sequence $x = (x_n)_{n \in \mathbb{Z}} \in \ell_2(\mathbb{Z})$ has a unique representation $x = \sum_{n \in \mathbb{Z}} x_n e^{(n)}$. We define the left shift operator $U : \ell_2(\mathbb{Z}) \to \ell_2(\mathbb{Z})$ by

$$U(x) = \sum_{n \in \mathbb{Z}} x_n e^{(n+1)}.$$

The relation

$$x - U(x) = \sum_{n \in \mathbb{Z}} (x_n - x_{n-1}) e^{(n)} = c \cdot e^{(0)}$$

implies $x_n = x_0$ for all $n > 0$ and $x_n = x_1$ for all $n < 0$. For a sequence in $\ell_2(\mathbb{Z})$, this is only possible if $x_0 = x_1 = 0$. So $x - U(x)$ is a multiple of $e^{(0)}$ if and only if $x = 0$. We define the map $f : \ell_2(\mathbb{Z}) \to \ell_2(\mathbb{Z})$ by

$$f(x) = (1 - \|x\|) e^{(0)} + U(x).$$

Then f maps $\overline{B}_{\ell_2(\mathbb{Z})}$ into $\overline{B}_{\ell_2(\mathbb{Z})}$, since we have for $\|x\| \leq 1$

$$\|f(x)\| \leq |1 - \|x\|| \cdot \|e^{(0)}\| + \|U(x)\| = (1 - \|x\|) + \|x\| = 1.$$

Finally, f is a fixed point free map. Indeed, if

$$x = f(x) = (1 - \|x\|) e^{(0)} + U(x),$$

then $x - U(x) = (1 - \|x\|)e^{(0)}$, which is clearly impossible if $x = 0$, and impossible if $x \neq 0$, as we have seen above. $\qquad \square$

Remark 7.3.13 ([201, Remark 2.4, p.12]) Although the map f in Example 7.3.12 is fixed-point-free, it is not difficult to check that

$$\inf\{\|x - f(x)\| : x \in \overline{B}_{\ell_2(\mathbb{Z})}\} = 0.$$

Indeed it suffices to consider the sequence $(x^{(n)})_{n=1}^{\infty}$ of sequences $x^{(n)} \in \ell_2(\mathbb{Z})$ with

$$x^{(n)} = \sum_{k=-n}^{n-1} \frac{1}{\sqrt{2n}} \cdot e^{(k)} \ (n = 1, 2, \dots).$$

Then we have

$$\|x^{(n)}\| = 1 \text{ and } \left\|x^{(n)} - f(x^{(n)})\right\| = \frac{1}{\sqrt{n}} \text{ for all } n \in \mathbb{N}.$$

The next example will show that $\inf\{\|x - f(x)\| : x \in \overline{B}_X\}$ can be greater than 0, where $\overline{B}_X$ denotes the closed unit ball in a Banach space X.

Example **7.3.14** ([201, Example 2.1, p. 12]) Let c_0 be the Banach space of all sequences that converge to 0 with the supremum norm defined by

$$\|x\|_{\infty} = \sup_k |x_k| \text{ for all } x \in c_0,$$

and $\overline{B}_{c_0}$ be the closed unit ball in c_0. We fix $k > 1$ and consider the map $g : [-1, 1] \to \mathbb{R}$ with $g(t) = \min\{1, k|t|\}$. We also define the map $f : \overline{B}_{c_0} \to \overline{B}_{c_0}$ by

$$f(x) = (1, g(x_1), g(x_2), \dots) \text{ for every sequence } x = (x_n)_{n=1}^{\infty} \in \overline{B}_{c_0}.$$

Obviously f is continuous. Moreover we have

$$\|x - f(x)\| > 1 - \frac{1}{k} \text{ for all } x \in \overline{B}_{c_0}.$$

If this were not the case, then we would have

$$\|x - f(x)\| \leq 1 - \frac{1}{k} \text{ for some } x \in \overline{B}_{c_0}.$$

This implies $|x_n| \geq 1/k$ for every $n \in \mathbb{N}$ which is a contradiction to $x \in c_0$.

7.4 Hausdorff distance

In this section, we consider the *Hausdorff distance*.

We need some notations.

We denote the classes of all bounded sets, and of all bounded closed sets in a metric space (X, d) by $\mathcal{M}_X$ and $\mathcal{M}_X^c$, respectively.

We recall the concept of the *Hausdorff distance* which is the maximum distance of a set to the nearest point in the other set. At a later stage, we will establish a relation between the Hausdorff distance and the *Hausdorff measure of noncompactness*.

Definition 7.4.1 Let (X, d) be a metric space. The function $d_H : \mathcal{M}_X \times \mathcal{M}_X \to \mathbb{R}$, defined by

$$d_H(S, \tilde{S}) = \max \left\{ \sup_{s \in S} \text{dist}(s, \tilde{S}), \sup_{\tilde{s} \in \tilde{S}} \text{dist}(\tilde{s}, S) \right\} \text{ for all } S, \tilde{S} \in \mathcal{M}_X,$$

is called the *Hausdorff distance*; the value $d_H(S, \tilde{S})$ is called the *Hausdorff distance* of S and $\tilde{S}$.

The following result shows that the Hausdorff distance is a metric on the bounded closed subsets of a metric space.

Theorem 7.4.2 *Let (X, d) be a metric space. Then we have*

(a) $(\mathcal{M}_X, d_H)$ *is a semimetric space;*

(b) $(\mathcal{M}_X^c, d_H)$ *is a metric space.*

Proof. Obviously, $0 \leq d_H(S, \tilde{S}) < \infty$ for all $S, \tilde{S} \in \mathcal{M}_X$, and $S = \tilde{S}$ implies $d_H(S, \tilde{S}) = 0$. It is also clear that $d_H(S, \tilde{S}) = d_H(\tilde{S}, S)$ for all $S, \tilde{S} \in \mathcal{M}_X$. Furthermore, if $S, \tilde{S} \in \mathcal{M}_X^c$ and $d_H(S, \tilde{S}) = 0$, then $S = \tilde{S}$.

To show that the triangle inequality holds, let $S, \tilde{S}, \tilde{\tilde{S}} \in \mathcal{M}_X$. It follows from the triangle inequality for the metric d

$$d(s, \tilde{\tilde{s}}) \leq d(s, \tilde{s}) + d(\tilde{s}, \tilde{\tilde{s}}) \text{ for all } s \in S, \tilde{s} \in \tilde{S} \text{ and } \tilde{\tilde{s}} \in \tilde{\tilde{S}}$$

that

$$\text{dist}(s, \tilde{\tilde{S}}) = \inf_{\tilde{\tilde{s}} \in \tilde{\tilde{S}}} d(s, \tilde{\tilde{s}}) \leq d(s, \tilde{s}) + d(\tilde{s}, \tilde{\tilde{s}}) \text{ for all } s \in S, \tilde{s} \in \tilde{S} \text{ and } \tilde{\tilde{s}} \in \tilde{\tilde{S}},$$

hence

$$\text{dist}(s, \tilde{\tilde{S}}) \leq d(s, \tilde{s}) + \inf_{\tilde{\tilde{s}} \in \tilde{\tilde{S}}} d(\tilde{s}, \tilde{\tilde{s}}) = d(s, \tilde{s}) + \text{dist}(\tilde{s}, \tilde{\tilde{S}})$$

$$\leq d(s, \tilde{s}) + \sup_{\tilde{s} \in \tilde{S}} \text{dist}(\tilde{s}, \tilde{\tilde{S}}) \leq d(s, \tilde{s}) + d_H(\tilde{S}, \tilde{\tilde{S}}) \text{ for all } s \in S \text{ and all } \tilde{s} \in \tilde{S}.$$

This implies that

$$\text{dist}(s, \tilde{\tilde{S}}) \leq \inf_{\tilde{s} \in \tilde{S}} d(s, \tilde{s}) + d_H(\tilde{S}, \tilde{\tilde{S}}) \leq \sup_{s \in S} \text{dist}(s, \tilde{S}) + d_H(\tilde{S}, \tilde{\tilde{S}})$$

$$\leq d_H(S, \tilde{S}) + d_H(\tilde{S}, \tilde{\tilde{S}}) \text{ for all } s \in S. \tag{7.9}$$

Replacing s and $\tilde{\tilde{S}}$ by $\tilde{\tilde{s}}$ and S in (7.9), we also obtain

$$\text{dist}(\tilde{\tilde{s}}, S) \leq d_H(\tilde{\tilde{S}}, \tilde{S}) + d_H(\tilde{S}, S) \text{ for all } \tilde{\tilde{s}} \in \tilde{\tilde{S}}. \tag{7.10}$$

Finally, (7.9) and (7.10) together imply that

$$d_H(S, \tilde{\tilde{S}}) = \max \left\{ \sup_{s \in S} \text{dist}(s, \tilde{\tilde{S}}), \sup_{\tilde{\tilde{s}} \in \tilde{\tilde{S}}} \text{dist}(\tilde{\tilde{s}}, S) \right\} \leq d_H(S, \tilde{S}) + d_H(\tilde{S}, \tilde{\tilde{S}}).$$

Thus we have shown that d_H satisfies the triangle inequality. $\square$

Remark 7.4.3 *Let (X, d) be a metric space and S be a nonempty subset of X. Then*

$$B_r(S) = \bigcup_{s \in S} B_r(s) = \{y \in X : \text{dist}(y, S) < r\}$$

is the open ball with centre in S and radius $r > 0$.
It can be shown that

$$d_H(S, \tilde{S}) = \inf \left\{ \varepsilon > 0 : S \subset B_\varepsilon(\tilde{S}) \text{ and } \tilde{S} \subset B_\varepsilon(S) \right\} \text{ for all } S, \tilde{S} \in \mathcal{M}_X.$$

7.5 The axioms of measures of noncompactness

In this section, we introduce the *axioms of a measure of noncompactness* on the class of *bounded subsets of a complete metric space*. It seems that the axiomatic approach is the best way of dealing with measures of noncompactness. It is possible to use several systems of axioms which are not necessarily equivalent. The set of axioms should satisfy two requirements

(i) it should have natural realizations and

(ii) it should provide useful tools for applications.

The axiomatic introduction of measures of noncompactness in Banach spaces in the books [4] and [12] differs from the approach we follow here.

The notion of a measure of noncompactness was originally introduced in metric spaces. So we are going to give our axiomatic definition in this class of spaces which can be found, for instance, in the book [201].

Definition 7.5.1 Let (X, d) be a complete metric space. A set function $\phi : \mathcal{M}_X \to [0, \infty)$ is called a *measure of noncompactness on* $\mathcal{M}_X$, if it satisfies the following conditions for all $Q, Q_1, Q_2 \in \mathcal{M}_X$

(MNC.1) $\phi(Q) = 0$ if and only if Q is relatively compact   *(regularity)*

(MNC.2) $\phi(Q) = \phi(\overline{Q})$ *(invariance under closure)*

(MNC.3) $\phi(Q_1 \cup Q_2) = \max\{\phi(Q_1), \phi(Q_2)\}$ *(semi–additivity).*

The number $\phi(Q)$ is called the *measure of noncompactness of the set* Q.

We start with a trivial example.

Example **7.5.2** *In every complete metric space, the map* $\phi : \mathcal{M}_X \to [0, \infty)$ *with*

$$\phi(Q) = \begin{cases} 0 & \text{if } Q \text{ is relatively compact} \\ 1 & \text{otherwise} \end{cases}$$

is a measure of noncompactness, the so–called trivial measure of noncompactness.

The following properties can easily be deduced from the axioms in Definition 7.5.1.

Proposition 7.5.3 *Let* ϕ *be a measure of noncompactness on a complete metric space* (X, d). *Then* ϕ *has the following properties*

$$Q \subset \tilde{Q} \text{ implies } \phi(Q) \leq \phi(\tilde{Q}) \quad \text{(monotonicity);} \tag{7.11}$$

$$\phi(Q_1 \cap Q_2) \leq \min\{\phi(Q_1), \phi(Q_2)\} \text{ for all } Q_1, Q_2 \in \mathcal{M}_X. \tag{7.12}$$

$$\text{If } Q \text{ is finite then } \phi(Q) = 0 \text{ (non–singularity).} \tag{7.13}$$

$$\begin{cases} \text{Generalized Cantor's intersection property} \\ \text{If } (Q_n) \text{ is a decreasing sequence of nonempty sets in } \mathcal{M}_X^c \text{ and} \\ \lim_{n \to \infty} \phi(Q_n) = 0, \text{ then the intersection} \\ \qquad\qquad Q_\infty = \bigcap Q_n \neq \emptyset \\ \text{is compact.} \end{cases} \tag{7.14}$$

Proof. Let $Q_1, Q_2 \in \mathcal{M}_X$ and $Q_1 \subset Q_2$. Then it follows from (MNC.3) that

$$\phi(Q_1) \leq \max\{\phi(Q_1), \phi(Q_2)\} = \phi(Q_1 \cup Q_2) = \phi(Q_2),$$

that is, (7.11) is satisfied.

Also, since $Q_1 \cap Q_2 \subset Q_1, Q_2$, we obtain from (7.11)

$$\phi(Q_1 \cap Q_2) \leq \phi(Q_1), \phi(Q_2),$$

that is, (7.12) is satisfied.

Also if Q is finite, then Q is compact, and so $\phi(Q) = 0$ by (MNC.1) and (MNC.2), that is, the condition in (7.13) holds.

To show the condition in (7.14), let $(x_n)_{n=1}^{\infty}$ be a sequence such that $x_n \in Q_n$ for $n = 1, 2, \ldots$. We consider the decreasing sequence $(C_n)_{n=1}^{\infty}$ of sets $C_n = \{x_k : k \geq n\}$. Obviously, we have $C_n \subset Q_n$ for all $n \in \mathbb{N}$, and by (MNC.3), (7.13) and (7.11)

$$\phi(C_1) = \phi\left(C_n \cup \{x_1, x_2, \ldots, x_{n-1}\}\right) = \max\left\{\phi(C_n), \phi\left(\{x_1, x_2, \ldots, x_{n-1}\}\right)\right\}$$
$$= \phi(C_n) \leq \phi(Q_n) \text{ for every } n.$$

Since $\lim_{n \to \infty} \phi(Q_n) = 0$, it follows that $\phi(C_1) = 0$, and so the set $\{x_n : n \in \mathbb{N}\}$ is relatively compact by (MNC.1). Let $\bar{x}$ be the limit of a subsequence of the sequence $(x_n)_{n=1}^{\infty}$. Since each set Q_n is closed, we have $\bar{x} \in Q_n$ for all $n \in \mathbb{N}$, and so $Q_{\infty} \neq \emptyset$. Finally, since $\phi(Q_{\infty}) \leq \phi(Q_n)$ for each $n \in \mathbb{N}$ by (7.11), we obtain $\phi(Q_{\infty}) = 0$, and so Q_{∞} is compact, since it is closed, being the intersection of the closed sets Q_n. $\square$

We compare Cantor's generalized intersection property (7.14) with Cantor's intersection theorem, Theorem 2.10.4, which we recall for the reader's convenience.

Remark 7.5.4 (Cantor's intersection theorem) (Theorem 2.10.4) *Let $(Q_n)_{n=1}^{\infty}$ be a decreasing sequence of nonempty sets in $\mathcal{M}_X^c$ and $\lim_{n \to \infty} \operatorname{diam}(Q_n) = 0$. Then we have*

$$Q_{\infty} = \bigcap_{n=1}^{\infty} Q_n = \{x\} \text{ for some } x \in X.$$

Remark 7.5.5 *If X is a Banach space then a measure of noncompactness ϕ may have some additional properties related to the linear structure of a normed space, for instance*

$$\phi(\lambda Q) = |\lambda|\phi(Q) \text{ for any scalar } \lambda \text{ and all } Q \in \mathcal{M}_X \quad \text{(homogeneity)}, \tag{7.15}$$

$$\phi(Q_1 + Q_2) \leq \phi(Q_1) + \phi(Q_2) \text{ for all } Q_1, Q_2 \in \mathcal{M}_X \quad \text{(subadditivity)}, \tag{7.16}$$

$$\phi(x + Q) = \phi(Q) \text{ for any } x \in X \text{ and all } Q \in \mathcal{M}_X \quad \text{(translation invariance)}. \tag{7.17}$$

$$\begin{cases} \text{For every } Q_0 \in \mathcal{M}_X \text{ and for all } \varepsilon > 0 \text{ there is } \delta > 0 \text{ such that} \\ |\phi(Q_0) - \phi(Q)| < \varepsilon \text{ for all } Q \in \mathcal{M}_X \text{ with } d_H(Q_0, Q) < \delta \\ \qquad\qquad\qquad \text{(continuity)}. \end{cases} \tag{7.18}$$

$$\begin{cases} \qquad\qquad \phi\left(\operatorname{co}(Q)\right) = \phi(Q) \text{ for all } Q \in \mathcal{M}_X \\ \text{(invariance under the passage to the convex hull)}. \end{cases} \tag{7.19}$$

$$\begin{cases} \qquad\qquad \text{There exists a constant } L_{\phi} \text{ such that} \\ |\phi(Q_1) - \phi(Q_2)| \leq L_{\phi} \cdot d_H(Q_1, Q_2) \text{ for all } Q_1, Q_2 \in \mathcal{M}_X \\ \qquad\qquad\qquad \text{(Lipschitzianity)}. \end{cases}$$

Remark 7.5.6 *Banaś and Goebel [12, Definition 3.1.1], [14] gave the following axioms for a measure of noncompactness on a Banach space X: A function $\psi : \mathcal{M}_X \to [0, \infty)$ is a measure of noncompactness if it satisfies the conditions (MNC.2), (7.11), (7.19),*

(i) *the family* $\ker\psi = \{Q \in \mathcal{M}_X : \psi(Q) = 0\}$ *is nonempty and contained in the family of all relatively compact subsets of* X*;*
(compare with (MNC.1)*)*

(ii) *if* (Q_n) *is a decreasing sequence of sets* $Q_n \in \mathcal{M}_X^c$ *with* $\lim_{n\to\infty} \psi(Q_n) = 0$*, then the intersection* $Q_\infty = \bigcap_{n=1}^\infty Q_n$ *satisfies* $\emptyset \neq Q_\infty \in \ker\psi$*;*
(compare with (7.14)*)*

and a convexity condition

$$\psi(\lambda Q + (1 - \lambda)\tilde{Q}) \leq \lambda\psi(Q) + (1 - \lambda)\psi(\tilde{Q}) \text{ for all } \lambda \in (0,1) \text{ and all } Q, \tilde{Q} \in \mathcal{M}_X.$$

The family $\ker(\mu)$ *described in* (i) *is called the* kernel of the measure of noncompactness μ. *A measure* μ *is said to be* sublinear *if it satisfies the conditions in* (7.15) *and* (7.16)*.*

Example 7.5.7 *It is well known that if* c_0 *has the natural norm* $\|\cdot\|_\infty$*, then a set* $Q \in \mathcal{M}_{c_0}$ *is relatively compact if and only if*

$$\lim_{n\to\infty} \left(\sup_{x \in Q} \left(\sup_{k \geq n} |x_k| \right) \right) = 0.$$

It is easy to show that the function $\mu : \mathcal{M}_{c_0} \to [0, \infty)$ *defined by*

$$\mu(Q) = \lim_{n\to\infty} \left(\sup_{x \in Q} \left(\sup_{k \geq n} |x_k| \right) \right) \text{ for all } Q \in \mathcal{M}_{c_0}$$

is a measure of noncompactness on c_0*. It will turn out in Example 7.9.6 that the function* μ *is the Hausdorff measure of noncompactness.*

In Remark 7.3.9 we mentioned the generalization of Peano's theorem to infinite dimensional Banach spaces as an application of a measure of noncompactness in the sense of Banaś and Goebel in Remark 7.5.6. The considerations of the following two paragraphs are taken from [12, Chapter 13.2, pp. 70/71].

Ambrosetti [6] was the first one who used a measure of noncompactness as a tool for generalizing Peano's theorem to arbitrary Banach spaces X. He proved the existence theorem under the assumption of uniform continuity of the function f in (7.7) assuming in addition that

$$\alpha(f(t, Q)) \leq k \cdot \alpha(Q) \text{ for all } t \in [0, T] \text{ and all } Q \in \mathcal{M}_X,$$

where k is an arbitrary constant and α denotes the Kuratowski measure of noncompactness. Extensions of Ambrosetti's result for uniformly continuous functions f were proved by Goebel and Rzymowski [73], Rzymowski [177] and Szufla [197].

We need the following concepts to be able to state the result mentioned in Remark 7.3.9. A real–valued nonnegative function $w : (0, t_0] \times [0, \infty)$ is said to be a *Kamke comparison function* if $w(t, 0) = 0$ for all t, $w(t, \cdot)$ is continuous on $(0, \infty)$ for each fixed $t \in (0, t_0]$ and $w(\cdot, u)$ is measurable on $(0, t_0)$ for each fixed $u \in [0, \infty)$, and the unique solution of the integral inequality

$$u(t) \leq \int_{s=0}^{t} w(s, u(s)) \, ds \text{ for } t \in (0, t_0],$$

which satisfies the conditions

$$\lim_{t \to 0^+} \frac{u(t)}{t} = \lim_{t \to 0^+} u(t) = 0,$$

is $u(t) = 0$ for all $t \in (0, t_0]$.

A function $f : (0, t_0] \times \mathcal{M}_X \to X$, where X is a Banach space, is said to satisfy the *Kamke comparison condition with respect to the measure of noncompactness μ on X*, if

$$\mu(f(t, Q)) \leq w(t, \mu(Q)) \text{ for all } Q \in \mathcal{M}_X \text{ and almost all } t \in (0, T], \qquad (7.20)$$

where w is a Kamke function.

The following result was proved by the use of *Bielecki's method of weighted norm [23, 22]*. We refer to its proof in [12, pp. 74–76].

Theorem 7.5.8 ([12, Theorem 7.13.1]) *Let X be a Banach space, μ be a sublinear measure of noncompactness and the function f be uniformly continuous on $[0, t_0] \times B_r(0)$ for some $r > 0$. If f satisfies $\|f(t, x)\| \leq c$ on $[0, t_0] \times B_r(0)$ for some constant c with $ct_0 \leq r$ and the Kamke comparison condition with respect to μ and the Kamke comparison function w being given by $w(t, u) = g(t)u$, where g is an integrable function on $[0, t_0]$ such that (7.20) holds almost everywhere on $[0, t_0]$, then the differential equation $x' = f(t, x)$ has at least one solution with $x(0) = 0$.*

There is an extensive literature on the application of measures of noncompactness in the sense of Remark 7.5.6 in the theory of differential equations. We refer the interested reader to the monographs [4, 12, 13, 66] and the references therein.

7.6   The Kuratowski measure of noncompactness

Now we define and study some properties of the measure of noncompactness introduced by Kuratowski in 1930 which historically is the oldest one, as mentioned at the beginning of the current chapter.

Definition 7.6.1 Let (X, d) be a complete metric space. The function

$$\alpha : \mathcal{M}_X \to [0, \infty)$$

with

$$\alpha(Q) = \inf \left\{ \varepsilon > 0 : Q \subset \bigcup_{k=1}^{n} S_k, \ S_k \subset X, \ \mathrm{diam}(S_k) < \varepsilon \ (k = 1, 2, \ldots, n \in \mathbb{N}) \right\}$$

is called the *Kuratowski measure of noncompactness (KMNC)*, and the real number $\alpha(Q)$ is called the *Kuratowski measure of noncompactness of Q*.

Remark 7.6.2 *Therefore $\alpha(Q)$ is the infimum of all positive real numbers ε such that the bounded set Q can be covered by a finite number of sets of diameters less than ε.*

It turns out that the Kuratowski measure of noncompactness is a measure of noncompactness in the sense of Definition 7.5.1.

Theorem 7.6.3 *Let (X, d) be a complete metric space. Then the Kuratowski measure of noncompactness α of Definition 7.6.1 is a measure of noncompactness in the sense of Definition 7.5.1.*

Proof. We have to show that the function α satisfies the conditions in (MNC.1), (MNC.2) and (MNC.3) of Definition 7.5.1.

(i) The condition in (MNC.1) is an immediate consequence of Definition 7.6.1.

(ii) Now we show the identity in (MNC.2).
Obviously $Q_1 \subset Q_2$ implies $\alpha(Q_1) \leq \alpha(Q_2)$, so α is monotone, that is, the condition in (7.11) holds. Since $Q \subset \overline{Q}$, this implies

$$\alpha(Q) \leq \alpha(\overline{Q}) \text{ for all } Q \in \mathcal{M}_X. \tag{7.21}$$

To prove the converse inequality of (7.21), let $\varepsilon > 0$ be given, and $S_k \in \mathcal{M}_X$ be sets with $\operatorname{diam}(S_k) < \varepsilon$ for $k = 1, 2, \ldots, n$ such that $Q \subset \bigcup_{k=1}^{n} S_k$. Then we have

$$\overline{Q} \subset \overline{\bigcup_{k=1}^{n} S_k} \subset \bigcup_{k=1}^{n} \overline{S_k}.$$

Since $\operatorname{diam}(S_k) = \operatorname{diam}(\overline{S_k})$ for $k = 1, 2, \ldots, n$, we conclude that $\alpha(\overline{Q}) \leq \alpha(Q)$. This and (7.21) together imply (MNC.2).

(iii) Finally we show the condition in (MNC.3).
It follows from (7.11) that

$$\alpha(Q_1) \leq \alpha(Q_1 \cup Q_2) \text{ and } \alpha(Q_2) \leq \alpha(Q_1 \cup Q_2),$$

hence

$$\max\{\alpha(Q_1), \alpha(Q_2)\} \leq \alpha(Q_1 \cup Q_2). \tag{7.22}$$

To show the converse inequality of (7.22), let $\rho = \max\{\alpha(Q_1), \alpha(Q_2)\}$ and $\varepsilon > 0$ be given. Then there are sets $S_k^{(1)}, S_j^{(2)} \in \mathcal{M}_X$ with $\operatorname{diam}(S_k^{(1)}) < \rho + \varepsilon$ for $k = 1, 2, \ldots, n(1)$, $\operatorname{diam}(S_j^{(2)}) < \rho + \varepsilon$ for $j = 1, 2, \ldots, n(2)$ and

$$Q_1 \subset \bigcup_{k=1}^{n(1)} S_k^{(1)} \text{ and } Q_2 \subset \bigcup_{j=1}^{n(2)} S_j^{(2)}.$$

It follows that

$$Q_1 \cup Q_2 = \bigcup_{k=1}^{n(1)} \bigcup_{j=1}^{n(2)} \left(S_k^{(1)} \cup S_j^{(2)} \right),$$

hence $Q_1 \cup Q_2$ is covered by a finite number of sets of diameter less than $\rho + \varepsilon$. Consequently we have

$$\alpha(Q_1 \cup Q_2) < \rho + \varepsilon.$$

Since $\varepsilon > 0$ was arbitrary, we have

$$\alpha(Q_1 \cup Q_2) \leq \rho = \max\{\alpha(Q_1), \alpha(Q_2)\}.$$

This and (7.22) together yield the condition in (MNC.3). $\qquad\square$

Remark 7.6.4 *By Theorem 7.6.3, the function α is a measure of noncompactness, and hence it satisfies the conditions in Proposition 7.5.3, in particular, α satisfies Cantor's generalized intersection property in (7.14). This result is due to Kuratowski.*

It will turn out that if X is a normed space, then the function α has some additional properties connected with the linear structures of a normed space, in particular, we will see that, in this case, α satisfies the conditions in (7.15), (7.16), (7.17) and (7.18) of Remark 7.5.5.

We need the following simple results.

Lemma 7.6.5 *Let X be a normed space and $Q \in \mathcal{M}_X$. Then we have*

$$\sup_{y \in co(Q)} \|x - y\| = \sup_{z \in Q} \|x - z\| \text{ for all } x \in X. \tag{7.23}$$

Proof. Since $Q \subset co(Q)$, it suffices to show that

$$\sup_{y \in co(Q)} \|x - y\| \le \sup_{z \in Q} \|x - z\| \text{ for all } x \in X. \tag{7.24}$$

Let $y \in co(Q)$ be given. Since $co(Q) = cvx(Q)$ by (7.2) in Theorem 7.2, there exist $x_k \in Q$ and $\lambda_k \ge 0$ for $k = 1, 2, \ldots, n$ such that $\sum_{k=1}^{n} \lambda_k = 1$ and $y = \sum_{k=1}^{n} \lambda_k x_k$. It follows from

$$x - y = \sum_{k=1}^{n} \lambda_k x - \sum_{k=1}^{n} \lambda_k x_k = \sum_{k=1}^{n} \lambda_k (x - x_k)$$

that

$$\|x - y\| \le \sum_{k=1}^{n} \|\lambda_k (x - x_k)\| = \sum_{k=1}^{n} |\lambda_k| \cdot \|x - x_k\|$$

$$\le \sup_{z \in Q} \|x - z\| \cdot \sum_{k=1}^{n} \lambda_k = \sup_{z \in Q} \|x - z\|.$$

Since $y \in co(Q)$ was arbitrary, we obtain (7.24). $\qquad \square$

We obtain the following corollary as a consequence of the equality (7.23) in Lemma 7.6.5.

Corollary 7.6.6 *Let X be a Banach space and $Q \in \mathcal{M}_X$. Then Q and $co(Q)$ have equal diameters, that is,*

$$\mathrm{diam}(Q) = \mathrm{diam}(co(Q)).$$

Let X be a normed space and $Q \in \mathcal{M}_X \setminus \{\emptyset\}$. Then the *convex closure of Q*, $\overline{co}(Q)$ is the smallest convex and closed subset of X that contains Q. It is easy to see that

$$\overline{co}(Q) = \overline{co(Q)}.$$

Theorem 7.6.7 (Darbo) *Let X be a normed space, and $Q, Q_1, Q_2 \in \mathcal{M}_X$. Then we have*

$$\alpha(Q_1 + Q_2) \le \alpha(Q_1) + \alpha(Q_2) \qquad \text{(subadditivity)}, \tag{7.25}$$
$$\alpha(Q + x) = \alpha(Q) \text{ for each } x \in X \qquad \text{(translation invariance)}, \tag{7.26}$$
$$\alpha(\lambda Q) = |\lambda| \alpha(Q) \text{ for each scalar } \lambda \qquad \text{(homogeneity)}, \tag{7.27}$$

and

$$\alpha(Q) = \alpha(co(Q)) \qquad \text{(invariance under the passage to the convex hull)}. \tag{7.28}$$

Proof.

(i) First we show the inequality in (7.25).

Let $S_k^{(1)} \in \mathcal{M}_X$ with $\mathrm{diam}(S_k^{(1)}) < \rho_1$ for each $k = 1, \ldots, n$ and

$$Q_1 \subset \bigcup_{k=1}^{n} S_k^{(1)}.$$

Also let $S_j^{(2)} \in \mathcal{M}_X$ with $\mathrm{diam}(S_j^{(2)}) < \rho_2$ for each $j = 1, \ldots, m$ and

$$Q_2 \subset \bigcup_{j=1}^{m} S_j^{(2)}.$$

Then we have

$$Q_1 + Q_2 \subset \bigcup_{k=1}^{n} \bigcup_{j=1}^{m} \left(S_k^{(1)} + S_j^{(2)} \right) \text{ and } \mathrm{diam}(S_k^{(1)} + S_j^{(2)}) < \rho_1 + \rho_2. \qquad (7.29)$$

It follows from (7.29) that $\alpha(Q_1 + Q_2) < \rho_1 + \rho_2$.
This shows the inequality in (7.25).

(ii) Now we show the equality in (7.26).

Let $x \in X$. It follows from (7.25) and the non–singularity of α ((7.13) in Proposition 7.5.3) that

$$\alpha(Q + x) \leq \alpha(Q) + \alpha(\{x\}) = \alpha(Q), \qquad (7.30)$$

and by the same argument we have

$$\alpha(Q) = \alpha((Q + x) + (-x)) \leq \alpha(Q + x) + \alpha(\{-x\}) = \alpha(Q + x). \qquad (7.31)$$

Now we obtain (7.26) from (7.30) and (7.31).

(iii) Now we show the equality in (7.27).

The equality in (7.27) is obvious for $\lambda = 0$.
So let $\lambda \neq 0$ and $S_k \in \mathcal{M}_X$ with $\mathrm{diam}(S_k) < \rho$ for $k = 1, \ldots, n$ and $Q \subset \cup_{k=1}^{n} S_k$.
Then we have for any scalar λ,

$$\lambda Q \subset \bigcup_{k=1}^{n} \lambda S_k \text{ and } \mathrm{diam}(\lambda S_k) = |\lambda| \, \mathrm{diam}\,(S_k).$$

Hence it follows that

$$\alpha(\lambda Q) \leq |\lambda| \alpha(Q).$$

Since $\lambda \neq 0$, we analogously obtain

$$\alpha(Q) = \alpha(\lambda^{-1}(\lambda Q)) \leq |\lambda^{-1}| \alpha(\lambda Q),$$

that is,

$$|\lambda| \alpha(Q) \leq \alpha(\lambda Q).$$

This completes the proof of (7.27).

(iv) Finally, we show the identity in (7.28).
We clearly have $\alpha(Q) \le \alpha(\text{co } Q)$, and it suffices to show that

$$\alpha(\text{co}(Q)) \le \alpha(Q). \tag{*}$$

Let $\alpha(Q) = \rho$ and $S_k \in \mathcal{M}_X$ be sets with $\text{diam}(S_k) < \rho$ for each $k = 1, 2, \ldots n$ and $Q = \cup_{k=1}^n S_k$. It follows by (7.2) in Theorem 7.2.2 that

$$\text{co}(Q) = \left\{ \sum_{k=1}^n \lambda_k x_k : \lambda_k \ge 0, \sum_{k=1}^n \lambda_k = 1, \ x_k \in \text{co}(S_k) \ (k = 1, 2, \ldots, n) \right\}. \tag{7.32}$$

Let $\varepsilon > 0$ be given, and

$$\mathbb{S} = \left\{ (\lambda_1, \lambda_2, \ldots, \lambda_n) : \sum_{k=1}^n \lambda_k = 1, \lambda_k \ge 0 \ (k = 1, 2, \ldots, n) \right\}.$$

Then $\mathbb{S}$ is a compact subset of $(\mathbb{R}^n, \|\cdot\|_\infty)$, where

$$\|(\lambda_1, \lambda_2, \ldots, \lambda_n)\|_\infty = \sup_{1 \le k \le n} |\lambda_k|.$$

We put $M = \sup\{\|x\| : x \in \cup_{k=1}^n \text{co}(S_k)\}$. Let

$$\mathbb{T} = \{(t_{j,1}, t_{j,2}, \ldots, t_{j,n}) : j = 1, 2, \ldots, m\} \subset \mathbb{S}$$

be a finite $\varepsilon/(Mn)$–net for $\mathbb{S}$, with respect to the $\|\cdot\|_\infty$–norm. Hence, if $\sum_{k=1}^n \lambda_k x_k$ is a convex combination of elements of Q, where we suppose that $x_k \in \text{co}(S_k)$ for $k = 1, 2, \ldots, n$, then there exists $(t_{j,1}, t_{j,2}, \ldots, t_{j,n}) \in \mathbb{T}$ such that

$$\|(\lambda_1, \lambda_2, \ldots, \lambda_n) - (t_{j,1}, t_{j,2}, \ldots, t_{j,n})\|_\infty < \frac{\varepsilon}{Mn}. \tag{7.33}$$

Since

$$\sum_{k=1}^n \lambda_k x_k = \sum_{k=1}^n t_{j,k} x_k + \sum_{k=1}^n (\lambda_k - t_{j,k}) x_k, \tag{7.34}$$

it follows from (7.32), (7.33) and (7.34) that

$$\text{co}(Q) \subset \bigcup_{j=1}^m \left\{ \sum_{k=1}^n t_{j,k} \text{co}(S_k) \right\} + \frac{\varepsilon}{Mn} \sum_{k=1}^n B_k, \tag{7.35}$$

where $B_k = \{x \in X : \|x\| \le M\}$ for $k = 1, 2, \ldots, n$. Now we have by (7.35), (7.1), (7.25), (MNC.3), (7.27) and Corollary 7.6.6

$$\alpha(\text{co}(Q)) \le \alpha\left(\bigcup_{j=1}^m \left\{ \sum_{k=1}^n t_{j,k} \text{co}(S_k) \right\} \right) + \alpha\left(\frac{\varepsilon}{Mn} \sum_{k=1}^n B_k \right)$$

$$\le \max_{1 \le j \le m} \alpha\left(\sum_{k=1}^n t_{j,k} \text{co}(S_k) \right) + \frac{\varepsilon}{Mn} \sum_{k=1}^n \alpha(B_k)$$

$$< \max_{1 \le j \le m} \sum_{k=1}^n t_{j,k} \alpha(\text{co}(S_k)) + \frac{\varepsilon}{Mn} 2nM$$

$$< \rho \max_{1 \le j \le m} \sum_{k=1}^n t_{j,k} + 2\varepsilon \le \rho + 2\varepsilon.$$

Since $\varepsilon > 0$ was arbitrary, we have $\alpha(\text{co}(Q)) \le \rho = \alpha(Q)$, that is, (*).
This completes the proof of Part (iv). $\qquad\square$

Remark 7.6.8 *The invariance of α under the passage to the convex hull in (7.28) is vital in the proof of Darbo's fixed point theorem which uses Schauder's fixed point theorem (Theorem 7.3.8).*

Theorem 7.6.9 (Darbo's fixed point theorem) *Let C be a non–empty bounded, closed and convex subset of a Banach space X and α be the Kuratowski measure of noncompactness on X. If $f : C \to C$ is a continuous map such that there exists a constant $c \in [0, 1)$ with*

$$\alpha(f(Q)) \leq c \cdot \alpha(Q) \text{ for every } Q \subset C, \tag{7.36}$$

then f has a fixed point in C.

Proof. We put $C_0 = C$ and define a decreasing sequence $(C_n)_{n=0}^{\infty}$ of sets

$$C_{n+1} = \overline{\text{co}}(f(C_n)) \text{ for } n = 0, 1, \ldots .$$

We use mathematical induction to show that the sequence $(C_n)_{n=0}^{\infty}$ is decreasing, that is,

$$C_{n+1} \subset C_n \text{ for all } n \geq 0. \tag{*}$$

We obtain for $n = 0$, since $f : C \to C$ and C is convex,

$$C_1 = \overline{\text{co}}(f(C_0)) = \overline{\text{co}}(f(C)) \subset \overline{\text{co}}(C) = C = C_0,$$

that is, $C_1 \subset C_0$. Now we assume that $(*)$ holds for some $n \geq 0$. Since C_{n+1} is convex, we have by definition

$$C_{n+2} = \overline{\text{co}}(f(C_{n+1})) \subset \overline{\text{co}}(f(C_n)) = \overline{\text{co}}(C_{n+1}) = C_{n+1}.$$

Thus it follows from mathematical induction that $(*)$ holds for all $n \geq 0$.
Since α is invariant under the passage to the convex hull by (7.28) in Theorem 7.6.7, we successively obtain from (7.36)

$$\alpha(C_{n+1}) = \alpha(\overline{\text{co}}(f(C_n))) = \alpha(f(C_n)) \leq c \cdot \alpha(C_n) \leq \cdots \leq c^{n+1} \cdot \alpha(C_0) = c^{n+1} \cdot \alpha(C).$$

Since $c < 1$, it follows that $\lim_{n \to \infty} \alpha(C_n) = 0$ and so $C_{\infty} = \bigcap_{n=1}^{\infty} C_n \neq \emptyset$ is compact by Cantor's generalized intersection property (7.14) in Proposition 7.5.3. Also C_{∞} is convex as the intersection of convex sets. So f has a fixed point by Schauder's fixed point theorem, Theorem 7.3.8. $\qquad \square$

We clearly have $\alpha(B_X) \leq 2$. But Furi and Vignoli, and Nussbaum have shown that $\alpha(B_X) = 2$ in any infinite dimensional Banach space. To prove this result we need the next theorem which we state without proof. A proof can be found in [60, pp. 303–307].

Theorem 7.6.10 (Ljusternik–Šnirleman-Borsuk) (see, for instance, [56, pp. 122 and 284]) *If S_n denotes the unit sphere of an n–dimensional real normed space E_n, F_k is a closed subset of E_n for each $k = 1, 2, \ldots, n$ and $S_n = \bigcup_{k=1}^{n} F_k$, then there exists a $k_0 \in \{1, 2, \ldots, n\}$ such that the set $S_n \cap F_{k_0}$ contains a pair of antipodal points, that is, there exists $x_0 \in S_n \cap F_{k_0}$ such that $\{x_0, -x_0\} \subset S_n \cap F_{k_0}$.*

Theorem 7.6.11 (Furi–Vignoli; Nussbaum) *Let X be an infinite dimensional Banach space. Then we have $\alpha(B_X) = 2$, where B_X denotes the unit ball in X.*

Proof. Obviously we have $\alpha(B_X) \leq 2$.

We assume $\alpha(B_X) < 2$. Then there exist bounded and closed subsets Q_k of X with $\mathrm{diam}(Q_k) < 2$ for $k = 1, 2, \ldots, n$ and

$$B_X \subset \bigcup_{k=1}^{n} Q_k.$$

Let $\{x_1, x_2, \ldots, x_n\}$ be a linearly independent subset of X, and

$$E_n = \mathrm{span}_{\mathbb{R}}(\{x_1, x_2, \ldots, x_n\}) = \left\{ \sum_{k=1}^{n} \lambda_k x_k : \lambda_k \in \mathbb{R} \ (k = 1, 2, \ldots, n) \right\}$$

be the set of all linear combinations of the set $\{x_1, x_2, \ldots, x_n\}$ with real coefficients. Clearly E_n is a real n-dimensional normed space, the norm of E_n, of course, being the restriction of the norm of X on E_n. We denote by $S_n = \{x \in E_n : \|x\| = 1\}$ the unit sphere in E_n. Then we have

$$S_n \subset \bigcup_{k=1}^{n} (S_n \cap Q_k), \ \mathrm{diam}(S_n \cap Q_k) < 2 \text{ for } k = 1, 2, \ldots,$$

and $S_n \cap Q_k$ is a closed subset of E_n for each $k = 1, 2, \ldots, n$. This is a contradiction to Theorem 7.6.10. $\qquad\square$

7.7   The Hausdorff measure of noncompactness

Now we define and study the Hausdorff or ball measure of noncompactness which was first introduced by Goldenštein, Goh'berg and Markus in 1957 [GGM1] and later studied by Goldenštein and Markus in 1965 [GGM2].

The definition of the Hausdorff measure of noncompactness is similar to that of the Kuratowski measure of noncompactness and the results are analogous.

Definition 7.7.1 Let (X, d) be a complete metric space. The function

$$\chi : \mathcal{M}_X \to [0, \infty)$$

with

$$\chi(Q) = \inf \left\{ \varepsilon > 0 : Q \subset \bigcup_{k=1}^{n} B_{r_k}(x_k), \ x_k \in X, \ r_k < \varepsilon \ (k = 1, 2, \ldots, n \in \mathbb{N}) \right\}$$

is called the *Hausdorff* or *ball measure of noncompactness*, and the real number $\chi(Q)$ is called the *Hausdorff* or *ball measure of noncompactness of Q*.

Remark 7.7.2 *It is not supposed in the definition of the Hausdorff measure of noncompactness of a set Q that the centres of the balls which cover Q belong to Q. Therefore we may equivalently define*

$$\chi(Q) = \inf\{\varepsilon > 0 : Q \text{ has a finite } \varepsilon\text{-net in } X\}.$$

As an example we compute the Hausdorff measure of noncompactness of bounded subsets of the Banach space $(C[a,b], \|\cdot\|)$ of continuous real–valued functions on the real interval $[a,b]$ where

$$\|f\| = \sup_{t\in[a,b]} |f(t)| \text{ for all } f \in C[a,b].$$

The next example can be found in [4].

Example **7.7.3** ([4, 1.1.10]) *The Hausdorff measure of noncompactness of any set* $Q \in \mathcal{M}_{C[a,b]}$ *is given by*

$$\chi(Q) = \frac{1}{2} \cdot \lim_{\delta\to 0} \sup_{f\in Q} \max_{0\le r\le \delta} \|f - f_\tau\|, \qquad (7.37)$$

where f_τ *denotes the* τ*–translate of the function* f*, that is,*

$$f_\tau(t) = \begin{cases} f(t + \tau) & (t \in [a, b - r]) \\ f(b) & (t \in (b - r, b]). \end{cases}$$

Proof. We write $\mathcal{M} = \mathcal{M}_{C[a,b]}$, for short.

Let $\varepsilon > 0$ and $Q \in \mathcal{M}$ be given and $\mathcal{B}$ be a finite $(\chi(Q) + \varepsilon)$–net of the set Q. Let $f \in Q$ and $g \in Q$ be such that $\|f - g\| < \chi(Q) + \varepsilon$. Furthermore, let $\delta > 0$ and $\tau \in [0, \delta]$. Then we have

$$\|f - f_\tau\| \le \|f - g\| + \|g - g_\tau\| + \|g_\tau - f_\tau\| \le 2 \cdot \|f - g\| + \|g - g_\tau\|$$
$$\le 2 \cdot \chi(Q) + 2\varepsilon + \max_{g\in Q} \max_{0\le\tau\le\delta} \|g - g_\tau\|,$$

hence

$$\sup_{f\in Q} \max_{0\le\tau\le\delta} \|f - f_\tau\| \le 2 \cdot \chi(Q) + 2\varepsilon + \max_{g\in Q} \max_{0\le r\le\delta} \|g - g_\tau\|.$$

Letting $\delta \to \infty$ and taking into account that the finite family $\mathcal{B}$ is equicontinuous, we obtain

$$\lim_{\delta\to 0} \sup_{f\in Q} \max_{0\le\tau\le\delta} \|f - f_\tau\| \le 2 \cdot \chi(Q) + 2\varepsilon,$$

and since $\varepsilon > 0$ was arbitrary, it follows that

$$\lim_{\delta\to 0} \sup_{f\in Q} \max_{0\le\tau\le\delta} \|f - f_\tau\| \le 2 \cdot \chi(Q) = 0. \qquad (7.38)$$

To prove the converse inequality of (7.38), we assume that the functions $f \in Q \in \mathcal{M}$ are extended from the interval $[a,b]$ to the whole real line by putting $f(t) = f(a)$ for $t < a$ and $f(t) = f(b)$ for $t > b$. We define the operators R_h, P_h for $h > 0$ by

$$(R_h f)(t) = \frac{1}{2} \cdot \max\{f(s) : s \in [t - h, t + h]\} + \frac{1}{2} \cdot \min\{f(s) : s \in [t - h, t + h]\}$$

and

$$(P_h f)(t) = \frac{1}{2h} \int_{t-h}^{t+h} f(s) \, ds.$$

We are going to show that the set $P_h R_h(Q)$ is a $(q_{2h}/2)$–net of the set Q, where

$$q_{2h} = \sup_{f\in Q} \max_{0\le\tau\le\delta} \|f - f_\tau\|.$$

We have

$$\|P_h R_h f - f\| = \max_{0 \le t \le b} \left| \frac{1}{2h} \int_{t-h}^{t+h} (R_h f)(s)\, ds - \frac{1}{2h} \int_{t-h}^{t+h} f(t)\, dt \right|$$

$$\le \frac{1}{2h} \max_{0 \le t \le b} \int_{t-h}^{t+h} |(R_h f)(s) - f(t)|\, ds. \qquad (7.39)$$

If $|t - s| \le h$, then clearly

$$\min\{f(\tau) : \tau \in [s - h, s + h]\} \le f(t) \le \max\{f(\tau) : \tau \in [s - h, s + h]\}.$$

Consequently

$$|(R_h f)(s) - f(t)| \le \frac{1}{2h} \cdot \max_{0 \le \tau \le 2h} \|f - f_\tau\|, \text{ which implies } |(R_h f)(s) - f(t)| \le \frac{q_{2h}}{2}.$$

Finally, we obtain from this and (7.39), $\chi(Q) \le q_{2h}/2$. Letting $h \to \infty$ we get

$$\chi(Q) \le \frac{1}{2} \cdot \limsup_{\delta \to 0} \max_{f \in Q} \max_{0 \le \tau \le \delta} \|f - f_\tau\|.$$

This completes the proof.   □

The next example is the analogue of Example 7.7.3 for the space $L^p[a, b]$; it can also be found in [4].

Example 7.7.4 ([4, 1.1.13]) *Let* $(L_p[a, b], \| \cdot \|)$ *for* $1 \le p < \infty$ *be the Banach space of equivalence classes of measurable real–valued functions defined on the interval* $[a, b]$ *with integrable pth power and the usual norm defined by*

$$\|f\| = \left(\int_a^b |f(t)|^p \, dt \right)^{1/p} \quad \text{for all } f \in L_p[a, b].$$

Then the Hausdorff measure of noncompactness of any set $Q \in \mathcal{M}_{L_p[a,b]}$ *is given by*

$$\chi(Q) = \frac{1}{2} \cdot \limsup_{\delta \to 0} \max_{f \in Q} \max_{0 \le \tau \le \delta} \|f - f_\tau\|,$$

where f_τ *denotes the* τ*–translate of the function* f *as in Example 7.7.3.*

Proof. The proof is left to the reader.   □

The Hausdorff measure of noncompactness has the following properties which are analogous to those of the Kuratowski measure of noncompactness in Theorem 7.6.3 and Remark 7.6.4, that is, the Hausdorff measure of noncompactness is a measure of noncompactness in the sense of Definition 7.5.1 and so satisfies the conditions in Proposition 7.5.3, and the analogous ones of the conditions in Theorem 7.6.7.

Theorem 7.7.5 *(a) Let* (X, d) *be a complete metric space and* $Q, Q_1, Q_2 \in \mathcal{M}_X$. *Then we have*

$$\chi(Q) = 0 \text{ if and only if } Q \text{ is relatively compact} \qquad (1)$$

$$\chi(Q) = \chi(\overline{Q}) \tag{2}$$

$$\chi(Q_1 \cup Q_2) = \max\{\chi(Q_1), \chi(Q_2)\} \tag{3}$$

$$Q_1 \subset Q_2 \text{ implies } \chi(Q_1) \le \chi(Q_2) \tag{4}$$

$$\chi(Q_1 \cap Q_2) \le \min\{\chi(Q_1), \chi(Q_2)\} \tag{5}$$

$$\text{If } Q \text{ is finite then } \chi(Q) = 0 \tag{6}$$

$$\left\{ \begin{array}{l} \text{Generalized Cantor's intersection property} \\ \textit{If } (Q_n) \textit{ is a decreasing sequence of nonempty sets in } \mathcal{M}_X^c \\ \textit{and } \lim_{n \to \infty} \chi(Q_n) = 0, \textit{ then the intersection} \\ \qquad Q_\infty = \bigcap Q_n \ne \emptyset \\ \textit{is compact.} \end{array} \right. \tag{7}$$

(b) If X is a Banach space, then we have for all $Q, Q_1, Q_2 \in \mathcal{M}_X$

$$\chi(Q_1 + Q_2) \le \chi(Q_1) + \chi(Q_2) \tag{8}$$

$$\chi(x + Q) = \chi(Q) \text{ for all } x \in X \tag{9}$$

$$\chi(\lambda Q) = |\lambda| \chi(Q) \text{ for each scalar } \lambda \tag{10}$$

$$\left\{ \begin{array}{l} \text{invariance under the passage to the convex hull} \\ \qquad \chi(co(Q)) = \chi(Q). \end{array} \right. \tag{11}$$

Proof. The proof, which is very similar to that of the corresponding results for the Kuratowski measure of noncompactness in Section 7.6, is left to the reader. $\square$

Remark 7.7.6 *(a) The results in (1), (2) and (3) show that the Hausdorff measure of noncompactness is a measure of noncompactness in the sense of Definition 7.5.1; the results in (4), (5), (6) and (7) are analogous of those of Proposition 7.5.3; and the conditions in (8), (9), (10) and (11) are analogous of those of Theorem 7.6.7 (b).*
(b) Theorems 7.6.3, 7.6.7, Remark 7.6.4 and Theorem 7.7.5 (a) and (b) show that the Kuratowski and Hausdorff measures of noncompactness are measures of noncompactness in the sense of Banaś and Goebel [12] mentioned in Remark 7.5.6.

The next result shows that the functions χ and α are in some sense equivalent.

Theorem 7.7.7 *Let (X, d) be a complete metric space and $Q \in \mathcal{M}_X$. Then we have*

$$\chi(Q) \le \alpha(Q) \le 2\chi(Q). \tag{7.40}$$

Proof. Let $\varepsilon > 0$ be given. If $\{x_1, x_2, \dots, x_n\}$ is an ε–net of Q, then $\{Q \cap B_\varepsilon(x_k) : k = 1, 2, \dots, n\}$ is a cover of Q with sets of diameter less than 2ε. This shows that

$$\alpha(Q) \le 2\chi(Q).$$

To prove the first inequality in (7.40), we assume that $\{S_k : k = 1, 2, \dots, n\}$ is a cover of Q with $\text{diam}(S_k) < \varepsilon$ for $k = 1, 2, \dots, n$, and let $y_k \in S_k$ for $k = 1, 2, \dots, n$. Then $\{y_1, y_2, \dots, y_n\}$ is an ε–net of Q. This shows that

$$\chi(Q) \le \alpha(Q). \qquad \square$$

Now we compute the Hausdorff measure of the closed unit ball $\overline{B_X}$ in an infinite dimensional Banach space X.

Theorem 7.7.8 *Let X be an infinite dimensional Banach space and $\overline{B_X}$ denote the closed unit ball in X. Then we have $\chi(\overline{B_X}) = 1$.*

Proof. Obviously we have $\chi(\overline{B_X}) \leq 1$.

If $\chi(\overline{B_X}) = q < 1$, then we can choose $\varepsilon > 0$ such that $q + \varepsilon < 1$. Now there exists a $(q+\varepsilon)$–net of $\overline{B_X}$, $\{x_1, x_2, \ldots, x_n\}$, say, and so we have

$$\overline{B_X} \subset \bigcup_{k=1}^{n} \left(x_k + (q+\varepsilon)\overline{B_X} \right).$$

It follows by Theorem 7.7.5 (4), (3), (9) and (10) that

$$q = \chi(\overline{B_X}) \leq \chi \left(\bigcup_{k=1}^{n} \left(x_k + (q+\varepsilon)\overline{B_X} \right) \right) = \max_{1 \leq k \leq n} \chi \left(x_k + (q+\varepsilon)\overline{B_X} \right)$$

$$= \chi \left((q+\varepsilon)\overline{B_X} \right) = (q+\varepsilon)\chi \left(\overline{B_X} \right) = q(q+\varepsilon).$$

Since $q + \varepsilon < 1$, it follows that $q = 0$, and so $\overline{B_X}$ is compact which is a contradiction to the infinite dimensionality of X. $\qquad\square$

Remark 7.7.9 *(a) Since $\overline{B_X} = \mathrm{co}(S_X)$, we have in infinite dimensional Banach space X*

$$\chi \left(\overline{B_X} \right) = \chi \left(B_X \right) = \chi \left(S_X \right)$$

where B_X and S_X denote the open unit ball and the unit sphere in X.
(b) Theorems 7.6.11 and 7.7.8 show that the second inequality in (7.40) of Theorem 7.7.7 is sharp.
(c) The first inequality in (7.40) of Theorem 7.7.7 is also sharp.
To see this, let $B = \{e^{(k)} : k \in \mathbb{N}\}$ be the set of the terms $e^{(k)}$ of the standard Schauder basis of c_0. Since

$$\left\| e^{(n)} - e^{(k)} \right\|_\infty = \sup_j \left| e_j^{(n)} - e_j^{(k)} \right| = 1 \text{ for all } k \neq n,$$

we have $\alpha(B) = 1$. On the other hand, $\chi(B) = 1$, because the distance from any finite subset of B to any element of c_0 cannot be smaller than 1.

Remark 7.7.10 *The Kuratowski and Hausdorff measures of noncompactness are closely related to the geometric properties of the space and it is possible to improve the inequality $\chi(Q) \leq \alpha(Q)$ in certain spaces.*
For instance, it can be shown that in Hilbert spaces

$$\sqrt{2}\chi(Q) \leq \alpha(Q) \leq 2\chi(Q),$$

and that we have in ℓ_p for $1 \leq p < \infty$

$$\sqrt[p]{2}\chi(Q) \leq \alpha(Q) \leq 2\chi(Q).$$

Remark 7.7.11 ([201, Example 2, p. 24]) Let $\ell_{\infty,\mathbb{R}}$ be the space of all real bounded sequences $x = (x_k)_{k=1}^\infty$ with the supremum norm defined by $\|x\|_\infty = \sup_k |x_k|$. Then we have

$$\alpha(Q) = 2\chi(Q) \text{ for all } Q \in \mathcal{M}_{\ell_{\infty,\mathbb{R}}}.$$

Proof. We know from Theorem 7.7.7 that

$$\alpha(Q) \leq 2\chi(Q) \text{ for all } Q \in \mathcal{M}_{\ell_{\infty,\mathbb{R}}}.$$

Let $\varepsilon > 0$ be given and $\rho = \alpha(Q)$. Then there are sets $S_1, S_2, \ldots, S_n \in \ell_{\infty,\mathbb{R}}$ with $\text{diam}(S_k) \leq \rho + \varepsilon$ for $k = 1, 2, \ldots, n$ and $Q \subset \bigcup_{k=1}^n S_k$. For each $j \in \mathbb{N}$, let

$$\alpha_{k,j} = \inf \left\{ x_j : x = (x_\ell)_{\ell \in \mathbb{N}} \in S_k \right\}, \quad \beta_{k,j} = \sup \left\{ x_j : x = (x_\ell)_{\ell \in \mathbb{N}} \in S_k \right\},$$

$$c_{k,j} = \frac{\alpha_{k,j} + \beta_{k,j}}{2}, \quad c_k = (c_{k,j})_{j=1}^\infty \text{ and } B_k = B_{\frac{\rho+\varepsilon}{2}}(c_k).$$

It can be seen that $S_k \subset B_k$ for each $k = 1, 2, \ldots, n$. Therefore we have $\chi(Q) \leq (\rho + \varepsilon)/2$, and letting $\varepsilon \to 0$, we obtain $2\chi(Q) \leq \alpha(Q)$. $\qquad\square$

To be able to prove one more useful property of χ, we need the following result.

Lemma 7.7.12 ([201, Lemma 2.9, p. 24]) Let A, B and C be subsets of a Banach space X, B be convex and closed, C be bounded and $A + C \subset B + C$. Then we have $A \subset B$.

Proof. Let $a \in A$. We have to show that $a \in B$. We know that given $c_1 \in C$, we have

$$a + c_1 \in A + C \subset B + C,$$

that is, $a + c_1 \in B + C$, and so there exist $b_1 \in B$ and $c_2 \in C$ such that

$$a + c_1 = b_1 + c_2.$$

For the same reason, since $c_2 \in C$, there exist $b_2 \in B$ and $c_3 \in C$ such that

$$a + c_2 = b_2 + c_3.$$

Repeating the process and summing the first n equations obtained, we get

$$na + \sum_{k=1}^n c_k = \sum_{k=1}^n b_k + \sum_{k=2}^{n+1} c_k,$$

or equivalently

$$a = \frac{1}{n} \sum_{k=1}^n b_k + \frac{c_{n+1}}{n} - \frac{c_1}{n}.$$

Since B is convex, it follows that

$$d_n = \frac{1}{n} \sum_{k=1}^n b_k \in B \text{ for all } n.$$

Moreover we have $c_1/n, c_{n+1}/n \to 0$ $(n \to \infty)$, since C is bounded. Therefore $d_n \to a$ $(n \to \infty)$, and since B is closed, it follows that $a \in B$, as required. $\qquad\square$

Theorem 7.7.13 *Let X be a Banach space and $Q \in \mathcal{M}_X$. Then we have*

$$\chi(B_r(Q)) = \chi(Q) + r,$$

where, by Remark 7.4.3,

$$B_r(Q) = \bigcup_{x \in Q} B_r(x) = \{y \in X : \text{dist}(y, Q) < r\}.$$

Proof. First, we observe that

$$B_r(Q) = Q + rB_0(1) \text{ for all } Q \in \mathcal{M}_X \text{ and all } r > 0. \qquad (7.41)$$

To see this, we first assume $y \in B_r(Q)$. Then there exist $x \in Q$ such that $y \in B_r(x)$, that is, $\|y - x\| < r$. We put $b = y - x$. Then $\|b\| = \|y - x\| < r$, hence $y = x + b$ with $x \in Q$ and $b \in rB_1(0)$, that is, $y \in Q + rB_1(0)$.

Conversely, if $y \in Q + rB_1(0)$, then there exist $x \in Q$ and $b \in rB_1(0)$ such that $y = x + b$, hence $\|y - x\| = \|b\| < r$, and so $y \in B_r(x)$ for some $x \in Q$, that is, $y \in B_r(Q)$. Thus we established (7.41).

Now we obtain by (7.41), Theorem 7.7.5 (8) and (10), and by Theorem 7.7.8

$$\chi(B_r(Q) = \chi(Q + rB_1(0)) \le \chi(Q) + r\chi(B_1(0)) = \chi(Q) + r.$$

In order to prove the converse inequality, we put

$$\rho = \chi(Q + rB_1(0)).$$

Let $r_1 > \rho$ be given. By the definition of the function χ, there exists a finite set H such that

$$Q + rB_1(0) \subset H + r_1 B_1(0).$$

This implies

$$Q + r\overline{B}_1(0) \subset \overline{\text{co}(H)} + (r_1 - r)\overline{B}_1(0) + r\overline{B}_1(0).$$

Since the set $\overline{\text{co}(H)} + (r_1 - r)\overline{B}_1(0)$ is convex and closed, Lemma 7.7.12 with $A = Q$ and $C = r\overline{B}_1(0)$ yields

$$Q \subset \overline{\text{co}(H)} + (r_1 - r)\overline{B}_1(0),$$

and so by Theorem 7.7.5 (8), (10) and (6), since the set H is finite, and by Theorem 7.7.8, we obtain

$$\chi(Q) \le \chi\left(\overline{\text{co}(H)}\right) + (r_1 - r)\chi\left(\overline{B}_1(0)\right) = 0 + (r_1 - r) \cdot 1,$$

hence $\chi(Q) + r \le r_1$. Finally, since $r_1 > \rho$ was arbitrary, we conclude

$$\chi(Q) + r \le \rho = \chi(Q + rB_1(0)) = \chi(B_r(Q)). \qquad \square$$

Now we establish a relation between the Hausdorff measure of noncompactness and the Hausdorff distance. In particular, (7.42) shows that if (X, d) is a complete metric space, then the map $\chi : (\mathcal{M}_X, d_H) \to \mathbb{R}$ is uniformly continuous.

Theorem 7.7.14 *Let (X, d) be a complete metric space. Then we have*

$$|\chi(Q_1) - \chi(Q_2)| \le d_H(Q_1, Q_2) \text{ for all } Q_1, Q_2 \in \mathcal{M}_X, \qquad (7.42)$$

and

$$\chi(Q) = d_H\left(Q, \mathcal{N}_X^c\right), \qquad (7.43)$$

where $\mathcal{N}_X^c$ denotes the class of all nonempty, bounded and compact subsets of (X, d).

Proof.

(i) First we show the inequality in (7.42).

Let $\varepsilon > 0$ be given and $d_H(Q_1, Q_2) = d$. Then it follows from the definition of the Hausdorff distance d_H (Definition 7.4.1) that there exists $y \in Q_2$ such that $\mathrm{dist}(y, Q_1) < d + \varepsilon$, hence

$$\mathrm{dist}(Q_1, Q_2) < d + \varepsilon, \text{ that is, } Q_1 \subset B_{d+\varepsilon}(Q_2).$$

Furthermore, by the definition of the Hausdorff measure of noncompactness χ, there exists a finite set S such that

$$Q_2 \subset B_{\chi(Q_2)+\varepsilon}(S).$$

Hence we have
$$Q_1 \subset B_{d+\varepsilon}(Q_2) \text{ and } Q_2 \subset B_{\chi(Q_2)+\varepsilon}(S), \tag{7.44}$$

and the inclusions in (7.44) imply

$$\mathrm{dist}(Q_1, S) \leq \mathrm{dist}(Q_1, Q_2) + \mathrm{dist}(Q_2, S) < \chi(Q_2) + d + 2\varepsilon,$$

hence
$$Q_1 \subset B_{d+\chi(Q_2)+2\varepsilon}(S),$$

and so
$$\chi(Q_1) \leq \chi(Q_2) + d + 2\varepsilon.$$

Interchanging the roles of Q_1 and Q_2, we also get

$$\chi(Q_2) \leq \chi(Q_1) + d + 2\varepsilon,$$

and consequently

$$|\chi(Q_1) - \chi(Q_2)| \leq d + 2\varepsilon = \mathrm{dist}(Q_1, Q_2) + 2\varepsilon.$$

Since $\varepsilon > 0$ was arbitrary, we obtain (7.42).

This concludes Part (i) of the proof.

(ii) Now we show the identity in (7.43).

We obtain from (7.42)
$$\chi(Q) \leq d_H(Q, \mathcal{N}_X^c). \tag{7.45}$$

To show the converse inequality, let $\varepsilon > 0$ be given. Then there exists a finite subset S of X such that
$$Q \subset B_{\chi(Q)+\varepsilon}(S) \text{ and } S \subset B_{\chi(Q)+\varepsilon}(Q). \tag{7.46}$$

Now the definition of the Hausdorff measure of noncompactness of a bounded set in a complete metric space and (7.46) together imply

$$d_H(Q, \mathcal{N}_X^c) \leq d_H(Q, S) \leq \chi(Q) + \varepsilon.$$

Since $\varepsilon > 0$ was arbitrary, we have $d_H(Q, \mathcal{N}_X^c) \leq \chi(Q)$, and this and (7.45) imply (7.43).

This concludes Part (ii) of the proof. $\qquad\square$

7.8 The inner Hausdorff and the separation measures of noncompactness

In this section, we introduce the *inner Hausdorff* and *separation measures of noncompactness*. Since we are not going to use them, we will not give any proofs for the stated properties and results.

First we take a short look at the *inner Hausdorff measure of noncompactness*.

We stated in Remark 7.7.2 that it is not supposed in the definition of the Hausdorff measure of noncompactness of a set Q that the centres of the balls which cover Q belong to Q; if, however, the centres of the balls which cover Q are required to belong to Q then the following is the definition of the so–called *inner Hausdorff measure of noncompactness*.

Definition 7.8.1 Let (X, d) be a complete metric space. The function

$$\chi_i : \mathcal{M}_X \to [0, \infty)$$

with

$$\chi_i(Q) = \inf \left\{ \varepsilon > 0 : Q \subset \bigcup_{k=1}^{n} B_{r_k}(x_k), \ x_k \in Q, \ r_k < \varepsilon \ (k = 1, 2, \dots, n \in \mathbb{N}) \right\}$$

is called the *inner Hausdorff*, or *inner ball measure of noncompactness*, and the real number $\chi_i(Q)$ is called the *inner Hausdorff*, or *inner ball measure of noncompactness of Q*.

Remark 7.8.2 *Similarly as in Remark 7.7.2, we can restate the definition of $\chi_i(Q)$ as follows*

$$\chi_i(Q) = \inf\{\varepsilon > 0 : Q \text{ has a finite } \varepsilon\text{–net in } Q\}.$$

There are certain properties of the Hausdorff measure of noncompactness that do not hold for the inner Hausdorff measure of noncompactness. We are going to list some of them without proof.

Remark 7.8.3 *If Q, Q_1 and Q_2 are bounded subsets of the metric space (X, d), then*

$$\chi_i(Q) = 0 \text{ if and only if } Q \text{ is totally bounded,}$$
$$\chi_i(Q) = \chi_i(\overline{Q}),$$

but in general,

$$Q_1 \subset Q_2 \text{ does not imply } \chi_i(Q_1) \leq \chi_i(Q_2),$$

and

$$\chi_i(Q_1 \cup Q_2) \neq \max\{\chi_i(Q_1), \chi_i(Q_2)\}.$$

Let Q, Q_1 and Q_2 be bounded subsets of the normed space X. Then

$$\chi_i(Q_1 + Q_2) \leq \chi_i(Q_1) + \chi_i(Q_2),$$
$$\chi_i(Q + x) = \chi_i(Q) \quad \text{for each } x \in X,$$
$$\chi_i(\lambda Q) = |\lambda| \chi_i(Q) \quad \text{for each } \lambda \in \mathbb{F},$$

but in general

$$\chi_i(Q) \neq \chi_i(\mathrm{co}(Q)).$$

It can be shown that if (X, d) is a complete metric space then

$$\chi(Q) \leq \chi_i(Q) \leq \alpha(Q) \text{ for all } Q \in \mathcal{M}_X.$$

In the fixed point theory in normed spaces (or more generally, in locally convex spaces) the relation $\alpha(Q) = \alpha(\mathrm{co}(Q))$ is of great importance. We remark that O. Hadžić [80], among other things, studied the inner Hausdorff measure of noncompactness in paranormed spaces. She proved under some additional conditions the inequality

$$\chi_i(\mathrm{co}(Q)) \leq \varphi[\chi_i(Q)], \text{ where } \varphi : [0, \infty) \mapsto [0, \infty),$$

and then obtained some fixed point theorems for multivalued mappings in general linear topological spaces.

Now we introduce one more measure of noncompactness, the *separation* or *Istrătescu measure of noncompactness* [95, 96] which also has many useful applications.

Definition 7.8.4 Let (X, d) be a complete metric space.
(a) A subset S of X is said to be *r–separated* or *r–discrete*, if $d(x, y) \geq r$ for all distinct elements x and y of S; the set S is called an *r–separation*. The function $\beta : \mathcal{M}_X \to [0, \infty)$ with

$$\beta(Q) = \sup\{r > 0 : Q \text{ has an infinite } r\text{–separation}\},$$

or equivalently,

$$\beta(Q) = \inf\{r > 0 : Q \text{ does not have an infinite } r\text{–separation}\},$$

is called the *separation* or *Istrătescu measure of noncompactness*; the real number $\beta(Q)$ is called the *separation* or *Istrătescu measure of noncompactness of Q*.

We state the following results without proof.

Remark 7.8.5 *(a) The function β is a measure of noncompactness. If X is a Banach space, then β is homogeneous and invariant under translations. It can also be checked that*

$$\chi(Q) \leq \beta(Q) \leq \alpha(Q) \text{ for all } Q \in \mathcal{M}_X \text{ ([201, II, Remark 3.2])}.$$

(b) Let X be a Banach space. Then β is invariant under the passage to the convex hull and algebraically semi–additive ([201, II, Theorems 3.6 and 3.4]).

Remark 7.8.6 *While $\alpha(B_X) = 2$ and $\chi(B_X) = 1$ in infinite dimensional Banach spaces by Theorems 7.6.11 (Furi–Vignoli; Nussbaum) and 7.7.8, it seems that the exact value of $\beta(B_X)$ is not known, in general ([50]).*

7.9 The Goldenštein–Goh'berg–Markus theorem

In this section, we prove the famous *Goldenštein–Goh'berg–Markus theorem* which gives an estimate for the Hausdorff measure of noncompactness of bounded sets in Banach spaces with Schauder bases.

We need the following result.

Lemma 7.9.1 *Let $(X, \|\cdot\|)$ be a Banach space with a Schauder basis $(b_n)_{n=1}^{\infty}$. We put*

$$\|x\| = \sup_n \left\| \sum_{k=1}^{n} \lambda_k b_k \right\| \text{ for all } x = \sum_{k=1}^{\infty} \lambda_k b_k \in X.$$

Then $(X, \|\cdot\|)$ is a Banach space.

Proof. It is easy to see that $\|\|\cdot\|\|$ is a norm on X.

So we have to prove that X is complete with respect to $\|\|\cdot\|\|$.

Let $\varepsilon > 0$ be given and $(x^{(n)})_{n=1}^{\infty}$ be a Cauchy sequence in $(X, \|\|\cdot\|\|)$, where

$$x^{(n)} = \sum_{k=1}^{\infty} \lambda_k^{(n)} b_k \text{ for all } n \in \mathbb{N}.$$

Then there exists $n_\varepsilon \in \mathbb{N}$ such that

$$\sup_n \left\|\sum_{k=1}^{n} \left(\lambda_k^{(p)} - \lambda_k^{(q)}\right) b_k\right\| = \left\|\left\| x^{(p)} - x^{(q)} \right\|\right\| < \varepsilon \text{ for all } p, q \geq n_\varepsilon. \tag{7.47}$$

We fix $m \in \mathbb{N}$ and obtain from (7.47)

$$\begin{aligned}
\left|\lambda_m^{(p)} - \lambda_m^{(q)}\right| \cdot \|b_m\| &= \left\|\left(\lambda_m^{(p)} - \lambda_m^{(q)}\right) b_m\right\| \\
&\leq \left\|\sum_{k=1}^{m} \left(\lambda_k^{(p)} - \lambda_k^{(q)}\right) b_k\right\| + \left\|\sum_{k=1}^{m-1} \left(\lambda_k^{(p)} - \lambda_k^{(q)}\right) b_k\right\| \\
&\leq \sup_n \left\|\sum_{k=1}^{n} \left(\lambda_k^{(p)} - \lambda_k^{(q)}\right) b_k\right\| + \sup_n \left\|\sum_{k=1}^{n} \left(\lambda_k^{(p)} - \lambda_k^{(q)}\right) b_k\right\| \\
&= 2 \cdot \left\|\left\| x^{(p)} - x^{(q)} \right\|\right\| < 2\varepsilon \text{ for all } p, q \geq n_\varepsilon.
\end{aligned}$$

Since $(b_n)_{n=1}^{\infty}$ is a Schauder basis, it follows that $b_n \neq 0$ for all $n \in \mathbb{N}$, for if $b_{n_0} = 0$ for some $n_0 \in \mathbb{N}$, then we would have

$$0 = 1 \cdot b_{n_0} = 0 \cdot b_{n_0},$$

and the expansion of 0 would not be unique. So we have

$$\left|\lambda_m^{(p)} - \lambda_m^{(q)}\right| < \frac{2\varepsilon}{\|b_m\|} \text{ for all } p, q \geq n_\varepsilon.$$

Therefore the sequence $(\lambda_k^{(p)})_{p=1}^{\infty}$ is a Cauchy sequence in $\mathbb{C}$ for each fixed $k \in \mathbb{N}$, hence convergent by the completeness of $\mathbb{C}$,

$$\lambda_k = \lim_{p \to \infty} \lambda_k^{(p)} \text{ for each } k \in \mathbb{N}, \text{ say.}$$

(i) We show that

$$x = \sum_{k=1}^{\infty} \lambda_k b_k \in X. \tag{7.48}$$

This means that we have to show that the series on the right converges. Since $(b_n)_{n=1}^{\infty}$ is a Schauder basis of X and $(X, \|\cdot\|)$ is a Banach space, $x \in X$ is the limit of the sequence

$$\left(x^{<m>}\right)_{m=1}^{\infty} = \left(\sum_{k=1}^{m} \lambda_k b_k\right)_{m=1}^{\infty}.$$

Let $\varepsilon > 0$ be given. Since $x^{(n_\varepsilon)} \in X$, there exists $m_\varepsilon \in \mathbb{N}$ such that

$$\left\|\sum_{k=m}^{l} \lambda_k^{(n_\varepsilon)} b_k\right\| < \varepsilon \text{ for all } l \geq m \geq m_\varepsilon.$$

Then we have for all $l \geq m \geq m_\varepsilon$ and all $p \geq n_\varepsilon$ by (7.47)

$$\left\| \sum_{k=m}^{l} \lambda_k^{(p)} b_k \right\| \leq \left\| \sum_{k=m}^{l} \left(\lambda_k^{(p)} - \lambda_k^{(n_\varepsilon)} \right) b_k \right\| + \left\| \sum_{k=m}^{l} \lambda_k^{(n_\varepsilon)} b_k \right\| \tag{7.49}$$

$$\leq \left\| \sum_{k=1}^{l} \left(\lambda_k^{(p)} - \lambda_k^{(n_\varepsilon)} \right) b_k \right\| + \left\| \sum_{k=1}^{m-1} \left(\lambda_k^{(p)} - \lambda_k^{(n_\varepsilon)} \right) b_k \right\| + \varepsilon$$

$$\leq \left\| \left| x^{(p)} - x^{(n_\varepsilon)} \right| \right\| + \left\| \left| x^{(p)} - x^{(n_\varepsilon)} \right| \right\| + \varepsilon < 3\varepsilon.$$

Letting $p \to \infty$ in (7.49), we obtain

$$\left\| \sum_{k=m}^{l} \lambda_k b_k \right\| \leq 3\varepsilon \text{ for all } l \geq m \geq m_\varepsilon. \tag{7.50}$$

So the sequence $(x^{<m>})_{m=1}^{\infty}$ is a Cauchy sequence in $(X, \|\cdot\|)$, and hence convergent by the completeness of X.

Thus we have shown that $\sum_{k-1}^{\infty} \lambda_k b_k$ converges.

(ii) Now we show that

$$\left\| \left| x^{(p)} - x \right| \right\| \to 0 \ (p \to \infty). \tag{7.51}$$

We have for all $p \geq n_\varepsilon$ by (7.49) and (7.50)

$$\left\| \sum_{k=1}^{n} \left(\lambda_k^{(p)} - \lambda_k \right) b_k \right\| \leq \left\| \sum_{k=1}^{m_\varepsilon} \left(\lambda_k^{(p)} - \lambda_k \right) b_k \right\| + \left\| \sum_{k=m_\varepsilon+1}^{n} \lambda_k^{(p)} b_k \right\| + \left\| \sum_{k=m_\varepsilon+1}^{n} \lambda_k b_k \right\|$$

$$< \left\| \sum_{k=1}^{m_\varepsilon} \left(\lambda_k^{(p)} - \lambda_k \right) b_k \right\| + 6\varepsilon,$$

and so

$$\left\| \left| x^{(p)} - x \right| \right\| = \sup_n \left\| \sum_{k=1}^{n} \left(\lambda_k^{(p)} - \lambda_k \right) b_k \right\|$$

$$\leq \left\| \sum_{k=1}^{m_\varepsilon} \left(\lambda_k^{(p)} - \lambda_k \right) b_k \right\| + 6\varepsilon \text{ for all } p \geq n_\varepsilon.$$

Hence we have shown (7.51).

Thus we have shown that $(X, \|\cdot\|)$ is a Banach space. $\square$

Let $(X, \|\cdot\|)$ be a Banach space with a Schauder basis $(b_n)_{n=1}^{\infty}$. We define the operators $\mathcal{P}_n, \mathcal{R}_n : X \to X$ by

$$\mathcal{P}_n(x) = \sum_{k=1}^{n} \lambda_k b_k \text{ and } \mathcal{R}_n(x) = \sum_{k=n+1}^{\infty} \lambda_k b_k \text{ for all } x = \sum_{k=1}^{\infty} \lambda_k x_k \in X.$$

Obviously $\mathcal{P}_n$ and $\mathcal{R}_n$ are linear operators, and $\mathcal{P}_n + \mathcal{R}_n = \mathcal{I}$ where $\mathcal{I}$ denotes the identity on X. By Lemma 7.9.1, $(X, \|\cdot\|)$ is a Banach space with

$$\|x\| = \sup_n \|\mathcal{P}_n(x)\| \text{ for all } x \in X.$$

Since obviously

$$\|x\| \le \|\|x\|\| \text{ for all } x \in X,$$

and X is complete with respect to both norms $\|\cdot\|$ and $\|\|\cdot\|\|$, the norms $\|\cdot\|$ and $\|\|\cdot\|\|$ are equivalent on X by Corollary 4.12.3, hence there is a constant $c > 0$ such that

$$\|x\| \le \|\|x\|\| \le c\|x\| \text{ for all } x \in X.$$

It also follows that

$$\|\mathcal{P}_n(x)\| \le \sup_m \|\mathcal{P}_m(x)\| = \|\|x\|\| \le c \cdot \|x\| \text{ for each } n \in \mathbb{N} \text{ and all } x \in X, \qquad (7.52)$$

hence $\mathcal{P}_n$ is continuous, and so is each $\mathcal{R}_n = \mathcal{I} - \mathcal{P}_n$, that is, we have

$$\mathcal{P}_n, \mathcal{R}_n \in \mathcal{B}(X) \text{ for all } n \in \mathbb{N}.$$

We also have by (7.52)

$$\|\mathcal{P}_n\| = \sup\left\{ \frac{\|\mathcal{P}_n(x)\|}{\|x\|} : \|x\| = 1 \right\} \le c \text{ for all } n \in \mathbb{N},$$

that is, the sequence $(\mathcal{P}_n)_{n=1}^{\infty}$ is uniformly bounded, hence

$$\mathcal{P} = \sup_n \|\mathcal{P}_n\| < \infty \text{ and } \mathcal{R} = \sup_n \|\mathcal{R}_n\| < \infty.$$

As a first result we show a generalization of Example 7.5.7; it is a characterization of bounded relatively compact sets in Banach spaces with a Schauder basis.

Theorem 7.9.2 *Let X be a Banach space with a Schauder basis $(b_n)_{n=1}^{\infty}$. Then the set $Q \in \mathcal{M}_X$ is relatively compact if and only if*

$$\lim_{n \to \infty} \left(\sup_{x \in Q} \|\mathcal{R}_n(x)\| \right) = 0. \qquad (7.53)$$

Proof.

(i) First we show that if $Q \in \mathcal{M}_X$ is relatively compact then (7.53) holds.
We assume that $Q \in \mathcal{M}_X$ is relatively compact.
Let $\varepsilon > 0$ be given. Then there exist $x_1, x_2, \ldots, x_p \in X$ such that

$$Q \subset \bigcup_{k=1}^{n} B_{\varepsilon}(x_k), \text{ where } x_k = \sum_{j=1}^{\infty} \lambda_j(x_k) b_j \text{ for } k = 1, 2, \ldots, p.$$

Since $\lim_{n \to \infty} \|\mathcal{R}_n(x_k)\| = 0$ for each $k = 1, 2, \ldots, p$, there exists an $N \in \mathbb{N}$ such that

$$\|\mathcal{R}_{N-1}(x_k)\| = \|x_k - \mathcal{P}_{N-1}(x_k)\| < \varepsilon \text{ for all } k = 1, 2, \ldots, p. \qquad (7.54)$$

Let $x = \sum_{j=1}^{\infty} \lambda_j(x_k) b_j \in Q$ be given. Then we have for all $n, m \in \mathbb{N}$ with $m > n > N$

$$\left\| \sum_{j=n+1}^{m} \lambda_j(x) b_j \right\| = \left\| \sum_{j=N}^{m} \lambda_j(x) b_j - \sum_{j=N}^{n} \lambda_j(x) b_j \right\|$$
$$= \|\mathcal{P}_m(\mathcal{R}_{N-1}(x)) - \mathcal{P}_n(\mathcal{R}_{N-1}(x))\|$$
$$\le \|\mathcal{P}_m - \mathcal{P}_n\| \cdot \|\mathcal{R}_{N-1}(x)\|$$

$$\leq 2 \cdot \sup_{n} \|\mathcal{P}_n\| \cdot \|\mathcal{R}_{N-1}(x)\| = 2 \cdot \mathcal{P} \cdot \|\mathcal{R}_{N-1}(x)\| \,.$$

If we fix $n > N$ and let $m \to \infty$, then we obtain

$$\|\mathcal{R}_n(x)\| \leq 2 \cdot \mathcal{P} \cdot \|\mathcal{R}_{N-1}(x)\| \,.$$

Let $n > N$ be given. We choose $x_i \in \{x_1, x_2, \ldots, x_p\}$ such that $\|x - x_i\| < \varepsilon$, and obtain by (7.54)

$$\begin{aligned}
\|\mathcal{R}_n(x)\| &\leq 2 \cdot \mathcal{P} \cdot \|\mathcal{R}_{N-1}(x)\| = 2 \cdot \mathcal{P} \cdot \|x - \mathcal{P}_{N-1}(x)\| \\
&\leq 2 \cdot \mathcal{P} \left(\|x - x_i\| + \|x_i - \mathcal{P}_{N-1}(x_i)\| + \|\mathcal{P}_{N-1}(x_i) - \mathcal{P}_{N-1}(x)\| \right) \\
&< 2 \cdot \mathcal{P} \left(\varepsilon + \|\mathcal{R}_{N-1}(x_i)\| + \|\mathcal{P}_{N-1}(x_i - x)\| \right) \\
&< 2 \cdot \mathcal{P} \left(\varepsilon + \varepsilon + \|\mathcal{P}_{N-1}\| \cdot \|x_i - x\| \right) \leq 2 \cdot \mathcal{P} (2\varepsilon + \mathcal{P}\varepsilon) \\
&= 2 \cdot \mathcal{P}\varepsilon(2 + \mathcal{P}).
\end{aligned}$$

Since $x \in Q$ was arbitrary, we have

$$\sup_{x \in Q} \|\mathcal{R}_n(x)\| \leq 2 \cdot \mathcal{P}\varepsilon(2 + \mathcal{P}) \text{ for all } n > N,$$

and so (7.53) holds.

(ii) Now we show that if (7.53) holds then the set $Q \in \mathcal{M}_X$ is relatively compact. We assume that (7.53) is satisfied. Let $\varepsilon > 0$ be given. Then there exists $N \in \mathbb{N}$ such that $\|\mathcal{R}_n(x)\| < \varepsilon/2$ for all $n \geq N$ and all $x \in Q$. We write

$$F = \operatorname{span}\left(\{b_1, b_2, \cdots, b_n\}\right) \text{ and } Q_0 = \left\{ x \in F : \|x\| \leq \mathcal{P} \cdot \sup_{y \in Q} \|y\| \right\}.$$

Obviously we have $\mathcal{P}_n(x) \in Q_0$ for every $x \in Q$. Since F is finite dimensional, Q_0 is relatively compact, and so there exist $y_1, y_2, \ldots, y_m$ with

$$Q_0 \subset \bigcup_{k=1}^{m} B_\varepsilon(y_k).$$

It follows that, for every $x \in Q$, there exists $k \in \{1, 2, \ldots, m\}$ such that

$$\|x - y_k\| \leq \|\mathcal{P}_N(x - y_k)\| + \|\mathcal{P}_N(x)\| < \varepsilon,$$

where $y_k \in F$ was chosen such that $\mathcal{P}_N(x) \in B_{\frac{\varepsilon}{2}}(y_k)$. Thus we have

$$Q \subset \bigcup_{j=1}^{m} B_\varepsilon(y_j),$$

and so Q is relatively compact. $\qquad\square$

Now we prove the following famous result, the second part of which gives an estimate for the Hausdorff measure of noncompactness of any bounded subset of a Banach space with a Schauder basis.

Theorem 7.9.3 (Goldenŝtein, Goh'berg, Markus)
Let X be a Banach space with a Schauder basis. Then the function $\mu : \mathcal{M}_X \to [0, \infty)$ with

$$\mu(Q) = \limsup_{n \to \infty} \left(\sup_{x \in Q} \|\mathcal{R}_n(x)\| \right) \tag{7.55}$$

is a measure of noncompactness on X which is invariant under the passage to the convex hull.

Moreover, the following inequalities hold for all $Q \in \mathcal{M}_X$

$$\frac{1}{L} \cdot \mu(Q) \leq \chi(Q) \leq \inf_n \left(\sup_{x \in Q} \|\mathcal{R}_n(x)\| \right) \leq \mu(Q), \tag{7.56}$$

where $L = \limsup_{n \to \infty} \|\mathcal{R}_n\|$ denotes the basis constant of the Schauder basis.

Proof. The property in (MNC.1) of Definition 7.5.1 follows from Theorem 7.9.2, the property in (MNC.3) can easily be shown, and the property in (MNC.2) follows from the continuity of the operator $\mathcal{R}_n$.
Furthermore $\mu(\mathrm{co}(Q)) = \mu(Q)$ follows from the linearity of $\mathcal{R}_n$.

(i) First we show the second inequality in (7.56).
We obviously have

$$Q \subset \mathcal{P}_n(Q) + \mathcal{R}_n(Q) \text{ for each } n \in \mathbb{N} \text{ and all } Q \in \mathcal{M}_X. \tag{7.57}$$

It follows from (7.57) and Theorem 7.7.5 (4) and (8) that

$$\chi(Q) \leq \chi(\mathcal{P}_n(Q) + \mathcal{R}_n(Q)) \leq \chi(\mathcal{P}_n(Q)) + \chi(\mathcal{R}_n(Q)) = \chi(\mathcal{R}_n(Q))$$
$$\leq \sup_{x \in Q} \|\mathcal{R}_n(x)\| \text{ for all } n \in \mathbb{N},$$

and we obtain

$$\chi(Q) \leq \inf_n \left(\sup_{x \in Q} \|\mathcal{R}_n(x)\| \right) \leq \limsup_{n \to \infty} \left(\sup_{x \in Q} \|\mathcal{R}_n(x)\| \right) = \mu(Q).$$

Thus we have shown the second inequality in (7.56).

(ii) Now we show the first inequality in (7.56).
Let $\rho = \chi(Q)$, and $\varepsilon > 0$ be given. Then there are $x_1, x_2, \ldots, x_m \in X$ such that

$$Q \subset \bigcup_{k=1}^m B_{\rho+\varepsilon}(x_k) \subset \{x_1, x_2, \ldots, x_m\} + (\rho + \varepsilon)\overline{B_X}. \tag{$*$}$$

To see the second inclusion in ($*$), we assume that $x \in \bigcup_{k=1}^m B_{\rho+\varepsilon}(x_k)$. Then $x \in B_{\rho+\varepsilon}(x_{k_0})$ for some $k_0 \in \{1, 2, \ldots, m\}$ and so $\|x - x_{k_0}\| < \rho + \varepsilon$. We put $y = x - x_{k_0}$. Then $\|y\| < \rho + \varepsilon$, that is, $y \in (\rho+\varepsilon)B_X \subset (\rho+\varepsilon)\overline{B_X}$, and consequently $x = x_{k_0} + y \in \{x_1, x_2, \ldots, x_m\} + (\rho + \varepsilon)\overline{B_X}$.
Now ($*$) implies that for every $x \in Q$ there exist $y \in Q_0 = \{x_1, x_2, \ldots, x_m\}$ and $z \in \overline{B_X}$ such that $x = y + (\rho + \varepsilon)z$, and so

$$\|\mathcal{R}_n(x)\| = \|\mathcal{R}_n(y + (\rho + \varepsilon)z)\| = \|\mathcal{R}_n(y) + (\rho + \varepsilon)\mathcal{R}_n(z)\|$$
$$\leq \|\mathcal{R}_n(y)\| + (\rho + \varepsilon)\|\mathcal{R}_n(z)\| \leq \max_{1 \leq k \leq m} \|\mathcal{R}_n(x_k)\| + (\rho + \varepsilon) \cdot \|\mathcal{R}_n\| \cdot \|z\|$$
$$\leq \max_{1 \leq k \leq m} \|\mathcal{R}_n(x_k)\| + (\rho + \varepsilon) \cdot \|\mathcal{R}_n\| \text{ for all } n.$$

This implies

$$\sup_{x \in Q} \|\mathcal{R}_n(x)\| \leq \|\mathcal{R}_n(Q_0)\| + (\rho + \varepsilon) \cdot \|\mathcal{R}_n\|,$$

and so we have $\lim_{n \to \infty} \|\mathcal{R}_n(Q_0)\| = 0$, since the finite set Q_0 is compact, hence

$$\mu(Q) = \limsup_{n \to \infty} \left(\sup_{x \in Q} \|\mathcal{R}_n(x)\| \right) \leq (\rho + \varepsilon) \limsup_{n \to \infty} \|\mathcal{R}_n\| = (\rho + \varepsilon) \cdot L.$$

Since $\varepsilon > 0$ was arbitrary, it follows that

$$\frac{1}{L} \cdot \mu(Q) \leq \rho = \chi(Q).$$

Thus we have established the first inequality in (7.56). $\qquad\square$

Remark 7.9.4 *If we write in Theorem 7.9.3*

$$\tilde{\mu}(Q) = \inf_n \left(\sup_{x \in Q} \|\mathcal{R}_n(x)\| \right),$$

then we obtain from the inequalities in (7.56)

$$\frac{1}{L} \tilde{\mu}(Q) \leq \chi(Q) \leq \tilde{\mu}(Q) \text{ for all } Q \in \mathcal{M}_X. \tag{7.58}$$

Remark 7.9.5 *The measure of noncompactness μ in Theorem 7.9.3 is equivalent to the Hausdorff measure of noncompactness by (7.56).*
If $L = 1$ then we obtain from (7.56)

$$\tilde{\mu}(Q) = \mu(Q) = \chi(Q) \text{ for all } Q \in \mathcal{M}_X.$$

The next example provides a proof for Example 7.5.7.

Example 7.9.6 *If $X = c_0$ or $X = \ell_p$ for $1 \leq p < \infty$ with the standard Schauder basis $(e^{(n)})_{n=1}^{\infty}$ and the natural norms, then we have*

$$\|\mathcal{R}_n(x)\| \leq \|x\| \text{ for all } n \in \mathbb{N} \text{ and all } x \in X,$$

hence

$$\|\mathcal{R}_n\| \leq 1 \text{ for all } n \in \mathbb{N}.$$

We also obtain for each $n \in \mathbb{N}$

$$\left\| \mathcal{R}_n(e^{(n+1)}) \right\| = \|e^{(n+1)}\| = 1,$$

hence $\|\mathcal{R}_n\| \geq 1$ for all $n \in \mathbb{N}$. Therefore $\|\mathcal{R}_n\| = 1$ for all $n \in \mathbb{N}$, and so $L = \lim_{n \to \infty} \|\mathcal{R}_n\| = 1$.
We put $\mu_n(Q) = \sup_{x \in Q} \|\mathcal{R}_n(x)\|$ for all $n \in \mathbb{N}$ and all $Q \in \mathcal{M}_X$. Since $0 \leq \|\mathcal{R}_{n+1}(x)\| \leq \|\mathcal{R}_n(x)\|$ for all n and all $x \in Q$, we obtain $0 \leq \mu_{n+1}(Q) \leq \mu_n(Q)$ for all n, and consequently $\mu(Q) = \lim_{n \to \infty} \mu_n(Q)$ exists for each $Q \in \mathcal{M}_X$. So we have by (7.56)

$$\chi(Q) = \mu(Q) = \lim_{n \to \infty} \left(\sup_{x \in Q} \|\mathcal{R}_n(x)\| \right)$$

for the Hausdorff measure of noncompactness of any $Q \in \mathcal{M}_{c_0}$ or $Q \in \mathcal{M}_{\ell_p}$ for $1 \leq p < \infty$.
If $Q \in \mathcal{M}_{c_0}$ then we obtain

$$\chi(Q) = \lim_{n \to \infty} \left(\sup_{x \in Q} \|\mathcal{R}_n(x)\|_\infty \right) = \lim_{n \to \infty} \left(\sup_{x \in Q} \left(\sup_{k \geq n+1} |x_k| \right) \right)$$

and $Q \in \mathcal{M}_{c_0}$ is relatively compact if and only if

$$\chi(Q) = \lim_{n \to \infty} \left(\sup_{x \in Q} \left(\sup_{k \geq n+1} |x_k| \right) \right) = 0,$$

as stated in Example 7.5.7.

Similarly, if $Q \in \mathcal{M}_{\ell_p}$ then we obtain

$$\chi(Q) = \lim_{n \to \infty} \left(\sup_{x \in Q} \|\mathcal{R}_n(x)\|_p \right) = \lim_{n \to \infty} \left(\sup_{x \in Q} \left(\sum_{k=n+1}^{\infty} |x_k|^p \right)^{1/p} \right)$$

and $Q \in \mathcal{M}_{\ell_p}$ is relatively compact if and only if

$$\lim_{n \to \infty} \left(\sup_{x \in Q} \sum_{k=n+1}^{\infty} |x_k|^p \right) = 0.$$

Now we apply Theorem 7.9.3 to give an estimate for the Hausdorff measure of noncompactness of bounded sets in the space of all convergent sequences.

Example 7.9.7 *In c with the natural supremum norm $\|\cdot\|_\infty$, $(b_n)_{n=0}^\infty$ with $b_0 = e = (1, 1, \dots)$ and $b_n = e^{(n)}$ for $n = 1, 2, \dots$ is a Schauder basis of c, and every sequence $x = (x_k)_{k=1}^\infty \in c$ has an expansion*

$$x = \xi \cdot e + \sum_{k=1}^{\infty} (x_k - \xi) e^{(k)}, \text{ where } \xi = \lim_{k \to \infty} x_k.$$

Then we have

$$\mathcal{R}_n(x) = \sum_{k=n+1}^{\infty} (x_k - \xi) e^{(k)} \text{ for all } n \in \mathbb{N}.$$

Since $|x_k - \xi| \leq \|x\|_\infty + |\xi| \leq 2 \cdot \|x\|_\infty$ for all k and all $x \in c$, we have

$$\|\mathcal{R}_n(x)\|_\infty \leq \sup_{k \geq n+1} |x_k - \xi| \leq 2 \cdot \|x\|_\infty,$$

that is,

$$\|\mathcal{R}_n\| \leq 2 \text{ for all } n \in \mathbb{N}. \tag{7.59}$$

On the other hand, if $n \in \mathbb{N}$ is given, then we have for $x = -e + 2 \cdot e^{(n+1)}$

$$\xi = -1, \ \|x\| = 1 \text{ and } \mathcal{R}_n(x) = 2 \cdot e^{(n+1)}, \text{ that is, } \|\mathcal{R}_n\| \geq 2.$$

This and (7.59) together imply $\|\mathcal{R}_n\| = 2$ for all $n \in \mathbb{N}$, and so

$$L = \lim_{n \to \infty} \|\mathcal{R}_n\| = 2.$$

Now we obtain from (7.56)

$$\frac{1}{2} \cdot \mu(Q) = \frac{1}{2} \cdot \lim_{n \to \infty} \left(\sup_{x \in Q} \|\mathcal{R}_n(x)\|_\infty \right) \leq \chi(Q) \leq \mu(Q) \text{ for every } Q \in \mathcal{M}_X,$$

where for each $x \in c$ with $\xi_x = \lim_{k \to \infty} x_k$

$$\|\mathcal{R}_n(x)\|_\infty = \sup_{k \geq n+1} |x_k - \xi_x|.$$

Moreover, the bounds are attained. Indeed, for $Q \in \overline{B}_c$, we have $\chi(Q) = 1$, but $\mu(Q) = 2$. On the other hand, for $Q = \overline{B}_c \cap c_0$, we obtain $\chi(Q) = \mu(Q) = 1$.

We close this section with defining one more measure of noncompactness which is also equivalent to the Hausdorff measure of noncompactness.

Theorem 7.9.8 ([201, Theorem 4.3, p. 37]) Let X be a Banach space with a Schauder basis. Then the function $\nu : \mathcal{M}_X \to [0, \infty)$ with

$$\nu(Q) = \liminf_{n \to \infty} \left(\sup_{x \in Q} \|\mathcal{R}_n(x)\| \right) \text{ for all } Q \in \mathcal{M}_X$$

is a measure of noncompactness on X which is invariant under the passage to the convex hull. Moreover, the following inequalities hold

$$\frac{1}{L} \cdot \nu(Q) \leq \chi(Q) \leq \nu(Q) \text{ for all } Q \in \mathcal{M}_X.$$

Proof. Let $\varepsilon > 0$ be given. Then we have for all sufficiently large $n \in \mathbb{N}$ and for all $k \geq n$

$$\|\mathcal{R}_k(x)\| = \|\mathcal{R}_k (\mathcal{R}_n(x))\| \leq \|\mathcal{R}_k\| \cdot \|\mathcal{R}_n(x)\| \leq (L + \varepsilon) \|\mathcal{R}_n(x)\|,$$

hence

$$\nu(Q) = \liminf_{n \to \infty} \left(\sup_{x \in Q} \|\mathcal{R}_n(x)\| \right) \leq \limsup_{n \to \infty} \left(\sup_{x \in Q} \|\mathcal{R}_n(x)\| \right) \leq$$

$$\leq (L + \varepsilon) \liminf_{n \to \infty} \left(\sup_{x \in Q} \|\mathcal{R}_n(x)\| \right) \text{ for all } Q \in \mathcal{M}_X.$$

Since $\varepsilon > 0$ was arbitrary, we obtain by (7.56)

$$\frac{1}{L} \cdot \nu(Q) \leq \frac{1}{L} \cdot \mu(Q) \leq \chi(Q) \leq \nu(Q).$$

The properties of ν now follow from those of μ, or can be proved similarly. $\square$

Remark 7.9.9 *It is obvious from Theorem 7.9.8 that $\nu = \mu$ if*

$$\lim_{n \to \infty} \left(\sup_{x \in Q} \|\mathcal{R}_n(x)\| \right)$$

exists for all $Q \in \mathcal{M}_X$, for instance, for $X = c_0$ or $X = \ell_p$ $(1 \leq p < \infty)$ with the standard basis by Example 7.9.6. When $X = c$ with the supremum norm and the standard basis, then we have $\nu \neq \mu$ by Example 7.9.7.

7.10 The Darbo–Sadovskiĭ theorem of operators

So far we studied *measures of noncompactness of bounded sets* in complete metric or Banach spaces. For the proof of the Darbo–Sadovskiĭ theorem which is a generalization of the Darbo theorem, Theorem 7.6.7, we need the concept of *condensing operators*.

Definition 7.10.1 ([201, Definition 5.1, p. 38]) Let X and Y be complete metric spaces, ϕ and ψ be measures of noncompactness on X and Y, and $f : D \subset X \to Y$ be an operator. Then

(a) f is said to be (ϕ, ψ)–*contractive with constant* $k > 0$ or simply (ϕ, ψ)–*contractive*, if f is continuous and satisfies

$$\psi(f(Q)) \leq k \cdot \phi(Q) \text{ for all bounded subsets } Q \text{ of } D.$$

In particular, if $X = Y$ and $\phi = \psi$ we simply say that f is $k - \phi$–*contractive*.

(b) f is said to be (ϕ, ψ)–*condensing with constant* $k > 0$ or simply (ϕ, ψ)–*condensing*, if f is continuous and satisfies

$$\psi(f(Q)) < k \cdot \phi(Q) \text{ for all bounded non–precompact subsets } Q \text{ of } D.$$

In particular, if $X = Y$ and $\phi = \psi$ we simply say that f is $k - \phi$–*condensing*. Moreover, if $k = 1$ then f is said to be ϕ–*condensing*.

Remark 7.10.2 ([201, Remark 5.2, p. 38])

(a) If $\phi = \alpha$, the Kuratowski measure of noncompactness, then the $k - \alpha$–contractive or $k - \alpha$–condensing operators are usually called k–*set contractive* or k–*set condensing*.

(b) If $\phi = \chi$, the Hausdorff measure of noncompactness, then the $k - \chi$–contractive or $k - \chi$–condensing operators are usually called k–*ball contractive* or k–*ball condensing*.

(c) Every compact operator is $k - (\phi, \psi)$–contractive and $k - (\phi, \psi)$–condensing for all $k > 0$.

(d) Every $k - (\phi, \psi)$–condensing operator is $k - (\phi, \psi)$–contractive, but the converse is not true, in general.

Now we state a few simple properties of contractive and condensing operators.

Proposition 7.10.3 ([201, Proposition 5.3, p. 39]) Let X, Y and Z be complete metric spaces, ϕ, ψ and κ be measures of noncompactness on X, Y and Z, respectively, and $f : X \to Y$ and $g : Y \to Z$.

(a) If f is $k - (\phi, \psi)$–contractive, then f is $k' - (\phi, \psi)$–condensing for all $k' > k$.

(b) If f is $k_1 - (\phi, \psi)$–contractive (condensing) and g is $k_2 - (\psi, \kappa)$–contractive (condensing), then $g \circ f$ is $k_1 k_2 - (\phi, \kappa)$–contractive (condensing).

(c) If X and Y are Banach spaces, ψ is subadditive, that is, ψ satisfies *(7.16)*, $f_1 : D \subset X \to Y$ is $k_1 - (\phi, \psi)$–contractive (condensing) and $f_2 : D \subset X \to Y$ is $k_2 - (\phi, \psi)$–contractive (condensing), then $f_1 + f_2$ is $(k_1 + k_2) - (\phi, \psi)$–contractive (condensing).

(d) If X and Y are Banach spaces and ψ is subadditive and homogeneous, that is, ψ satisfies *(7.16)* and *(7.15)*, then the set of $k - (\phi, \psi)$–contractive (condensing) operators is convex.

(e) If X and Y are Banach spaces and ψ is invariant under passage to the convex hull, then the set of $k - (\phi, \psi)$–contractive (condensing) operators is convex.

Proof. Since Parts (a), (b), (c) and (d) are easy to verify, we only prove Part (e). Let f_1 and f_2 be two $k - (\phi, \psi)$–contractive operators, $\lambda \in [0, 1]$, and $f_\lambda = \lambda f_1 + (1 - \lambda) f_2$. Then we have $f_\lambda(Q) \subset \text{co}(f_1(Q) \cup f_2(Q))$ for all bounded subsets Q of D, and therefore it follows that

$$\psi(f_\lambda(Q)) \leq \psi\left(\text{co}\left(f_1(Q) \cup f_2(Q)\right)\right) = \psi\left(f_1(Q) \cup f_2(Q)\right)$$

$$= \max \left\{ \psi \left(f_1(Q) \right), \psi \left(f_2(Q) \right) \right\}.$$

If $\max\{\psi(f_1(Q)), \psi(f_2(Q))\} = \psi(f_l(Q))$ for $l \in \{1, 2\}$, then we obtain

$$\psi \left(f_\lambda(Q) \right) \leq \psi \left(f_l(Q) \right) \leq k \cdot \phi(Q).$$

Thus we have $\psi(f_\lambda(Q)) \leq k \cdot \phi(Q)$ for every bounded subset Q of D. As f_λ obviously is continuous, we conclude that f_λ is $k - (\phi, \psi)$–contractive.
The proof is analogous for $k - (\phi, \psi)$–condensing operators. □

Example 7.10.4 ([201, Example 5, p. 39]) Let X be a Banach space, $D \subset X$ and $f, g : D \to X$ be operators where f is compact and g is k–contractive, that is, there is $k \in [0, 1)$ such that $\|g(x) - g(y)\| \leq k\|x - y\|$ for all $x, y \in D$. Then $f + g$ is a k–set contractive operator.

Proof. Let Q be a bounded subset of D. We obtain immediately from the definition of the Kuratowski measure of noncompactness that $\alpha(g(Q)) \leq k\alpha(Q)$. Since f is compact, we have $\alpha(f(Q)) = 0$. Therefore it follows that

$$\alpha((f + g)(Q)) = \alpha(f(Q) + g(Q)) \leq \alpha(f(Q)) + \alpha(g(Q)) \leq k\alpha(Q).$$

Since $f + g$ is continuous, we obtain that it is a k–set contractive operator. □

Now we give an example of a set–condensing operator which is not k–set–contractive for any $k \in [0, 1)$.

Example 7.10.5 ([201, Example 6, p. 40]) Let $h : [0, 1] \to \mathbb{R}^+$ be a strictly decreasing function with the $h(0) = 1$, and $\overline{B} = \overline{B}_1(0)$ denote the closed unit ball of an infinite dimensional Banach space X. We define the operator $f : \overline{B} \to \overline{B}$ by $f(x) = h(\|x\|)x$ for all $x \in \overline{B}$. Then f is set–condensing, but not k–set–contractive for any $k \in [0, 1)$.

Proof.

(i) First we show that f is not k–set–contractive for any $k \in [0, 1)$.
   Let $y \in B_{h(r)r}(0)$, that is, $\|y\| < h(r)r = \|f(x)\|$ for some x with $\|x\| = r$, and so $y \in f(\overline{B}_r(0))$. Hence we have $B_{h(r)r}(0) \subset f(\overline{B}_r(0))$, and we obtain

$$\alpha \left(f(\overline{B}_r(0)) \right) \geq \alpha \left(B_{h(r)r}(0) \right) = 2 \cdot h(r)r = 2 \cdot h(r)\alpha \left(\overline{B}_r(0) \right).$$

   Since $\lim_{r \to 0} h(r) = 1$, it follows that f cannot be k–contractive for any $k \in [0, 1)$.

(ii) Now we show that f is set–condensing.
    Let $Q \subset \overline{B}$ be a set which is not precompact. Then $\alpha(Q) = \rho > 0$. We choose $r \in (0, \rho/2)$ and define the sets $Q_1 = Q \cap \overline{B}_r(0)$ and $Q_2 = Q \setminus \overline{B}_r(0)$. We obviously have $f(Q) = f(Q_1) \cup f(Q_2)$. Moreover, since $f(Q_1) \subset \overline{co}(Q_1 \cup \{0\})$, we obtain

$$\alpha(f(Q_1)) \leq \alpha \left(\overline{co}(Q_1 \cup \{0\}) \right) = \alpha(Q_1 \cup \{0\}) = \alpha(Q_1) \leq 2r < \rho = \alpha(Q).$$

On the other hand, the function h is strictly decreasing, and so

$$f(Q_2) \subset \{ta : 0 \leq t \leq h(r), \ a \in Q_2\} \subset \overline{co} \left(\{0\} \cup h(r)Q_2 \right),$$

which implies

$$\alpha(f(Q_2)) \leq h(r)\alpha(Q_2) \leq h(r)\alpha(Q) \leq \alpha(Q).$$

Hence we have

$$\alpha(f(Q)) = \alpha \left(f(Q_1) \cup f(Q_2) \right) = \max \left\{ \alpha(f(Q_1)), \alpha(f(Q_2)) \right\} < \alpha(Q)$$

and consequently f is set–condensing. □

Now we generalize Darbo's fixed point theorem, Theorem 7.6.9.

Theorem 7.10.6 (Darbo–Sadovskiĭ) ([201, Theorem 5.4, p. 40]) Let X be a Banach space, ϕ be a measure of noncompactness which is invariant under passage to the convex hull, $C \neq \emptyset$ be a bounded, closed and convex subset of X and $f : C \to C$ be a $\phi-$ condensing operator. Then f has a fixed point.

Proof. We choose a point $c \in C$ and denote by Σ the class of all closed and convex subsets K of C such that $c \in K$ and $f(K) \subset K$. Furthermore, we put

$$S_1 = \bigcap_{K \in \Sigma} K \text{ and } S_2 = \overline{co}\,(f(S_1) \cup \{c\}).$$

Obviously $\Sigma \neq \emptyset$, since $C \in \Sigma$, and $S_1 \neq \emptyset$, since $c \in S_1$. We also have

$$f(S_1) = f\left(\bigcap_{K \in \Sigma} K\right) \subset \bigcap_{K \in \Sigma} f(K) \subset \bigcap_{K \in \Sigma} K = S_1,$$

and consequently $f : S_1 \to S_1$.
Moreover, we have $S_1 = S_2$. Indeed, since $c \in S_1$ and $f(S_1) \subset S_2$, it follows that $S_2 \subset S_1$. This implies $f(S_2) \subset f(S_1) \subset S_2$, and so $S_2 \in \Sigma$, and hence $S_1 \subset S_2$. Therefore the properties of ϕ now imply

$$\phi(S_1) = \phi(S_2) = \phi\,(\overline{co}\,(f(S_1) \cup \{c\})) = \phi\,(\text{co}\,(f(S_1) \cup \{c\})) = \phi\,(f(S_1) \cup \{c\})$$
$$= \max\,\{\phi(f(S_1)), \phi(\{c\})\} = \phi(f(S_1)).$$

Since f is ϕ–condensing, it follows that $\phi(S_1) = 0$, and so S_1 is compact. Obviously S_1 is also convex. Thus it follows from Schauder's fixed point theorem, Theorem 7.3.8, that the operator $f : C \to C$ has a fixed point.   □

The following example will show that the theorem of Darbo and Sadovskiĭ fails to be true, if we assume that f is a k–contractive operator with constant $k = 1$.

Example **7.10.7** ([201, Example 7, p. 41]) Let $\overline{B}_{\ell_2}$ be the closed unit ball in ℓ_2. We define the operator $f : \overline{B}_{\ell_2} \to \overline{B}_{\ell_2}$ by

$$f(x) = f\,((x_k)_{k=1}^\infty) = \left(\sqrt{1 - \|x\|_2^2}, x_1, x_2, \dots\right).$$

Then we can write $f = g + h$ where g is the one-dimensional mapping

$$g(x) = g\,((x_k)_{k=1}^\infty) = \sqrt{1 - \|x\|_2^2}\,e^{(1)} \text{ and } h(x) = (0, x_1, x_2, \dots)$$

is an isometry. Hence f is a well–defined, continuous operator, and for every bounded subset Q of $\overline{B}_{\ell_2}$, we have

$$\alpha(f(Q)) \leq \alpha(g(Q) + h(Q)) \leq \alpha(g(Q)) + \alpha(h(Q)) = 0 + \alpha(Q).$$

So f is a k–set–contractive operator with constant $k = 1$. But f has no fixed points. If f had a fixed point $x \in \overline{B}_{\ell_2}$, then we would have $x_k = x_{k+1}$ for all $k \in \mathbb{N}$. Since $x \in \ell_2$, this implies $x_k = 0$ for all $k \in \mathbb{N}$, and then $f(x) = \sqrt{1 - \|x\|_2^2}\,e^{(0)} = e^{(0)} = (0, 0, 0, \dots)$, a contradiction.

7.11   Measures of noncompactness of operators

In this section we are going to define and study *measures of noncompactness of operators* between Banach spaces. It will turn out that the concept of the measure of noncompactness of a linear operator between Banach spaces is somehow similar to that of the norm of bounded linear operators between Banach spaces.

Definition 7.11.1 Let ϕ and ψ be measures of noncompactness on the Banach spaces X and Y, respectively.

(a)   An operator $A : X \to Y$ is said to be (ϕ, ψ)–*bounded*, if

$$A(Q) \in \mathcal{M}_Y \text{ for all } Q \in \mathcal{M}_X,$$

and if there is a nonnegative real number c such that

$$\psi(A(Q)) \le c \cdot \phi(Q) \text{ for all } Q \in \mathcal{M}_X. \tag{7.60}$$

(b)   If an operator A is (ϕ, ψ)–bounded, then the number

$$\|A\|_{(\phi,\psi)} = \inf\{c \ge 0 : (7.60) \text{ holds}\} \tag{7.61}$$

is called the (ϕ, ψ)–*operator norm of A*, or (ϕ, ψ)–*measure of noncompactness of A*. If $\psi = \phi$, we write $\|A\|_\phi = \|A\|_{(\phi,\phi)}$, for short.

Remark 7.11.2 *By Definition 7.10.1 (a) and Definition 7.11.1, a (ϕ, ψ)–bounded operator is a contractive operator between Banach spaces.*

Remark 7.11.3 *The definition of the (ϕ, ψ)–operator norm is similar to that of the norm of an operator in $\mathcal{B}(X,Y)$ where X and Y are Banach spaces. We recall that a linear operator $A : X \to Y$ is said to be bounded, if there exists a constant c such that*

$$\|A(x)\| \le c \cdot \|x\| \text{ for all } x \in X. \tag{7.62}$$

If a linear operator A is bounded, that is, if $A \in \mathcal{B}(X,Y)$, then the number

$$\|A\| = \inf\{c \ge 0 : (7.62) \text{ holds}\}$$

is the operator norm of A denoted by $\|A\|$.

Now we compute the Hausdorff measure of noncompactness of a bounded linear operator between infinite dimensional Banach spaces.

Theorem 7.11.4 *Let X and Y be infinite dimensional Banach spaces and $A : X \to Y$ be a bounded linear operator. Then we have*

$$\|A\|_\chi = \chi(A(S_X)) = \chi(A(\overline{B}_X)) = \chi(A(B_X)) \tag{7.63}$$

where B_X and $\overline{B}_X$ and S_X denote the open and closed unit balls, and the unit sphere in X.

Proof. We write $B = B_X$, $\overline{B} = \overline{B}_X$ and $S = S_X$, for short.
Since $\mathrm{co}(S) = \overline{B}$ and $A(\mathrm{co}(S)) = \mathrm{co}(A(S))$, we have by the invariance of χ under passage to the convex hull in Theorem 7.7.5 (b) (11)

$$\chi(A(\overline{B})) = \chi(A(\mathrm{co}(S))) = \chi(\mathrm{co}(A(S))) = \chi(A(S)).$$

We show that

$$\|A\|_\chi = \chi(A(B)). \tag{7.64}$$

(i) First (7.61), Theorem 7.7.8 and Remark 7.7.9 imply

$$\chi(A(B)) \leq \|A\|_\chi \cdot \chi(B) = \|A\|_\chi \cdot 1 = \|A\|_\chi. \tag{7.65}$$

(ii) Now we show the converse inequality of (7.65), that is,

$$\|A\|_\chi \leq \chi(A(B)). \tag{7.66}$$

Let $Q \in \mathcal{M}_X$ and $\{x_k : 1 \leq k \leq n\}$ be a finite r–net of Q. Then we have $Q \subset \bigcup_{k=1}^n B_r(x_k)$, and so

$$A(Q) \subset \bigcup_{k=1}^n A(B_r(x_k)). \tag{7.67}$$

Now it follows from (7.67), Theorem 7.7.5 (a) (4), (3), the additivity of A (Theorem 7.7.5 (b) (8)), and from Theorem 7.7.5 (b) (6) and (10) that

$$
\begin{aligned}
\chi(A(Q)) &\leq \chi\left(\bigcup_{k=1}^n A(B_r(x_k))\right) \leq \max_{1 \leq k \leq n} \chi\left(A(B_r(x_k))\right) \\
&= \max_{1 \leq k \leq n} \chi\left(A\left(\{x_k\} + B_r(0)\right)\right) = \max_{1 \leq k \leq n} \chi\left(A\left(\{x_k\}\right) + A(B_r(0))\right) \\
&\leq \max_{1 \leq k \leq n} \chi\left(A\left(\{x_k\}\right)\right) + \chi(A(B_r(0))) = \chi(A(rB_1(0))) = \chi(rA(B)) \\
&= r\chi(A(B)).
\end{aligned}
$$

This implies

$$\chi(A(Q)) \leq \chi(Q) \cdot \chi(A(B)),$$

and hence (7.66) holds.

Finally (7.66) and (7.65) imply (7.63).   □

Theorem 7.11.5 *Let X, Y and Z be Banach spaces, $A \in \mathcal{B}(X,Y)$ and $\tilde{A} \in \mathcal{B}(Y,Z)$. Then $\|A\|_\chi$ is a seminorm on $\mathcal{B}(X,Y)$ and*

$$\|A\|_\chi = 0 \text{ if and only if } L \in \mathcal{K}(X,Y); \tag{7.68}$$

$$\|A\|_\chi \leq \|A\|; \tag{7.69}$$

$$\|\tilde{A} \circ A\|_\chi \leq \|\tilde{A}\|_\chi \cdot \|A\|_\chi. \tag{7.70}$$

Proof.

(i) First we show that $\|\cdot\|_\chi$ is a seminorm.

(α) Obviously we have $\|A\|_\chi \in [0,\infty)$ for all $A \in \mathcal{B}(X,Y)$.

(β) Now we show the homogeneity of $\|\cdot\|_\chi$, that is,

$$\|\lambda A\|_\chi = |\lambda|\,\|A\|_\chi \text{ for all scalars } \lambda \text{ and all } A \in \mathcal{B}(X,Y). \tag{$*$}$$

Since trivially $\|\lambda A\|_\chi = |\lambda|\,\|A\|_\chi$ for $\lambda = 0$, we may assume that $\lambda \neq 0$. Let $\rho = \|A\|_\chi$, and $\varepsilon > 0$ be given. Then we have

$$\chi(A(Q)) \leq (\rho + \varepsilon)\chi(Q),$$

and so by Theorem 7.7.5 (b) (10)

$$\chi(\lambda A(Q)) = |\lambda| \chi(A(Q)) \le |\lambda|(\rho + \varepsilon)\chi(Q) \text{ for all } Q \in \mathcal{M}_X.$$

Since $\varepsilon > 0$ was arbitrary, it follows that

$$\chi(\lambda A(Q)) \le |\lambda|\rho \, \chi(Q) \text{ for all } Q \in \mathcal{M}_X,$$

hence

$$\|\lambda A\|_\chi \le |\lambda|\rho = |\lambda| \, \|A\|_\chi. \tag{**}$$

It also follows by what we have just shown that

$$\|A\|_\chi = \|\lambda^{-1}(\lambda A)\|_\chi \le |\lambda^{-1}| \, \|\lambda A\|_\chi$$

$$|\lambda| \, \|A\|_\chi \le \|\lambda A\|_\chi. \tag{***}$$

Now (*) follows from (**) and (***).

(γ) Finally, we prove that $\|\cdot\|_\chi$ satisfies the triangle inequality.
Let $\rho_k = \|A_k\|_\chi$ for $k = 1, 2$, and $\varepsilon > 0$ be given. Then we have

$$\chi(A_k(Q)) \le (\rho_k + \varepsilon/2)\chi(Q) \text{ for all } Q \in \mathcal{M}_X,$$

and so by Theorem 7.7.5 (b) (8)

$$\chi\left((A_1 + A_2)(Q)\right) = \chi\left(A_1(Q) + A_2(Q)\right) \le \chi(A_1(Q)) + \chi(A_2(Q))$$
$$\le (\rho_1 + \rho_2 + \varepsilon)\chi(Q) \text{ for all } Q \in \mathcal{M}_X.$$

Since $\varepsilon > 0$ was arbitrary, this implies

$$\chi((A_1 + A_2)(Q)) \le (\rho_1 + \rho_2)\chi(Q) \text{ for all } Q \in \mathcal{M}_X,$$

hence

$$\|A_1 + A_2\|_\chi \le \rho_1 + \rho_2 = \|A_1\|_\chi + \|A_2\|_\chi$$

which is the triangle inequality.
Thus we have shown that $\|\cdot\|_\chi$ is a seminorm.

(ii) The statement in (7.68) follows from the observation that an operator $A : X \to Y$ is compact if and only if it is continuous and maps bounded sets into relatively compact sets.

(iii) Now we show the inequality in (7.69).
We have by (7.63) in Theorem 7.11.4 and by the definition of the function χ

$$\|A\|_\chi = \chi(A(S_X)) \le \sup_{x \in S_X} \|A(x)\| = \|A\|,$$

that is, the inequality in (7.69) holds.

(iv) Finally, we show the inequality in (7.70).
We have for all $Q \in \mathcal{M}_X$ by Definition 7.11.1

$$\chi\left((\tilde{A} \circ A)(Q)\right) = \chi(\tilde{A}(A(Q))) \le \|\tilde{A}\|_\chi \cdot \chi(A(Q)) \le \|\tilde{A}\|_\chi \cdot \|A\|_\chi \cdot \chi(Q),$$

whence the inequality in (7.70) follows. $\qquad\square$

We are going to prove that, in Banach spaces, there is a connection between the Hausdorff measure of noncompactness of an operator A and the Hausdorff measure of noncompactness of its adjoint operator A'. This will be shown by the use of the next theorem by Harte [86] of 1984.

Theorem 7.11.6 (Harte) ([86, Theorem 1]) *Let X, Y, Z, W and E be Banach spaces, $T \in \mathcal{B}(X,Y)$, $S \in \mathcal{B}(W,Z)$, $U \in \mathcal{B}(E, \mathcal{B}(Y,Z))$ and $V \in \mathcal{B}(E, \mathcal{B}(X,W))$. We denote by $\mathcal{L}_S : \mathcal{B}(X,W) \to \mathcal{B}(X,Z)$ and $\mathcal{R}_T : \mathcal{B}(Y,Z) \to \mathcal{B}(X,Z)$ the operators defined by*

$$\mathcal{L}_S(A) = S \circ A \text{ for all } A \in \mathcal{B}(X,W)$$
$$\mathcal{R}_T(B) = B \circ T \text{ for all } B \in \mathcal{B}(Y,Z).$$

If $T_1, T_2 \in \mathcal{B}(X,Y)$, then we denote by $T_1 \star T_2$ the operator from $N(T_1 - T_2)$ with

$$(T_1 \star T_2)x = T_1 x = T_2 x \text{ for all } x \in N(T_1 - T_2).$$

Then the following inequality holds

$$\|(\mathcal{R}_T \circ U) \star (\mathcal{L}_S \circ V)\|_\chi \leq 4\|U\| \, \|T\|_\chi + 2\|V\| \, \|S\|_\chi, \tag{7.71}$$

and so, if T and S are compact operators then $(\mathcal{R}_T \circ U) \star (\mathcal{L}_S \circ V)$ is compact.

Proof. We put

$$F = N[(\mathcal{R}_T \circ U) \star (\mathcal{L}_S \circ V)] = \{e \in E : U(e)T = SV(e)\}. \tag{7.72}$$

If $\|T\|_\chi < \delta$, then there exists a finite subset $X_\delta = \{x_1, \ldots, x_n\} \subset \overline{B}_X$ such that

$$T(\overline{B}_X) \subset \bigcup_{i=1}^{n} B_{2\delta}(Tx_i). \tag{7.73}$$

Analogously, if $\|S\|_\chi < \epsilon$, then, for each $x_i \in X_\delta$, there exists a finite subset $F_\epsilon(x_i) = \{e_{1,i}, \ldots, e_{m_i,i}\} \subset \overline{B}_F = \{x \in F : \|x\| \leq 1\}$ such that

$$\{SV(e)x_i : e \in \overline{B}_F\} \subset \|V\| S(\overline{B}_W) \subset \bigcup_{j=1}^{m_i} B_{2\|V\|\delta}(SV(e_{j,i})x_i), \tag{7.74}$$

where $\overline{B}_W$ is the closed unit ball in W.

We consider the finite set

$$F_{\epsilon,\delta} = \bigcup_{i=1}^{n} F_\epsilon(x_i) \subset \overline{B}_F.$$

If $e \in \overline{B}_F$ and $x \in \overline{B}_X$, then (7.73) implies that there exists $x_{i_0} \in X_\delta$ such that $\|Tx - Tx_{i_0}\| \leq 2\delta$, and (7.74) implies that there exists $e_0 \in F_\epsilon(x_{i_0})$ such that $\|SV(e)x_{i_0} - SV(e_0)x_{i_0}\| \leq 2\|V\|\epsilon$. It follows from the definition of F in (7.72) that

$$\|[(\mathcal{R}_T \circ U) \star (\mathcal{L}_S \circ V)(e) - (\mathcal{R}_T \circ U) \star (\mathcal{L}_S \circ V)(e_0)]x\|$$
$$\leq \|U(e)(Tx - Tx_{i_0})\| + \|S(V(e) - V(e_0))x_{i_0}\| + \|U(e_0)(Tx_{i_0} - Tx)\|$$
$$\leq 2\|U\|\delta + 2\|V\|\epsilon + 2\|U\|\delta. \tag{7.75}$$

Obviously (7.75) implies

$$\|(\mathcal{R}_T \circ U) \star (\mathcal{L}_S \circ V)(e) - (\mathcal{R}_T \circ U) \star (\mathcal{L}_S \circ V)(e_0)\| \leq 4\|U\|\delta + 2\|V\|\epsilon. \qquad \square$$

Corollary 7.11.7 ([86, Corollary 1]) *If X and Y are Banach spaces and $A \in \mathcal{B}(X,Y)$, then we have*

$$\frac{1}{4}\|A\|_\chi \leq \|A'\|_\chi \leq 4\|A\|_\chi. \tag{7.76}$$

Proof. If we put $W = Z = \mathbb{F}$, $E = Y^*$, $S = I_\mathbb{F}$ and $V = A^*$ in Theorem 7.11.6, then we have $F = E$, $\mathcal{R}_T \circ U = \mathcal{L}_S \circ V = T'$ and $\|S\|_\chi = 0$. Hence the right-hand inequality in (7.76) follows from (7.71).
The left-hand side follows from the right-hand inequality and Corollary 4.9.9. $\quad\square$

Corollary 7.11.8 (Schauder theorem) ([182, Satz 1 and Satz 2])
If X and Y are Banach spaces and $A \in \mathcal{B}(X,Y)$, then the operator A is compact if and only if A' is compact.

Proof. The proof follows by Corollary 7.11.7. $\quad\square$

Remark 7.11.9 *A sharper inequality than (7.76) is known. Namely (see the proof in [3, Theorem 2.5.1] or in [86]), if X and Y are Banach spaces and $A \in \mathcal{B}(X,Y)$ then we have*

$$\frac{1}{2}\|A\|_\chi \leq \|A'\|_\chi \leq 2\|A\|_\chi \text{ and } \frac{1}{2}\|A\|_\alpha \leq \|A'\|_\alpha \leq 2\|A\|_\alpha.$$

We are going to prove that the adjoint operator of a finite rank operator is also a finite rank operator, and that the converse implication also holds. To prove this we need the following characterization of finite rank operators.

Theorem 7.11.10 *Let X and Y be normed spaces and $A \in \mathcal{B}(X,Y)$. Then A is a finite rank operator, $\dim R(A) = n < \infty$ say, if and only if there exist linearly independent subsets $\{y_1,\ldots,y_n\}$ in Y and $\{f_1,\ldots,f_n\}$ in X^* such that*

$$Ax = \sum_{k=1}^{n} f_k(x)y_k \text{ for all } x \in X.$$

Moreover, there exists $\{g_1,\ldots,g_n\} \subset Y^*$ such that

$$f_k(x) = g_k(Ax) \text{ for all } x \in X, \text{ that is, } f_k = A'g_k \text{ for all } k = 1,\ldots,n.$$

Proof. We assume that A is a finite rank operator and $\dim R(A) = n < \infty$. Then there exists a linearly independent subset $\{y_1,\ldots,y_n\}$ of Y such that

$$Ax = \sum_{k=1}^{n} f_k(x)y_k \text{ for all } x \in X, \tag{7.77}$$

where $f_k : X \mapsto \mathbb{F}$ for $k = 1,\ldots,n$. It is easy to show that f_k for $k = 1,\ldots,n$ are linear functionals. We are going to prove that these functionals are continuous. Since the linear space $Y_1 = \text{span}\{y_2,\ldots,y_n\}$ is a closed subspace of Y and $y_1 \notin Y_1$, by Corollary 4.5.4, there exists a functional $g_1 \in Y^*$ such that

$$g_1(y_1) = 1 \text{ and } g_1(y) = 0 \text{ for all } y \in Y_1. \tag{7.78}$$

Hence, it follows from (7.77) and (7.78) that

$$g_1(Ax) = \sum_{k=1}^{n} f_k(x)g_1(y_k) = f_1(x) \text{ for all } x \in X,$$

and so $A'g_1 = f_1$. Therefore we have shown that $f_1 \in X^*$. Analogously it can be shown that $\{f_2,\ldots,f_n\} \subset X^*$ and that there exists $\{g_2,\ldots,g_n\} \subset Y^*$ such that $f_k = A'g_k$ for $k = 2,\ldots,n$. This completes the nontrivial part of the proof. $\quad\square$

Theorem 7.11.11 *If X and Y are normed spaces and $A \in \mathcal{B}(X, Y)$, then A is a finite rank operator if and only if A' is a finite rank operator.*

Proof.

(i)   First we show that if A is a finite rank operator then A' is a finite rank operator. We assume that A is a finite rank operator and $\dim R(A) = n < \infty$. Then, by Theorem 7.11.10, there exist a linearly independent subset $\{y_1, \ldots, y_n\}$ in Y and a subset $\{f_1, \ldots, f_n\}$ of X^* such that

$$Ax = \sum_{k=1}^{n} f_k(x) y_k \text{ for all } x \in X.$$

Hence we have

$$(A'g)x = g(Ax) = \sum_{k=1}^{n} f_k(x) g(y_k) \text{ for all } g \in Y^* \text{ and all } x \in X,$$

that is,

$$A'g = \sum_{k=1}^{n} J_Y(y_k)(g) f_k \text{ for all } g \in Y^*,$$

and, by Theorem 7.11.10, A' is a finite rank operator.

(ii) Now we show that if A' is a finite rank operator then A is a finite rank operator. If A' is a finite rank operator, then we know by what we have just proved that A'' is a finite rank operator and finally Lemma 4.9.10 implies that A is a finite rank operator. □

7.12   Ascent and descent of operators

The notions and properties of the ascent and the descent of operators play important roles in operator theory. Many results are valid for linear operators on linear spaces without norm.

The standard results of the theory presented in this section can be found in several textbooks, for instance, in [92, 199, 58, 142, 2, 3, 38, 88].

Let $A \in \mathcal{L}(X)$. The following inclusions are obvious:

$$N(A^n) \subset N(A^{n+1}) \text{ for } n = 0, 1, 2, \ldots, \text{ where } A^0 = I.$$

If $N(A^n) = N(A^{n+1})$ for some $n \in \{0, 1, 2, \ldots\}$, then $N(A^k) = N(A^n)$ for every $k = n+1, n+2, \ldots$

Definition 7.12.1 Let $A \in \mathcal{L}(X)$. If there exists some $n \in \mathbb{N}$ such that the following holds

$$N(A^n) = N(A^{n+1}), \tag{7.79}$$

then A has a *finite ascent*. The smallest n such that (7.79) holds is the ascent of A, denoted by $asc(A)$.

If (7.79) does not hold for any n, then the ascent of A is infinite, noted by $asc(A) = \infty$.

Dually, we have

$$R(A^n) \supset R(A^{n+1}) \text{ for } n = 0, 1, 2, \ldots, \text{ where } A^0 = I.$$

If $R(A^n) = R(A^{n+1})$ for some $n \in \mathbb{N}$, then $R(A^k) = R(A^n)$ for every $k = n+1, n+2, \ldots$

Definition 7.12.2 Let $A \in \mathcal{L}(X)$. If there exists some $n \in \mathbb{N}$ such that

$$R(A^n) = R(A^{n+1}), \tag{7.80}$$

then A has a *finite descent*. In this case the descent of A is the smallest n such that (7.80) holds, and $dsc(A) = n$.
If (7.80) does not hold for any $n \in \mathbb{N}$, then A has an infinite descent, and we write $dsc(A) = \infty$.

Lemma 7.12.3 *The following conditions are equivalent for $A \in \mathcal{L}(X)$:*

$$\text{asc}(A) \leq m < \infty, \tag{7.81}$$

$$N(A^n) \cap R(A^m) = \{0\} \text{ for every } n \in \mathbb{N} \tag{7.82}$$

$$N(A^n) \cap R(A^m) = \{0\} \text{ for some } n \in \mathbb{N} \tag{7.83}$$

and

$$N(A) \cap R(A^m) = \{0\}. \tag{7.84}$$

Proof.

(i) First we show that (7.81) implies (7.82).
If $\text{asc}(A) \leq m < \infty$, $n \in \mathbb{N}$ and $y \in N(A^n) \cap R(A^m)$, then $y = A^m x$, $x \in X$ and $A^n y = 0$. Consequently we have $A^{m+n} x = 0$ and $x \in N(A^{m+n}) = N(A^m)$, and hence $y = A^m x = 0$.

(ii) It is obvious that (7.82) implies (7.83) and (7.83) implies (7.84).

(iii) Finally, we show that (7.84) implies (7.81).
We assume that $N(A) \cap R(A^m) = \{0\}$. If $x \in N(A^{m+1})$, that is, $A(A^m x) = 0$, then we have $A^m x \in N(A) \cap R(A^m) = \{0\}$, or $x \in N(A^m)$. Therefore we obtain $N(A^m) = N(A^{m+1})$ and $\text{asc}(A) \leq m$. $\qquad \square$

Lemma 7.12.4 *The following conditions are equivalent for $A \in \mathcal{L}(X)$:*

$$dsc(A) \leq m < \infty. \tag{7.85}$$

For every $n \in \mathbb{N}$, $R(A^n)$ has a direct complement M_n in X such that $M_n \subset N(A^m)$ $\tag{7.86}$

$$N(A^m) + R(A^n) = X \text{ for every } n \in \mathbb{N}, \tag{7.87}$$

$$N(A^m) + R(A^n) = X \text{ for some } n \in \mathbb{N} \tag{7.88}$$

and

$$N(A^m) + R(A) = X. \tag{7.89}$$

Proof.

(i) First we show that (7.85) implies (7.86).

Let $\mathrm{dsc}(A) \leq m$, $n \in \mathbb{N}$, and let M be a direct complement of $R(A^n)$ in X, that is,

$$X = M \oplus R(A^n). \tag{7.90}$$

Let $\{x_i : i \in I\}$ be an algebraic basis in M. Since $A^m(M) \subset A^m(X) = A^{m+n}(X)$, for every $i \in I$, there exists some $y_i \in X$ such that $A^m x_i = A^{m+n} y_i$. Let $z_i = x_i - A^n y_i$. It follows from $A^m z_i = A^m x_i - A^{m+n} y_i = 0$ that $M_n = \mathrm{span}\{z_i : i \in I\} \subset N(A^m)$. According to (7.90), for every $x \in X$, there exist scalars α_i, $i \in I$ and $y \in X$ such that

$$x = \sum \alpha_i x_i + A^n y = \sum \alpha_i (z_i + A^n y_i) + A^n y = \sum \alpha_i z_i + A^n z.$$

Hence, we have $X = M_n + R(A^n)$. To prove that this sum is direct, we assume $x \in M_n \cap R(A^n)$. Then we have $x = \sum \beta_i z_i = A^n v$ and hence

$$\sum \beta_i x_i = \sum \beta_i A^n y_i + A^n v \in R(A^n).$$

It follows from (7.90) that $\sum \beta_i x_i = 0$. We get $\beta_i = 0$ for every $i \in I$, and consequently $x = 0$, hence $X = M_n \oplus R(A^n)$.

(ii) Obviously (7.86) implies (7.87), (7.87) implies (7.88) and (7.88) implies (7.89).

(iii) Finally, we show that (7.89) implies (7.85).

We assume that $N(A^m) + R(A) = X$. Then we have $A^m(X) = A^m(R(A)) = A^{m+1}(X)$, so $\mathrm{dsc}(A) \leq m$. $\square$

The results of Lemmas 7.12.3 and 7.12.4 would also follow from the following characterizations of finite ascent and descent due to Taylor and Lay [199]. The proofs of their results are purely algebraic.

Theorem 7.12.5 (Taylor and Lay) ([199, Theorem 6.3]) *Let X be a linear space and $A \in \mathcal{L}(X)$. For $i, j = 0, 1, \ldots$, there is an algebraic isomorphism, denoted by $\cong$ such that*

$$N(A^{i+j})/N(A^i) \cong R(A^i) \cap N(A^j). \tag{7.91}$$

Thus, for $j = 1, 2, \ldots$, $\mathrm{asc}(A) \leq p$ if and only in $R(A^p) \cap N(A^j) = \{0\}$.

Proof. We fix i and j, and observe that $A^i : N(A^{i+j}) \to R(A^i) \cap N(A^j)$. If $y \in R(A^i) \cap N(A^j)$, then there exists x in the domain of A^i such that $y = A^i x$. In fact, $x \in N(A^{i+j})$, since $0 = A^j x = A^j(A^i x)$. Thus A^i maps $N(A^{i+j})$ onto $R(A^i) \cap N(A^j)$. If we define $\phi : N(A^{i+j})/N(A^i) \to R(A^i) \cap N(A^j)$ by $\phi([x]) = A^j x$ for $[x] = x + N(A^i) \in N(A^{i+j})/N(A^i)$, then ϕ is the desired isomorphism, by Theorem 2.5.9.

The second statement follows from (7.91) and the fact that, for any $j = 1, 2, \ldots$, $\mathrm{asc}(A) \leq p$ if and only if $N(A^{p+j})/N(A^p) = \{0\}$. $\square$

The following result is the special case of a theorem by Taylor and Lay [199, Theorem 6.4], when the domain of the operator is the whole space.

Theorem 7.12.6 (Taylor and Lay) *Let X be a linear space and $A \in \mathcal{L}(X)$ with domain $D(A) = X$. Then, for $i, j = 0, 1, \ldots$, there is an algebraic isomorphism $\cong$ such that*

$$R(A^i)/R(A^{i+j}) \cong X/[R(A^j) + N(A^i)]. \tag{7.92}$$

Thus, for $j = 1, 2, \ldots$, $\mathrm{dsc}(A) \leq q$ if and only if $X = R(A^j) + N(A^q)$.

Proof. Let φ be the natural homomorphism from $R(A^i)$ onto $R(A^i)/R(A^{i+j})$. We put $f = \varphi A^i$. Obviously $R(f) = R(A^i)/R(A^{i+j})$.

Now we determine $N(f)$.

If $f(x) = 0$, then $A^i x \in R(A^{i+j})$; that is, $A^i x = A^{i+j} w$ for some $w \in X$. Hence $A^i(x - A^j w) = 0$, and

$$x = A^j w + (x - A^j w) \in R(A^j) + N(A^i).$$

Now we suppose $x \in R(A^j)$. Then $A^i x \in R(A^{i+j})$, and so $f(x) = 0$. Also if $x \in N(A^i)$, then $f(x) = \varphi A^i x = \varphi(0) = 0$. We conclude $N(f) = R(A^j) + N(A^i)$. The one–to–one mapping

$$\hat{f} : X/[R(A^j) + N(A^i)] \to R(A^i)/R(A^{i+j})$$

induced by f is the desired isomorphism, by Theorem 2.5.9.

Finally, the second statement of the theorem easily follows from (7.92). $\qquad \square$

Theorem 7.12.7 *If $A \in \mathcal{L}(X)$, $\mathrm{asc}(A) < \infty$ and $\mathrm{dsc}(A) < \infty$, then we have*

$$\mathrm{asc}(A) = \mathrm{dsc}(A).$$

Proof. Let $p = \mathrm{asc}(A)$ and $q = \mathrm{dsc}(A)$.

(i) First we assume $p \leq q$.

Then $R(A^q) \subset R(A^p)$. Let $q > 0$, because otherwise the proof is finished. Let $y \in R(A^p)$. We conclude from Lemma 7.12.4 that $X = N(A^q) + R(A^q)$, so there exist $z \in N(A^q)$ and $w \in X$ such that $y = z + A^q w$. Since $z = y - A^q w \in R(A^p)$, we have $z \in N(A^q) \cap R(A^p)$. It follows from Lemma 7.12.3 that $N(A^q) \cap R(A^p) = \{0\}$, so $z = 0$ and therefore $y \in R(A^q)$. We have just proved that $R(A^q) = R(A^p)$, so $q \leq p$, and hence $p = q$.

(ii) Now we assume $q \leq p$ and $p > 0$.

We get $N(A^q) \subset N(A^p)$. Let $x \in N(A^p)$. We have from Lemma 7.12.4 $X = N(A^q) + R(A^p)$, so there exist $u \in N(A^q)$ and $v \in X$ such that $x = u + A^p v$. Since $A^p x = 0$ and $A^p u = 0$, we have $A^{2p} v = 0$. Consequently it holds that $v \in N(A^{2p}) = N(A^p)$ and so $x = u \in N(A^q)$. We have just proved $N(A^q) = N(A^p)$ and so $p \leq q$, hence, $p = q$. $\qquad \square$

Theorem 7.12.8 *Let $A \in \mathcal{L}(X)$, $\mathrm{asc}(A) < \infty$ and $\mathrm{dsc}(A) < \infty$. Then we have $\mathrm{asc}(A) = \mathrm{dsc}(A) = p$ and*

$$X = R(A^p) \oplus N(A^p). \tag{7.93}$$

Moreover, $R(A^p)$ and $N(A^p)$ are invariant subspaces of A, the operator $A_1 = A|_{R(A^p)} : R(A^p) \to R(A^p)$ is invertible, and the operator $A_2 = A|_{N(A^p)} : N(A^p) \to N(A^p)$ is nilpotent. If $A \in \mathcal{B}(X)$, then $R(A^p)$ and $N(A^p)$ are closed.

Conversely if the following holds for some $m \in \mathbb{N}$

$$X = R(A^m) \oplus N(A^m), \tag{7.94}$$

then we have $\mathrm{asc}(A) = \mathrm{dsc}(A) \leq m$.

Proof. If $\mathrm{asc}(A) = \mathrm{dsc}(A) = p < \infty$, then (7.93) follows from Lemmas 7.12.3 and 7.12.4. On the other hand, it follows from $\mathrm{dsc}(A) = p$ that $A(R(A^p)) = R(A^p)$ and therefore A_1 is surjective. Since $\mathrm{asc}(A) < \infty$, A_1 is one to one by Lemma 7.12.3. If $x \in N(A^p)$, then $A_2^p x = A^p x = 0$ and so A_2 is nilpotent.

If $A \in \mathcal{B}(X)$, then $N(A^p)$ is closed and it follows from Kato's theorem, Theorem 4.15.5, and (7.93) that $R(A^p)$ is closed.

If (7.94) holds, then we conclude from Lemmas 7.12.3 and 7.12.4 that $\mathrm{asc}(A) \leq m$ and $\mathrm{dsc}(A) \leq m$. According to Theorem 7.12.7, we have $\mathrm{asc}(A) = \mathrm{dsc}(A) \leq m$. $\qquad \square$

Lemma 7.12.9 *If $A \in \mathcal{B}(X)$, then $\operatorname{asc}(A) = \operatorname{dsc}(A) < \infty$ if and only if $\operatorname{asc}(A') = \operatorname{dsc}(A') < \infty$. Moreover, we have*

$$\operatorname{asc}(A) = \operatorname{dsc}(A) = \operatorname{asc}(A') = \operatorname{dsc}(A').$$

Proof. Let $\operatorname{asc}(A) = \operatorname{dsc}(A) = p < \infty$. Then we have $N(A^p) = N(A^{p+1})$ and $R(A^p) = R(A^{p+1})$. It follows from Theorem 7.12.8 that $R(A^p)$ is closed, so we get by (4.113)

$$R((A')^p) = R((A^p)') = N(A^p)^\circ = N(A^{p+1})^\circ = R((A^{p+1})') = R((A')^{p+1}).$$

It follows from (4.77) that

$$N((A')^p) = N((A^p)') = R(A^p)^\circ = R(A^{p+1})^\circ = N((A^{p+1})') = N((A')^{p+1}).$$

Consequently we have $\operatorname{asc}(A') \le p$ and $\operatorname{dsc}(A') \le p$.
Now we assume $\operatorname{asc}(A') = \operatorname{dsc}(A') = q < \infty$. Then we have $N((A')^q) = N((A')^{q+1})$ and $R((A')^q) = R((A')^{q+1})$ is closed by Theorem 7.12.8. We get from Lemma 4.9.6, (4.113) and (4.77)

$$N(A^q) = {}^\circ(N(A^q)^\circ) = {}^\circ(R((A')^q)) = {}^\circ(R((A')^{q+1})) = {}^\circ(N(A^{q+1})^\circ) = N(A^{q+1}),$$
$$R(A^q) = {}^\circ N((A')^q) = {}^\circ N((A')^{q+1}) = R(A^{q+1}).$$

It follows that $\operatorname{asc}(A) \le q$ and $\operatorname{dsc}(A) \le q$. Finally, we have $p = q$. $\qquad\square$

Lemma 7.12.10 *Let $A \in \mathcal{B}(X)$ and $R(A^n)$ be closed for $n = 1, 2, \ldots$. Then we have $\operatorname{asc}(A) = \operatorname{dsc}(A')$ and $\operatorname{dsc}(A) = \operatorname{asc}(A')$.*

Proof. As in the proof of Lemma 7.12.9, we obtain that $\operatorname{asc}(A) < \infty$ if and only if $\operatorname{dsc}(A') < \infty$, and in this case $\operatorname{asc}(A) = \operatorname{dsc}(A')$. It follows that $\operatorname{asc}(A) = \infty$ if and only if $\operatorname{dsc}(A') = \infty$. Therefore we have in both cases $\operatorname{asc}(A) = \operatorname{dsc}(A')$. Analogously we obtain $\operatorname{dsc}(A) = \operatorname{asc}(A')$. $\qquad\square$

7.13 Fredholm theorems for operators with $\|\cdot\|_\chi < 1$

If X is a Banach, $A \in \mathcal{B}(X)$ and λ is a scalar, then we write $A - \lambda$ instead of $A - \lambda I$, where I denotes the identity map on X.

Let $k(t, s)$ be a continuous function on the square $a \le t, s \le b$, and

$$x(t) - \int_a^b k(t, s) x(s)\, ds = y(t), \ (a \le t \le b),$$

be the well–known *Fredholm integral equation of the second kind*; it is assumed that $y \in C[a, b]$ is a given function, and $x \in C[a, b]$ is the unknown. The operator $F : C[a, b] \to C[a, b]$, defined by

$$(Fx)(t) = \int_a^b k(t, s) x(s)\, ds, \ x \in C[a, b] \qquad (7.95)$$

is compact (this is not trivial but not hard to prove). Hence the equation in (7.95) could be written in operator form as

$$(I - F)x = y.$$

If X is a Banach space, $K \in \mathcal{K}(X)$, then theorems related to the operator $I - K$, are usually called *Fredholm theorems*. In this section, we mainly present results from [GGM2]. We consider the operator $I - A$ with $A \in \mathcal{B}(X)$ and $\|A\|_\chi < 1$. We recall that if $\|A\|_\chi = 0$ if and only if $A \in \mathcal{K}(X)$ by (7.68) in Theorem 7.11.5.

The following notations are very useful in Fredholm theory and will frequently be used in the remainder of this chapter and also in Chapter 8. We recall that if $A \in \mathcal{B}(X)$, then $R(A)$ and $N(A)$ denote, respectively, the image and null spaces of the operator A. Let $\alpha(A) = \dim N(A)$ and $\beta(A) = \dim X/R(A)$. Usually $\alpha(A)$ and $\beta(A)$ are called, respectively, the *nullity* and the *defect of* A. If at least one of the values $\alpha(A)$ and $\beta(A)$ is finite, then the difference $i(A) = \alpha(A) - \beta(A)$ is called the *index of the operator* A.

Remark 7.13.1 *The use of the Greek letter α for the Kuratowski measure of noncompactness on the one hand, and for the nullity on the other hand is standard. So we did not introduce different notations. The meaning of α will be clear from the context in each case.*

Theorem 7.13.2 *Let X be a Banach space, $A \in \mathcal{B}(X)$ and $\|A\|_\chi < 1$. Then we have $\alpha(I - A) < \infty$.*

Proof. We denote by S the unit sphere in the subspace $N(I - A)$. Let $\varepsilon > 0$ and k be a natural number such that $\|A^k\|_\chi < \epsilon$ (the existence of the number k follows from $\|A\|_\chi < 1$ and (7.70) in Theorem 7.11.5). Hence for the set $A^k(S)$ there exists a finite ε-net which is also an ε-net for the set S (since $A^k(S) = S$). We have proved that S is a compact subset of X and by Corollary 4.13.4 this implies $\alpha(I - A) < \infty$. $\qquad\square$

Theorem 7.13.3 *Let X be a Banach space, $A \in \mathcal{B}(X)$ and $\|A\|_\chi < 1$. Then we have $R(I - A) = \overline{R(I - A)}$.*

Proof. It follows from Theorem 7.13.2 and Lemma 4.5.10 that there exists a closed subspace M in X such that $M \oplus N(I - A) = X$. Obviously $x \in M$ and $(I - A)x = 0$ imply $x = 0$. Hence $R(I - A)$ is a closed subspace in X if and only if

$$\inf_{x \in M, \, \|x\|=1} \|(I - A)x\| > 0.$$

We assume that $R(I - A) \neq \overline{R(I - A)}$. Then there exists a sequence (x_n) in M such that $\|x_n\| = 1$ and $\lim_{n \to \infty}(I - A)x_n = 0$. Let us remark that for each natural number m we have $\lim_{n \to \infty}(I - A^m)x_n = 0$. Let $\varepsilon > 0$ and k be a natural number such that $\|A^k\|_\chi < \varepsilon$. There exists a natural number n_0 such that for each $n > n_0$ we have $\|x_n - A^k x_n\| < \delta$, where $\delta = (\varepsilon - \|A^k\|_\chi)/2$. From $\varepsilon - \delta > \|A^k\|_\chi$, it follows that there exists $Q \subset X$ such that Q is a finite $\varepsilon - \delta$-net for $\{A^k x_n : n = 1, 2, \ldots\}$. It is easy to prove that Q is a finite ε-net for $\{x_{n_0+m} : m = 1, 2, \ldots\}$. Since ε was arbitrary, it follows that the sequence (x_n) has a subsequence (x_{n_k}) that converges to $x_0 \in X$. Obviously we have $\|x_0\| = 1$, $x_0 - Ax_0 = 0$ and $x_0 \in M$, which is a contradiction. $\qquad\square$

Lemma 7.13.4 *Let X be a Banach space, $A \in \mathcal{B}(X)$ and $\|A\|_\chi < 1$. If the equation $x - Ax = y$ has a solution for each $y \in X$, then the homogeneous equation $x - Ax = 0$ only has the trivial solution, that is,*

$$R(I - A) = X \text{ implies } N(I - A) = \{0\}.$$

Proof. It follows from $R(I - A) = X$, that the descent $\mathrm{dsc}(I - A)$ of the operator $I - A$ is equal to zero. Now if $N(I - A) \neq \{0\}$, then the ascent $\mathrm{asc}(I - A)$ of the operator $I - A$

is infinite by Theorem 7.12.7. By the Riesz lemma, Lemma 4.13.2, there exists a sequence (y_n) in X with the following property

$$\|y_n\| = 1, \ y_n \in N((I-A)^n) \text{ and } \|y_n - x\| > \frac{1}{2} \text{ for each } x \in N((I-A)^{n-1}). \qquad (7.96)$$

Let us remark that $(I - A^k)(N((I-A)^n)) \subset N((I-A)^{n-1})$ for $n = 1, 2, \ldots$. Let k be a natural number such that $\|A^k\|_\chi < 1/4$. Hence, for $\{A^k y_n : n = 1, 2, \ldots\}$ there exist a finite ε-net for $\epsilon = 1/4$ and an element z in this net such that $\|A^k y_p - z\|, \|A^k y_m - z\| < 1/4$ for some $p > m$ hence

$$\|A^k y_p - A^k y_m\| \leq \|A^k y_p - z\| + \|z - A^k y_m\| < \frac{1}{2}. \qquad (7.97)$$

From

$$A^k y_p - A^k y_m = y_p - x, \text{ where } x = y_m + (I-A^k)y_p - (I-A^k)y_m \in N((I-A)^{p-1}),$$

and (7.96) it follows that $\|y_p - x\| > 1/2$, that is, $\|A^k y_p - A^k y_m\| > 1/2$. This is a contradiction to (7.97). $\qquad \square$

Remark 7.13.5 *We remark that Theorems 7.13.2 and 7.13.3, and Lemma 7.13.4 hold true in the case when the condition $\|A\|_\chi < 1$ is replaced by the weaker condition $\|A^k\|_\chi < 1$ for some natural number k.*

The following two lemmas are necessary for further results.

Lemma 7.13.6 *Let M be a closed subspace of a normed space X and E be a finite dimensional subspace of X. Then $M + E$ is a closed subspace of X.*

Proof. It is sufficient to prove the lemma in the case of $\dim E = 1$. Let $e_1 \in E$ and $e_1 \neq 0$. If $e_1 \in M$, then $M + E = M$ is a closed subspace of X. If $e_1 \notin M$, then we have $d = \text{dist}(e_1, M) > 0$. We remark that $M + E = \{\lambda e_1 + v : \lambda \in \mathbb{F}, \ v \in M\}$, and we obtain for $\lambda \neq 0$

$$\frac{\|\lambda e_1 + v\|}{|\lambda|} = \|e_1 + \lambda^{-1}v\| \geq d,$$

that is,

$$|\lambda| \leq \frac{\|\lambda e_1 + v\|}{d} \text{ for } \lambda \in \mathbb{F} \text{ and } v \in M. \qquad (7.98)$$

To prove that $M + E$ is a closed subspace of X, we assume that the sequence (x_n) in $M + E$ converges to $x \in X$. We have to prove that $x \in M + E$. Let $x_n = \lambda_n e_1 + v_n$, $\lambda_n \in \mathbb{F}$, $v_n \in M$ for $n = 1, 2, \ldots$. It follows from (7.98) that

$$|\lambda_n - \lambda_m| \leq \frac{\|x_n - x_m\|}{d},$$

and there is $\lambda \in \mathbb{F}$ such that $\lim_{n\to\infty} \lambda_n = \lambda$. Obviously, $\lim_{n\to\infty} v_n = x - \lambda e_1 \in M$, hence $x = (x - \lambda e_1) + \lambda e_1 \in M + E$. $\qquad \square$

Lemma 7.13.7 *Let M be a closed subspace of a normed space X. Then we have*

$$\text{codim} M = \dim M^0 \text{ and } \text{codim} M^0 = \dim M. \qquad (7.99)$$

Proof.

(i) First we show the first equality in (7.99).

(α) We assume that $\mathrm{codim} M = m < \infty$.

Let $\{\tilde{x}_1, \ldots, \tilde{x}_m\}$ be a basis for the quotient space $\tilde{X} = X/M$ and $x_k \in \tilde{x}_k$ for $k = 1, \ldots, m$. For each $x \in X$, $\tilde{x} = x + M$ is uniquely represented as $\tilde{x} = \lambda_1 \tilde{x}_1 + \ldots + \lambda_m \tilde{x}_m$, hence the representation

$$x = \lambda_1 x_1 + \ldots + \lambda_m x_m + v \text{ for all } v \in M$$

is unique. Let $M_k = M \oplus \mathrm{span}\{x_1, \ldots, x_{k-1}, x_{k+1}, \ldots, x_m\}$ for $k = 1, \ldots, m$. By Lemma 7.13.6, M_k is a closed subspace of X. We remark that $x_k \notin M_k$. It follows that there exists $f_k \in X^*$ such that $f_k \in M_k^0$ and $f_k(x_k) = 1$ by Corollary 4.5.4. It is easy to prove that the functionals f_k for $k = 1, \ldots, m$ are linearly independent vectors. Thus we have proved that $\dim M^0 \geq \mathrm{codim} M$. On the other hand, if $f \in M^0$ and $f(x_k) = \alpha_k$ for $k = 1, \ldots, m$, then $(f - \alpha_1 f_1 - \cdots - \alpha_m f_m)(x) = 0$ for each $x \in X$. Hence we have $f = \alpha_1 f_1 + \cdots + \alpha_m f_m$, and so $\dim M^0 \leq m$. Thus we have shown that $\dim M^0 = \mathrm{codim} M$.

(β) If $\mathrm{codim} M = \infty$, then it is easy to prove that there exists a sequence (M_n) of closed subspaces in X with the property

$$M \subsetneq M_1 \subsetneq M_2 \subsetneq \cdots .$$

Hence it follows by Corollary 4.5.4 that

$$M^0 \supsetneq M_1^0 \supsetneq M_2^0 \supsetneq \cdots , \tag{7.100}$$

and so $\dim M^0 = \infty$.

Thus we have shown the first equality in (7.99).

(ii) Now prove the second equality in (7.99).

(γ) We assume $\dim M = n < \infty$.

If $\{x_1, \ldots, x_n\}$ is a basis of M, then there exist n functionals $f_1, \ldots, f_n \in X^*$ such that $f_k(x_j) = \delta_{kj}$ by Corollary 4.5.12. Let $f \in X^*$ and $f_0 = f - \sum_{k=1}^{n} f(x_k) f_k$. Obviously $f_0(x_k) = 0$ for $k = 1, \ldots, n$ and it follows that $f_0 \in M^0$. Hence $\tilde{f}$ is a linear combination of elements $\tilde{f}_k$, where $\tilde{f}$ and $\tilde{f}_k$ are elements in X^*/M^0. Since the vectors $\tilde{f}_k$ are linearly independent, we have $\mathrm{codim} M^0 = n$.

(δ) If $\dim M = \infty$, then there exists a sequence (E_n) of finite dimensional subspaces in X such that

$$E_1 \subsetneq E_2 \subsetneq \cdots \subsetneq M.$$

Hence it follows from Corollary 4.5.4 that

$$E_1^0 \supsetneq E_2^0 \supsetneq \cdots \supsetneq M^0,$$

and so $\mathrm{codim} M^0 = \infty$.

Thus we have shown the second equality in (7.99). $\square$

Theorem 7.13.8 (Fredholm's alternative) *Let X be a Banach space and $K \in \mathcal{K}(X)$. Then $R(I - K)$ is a closed subspace of X and*

$$\alpha(I - K) = \alpha(I' - K') < \infty.$$

In particular, we either have $R(I - K) = X$ and $N(I - K) = \{0\}$, or we have $R(I - K) \neq X$ and $N(I - K) \neq \{0\}$.

Proof. By Corollary 7.11.8, and Theorems 7.13.2 and 7.13.3 it follows that $R(I - K)$ is a closed subspace of X, and $\alpha(I - K) < \infty$ and $\alpha(I' - K') < \infty$. We assume that

$$\alpha(I - K) = n \text{ and } \alpha(I' - K') = m.$$

We prove $n = m$.

If $n = 0$, it follows from (4.113) in Theorem 4.15.13 that $R(I' - K') = X'$, and then Corollary 7.11.8 and Lemma 7.13.4 imply $m = 0$. If $m = 0$, Lemmas 4.9.8, 4.9.6 and 7.13.4 together imply $n = 0$.

Now we assume that $n > 0$ and $m > 0$. Let $\{x_1, \ldots, x_n\}$ be a basis of the space $N(I - K)$, and $\{f_1, \ldots, f_m\}$ be a basis of the space $N(I' - K')$. By Corollary 4.5.12 there exist n functionals $g_1, \ldots, g_n \in X^*$ such that

$$g_k(x_j) = 0 \text{ for } k \neq j \text{ and } g_k(x_i) = 1. \tag{7.101}$$

And by Lemma 4.5.13, there exist m vectors $z_1, \ldots, z_m \in X$ such that

$$f_k(z_j) = 0 \text{ for } k \neq j \text{ and } f_k(z_i) = 1.$$

We are going to prove that $n < m$ is not possible. We assume that $n < m$ and $F \in \mathcal{B}(X)$ is an operator defined by

$$F(x) = \sum_{k=1}^{n} g_k(x) z_k \text{ for all } x \in X.$$

Clearly $K_1 = K + F \in \mathcal{K}(X)$, and we show that

$$N(I - K_1) = \{0\}. \tag{7.102}$$

From $x_0 \in N(I - K_1)$, it follows for each $k = 1, \ldots, n$ that

$$0 = f_k((I - K_1)x_0) = f_k((I - K)x_0) - f_k(Fx_0)$$

$$= ((I' - K_1')f_k)x_0 - \sum_{i=1}^{n} g_i(x_0) f_k(z_i) = -g_k(x_0),$$

and so

$$g_k(x_0) = 0 \text{ for each } k = 1, \ldots, n. \tag{7.103}$$

Hence $x_0 \in N(I - K)$, and there exist scalars $\alpha_1, \ldots, \alpha_n$ such that

$$x_0 = \sum_{j=1}^{n} \alpha_j x_j. \tag{7.104}$$

It follows from (7.101), (7.103) and (7.104) that

$$0 = g_k(x_0) = \sum_{j=1}^{n} \alpha_j g_k(x_j) = \alpha_k, \tag{7.105}$$

and so $x_0 = 0$. Thus we have proved (7.102). Now it follows from the already proved part of the theorem that $N(I' - K_1') = \{0\}$. We remark that we have for each $x \in X$

$$((I' - K_1')f_{n+1}) x = ((I' - K')f_{n+1}) x - (F'f_{n+1})x = -f_{n+1}(Fx)$$

$$= -f_{n+1}\left(\sum_{i=1}^{n} g_i(x)z_i\right) = -\sum_{i=1}^{n} g_i(x)f_{n+1}(z_i) = 0,$$

hence $f_{n+1} \in N(I' - K_1')$. This is a contradiction and so we must have $n \geq m$. On the other hand if instead of K we consider the operator K', then it follows from the already proved part of this theorem, Lemma 4.9.7 and Theorem 4.15.13 that

$$\alpha(I' - K') \geq \alpha(I'' - K'') = \beta(I' - K') = \alpha(I - K). \qquad \square$$

Theorem 7.13.9 *Let X be a Banach space, $A \in \mathcal{B}(X)$ and $\|A\|_\chi < 1$. Then the equations $x - Ax = 0$ and $f - A'f = 0$ have the same number of maximal linearly independent solutions, that is, $\alpha(I - A) = \alpha(I' - A')$.*

Proof. Theorem 7.13.2 implies $\alpha(I - A) < \infty$. We remark that it follows from $\|A\|_\chi < 1$, (7.70) and Corollary 7.11.7 that there exists a natural number k, such that $\|A'^k\|_\chi < 1$. Furthermore, it follows from Remark 7.13.5 and Theorem 7.13.2 that $n(I' - A') < \infty$. We assume that $\alpha(I - A) > \alpha(I' - A')$. Let $x_1, \ldots, x_n$ for $n = \alpha(I' - A')$ be linearly independent elements in $N(I - A)$, and $\phi_1, \ldots, \phi_n$ be linearly independent functionals in X' such that $\phi_i(x_j) = \delta_{ij}$ (Corollary 4.5.12). Let $\{y_1, \ldots, y_n\}$ be a basis of direct complements of $R(A)$ in X, and $F \in \mathcal{B}(X)$ be a finite dimensional operator defined by $F(x) = \sum_{k=1}^{n} \phi_k y_k$ for $x \in X$. Let us remark that $\|A + F\|_\chi < 1$ and $R(I - (A + F)) = X$. Lemma 7.13.4 implies $n(I - (A + F)) = 0$, and by Banach's theorem of the bounded inverse, Theorem 4.6.3, we have $(I - (A + F))^{-1} \in \mathcal{B}(X)$. The operator $K = (I - (A + F))^{-1}F$ is in $\mathcal{K}(X)$ by Corollary 4.14.9, and obviously

$$\alpha(I - A) = \alpha[(I - (A + F))(I + K)] = \alpha(I + K)$$

and

$$\alpha(I' - A') = \alpha[(I + K)'(I - (A + F))'] = \alpha(I' + K').$$

It follows from Theorem 7.13.8 that $\alpha(I - K) = \alpha(I' - K')$, which is a contradiction to the assumption $\alpha(I - A) > \alpha(I' - A')$. Hence we have proved that $\alpha(I - A) \leq \alpha(I' - A')$. On the other hand, if we consider the operator A' instead of A, it follows from the already proved part of the theorem that $\alpha(I' - A') \leq \alpha(I'' - A'') = \beta(I' - A') = \alpha(I - A)$. $\square$

If X is finite dimensional and $A \in \mathcal{L}(X)$, then $\mathrm{asc}(A) = \mathrm{dsc}(A) < \infty$. If X is infinite dimensional, then we cannot conclude from $A \in \mathcal{L}(X)$ that $\mathrm{asc}(A) = \mathrm{dsc}(A)$.

In the following theorem, we study the ascent and descent of the operator $A - \lambda$ where $A \in \mathcal{K}(X)$ and $\lambda \neq 0$.

Theorem 7.13.10 *Let X be a Banach space, $A \in \mathcal{B}(X)$ be a compact operator and $\lambda \neq 0$. Then we have*

$$\mathrm{asc}(A - \lambda) = \mathrm{dsc}(A - \lambda) < \infty, \qquad (7.106)$$

that is, the operator $A - \lambda$ has a finite ascent and descent.

Proof. We show that $\mathrm{asc}(A - \lambda) < \infty$. We may assume that $\lambda = 1$. We note $(I - A)^n = I - K_n$, where K_n is a compact operator for $n = 1, 2, \ldots$. It follows from Theorem 7.13.8 that

$$\alpha(I - A)^n < \infty \text{ for each } n = 1, 2, \ldots.$$

We assume

$$N(I - A)^n \neq N(I - A)^{n+1} \text{ for each } n = 0, 1, 2, \ldots.$$

The Riesz lemma, Lemma 4.13.2, implies that, for each n, there exists $x_n \in N(I - A)^n$ such that

$$\|x_n\| = 1 \text{ and } \inf\{\|x_n - y\| : y \in N(I - A)^n\} \geq \frac{1}{2}.$$

Hence we have for $n > m$

$$\|Ax_n - Ax_m\| = \|x_n - [(I - A)x_n + x_m - (I - A)x_m]\| \geq \frac{1}{2},$$

and the sequence (Ax_n) has no convergent subsequence. This is a contradiction to the assumption that A is a compact operator. It follows that $\operatorname{asc}(A - I) < \infty$.

We show that $\operatorname{dsc}(A - \lambda) < \infty$. Again it suffices to assume $\lambda = 1$. Since $(I - A)^n = I - K_n$, where K_n is a compact operator for $n = 1, 2, \ldots$, Theorem 7.13.3 implies

$$R(I - A)^n = \overline{R(I - A)^n} \text{ for each } n = 1, 2, \ldots.$$

Hence, we have by Corollary 4.9.9 and Theorem 4.9.3

$$R(I - A)^n = {}^o[N(I' - A')^n] \text{ for each } n = 1, 2, \ldots. \tag{7.107}$$

Since A' is a compact operator by Corollary 7.11.8, by the already proved part of the theorem, there exists a natural number p such that

$$N(I' - A')^p = N(I' - A')^{p+1}. \tag{7.108}$$

It follows from (7.107) and (7.108) that

$$[R(I - A)^p]^o = [R(I - A)^{p+1}]^o,$$

and so we obtain from Lemma 4.9.6

$$R(I - A)^p = {}^o[(R(I - A)^p)^o] = {}^o[(R(I - A)^{p+1})^o] = R(I - A)^{p+1}.$$

Hence, $\operatorname{dsc}(I - A) \leq p$, and (7.106) follows from Theorem 7.12.7. $\qquad\square$

Chapter 8

Fredholm Theory

Chapter 8 is focused on the Fredholm theory and Fredholm operators which are generalizations of operators that are the difference of the identity and a compact linear operator on a Banach space. Fredholm operators play a very important role in the spectral theory of operators. The chapter presents a study of Fredholm and semi–Fredholm operators, the index and Atkinson's theorems, and Yood's results for all and Yood's results for all upper semi–Fredholm operators with nonpositive index, lower semi–Fredholm operators with nonnegative index, and properties of right and left Fredholm operators. It also establishes the openness of the set of proper semi–Fredholm operators in the space of bounded linear operators between Banach spaces, and gives the proofs of the punctured neighborhood theorem and the Kato decomposition theorem, for which West's proof is used. Finally it provides detailed studies of the ascent and descent of operators, and of the generalized kernel and range of bounded linear operators, the properties of Browder and semi–Browder operators, essential spectra and essential type subsets of the spectrum.

8.1 Introduction

Fredholm operators, the basic objects in Fredholm operator theory, are generalizations of the operators of the form $A = I - K$, where $K \in \mathcal{K}(X)$ and X is a Banach space. They are very important for the spectral theory of operators. In this chapter we present some elements from Fredholm theory. We refer the interested readers to the textbooks and monographs [3, 2, 15, 38, 101, 112, 142] for further reading.

We will only present results on bounded linear operators between Banach spaces. The theory, however, has also been generalized to unbounded closed operators on Banach spaces, for instance in [74, 111, 112, 187, 188, 183, 185], and has important applications in the fields of differential and integral equations. Schechter in [183, 185] was the first to simplify many results in Fredholm theory, using only the closed graph theorem, and was able to completely avoid the concept of the opening between two subspaces.

This chapter presents a study of Fredholm and semi–Fredholm operators, the index and Atkinson's theorems, Theorems 8.5.1 and 8.6.1, and Yood's results Theorems 8.4.4 for all upper semi–Fredholm operators with nonpositive index, lower semi–Fredholm operators with nonnegative index, and properties of right and left Fredholm operators. It also establishes the openness of the set of proper semi–Fredholm operators in the space of bounded linear operators between Banach spaces, and gives the proofs of the punctured neighborhood theorem and the Kato decomposition theorem, Theorems 8.10.2 and 8.11.5. Finally, it provides detailed studies of the ascent and descent of operators, the properties of Browder and semi–Browder operators, essential spectra and essential type subsets of the spectrum in Sections 8.12–8.14.

Throughout this section we use the term *complement* to mean *topological complement*. If X is a Banach space, $A \in \mathcal{B}(X)$ and λ is a scalar, then we write $A - \lambda$ instead of $A - \lambda I$, where I denotes the identity map on X.

8.2 Preliminary results

Let $A \in \mathcal{L}(X, Y)$. Then $N(A)$ and $R(A)$, respectively, denote the null–space and the range of A. Let $\alpha(A) = \dim N(A)$ if $N(A)$ is finite dimensional, and let $\alpha(A) = \infty$ if $N(A)$ is infinite dimensional. Similarly, let $\beta(A) = \dim Y/R(A) = \operatorname{codim} R(A)$ if $Y/R(A)$ is finite dimensional, and let $\beta(A) = \infty$ if $Y/R(A)$ is infinite dimensional. The quantity $\alpha(A)$ is the *nullity*, and $\beta(A)$ is the *defect* of $A \in \mathcal{L}(X, Y)$.

If $A \in \mathcal{B}(X, Y)$, then $A' \in \mathcal{B}(Y^*, X^*)$ is its adjoint operator.

Let M and N be closed subspaces of Y. Then $J_M : M \to Y$ is the natural inclusion, and $Q_N : Y \to Y/N$ is the natural epimorphism. Obviously we have $J_M \in \mathcal{B}(M, Y)$ and $Q_N \in \mathcal{B}(Y, Y/N)$.

We recall that if $U \subset X$ and $W \subset X^*$ then the *annihilators of* U and W are the sets

$$U^\circ = \{x^* \in X^* : x^*(u) = 0 \text{ for every } u \in U\}$$

and

$$^\circ W = \{x \in X : w(x) = 0 \text{ for every } w \in W\}.$$

The set U° is a closed subspace of X^*, and $^\circ W$ is a closed subspace of X. If $A \in \mathcal{B}(X, Y)$, then we have by (4.77)

$$R(A)^\circ = N(A') \text{ and } \overline{R(A)} = {}^\circ[R(A)^\circ] = {}^\circ[N(A')]. \tag{8.1}$$

Moreover, $R(A)$ is closed if and only if $R(A')$ is closed, and in this case, it follows from (4.113) and (4.77) that
in this case and

$$R(A') = N(A)^\circ \text{ and } R(A) = {}^\circ N(A'). \tag{8.2}$$

We need the following auxiliary results.

Lemma 8.2.1 *Let U and V be subspaces of X, such that V is closed and $V \subset U$. Then we have*

$$\dim U/V = \dim V^\circ/U^\circ. \tag{8.3}$$

Proof.

(i) First we assume $\dim U/V = k < \infty$.

Let $\{\tilde{x}_1, \ldots, \tilde{x}_k\}$ denote a basis of U/V, where $x_i \in U$, $\tilde{x}_i = x_i + V$ for $i = 1, \ldots, k$, and let $V_i = V \oplus \operatorname{span}(\{x_1, \ldots, x_{i-1}, x_{i+1}, \ldots, x_k\})$ for $i = 2, \ldots, k-1$. Now Lemma 7.13.6 implies that V_i is closed in X, and obviously $x_i \notin V_i$. By the Hahn-Banach theorem, there exists some $f_i \in X^*$ satisfying $f_i \in V_i^\circ$ and $f_i(x_i) = 1$. It is easy to see that

$\tilde{f}_i = f_i + U^\circ$ for $i = 1, \ldots, k$ are linearly independent. Thus we have $\dim V^\circ/U^\circ \geq k$. On the other hand, if $f \in V^\circ$, let $f(x_i) = \alpha_i$ for $i = 1, \ldots, k$. Since $\{\tilde{x}_1, \ldots, \tilde{x}_k\}$ is a basis of U/V, it follows that, for every $x \in U$, the element $\tilde{x} = x + V$ has the unique representation $\tilde{x} = \lambda_1 \tilde{x}_1 + \ldots + \lambda_k \tilde{x}_k$ ($\lambda_i \in \mathbb{C}$), and consequently the representation

$$x = \lambda_1 x_1 + \ldots + \lambda_k x_k + v \text{ for } v \in V, \tag{8.4}$$

is unique in terms of $\lambda_1, \ldots, \lambda_k$. From (8.4) we get $(f - \alpha_1 f_1 - \cdots - \alpha_k f_k)(x) = 0$ for every $x \in U$, that is, $f - (\alpha_1 f_1 + \cdots + \alpha_k f_k) \in U^\circ$. In particular, for $\tilde{f} = f + U^\circ$ we have $\tilde{f} = \alpha_1 \tilde{f}_1 + \cdots + \alpha_k \tilde{f}_k$, and consequently $\dim V^\circ/U^\circ \leq k$.

(ii) Now we assume $\dim U/V = \infty$.
Then there exists a sequence (V_n) of closed subspaces of X satisfying

$$V \subsetneq V_1 \subsetneq V_2 \subsetneq \cdots \subsetneq U.$$

It follows from the Hahn-Banach theorem that

$$V^\circ \supsetneq V_1^\circ \supsetneq V_2^\circ \supsetneq \cdots \supsetneq U^\circ$$

is satisfied, and consequently $\dim(V^\circ/U^\circ) = \infty$. □

As a corollary of Lemma 8.2.1, we obtain the following stronger result than Lemma 7.13.7.

Corollary 8.2.2 *Let M be a subspace of X. Then we have*

$$\operatorname{codim} M^\circ = \dim M. \tag{8.5}$$

If M is closed, then we have

$$\operatorname{codim} M = \dim M^\circ. \tag{8.6}$$

Proof. We take $U = M$ and $V = \{0\}$ in Lemma 8.2.1 to obtain (8.5). Moreover, we take $U = X$ and $V = M$ to obtain (8.6). □

Now we prove some results from linear algebra which are needed later in this chapter.

Lemma 8.2.3 ([105, Lemma 2.2]) *Let S_1 and S_2 be subspaces of a linear space X. Then the quotient spaces $S_1/(S_1 \cap S_2)$ and $(S_1 + S_2)/S_2$ are isomorphic.*

Proof. Let $[x] = s_1 + s_2 + S_2 = s_1 + S_2$ with $s_1 \in S_1$ and $s_2 \in S_2$ be an element in the quotient space $(S_1 + S_2)/S_2$. We define

$$\psi(s_1) = [s_1] \text{ for each } s_1 \in S_1.$$

It follows that $\psi \in \mathcal{L}(S_1, (S_1 + S_2)/S_2)$.
If $[x] \in (S_1 + S_2)/S_2$, then $x = s_1 + s_2$ with $s_1 \in S_2$ and $s_2 \in S_2$, and so $[x] = [s_1]$. This shows that ψ is onto $(S_1 + S_2)/S_2$. Combining this with the fact that $N(\psi) = S_1 \cap S_2$, we obtain the desired result. □

Lemma 8.2.4 ([198, Lemma 2.1]) *Let S_1 and S_2 be subspaces of a linear space X and $S_1 \subset S_2$. Then we have*

$$\dim X/S_1 = \dim X/S_2 + \dim S_2/S_1.$$

(We use the convention that if a dimension is infinite then $\infty + p = p + \infty = \infty$ if $p \in \mathbb{N}_0 \cup \{\infty\}$.)

Proof. Let T_1 and T_2 be the complements of S_1 in S_2, and of S_2 in X, respectively, that is, $S_2 = S_1 \oplus T_1$, and $X = S_2 \oplus T_2$. Then it is easy to see that $T_1 \oplus T_2$ is a complement of S_1 in X. Hence X/S_1 and $T_1 \oplus T_2$ are isomorphic, and S_2/S_1 and T are isomorphic. Clearly

$$\dim(T_1 \oplus T_2) = \dim T_1 + \dim T_2,$$

(with the convention on the infinity of a dimension). □

Lemma 8.2.5 ([198, Lemma 2.2]) *Let S and T be subspaces of a linear space X such that $S \cap T = \{0\}$ and $\dim X/T \le \dim S < \infty$. Then we have $X = S \oplus T$.*

Proof. Let $W = S \oplus T$. Then it follows that $T \subset W \subset X$, hence by Lemma 8.2.4

$$\dim X/T = \dim X/W + \dim W/T. \tag{8.7}$$

Also $\dim S = \dim W/T$ and it follows from (8.7) that $\dim W/T \le \dim X/T$. We also see that

$$\dim X/T = \dim S = \dim W/T,$$

whence $\dim X/W = 0$, and so $X = W$. □

8.3 Generalized kernel and generalized range of operators

In this section, we study two special subspaces of a linear space X which are related to a given operator $A \in \mathcal{L}(X)$. The results are very important in the spectral theory of operators.

Definition 8.3.1 For $A \in \mathcal{L}(X)$ we define the *generalized kernel* and *generalized range* of A by

$$N^\infty(A) = \bigcup_{n=1}^\infty N(A^n) \text{ and } R^\infty(A) = \bigcap_{n=1}^\infty R(A^n), \text{ respectively.}$$

Unfortunately the terminology of Definition 8.3.1 is not unique. Some authors refer to generalized kernels and generalized ranges as *hyper–kernels* and *hyper–ranges*.

Both $R^\infty(A)$ and $N^\infty(A)$ are linear subspaces of X. Moreover, these subspaces are A–invariant.

Lemma 8.3.2 *Let $A \in \mathcal{L}(X)$. If there exists $m \in \mathbb{N}$ such that*

$$N(A) \cap A^m(X) = N(A) \cap A^{m+k}(X) \text{ for all } k \in \mathbb{N}, \tag{8.8}$$

then $A(R^\infty(A)) = R^\infty(A)$.

Proof. It follows from $R^\infty(A) \subset A^n(X)$ that $A(R^\infty(A)) \subset A^{n+1}(X)$ for $n = 1, 2, \ldots$, so $A(R^\infty(A)) \subset \cap_{n=1}^\infty A^{n+1}(X) = R^\infty(A)$.
To prove the converse inclusion, we assume that $y \in R^\infty(A)$. Then, for every $k \in \mathbb{N}$, there exists $x_k \in X$ such that $y = A^{m+k}(x_k)$. Let $z_k = A^m(x_1) - A^{m+k-1}(x_k)$ for $k \in \mathbb{N}$. Since $z_k \in N(A) \cap A^m(X)$, from (8.8), we get $z_k \in A^{m+k-1}(X)$ for every $k \in \mathbb{N}$. Consequently, $A^m(x_1) \in A^{m+k-1}(X)$ for every $k \in \mathbb{N}$, so $A^m(x_1) \in R(A^\infty)$. We obtain $y = A^{m+1}(x_1) \in A(R^\infty(A))$. □

Theorem 8.3.3 *Let $A \in \mathcal{L}(X)$. If $\alpha(A) < \infty$ or $\beta(A) < \infty$, then there exists some $m \in \mathbb{N}$ such that the statement in (8.8) holds. Hence we have*

$$A(R^\infty(A)) = R^\infty(A).$$

Proof.

(i) First we assume $\alpha(A) < \infty$, and prove that there exists some $m \in \mathbb{N}$ such that the statement in (8.8) holds.

We assume that the statement in (8.8) is not satisfied. Then there exists a subsequence $(n_k)_k$ of $\mathbb{N}$ such that

$$N(A) \cap A^{n_1}(X) \neq N(A) \cap A^{n_1+1}(X) = \cdots = N(A) \cap A^{n_2}(X) \neq$$
$$\neq N(A) \cap A^{n_2+1}(X) = \cdots \tag{8.9}$$

Hence, for every $k = 1, 2, \ldots$, there exists

$$x_k \in N(A) \cap (A^{n_k}(X) \backslash A^{n_k+1}(X)).$$

Since $\dim N(A) < \infty$, this cannot hold.

Now applying Lemma 8.3.2 we get $A(R^\infty(A)) = R^\infty(A)$.

(ii) Now we assume that $\beta(A) < \infty$.

Then there exists a subspace V of X such that $\dim V < \infty$ and

$$X = V \oplus A(X). \tag{8.10}$$

Again we prove that (8.8) holds. We assume that this statement is not true. Then there exists a subsequence $(n_k)_k$ of $\mathbb{N}$ such that (8.9) holds. Therefore, for every $k \in \mathbb{N}$, there exists $x_k \in X$ such that $A^{n_k} x_k \in N(A) \cap A^{n_k}(X)$ and $A^{n_k} x_k \notin N(A) \cap A^{n_k+1}(X)$. Because of (8.10) there exist $v_k \in V$ and $w_k \in A(X)$ such that $x_k = v_k + w_k$. Since $\dim V < \infty$, the set $\{v_k : k \in \mathbb{N}\}$ is linearly dependent and so there exist $k_1, \ldots, k_j \in \mathbb{N}$ $(k_1 < \cdots < k_j)$ and scalars $\alpha_1, \ldots, \alpha_{j-1}$ for $j \in \mathbb{N}$ and $j \geq 2$ such that

$$v_{k_j} = \alpha_1 v_{k_1} + \cdots + \alpha_{j-1} v_{k_{j-1}}.$$

Therefore we have

$$x_{k_j} - \alpha_1 x_{k_1} - \cdots - \alpha_{j-1} x_{k_{j-1}} = w_{k_j} - \alpha_1 w_{k_1} - \cdots - \alpha_{j-1} w_{k_{j-1}}.$$

Since $A^{n_{k_j}} x_{k_1} = \cdots = A^{n_{k_j}} x_{k_{j-1}} = 0$, it follows that

$$A^{n_{k_j}} x_{k_j} = A^{n_{k_j}} (w_{k_j} - \alpha_1 w_{k_1} - \cdots - \alpha_{j-1} w_{k_{j-1}})$$
$$\in A^{n_{k_j}} (A(X)) = A^{n_{k_j}+1}(X),$$

which is a contradiction.

So we again apply Lemma 8.3.2 to get the assertion. $\qquad\qquad \square$

Lemma 8.3.4 *Let $A \in \mathcal{L}(X)$. Then we have:*

$$\text{If } \operatorname{asc}(A) < \infty, \text{ then } N^\infty(A) \cap R^\infty(A) = \{0\}. \tag{8.11}$$

$$\text{If } \operatorname{dsc}(A) < \infty, \text{ then } N^\infty(A) + R^\infty(A) = X. \tag{8.12}$$

Proof.

(i) First we show (8.11).

Let $m = \mathrm{asc}(A) < \infty$. Then we have $N^\infty(A) = N(A^m)$. Since $R^\infty(A) \subset R(A^m)$, we conclude that $N^\infty(A) \cap R^\infty(A) \subset N(A^m) \cap R(A^m)$. By Lemma 7.12.3, we obtain $N(A^m) \cap R(A^m) = \{0\}$ and hence $N^\infty(A) \cap R^\infty(A) = \{0\}$.

(ii) Finally, we prove (8.12).

Let $m = \mathrm{dsc}(A) < \infty$. Then we have $R^\infty(A) = R(A^m)$. Since $N(A^m) \subset N^\infty(A)$, we conclude that $N(A^m) + R(A^m) \subset N^\infty(A) + R^\infty(A)$. By Lemma 7.12.4, we obtain $N(A^m) + R(A^m) = X$ and hence $N^\infty(A) + R^\infty(A) = X$. □

Remark 8.3.5 *We remark that the condition* $N^\infty(A) \cap R^\infty(A) = \{0\}$ *is equivalent to the condition* $N(A) \cap R^\infty(A) = \{0\}$, *that is, to the fact that A is one to one on* $R^\infty(A)$. *In fact, if* $N^\infty(A) \cap R^\infty(A) = \{0\}$, *because of* $N(A) \cap R^\infty(A) \subset N^\infty(A) \cap R^\infty(A)$, *it follows that* $N(A) \cap R^\infty(A) = \{0\}$. *On the other hand, if A is one to one on* $R^\infty(A)$, *then A^n is also one to one on* $R^\infty(A)$ *for every* $n \in \mathbb{N}$, *that is,* $N(A^n) \cap R^\infty(A) = \{0\}$ *for every* $n \in \mathbb{N}$ *and hence* $N^\infty(A) \cap R^\infty(A) = \{0\}$.

Corollary 8.3.6 *Let $A \in \mathcal{L}(X)$ and* $\mathrm{asc}(A) < \infty$. *Then we have:*

$$A \text{ is one to one on } R^\infty(A). \tag{8.13}$$

$$\text{For every subspace } M \text{ of } X \text{ such that } A(M) = M, \\ \text{it follows that } A \text{ is one to one on } M. \tag{8.14}$$

Proof.

(i) The statement in (8.13) follows from (8.11).

(ii) Now we show (8.14).

We assume that M is a subspace of X such that $A(M) = M$. Then $A^n(M) = M$ for every $n \in \mathbb{N}$, and hence $M \subset R(A^n)$ for every $n \in \mathbb{N}$. Thus we have $M \subset R^\infty(A)$ and A is one to one on M by (8.13). □

Corollary 8.3.7 *Let $A \in \mathcal{L}(X)$. If* $\mathrm{asc}(A) < \infty$ *or* $\mathrm{dsc}(A) < \infty$, *then we have* $A(R^\infty(A)) = R^\infty(A)$.

Proof. We assume $p = \mathrm{asc}(A) < \infty$. Then it follows by Lemma 7.12.3 that $N(A) \cap A^p(X) = \{0\}$. Since $N(A) \cap A^p(X) \supset N(A) \cap A^{p+k}(X)$ for every $k = 1, 2, \ldots$, we have

$$N(A) \cap A^p(X) = N(A) \cap A^{p+k}(X) = \{0\} \text{ for all } k \in \mathbb{N}.$$

If $q = \mathrm{dsc}(A) < \infty$, then we have

$$N(A) \cap A^q(X) = N(A) \cap A^{q+k}(X) \text{ for all } k \in \mathbb{N}.$$

Now the assertion follows from Lemma 8.3.2. □

Theorem 8.3.8 *Let $A \in \mathcal{L}(X)$. We assume that* $\alpha(A) < \infty$ *or* $\beta(A) < \infty$. *Then*

$$\mathrm{asc}(A) < \infty \text{ if and only if } N^\infty(A) \cap R^\infty(A) = \{0\}. \tag{8.15}$$

Proof.

(i) The sufficiency of $\mathrm{asc}(A)$ in the equivalence in (8.15) is the assertion in (8.11).

(ii) Conversely, we assume that $\alpha(A) < \infty$ or $\beta(A) < \infty$ and that $N^\infty(A) \cap R^\infty(A) = \{0\}$. Then we have $N(A) \cap R^\infty(A) = \{0\}$ and, by the proof of Theorem 8.3.3, there exists $m \in \mathbb{N}$ such that

$$N(A) \cap A^m(X) = N(A) \cap A^{m+k}(X) \text{ for all } k \in \mathbb{N}.$$

Therefore we have $N(A) \cap A^m(X) = N(A) \cap R^\infty(A) = \{0\}$ and it follows from Lemma 7.12.3 that $\mathrm{asc}(A) \leq m < \infty$. □

Theorem 8.3.9 *Let $A \in \mathcal{L}(X)$. We assume $\beta(A) < \infty$. Then*

$$\mathrm{dsc}(A) < \infty \text{ if and only if } N^\infty(A) + R^\infty(A) = X. \tag{8.16}$$

Proof.

(i) The sufficiency of $\mathrm{dsc}(A)$ in the equivalence in (8.16) is the assertion in (8.12).

(ii) Conversely, we assume that $\beta(A) < \infty$ and $N^\infty(A) + R^\infty(A) = X$. Since

$$R(A) \subseteq N(A) + R(A) \subseteq N(A^2) + R(A) \subseteq \cdots \subseteq N(A^n) + R(A) \subseteq \ldots \text{ for } n \in \mathbb{N},$$

and since $R(A)$ has finite codimension in X, we conclude that there exists $m \in \mathbb{N}$ such that
$$N(A^m) + R(A) = N(A^{m+k}) + R(A) \text{ for all k } \in \mathbb{N}.$$

Consequently, we have $N^\infty(A) + R(A) = N(A^m) + R(A)$. Since $X = N^\infty(A) + R^\infty(A) \subset N^\infty(A) + R(A)$, we get $N(A^m) + R(A) = X$. It follows from Lemma 7.12.4 that $\mathrm{dsc}(A) \leq m < \infty$. □

Lemma 8.3.10 *For $A \in \mathcal{L}(X)$ the following conditions are equivalent:*

(1) $N(A) \subset R(A^m)$ for every $m \in \mathbb{N}$.

(2) $N(A^n) \subset R(A^m)$ for every $n \in \mathbb{N}$ and every $m \in \mathbb{N}$.

(3) $N(A^n) \subset R(A)$ for every $n \in \mathbb{N}$.

Proof.

(i) First we show that the statement in (1) implies that in (2).
We use mathematical induction with respect to n.
If $n = 1$ then the statement obviously holds.
We suppose that the statement holds for some $n \geq 1$, and prove it for $n + 1$. Let $x \in N(A^{n+1})$. From $A^n(Ax) = 0$ it follows that $Ax \in N(A^n)$. According to the induction assumption, we have $Ax = A^{m+1}x_1$ for some $x_1 \in X$. Hence, $x - A^m(x_1) \in N(A)$, and from (1) it follows that $x - A^m x_1 \in R(A^m)$, and thus $x \in R(A^m)$.

(ii) It is obvious that the statement in (2) implies that in (1).

(iii) Now we show that the statement in (3) implies that in (2).
We use mathematical induction with respect to m.
If $m = 1$ then the statement holds.
We suppose that the statement holds for some m, and prove it for $m + 1$. Let $x \in N(A^n)$. By the induction assumption, we have $x \in R(A^m)$, and there exists some $x_1 \in X$ such that $x = A^m x_1$. From $0 = A^n x = A^{n+m}x_1$, we get $x_1 \in N(A^{n+m})$ and from (3), we obtain $x_1 \in R(A)$, and consequently $x \in R(A^{m+1})$.

(iv)   Finally, it is obvious that the statement in (2) implies that in (3). $\square$

Definition 8.3.11 For $A \in \mathcal{L}(X)$ we define

$$k(A) = \dim[N(A)/(N(A) \cap R^{\infty}(A))]. \tag{8.17}$$

We note that $k(A) = k$, when A has Kaashoek's property $P(I : k)$ ([105], [104, pp. 452–453]), and $k(A) = 0$ when in Kato's notation $v(A : I) = \infty$ ([111, pp. 289–290]).

The following result obviously holds.

Lemma 8.3.12 *For $A \in \mathcal{L}(X)$ any of the following conditions is equivalent to any condition in Lemma 8.3.10:*

(1) $N(A) \subset R^{\infty}(A)$.

(2) $N^{\infty}(A) \subset R(A)$.

(3) $N^{\infty}(A) \subset R^{\infty}(A)$.

(4) $k(A) = 0$.

8.4   Fredholm and semi-Fredholm operators

Now we study Fredholm and semi-Fredholm operators on Banach spaces.

Definition 8.4.1 Let

$$\Phi_+(X, Y) = \{A \in \mathcal{B}(X, Y) : \alpha(A) < \infty \text{ and } R(A) \text{ is closed}\},$$
$$\Phi_-(X, Y) = \{A \in \mathcal{B}(X, Y) : \beta(A) < \infty\},$$
$$\Phi_\pm(X, Y) = \Phi_+(X, Y) \cup \Phi_-(X, Y),$$
$$\Phi(X, Y) = \Phi_+(X, Y) \cap \Phi_-(X, Y).$$

If $\beta(A) < \infty$ then $R(A)$ is closed in Y by Kato's theorem, Theorem 4.15.5.
The classes $\Phi_+(X, Y)$, $\Phi_-(X, Y)$, $\Phi_\pm(X, Y)$ and $\Phi(X, Y)$, respectively, are called *upper semi–Fredholm, lower semi–Fredholm, semi–Fredholm* and *Fredholm operators*.
We write $\Phi(X) = \Phi(X, X)$ etc., for short.
If $A \in \Phi_\pm(X, Y)$, then the *index of A* is defined as $i(A) = \alpha(A) - \beta(A)$.
Let $\Phi_0(X, Y) = \{A \in \Phi(X, Y) : i(A) = 0\}$,

$$\Phi_+^-(X, Y) = \{A \in \Phi_+(X, Y) : i(A) \leq 0\}$$

and

$$\Phi_-^+(X, Y) = \{A \in \Phi_-(X, Y) : i(A) \geq 0\}.$$

The members of the class $\Phi_0(X, Y)$ are referred to as *Weyl operators*.
If the reference to the spaces X and Y is not necessary, we also write $\Phi_+ = \Phi_+(X, Y)$, and similarly in each of the cases above, for short.

Lemma 8.4.2 *If $A \in \mathcal{B}(X, Y)$ and $R(A)$ is closed, then we have $\alpha(A') = \beta(A)$, $\beta(A') = \alpha(A)$ and $\alpha(A) = \alpha(A'')$.*

Proof. We get from (8.1) and (8.6)

$$\alpha(A') = \dim N(A') = \dim R(A)^\circ = \operatorname{co} \dim R(A) = \beta(A).$$

It follows from (8.2) and (8.5) that

$$\beta(A') = \operatorname{co} \dim R(A') = \operatorname{co} \dim N(A)^\circ = \dim N(A) = \alpha(A). \qquad \square$$

Lemma 8.4.3 *Let $A \in \mathcal{B}(X,Y)$. Then we have $A \in \Phi_+(X,Y)$ if and only if $A' \in \Phi_-(Y^*, X^*)$. Also, $A \in \Phi_-(X,Y)$ if and only if $A' \in \Phi_+(Y^*, X^*)$. We have $i(A) = -i(A')$ in both cases.*

Proof. The result follows from Lemma 8.4.2 and Theorem 4.15.13. $\qquad \square$

Now, we characterize the Φ_+–operators; the result is due to Yood.

Theorem 8.4.4 ([211]) *For $A \in \mathcal{B}(X,Y)$ the following conditions are equivalent:*

(1) $A \in \Phi_+(X,Y)$.

(2) *For every bounded sequence $(x_n)_n$ in X for which $(Ax_n)_n$ has a convergent subsequence it follows that $(x_n)_n$ has a convergent subsequence.*

(3) *For every bounded set $Q \subset X$ for which $A(Q)$ is totally bounded it follows that Q itself is totally bounded.*

Proof.

(i) First we show that the statement in (1) implies that in (2).
Let $A \in \Phi_+(X,Y)$ and $(x_n)_n$ be a bounded sequence in X which has no convergent subsequence. Then there exists a closed subspace X_1 of X such that $X = N(A) \oplus X_1$. Since A is one to one on X_1, and $A(X_1) = R(A)$ is closed, we get that the restriction $A|_{X_1} : X_1 \to R(A)$ of A on X_1 has a bounded inverse. There exist $u_n \in N(A)$ and $v_n \in X_1$ such that $x_n = u_n + v_n$ for $n = 1, 2, \ldots$. We assume that the sequence (Ax_n) has a convergent subsequence, (Ax'_n) say. It follows that (Av'_n) is a convergent sequence and consequently $(A|_{X_1})^{-1}(Av'_n)) = (v'_n)$ is a convergent sequence. Thus, the sequence (v'_n) is bounded. Since the sequence (x'_n) is bounded, it follows that (u'_n) is also bounded. However, (u'_n) is a sequence in the finite dimensional space $N(A)$, so it contains a convergent subsequence (u''_n). From $x''_n = u''_n + v''_n$ it follows that (x''_n) is a convergent sequence, which is not possible.

(ii) It is obvious that the statement in (2) implies that in (3).

(iii) Finally, we show that the statement in (3) implies that in (1).
We assume that, for every bounded set Q, which is not totally bounded, the set $A(Q)$ is not totally bounded. Let $\overline{B}$ be the closed unit ball of $N(A)$, then $A(\overline{B}) = \{0\}$ is totally bounded. By hypothesis, $\overline{B}$ must also be totally bounded, and hence $N(A)$ is finite dimensional. Consequently, $\alpha(A) < \infty$. It follows that there exists a closed subspace X_1 of X such that $X = N(A) \oplus X_1$. Suppose that the restriction $A|_{X_1}$ is not bounded below, that is,

$$\inf_{x \in X_1, \|x\|=1} \|Ax\| = 0.$$

There exists a sequence $(x_n)_n$ in X_1 such that $\|x_n\| = 1$ and $Ax_n \to 0$ as $n \to \infty$. It follows that $\{Ax_n : n = 1, 2, \ldots\}$ is totally bounded, so $\{x_n : n = 1, 2, \ldots\}$ is totally bounded. There exists a subsequence $(x_{n(k)})_k$ of $(x_n)_n$ such that $x_{n(k)} \to x \in X_1$ as $k \to \infty$. From $A(x_{n(k)}) \to A(x) = 0$ as $k \to \infty$, we obtain $x \in X_1 \cap N(A)$. It follows that $x = 0$, which is not possible since $\|x\| = 1$. We have just proved that $A_{|X_1}$ is bounded below, so $R(A) = A(X_1)$ is closed. Thus we have $A \in \Phi_+(X, Y)$. $\qquad\square$

Theorem 8.4.5 ([211]) *Let $A \in \mathcal{B}(X, Y)$ and $B \in \mathcal{B}(Y, Z)$.*

(1) *If BA belongs to the class Φ_+, then A and $BJ_{R(A)}$ belong to the class Φ_+.*

(2) *If BA belongs to the class Φ_-, then B and $Q_{N(B)}A$ also belong to the class Φ_-.*

Proof.

(i)   First we show the implication in (1).
Let Q be a bounded subset of X, such that Q is not totally bounded. It follows from Theorem 8.4.4 that $B(A(Q))$ is not totally bounded. Consequently, $A(Q)$ is not totally bounded. It follows from Theorem 8.4.4 that $A \in \Phi_+(X, Y)$.
Let $(y_n)_n$ be a bounded sequence in $R(A)$, which has no convergent subsequence. It follows from $A \in \Phi_+(X, Y)$ that there exists a closed subspace X_1 of X such that $X = N(A) \oplus X_1$. Then the restriction $A_1 = A|_{X_1} : X_1 \to R(A)$ of A on X_1 has a bounded inverse from $R(A)$ to X_1. There exists some $x_n \in X_1$ such that $Ax_n = y_n$, that is, $x_n = A_1^{-1}y_n$. It follows that $(x_n)_n$ is bounded and has no convergent subsequence. Since BA is a Φ_+-operator, it follows from Part (2) of Theorem 8.4.4 that $(BAx_n)_n = (By_n)_n$ has no convergent subsequence. Now we conclude by Part (1) of Theorem 8.4.4 that $BJ_{R(A)}$ is a Φ_+-operator.

(ii)   Finally, we show the implication in (2).
It follows from $BA \in \Phi_-(X, Z)$ that $A'B' \in \Phi_+(Z^*, X^*)$ (see Lemma 8.4.3). We get from (1) that B' and $A'J_{R(B')} = A'J_{N(B)^\circ}$ are Φ_+-operators. We get from (8.2) that $R(Q'_{N(B)}) = N(Q_{N(B)})^\circ = N(B)^\circ$ and it follows from (8.1) that $N(Q'_{N(B)}) = R(Q_{N(B)})^\circ = \{0\}$. Therefore $Q'_{N(B)}$ is an isomorphism from $(Y/N(B))^*$ onto $N(B)^\circ$, so $Q'_{N(B)}$ is a Φ_+-operator. It follows from (1) that $(Q_{N(B)}A)' = A'J_{N(B)^\circ}Q'_{N(B)}$ is a Φ_+-operator. According to Lemma 8.4.3 we get that B and $Q_{N(B)}A$ are Φ_--operators.
$\square$

Theorem 8.4.6 (1) *If $A \in \Phi_+(X, Y)$ and $K \in \mathcal{K}(X, Y)$, then $A + K \in \Phi_+(X, Y)$.*

(2) *If $A \in \Phi_-(X, Y)$ and $K \in \mathcal{K}(X, Y)$, then $A + K \in \Phi_-(X, Y)$.*

(3) *If $A \in \Phi(X, Y)$ and $K \in \mathcal{K}(X, Y)$, then $A + K \in \Phi(X, Y)$.*

Proof.

(i)   First we show the implication in (1).
Let $A \in \Phi_+(X, Y)$ and $K \in \mathcal{K}(X, Y)$. Let (x_n) be a bounded sequence in X such that $((A + K)x_n)$ has a convergent subsequence $((A + K)x'_n)$. Since (x'_n) is bounded in X and $K \in \mathcal{K}(X, Y)$, there exists a subsequence (x''_n) of (x'_n) for which (Kx''_n) is convergent. Hence (Ax''_n) is also convergent. Since $A \in \Phi_+(X, Y)$, the sequence (x''_n) must have a convergent subsequence by Part (2) of Theorem 8.4.4. Therefore the sequence (x_n) has a convergent subsequence. Applying Part (2) of Theorem 8.4.4 to $A + K$, we conclude that $A + K \in \Phi_+(X, Y)$.

(ii)   Now we show the implication in (2).
Let $A \in \Phi_-(X, Y)$ and $K \in \mathcal{K}(X, Y)$. Then $A' \in \Phi_+(Y^*, X^*)$ and $K' \in \mathcal{K}(Y^*, X^*)$. It follows from (1) that $A' + K' \in \Phi_+(Y^*, X^*)$, and so $A + K \in \Phi_-(X, Y)$.

(iii) Finally, the implication (3) follows from those in (1) and (2). $\qquad\square$

We close this section with an interesting application of the Hausdorff measure of non-compactness in Fredholm theory concerning the characterization of the class $\Phi_+(X,Y)$. The following result, which we are going to state without proof, is due to A. Lebow and M. Schechter in 1971 [130]. For further results we refer the reader to [187, Chapter 14].

Theorem 8.4.7 (Lebow and Schechter) ([130, Theorem 4.10])
An operator $A \in \mathcal{B}(X,Y)$ is in $\Phi_+(X,Y)$ if and only there exists a constant c such that

$$\chi(Q) \leq c \cdot \chi(A(Q)) \text{ for all } Q \in \mathcal{M}_X.$$

8.5   The index theorem

Now we prove the very important *index theorem* for semi-Fredholm operators ([111, 187, 38].

Theorem 8.5.1 (The index theorem) *Let $A \in \mathcal{B}(X,Y)$ and $B \in \mathcal{B}(Y,Z)$.*

(1)   If $A \in \Phi_+(X,Y)$ and $B \in \Phi_+(Y,Z)$, then $BA \in \Phi_+(X,Z)$ and

$$\alpha(BA) \leq \alpha(A) + \alpha(B), \ \beta(BA) \leq \beta(A) + \beta(B) \text{ and } i(BA) = i(A) + i(B).$$

(2)   The statement of (1) is also valid for the class Φ_-.

(3)   The statement of (1) is also valid for the class Φ.

Proof.

(i)   First we show the statement in (1).
Let $A \in \Phi_+(X,Y)$, $B \in \Phi_+(Y,Z)$, and let Q be a bounded but not totally bounded subset of X. By Theorem 8.4.4, $A(Q)$ is bounded, but not totally bounded. Since $A(Q)$ is bounded, $B(A(Q))$ is not totally bounded for the same reason. It follows from Theorem 8.4.4 that $BA \in \Phi_+(X,Z)$.
Let X_1 be a complement of $N(A)$ in X, that is,

$$X = N(A) \oplus X_1. \tag{8.18}$$

Then the restriction $A|_{X_1} : X_1 \to R(A)$ of A on X_1 has the inverse $(A|_{X_1})^{-1} : R(A) \to X_1$. Let $W = (A|_{X_1})^{-1}(N(B) \cap R(A))$. We shall prove

$$N(BA) = N(A) \oplus W. \tag{8.19}$$

We obviously have $W \subset N(BA)$ and $N(A) \subset N(BA)$, so $N(A) + W \subset N(BA)$. Let $x \in N(BA)$. It follows from (8.18) that there exist $z \in N(A)$ and $x_1 \in X_1$ such that $x = z + x_1$. Since $BAx_1 = BAx = 0$, it follows that $A|_{X_1}x_1 = Ax_1 \in R(A) \cap N(B)$, and so $x_1 \in W$. Therefore we have $N(BA) = N(A) + W$. If $w \in N(A) \cap W$, then $w \in X_1 \cap N(A) = \{0\}$. Hence, $N(A) \cap W = \{0\}$. We obtain from (8.19)

$$\dim N(BA) = \dim N(A) + \dim(N(B) \cap R(A)), \tag{8.20}$$

and consequently,
$$\dim N(BA) \le \dim N(A) + \dim N(B),$$

that is, $\alpha(BA) \le \alpha(A) + \alpha(B)$. Thus $\alpha(BA) < \infty$ and $BA \in \Phi_+(X, Z)$. Let $Y_0 = R(A) \cap N(B)$. There exist subspaces Y_1 and Y_2 of Y such that

$$R(A) = Y_0 \oplus Y_1, \tag{8.21}$$

$$N(B) = Y_0 \oplus Y_2. \tag{8.22}$$

Let Y_3 be the complement of $R(A) \oplus Y_2$ in Y, that is,

$$Y = R(A) \oplus Y_2 \oplus Y_3. \tag{8.23}$$

We get

$$R(B) = R(BA) + B(Y_3). \tag{8.24}$$

It follows from (8.21) that $R(BA) = B(Y_1)$. Since

$$(Y_1 \oplus Y_3) \cap N(B) = (Y_1 \oplus Y_3) \cap (Y_0 \oplus Y_2) = \{0\},$$

we get that B is one to one on $Y_1 \oplus Y_3$. Hence, $\dim B(Y_3) = \dim Y_3$, and it follows from $Y_1 \cap Y_3 = \{0\}$ that $B(Y_1) \cap B(Y_3) = \{0\}$, so $R(BA) \cap B(Y_3) = \{0\}$. We obtain from (8.24)

$$R(B) = R(BA) \oplus B(Y_3),$$

and consequently

$$\beta(BA) = \beta(B) + \dim B(Y_3) = \beta(B) + \dim Y_3. \tag{8.25}$$

It follows from (8.23) that $\dim Y_3 \le \beta(A)$, so we get $\beta(BA) \le \beta(B) + \beta(A)$ from (8.25).

We obtain from (8.20)

$$\alpha(BA) = \alpha(A) + \dim Y_0. \tag{8.26}$$

We recall that it follows from (8.22) that the subspaces Y_0 and Y_2 are finite dimensional.

We get from (8.26), (8.25), (8.22) and (8.23) that

$$\begin{aligned}
i(BA) &= \alpha(BA) - \beta(BA) = \alpha(A) + \dim Y_0 - \beta(B) - \dim Y_3 \\
&= \alpha(A) + (\dim Y_0 + \dim Y_2) - \beta(B) - (\dim Y_2 + \dim Y_3) \\
&= \alpha(A) + \alpha(B) - \beta(B) - \beta(A) \\
&= i(A) + i(B).
\end{aligned}$$

(ii) Now we prove the statement in (2).

Let $A \in \Phi_-(X, Y)$ and $B \in \Phi_-(Y, Z)$. Then $A' \in \Phi_+(Y^*, X^*)$ and $B' \in \Phi_+(Z^*, Y^*)$ by Lemma 8.4.3, and we have $A'B' \in \Phi_+(Z^*, X^*)$ according to the previous consideration. It follows that $BA \in \Phi_-(X, Z)$. Using Lemma 8.4.2 and (1), we get

$$\begin{aligned}
\alpha(BA) &= \beta[(BA)'] = \beta(A'B') \\
&\le \beta(A') + \beta(B') = \alpha(A) + \alpha(B).
\end{aligned}$$

Analogously, $\beta(BA) \le \beta(A) + \beta(B)$ and, by Lemma 8.4.3, it follows that

$$i(BA) = -i((BA)') = -i(A'B') = -i(A') - i(B') = i(A) + i(B).$$

(iii)   Finally, the statement in (3) follows from those in (1) and (2). □

Remark 8.5.2 *We remark that a short proof is given for the index of Fredholm operators in [210, Theorem].*

Remark 8.5.3 *Let us point out that from the proof of Theorem 8.5.1, if $A \in \mathcal{L}(X, Y)$ and $B \in \mathcal{L}(Y, Z)$, then*

$$\alpha(BA) \leq \alpha(A) + \alpha(B) \text{ and } \beta(BA) \leq \beta(A) + \beta(B).$$

Moreover, if $\alpha(A) < \infty$ and $\alpha(B) < \infty$ (or, $\beta(A) < \infty$ and $\beta(B) < \infty$), then

$$i(BA) = i(A) + i(B).$$

Therefore, $\alpha(A^n) \leq n\alpha(A)$ and $\beta(A^n) \leq n\beta(A)$ for every $n \in \mathbb{N}$. Moreover, if $\alpha(A) < \infty$ or $\beta(A) < \infty$, then $i(A^n) = ni(A)$ for every $n \in \mathbb{N}$.
If $N(B) \subset R(A)$, we get from (8.20)

$$\dim N(BA) = \dim N(A) + \dim N(B),$$

so $\alpha(BA) = \alpha(A) + \alpha(B)$. We obtain from (8.22), (8.23) and (8.25) $\beta(BA) = \beta(A) + \beta(B)$. If $N(A^n) \subset R(A)$ or $N(A) \subset R(A^n)$ for every $n \in \mathbb{N}$ (these two conditions are equivalent according to Lemma 8.3.10, we prove by mathematical induction with respect to n that

$$\alpha(A^n) = n\alpha(A) \quad \text{and} \quad \beta(A^n) = n\beta(A).$$

Corollary 8.5.4   *(1) Let $A \in \Phi_+(X, Y)$ and $B \in \Phi_+(Y, Z)$. If $A \notin \Phi_-(X, Y)$ or $B \notin \Phi_-(Y, Z)$, then $BA \in \Phi_+(X, Z) \setminus \Phi_-(X, Z)$.*

(2) Let $A \in \Phi_-(X, Y)$ and $B \in \Phi_-(Y, Z)$. If $A \notin \Phi_+(X, Y)$ or $B \notin \Phi_+(Y, Z)$, then $BA \in \Phi_-(X, Z) \setminus \Phi_+(X, Z)$.

   Proof.

(i)   First we show the statement in (1).
We assume that A and B are Φ_+–operators and at least one of them is not Fredholm. It follows from Theorem 8.5.1 that $BA \in \Phi_+(X, Z)$ and $i(BA) = i(B) + i(A) = -\infty$, so $BA \notin \Phi_-(X, Z)$.

(ii)   The proof of the statement in (2) is analogous. □

   We see that $\Phi_+(X)$, $\Phi_-(X)$, $\Phi_+(X) \setminus \Phi_-(X)$, and $\Phi_-(X) \setminus \Phi_+(X)$ are semigroups in $\mathcal{B}(X)$.
   The following two results are obtained from the index theorem, Theorem 8.5.1.

Theorem 8.5.5 *Let $A \in \mathcal{L}(X)$. Then we have:*

$$\text{If asc}(A) < \infty, \text{ then } \alpha(A) \leq \beta(A). \tag{8.27}$$

$$\text{If dsc}(A) < \infty, \text{ then } \beta(A) \leq \alpha(A). \tag{8.28}$$

Proof.

(i) First we show (8.27).

Let $p = \mathrm{asc}(A) < \infty$. If $\beta(A) = \infty$, then it follows that $\alpha(A) \leq \beta(A)$.

We assume $\beta(A) < \infty$. Then we get $N(A) \cap R(A^p) = \{0\}$ from Lemma 7.12.3, so $\alpha(A) \leq \beta(A^p) \leq p\beta(A) < \infty$ by Remark 8.5.3. We also have for every $n \geq p$ by the index theorem, Theorem 8.5.1,

$$ni(A) = i(A^n) = \alpha(A^n) - \beta(A^n)$$
$$= \alpha(A^p) - \beta(A^n). \qquad (8.29)$$

If $\mathrm{dsc}(A) < \infty$, then we have $\mathrm{dsc}(A) = \mathrm{asc}(A) = p$ and, for every $n \geq p$, we get $ni(A) = \alpha(A^p) - \beta(A^p)$. Since $\alpha(A^p) - \beta(A^p)$ is constant, it follows that $i(A) = 0$, that is, $\alpha(A) = \beta(A)$. If $\mathrm{dsc}(A) = \infty$, then we obtain $\lim_{n\to\infty} \beta(A^n) = \infty$, and consequently $\alpha(A^p) - \beta(A^n) < 0$ for a suitable value of n. From (8.29), we obtain $i(A) < 0$ and so $\alpha(A) < \beta(A)$.

(ii) Now we show (8.28).

Let $q = \mathrm{dsc}(A) < \infty$. If $\alpha(A) = \infty$, then we have $\beta(A) \leq \alpha(A)$.

Let $\alpha(A) < \infty$. Then it follows from Lemma 7.12.4 that there exists a subspace M of X such that $X = M \oplus R(A)$ and $M \subset N(A^q)$, so $\beta(A) = \dim M \leq \alpha(A^q) \leq q\alpha(A) < \infty$. Now, as in the proof of (8.27), we obtain $\beta(A) = \alpha(A)$ if $\mathrm{asc}(A) < \infty$ holds, and $\beta(A) < \alpha(A)$ if $\mathrm{asc}(A) = \infty$ holds. $\qquad \square$

Corollary 8.5.6 *Let $A \in \mathcal{L}(X)$. Then we have:*

$$\text{If } \mathrm{asc}(A) < \infty \text{ and } \mathrm{dsc}(A) < \infty, \text{ then } \alpha(A) = \beta(A). \qquad (8.30)$$

$$\text{If } \alpha(A) = \beta(A) < \infty, \text{ then } \mathrm{asc}(A) < \infty \text{ if and only if } \mathrm{dsc}(A) < \infty. \qquad (8.31)$$

Proof.

(i) The implication in (8.30) follows from Theorem 8.5.5.

(ii) Now we show the equivalence in (8.31).

Let $\alpha(A) = \beta(A) < \infty$ and $p = \mathrm{asc}(A) < \infty$. Then we have $\alpha(A^n) - \beta(A^n) = i(A^n) = ni(A) = 0$ for $n = 0, 1, 2, \ldots$ (see Remark 8.5.3). Hence we obtain

$$\beta(A^p) = \alpha(A^p) = \alpha(A^{p+1}) = \beta(A^{p+1}) \leq (p+1)\beta(A) < \infty.$$

It follows that $R(A^p) = R(A^{p+1})$, that is, $\mathrm{dsc}(A) \leq p$. If $\mathrm{dsc}(A) < \infty$, we analogously prove $\mathrm{asc}(A) < \infty$. $\qquad \square$

8.6   Atkinson's theorem

If M and N are closed subspaces of a Banach space X and $X = M \oplus N$, then we use $P_{M,N}$ to denote the projection from X onto M parallel to N.

In 1951, *Atkinson* proved the next very important theorem ([Atk] or [8]).

Theorem 8.6.1 (Atkinson) *Let $A \in \mathcal{B}(X, Y)$ and I denote the identity on X or Y depending on context. Then the following conditions are equivalent:*

(1) $A \in \Phi(X, Y)$.

(2) There exist $A_1, A_2 \in \mathcal{B}(Y, X)$, $F_1 \in \mathcal{F}(X)$ and $F_2 \in \mathcal{F}(Y)$ such that

$$A_1 A = I + F_1 \text{ and } AA_2 = I + F_2.$$

(3) There exist $A_1, A_2 \in \mathcal{B}(Y, X)$, $K_1 \in \mathcal{K}(X)$ and $K_2 \in \mathcal{K}(Y)$ such that

$$A_1 A = I + K_1 \text{ and } AA_2 = I + K_2.$$

Proof.

(i) First we show that the statement in (1) implies that in (2).
Let $A \in \Phi(X, Y)$. Then there exist a closed subspace X_1 in X, and a finite dimensional subspace Y_1 in Y, such that

$$X = N(A) \oplus X_1 \text{ and } Y = R(A) \oplus Y_1.$$

The operator $A|_{X_1} : X_1 \to R(A)$ has a bounded inverse $A|_{X_1}^{-1} \in \mathcal{B}(R(A), X_1)$. We consider the projections $P_{R(A),Y_1} \in \mathcal{B}(Y)$ and $P_{N(A),X_1} \in \mathcal{B}(X)$. Let $A_1 = A_2 = A|_{X_1}^{-1} P_{R(A),Y_1}$. Obviously, $A_1 = A_2 \in \mathcal{B}(Y, X)$ holds, and it is easy to see that $A_1 A = P_{X_1,N(A)} = I - P_{N(A),X_1}$ and $AA_2 = P_{R(A),Y_1} = I - P_{Y_1,R(A)}$. For $F_1 = -P_{N(A),X_1}$ and $F_2 = -P_{Y_1,R(A)}$, we get $A_1 A = I + F_1$, $AA_2 = I + F_2$, and $F_1 \in \mathcal{F}(X)$ and $F_2 \in \mathcal{F}(Y)$.

(ii) It is obvious that the statement in (2) implies that in (3).

(iii) Finally, we show that the statement in (3) implies that in (1).
Let $A_1 A = I + K_1$, $AA_2 = I + K_2$ for $A_1, A_2 \in \mathcal{B}(Y, X)$, $K_1 \in \mathcal{K}(X)$ and $K_2 \in \mathcal{K}(Y)$. It follows from (3) of Theorem 8.4.6 that $I + K_1 \in \Phi(X)$ and $I + K_2 \in \Phi(Y)$. Thus we have $A_1 A \in \Phi(X)$ and $AA_2 \in \Phi(Y)$, and according to Theorem 8.4.5 we obtain $A \in \Phi(X, Y)$. $\square$

Remark 8.6.2 *We note that it follows from the proof of Theorem 8.6.1 that we can choose Fredholm operators A_1 and A_2.*

The following result shows that the set of Fredholm operators is an open set in $\mathcal{B}(X)$.

Theorem 8.6.3 *Let $A \in \Phi(X, Y)$. Then there exists some $\varepsilon > 0$ such that for every $B \in \mathcal{B}(X, Y)$, if $\|B\| < \epsilon$ then $A + B \in \Phi(X, Y)$ and*

(1) $\alpha(A + B) \le \alpha(A)$,

(2) $\beta(A + B) \le \beta(A)$,

(3) $i(A + B) = i(A)$.

Proof.

(i) First we prove the equality in (3).
It follows from Atkinson's theorem, Theorem 8.6.1, that there exist $A_1 \in \mathcal{B}(Y, X)$, $K_1 \in \mathcal{K}(X)$ and $K_2 \in \mathcal{K}(Y)$ such that

$$A_1 A = I + K_1, \quad AA_1 = I + K_2. \tag{8.32}$$

Let $\varepsilon = \|A_1\|^{-1}$, $B \in \mathcal{B}(X, Y)$ and $\|B\| < \varepsilon$. It follows from (8.32) that

$$A_1(A + B) = (I + A_1 B) + K_1, (A + B)A_1 = (I + BA_1) + K_2. \tag{8.33}$$

Since $\|A_1 B\| \leq \|A_1\| \|B\| < 1$, there exists $(I + A_1 B)^{-1} \in \mathcal{B}(X)$. Analogously, there exists $(I + B A_1)^{-1} \in \mathcal{B}(Y)$. We obtain from (8.33)

$$(I + A_1 B)^{-1} A_1 (A + B) = I + (I + A_1 B)^{-1} K_1 \tag{8.34}$$

and

$$(A + B) A_1 (I + B A_1)^{-1} = I + K_2 (I + B A_1)^{-1}. \tag{8.35}$$

Since $(I + A_1 B)^{-1} K_1 \in \mathcal{K}(X)$ and $K_2 (I + B A_1)^{-1} \in \mathcal{K}(Y)$, Theorem 8.6.1 implies $A + B \in \Phi(X, Y)$. It follows from Theorems 8.5.1 and 7.13.8, and (8.32) and (8.34) that

$$i(A_1) + i(A) = 0, \quad i(A_1) + i(A + B) = 0.$$

Consequently we have $i(A + B) = i(A)$, that is, that the equality in (3) holds.

(ii) Now we prove the inequality in (1).
It follows from (8.34) that

$$N(A + B) \subset N((I + A_1 B)^{-1} A_1 (A + B))$$

implies

$$\alpha(A + B) \leq \alpha(I + (I + A_1 B)^{-1} K_1).$$

Also it follows from Theorem 7.13.8 that

$$\alpha(I + (I + A_1 B)^{-1} K_1) = \beta(I + (I + A_1 B)^{-1} K_1).$$

Since

$$X = R(I + (I + A_1 B)^{-1} K_1) + R((I + A_1 B)^{-1} K_1),$$

we get

$$\beta(I + (I + A_1 B)^{-1} K_1) \leq \dim R((I + A_1 B)^{-1} K_1) = \dim R(K_1).$$

It follows from $R(K_1) = N(A)$ (see the proof of Theorem 8.6.1) that $\alpha(A+B) \leq \alpha(A)$, that is, the inequality in (1) holds.

(iii) Finally, we prove the inequality in (2).
The identity in (8.35) implies $R(A + B) \supset R(I + K_2 (I + B A_1)^{-1})$, and consequently

$$\beta(A + B) \leq \beta(I + K_2 (I + B A_1)^{-1}) \leq \dim R(K_2 (I + B A_1)^{-1})$$
$$= \dim R(K_2) = \beta(A).$$

$$\square$$

The following result is a corollary of Atkinson's theorem and the Fredholm alternative (Theorems 8.6.1 and 7.13.8).

Theorem 8.6.4 *If $A \in \Phi(X, Y)$ and $K \in \mathcal{K}(X, Y)$, then $A + K \in \Phi(X, Y)$ and $i(A + K) = i(A)$.*

Proof. According to Theorem 8.6.1 there exist $B \in \Phi(Y, X)$ and $K_1 \in \mathcal{K}(X)$ such that $BA = I + K_1$. For $K \in \mathcal{K}(X, Y)$, according to Theorem 8.4.6 (3), we get $A + K \in \Phi(X, Y)$ and

$$B(A + K) = BA + BK = I + (K_1 + BK) = I + K_2 \text{ and } K_2 \in \mathcal{K}(X).$$

Now, from Theorem 7.13.8 we have

$$i(B(A+K)) = i(I+K_2) = 0 \text{ and } i(BA) = i(I+K_1) = 0,$$

and it follows from Theorem 8.5.1 that

$$i(B) + i(A+K) = 0 \text{ and } i(B) + i(A) = 0.$$

Consequently we obtain $i(A+K) = i(A)$. □

Corollary 8.6.5 *If $A \in \Phi_+(X,Y)$ ($A \in \Phi_-(X,Y)$) and $K \in \mathcal{K}(X,Y)$, then $A + K \in \Phi_+(X,Y)$ ($A + K \in \Phi_-(X,Y)$) and $i(A+K) = i(A)$.*

Proof. If $A \in \Phi(X,Y)$, then the result follows from Theorem 8.6.4. If $A \in \Phi_+(X,Y)$ and $A \notin \Phi(X,Y)$, it follows from Theorem 8.4.6 that $A + K \in \Phi_+(X,Y) \setminus \Phi_-(X,Y)$, so $i(A+K) = i(A) = -\infty$. Analogously, if $A \in \Phi_-(X,Y) \setminus \Phi_+(X,Y)$, then $A + K \in \Phi_-(X,Y) \setminus \Phi_+(X,Y)$ and $i(A+K) = i(A) = \infty$. □

There are some more corollaries of Atkinson's theorem, Theorem 8.6.1.

Theorem 8.6.6 *Let $A \in \mathcal{B}(X,Y)$, and $B \in \mathcal{B}(Y,Z)$. Then we have:*

(1) $BA \in \Phi_+(X,Z)$ and $A \in \Phi(X,Y)$ imply $B \in \Phi_+(Y,Z)$.

(2) $BA \in \Phi_-(X,Z)$ and $B \in \Phi(Y,Z)$ imply $A \in \Phi_-(X,Y)$.

(3) If $BA \in \Phi(X,Z)$, then $A \in \Phi(X,Y)$ is equivalent to $B \in \Phi(Y,Z)$.

Proof.

(i) First we show the implication in (1).
We assume $A \in \Phi(X,Y)$. Then there exist $A_2 \in \Phi(Y,X)$ and $K_2 \in \mathcal{K}(Y)$ such that $AA_2 = I + K_2$ by Theorem 8.6.1. Hence we get $B = BAA_2 - BK_2$. Since $A_2 \in \Phi(Y,X)$ and $BA \in \Phi_+(X,Z)$, it follows that $BAA_2 \in \Phi_+(Y,Z)$ by Theorem 8.5.1. We conclude from $BK_2 \in \mathcal{K}(Y,Z)$ that $B \in \Phi_+(Y,Z)$ by Part (1) of Theorem 8.4.6.

(ii) The implication in (2) is shown analogously.

(iii) Finally, we observe that the statement in (3) follows from the implications in (1) and (2), and Theorem 8.4.5. □

Corollary 8.6.7 *Let $A \in \mathcal{B}(X,Y)$ and $B \in \mathcal{B}(Y,Z)$. Then we have:*

(1) $BA \in \Phi(X,Z)$ and $\alpha(B) < \infty$ imply $A \in \Phi(X,Y)$ and $B \in \Phi(Y,Z)$.

(2) $BA \in \Phi(X,Z)$ and $\beta(A) < \infty$ imply $A \in \Phi(X,Y)$ and $B \in \Phi(Y,Z)$.

Proof.

(i) First we show the implication in (1).
It follows from $R(B) \supset R(BA)$ that $\beta(B) \leq \beta(BA)$, and consequently we have $B \in \Phi(Y,Z)$. From the statement in (3) of Theorem 8.6.6, we get $A \in \Phi(X,Y)$.

(ii) Finally we show the implication in (2).
It follows from $N(A) \subset N(BA)$ that $\alpha(A) \leq \alpha(BA)$, so $A \in \Phi(X,Y)$ and we obtain $B \in \Phi(Y,Z)$ from the statement in (3) of Theorem 8.6.6. □

Corollary 8.6.8 *Let $A \in \mathcal{B}(X,Y)$. Then we have:*

(1)   *If M is a closed subspace of X, $\operatorname{codim} M < \infty$ and $AJ_M \in \Phi_+(M,Y)$, then $A \in \Phi_+(X,Y)$.*

(2)   *If V is a subspace of Y, $\dim V < \infty$ and $Q_V A \in \Phi_-(X,Y/V)$, then $A \in \Phi_-(X,Y)$.*

    Proof.

(i)   First we show the statement in (2).
Since $N(Q_V) = V$ and Q_V is surjective, we get that Q_V is Fredholm. It follows from the statement (2) in Theorem 8.6.6 that $A \in \Phi_-(X,Y)$.

(ii)   The statement in (1) is shown analogously. $\square$

8.7   Yood results for Φ_+^- and Φ_-^+ operators

Every bounded-below operator is also a Φ_+^--operator, and every surjective operator is a Φ_-^+-operator. We start with the following results of Yood [211], which show that every Φ_+^--operator can be obtained by a compact perturbation of a bounded-below operator, and every Φ_-^+-operator can be obtained by a compact perturbation of a surjective operator.

We recall that $j(A)$ is the minimum modulus of the operator $A \in \mathcal{B}(X,Y)$ (Definition 4.15.1).

Theorem 8.7.1 *Let $A \in \mathcal{B}(X,Y)$. Then $A \in \Phi_+^-(X,Y)$ if and only if there exists $F \in \mathcal{F}(X,Y)$ such that $\dim R(F) = \alpha(A)$ and $j(A+F) > 0$.*

    Proof. We assume that $A \in \Phi_+^-(X,Y)$. Let $\{x_1, x_2, \ldots, x_n\}$ be a basis in $N(A)$. There exist $f_1, f_2, \ldots, f_n \in X'$ such that $f_i(x_j) = \delta_{ij}$ for $i,j = 1,2,\ldots,n$. From $i(A) \leq 0$, we obtain $\beta(A) \geq \alpha(A)$, so $\dim N(A') \geq n$. Let $\{g_1, g_2, \ldots, g_n\}$ be a linearly independent subset of $N(A')$. Then there exist $y_1, y_2, \ldots, y_n \in Y$, such that $g_i(y_j) = \delta_{ij}$ for $i,j = 1,2,\ldots,n$. We define the operator $F \in \mathcal{B}(X,Y)$ as follows

$$F(x) = \sum_{k=1}^{n} f_k(x)y_k \text{ for } x \in X.$$

Obviously we have $\dim R(F) = n$ and $F \in \mathcal{F}(X,Y)$. We prove that $j(A+F) > 0$. It follows from Theorem 8.4.6 (1) that $R(A+F)$ is closed. If $x_0 \in X$ and $(A+F)x_0 = 0$, then

$$Ax_0 + \sum_{k=1}^{n} f_k(x_0)y_k = 0 \tag{8.36}$$

and

$$0 = g_j\left(Ax_0 + \sum_{k=1}^{n} f_k(x_0)y_k\right) = g_j(Ax_0) + \sum_{k=1}^{n} f_k(x_0)g_j(y_k)$$
$$= g_j(Ax_0) + f_j(x_0) \text{ for } j = 1,2,\ldots,n.$$

Since $g_j(Ax_0) = (A'g_j)x_0 = 0$, we have $f_j(x_0) = 0$ for $j = 1,2,\ldots,n$. It follows from

(8.36) that $Ax_0 = 0$, that is, $x_0 \in N(A)$. Hence there exist scalars $\lambda_1, \lambda_2, \ldots, \lambda_n$ such that $x_0 = \sum_{k=1}^{n} \lambda_k x_k$. We obtain

$$0 = f_j(x_0) = f_j\left(\sum_{k=1}^{n} \lambda_k x_k\right) = \lambda_j$$

that is, $x_0 = 0$. Consequently we have $j(A + F) > 0$. $\qquad\square$

We recall that $q(A)$ is the modulus of surjectivity of $A \in \mathcal{B}(X, Y)$ (Definition 4.15.7).

Theorem 8.7.2 *Let $A \in \mathcal{B}(X, Y)$. Then $A \in \Phi_-^{+}(X, Y)$ if and only if there exists $F \in \mathcal{F}(X, Y)$ such that $\dim R(F) = \beta(A)$ and $q(A + F) > 0$.*

Proof. We assume that $A \in \Phi_-^{+}(X, Y)$. Let $\beta(A) = n$. Since $i(A) \geq 0$, there exist linearly independent vectors $x_1, x_2, \ldots, x_n$ in $N(A)$. Also, there exist functionals $f_1, f_2 \ldots, f_n \in X^*$ such that $f_i(x_j) = \delta_{ij}$ for all $i, j \in \{1, 2, \ldots, n\}$. Let $\{y_1, y_2, \ldots, y_n\}$ be a basis of an algebraic complement of $R(A)$ in Y, and

$$F(x) = \sum_{k=1}^{n} f_k(x) y_k.$$

Obviously, we have $\dim R(F) = n$ and $F \in \mathcal{F}(X, Y)$. We prove that $A + F$ is surjective, so $q(A + F) > 0$. Let $y \in Y$. Since $Y = R(A) \oplus \mathrm{span}(\{y_1, y_2, \ldots, y_n\})$, there exist $x_0 \in X$ and scalars $\lambda_1, \lambda_2, \ldots, \lambda_n$ such that $y = Ax_0 + \lambda_1 y_1 + \lambda_2 y_2 + \cdots + \lambda_n y_n$. It follows from $X = \cap_{k=1}^{n} N(f_k) \oplus \mathrm{span}(\{x_1, x_2, \ldots, x_n\})$ that there exist $z_1 \in \cap_{k=1}^{n} N(f_k)$ and $z_2 \in \mathrm{span}(\{x_1, x_2, \ldots, x_n\})$ such that $x_0 = z_1 + z_2$. From $x_1, x_2, \ldots, x_n \in N(A)$, we obtain $Az_2 = 0$, so $Ax_0 = Az_1$. Since $z_1 \in \cap_{k=1}^{n} N(f_k)$, it follows that $F(z_1) = \sum_{k=1}^{n} f_k(z_1) y_k = 0$. Consequently we have $(A + F)z_1 = Ax_0$ and

$$(A + F)\left(z_1 + \sum_{k=1}^{n} \lambda_k x_k\right)$$

$$= Ax_0 + A\left(\sum_{k=1}^{n} \lambda_k x_k\right) + F\left(\sum_{k=1}^{n} \lambda_k x_k\right)$$

$$= Ax_0 + \sum_{k=1}^{n} \lambda_k y_k = y. \qquad\square$$

Theorem 8.7.3 *Let $A \in \mathcal{B}(X, Y)$. The following conditions are equivalent:*

$$A \in \Phi_0(X, Y). \tag{8.37}$$

There exist $B \in \mathcal{B}(X, Y))$ and $F \in \mathcal{F}(X, Y)$ such that B is an invertible operator, $\dim R(F) = \alpha(A)(= \beta(A))$ and $A = B + F$, (8.38)

Proof.

(i) First we show that (8.37) implies (8.38).

If $A \in \Phi_0(X, Y)$, then it follows from Theorem 8.7.1 that there exists $F_0 \in \mathcal{F}(X, Y)$ such that $\dim R(F_0) = \alpha(A)$ and $j(A + F_0) > 0$. Thus we have $\alpha(A + F_0) = 0$. From Theorem 8.6.4, we obtain $A + F_0 \in \Phi_0(X, Y)$. Consequently $\beta(A + F_0) = \alpha(A + F_0) = 0$, so $A + F_0$ is invertible. For $B = A + F_0$ and $F = -F_0$ we have $A = B + F$, B is invertible, $F \in \mathcal{F}(X, Y)$ and $\dim R(F) = \dim R(F_0) = \alpha(A)$.

(ii) The converse implication follows from Theorem 8.6.4.   □

Remark 8.7.4 *We think that Theorem 8.7.3 was first proved by Nikolsky [146, Theorem 1].*

Remark 8.7.5 *In [143], Murphy and West proved that an index–zero operator A on a Banach space has a decomposition $A = V + F$ where V is invertible, F is of finite rank and the commutator $[V, F] = VF - FV$ of V and F satisfies $[V, F]^3 = 0$.*

Remark 8.7.6 *The implication (8.37) implies that (8.38) can be proved directly, without using Theorem 8.7.1 ([15, Theorem 0.2.8]):*
It follows from $A \in \Phi_0(X, Y)$ that $\dim N(A) = \operatorname{codim}(R(A)) = n < \infty$ and there exists a closed subspace M of X and a finite dimensional subspace W of Y with $\dim W = n$ such that

$$X = N(A) \oplus M \ and \ Y = W \oplus R(A). \tag{8.39}$$

There exists an isomorphism $C : N(A) \to W$ and a projection $P \in \mathcal{B}(X)$ such that $R(P) = N(A)$ and $N(P) = M$. Let $F_0 = CP$. Obviously we have $R(F_0) = W$, so $F_0 \in \mathcal{F}(X, Y)$ and $\dim R(F_0) = \alpha(A)$. The operator $A + F_0$ is bijective from M onto $R(A)$, and from $N(A)$ onto W. Now it follows from (8.39) that $A + F_0$ is bijective on X, so $A + F_0$ is invertible. For $B = A + F_0$ and $F = -F_0$ we obtain that (8.38) holds.

8.8 Φ, Φ_l and Φ_r operators and Calkin algebras

We define the following subsets of lower and upper semi–Fredholm operators in $\mathcal{B}(X, Y)$ by

$$\Phi_r(X, Y) = \{A \in \Phi_-(X, Y) : \text{there exists a bounded projection from } X \text{ onto } N(A)\},$$

and

$$\Phi_l(X, Y) = \{A \in \Phi_+(X, Y) : \text{there exists a bounded projection from } Y \text{ onto } R(A)\}.$$

Operators in $\Phi_r(X, Y)$ ($\Phi_l(X, Y)$) are called *right (left) Fredholm operators* or *right (left) essentially invertible operators.* Obviously, $\Phi(X, Y) = \Phi_r(X, Y) \cap \Phi_l(X, Y)$.
Similarly, as in Definition 8.4.1, we also write $\Phi_r = \Phi_r(X, Y)$ and $\Phi_l = \Phi_l(X, Y)$, for short, if no reference to the spaces X and Y is needed.
We need the following auxiliary result.

Lemma 8.8.1 *Let $A \in \Phi_+(X, Y)$ and M be a closed subspace of X. Then $A(M)$ is closed in Y.*

Proof. Since $A \in \Phi_+(X, Y)$, it follows that there exists a closed subspace X_1 of X such that $X = N(A) \oplus X_1$. Then the restriction $A|_{X_1} : X_1 \to Y$ is bounded below and $M = (M \cap X_1) \oplus M_1$ for some subspace M_1 of M. Since $M_1 \cap X_1 = (M_1 \cap M) \cap X_1 = M_1 \cap (M \cap X_1) = \{0\}$, it follows that $\dim M_1 \leq \operatorname{co} \dim X_1 < \infty$. Hence $A(M) = A(M \cap X_1) + A(M_1)$, $\dim A(M_1) < \infty$, and $A(M \cap X_1)$ is closed, since $M \cap X_1$ is closed and $A|_{X_1}$ is bounded below. It follows from Lemma 7.13.6 that $A(M)$ is closed.   □

Theorem 8.8.2 *Let $A \in \mathcal{B}(X, Y)$. Then the following conditions are equivalent:*

(1) $A \in \Phi_r(X, Y)$.

(2) *There exist $A_1 \in \mathcal{B}(Y, X)$ and $F \in \mathcal{F}(Y)$ such that $AA_1 = I + F$.*

(3) *There exist $A_1 \in \mathcal{B}(Y, X)$ and $K \in \mathcal{K}(Y)$ such that $AA_1 = I + K$.*

Proof.

(i) First we show that (1) implies (2).
We assume $A \in \Phi_-(X, Y)$, and suppose that there exists a bounded projection from X onto $N(A)$. It follows that $X = N(A) \oplus X_1$, where X_1 is closed. From $A \in \Phi_-(X, Y)$ it follows that $Y = R(A) \oplus Y_1$, where Y_1 is finite dimensional in Y. Now, the restriction $A|_{X_1} : X_1 \to R(A)$ of A on X_1 has a bounded inverse and we conclude that $A_1 = A|_{X_1}^{-1} P_{R(A), Y_1} \in \mathcal{B}(Y, X)$ and $AA_1 = P_{R(A), Y_1} = I - P_{Y_1, R(A)}$, that is, $AA_1 = I + F$, where $F = -P_{Y_1, R(A)} \in \mathcal{F}(Y)$.

(ii) It is obvious that the statement in (2) implies that in (3).

(iii) Finally, we show that the statement in (3) implies that in (1).
Let $AA_1 = I + K$ for some $A_1 \in \mathcal{B}(Y, X)$ and $K \in \mathcal{K}(Y)$. Since $I + K \in \Phi(Y)$, it follows that $A_1 \in \Phi_+(Y, X)$ and $A \in \Phi_-(X, Y)$ by Theorem 8.4.5, and there exists a closed subspace V of Y such that $Y = N(I + K) \oplus V$. It follows from Lemma 8.8.1 that $A_1(V)$ is closed in X.
Let $x \in N(A) \cap A_1(V)$. Then we have $Ax = 0$ and $x = A_1 y$ for $y \in V$. It follows that $AA_1 y = 0$, so $(I + K)y = 0$. We get $y \in N(I + K) \cap V$. It follows that $y = 0$ and $x = 0$. We shall prove that $\dim X/(N(A) \oplus A_1(V)) < \infty$. If $\tilde{x}_1, \ldots, \tilde{x}_n$ are linearly independent in $X/(N(A) \oplus A_1(V))$, $x_i \in X$ and $\tilde{x}_i = x_i + N(A) \oplus A_1(V)$ for $i = 1, \ldots n$, then $A\tilde{x}_i = Ax_i + AA_1(V)$ for $i = 1, \ldots, n$ and $A\tilde{x}_1, \ldots, A\tilde{x}_n$ are linearly independent in $Y/AA_1(V)$.
Since $AA_1(V) = (I + K)V = R(I + K)$, we get

$$\dim X/(N(A) \oplus A_1(V)) \leq \dim Y/R(I + K) = \beta(I + K) < \infty.$$

Consequently, there exists a finite dimensional subspace M of X such that $X = N(A) \oplus (A_1(V) \oplus M)$. Since $A_1(V) \oplus M$ is closed in X by Lemma 7.13.6, there exists a bounded projector from X onto $N(A)$. $\qquad \square$

Theorem 8.8.3 *Let $A \in \mathcal{B}(X, Y)$. Then the following conditions are equivalent:*

(1) $A \in \Phi_l(X, Y)$.

(2) *There exist $A_1 \in \mathcal{B}(Y, X)$ and $F \in \mathcal{F}(X)$ such that $A_1 A = I + F$.*

(3) *There exist $A_1 \in \mathcal{B}(Y, X)$ and $K \in \mathcal{K}(X)$ such that $A_1 A = I + K$.*

Proof.

(i) First we show that (1) implies (2).
We assume $A \in \Phi_+(X, Y)$, and suppose that there exists a bounded projection P from Y onto $R(A)$. Then $\dim N(A) < \infty$ and there exists a closed subspace X_1 of X such that $X = N(A) \oplus X_1$. The operator $A|_{X_1} : X_1 \to R(A)$ has a bounded inverse and, for $A_1 = A|_{X_1}^{-1} P$, it is easy to see that $A_1 A = P_{X_1, N(A)} = I - P_{N(A), X_1}$, that is, there exist $A_1 \in \mathcal{B}(Y, X)$ and $F = -P_{N(A), X_1} \in \mathcal{F}(X)$ such that $A_1 A = I + F$.

(ii) It is obvious that the statement in (2) implies that in (3).

(iii) Finally, we show that the statement in (3) implies that in (1).

Let $A_1 A = I + K$ for some $A_1 \in \mathcal{B}(Y, X)$ and $K \in \mathcal{K}(X)$. Since $I + K \in \Phi(X)$, it follows that $A \in \Phi_+(X, Y)$ and $A_1 \in \Phi_-(Y, X)$ by Theorem 8.4.5. We conclude in the same way as in the proof of Theorem 8.8.2 that there exists a closed subspace V of X such that

$$X = N(I + K) \oplus V. \tag{8.40}$$

Now we have $\dim Y / (N(A_1) \oplus A(V)) < \infty$. It follows from (8.40) that $R(A) = A(N(I + K)) + A(V)$. Hence we obtain

$$R(A) + N(A_1) = A(N(I + K)) + A(V) + N(A_1). \tag{8.41}$$

Since $\mathrm{codim}((N(A_1) \oplus A(V)) < \infty$, we get that $\mathrm{codim}(R(A) + N(A_1)) < \infty$.
From the fact $\dim(N(A_1) \cap A(N(I + K))) < \infty$, and knowing that $N(A_1)$ is closed, we conclude that there exists a closed subspace Z of Y such that

$$N(A_1) = (N(A_1) \cap A(N(I + K))) \oplus Z. \tag{8.42}$$

From (8.42) and (8.41), we get $R(A) + N(A_1) = R(A) + Z$. We are going to prove that the sum $R(A) + Z$ is direct. Let $y \in R(A) \cap Z$ and $y = Ax$ for $x \in X$. Since $Z \subset N(A_1)$, we conclude $0 = A_1 y = A_1 A x = (I + K)x$, that is, $x \in N(I + K)$, and therefore $y \in N(A_1) \cap A(N(I + K))$. From $y \in Z$ and (8.42), we obtain $y = 0$. Consequently, we have $\mathrm{codim}(R(A) \oplus Z) < \infty$, and there exists a finite dimensional subspace W of Y such that $Y = R(A) \oplus Z \oplus W$. Since $Z \oplus W$ is closed, there exists a bounded projection from Y onto $R(A)$. $\qquad \square$

Corollary 8.8.4 *Let $A \in \mathcal{B}(X, Y)$ and $B \in \mathcal{B}(Y, Z)$. Then we have:*

(1) If $A \in \Phi_l(X, Y)$ and $B \in \Phi_l(Y, Z)$, then $BA \in \Phi_l(X, Z)$.

(2) If $A \in \Phi_r(X, Y)$ and $B \in \Phi_r(Y, Z)$, then $BA \in \Phi_r(X, Z)$.

Proof.

(i) First we show the implication in (1).

Let $A \in \Phi_l(X, Y)$ and $B \in \Phi_l(Y, Z)$. By Theorem 8.8.3, there exist $A_1 \in \mathcal{B}(Y, X)$, $K_1 \in \mathcal{K}(X)$, $B_1 \in \mathcal{B}(Z, Y)$ and $K_2 \in \mathcal{K}(Y)$ such that $A_1 A = I + K_1$ and $B_1 B = I + K_2$. Therefore we have $A_1 B_1 B A = A_1 (I + K_2)A = A_1 A + A_1 K_2 A = I + K_1 + A_1 K_2 A$. Since $K_1 + A_1 K_2 A \in \mathcal{K}(X)$, it follows from Theorem 8.8.3 that $BA \in \Phi_l(X, Z)$.

(ii) The implication in (2) is proved analogously. $\qquad \square$

Corollary 8.8.5 *Let $A \in \mathcal{B}(X, Y)$ and $B \in \mathcal{B}(Y, Z)$. Then we have:*

(1) If $BA \in \Phi_l(X, Z)$, then $A \in \Phi_l(X, Y)$.

(2) If $BA \in \Phi_r(X, Z)$, then $B \in \Phi_r(Y, Z)$.

(3) If $BA \in \Phi(X, Z)$, then $A \in \Phi_l(X, Y)$ and $B \in \Phi_r(Y, Z)$.

Proof.

(i) First we show the implication in (1).

Let $BA \in \Phi_l(X, Z)$. By Theorem 8.8.3, there exist $B_1 \in \mathcal{B}(Z, X)$ and $K \in \mathcal{K}(X)$ such that $B_1 B A = I + K$, which implies $A \in \Phi_l(X, Y)$.

(ii) The implication in (2) is shown analogously.

(iii) Finally, the implication in (3) follows from those in (1) and (2), since $\Phi(X, Z) = \Phi_l(X, Z) \cap \Phi_r(X, Z)$. $\square$

Corollary 8.8.6 *(1) If $A \in \Phi_l(X, Y)$ and $K \in \mathcal{K}(X, Y)$, then $A + K \in \Phi_l(X, Y)$.*

(2) If $A \in \Phi_r(X, Y)$ and $K \in \mathcal{K}(X, Y)$, then $A + K \in \Phi_r(X, Y)$.

Proof.

(i) First we show the implication in (1).
Let $A \in \Phi_l(X, Y)$. Then there exist $A_1 \in \mathcal{B}(Y, X)$ and $K_1 \in \mathcal{K}(X)$ such that $A_1 A = I + K_1$. Therefore we have $A_1(A + K) = I + K_1 + A_1 K$, and since $K_1 + A_1 K \in \mathcal{K}(X)$, we conclude that $A + K \in \Phi_l(X, Y)$.

(ii) The implication in (2) is proved analogously. $\square$

Corollary 8.8.7 *Let $A \in \mathcal{B}(X, Y)$ and $B \in \mathcal{B}(Y, Z)$. Then we have:*

(1) $BA \in \Phi_l(X, Z)$ and $A \in \Phi(X, Y)$ imply $B \in \Phi_l(Y, Z)$.

(2) $BA \in \Phi_r(X, Z)$ and $B \in \Phi(Y, Z)$ imply $A \in \Phi_r(X, Y)$.

Proof. The statements follow from Theorem 8.6.1 and Corollaries 8.8.4 and 8.8.6, analogously as in the proof of Theorem 8.6.6. $\square$

The set of all compact operators $\mathcal{K}(X)$ is a closed two–sided ideal of the Banach algebra $\mathcal{B}(X)$. Thus we have:

Definition 8.8.8 The *Calkin algebra* $\mathcal{C}(X) = \mathcal{B}(X)/\mathcal{K}(X)$ is a Banach algebra with the norm

$$\|A + \mathcal{K}(X)\| = \inf_{K \in \mathcal{K}(X)} \|A - K\| = \|A\|_K, \text{ for all } A \in \mathcal{B}(X),$$

where $\pi : \mathcal{B}(X) \to \mathcal{C}(X)$ denotes the natural homomorphism. We write $r_e(A)$ for the *spectral radius of $\pi(A)$* in $\mathcal{C}(X)$, and call $r_e(A)$ the *essential spectral radius of A*.

The Calkin algebra plays an important role in the theory of Fredholm operators. As can be seen from Atkinson's theorem, the set of all Fredholm operators in $\mathcal{B}(X)$ is the inverse image of π of the set of all invertible elements in the Calkin algebra.

Corollary 8.8.9 *The operator $A \in \mathcal{B}(X)$ is Fredholm if and only if $\pi(A)$ is invertible in the Calkin algebra $\mathcal{C}(X)$.*

Proof. The corollary follows from Theorem 8.6.1. $\square$

Corollary 8.8.10 *The operator $A \in \mathcal{B}(X)$ is in $\Phi_l(X)$ if and only if $\pi(A)$ is left invertible in the Calkin algebra $\mathcal{C}(X)$.*

Furthermore, it follows from Theorem 8.8.3, and Theorem 8.8.2, respectively that:

Corollary 8.8.11 *The operator $A \in \mathcal{B}(X)$ is in $\Phi_r(X)$ if and only if $\pi(A)$ is right invertible in the Calkin algebra $\mathcal{C}(X)$.*

Corollary 8.8.12 *Let $A \in \mathcal{B}(X)$. Then X is infinite dimensional if and only if $\sigma(\pi(A)) \neq \emptyset$.*

Proof. We assume $\dim X = \infty$. Then $\mathcal{K}(X)$ is a proper subspace of $\mathcal{B}(X)$ and $\mathcal{C}(X)$ is non–trivial. Hence, $\sigma(\pi(A)) \neq \emptyset$.

To prove the converse implication, we assume $\dim X < \infty$. Then, for every $\lambda \in \mathbb{C}$, $\alpha(A-\lambda) < \infty$ and $\beta(A - \lambda) < \infty$, and hence, $A - \lambda$ is Fredholm. So, $\sigma(\pi(A)) = \emptyset$. $\qquad\square$

We close this section with the following results due to Pfaffenberger [153]. They are related to the study of the relationship between the semi–Fredholm operators on a Banach space and the right and left null divisors in the quotient algebra of all the bounded operators modulo the ideal of compact operators.

Theorem 8.8.13 ([153, Lemma 3.1]) *Let X be a Banach space, and N^r denote the right null divisor in in the Calkin algebra $\mathcal{C}(X) = \mathcal{B}(X)/\mathcal{K}(X)$ (Definition 8.8.8). Then we have*

$$\pi^{-1}(N^r) \subset [\Phi_-(X)]^c = \mathcal{B}(X) \setminus \Phi_-(X).$$

Proof. We assume $A \in \pi^{-1}(N^r) \cap \Phi_-(X)$. Then $A \in \pi^{-1}(N^r)$ if and only if there exists a $B \notin \mathcal{K}(X)$ with $BA \in \mathcal{K}(X)$, and $A \in \Phi_-(X)$ implies $\text{codim}_X(R(A)) < \infty$, so $X = R(A) \oplus M$, where $\dim M < \infty$.

Clearly B is not compact on $R(A)$, since if B were compact on $R(A)$ this would imply that B is compact on $X = R(A) \oplus M$, contrary to the choice of B. But if the restriction $B|_{R(A)}$ of B on $R(A)$ is not compact, then $BA \notin \mathcal{K}(X)$, which is a contradiction. So we must have $\pi^{-1}(N^r) \cap \Phi_-(X) = \emptyset$ which implies $\pi^{-1}(N^r) \subset [\Phi_-(X)]^c$. $\qquad\square$

Theorem 8.8.14 ([153, Lemma 3.2]) *Let X be a Banach space, and N^l denote the left null divisor in the Calkin algebra $\mathcal{C}(X) = \mathcal{B}(X)/\mathcal{K}(X)$. Then we have*

$$\pi^{-1}(N^l) \subset [\Phi_+(X)]^c = \mathcal{B}(X) \setminus \Phi_+(X).$$

Proof. We assume $A \in \pi^{-1}(N^l) \cap \Phi_+(X)$. Then $A \in \pi^{-1}(N^l)$ if and only if there exists a $B \notin \mathcal{K}(X)$ with $AB \in \mathcal{K}(X)$, and $A \in \Phi_+(X)$. Also $A \in \Phi_+(X)$. The set $B(S_X)$ is bounded, but not totally bounded, so $AB(S_X)$ is not totally bounded by Part (3) of Theorem 8.4.4. But this implies $AB \notin \mathcal{K}(X)$, a contradiction. So $\pi^{-1}(N^l) \cap \Phi_+(X) = \emptyset$ which implies $\pi^{-1}(N^l) \subset [\Phi_+(X)]^c$. $\qquad\square$

Remark 8.8.15 *It was shown in [153, Corollary 3.7] that if $X = \ell_p$ for $1 < p < \infty$ or any Hilbert space, then*
(a) $[\Phi_+(X)]^c = \pi^{-1}(N^l)$,
(b) $[\Phi_-(X)]^c = \pi^{-1}(N^r)$, and
(c) $[\Phi_+ \cup \Phi_-(X)]^c = \pi^{-1}(N)$, where $N = N^l \cap N^r$ is the set of two–sided null divisors of $\mathcal{B}(X)/\mathcal{K}(X)$.

Remark 8.8.16 *In [162], the author studied the set of regular elements in the Calkin algbera $\mathcal{C}(X)$. Among other results it was proved that*

$$B(X)^\sqcap + K(X) = \pi^{-1}(C(X)^\sqcap),$$

$$B(X)^\bullet\Phi(X) + K(X) = \pi^{-1}(C(X)^\bullet C(X)^{-1})$$

and

$$B(X)^\sqcap \cap \overline{\Phi(X)} + K(X) = \pi^{-1}(C(X)^\sqcap \cap \overline{C(X)^{-1}}).$$

Remark 8.8.17 *In Section 6.14, we studied the invertibility of the differences of projections. We remark that, in [120], the authors studied the Fredholm properties of the difference of orthogonal projections in a Hilbert space. Among other things, the following results were obtained.*

Theorem 8.8.18 ([118, Theorem 3.1]) *Let R and K be closed subspaces of a Hilbert space X and let P and Q be the orthogonal projections with the ranges R and K, respectively. The following statements are equivalent*

1. $P - Q \in \Phi(X)$;

2. $I - PQ \in \Phi(X)$ and $I - (I - P)(I - Q) = P + Q - PQ \in \Phi(X)$;

3. $R + K$ is closed in X and $\dim[(R \cap K) \oplus (R^{\perp} \cap K^{\perp})] < \infty$;

4. $\|P + Q - I\|_K < 1$;

5. $P + Q \in \Phi(X)$ and $I - PQ \in \Phi(X)$.

8.9 Φ_+ and Φ_- are open sets

The early proofs that the set of semi–Fredholm operators is open used the Borsuk antipodal theorem [28]. Searcóid [190], Labuschagne and Swart [129], Kroh and Volkman [120] gave a proof of the openness of the set of proper semi-Fredholm operators in $\mathcal{B}(X,Y)$ which does not depend on the Borsuk antipodal theorem, unlike the proof in [38] or [75].

We need the following auxiliary results [154].

Lemma 8.9.1 *Let M be a closed subspace of X such that $\operatorname{codim} M = m$ for some $m \in \mathbb{N}$. Then, for every $\varepsilon > 0$, there exists a projection $P \in F(X)$ such that $N(P) = M$ and $\|P\| \leq (1 + \varepsilon)m$.*

Proof. Let $Q_M : X \to X/M$ be the natural quotient mapping. Since $\dim X/M = m$, from Corollary 4.5.12, there exist $\tilde{x}_1, \tilde{x}_2 \ldots, \tilde{x}_m \in X/M$ and $\tilde{f}_1, \tilde{f}_2 \ldots, \tilde{f}_m \in (X/M)^*$ such that $\|\tilde{x}_i\| = 1$, $\|\tilde{f}_i\| = 1$ and $\tilde{f}_i(\tilde{x}_j) = \delta_{ij}$ for all $i, j = 1, 2, \ldots, m$. Moreover, there exist elements $x_i \in X$ such that $Q_M x_i = \tilde{x}_i$ and $\|x_i\| < 1 + \epsilon$. Let $f_i = Q_M^* \tilde{f}_i$ for $i = 1, 2, \ldots, m$. Then $P(x) = \sum_{k=1}^m f_k(x)x_k$ is the requested projection. $\square$

We recall that the sets $\mathcal{J}(X,Y)$ and $\mathcal{Q}(X,Y)$ of all bounded-below and surjective operators in $\mathcal{B}(X,Y)$ are open subsets of $\mathcal{B}(X,Y)$ by Corollary 4.15.12.

Lemma 8.9.2 *Let $A \in \mathcal{J}(X,Y)$. If there exists a sequence of Fredholm operators $(A_n)_n$ of the same index k such that $A_n \to A$ $(n \to \infty)$, then $i(A) = k$.*

Proof. By Corollary 4.15.12, there exists $\varepsilon > 0$ with $\varepsilon < j(A)$ such that it follows from $U \in \mathcal{B}(X,Y)$ and $\|A - U\| < \varepsilon$ that U is bounded below. We can assume that $\|A - A_n\| < \varepsilon$ for every n. Then, for every $n \in \mathbb{N}$, we have $\alpha(A_n) = 0$, so $\beta(A_n) = -i(A_n) = -k$ and $k < 0$. By Lemma 8.9.1 with εm and m replaced by ε and $-k$, respectively, there exist projections $Q_n \in \mathcal{F}(Y)$ such that $N(Q_n) = R(A_n)$ and $\|Q_n\| \leq -k + \varepsilon$ for every $n \in \mathbb{N}$. Let the operator $\tilde{A}_n : X \to R(A_n)$ be defined as the reduction of A_n. Then $\tilde{A}_n$ is bijective and hence, $\tilde{A}_n$ has a bounded inverse defined on $R(A_n)$. Let $B_n = \tilde{A}_n^{-1}(I - Q_n)$. Obviously, $B_n \in B(Y,X)$ and

$$B_n A_n = I \quad \text{and} \quad A_n B_n = I - Q_n. \tag{8.43}$$

From (8.43), using Theorem 8.6.1, we conclude that $B_n \in \Phi(Y,X)$. Using Theorem 8.5.1 we obtain $i(B_n) = -i(A_n) = -k$. We assume $y \in Y$ and $B_n y \neq 0$ for $n \in \mathbb{N}$. Then we have

$$j(A) \leq \frac{\|AB_n y\|}{\|B_n y\|} = \frac{\|((A - A_n)B_n + (I - Q_n))y\|}{\|B_n y\|}$$

$$\leq \frac{(\|A - A_n\|\|B_n\| + \|I - Q_n\|)\|y\|}{\|B_n y\|}$$

$$\leq \frac{(\varepsilon\|B_n\| + 1 - k + \varepsilon)\|y\|}{\|B_n y\|}.$$

Consequently, we have

$$\|B_n\| \leq (j(A))^{-1}(\varepsilon\|B_n\| + 1 - k + \varepsilon),$$

so

$$\|B_n\| \leq \frac{(1 - k + \varepsilon)}{j(A) - \varepsilon}.$$

We obtain that the sequence $(B_n)_n$ is uniformly bounded. Because of this there exists $n_0 \in \mathbb{N}$ such that $\|B_n(A - A_n)\| < 1$ for all $n \geq n_0$. Knowing that $B_n A = I + B_n(A - A_n)$, we conclude that $B_n A$ is invertible in $\mathcal{B}(X)$. Consequently, for $B = (B_n A)^{-1}B_n$, we have $B \in \Phi(Y)$ and $BA = I$. According to Theorem 8.5.1, we conclude

$$i(B) = i((B_n A)^{-1}) + i(B_n) = i(B_n) = -k,$$

and from $BA = I$ we get $i(A) = -i(B) = k$. ☐

Lemma 8.9.3 *The sets*

$$\mathcal{J}(X, Y) \setminus \Phi_-(X, Y) = \{A \in \mathcal{B}(X, Y) : A \text{ is bounded below and } i(A) = -\infty\}$$

and

$$\mathcal{Q}(X, Y) \setminus \Phi_+(X, Y) = \{A \in \mathcal{B}(X, Y) : T \text{ is surjective and } i(A) = \infty\}$$

are open in $\mathcal{B}(X, Y)$.

Proof. Let $A \in \mathcal{B}(X, Y)$ be bounded below and $i(A) = -\infty$. Then there exists some $\varepsilon > 0$ such that if $B \in \mathcal{B}(X, Y)$ and $\|B - A\| < \varepsilon$, then it follows that B is bounded below. We prove $i(B) = -\infty$.

We assume that the opposite holds, that is, there exists a sequence $(A_n)_n$ in $\mathcal{B}(X, Y)$ such that $A_n \to A$ as $n \to \infty$, $\|A - A_n\| < \varepsilon$, and $i(A_n) > -\infty$ for every $n \in \mathbb{N}$. It follows that $A_n \in \Phi(X, Y)$ for every $n \in \mathbb{N}$.

We prove that all A_n have the same index.

For every $m, n \in \mathbb{N}$, let

$$b = \sup\{a \in [0, 1] : i(aA_n + (1 - a)A_m) = i(A_m)\}$$

and $C = bA_n + (1 - b)A_m$. Then C is bounded below (because $\|A - C\| < \varepsilon$) and C is the limit of Fredholm operators each having the same index $i(A_m)$. It follows from Lemma 8.9.2 that $i(C) = i(A_m)$, so $b \in \{a \in [0, 1] : i(aA_n + (1 - a)A_m) = i(A_m)\}$. It follows from Theorem 8.6.3 that $b = 1$, that is, $i(A_n) = i(A_m)$. Consequently, for every $n \in \mathbb{N}$, we have $i(A_n) = k$, where $k \in \mathbb{Z}$ and $k < 0$. Now, we apply Lemma 8.9.2 to the sequence $(A_n)_n$ and A, to obtain $i(A) = k \neq -\infty$, which is not possible.

If $A \in \mathcal{B}(X, Y)$ is surjective and $i(A) = +\infty$, then $A' \in \mathcal{B}(Y^*, X^*)$ is bounded below (because $j(A') = q(A) > 0$), and it follows from Lemma 8.4.3 that $i(T') = -i(T) = -\infty$. The openness of the set $\mathcal{Q}(X, Y) \setminus \Phi_+(X, Y)$ follows from the openness of the set $\mathcal{J}(Y^*, X^*) \setminus \Phi_-(Y^*, X^*)$. ☐

The following theorem shows that the set of proper semi–Fredholm operators, that is, the set of semi–Fredholm operators which are not Fredholm, is open.

Theorem 8.9.4 *Let $A \in \mathcal{B}(X,Y)$.*

(1) If $A \in \Phi_+(X,Y)$ and $\beta(A) = \infty$, then there exists some $\varepsilon > 0$ such that, for $B \in \mathcal{B}(X,Y)$ and $\|B\| < \varepsilon$, it follows that $A + B \in \Phi_+(X,Y)$, $\alpha(A + B) \leq \alpha(A)$ and $\beta(A + B) = \beta(A) = \infty$.

(2) If $A \in \Phi_-(X,Y)$ and $\alpha(A) = \infty$, then there exists some $\varepsilon > 0$ such that, for $B \in \mathcal{B}(X,Y)$ and $\|B\| < \epsilon$, it follows that $A + B \in \Phi_-(X,Y)$, $\beta(A + B) \leq \beta(A)$ and $\alpha(A + B) = \alpha(A) = \infty$.

Proof.

(i) First we prove the statement in (1).

Let $A \in \Phi_+(X,Y)$ and $\beta(A) = \infty$. Then $i(A) = -\infty$. Also there exists a closed subspace M of X such that $X = N(A) \oplus M$. It follows that the restriction $A|_M : M \to Y$ is bounded below and $i(A|_M) = -\infty$. We get from Lemma 8.9.3 that there exists $\varepsilon > 0$ such that, for every $B \in \mathcal{B}(X,Y)$, if $\|B\| < \varepsilon$, then $(A+B)|_M \in B(M,Y)$ is bounded below and $i((A + B)|_M) = -\infty$. By Corollary 8.6.8, we have $A + B \in \Phi_+(X,Y)$. Since $N(A + B) \cap M = \{0\}$, it follows that $\alpha(A + B) \leq \text{codim } M = \alpha(A)$. Now, $N(A)$ is finite dimensional, $R(A + B) = R((A + B)|_M) + (A + B)(N(A))$ and codim $R((A + B)|_M) = \infty$, so $\beta(A + B) = \infty$.

(ii) Finally the statement in (2) follows dually from that in (1). $\qquad\square$

Theorem 8.9.5 *If $A \in \Phi_+(X,Y)$ (respectively $\Phi_-(X,Y), \Phi(X,Y)$), then there exists some $\varepsilon > 0$ such that if $B \in \mathcal{B}(X,Y)$ and $\|B\| < \varepsilon$, then it follows that $A + B \in \Phi_+(X,Y)$ (respectively $\Phi_-(X,Y), \Phi(X,Y)$) and*

$$\alpha(A + B) \leq \alpha(A), \tag{1}$$
$$\beta(A + B) \leq \beta(A), \tag{2}$$
$$i(A + B) = i(A). \tag{3}$$

Proof. The theorem follows from Theorems 8.6.3 and 8.9.4. $\qquad\square$

Corollary 8.9.6 *The sets*

$$\Phi_+(X,Y), \ \Phi_-(X,Y), \ \Phi_+(X,Y) \setminus \Phi_-(X,Y), \ \Phi_-(X,Y) \setminus \Phi_+(X,Y),$$

$$\Phi(X,Y), \ \Phi_0(X,Y), \ \Phi_+^-(X,Y) \text{ and } \Phi_-^+(X,Y)$$

are open in $\mathcal{B}(X,Y)$.
The sets $\Phi_+(X), \ \Phi_-(X), \ \Phi_+(X) \setminus \Phi_-(X), \ \Phi_-(X) \setminus \Phi_+(X), \ \Phi(X), \ \Phi_0(X), \ \Phi_+^-(X)$ and $\Phi_-^+(X)$ are open semigroups in $B(X)$.

Proof. The corollary follows from Theorems 8.6.3, 8.9.4, and 8.5.1, and Corollary 8.5.4. $\square$

Corollary 8.9.7 *If $A \in \mathcal{B}(X,Y)$ belongs to the boundary of the set $\Phi(X,Y)$*

(respectively $\Phi_+(X,Y), \ \Phi_-(X,Y), \ \Phi_+(X,Y) \setminus \Phi_-(X,Y), \ \Phi_-(X,Y) \setminus \Phi_+(X,Y),$
$\Phi_+^-(X,Y), \ \Phi_-^+(X,Y), \ \Phi_0(X,Y)),$

then $A \notin \Phi_+(X,Y) \bigcup \Phi_-(X,Y)$.

Proof. Let $A \in \partial\Phi(X,Y)$. Since $\Phi(X,Y)$ is open, it follows that $A \notin \Phi(X,Y)$. If $A \in \Phi_\pm(X,Y)$, then $A \in \Phi_\pm(X,Y) \setminus \Phi(X,Y)$. As the set $\Phi_\pm(X,Y) \setminus \Phi(X,Y) = (\Phi_+(X,Y) \setminus \Phi_-(X,Y)) \cup (\Phi_-(X,Y) \setminus \Phi_+(X,Y))$ is open by Corollary 8.9.6, there exists a neighborhood $B_\varepsilon(A)$ of A such that $B_\varepsilon(A) \subset \Phi_\pm(X,Y) \setminus \Phi(X,Y)$. Hence we have $B_\varepsilon(A) \cap \Phi(X,Y) = \emptyset$, which is impossible because of $A \in \partial\Phi(X,Y)$. $\qquad\square$

8.10   The punctured neighborhood theorem

Let $A \in \mathcal{L}(X)$. We recall the notions of $N^\infty(A) = \bigcup_{n=1}^\infty N(A^n)$, $R^\infty(A) = \bigcap_{n=1}^\infty R(A^n)$ in Definition 8.3.1 and

$$k(A) = \dim \left(N(A)/[N(A) \cap R^\infty(A)]\right)$$

in Definition 8.3.11. We define for each $n \in \mathbb{N}$

$$a_n(A) = \dim \left(N(A)/[N(A) \cap R(A^n)]\right)$$

and

$$b_n(A) = \dim \left([R(A) + N(A^n)]/R(A)\right).$$

We note that, by Theorem 8.3.3, if $\alpha(A) < \infty$ or $\beta(A) < \infty$ then $k(A) = a_n(A)$ for some $n \in \mathbb{N}$.

We need the following result.

Lemma 8.10.1 *Let* $A \in \mathcal{B}(X)$ *be a semi–Fredholm operator. Then we have* $a_n(A) = b_n(A) = a_n(A')$ *for each* $n \in \mathbb{N}$.

Proof. We use mathematical induction with respect to n. We have for $n = 1$ by Lemma 8.2.3

$$N(A)/N(A) \cap R(A) \text{ is isomorphic to } (R(A) + N(A))/R(A),$$

and so $a_1(A) = b_1(A)$.
Now we assume $a_k(A) = b_k(A)$, and prove $a_{k+1}(A) = b_{k+1}(A)$. We have by Lemma 8.2.4

$$a_{k+1} = a_k + \dim[(N(A) \cap R(A^n))/(N(A) \cap R(A^{n+1}))].$$

Since $(N(A) \cap R(A^n))/(N(A) \cap R(A^{n+1}))$ and $(R(A) + N(A^{n+1}))/(R(A) + N(A^n))$ are isomorphic for each $n \in \mathbb{N}$ by Lemma 8.2.3 (see also [77]), we obtain that $a_k(A) = b_k(A)$ implies $a_{k+1}(A) = b_{k+1}(A)$.
Using Lemma 8.2.1 we obtain

$$\begin{aligned}
a_n(A') &= N(A')/(N(A') \cap R(A'^n)) = R(A)^\circ/(R(A)^\circ \cap N(A^n)^\circ) \\
&= R(A)^\circ/[R(A) \cup N(A^n)]^\circ = R(A)^\circ/[R(A) + N(A^n)]^\circ \\
&= (R(A) + N(A^n))/R(A) = b_n(A) \qquad \square
\end{aligned}$$

Now we are going to prove the *punctured neighborhood theorem.*

Theorem 8.10.2 (The punctured neighborhood theorem) ([111])
Let $A \in \mathcal{B}(X)$. *If* A *is a semi–Fredholm operator, then there exists* $\varepsilon > 0$ *such that, for every* λ *with* $0 < |\lambda| < \varepsilon$, *it follows that* $A - \lambda$ *is semi–Fredholm*

(1) $\alpha(A - \lambda) = \alpha(A) - k(A)$,

(2) $\beta(A - \lambda) = \beta(A) - k(A)$,

(3) $i(A - \lambda) = i(A)$.

Proof.

(i) We assume $A \in \Phi_+(X)$. It follows from the definition of $k(A)$ that $k(A) \leq \alpha(A)$. Also Theorem 8.9.5 implies that there exists some $\varepsilon_1 > 0$ such that if $|\lambda| < \varepsilon_1$ then $A - \lambda \in \Phi_+(X)$, $\alpha(A - \lambda) \leq \alpha(A)$, $\beta(A - \lambda) \leq \beta(A)$ and $i(A - \lambda) = i(A)$. Let $x \in N(A - \lambda)$, that is, $Ax = \lambda x$. It follows that $A^2 x = \lambda Ax = \lambda^2 x$ and finally $A^n x = \lambda^n x$ for $n = 1, 2, \ldots$. If $\lambda \neq 0$, then

$$N(A - \lambda) \subset \bigcap_{n=1}^{\infty} R(A^n) = R(A^\infty). \tag{8.44}$$

We get from Theorem 8.5.1 (1) that $A^n \in \Phi_+(X)$, so $R(A^n)$ is closed for every $n \in \mathbb{N}$. Hence $R^\infty(A)$ is closed in X. We put $X_1 = R^\infty(A)$. It follows from Theorem 8.3.3 that $A(X_1) = X_1$. Let A_1 be the restriction of A on X_1. Since $A_1 \in \Phi(X_1)$, according to Theorem 8.6.3 there exists some $\varepsilon_2 > 0$ such that if $0 < |\lambda| < \varepsilon_2$ then $A_1 - \lambda \in \Phi(X_1)$, $\alpha(A_1 - \lambda) \leq \alpha(A_1)$, $\beta(A_1 - \lambda) \leq \beta(A_1)$ and $i(A_1 - \lambda) = i(A_1)$. Since $\beta(A_1) = 0$, it follows that $\beta(A_1 - \lambda) = 0$. We have from (8.44)

$$\alpha(A - \lambda) = \alpha(A_1 - \lambda) = i(A_1 - \lambda) = i(A_1) = \alpha(A_1).$$

From

$$\begin{aligned}
\alpha(A_1) &= \dim\left[N(A) \cap R^\infty(A)\right] \\
&= \dim N(A) - \dim\left(N(A)/[N(A) \cap R^\infty(A)]\right) = \alpha(A) - k(A),
\end{aligned}$$

we get that $\alpha(A - \lambda)$ has a constant value $\alpha(A) - k(A)$ in a neighborhood $0 < |\lambda| < \varepsilon_2$. Let $\varepsilon = \min\{\varepsilon_1, \varepsilon_2\}$. For $0 < |\lambda| < \varepsilon$, we have $i(A - \lambda) = i(A)$ and $\alpha(A - \lambda) = \alpha(A) - k(A)$, so $\beta(A - \lambda)$ has a constant value $\beta(A) - k(A)$.

If $A \in \Phi_-(X)$, then $A' \in \Phi_+(X^*)$, and it follows from the previously proved part that there exists some $\varepsilon > 0$ such that if $0 < |\lambda| < \varepsilon$ then $\alpha(A' - \lambda)$ has the constant value $\alpha(A') - k(A')$, $\beta(A' - \lambda)$ has the constant value $\beta(A') - k(A')$ and $i(A' - \lambda) = i(A')$. Since $A^n \in \Phi_-(X)$, it follows that $R(A^n)$ is closed for every $n \in \mathbb{N}$, and by Lemma 8.10.1, we have $k(A') = k(A)$. Now it follows from Lemma 8.4.2 that $i(A - \lambda) = -i(A' - \lambda) = -i(A') = i(A)$, and

$$\beta(A - \lambda) = \alpha(A' - \lambda) = \alpha(A') - k(A') = \beta(A) - k(A),$$

that is, (2) holds.

(ii) Analogously, it can be proved that (1) holds. Then (3) is an immediate consequence of (1) and (2). □

Definition 8.10.3 The *semi–Fredholm domain* of $A \in \mathcal{B}(X)$ is defined as

$$\Phi_\pm(A) = \{\lambda \in \mathbb{C} : A - \lambda \in \Phi_+(X) \cup \Phi_-(X)\}.$$

The *Fredholm domain* of $A \in \mathcal{B}(X)$ is defined as

$$\Phi(A) = \{\lambda \in \mathbb{C} : A - \lambda \in \Phi(X)\}.$$

The sets $\Phi_\pm(A)$ and $\Phi(A)$ in Definition 8.10.3 are open and the following result follows from the punctured neighborhood theorem, Theorem 8.10.2.

Corollary 8.10.4 *Let $A \in \mathcal{B}(X)$. Then the function $\lambda \to i(A - \lambda)$ is constant on every connected component of the semi–Fredholm domain of A.*

We continue with the following result which is obtained from the punctured neighborhood theorem, Theorem 8.10.2.

Theorem 8.10.5 *Let $A \in \Phi_+(X)$ and $0 < \alpha(A) < \infty$. Then the following conditions are equivalent:*

(1) $\mathrm{asc}(A) < \infty$;

(2) 0 *is an isolated point in* $\sigma_a(A)$.

   Proof.

(i) First we show that (1) implies (2).
    We conclude from (1), Theorem 8.10.2 and Lemma 7.12.3 that $k(A) = \alpha(A)$ and $\alpha(A - \lambda) = 0$ for $0 < |\lambda| < \varepsilon$. Furthermore, since $R(A - \lambda)$ is closed, we obtain (2).

(ii) Finally, we show that (2) implies (1).
    We note that (2) implies $k(A) = \alpha(A)$, and so $N(A) \cap R(A^\infty) = \{0\}$. By Theorem 8.3.3, we obtain (1). $\square$.

   Similarly as in the previous theorem, we obtain the following result using Theorem 8.10.2 and Lemma 7.12.3.

Theorem 8.10.6 *Let $A \in \Phi_-(X)$ and $0 < \beta(A) < \infty$. Then the following conditions are equivalent:*

(1) $\mathrm{dsc}(A) < \infty$;

(2) 0 *is an isolated point in* $\sigma_d(A)$.

8.11   The Kato decomposition theorem

   In 1956, Kato [111] proved a famous decomposition theorem for the bundle of operators. In 1987, T. T. West [204] proved, in an elementary way, the following version of the Kato decomposition theorem. It shows that, for any semi–Fredholm operator $A \in \mathcal{B}(X)$, it is possible to decompose X into two A–invariant subspaces X_1 and X_2 of X such that $\dim X_1 < \infty$ and such that the restriction A_1 of A onto X_1 is nilpotent and if A_2 is the restriction of A onto X_2, then the nullity or defect of $A_2 - \lambda$ is constant for small values of λ.

   In this section we present some results of West (see also [205]).

   It is easy to prove the following lemma.

Lemma 8.11.1 *Let $A \in \mathcal{L}(X)$ and $n \in \mathbb{N}$. Then we have*

$N(A) \subset R(A^n)$ *if and only if* $N(A^2) \subset R(A^{n-1})$ *if and only if* ...

$$\text{if and only if } N(A^n) \subset R(A). \quad (8.45)$$

The Kato decomposition is non-trivial if $N^\infty(A) \nsubseteq R^\infty(A)$. In this case, there exists a smallest integer ν such that

$$N(A^{\nu-1}) \subset R(A) \text{ and } N(A^\nu) \nsubseteq R(A). \tag{8.46}$$

Hence we can choose $y \in N(A^\nu) \nsubseteq R(A)$, and we consider $A^k y$ for $k = 1, 2, \ldots, \nu - 1$.

Lemma 8.11.2 *With these hypotheses*

$$y \in N(A^\nu) \setminus R(A),$$

$$Ay \in N(A^{\nu-1}) \setminus R(A^2),$$

$$A^2 y \in N(A^{\nu-2}) \setminus R(A^3),$$

$$\cdots\cdots\cdots\cdots\cdots\cdots\cdots\cdots$$

$$A^{\nu-2} y \in N(A^2) \setminus R(A^{\nu-1}),$$

$$A^{\nu-1} y \in N(A) \setminus R(A^\nu);$$

furthermore, the elements $y, Ay, \ldots, A^{\nu-1}y$ *are linearly independent modulo the subspace* $R(A^\nu)$.

Proof. The subset inclusions follow from the hypotheses and Lemma 8.11.1. The lemma is true for $\nu = 1$. Hence we assume $\nu \geq 2$ and $y \in N(A^\nu) \setminus R(A)$. Then we have $Ay \in A(N(A^\nu)) \subset N(A^{\nu-1})$. If $Ay \in R(A^2)$, then

$$A^{-1}(Ay) \bigcap R(A) = \{y + N(A)\} \bigcap R(A) \neq \emptyset.$$

Hence there exists $z \in N(A)$ such that $y + z \in R(A)$. Now we have

$$z \in N(A) \subset N(N^{\nu-1}) \subset R(A)$$

by Lemma 8.11.1, hence $y \in R(A)$, which is a contradiction. So we have

$$Ay \in N(A^{\nu-1}) \setminus R(A^2),$$

and the process can be continued for $A^2 y, \ldots, A^{\nu-1}y$. Clearly we have $A^\nu y = 0$. Now we assume that there exist $\alpha_k \in \mathbb{C}$ for $k = 0, 1, \ldots, \nu - 1$ such that

$$\sum_{k=0}^{\nu-1} \alpha_k A^k y \in R(A^\nu).$$

Applying $A^{\nu-1}$ to this inclusion, we obtain

$$\alpha_0 A^{\nu-1} y \in R(A^{2\nu-1}) \subset R(A^\nu).$$

This gives $\alpha_0 = 0$, and in a similar way we obtain

$$\alpha_k = 0 \text{ for } k = 0, 1, \ldots, \nu - 1. \qquad \square$$

To consider our special case of Kato's decomposition theorem [111, Theorem 4] we define the *jump of a semi-Fredholm operator* to be

$$ju(A) = \alpha(A) - \alpha(A + \lambda) \text{ for } 0 < |\lambda| < \varepsilon \text{ and } A \in \Phi_+(X);$$

$$ju(A) = \beta(A) - \beta(A + \lambda) \text{ for } 0 < |\lambda| < \epsilon \text{ and } A \in \Phi_-(X).$$

We note that these values are well defined by the punctured neighborhood theorem, Theorem 8.10.2.

Now we turn our attention to the restriction of A to its invariant subspace $R^\infty(A)$.

Proposition 8.11.3 *Under the same hypotheses we can choose* $f \in N(A'^\nu)$ *such that*

$$A'^i f(A^{\nu-j-1}y) = \delta_{ij} \text{ for } 0 \le i,j \le \nu - 1.$$

Proof. Since $y, Ay, \ldots, A^{\nu-1}y$ are linearly independent modulo the closed subspace $R(A)$, by the Hahn-Banach theorem, there exists

$$f \in R(A^\nu)^\circ = N(A'^\nu)$$

such that

$$f(A^{\nu-1}y) = 1$$

and

$$f(A^j y) = 0 \text{ for } 0 \le j \le \nu - 2.$$

Now we have $A'f(A^{\nu-1}y) = A'^\nu f(y) = 0$ and, for $0 \le j \le \nu - 2$, we have

$$A'f(A^j y) = f(A^{j+1}y),$$

thus $A'f(A^{\nu-2}y) = 1$ and $A'f(A^j y) = 0$ for $0 \le j \le \nu - 3$. Continuing the process proves the proposition. We observe that

$$f \in N(A'^\nu) \setminus R(A'),$$

$$A'f \in N(A'^{(\nu-1)}) \setminus R(A'^2),$$
$$A'^2 f \in N(A'^{(\nu-2)}) \setminus R(A'^3),$$

$$\cdots\cdots\cdots\cdots\cdots\cdots\cdots\cdots$$

$$A'^{(\nu-2)} f \in N(A^2) \setminus R(A'^{(\nu-1)}),$$
$$A'^{(\nu-1)} f \in N(A') \setminus R(A'^\nu);$$
$$A'^\nu f = 0. \qquad \square$$

Let Y denote the subspace spanned by $y, Ty, \ldots, T^{\nu-1}y$ of Lemma 8.46.

Proposition 8.11.4 *The map*

$$P = \sum_{j=0}^{\nu-1} A'^j f \bigotimes A^{\nu-j-1}y$$

is a projector in $\mathcal{B}(X)$ *with range* Y *which commutes with* A; $A_{|Y}$ *is nilpotent and* $ju(A_{|Y}) = 1$.

Proof. The map P is a projection in $\mathcal{B}(X)$ by Proposition 8.11.3. Furthermore we have

$$AP = \sum_{j=0}^{\nu-1} A'^j f \bigotimes A^{\nu-j}y = \sum_{j=1}^{\nu-1} A'^j f \bigotimes A^{\nu-j}y$$

and

$$PA = \sum_{j=0}^{\nu-1} A'^{(j+1)} f \bigotimes A^{\nu-j-1}y = \sum_{j=0}^{\nu-2} A'^{(j+1)} f \bigotimes A^{\nu-j-1}y,$$

hence $PA = AP$. It is clear that $A_{|Y}$ is nilpotent, $\alpha(A_{|Y}) = 1$, and Y being finite dimensional, $ju(A_{|Y}) = 1$. $\qquad \square$

The next result in this section is a special case of Kato's decomposition theorem.

Theorem 8.11.5 (Kato's decomposition theorem)
If $A \in \Phi_{\pm}(X)$, then $A = A_1 \oplus A_2$, where A_1 is a finite dimensional nilpotent direct summand, $A_2 \in \Phi_{\pm}(X_2)$, and $ju(A_2) = 0$.

Proof. Let $A \in \Phi_{\pm}(X)$. If $ju(A) = 0$ then there is nothing to prove.
If $ju(A) > 0$, let P be the non–zero finite rank projection of Proposition 8.11.4 commuting with A. We note that $A|_{N(A)}$ is semi–Fredholm and $ju(A|_{N(P)}) = ju(A) - 1$. Continuing the process a finite number of times reduces the jump of the residual operator to zero. $\square$

Remark 8.11.6 *Some result which generalizes Kato's decomposition theorem, Theorem 8.11.5, was obtained in [163, 142, 141, 3, 2]. The generalization was obtained for operators $A \in \mathcal{B}(X)$ with a closed range and with $k(A) < \infty$.*

Remark 8.11.7 *If λ is in a semi–Fredholm domain of $A \in \mathcal{B}(X)$ and $ju(A - \lambda) > 0$, then λ is called a jumping point. Using Kato's decomposition theorem, Theorem 8.11.5, in [167, Theorem 1.1], an asymptotic formula for the jumps in semi–Fredholm domains was obtained.*

8.12   Browder and semi-Browder operators

Browder operators form a very important subset of the set of Fredholm operators $\Phi(X)$. The definition of Browder operators involves the notions of the ascent and descent of operators.

Definition 8.12.1 The operator $A \in \mathcal{B}(X)$ is a *Browder operator* (or *Riesz–Schauder operator*, if $A \in \Phi(X)$, $asc(A) < \infty$ and $dsc(A) < \infty$. We use $\mathfrak{B}(X)$ to denote the set of all Browder operators on X.

We note that if A is Browder, then we get from (8.30) $\alpha(A) = \beta(A) < \infty$, that is, $A \in \Phi_0(X)$.

Remark 8.12.2 *Let us point out that if, in the definition of a Browder operator, we omit the condition $A \in \Phi(X)$, then A is a Drazin invertible element in $\mathcal{B}(X)$ ([59]), that is, there is a unique $B \in \mathcal{B}(X)$ and a natural number n such that*

$$A^n BA = A, \quad BAB = B \text{ and } AB = BA.$$

Usually B is called a Drazin *inverse of A, denoted by A^d. Drazin inverses are of great importance in the theory of generalized inverses (see [17, 39, 115]).*

We state the following auxiliary result.

Lemma 8.12.3 *If $A, P \in \mathcal{L}(X)$ and $P^2 = P$, then $AP = PA$ if and only if $R(P)$ and $N(P)$ are invariant for A.*

Proof. If $AP = PA$ and $x \in R(P)$, then $Ax = APx = PAx \in R(P)$. If $x \in N(P)$, then $Ax = A(I - P)x = (I - P)Ax \in R(I - P) = N(P)$. Hence $R(P)$ and $N(P)$ are A–invariant subspaces.
Now we suppose that if $R(P)$ and $N(P)$ are A–invariant subspaces, then it follows from $X = R(P) \oplus N(P)$ that, for any $x \in X$, there exist $x_1 \in R(P)$ and $x_2 \in N(P)$ such that $x = x_1 + x_2$. Hence we have $APx = Ax_1$ and $PAx = PAx_1 + PAx_2 = Ax_1$, and so $AP = PA$. $\square$

Now, we characterize Browder operators.

Theorem 8.12.4 *Let $A \in \mathcal{B}(X)$. Then the following conditions are equivalent:*

(1) $A \in \mathfrak{B}(X)$.

(2) *There exist $F \in \mathcal{F}(X)$ and an invertible operator $B \in \mathcal{B}(X)$ such that $FB = BF$ and $A = B + F$.*

(3) *There exist $K \in \mathcal{K}(X)$ and an invertible operator $B \in \mathcal{B}(X)$ such that $KB = BK$ and $A = B + K$.*

Proof.

(i) First we show that the condition in (1) implies the statement in (2).

We assume that $A \in \Phi(X)$, $\mathrm{asc}(A) < \infty$ and $\mathrm{dsc}(A) < \infty$. By Theorem 7.12.8, we have $\mathrm{asc}(A) = \mathrm{dsc}(A) = p$ and $X = R(A^p) \oplus N(A^p)$. There exists a projection $F \in \mathcal{B}(X)$ such that $R(F) = N(A^p)$ and $N(F) = R(A^p)$. It follows from Theorem 8.5.1 (3) that $A^p \in \Phi(X)$, so $\dim R(F) < \infty$, that is, $F \in \mathcal{F}(X)$. It is easy to see that $A^p - F$ is a bijection from X onto X, so $A^p - F$ is invertible. Knowing that the subspaces $N(A^p)$ and $R(A^p)$ are invariant for A by Theorem 7.12.8, it follows from Lemma 8.12.3 that $FA = AF$. Hence we have

$$A^p - F = (A - F)(A^{p-1} + A^{p-2}F + A^{p-3}F + \cdots + F)$$
$$= (A^{p-1} + A^{p-2}F + A^{p-3}F + \cdots + F)(A - F).$$

Since $A^p - F$ is invertible and A and F commute, it follows that $A - F$ is also invertible. Consequently, for $B = A - F$, we have $A = B + F$, $B \in (\mathcal{B}(X))^{-1}$ and $BF = FB$.

(ii) It is obvious that the statement in (2) implies that in (3).

(iii) Finally, we show that the statement in (3) implies the condition in (1).

Let $A = B + K$ where $B \in (\mathcal{B}(X))^{-1}$, $K \in \mathcal{K}(X)$ and $BK = KB$. It follows from Corollary 8.6.5 that $A \in \Phi(X)$ and $i(A) = 0$, so $\alpha(A) = \beta(A) < \infty$. Since $A = B(I + B^{-1}K)$, and B commutes with K, we get $A^n = B^n(I + B^{-1}K)^n$. Consequently we have $N(A^n) = N((I + B^{-1}K)^n)$ for $n = 1, 2, \ldots$. Since $\mathrm{asc}(I + B^{-1}K) < \infty$ by Theorem 7.13.10, we get $\mathrm{asc}(A) < \infty$. It follows from (8.31) that $\mathrm{dsc}(A) < \infty$. □

The following two examples show that the commutativity condition in Theorem 8.12.4 is essential.

Example 8.12.5 (Yood) ([211]) *Let $X = \ell_1$. For $x = (\xi_1, \xi_2, \ldots)$ and $y = (\eta_1, \eta_2, \ldots)$, we define $Bx = y$ as $\eta_1 = \xi_2$, $\eta_{2n} = \xi_{2n+2}$ and $\eta_{2n+1} = \xi_{2n-1}$ for $n = 1, 2, \ldots$, so*

$$Bx = (\xi_2, \xi_4, \xi_1, \xi_6, \xi_3, \xi_8, \ldots).$$

Obviously B is an isometric isomorphism on X. We define K by $Kx = y$, where $\eta_1 = \xi_2$, and $\eta_i = 0$ for $i \neq 1$. Clearly $K \in \mathcal{K}(X)$. Let $A = B - K$, and $x_n = (\xi_i^n)_i$ be defined by

$$\xi_{2n}^n = 1, \text{ and } \xi_i^n = 0 \text{ for } i \neq 2n.$$

By mathematical induction with respect to n, we prove that $A^n(x_n) = 0$ and $A^{n-1}(x_n) \neq 0$ for $n \in \mathbb{N}$. The statement is satisfied for $n = 1$, so we suppose that it is valid for some $m \geq 1$. Then we have

$$A^{m+1}(x_{m+1}) = A^m A(x_{m+1}) = A^m(x_m) = 0,$$
$$A^m(x_{m+1}) = A^{m-1}A(x_{m+1}) = A^{m-1}(x_m) \neq 0.$$

We obtain $\mathrm{asc}(A) = \infty$. Since $A \in \Phi_0(X)$ by Corollary 8.6.5, it follows from (8.31) that $\mathrm{dsc}(A) = \infty$.

Example **8.12.6 (Kaashoek and Lay)** ([106]) *Let*

$$X = \ell_1(\mathbb{Z}) = \left\{ (\dots, \xi_{-1}, \xi_0, \xi_1, \dots) : \xi_i \in \mathbb{C}, \sum_{i=-\infty}^{\infty} |\xi_i| < \infty \right\},$$

and let $B, K \in \mathcal{B}(X)$ *be defined as follows: for* $x = (\dots, \xi_{-1}, \xi_0, \xi_1, \dots) \in \ell_1(\mathbb{Z})$, *we have* $B(x) = y$, *where* $y = (\eta_n)$ *is given by* $\eta_n = \xi_{n-1}$ *for* $n = 0, \pm 1, \pm 2, \dots$, *and*

$$(Kx)_n = \begin{cases} 0, & \text{for } n \neq 1, \\ \xi_0 & \text{for } n = 1. \end{cases}$$

Then B *is an isometric isomorphism on* X, K *has finite rank (so it is also compact),* $BK \neq KB$, *and* $\mathrm{asc}(B + K) = \mathrm{dsc}(B + K) = \infty$.

Theorem 8.12.7 *Let* $A_1, A_2 \in \mathcal{B}(X)$ *be Browder operators and* $A_1 A_2 = A_2 A_1$. *Then* $A_1 A_2$ *is also a Browder operator.*

Proof. Let $A = A_1 A_2$. It follows from Theorem 8.5.1 that $A \in \Phi_0(X)$. We have for $n \in \mathbb{N}$ that A^n, A_1^n and A_2^n are Fredholm operators and, consequently, have finite dimensional null–spaces. It follows from (8.20) that

$$\alpha(A^n) = \alpha(A_2^n) + \dim(R(A_2^n) \cap N(A_1^n)).$$

Since $\mathrm{asc}(A_1) = \mathrm{dsc}(A_1) < \infty$ and $\mathrm{asc}(A_2) = \mathrm{dsc}(A_2) < \infty$, we have that $\alpha(A^n)$ has the same value for all $n \geq \max(\mathrm{asc}(A_1), \mathrm{asc}(A_2))$. Therefore, $\mathrm{asc}(A) < \infty$. Now it follows from (8.31) that $\mathrm{dsc}(A) < \infty$. $\square$

Analogously as in the case of semi–Fredholm operators, there are two class of semi–Browder operators.

Definition 8.12.8 Let $A \in \mathcal{B}(X)$. Then A is an *upper semi–Browder operator*, if $A \in \Phi_+(X)$ and $\mathrm{asc}(A) < \infty$; A is a *lower semi–Browder operator*, if $A \in \Phi_-(X)$ and $\mathrm{dsc}(A) < \infty$. Let $\mathfrak{B}_+(X)$ ($\mathfrak{B}_-(X)$) denote the set of all upper (lower) semi-Browder operators. An operator A is a *semi–Browder operator* if it is an upper or lower semi–Browder operator.

If follows from Theorem 8.5.5 that $A \in \mathfrak{B}_+(X)$ if and only if $A \in \Phi_+^-(X)$ and $\mathrm{asc}(A) < \infty$, and $A \in \mathfrak{B}_-(X)$ if and only if $A \in \Phi_-^+(X)$ and $\mathrm{dsc}(A) < \infty$.

Clearly A is a Browder operator if it is both an upper and lower semi–Browder operator, that is, $\mathfrak{B}(X) = \mathfrak{B}_+(X) \cap \mathfrak{B}_-(X)$.

Corollary 8.12.9 *Let* $A \in \mathcal{B}(X)$. *Then* $A \in \mathfrak{B}_+(X)$ *if and only if* $A' \in \mathfrak{B}_-(X^*)$; *moreover,* $A \in \mathfrak{B}_-(X)$ *if and only if* $A' \in \mathfrak{B}_+(X^*)$.

Proof. The corollary follows from Lemmas 7.12.10 and 8.4.3. $\square$

Theorem 8.12.10 *Let* $A, B \in \mathcal{B}(X)$ *and* $AB = BA$. *Then we have*

(1) $A, B \in \mathfrak{B}_+(X)$ *if and only if* $AB \in \mathfrak{B}_+(X)$,

(2) $A, B \in \mathfrak{B}_-(X)$ *if and only if* $AB \in \mathfrak{B}_-(X)$.

Proof.

(2) We prove the equivalence in (2).

(i)   First we assume $A, B \in \mathfrak{B}_-(X)$. Then we have $AB \in \Phi_-(X)$ by Theorem 8.5.1 (2). Let $p = \max\{\mathrm{dsc}(A), \mathrm{dsc}(B)\}$. Then we have $A^p(X) = A^{p+1}(X)$ and $B^p(X) = B^{p+1}(X)$, hence,

$$(AB)^p(X) = A^p B^p(X) = A^p B^{p+1}(X) = B^{p+1} A^p(X)$$
$$= B^{p+1} A^{p+1}(X) = (AB)^{p+1}(X).$$

It follows that $\mathrm{dsc}(AB) < \infty$, so $AB \in \mathfrak{B}_-(X)$.

(ii) To show the converse implication we assume $AB \in \mathfrak{B}_-(X)$.
Using Theorem 8.4.5 (2) it follows from $AB = BA \in \Phi_-(X)$ that $A, B \in \Phi_-(X)$. If $p = \mathrm{dsc}(AB) < \infty$, then $R^\infty(AB) = (AB)^p(X)$, and since $(AB)^p \in \Phi_-(X)$, we obtain $\mathrm{codim}(R^\infty(AB)) < \infty$. We have for every $n \in \mathbb{N}$

$$A^n(X) \supset A^n B^n(X) = (AB)^n(X) \supset R^\infty(AB),$$

so $dsc(A) < \infty$. Analogously, we obtain $\mathrm{dsc}(B) < \infty$ and $A, B \in \mathfrak{B}_-(X)$.

(1)   The equivalence in (1) follows from (2) and Corollary 8.12.9.   □

Lemma 8.12.11 *Let $A \in \mathcal{B}(X)$. Then we have:*

(1) $A \in \mathcal{B}_+(X)$ if and only if $R(A)$ is closed and $\dim N^\infty(A) < \infty$.

(2) $A \in \mathcal{B}_-(X)$ if and only if $\mathrm{codim}\, R^\infty(A) < \infty$.

   Proof.

(1)   We prove the equivalence in (1).

   (i)   First we assume $A \in \mathcal{B}_+(X)$.
   Then $R(A)$ is closed and $N^\infty(A) = N(A^p)$, where $p = \mathrm{asc}(A) < \infty$. Since $A^p \in \Phi_+(X)$, we conclude that $\dim N^\infty(A) < \infty$.

   (ii) To show the converse implication, we assume that $R(A)$ is closed and

   $$\dim N^\infty(A) < \infty.$$

   Then $A \in \Phi_+(X)$ and $\mathrm{asc}(A) < \infty$, that is, $A \in \mathcal{B}_+(X)$.

(2)   The equivalence in (2) can be proved similarly.   □

Lemma 8.12.12 *Let $A \in \mathcal{B}(X)$. Then we have:*

(1) $A \in \mathcal{B}_+(X)$ and $k(A) = 0$ if and only if A is bounded below;

(2) $A \in \mathcal{B}_-(X)$ and $k(A) = 0$ if and only if A is onto.

   Proof.

(1)   We prove the equivalence in (1).

   (i)   First we assume $A \in \mathcal{B}_+(X)$ and $k(a) = 0$.
   Since $R(A)$ is closed, in order to prove that A is bounded below, it is enough to show that A is injective. We assume to the contrary that there exists $x_0 \in N(A)$ such that $x_0 \neq 0$. Since $k(A) = 0$, it follows from Lemma 8.3.12 that $N(A^n) \subset R(A)$, for every $n \in \mathbb{N}$. Hence there exists $x_1 \in X$ such that $x_0 = Ax_1$. Since $x_1 \in N(A^2)$, there exists $x_2 \in X$ such that $x_1 = Ax_2$. So we find inductively elements $x_i \in X$ such that $x_{i-1} = Ax_i$, $i \in \mathbb{N}$. Using the induction we can easily prove that the elements x_i are linearly independent. Since $x_i \in N^\infty(A)$, we get that $\dim N^\infty(A) = \infty$. This is a contradiction to (1).

(ii) Now we show the converse implication in (1).
If A is bounded below, then $R(A)$ is closed and $N(A) = \{0\}$, which implies that $A \in \Phi_+(X)$, $\mathrm{asc}(A) = 0$ and $k(A) = 0$.

(2) The equivalence in (2) follows from Corollary 8.12.9, Lemma 8.10.1 and the equivalence in (1). □

Now we characterize upper semi–Browder operators.

Theorem 8.12.13 *Let $A \in \mathcal{B}(X)$. Then*

(1) $A \in \mathfrak{B}_+(X)$,

implies that

(2) there exist $F \in \mathcal{F}(X)$ and $B \in \mathcal{B}(X)$, such that B is bounded below, $BF = FB$ and $A = B + F$.

Proof. Let $A \in \mathfrak{B}_+(X)$. It follows from the Kato decomposition theorem, Theorem 8.11.5, that the operator A is decomposable in the sense of Kato, that is, X is a direct sum of closed subspaces X_0 and X_1, which are A–invariant, and $\dim X_1 < \infty$. Furthermore, if $A_0 = A|_{X_0} : X_0 \to X_0$ and $A_1 = A|_{X_1} : X_1 \to X_1$, then A_1 is nilpotent on X_1, $R(A_0)$ is closed and $k(A_0) = 0$. Since $N^\infty(A_0) \subset N^\infty(A)$, Lemma 8.12.11 (1) yields that A_0 is an upper semi–Browder operator. It follows from Lemma 8.12.12 (1) that A_0 is bounded below. Thus we have $N(A^n) = N(A_0^n) \oplus N(A_1^n) = N(A_1^n) \subset X_1$ for every $n \in \mathbb{N}$, and so $N^\infty(A) \subset X_1$. As $X_1 \subset N^\infty(A)$, we get $X_1 = N^\infty(A) = N(A^p)$, where $p = \mathrm{asc}(A)$.
There exists a projection $F \in \mathcal{B}(X)$ such that $R(F) = X_1$ and $N(F) = X_0$. Clearly, $F \in F(X)$. By Lemma 8.12.3, A and F commute. It is easy to see that $A^p - F$ is injective. Since $A^p - F = (A^{p-1} + A^{p-2}F + \cdots + F)(A - F)$, we conclude that $N(A - F) \subset N(A^p - F)$. Thus $A - F$ is injective. Since $R(A - F) = (A - F)X_0 + (A - F)X_1 = R(A_0) + (A - F)X_1$, it follows by Lemma 7.13.6 that $R(A - F)$ is closed. Therefore, the operator $B = A - F$ is bounded below and $BF = FB$. □

A characterization of lower semi–Browder operators is given in the next theorem.

Theorem 8.12.14 *Let $A \in \mathcal{B}(X)$. Then*

(1) $A \in \mathfrak{B}_-(X)$

implies that

(2) there exist $F \in \mathcal{F}(X)$ and $B \in \mathcal{B}(X)$ such that B is onto, $BF = FB$ and $A = B + F$.

Proof. Let $A \in \mathfrak{B}_-(X)$. Then A is decomposable in the sense of Kato, that is, X is a direct sum of closed subspaces X_0 and X_1, which are A–invariant, and $\dim X_1 < \infty$. The restrictions $A_0 = A|_{X_0} : X_0 \to X_0$ and $A_1 = A|_{X_1} : X_1 \to X_1$ have the following properties: A_1 is nilpotent on X_1, $R(A_0)$ is closed and $k(A_0) = 0$. Hence there exists $p \in \mathbb{N}$ such that $A_1^{p-1} \neq 0 = A_1^p$. We obtain for $n \geq p$

$$R(A^n) = R(A_0^n) \oplus R(A_1^n) = R(A_0^n),$$

which implies that $R^\infty(A) = R^\infty(A_0)$. It follows from Lemma 8.12.11 (2) that $A_0 \in \mathfrak{B}_-(X)$. Now we conclude by Lemma 8.12.12 (2) that A_0 is onto, which implies that $R(A_0^n) = X_0$ for every $n \in \mathbb{N}$. We get by Theorem 8.11.5 that $R(A^n) = X_0$ for all $n \geq p$, and so $R^\infty(A) = R(A^p) = X_0$. Let $F \in \mathcal{B}(X)$ be the projection such that $R(F) = X_1$ and $N(F) = X_0$. We see that $F \in \mathcal{F}(X)$ and $AF = FA$. It is easy to see that $A^p - F$ is onto.

Since $A^p - F = (A - F)(A^{p-1} + A^{p-2}F + \cdots + F)$, we conclude that $A - F$ is onto. For $B = A - F$ we have $A = B + F$, B is onto and $BF = FB$. $\qquad\square$

We remark that the commutativity condition in Theorems 8.12.13 and 8.12.14 is essential (Examples 8.12.5 and 8.12.6).

In 1978, Grabiner [76] proved that semi–Browder operators are stable under commuting compact perturbations. First he proved the following auxiliary result.

Lemma 8.12.15 *We assume that $B \in \mathcal{B}(X)$ is onto, and $K \in \mathcal{K}(X)$ commutes with B. Then we have*
$$C = B - K \in \mathfrak{B}_-(X).$$

Proof. Since $B \in \Phi_-(X)$, it follows from Corollary 8.6.5 that $C \in \Phi_-(X)$. Now it follows by the index theorem, Theorem 8.5.1 (2), that $V^k \in \Phi_-(X)$ for each $k \in \mathbb{N}$. Hence $\dim X/R(C^k) < \infty$ for each $k \in \mathbb{N}$, and the map $\tilde{B}$ induced by B on $X/R(C^k)$ is onto. Thus $\tilde{B}$ is one to one, and so $N(B) \subset R(B^k)$ for each $k \in \mathbb{N}$. Since B is onto, there exists a positive number γ such that
$$\|Bx\| \geq \gamma \cdot \mathrm{dist}(x, N(B)) \text{ for all } x \in X.$$

Clearly we have
$$B(R(C^k)) = R(C^k B) = R(C^k).$$

Now if $x \in X$ and $z \in R(C^k)$, there exists $y \in R(C^k)$ such that $By = z$. Hence $N(B) \subset R(B^k)$ implies
$$\|Bx - z\| = \|B(x - y)\| \geq \gamma \cdot \mathrm{dist}(x - y, N(B)) \geq \gamma \cdot \mathrm{dist}(x - y, R(C^k)).$$

Since this holds for all $z \in R(C^k)$ it implies
$$\mathrm{dist}(Bx, R(C^k)) \geq \gamma \cdot \mathrm{dist}(x, R(C^k)).$$

We suppose $\mathrm{dsc}(V) = \infty$ and $\varepsilon \in (0, 1)$. Now, by the Riesz lemma, there exists $x_n \in R(C^n)$ with $\|x_n\| = 1$ and $\mathrm{dist}(x_n, R(C^k)) > \varepsilon$ for each $n \in \mathbb{N}$. If $m > n$, then $Kx_m - Kx_n = (Kx_m - (B - K)x_n) - Bx_n$, and so
$$\|Kx_m - Kx_n\| \geq \mathrm{dist}(Bx_n, R(C^{n+1})) \geq \gamma \cdot \mathrm{dist}(x_n, R(C^{n+1})) > \gamma \cdot \varepsilon.$$

This is in a contradiction to the compactness of K, so we must have $\mathrm{dsc}(C) < \infty$. $\qquad\square$

Theorem 8.12.16 *Let $A \in \mathcal{B}(X)$, $K \in \mathcal{K}(X)$ and $AK = KA$. Then*
$$A \in \mathfrak{B}_-(X) \text{ implies } A - K \in \mathfrak{B}_-(X). \tag{8.47}$$

and
$$A \in \mathfrak{B}_+(X) \text{ implies } A - K \in \mathfrak{B}_+(X). \tag{8.48}$$

Proof.

(i) First we show the implication in (8.47).

It is known that $B = A - K \in \Phi_-(X)$. If $\mathrm{dsc}(A) = p$, then $R(A^p)$ is a closed subspace of X and the restriction of A on $R(A^p)$ is onto. By Lemma 8.12.15, the restriction of B on $R(T^p)$ has finite descent, and so there exists $k \in \mathbb{N}$ such that
$$R(B^m) \supset R(B^m A^p) \supset R(A^k A^p) \text{ for any } m > k.$$

Since $R(A^k A^p)$ has finite codimension, it follows that N has finite descent. This proves (8.47).

(ii) Now we prove the implication in (8.48).

  We note that $A \in \mathfrak{B}_+(X)$ implies $A' \in \mathfrak{B}_-(X^*)$. Now the implication in (8.48) follows from the implication in (8.47), that is, $\text{asc}(A - K) = \text{dsc}(A' - K') < \infty$. $\square$

Remark 8.12.17 *Theorem 8.12.16 implies the converse of Theorems 8.12.13 and 8.12.14.*

Remark 8.12.18 *Theorem 8.12.16 was generalized in [164, 165].*

8.13 Essential spectra

There are several definitions of the *essential spectrum* of an operator $A \in \mathcal{B}(X)$ which coincide if X is a Hilbert space and A is self–adjoint.

Definition 8.13.1 Let $A \in \mathcal{B}(X)$. The *Fredholm (Wolf, or Calkin) essential spectrum* of A, denoted by $\sigma_{ef}(A)$, is the set of all complex λ, such that $A - \lambda$ is not Fredholm, that is,

$$\sigma_{ef}(A) = \{\lambda \in \mathbb{C} : A - \lambda \notin \Phi(X)\}.$$

Let $\pi : \mathcal{B}(X) \to \mathcal{B}(X)/\mathcal{K}(X)$ be the natural homomorphism from $\mathcal{B}(X)$ to the Calkin algebra $\mathcal{C}(X) \equiv \mathcal{B}(X)/\mathcal{K}(X)$. Then the Fredholm essential spectrum of A is equal to $\sigma(\pi(A))$, the spectrum of $\pi(A)$ in $\mathcal{C}(X)$.

Definition 8.13.2 (Schechter) ([184, 186]) The *Schechter essential spectrum* of $A \in B(X)$ is defined as

$$\sigma_{em}(A) = \{\lambda \in \mathbb{C} : A - \lambda \notin \Phi_0(X)\}. \tag{8.49}$$

Remark 8.13.3 *Some authors refer to the Schechter essential spectrum as the* Weyl *essential spectrum, or simply,* Weyl *spectrum.*

It follows from (8.49) and Theorem 8.7.3 that

$$\sigma_{em}(A) = \bigcap_{F \in \mathcal{F}(X)} \sigma(A + F) = \bigcap_{E \in \mathcal{K}(X)} \sigma(A + E).$$

We note that

$$\sigma(A) = \sigma_{em}(A) \cup \sigma_p(A), \tag{8.50}$$

where $\sigma_p(A)$ denotes the point spectrum of A.

Theorem 8.13.4 *Let $A \in \mathcal{B}(X)$. Then we have*

$$\partial \sigma(A) \subset \sigma_{ef}(A) \cup iso\,\sigma(A),$$

Proof. We assume that $\lambda \in \partial \sigma(A)$ and $\lambda \notin \sigma_{ef}(A)$. Then $A - \lambda \in \Phi(X)$ and it follows from Theorem 8.10.2 that λ is an isolated point of $\sigma(A)$. $\square$

Example **8.13.5** *Let $U : \ell_2 \to \ell_2$ be the right shift, that is,*

$$U(x_1, x_2, \dots) = (0, x_1, x_2, \dots) \text{ for all } (x_1, x_2, \dots) \in \ell_2.$$

Then $\sigma_{em}(U) = \sigma(U) = \{\lambda : |\lambda| \leq 1\}$ and $\sigma_{ef}(U) = \{\lambda : |\lambda| = 1\}$.

Proof. The set ℓ_2 is a Hilbert space endowed with the scalar product $\langle x, y \rangle = \sum_{i=1}^{\infty} x_i \overline{y_i}$ where $x = (x_i), y = (y_i) \in \ell_2$. Since U does not have eigenvalues, it follows from (8.50) that $\sigma_{em}(U) = \sigma(U)$. Furthermore, we have $\sigma(U) = \{\lambda : |\lambda| \leq 1\}$. Since $U^*U = I$ and $UU^* = I - P_{\mathrm{span}(\{e_1\})}$, we conclude that $\pi(U)$ is unitary in the Calkin algebra $\mathcal{C}(\ell_2)$, so $\sigma_{ef}(U) = \sigma(\pi(U)) \subset \{\lambda : |\lambda| = 1\}$. As $\mathrm{iso}\,\sigma(U) = \emptyset$, it follows from Theorem 8.13.4 that $\partial\sigma(U) \subset \sigma_{ef}(U)$, and so $\{\lambda : |\lambda| = 1\} = \sigma_{ef}(U)$. Consequently, we have $\sigma_{ef}(U) = \{\lambda : |\lambda| = 1\}$. $\square$

The Schechter essential spectrum is the greatest subset of $\sigma(A)$, which is invariant for compact perturbations. The Fredholm essential spectrum does not have this nice property. The undesirable property is that the Schechter essential spectrum can exclude open subsets of the spectrum, meaning that a significant part of the spectrum does not belong to the Schechter essential spectrum. The Schechter essential spectrum can be enlarged using the Browder essential spectrum.

Definition 8.13.6 The *Browder essential spectrum* of $A \in \mathcal{B}(X)$, denoted by $\sigma_{eb}(A)$, is the set of all complex λ, such that $A - \lambda$ is not a Browder operator, that is,

$$\sigma_{eb}(A) = \{\lambda \in \mathbb{C} : A - \lambda \notin \mathfrak{B}(X)\}.$$

It follows from Theorem 8.12.4 that

$$\sigma_{eb}(A) = \bigcap_{F \in \mathcal{F}(X), AF=FA} \sigma(A + F) = \bigcap_{K \in \mathcal{K}(X), AK=KA} \sigma(A + K).$$

We note that the set $\sigma(A) \backslash \sigma_{eb}(A)$ is at most countable (it consists of isolated points).

It follows from Theorems 8.10.5 and 8.10.6 that the Browder essential spectrum can be described in the following way.

Corollary 8.13.7 *The Browder essential spectrum of $A \in \mathcal{B}(X)$ is equal to the union of $\sigma_{ef}(A)$ ($\sigma_{em}(A)$) and the set of all points of $\sigma(A)$ which are not isolated in $\sigma(A)$, that is,*

$$\sigma_{eb}(A) = \sigma_{ef}(A) \cup \mathrm{acc}(\sigma(A)) = \sigma_{em}(A) \cup \mathrm{acc}(\sigma(A)).$$

The statement of the following theorem is actually Browder's original definition of the essential spectrum for closed densely defined operators on Banach spaces, which appeared in [32].

Theorem 8.13.8 *The Browder essential spectrum of $A \in \mathcal{B}(X)$ is equal to the set of all $\lambda \in \sigma(A)$ such that at least one of the following conditions holds:*

$$R(A - \lambda) \text{ is not closed in } X;$$
$$\dim \bigcup_{n \geq 1} N((A - \lambda)^n) = \infty;$$
$$\lambda \text{ is the accumulation point of the set } \sigma(A).$$

Proof. The proof follows from Theorems 8.10.5 and 8.10.6 and the definition of Browder operators. $\square$

Definition 8.13.9 ([111]) The *Kato essential spectrum* of $A \in \mathcal{B}(X)$, denoted by $\sigma_{ek}(A)$, is the set of all complex λ, such that $A - \lambda$ is not semi–Fredholm, that is,

$$\sigma_{ek}(A) = \{\lambda \in \mathbb{C} : A - \lambda \notin \Phi_+(X) \cup \Phi_-(X)\}.$$

Definition 8.13.10 ([79]) Let $A \in \mathcal{B}(X)$. Then the sets

$$\sigma_{e\alpha}(A) = \{\lambda \in \mathbb{C} : A - \lambda \notin \Phi_+(X)\}$$

and

$$\sigma_{e\beta}(A) = \{\lambda \in \mathbb{C} : A - \lambda \notin \Phi_-(X)\}$$

are called the *upper semi-Fredholm (Gustafson) essential spectrum*, and the *lower semi-Fredholm (Weidmann) essential spectrum* of A.

We note that

$$\sigma_{ek}(A) = \sigma_{e\alpha}(A) \cap \sigma_{e\beta}(A) \text{ and } \sigma_{ef}(A) = \sigma_{e\alpha}(A) \cup \sigma_{e\beta}(A).$$

Lemma 8.13.11 *Let $A \in \mathcal{B}(X)$. Then we have*

$$\partial \sigma_{ef}(A) \subset \sigma_{ek}(A). \tag{8.51}$$

Proof. Let $\lambda \in \partial \sigma_{ef}(A)$. It follows that $A - \lambda \in \partial \Phi(X)$. From Corollary 8.9.7, we have $A - \lambda \notin \Phi_\pm(X)$, that is, $\lambda \in \sigma_{ek}(A)$. □

Corollary 8.13.12 *The Kato essential spectrum of $A \in \mathcal{B}(X)$ is non–empty.*

Proof. Since $\sigma_{ef}(A) \neq \emptyset$, it follows that $\partial \sigma_{ef}(A) \neq \emptyset$. Now (8.51) implies $\sigma_{ek}(A) \neq \emptyset$. □

The following inclusions hold between the introduced spectra of $A \in \mathcal{B}(X)$.

$$\emptyset \neq \sigma_{ek}(A) \subset \begin{matrix} \sigma_{e\beta}(A) \\ \sigma_{e\alpha}(A) \end{matrix} \subset \sigma_{ef}(A) \subset \sigma_{em}(A) \subset \sigma_{eb}(A) \subset \sigma(A). \tag{8.52}$$

Theorem 8.13.13 *If $\lambda \in \partial \sigma(A)$ is not an isolated point of $\sigma(A)$, then $\lambda \in \sigma_{ek}(A)$.*

Proof. If $\lambda \notin \sigma_{ek}(A)$, then $A - \lambda \in \Phi_\pm(X)$. According to Theorem 8.10.2, it follows that there exists some $\varepsilon > 0$ such that if $0 < |\mu - \lambda| < \varepsilon$ then $\mu \in \rho(A)$. We conclude that λ is an isolated point of $\sigma(A)$. □

Remark 8.13.14 *We point out that if E and F are subsets of $\mathbb{C}$ that satisfy $E \subset F$, then it follows from $\partial F \subset E$ that $\partial F \subset \partial E$. Thus, from the inclusions (8.52) and Corollary 8.9.7, the following inclusions hold.*

Corollary 8.13.15 *Let $A, B \in \mathcal{B}(X)$. Then we have*

$$\partial \sigma_{eb}(A) \subset \partial \sigma_{em}(A) \subset \partial \sigma_{ef}(A) \subset \begin{matrix} \partial \sigma_{e\beta}(A) \\ \partial \sigma_{e\alpha}(A) \end{matrix} \subset \partial \sigma_{ek}(A) \tag{8.53}$$

and

$$\hat{\sigma}_{ek}(A) = \hat{\sigma}_{e\alpha}(A) = \hat{\sigma}_{e\beta}(A) = \hat{\sigma}_{ef}(A) = \hat{\sigma}_{em}(A) = \hat{\sigma}_{eb}(A),$$

where $\hat{\sigma}_{ek}(A)$ is the polynomial convex hull of $\sigma_{ek}(A)$, etc.

Corollary 8.13.16 *Let H be a Hilbert space and $A \in \mathcal{B}(H)$ be self-adjoint. Then all the essential spectra of A are equal.*

Proof. Since $\sigma(A)$ is real, all the essential spectra of A are real, so they do not have holes, and so they are equal. □

We also have the next result for the essential spectral radius of A.

Corollary 8.13.17 *Let $A \in \mathcal{B}(X)$. Then we have*

$$r_e(A) = \lim_{n \to \infty} \|A^n\|_K^{\frac{1}{n}}$$

and

$$\max_{\lambda \in \sigma_{ek}(A)} |\lambda| = \max_{\lambda \in \sigma_{ea}(A)} |\lambda| = \max_{\lambda \in \sigma_{e\beta}(A)} |\lambda| = \max_{\lambda \in \sigma_{ef}(A)} |\lambda|$$
$$= \max_{\lambda \in \sigma_{em}(A)} |\lambda| = \max_{\lambda \in \sigma_{eb}(A)} |\lambda| \equiv r_e(A).$$

Remark 8.13.18 *There is a nice application of the Kuratowski and Hausdorff measures of noncompactness due to Nussbaum [147, Theorem 1].*
Let $A \in \mathcal{B}(X)$. Then the limits

$$\lim_{n \to \infty} (\|A^n\|_\alpha)^{1/n} \quad \lim_{n \to \infty} (\|A^n\|_\chi)^{1/n} \quad and \quad \lim_{n \to \infty} (\|A^n\|_K)^{1/n}$$

exist, and are equal to $r_e(A)$.

8.14   Essential type subsets of the spectrum

We investigate the following subsets of the Schechter essential spectrum of an operator $A \in \mathcal{B}(X)$.

Definition 8.14.1 ([158, 159]) *The essential approximate point spectrum of $A \in \mathcal{B}(X)$ is defined as*

$$\sigma_{ea}(A) = \{\lambda \in \mathbb{C} : A - \lambda \notin \Phi_+^-(X)\}$$

and the *essential approximate defect spectrum* of A is defined as

$$\sigma_{ed}(A) = \{\lambda \in \mathbb{C} : A - \lambda \notin \Phi_-^+(X)\}.$$

It follows from Definition 8.14.1 that

$$\sigma_{ea}(A) = \sigma_{e\alpha}(A) \cup \{\lambda \in \mathbb{C} : A - \lambda \in \Phi_+(X) \text{ and } i(A - \lambda) > 0\},$$

and

$$\sigma_{ed}(A) = \sigma_{e\beta}(A) \cup \{\lambda \in \mathbb{C} : A - \lambda \in \Phi_-(X) \text{ and } i(A - \lambda) < 0\}, \qquad (8.54)$$

so $\sigma_{ea}(A) \neq \emptyset$ and $\sigma_{ed}(A) \neq \emptyset$.

For $A \in \mathcal{B}(X)$, we denote by $\sigma_a(A)$, the approximate point spectrum of A and by $\sigma_d(A)$, the approximate defect spectrum of A.

It follows from Theorems 8.7.1 and 8.7.2 that

$$\sigma_{ea}(A) = \bigcap_{K \in \mathcal{K}(X)} \sigma_a(A + K) \qquad (8.55)$$

and

$$\sigma_{ed}(A) = \bigcap_{K \in \mathcal{K}(X)} \sigma_d(A + K).$$

Theorem 8.14.2 *Let $A \in \mathcal{B}(X)$. Then*

(1) $\sigma_{e\alpha}(A) \subset \sigma_{ea}(A) \subset \sigma_{em}(A)$,

(2) $\partial \sigma_{em}(A) \subset \partial \sigma_{ea}(A) \subset \partial \sigma_{e\alpha}(A)$,

(3) $\hat{\sigma}_{e\alpha}(A) = \hat{\sigma}_{ea}(A) = \hat{\sigma}_{em}(A)$,

(4) $\sigma_{ea}(A)$ $(\sigma_{em}(A))$ *consists of* $\sigma_{e\alpha}(A)$ $(\sigma_{ea}(A))$ *including some holes in* $\sigma_{e\alpha}(A)$ $(\sigma_{ea}(A))$,

(5) *if* $\sigma_{e\alpha}(A)$ *is connected, then* $\sigma_{ea}(A)$ *is connected, and if* $\sigma_{ea}(A)$ *is connected, then* $\sigma_{em}(A)$ *is connected.*

Proof. It is sufficient to prove the inclusions in (2).
In order to do so, it is enough, because of Remark 8.13.14 and (1), to prove the inclusions $\partial \sigma_{em}(A) \subset \sigma_{ea}(A)$ and $\partial \sigma_{ea}(A) \subset \sigma_{e\alpha}(A)$.
By Corollary 8.13.15, $\partial \sigma_{em}(A) \subset \partial \sigma_{ek}(A) \subset \sigma_{ek}(A)$, and since $\sigma_{ek}(A) \subset \sigma_{ea}(A)$, we obtain $\partial \sigma_{em}(A) \subset \sigma_{ea}(A)$.
We assume $\lambda \in \partial \sigma_{ea}(A)$. If $\lambda \notin \sigma_{e\alpha}(A)$, then $A - \lambda \in \Phi_+(X)$ and $i(A - \lambda) > 0$. It follows from the continuity of the index that there exists $\varepsilon > 0$ such that $|\lambda - \lambda'| < \epsilon$ implies $A - \lambda' \in \Phi_+(X)$ and $i(A - \lambda') > 0$, that is, $\lambda' \in \sigma_{ea}(A)$. This contradicts $\lambda \in \partial \sigma_{ea}(A)$. Hence we must have $\partial \sigma_{ea}(A) \subset \sigma_{e\alpha}(A)$. $\square$

The following theorem can be proved similarly.

Theorem 8.14.3 *Let $A \in \mathcal{B}(X)$. Then*

(1) $\sigma_{e\beta}(A) \subset \sigma_{ed}(A) \subset \sigma_{em}(A)$,

(2) $\partial \sigma_{em}(A) \subset \partial \sigma_{ed}(A) \subset \partial \sigma_{e\beta}(A)$,

(3) $\hat{\sigma}_{e\beta}(A) = \hat{\sigma}_{ed}(A) = \hat{\sigma}_{em}(A)$,

(4) $\sigma_{ed}(A)$ $(\sigma_{em}(A))$ *consists of* $\sigma_{e\beta}(A)$ $(\sigma_{ed}(A))$ *and possibly some holes in* $\sigma_{e\beta}(A)$ $(\sigma_{ed}(A))$,

(5) *if* $\sigma_{e\beta}(A)$ *is connected, then* $\sigma_{ed}(A)$ *is also connected; if* $\sigma_{ed}(A)$ *is connected, then* $\sigma_{em}(A)$ *is also connected.*

Example 8.14.4 *Let A be a bounded self–adjoint operator on a Hilbert space. It follows from Theorem 8.14.2 and the fact that $\sigma(A)$ is real that $\sigma_{e\alpha}(A) = \sigma_{ea}(A) = \sigma_{em}(A)$. Similarly we obtain $\sigma_{e\beta}(A) = \sigma_{ed}(A) = \sigma_{em}(A)$.*

Example 8.14.5 *Let U be the right shift operator on ℓ_2. Then we have $\sigma_{ea}(U) = \{\lambda \in \mathbb{C} : |\lambda| = 1\} \equiv S$ and $\sigma_{ed}(U) = \{\lambda \in \mathbb{C} : |\lambda| \le 1\} \equiv D$.*

Proof. It is well known ([82, Solution 67]) that $\sigma_a(U) = S$. By (8.55), we see that $\sigma_{ea}(U) \subset \sigma_a(U)$ and so $\sigma_{ea}(U) \subset S$. Since $\sigma_{em}(U) = D$ by Example 8.13.6, we get $S \subset \sigma_{ea}(U)$ from Theorem 8.14.2 (2).
Using the inclusion $\sigma_{ed}(U) \subset \sigma_{em}(U)$, we conclude that $\sigma_{ed}(U) \subset D$. Since $\sigma_{ef}(U) = S$ by Example 8.13.6, it follows that $U - \lambda \in \Phi(X)$ for $\lambda \in \mathbb{C}$ such that $|\lambda| < 1$. We obtain from $\alpha(U) = 0$ and $\beta(U) = -1$ that $i(U) = -1$, and it follows by Corollary 8.10.4 that $i(U - \lambda) = -1$ for $\lambda \in \mathbb{C}$ such that $|\lambda| < 1$. By (8.54), we conclude that $\{\lambda \in \mathbb{C} : |\lambda| < 1\} \subset \sigma_{ed}(U)$, which implies $D \subset \sigma_{ed}(U)$. $\square$

Example 8.14.6 *Let U be the right shift operator on ℓ_2. Then we have $\sigma_{ea}(U^*) = \{\lambda \in \mathbb{C} : |\lambda| \leq 1\} \equiv D$ and $\sigma_{ed}(U^*) = \{\lambda \in \mathbb{C} : |\lambda| = 1\} \equiv S$.*

Proof. Since $\sigma_{em}(U^*) = D$ by [21, Example 1.2 and Corollary 2.6], it follows from Theorem 8.14.2 (2) that $\sigma_{ea}(U^*) \supset S$. Since $\sigma_{ef}(U^*) = \sigma_{ef}(U) = S$ by Example 8.13.6, for every $|\lambda| < 1$ we get that $U^* - \lambda$ is Fredholm and $i(U^* - \lambda) = 1$ (this follows from the fact that $i(U^*) = 1$ and Corollary 8.10.4). It follows from (8.54) that $\sigma_{ea}(U^*) \supset D \setminus S$.
It follows from $\sigma_{em}(U^*) = D$ that $S \subset \sigma_{ed}(U^*)$ by Theorem 8.14.3 (2). Since for every $|\lambda| < 1$, $U^* - \lambda \in \Phi_-^+(X)$, that is, $\lambda \notin \sigma_{ed}(U^*)$, we obtain $\sigma_{ed}(U^*) \subset \sigma_{em}(U^*) \setminus \{\lambda \in \mathbb{C} : |\lambda| < 1\} = S$. $\qquad\square$

Theorem 8.14.7 *Let $A \in \mathcal{B}(X)$ and p be a polynomial. Then we have*

$$\sigma_{ea}(p(A)) \subset p(\sigma_{ea}(A)) \tag{8.56}$$

$$\sigma_{ed}(p(A)) \subset p(\sigma_{ed}(A)). \tag{8.57}$$

Proof.

(i) First we show the inclusion in (8.56).
We can suppose that p is not a constant. Let $\lambda \notin p(\sigma_{ea}(A))$. Since

$$p(\mu) - \lambda = c(\mu - \mu_1) \ldots (\mu - \mu_n),$$

we get

$$p(A) - \lambda = c(A - \mu_1) \ldots (A - \mu_n). \tag{8.58}$$

For every $j = 1, \ldots, n$, $p(\mu_j) = \lambda \notin p\{\sigma_{ea}(A)\}$ holds, so $\mu_j \notin \sigma_{ea}(A)$, that is, $A - \mu_j \in \Phi_+^-(X)$. Then it follows from (8.58) and the index theorem, Theorem 8.5.1 (1), that $p(A) - \lambda \in \Phi_+^-(X)$, that is, $\lambda \notin \sigma_{ea}(p(A))$.

(i) The proof of the inclusion in (8.57) is similar to the proof of (8.56). $\qquad\square$

The following example shows that the inclusion (8.56) can be proper.

Example 8.14.8 *Let $A = U^* \oplus (U + 2)$, where U is the right shift operator, and $p(\lambda) = \lambda(\lambda - 2)$. Then $0 \in p(\sigma_{ea}(A))$ and $0 \notin \sigma_{ea}(p(A))$.*

Proof. We have $p(A) = A(A-2) = (U^* \oplus (U+2))((U^* - 2) \oplus U)$; since U^* is a Fredholm operator of index 1, and since $U + 2$ and $U^* - 2$ are invertible, it follows that A and $A - 2$, respectively, are Fredholm operators of index 1 and -1. Thus $p(A)$ is a Fredholm operator of index 0, and $0 \notin \sigma_{ea}(p(A))$. Since A is a Fredholm operator of index 1, it follows that $0 \in \sigma_{ea}(A)$. Hence we have $0 = p(0) \in p\{\sigma_{ea}(A)\}$.
$\square$

Also the inclusion in (8.57) can be proper; the proof is similar to that of Example 8.14.8.

Example 8.14.9 *Let $A = U \oplus (U^* + 2)$, where U is the right shift operator, and $p(\lambda) = \lambda(\lambda - 2)$. Then we have $0 \in p(\sigma_{ed}(A))$ and $0 \notin \sigma_{ed}(p(A))$.*

The set of all upper (lower) semi–Browder operators induces the corresponding spectrum.

Definition 8.14.10 ([160]) Let $A \in \mathcal{B}(X)$. The *Browder essential approximate point spectrum* of A is defined as

$$\sigma_{ab}(A) = \{\lambda \in \mathbb{C} : A - \lambda \notin \mathfrak{B}_+(X)\},$$

and the *Browder essential defect spectrum* of A is defined as

$$\sigma_{db}(A) = \{\lambda \in \mathbb{C} : A - \lambda \notin \mathfrak{B}_-(X)\}.$$

Obviously we have $\sigma_{eb}(A) = \{\lambda \in \mathbb{C} : A - \lambda \notin \mathcal{B}(X)\} = \sigma_{ab}(A) \cup \sigma_{db}(A)$.
It follows from Theorems 8.12.13, 8.12.14 and 8.12.16 that

$$\sigma_{ab}(A) = \bigcap_{TF=FT, F \in \mathcal{F}(X)} \sigma_a(A+F) = \bigcap_{TK=KT, K \in \mathcal{K}(X)} \sigma_a(A+K)$$

and

$$\sigma_{db}(A) = \bigcap_{TF=FT, F \in \mathcal{F}(X)} \sigma_d(A+F) = \bigcap_{TK=KT, K \in \mathcal{K}(X)} \sigma_d(A+K).$$

Corollary 8.14.11 *Let $A \in \mathcal{B}(X)$. Then we have*

(1) $\lambda \in \sigma_a(A) \backslash \sigma_{ab}(A)$ *if and only if λ is isolated in $\sigma_a(A)$, $0 < \alpha(A-\lambda) < \infty$, $\mathrm{asc}(A-\lambda) < \infty$ and $R(A - \lambda)$ is closed;*

(2) $\lambda \in \sigma_d(A) \backslash \sigma_{db}(A)$ *if and only if λ is isolated in $\sigma_d(A)$, $0 < \beta(A-\lambda) < \infty$, $\mathrm{dsc}(A-\lambda) < \infty$ and $R(A - \lambda)$ is closed.*

Proof.

(i) First we prove the statement in (1).
We assume $\lambda \in \sigma_a(A) \setminus \sigma_{ab}(A)$. Then $A - \lambda \in \Phi_+(X)$, $\mathrm{asc}(A - \lambda) < \infty$ and $A - \lambda$ is not bounded below. Since $R(A - \lambda)$ is closed, it follows that $A - \lambda$ is not injective. Thus $0 < \alpha(A - \lambda) < \infty$. It follows from Theorem 8.10.5 that λ_0 is an isolated point of $\sigma_a(A)$.
Conversely, if λ is isolated in $\sigma_a(A)$ such that $0 < \alpha(A - \lambda) < \infty$, $\mathrm{asc}(A - \lambda) < \infty$ and $R(A - \lambda)$ is closed, then clearly $\lambda \in \sigma_a(A) \setminus \sigma_{ab}(A)$.

(ii) The statement in (2) can be proved by duality using Theorem 8.10.6. $\qquad \square$

Theorem 8.14.12 *Let $A \in \mathcal{B}(X)$.*

(1) *If $\lambda \in \sigma_a(A)$ is isolated in $\sigma_a(A)$ and $\mathrm{asc}(A - \lambda) = \infty$, then $\lambda \in \sigma_{e\alpha}(A)$.*

(2) *If $\lambda \in \sigma_d(A)$ is isolated in $\sigma_d(A)$ and $\mathrm{dsc}(A - \lambda) = \infty$, then $\lambda \in \sigma_{e\beta}(A)$.*

Proof.

(i) First we show the implication in (1).
We assume that λ is isolated in $\sigma_a(A)$, $\mathrm{asc}(A - \lambda) = \infty$ and $\lambda \notin \sigma_{e\alpha}(A)$. Then we have $A - \lambda \in \Phi_+(X)$. It follows from Kato's decomposition theorem, Theorem 8.11.5, that the operator $A - \lambda$ is decomposable in the sense of Kato, that is, X is a direct sum of closed subspaces X_0 and X_1 which are $(A - \lambda)$-invariant. Consequently, they are A-invariant, and they have the following properties: $\dim X_1 < \infty$ and $A_1 - \lambda$ is nilpotent on X_1 where $A_1 = A|_{X_1} : X_1 \to X_1$; if $A_0 = A|_{X_0} : X_0 \to X_0$, then $R(A_0 - \lambda)$ is closed and $k(A_0 - \lambda) = 0$. Since $N(A - \lambda) = N(A_0 - \lambda) \oplus N(A_1 - \lambda)$, we conclude that $\alpha(A_0 - \lambda) \leq \alpha(A - \lambda) < \infty$, so $A_0 - \lambda \in \Phi_+(X_0)$. As $k(A_0 - \lambda) = 0$,

it follows from Theorem 8.10.2 that $\alpha(A_0 - \mu)$ is constant in a neighborhood of λ. Since λ is isolated in $\sigma_a(A)$, we get that $\alpha(A_0 - \mu) = 0$ for μ in a neighborhood of λ. Thus we have $\alpha(A_0 - \lambda) = 0$. Let F be the operator defined by $F = 0$ on X_0 and $F = I$ on X_1. Obviously we have $F \in \mathcal{F}(X)$ and $AF = FA$. From the nilpotency of $A_1 - \lambda$ we get $-1 \notin \sigma(A_1 - \lambda) = \{0\}$, so $A_1 + I - \lambda$ is a bijection which implies that the restriction $(A + F - \lambda)|_{X_1} : X_1 \to X_1$ is an injection. Since the restriction $(A + F - \lambda)|_{X_0} = A_0 - \lambda$ is also an injection, it follows that $A + F - \lambda$ is an injection, that is $\mathrm{asc}(A + F - \lambda) = 0$. From Theorem 8.4.6 (1), we obtain $A + F - \lambda \in \Phi_+(X)$. Thus we have $A + F - \lambda \in \mathcal{B}_+(X)$. It follows by (8.48) that $\mathrm{asc}(A - \lambda) < \infty$, which is not possible.

(ii)   The implication in (2) can be proved by duality.   □

Corollary 8.14.13 *Let* $A \in \mathcal{B}(X)$*. Then we have*

(1) $\sigma_{ab}(A) = \sigma_{e\alpha}(A) \cup \mathrm{acc}(\sigma_a(A)) = \sigma_{ea}(A) \cup \mathrm{acc}(\sigma_a(A))$.

(2) $\sigma_{ad}(A) = \sigma_{e\beta}(A) \cup \mathrm{acc}(\sigma_d(A)) = \sigma_{ed}(A) \cup \mathrm{acc}(\sigma_d(A))$.

Proof. The corollary follows from Corollary 8.14.11 and Theorem 8.14.12.   □

Corollary 8.14.14 *Let* $A \in \mathcal{B}(X)$*. Then we have*

(1) $\sigma_{ea}(A) \subset \sigma_{ab}(A) \subset \sigma_{eb}(A)$;

(2) $\partial\sigma_{eb}(A) \subset \partial\sigma_{ab}(A) \subset \partial\sigma_{ea}(A)$;

(3) $\hat{\sigma}_{ea}(A = \hat{\sigma}_{ab}(A) = \hat{\sigma}_{eb}(A)$;

(4) $\sigma_{ab}(A)$ $(\sigma_{eb}(A))$ *consists of* $\sigma_{ea}(A)$ $(\sigma_{ab}(A))$ *and possibly some holes in* $\sigma_{ea}(A)$ $(\sigma_{ab}(A))$.

(5) *If* $\sigma_{ea}(A)$ *is connected then* $\sigma_{ab}(A)$ *is connected; if* $\sigma_{ab}(A)$ *is connected, then* $\sigma_{eb}(A)$ *is connected.*

Proof. It is enough to prove (2).
Since $\partial\sigma_{eb}(A) \subset \partial\sigma_{em}(A)$ by (8.53) in Corollary 8.13.15 and $\partial\sigma_{em}(A) \subset \partial\sigma_{ea}(A)$ by Theorem 8.14.2 (2), we get $\partial\sigma_{eb}(A) \subset \sigma_{ea}(A) \subset \sigma_{ab}(A)$. It follows from Remark 8.13.14 that $\partial\sigma_{eb}(A) \subset \partial\sigma_{ab}(A)$. Let $\lambda_0 \in \partial\sigma_{ab}(A) \setminus \sigma_{eb}(A)$. Then $A - \lambda_0 \in \Phi_+(X)$ and $\mathrm{asc}(A - \lambda_0) = \infty$. Then it follows from the punctured neighborhood theorem, Theorem 8.10.2, that there exists $\varepsilon > 0$ such that if $0 < |\lambda - \lambda_0| < \varepsilon$ then $R(A - \lambda)$ is closed and $\alpha(A - \lambda)$ is constant. Since $\lambda_0 \in \partial\sigma_{ab}(A)$, we conclude that there exists μ with $|\mu - \lambda_0| < \varepsilon$ such that $A - \mu \in \mathcal{B}_+(X)$. We conclude from Theorem 8.10.5 that $\alpha(A - \lambda) = 0$ for every λ with $0 < |\lambda - \lambda_0| < \varepsilon$. So λ_0 is an isolated point of $\sigma_a(A)$ and it follows by Theorem 8.14.12 that $\lambda_0 \in \sigma_{ea}(A)$ which is not possible.   □

The following result can be proved similarly.

Corollary 8.14.15 *Let* $A \in \mathcal{B}(X)$*. Then we have*

(1) $\sigma_{ed}(A) \subset \sigma_{db}(A) \subset \sigma_{eb}(A)$;

(2) $\partial\sigma_{eb}(A) \subset \partial\sigma_{db}(A) \subset \partial\sigma_{ed}(A)$;

(3) $\hat{\sigma}_{ed}(A) = \hat{\sigma}_{db}(A) = \hat{\sigma}_{eb}(A)$;

(4) $\sigma_{db}(A)$ $(\sigma_{eb}(A))$ *consists of* $\sigma_{ed}(A)$ $(\sigma_{db}(A))$ *and possibly some holes in* $\sigma_{ed}(A)$ $(\sigma_{db}(A))$.

(5) *If* $\sigma_{ed}(A)$ *is connected then* $\sigma_{db}(A)$ *is connected; if* $\sigma_{db}(A)$ *is connected then* $\sigma_{eb}(A)$ *is connected.*

The following example shows that, in general, $\sigma_{ea}(A) \neq \sigma_{ab}(A)$ and $\sigma_{ed}(A) \neq \sigma_{db}(A)$.

Example 8.14.16 *Let X be a separable Hilbert space, V be the right shift operator on X and $N \in \mathcal{B}(X)$ be quasi–nilpotent. If $A = V \oplus V^* \oplus N$ then we have $\sigma_{ab}(A) = \sigma_{db}(A) = D$ and $\sigma_{ea}(A) = \sigma_{ed}(A) = \partial D \cup \{0\}$, where D is the closed unit ball.*

Proof. Salinas in [178] proved that $\sigma_{em}(A) = \partial D \cup \{0\}$ and $\sigma_{eb}(A) = D$ [178]. It follows from Theorems 8.14.2 and 8.14.3 that $\sigma_{ea}(A) = \sigma_{ed}(A) = \partial D \cup \{0\}$. Let $0 < |\lambda| < 1$. Since $\sigma_{em}(A) = \partial D \cup \{0\}$, we get $A - \lambda \in \Phi_0(X)$. If $\lambda \notin \sigma_{ab}(A)$ ($\lambda \notin \sigma_{db}(A)$), then $\mathrm{asc}(A - \lambda) < \infty$ ($\mathrm{dsc}(A - \lambda) < \infty$). It follows from (8.31) in Corollary 8.5.6 that $A - \lambda$ is a Browder operator, that is, $\lambda \notin \sigma_{eb}(A)$. This is a contradiction. $\square$

We obtain the following two corollaries from Theorems 8.12.13, 8.12.14 and 8.14.12, and their proofs.

Corollary 8.14.17 *Let $A \in \Phi_+^-(X)$. Then the following conditions are equivalent:*

(1) $A = B + F$ *where* $\alpha(B) = 0$, $F \in \mathcal{F}(X)$ *and* $BF = FB$.

(2) *There exists a finite rank projection P such that $AP = PA$ and $\alpha(A|_{N(P)}) = 0$.*

(3) *There exists $\varepsilon > 0$ such that $\alpha(A + \lambda) = 0$ holds for $0 < |\lambda| < \varepsilon$.*

(4) $\mathrm{asc}(A) < \infty$.

Corollary 8.14.18 *Let $A \in \Phi_-^+(X)$. Then the following statements are equivalent:*

(1) $A = B + F$ *where* $\beta(B) = 0$, $F \in \mathcal{F}(X)$ *and* $BF = FB$.

(2) *There exists a finite rank projection P such that $AP = PA$ and $\beta(A|_{N(P)}) = 0$.*

(3) *There exists $\varepsilon > 0$ such that $\beta(A + \lambda) = 0$ for $0 < |\lambda| < \varepsilon$.*

(4) $\mathrm{dsc}(A) < \infty$.

Theorem 8.14.19 *Let $A \in \mathcal{B}(X)$ and p be a polynomial. Then we have*

$$\sigma_{ab}(p(A)) = p(\sigma_{ab}(A)), \tag{8.59}$$

$$\sigma_{db}(p(A)) = p(\sigma_{db}(A)). \tag{8.60}$$

Proof.

(i) First we show (8.59).

Let $p(t) - \lambda = c(t - \lambda_1) \ldots (t - \lambda_n)$ where $c \neq 0$. Then we have $p(A) - \lambda = c(A - \lambda_1) \ldots (A - \lambda_n)$.

If $\lambda \in \sigma_{ab}(p(A))$, then $p(A) - \lambda \notin \mathfrak{B}_+(X)$ and it follows by Theorem 8.12.10 (1) that there exists $i \in \{1, 2, \ldots, n\}$ such that $A - \lambda_i \notin \mathfrak{B}_+(X)$, that is, $\lambda_i \in \sigma_{ab}(A)$. Since $\lambda = p(\lambda_i)$, it follows that $\lambda \in p(\sigma_{ab}(A))$.

Now we prove the opposite inclusion. We assume $\lambda \in p(\sigma_{ab}(A))$. Hence we have $\lambda_i \in \sigma_{ab}(A)$ for some $i \in \{1, 2, \ldots, n\}$, that is, $A - \lambda_i \notin \mathfrak{B}_+(X)$. We conclude from Theorem 8.12.10 (1) that $p(A) - \lambda \notin \mathfrak{B}_+(X)$, and so $\lambda \in \sigma_{ab}(p(A))$.

(ii) Finally we show (8.60).

Using (8.59) we get $\sigma_{db}(p(A)) = \sigma_{ab}(p(A)') = \sigma_{ab}(p(A')) = p(\sigma_{ab}(A')) = p(\sigma_{db}(A))$. $\square$

We close this chapter with the following remark.

Remark 8.14.20 *In the paper [Nikol], the author studied the perturbation of the point spectra of bounded linear operators by compact operators.*

Chapter 9

Sequence Spaces

Chapter 9 presents interesting applications of the results of the preceding chapters to modern topics in the large field of summability, in particular, in the characterizations of (infinite) matrix transformations and of compact linear operators between sequence spaces. The chapter contains an introduction to the theory of FK and BK spaces, which is a very powerful tool for the characterization of classes of matrix transformations because of the fundamental result that matrix maps between FK spaces are continuous. Additional topics are the studies and determinations of multiplier spaces and β–, γ–, functional and continuous duals, the properties of transposes of matrices and the complete known characterizations of matrix transformation between the classical sequence spaces of bounded, convergent and null sequences and of p–summable series for $1 \leq p < \infty$ in terms of conditions on the entries of the matrices. Finally, the focus is on the study of compact matrix operators, the representation of compact operators on the space of convergent sequences, an estimate of the Hausdorff measure of noncompactness and the characterization of such operators, and the characterizations of compact matrix operators between the classical sequence spaces by the application of the Hausdorff measure of noncompactness.

9.1   Introduction

This is the first of two chapters mainly dedicated to applications.

In this chapter, we apply our results of the preceding sections to modern topics in the large field of *summability*, and obtain many recent research results. We also study *measures of noncompactness* with applications to the characterizations of classes of compact operators between the *classical sequence spaces*.

In particular, Chapter 9 contains an introduction to the theory of FK and BK spaces in Section 9.2, which is a very powerful tool for the characterization of classes of matrix transformations because of the fundamental result that matrix maps between FK spaces are continuous, and some useful applications to the characterizations of matrix mappings from FK spaces into the spaces of bounded, convergent and null sequences are discussed in Section 9.3. Further topics are the studies and determinations of multiplier spaces and β–, γ–, functional and continuous duals, and relations between them in Sections 9.4 and 9.5; of properties of transposes of matrices in Section 9.6; and in Section 9.7, the complete known characterizations of matrix transformation between the classical sequence spaces of bounded, convergent and null sequences and of p–summable series for $1 \leq p < \infty$ in terms of conditions on the entries of the matrices. Finally, in Sections 9.8–9.10, the focus is on the study of compact matrix operators, the representation of compact operators on the space of convergent sequences, an estimate of the Hausdorff measure of noncompactness and the characterization of such operators, and the characterizations of compact matrix operators between the classical sequence spaces by the application of the Hausdorff measure of noncompactness.

In this chapter, all indices start from 0. We will always use the symbols $A, B, \ldots$ for infinite matrices of complex numbers; this is standard in the theory of sequence spaces. Therefore, from now on we are going to denote bounded linear operators between Banach sequence space by the symbol L to distinguish between matrices and operators.

Throughout this chapter, we use the following standard notations from summability theory.

Let $A = (a_{nk})_{n,k=0}^{\infty}$ be an infinite matrix of complex numbers, X and Y be subsets of ω, and $x = (x_k)_{k=0}^{\infty}$ be a sequence in ω. We write $A_n = (a_{nk})_{k=0}^{\infty}$ for the sequence in the nth row of the matrix A,

$$A_n x = \sum_{k=0}^{\infty} a_{nk} x_k \text{ and } Ax = (A_n x)_{n=0}^{\infty} \text{ for the } A \text{ transform of } x$$

(provided all the series converge),

$$X_A = \{x \in \omega : Ax \in X\} \text{ for the } matrix \ domain \ of \ A \ in \ X,$$

in particular, c_A for the matrix domain of A in c, the so–called *convergence domain of A*, and $(X, Y) = \{A : X \subset Y_A\}$ for the class of all matrix transformations from X into Y, that is, $A \in (X, Y)$ if and only if the series $A_n x = \sum_{k=0}^{\infty} a_{nk} x_k$ converge for all n and all $x \in X$, and $Ax = (A_n x)_{n=0}^{\infty} \in Y$ for all $x \in X$.

By classical sequence spaces we mean c, c_0 and ℓ_p for $1 \leq p \leq \infty$.

9.2   FK and BK spaces

The *classical summability theory* deals with a generalization of the concept of the *convergence of sequences* and *series of real* or *complex numbers*. One of the original ideas was to assign, in some way, a *limit to divergent sequences* or *series*. Classical methods of summability were also introduced for applications to problems in analysis such as the *analytic continuation of power series* and *improvement of the rate of convergence of numerical series*. These goals were achieved by considering a *transform* rather than the original sequence or series. This can be done in various different ways, the most important one being the use of transformations by infinite matrices, in particular, by the most popular *Hausdorff matrices* and their special cases, the *Cesàro matrices of order* $\alpha > -1$, the *Hölder* and *Euler matrices*, and *Nörlund matrices*.

The *Abel* and *Borel methods* are two popular methods not given by a matrix.

Rather than studying given matrix methods, we will focus on general properties of matrix transformations. We refer to [27, 48, 84, 133, 151, 218, 1] for further reading on summability methods.

A wide field of modern summability is the study of sequence spaces and matrix transformations, in particular, the characterizations of classes of matrix mappings that map a given sequence space X into another one Y. By characterization of such a class we mean necessary and sufficient conditions on the entries of an infinite matrix $A = (a_{nk})_{n,k=0}^{\infty}$ to map X into Y. The first such result was the famous *Toeplitz theorem* of 1911 [200] which characterizes the class of all so–called *regular* infinite matrices, that is, all matrix maps that map every convergent sequence into a convergent sequence and preserve the limits. The theorem states that an infinite matrix A is regular if and only if

$$\sup_{n} \sum_{k=0}^{\infty} |a_{nk}| < \infty \quad \text{(row norm condition)} \tag{i}$$

$$\lim_{n\to\infty} \sum_{k=0}^{\infty} a_{nk} = 1 \tag{ii}$$

$$\lim_{n\to\infty} a_{nk} = 0 \text{ for all } k. \tag{iii}$$

The original proof of Toeplitz's theorem was purely analytical and used the method of the *gliding hump*.

Modern summability uses functional analytical methods, in particular, the theory of FK and BK *spaces* which can be applied, among other things, for the characterizations of all known classes of matrix transformations between the classical sequence spaces with the exception of the classes (ℓ_∞, c) and (ℓ_∞, c_0) of matrix maps from ℓ_∞ to c and c_0, for which no functional analytic proof seems to be known. The class (ℓ_∞, c) of the so–called *coercive matrices* was first characterized by Schur in 1920 (for instance, [194, **10.**], and [194, **21. (20.1)**] for (ℓ_∞, c_0)).

For a comprehensive survey of results on matrix transformations we recommend [194]. The reader is also referred to [209, 27, 175, 108, 218, 133, 216, 215, 217, 13].

Now we give a short introduction to the theory of FK spaces which is the most powerful tool in the characterization of matrix transformations between sequence spaces. We remark that *K. Zeller* significantly influenced the development of the theory of FK spaces and its applications in summability.

The fundamental result of this section is Theorem 9.2.7 which states that matrix maps between FK spaces are continuous. Some of the results in this section can also be found in [134, 207, 209]. We also refer the reader to [108, 215].

We start with a slightly more general definition.

Definition 9.2.1 Let H be a linear space and a Hausdorff space. A subspace X of H is called an *FH space* if it is a Fréchet space X such that the topology of X is stronger than the restriction of the topology of H on X.

If $H = \omega$ with its topology given by the metric d_ω of Example 1.9.2, then an FH space is called an *FK space*.

A *BH space* or a *BK space* is an FH or FK space which is a Banach space.

Remark 9.2.2 *(a) If X is an FH space, then the inclusion map $\iota : X \to H$ with $\iota(x) = x$ for all $x \in X$ is continuous. Therefore X is continuously embedded in H.*

(b) Since convergence in (ω, d_ω) and coordinatewise convergence are equivalent by Theorem 3.3.5, convergence in an FK space implies coordinatewise convergence.

(c) The letters **F**, **H**, **K** *and* **B** *stand for* **F**réchet, **H**ausdorff, **K**oordinate, *the German word for coordinate, and* **B**anach.

(d) Some authors include local convexity in the definition of an FH space. Since most of the theory presented here can be developed without local convexity, we follow Wilansky [209] and do not include it in our definition.

Example 9.2.3 *(a) By Example 3.2.4, ω is an FK space with its natural metric d_ω of*

Example 1.9.2.

(b) The spaces ℓ_∞, c and c_0 and ℓ_p $(1 \le p < \infty)$ are Banach spaces by Examples 2.10.2 (d) and 3.2.4. If $\|\cdot\|$ denotes the natural norm on each of these spaces, then $|x_k| \le \|x\|$ for all k implies that the spaces are BK spaces.

The following results are fundamental.

Theorem 9.2.4 *Let X be a Fréchet space, Y be an FH space, $\mathcal{T}_Y$ and $\mathcal{T}_H|_Y$ denote the topologies on Y and of H on Y, respectively, and $f : X \to Y$ be a linear map. Then $f : X \to (Y, \mathcal{T}_H|_Y)$ is continuous, if and only if $f : X \to (Y, \mathcal{T}_Y)$ is continuous.*

Proof.

(i) First, we assume that $f : X \to (Y, \mathcal{T}_Y)$ is continuous. Since Y is an FH space, we have $\mathcal{T}_H|_Y \subset \mathcal{T}_Y$, and so $f : X \to (Y, \mathcal{T}_H|_Y)$ is continuous.

(ii) Conversely, we assume that $f : X \to (Y, \mathcal{T}_H|_Y)$ is continuous. Then it has closed graph by Theorem 2.9.3. Since Y is an FH space, we again have $\mathcal{T}_H|_Y \subset \mathcal{T}_Y$, and so $f : X \to (Y, \mathcal{T}_Y)$ has closed graph. Consequently $f : X \to (Y, \mathcal{T}_Y)$ is continuous by the closed graph theorem, Theorem 3.6.1. $\qquad\square$

We obtain the following as an immediate consequence of Theorem 9.2.4.

Corollary 9.2.5 *Let X be a Fréchet space, Y be an FK space, $f : X \to Y$ be linear, and the coordinates $P_n : X \to \mathbb{C}$ for $n = 0, 1, \dots$ be defined by $P_n(x) = x_n$ for all $x \in X$. If $P_n \circ f : X \to \mathbb{C}$ is continuous for every n, then $f : X \to Y$ is continuous.*

Proof. Since convergence and coordinatewise convergence are equivalent in ω by Theorem 3.3.5, the continuity of $P_n \circ f : X \to \mathbb{C}$ for all n implies the continuity of $f : X \to \omega$, and so $f : X \to Y$ is continuous by Theorem 9.2.4. $\qquad\square$

We recall that ϕ denotes the set of all finite sequences. Thus $x = (x_k)_{k=0}^\infty \in \phi$ if and only if there is an integer k such that $x_j = 0$ for all $j > k$.

Theorem 9.2.6 *Let $X \supset \phi$ be an FK space. If the series $\sum_{k=0}^\infty a_k x_k$ converge for all $x \in X$, then the linear functional f_a defined by $f_a(x) = \sum_{k=0}^\infty a_k x_k$ for all $x \in X$ is continuous.*

Proof. We define the functionals $f_a^{[n]}$ for all $n \in \mathbb{N}_0$ by $f_a^{[n]}(x) = \sum_{k=0}^n a_k x_k$ for all $x \in X$. Since X is an FK space and $f_a^{[n]} = \sum_{k=0}^n a_k P_k$ is a finite linear combination of the continuous coordinates P_k $(k = 0, 1, \dots)$, we have $f_a^{[n]} \in X'$ for all n. By hypothesis, the limits $f_a(x) = \lim_{n \to \infty} f_a^{[n]}(x)$ exist for all $x \in X$, hence $f_a \in X'$ by the Banach–Steinhaus theorem, Theorem 3.7.5. $\qquad\square$

The next result is one of the most important ones in the theory of matrix transformations between sequence spaces.

Theorem 9.2.7 *Matrix transformations between FK spaces are continuous.*

Proof. Let X and Y be FK spaces, $A \in (X, Y)$ and $L_A : X \to Y$ be defined by $L_A(x) = Ax$ for all $x \in X$. Since the maps $P_n \circ L_A : X \to \mathbb{C}$ are continuous for all n by Theorem 9.2.6, $L_A : X \to Y$ is continuous by Corollary 9.2.5. $\qquad\square$

It turns out that the FH topology of an FH space is unique, up to homeomorphisms.

Theorem 9.2.8 *Let $(X, \mathcal{T}_X)$ and $(Y, \mathcal{T}_Y)$ be FH spaces with $X \subset Y$, and $\mathcal{T}_Y|_X$ denote the topology of Y on X. Then*

$$\mathcal{T}_X \supset \mathcal{T}_Y|_X \, ; \tag{9.1}$$

$$\mathcal{T}_X = \mathcal{T}_Y|_X \quad \text{if and only if } X \text{ is a closed subspace of } Y. \tag{9.2}$$

In particular, the topology of an FH space is unique.

Proof.

(i) First we show the inclusion in (9.1).

Since X is an FH space, the inclusion map $\iota : (X, \mathcal{T}_X) \to (H, \mathcal{T}_H)$ is continuous by Remark 9.2.2 (a). Therefore $\iota : (X, \mathcal{T}_X) \to (Y, \mathcal{T}_Y)$ is continuous by Theorem 9.2.4. Thus the inclusion in (9.1) holds.

(ii) Now we show the identity in (9.2).

Let $\mathcal{T}$ and $\mathcal{T}'$ be FH topologies for an FH space. Then it follows by what we have just shown in Part (i) that $\mathcal{T} \subset \mathcal{T}' \subset \mathcal{T}$.

(α) If X is closed in Y, then X becomes an FH space with $\mathcal{T}_Y|_X$. It follows from the uniqueness that $\mathcal{T}_X = \mathcal{T}_Y|_X$.

(β) Conversely, if $\mathcal{T}_X = \mathcal{T}_Y|_X$, then X is a complete subspace of Y, and so closed in Y.

$\square$

Theorem 9.2.9 *Let X, Y and Z be FH spaces with $X \subset Y \subset Z$. If X is closed in Z, then X is closed in Y.*

Proof. Since X is closed in $(Y, \mathcal{T}_Z|_Y)$, it is closed in $(Y, \mathcal{T}_Y)$ by Theorem 9.2.8. $\square$

Let Y be a topological space, and $E \subset Y$. Then we write $\mathrm{cl}_Y(E)$ for the closure of E in Y.

Theorem 9.2.10 *Let X and Y be FH spaces with $X \subset Y$, and E be a subset of X. Then we have*

$$\mathrm{cl}_Y(E) = \mathrm{cl}_Y(\mathrm{cl}_X(E)), \quad \text{in particular } \mathrm{cl}_X(E) \subset \mathrm{cl}_Y(E).$$

Proof. Since $\mathcal{T}_Y|_X \subset \mathcal{T}_X$ by Theorem 9.2.8, it follows that $\mathrm{cl}_X(E) \subset \mathrm{cl}_Y(E)$. This implies

$$\mathrm{cl}_Y(\mathrm{cl}_X(E)) \subset \mathrm{cl}_Y(\mathrm{cl}_Y(E)) = \mathrm{cl}_Y(E).$$

Conversely, $E \subset \mathrm{cl}_X(E)$ implies $\mathrm{cl}_Y(E) \subset \mathrm{cl}_Y(\mathrm{cl}_X(E))$. $\square$

Example 9.2.11 *(a) Since c_0 and c are closed in ℓ_∞, their BK topologies are the same; since ℓ_1 is not closed in ℓ_∞, its BK topology is strictly stronger than that of ℓ_∞ on ℓ_1 by Theorem 9.2.8.*
(b) If c is not closed in an FK space X, then X must contain unbounded sequences by Theorem 9.2.9, for if X did not contain an unbounded sequence, then $X \subset \ell_\infty$, and since c is closed in ℓ_∞ by Part (a), it would be closed in X, a contradiction.

Now we introduce two important properties of FK and BK spaces.

Definition 9.2.12 *Let $X \supset \phi$ be an FK space. Then X is said to have*
(a) AD if $\mathrm{cl}_X(\phi) = X$;
(b) AK if each sequence $x = (x_k)_{k=0}^\infty \in X$ has a unique representation $x = \sum_{k=0}^\infty x_k e^{(k)}$, that is, every sequence x is the limit of its m–sections $x^{[m]} = \sum_{k=0}^m x_k e^{(k)}$.

Remark 9.2.13 *(a) The letters* **AD** *and* **AK** *stand for* **a***bschnitts***d***icht and* **A***bschnitts***k***onvergenz, the German words for sectionally dense, and sectional convergence, respectively.*
(b) The AK property of a BK space X means that the sequence $(e^{(n)})_{n=0}^{\infty}$ is a Schauder basis of X and the coefficients of the expansion of every sequence $x \in X$ with respect to this basis are equal to the terms of the sequence.

Example 9.2.14 *(a) Every FK space with AK obviously has AD.*
(b) An example of an FK space with AD which does not have AK can be found in [209, Example 5.2.14].
(c) The spaces ω, c_0 and ℓ_p $(1 \leq p < \infty)$ have AK by Example 3.4.3 (c).
(d) The space c does not have AK; every sequence $x = (x_k)_{k=0}^{\infty} \in c$ has a unique representation $x = \ell e + \sum_{k=0}^{\infty}(x_k - \xi)e^{(k)}$ where $\xi = \lim_{k \to \infty} x_k$ by Example 3.4.3 (c).

We close this section with two more examples.

Example 9.2.15 *Let $H = \mathcal{F} = \{f : [0,1] \to \mathbb{R}\}$ be the space of all real–valued functions defined on the unit interval $[0,1]$, and, for every $t \in [0,1]$, let $\hat{t} : \mathcal{F} \to \mathbb{R}$ be the function with $\hat{t}(f) = f(t)$. We assume that $\mathcal{F}$ has the weak topology by $\Phi = \{\hat{t} : t \in [0,1]\}$. Then the space $C[0,1]$ of all continuous real–valued functions on $[0,1]$ is a BH space with $\|f\| = \sup_{t \in [0,1]} |f(t)|$.*

Proof. Let $(f_k)_{k=0}^{\infty}$ be a sequence in $C[0,1]$ with $f_k \to 0$ $(k \to \infty)$. Then it follows that $\hat{t}(f_k) = f_k(t) \to 0$ $(k \to \infty)$ for all $\hat{t} \in \Phi$, that is $f_k \to 0$ $(k \to \infty)$ in $\mathcal{F}$. □

The class of FK spaces is fairly large. In fact, it is not quite simple to find a Fréchet sequence space which is not an FK space.

Example 9.2.16 (A Banach sequence space which is not a BK space)
We consider the spaces $(c_0, \|\cdot\|_{\infty})$ and $(\ell_2, \|\cdot\|_2)$. Since they have the same algebraic dimension, there is an isomorphism $f : c_0 \to \ell_2$. We define a second norm $\|\cdot\|$ on c_0 by $\|x\| = \|f(x)\|_2$ for all $x \in c_0$. Then $(c_0, \|\cdot\|)$ becomes a Banach space. But c_0 and ℓ_2 are not linearly homeomorphic, since ℓ_2 is reflexive, and c_0 is not. Therefore the two norms on c_0 are incomparable. By Example 9.2.3 and Theorem 9.2.8, $(c_0, \|\cdot\|)$ is a Banach sequence space which is not a BK space.

9.3   Matrix transformations into ℓ_{∞}, c, c_0, and ℓ_1

Now we apply the results of the previous section to characterize the classes matrix transformations from an arbitrary FK space into the spaces ℓ_{∞}, c, c_0 and ℓ_1.

If $X \subset \omega$ is a linear metric space and $a \in \omega$, then we write for $\delta > 0$

$$\|a\|_{\delta}^* = \|a\|_{X;\delta}^* = \sup_{x \in \overline{B}_{\delta}(0)} \left| \sum_{k=0}^{\infty} a_k x_k \right|,$$

provided the expression on the right-hand side exists and is finite, which is the case whenever the series $\sum_{k=0}^{\infty} a_k x_k$ converge for all $x \in X$ (Theorem 9.2.6); if X is a normed space, then we write $\|\cdot\|^* = \|\cdot\|_{X;1}^* = \|\cdot\|_1^*$, for short.

First, we characterize the classes (X, ℓ_{∞}) and (X, ℓ_1). We need the following well–known result for the characterization of the class (X, ℓ_1).

Lemma 9.3.1 (Peyerimhoff) ([152]) *Let* $a_0, a_1, \ldots, a_n \in \mathbb{C}$. *Then the following inequality holds*

$$\sum_{k=0}^{n} |a_k| \leq 4 \cdot \max_{N \subset \{0,\ldots,n\}} \left| \sum_{k \in N} a_k \right|. \tag{9.3}$$

Proof. The proof is easy and left to the reader. $\qquad\square$

Throughout, we write $\sup_N$ for the supremum taken over all finite subsets of N of $\mathbb{N}_0$.

Theorem 9.3.2 *Let X be an FK space. Then we have*

(a) $A \in (X, \ell_\infty)$ *if and only if*

$$\|A\|_{(X,\ell_\infty);\delta} = \sup_n \|A_n\|_{X;\delta}^* < \infty \text{ for some } \delta > 0; \tag{9.4}$$

(b) $A \in (X, \ell_1)$ *if and only if*

$$\|A\|_{(X,\ell_1);\delta} = \sup_N \left\| \left(\sum_{n \in N} a_{nk} \right)_{k=0}^{\infty} \right\|_{X;\delta}^* < \infty \text{ for some } \delta > 0. \tag{9.5}$$

Proof. We write $\overline{B}_\delta = \overline{B}_\delta(0)$, for short.

(a) First, we assume that (9.4) is satisfied. Then the series $A_n x$ converge for all $x \in \overline{B}_\delta$ and for all n, and $Ax \in \ell_\infty$ for all $x \in \overline{B}_\delta$. Since the set $\overline{B}_\delta$ is absorbing by Remark 3.3.4 (j), we conclude that the series $A_n x$ converge for all n and all $x \in X$, and $Ax \in \ell_\infty$ for all x.
Conversely, we assume $A \in (X, \ell_\infty)$. Then the map $L_A : X \to \ell_\infty$ defined by

$$L_A(x) = Ax \text{ for all } x \in X \tag{9.6}$$

is continuous by Theorem 9.2.7. Hence there exists a neighborhood N of 0 in X and a real $\delta > 0$ such that $\overline{B}_\delta \subset N$ and $\|L_A(x)\|_\infty < 1$ for all $x \in N$. This implies (9.4). Thus we have proved Part (a) of the theorem.

(b) We write $\|A\|_\delta = \|A\|_{(X,\ell_1);\delta}$, for short, and

$$b^{(N)} = \sum_{n \in N} A_n \text{ for any subset } N \text{ of } \mathbb{N}_0,$$

that is, $b^{(N)} = (b_k^{(N)})_{k=0}^{\infty}$ is the sequence with $b_k^{(N)} = \sum_{n \in N} a_{nk}$ for all k.
First, we assume that (9.5) is satisfied. In particular, this means that the series $\sum_{k=0}^{\infty} b_k^{(N)} x_k$ converges for each finite subset N of $\mathbb{N}_0$ and for each $x \in \overline{B}_\delta$, hence the series $A_n x = \sum_{k=0}^{\infty} b_k^{(\{n\})} x_k$ converges for each $n \in \mathbb{N}_0$ and for each $x \in \overline{B}_\delta$. Since the set $\overline{B}_\delta$ is absorbing by Remark 3.3.4 (j), the series $A_n x$ converge for all $n \in \mathbb{N}_0$ and all $x \in X$.
Let $m \in \mathbb{N}_0$ be given. Then we have by Lemma 9.3.1

$$\sum_{n=0}^{m} |A_n x| \leq 4 \cdot \max_{N_m \subset \{0,\cdots,m\}} \left| \sum_{n \in N_m} \sum_{k=0}^{\infty} a_{nk} x_k \right|$$

$$= 4 \cdot \max_{N_m \subset \{0,\cdots,m\}} \left| \sum_{k=0}^{\infty} \left(\sum_{n \in N_m} a_{nk} \right) x_k \right|$$

$$= 4 \cdot \max_{N_m \subset \{0,\cdots,m\}} \left| \sum_{k=0}^{\infty} b_k^{(N_m)} x_k \right| \leq 4 \cdot \max_{N_m \subset \{0,\cdots,m\}} \left\| b^{(N_m)} \right\|_{X;\delta}^{*}$$

$$\leq 4 \cdot \|A\|_{\delta} < \infty,$$

for all $x \in \overline{B}_{\delta}$. Since $m \in \mathbb{N}_0$ was arbitrary, this implies

$$\sum_{n=0}^{\infty} |A_n x| \leq 4 \cdot \|A\|_{\delta} < \infty \text{ for all } x \in \overline{B}_{\delta}. \tag{9.7}$$

Also, since the set $\overline{B}_{\delta}$ is absorbing, given $x \in X$ there is $\lambda > 0$ such that $\tilde{x} = \lambda x \in \overline{B}_{\delta}$ and it follows from (9.7) that

$$\sum_{n=0}^{\infty} |A_n x| = \sum_{n=0}^{\infty} \left| \frac{1}{\lambda} \cdot A_n \tilde{x} \right| \leq \frac{4}{\lambda} \cdot \|A\|_{\delta} < \infty,$$

hence $Ax \in \ell_1$. Thus we have shown the sufficiency of (9.5) for $A \in (X, \ell_1)$.

Now we prove the necessity of the condition in (9.5).

We assume $A \in (X, \ell_1)$. Then the series $A_n x$ converge for all $n \in \mathbb{N}_0$ and all $x \in X$, hence the series

$$\sum_{k=0}^{\infty} b_k^{(N)} x_k = \sum_{n \in N} A_n x$$

converge for all finite subsets N of $\mathbb{N}_0$ and all $x \in X$. Since X is an FK space, $\|b^{(N)}\|_{X,\delta}^{*}$ is defined and finite for all finite subsets N of $\mathbb{N}_0$ by Theorem 9.2.6. Also the map $L_A : X \to \ell_1$ defined as in (9.6) is continuous by Theorem 9.2.7. Hence there exist a neighborhood U of 0 in X and a positive real number δ such that $\overline{B}_{\delta} \subset U$ and $\|L_A(x)\|_1 < 1$ for all $x \in \overline{B}_{\delta}$. This implies

$$\left| \sum_{n \in N} A_n x \right| = \left| \sum_{n \in N} \left(\sum_{k=0}^{\infty} a_{nk} x_k \right) \right| = \left| \sum_{k=0}^{\infty} b_k^{(N)} x_k \right|$$

$$\leq \sum_{n=0}^{\infty} \left| \sum_{k=0}^{\infty} a_{nk} x_k \right| = \sum_{n=0}^{\infty} \|A_n x\|_1 = \|L_A(x)\|_1 < 1, \tag{9.8}$$

for all finite subsets N of $\mathbb{N}_0$. We obtain from (9.8)

$$\|A\|_{\delta} = \sup_N \left\| b^{(N)} \right\|_{X;\delta}^{*} \leq 1.$$

Thus we have shown that $A \in (X, \ell_1)$ implies (9.5).

This completes the proof of Part (b) of the theorem.   □

We obtain the following result for BK spaces.

Theorem 9.3.3 *Let X and Y be BK spaces.*

(a) *Then $(X, Y) \subset \mathcal{B}(X, Y)$, that is, every $A \in (X, Y)$ defines an operator $L_A \in \mathcal{B}(X, Y)$ by (9.6).*

(b) *If X has AK then $\mathcal{B}(X, Y) \subset (X, Y)$, that is, for each $L \in \mathcal{B}(X, Y)$ there exists $A \in (X, Y)$ such that*

$$Ax = L(x) \text{ for all } x \in X \tag{9.9}$$

holds. In this case we say that the operator L is given by the matrix A.

(c) We have $A \in (X, \ell_\infty)$ if and only if

$$\|A\|_{(X,\ell_\infty)} = \sup_n \|A_n\|^* < \infty; \tag{9.10}$$

if $A \in (X, Y)$, where $Y = c_0$, c or ℓ_∞, then

$$\|L_A\| = \|A\|_{(X,\ell_\infty)}. \tag{9.11}$$

(d) We have $A \in (X, \ell_1)$ if and only if

$$\|A\|_{(X,\ell_1)} = \sup_N \left\|\left(\sum_{n \in N} a_{nk}\right)_{k=0}^\infty\right\|^* < \infty \text{ for some } \delta > 0; \tag{9.12}$$

if $A \in (X, \ell_1)$, then

$$\|A\|_{(X,\ell_1)} \leq \|L_A\| \leq 4 \cdot \|A\|_{(X,\ell_1)}. \tag{9.13}$$

Proof.

(a) This is Theorem 9.2.7.

(b) Let $L \in \mathcal{B}(X, Y)$ be given. We write $L_n = P_n \circ L$ for all n, and put $a_{nk} = L_n(e^{(k)})$ for all n and k. Let $x = (x_k)_{k=0}^\infty \in X$ be given. Since X has AK, we have $x = \sum_{k=0}^\infty x_k e^{(k)}$, and since Y is a BK space, it follows that $L_n \in X^*$ for all n. Hence we obtain $L_n(x) = \sum_{k=0}^\infty x_k L_n(e^{(k)}) = \sum_{k=0}^\infty a_{nk} x_k = A_n x$ for all n, and so $L(x) = Ax$.

(c) This follows from Theorem 9.3.2 (a) and the definition of $\|A\|_{(X,\ell_\infty)}$.

(d) This follows from Theorem 9.3.2 (b) and the definition of $\|A\|_{(X,\ell_1)}$. $\qquad\square$

As a simple application of the previous results, we characterize the class $\mathcal{B}(\ell_1)$ and obtain the operator norm of any $L \in \mathcal{B}(\ell_1)$.

Example 9.3.4 We have $L \in \mathcal{B}(\ell_1)$ if and only if

$$\|L\| = \sup_k \sum_{n=0}^\infty |a_{nk}| < \infty, \tag{9.14}$$

where $A = (a_{nk})_{n,k=0}^\infty$ is the matrix that represents L as in (9.9).

Proof. Since ℓ_1 is a BK space with AK by Example 9.2.14 (c), it follows by Theorem 9.3.3 (a) and (b) that $L \in \mathcal{B}(\ell_1)$ if and only if $A \in (\ell_1, \ell_1)$, where $L(x) = Ax$ for all $x \in \ell_1$. Let $L \in \mathcal{B}(\ell_1)$ and $k \in \mathbb{N}_0$ be given. Then we have

$$\|L(e^{(k)})\|_1 = \sum_{n=0}^\infty |A_n e^{(k)}| = \sum_{n=0}^\infty |a_{nk}| \leq \|L\| \cdot \|e^{(k)}\|_1 = \|L\|.$$

Since $k \in \mathbb{N}_0$ was arbitrary, we have

$$\|L\| \geq \sup_k \sum_{n=0}^\infty |a_{nk}| < \infty. \tag{9.15}$$

Conversely, we assume that $\sup_k \sum_{n=0}^\infty |a_{nk}| < \infty$. Then we obtain for all $x \in \ell_1$

$$\|L(x)\|_1 = \sum_{n=0}^\infty |A_n x| \leq \sum_{n=0}^\infty \sum_{k=0}^\infty |a_{nk} x_k| = \sum_{k=0}^\infty \left(|x_k| \sum_{n=0}^\infty |a_{nk}|\right)$$

$$\leq \left(\sup_k \sum_{n=0}^{\infty} |a_{nk}| \right) \cdot \|x\|_1 < \infty,$$

hence $L \in \mathcal{B}(\ell_1)$ and

$$\|L\| \leq \sup_k \sum_{n=0}^{\infty} |a_{nk}|. \tag{9.16}$$

Finally, (9.15) and (9.16) yield (9.14). □

The following results can be applied to obtain, for instance, the characterizations of the classes (c_0, c_0) and (c_0, c) from that of (c_0, ℓ_∞) and then the characterizations of (c, c_0) and (c, c).

Theorem 9.3.5 *Let X be an FK space with AD, and Y and Y_1 be FK spaces, where Y_1 is a closed subspace of Y. Then $A \in (X, Y_1)$ if and only if $A \in (X, Y)$ and $Ae^{(k)} \in Y_1$ for all k.*

Proof. First, we assume $A \in (X, Y_1)$. Then $Y_1 \subset Y$ implies $A \in (X, Y)$, and $e^{(k)} \in X$ for all k implies $Ae^{(k)} \in Y_1$ for all k.
Conversely, we assume $A \in (X, Y)$ and $Ae^{(k)} \in Y_1$ for all k. We define the map $L_A : X \to Y$ by (9.6). Then $Ae^{(k)} \in Y_1$ implies $L_A(\phi) \subset Y_1$. By Theorem 9.2.7, L_A is continuous, hence $L_A(\mathrm{cl}_X(\phi)) \subset \mathrm{cl}_Y(L_A(\phi))$. Since Y_1 is closed in Y, and ϕ is dense in the AD space X, we have $L_A(X) = L_A(\mathrm{cl}_X(\phi)) \subset \mathrm{cl}_Y(L_A(\phi)) \subset \mathrm{cl}_Y(Y_1) = \mathrm{cl}_{Y_1}(Y_1) = Y_1$ by Theorem 9.2.8. □

Theorem 9.3.6 *Let X be an FK space, $X_1 = X \oplus e = \{x_1 = x + \lambda e : x \in X, \lambda \in \mathbb{C}\}$, and Y be a linear subspace of ω. Then $A \in (X_1, Y)$ if and only if $A \in (X, Y)$ and $Ae \in Y$.*

Proof. First, we assume $A \in (X_1, Y)$. Then $X \subset X_1$ implies $A \in (X, Y)$, and $e \in X_1$ implies $Ae \in Y$.
Conversely, we assume $A \in (X, Y)$ and $Ae \in Y$. Let $x_1 \in X_1$ be given. Then there are $x \in X$ and $\lambda \in \mathbb{C}$ such that $x_1 = x + \lambda e$, and it follows that $Ax_1 = A(x + \lambda e) = Ax + \lambda Ae \in Y$. □

9.4   Dual spaces

The so–called β–*duals* of sequence spaces are of greater interest in summability than the continuous duals. They naturally arise in the characterization of matrix transformations between sequence spaces in connection with the convergence of the series $A_n x$ for all sequences x in a given sequence space X. Their study is also needed to establish necessary and sufficient conditions on the entries of an infinite matrix to map a given sequence space X into a given sequence space Y.

The β–duals are special cases of *multiplier spaces*.

We write cs and bs for the sets of all *convergent* and *bounded series*.

Definition 9.4.1 Let X and Y be subsets of ω. The set

$$M(X, Y) = \{a \in \omega : a \cdot x = (a_k x_k)_{k=0}^{\infty} \in Y \text{ for all } x \in X\}$$

is called the *multiplier space of X in Y*.
Special cases are $X^\alpha = M(X, \ell_1)$, $X^\beta = M(X, cs)$ and $X^\gamma = M(X, bs)$, the α–, β– and γ–*duals of X*.

The following simple results are fundamental, and will frequently be used in the sequel.

Proposition 9.4.2 *Let X, $\tilde{X}$, Y and $\tilde{Y}$ be subsets of ω, and $\{X_\delta : \delta \in I\}$ be a collection of subsets X_δ of ω, where I is an indexing set. Then we have*

> (i) $Y \subset \tilde{Y}$ implies $M(X,Y) \subset M(X,\tilde{Y})$;

> (ii) $X \subset \tilde{X}$ implies $M(\tilde{X},Y) \subset M(X,Y)$;

> (iii) $X \subset M(M(X,Y), Y)$;

> (iv) $M(X,Y) = M\left(M(M(X,Y), Y), Y\right)$;

> (v) $M\left(\bigcup_{\delta \in I} X_\delta, Y\right) = \bigcap_{\delta \in I} M(X_\delta, Y)$.

Proof.

(i), (ii) Parts (i) and (ii) are trivial.

(iii) If $x \in X$, then $a \cdot x \in Y$ for all $a \in M(X,Y)$, and so $x \in M(M(X,Y),Y)$.

(iv) We replace X by $M(X,Y)$ in Part (iii) to obtain

$$M(X,Y) \subset M(M(M(X,Y),Y),Y). \qquad (9.17)$$

Conversely we have $X \subset M(M(X,Y))$ by Part (iii), and so by Part (ii)

$$M(M(M(X,Y),Y),Y) \subset M(X,Y). \qquad (9.18)$$

The identity in Part (iv) now follows from the inclusions in (9.17) and (9.18).

(v) First $X_\delta \subset \bigcup_{\delta \in I} X_\delta$ for all $\delta \in I$ implies $M(\bigcup_{\delta \in I} X_\delta, Y) \subset \bigcap_{\delta \in I} M(X_\delta, Y)$ by Part (ii).
Conversely, if $a \in \bigcap_{\delta \in I} M(X_\delta, Y)$, then $a \in M(X_\delta, Y)$ for all $\delta \in I$, and so we have $a \cdot x \in Y$ for all $x \in X_\delta$ and all $\delta \in I$. This implies $a \cdot x \in Y$ for all $x \in \bigcup_{\delta \in I} X_\delta$, hence $a \in M(\bigcup_{\delta \in I} X_\delta, Y)$.   □

We obtain the following result as an immediate consequence of Proposition 9.4.2.

Corollary 9.4.3 *Let X and $\tilde{X}$ be subsets of ω, $\{X_\delta : \delta \in I\}$ be a collection of subsets of ω, where I is an indexing set, and † denote any of the symbols α, β or γ. Then we have*

> (i) $X^\alpha \subset X^\beta \subset X^\gamma$;   (ii) $X \subset \tilde{X}$ implies $\tilde{X}^\dagger \subset X^\dagger$;
> (iii) $X \subset X^{\dagger\dagger} = (X^\dagger)^\dagger$;   (iv) $X^\dagger = X^{\dagger\dagger\dagger}$;

$$(v) \ \left(\bigcup_{\delta \in I} X_\delta\right)^\dagger = \bigcap_{\delta \in I} X_\delta^\dagger.$$

Example **9.4.4** *We have*

> (i) $M(c_0, c) = \ell_\infty$; (ii) $M(c, c) = c$; (iii) $M(\ell_\infty, c) = c_0$.

Proof.

(i) If $a \in \ell_\infty$, then $a \cdot x \in c$ for all $x \in c_0$, and so $\ell_\infty \subset M(c_0, c)$.
     Conversely, we assume $a \notin \ell_\infty$. Then there is a subsequence $(a_{k(j)})_{j=0}^\infty$ of the sequence a such that $|a_{k(j)}| > j + 1$ for all $j = 0, 1, \ldots$. We define the sequence x by

$$x_k = \begin{cases} \dfrac{(-1)^j}{a_{k(j)}} & (k = k(j)) \\ 0 & (k \neq k(j)) \end{cases} \quad (j = 0, 1, \ldots). \tag{9.19}$$

     Then we have $x \in c_0$ and $a_{k(j)} x_{k(j)} = (-1)^j$ for all $j = 0, 1, \ldots$, hence $a \cdot x \notin c$. This shows that $M(c_0, c) \subset \ell_\infty$.

(ii) If $a \in c$, then $a \cdot x \in c$ for all $x \in c$, and so $c \subset M(c, c)$.
      Conversely, if $a \in M(c, c)$, then $a \cdot x \in c$ for all $x \in c$, in particular, for $x = e \in c$, $a \cdot e = a \in c$, and so $M(c, c) \subset c$.

(iii) If $a \in c_0$, then $a \cdot x \in c$ for all $x \in \ell_\infty$, and so $c_0 \subset M(\ell_\infty, c_0)$.
       Conversely, we assume $a \notin c_0$. Then there are a real $b > 0$ and a subsequence $(a_{k(j)})_{j=0}^\infty$ of the sequence a such that $|a_{k(j)}| > b$ for all $j = 0, 1, \ldots$. We define the sequence x as in (9.19). Then we have $x \in \ell_\infty$ and $a_{k(j)} x_{k(j)} = (-1)^j$ for all $j = 0, 1, \ldots$, hence $a \notin M(\ell_\infty, c)$. This shows that $M(\ell_\infty, c) = c_0$. $\square$

Example 9.4.5 *Let* $\dagger$ *denote any of the symbols* α, β *or* γ. *Then we have* $\omega^\dagger = \phi$, $\phi^\dagger = \omega$, $c_0^\dagger = c^\dagger = \ell_\infty^\dagger = \ell_1$, $\ell_1^\dagger = \ell_\infty$, *and* $\ell_p^\dagger = \ell_q$ $(1 < p < \infty; q = p/(p-1))$.

It turns out that the multiplier spaces of BK spaces are again BK spaces. This result does not extend to FK spaces, in general.

Theorem 9.4.6 ([209, Theorem 4.3.15]) *Let* $X \supset \phi$ *and* $Y \supset \phi$ *be BK spaces with the norms* $\|\cdot\|_X$ *and* $\|\cdot\|_Y$. *Then* $Z = M(X, Y)$ *is a BK space with respect to the norm* $\|\cdot\|_Z$, *where*

$$\|z\|_Z = \sup_{\|x\|_X = 1} \|z \cdot x\|_Y \quad \text{for all } z \in Z.$$

Proof. Every $z \in Z$ defines a diagonal matrix map $\tilde{Z} \in (X, Y)$ by

$$\tilde{Z}x = z \cdot x = (z_k x_k)_{k=0}^\infty \quad \text{for all } x \in X.$$

It follows by Theorem 9.3.3 (a) that $L_{\tilde{Z}} \in \mathcal{B}(X, Y)$ with $L_{\tilde{Z}} = \tilde{Z}x$ for all $x \in X$.

(i) We show that the coordinates are continuous.
     We fix n and put $\lambda = 1/\|e^{(n)}\|_X$ and $\mu = \|e^{(n)}\|_Y$. Then we have $\|\lambda e^{(n)}\|_X = 1$ and

$$\lambda \mu |z_n| = \lambda \|z_n e^{(n)}\|_Y = \lambda \|e^{(n)} \cdot z\|_Y = \|(\lambda e^{(n)}) \cdot z\|_Y \leq \|z\|_Z,$$

     hence $|z_n| \leq (\lambda \mu)^{-1} \cdot \|z\|_Z$ for all n.
     This shows that the coordinates are continuous.

(ii) Now we show that Z is a closed subspace of the Banach space $\mathcal{B}(X, Y)$.
      Let $(\tilde{Z}^{(m)})_{m=0}^\infty$ be a sequence in (X, Y) such that

$$L_{\tilde{Z}^{(m)}} \to L_T \in \mathcal{B}(X, Y) \quad (m \to \infty).$$

      Then we obtain for each fixed $x \in X$

$$\tilde{Z}^{(m)}x = L_{\tilde{Z}^{(m)}}(x) \to L_T(x) \in Y \quad (m \to \infty).$$

Since X is a BK space, this implies

$$\tilde{Z}_k^{(m)} x = z_k^{(m)} x_k \to (L_T(x))_k \ (m \to \infty) \text{ for each fixed } k.$$

If we choose $x = e^{(k)}$, then we obtain

$$\tilde{Z}_k^{(m)} e^{(k)} = z_k^{(m)} \to t_k = \left(L_T(e^{(k)}) \right)_k \ (m \to \infty).$$

Therefore we have for all $x \in X$ and each fixed k

$$\left(\tilde{Z}^{(m)} x \right)_k \to t_k x_k \text{ and } \left(\tilde{Z}^{(m)} x \right)_k \to (L_T(x))_k \ (m \to \infty),$$

hence $L_T(x) = t \cdot x = \tilde{T} x$ for all $x \in X$. This shows that $L_T = \tilde{T}$ and so Z is closed. This completes Part (ii) of the proof. $\qquad \square$

It follows from Theorem 9.4.6 that the α–, β– and γ–duals of BK spaces are BK spaces.

Corollary 9.4.7 *The α–, β– and γ–duals of a BK space X are BK spaces with*

$$\|a\|_\alpha = \sup_{x \in S_X} \|ax\|_1 = \sup_{x \in S_X} \left(\sum_{k=0}^\infty |a_k x_k| \right) \text{ for all } a \in X^\alpha,$$

and

$$\|a\|_\beta = \sup_{x \in S_X} \|a\|_{bs} = \sup_{x \in S_X} \left(\sup_n \left| \sum_{k=0}^n a_k x_k \right| \right) \text{ for all } a \in X^\beta, X^\gamma.$$

Furthermore, X^β is a closed subspace of X^γ.

Proof. The first part is an immediate consequence of Theorem 9.4.6.
Since the BK norms on X^β and X^γ are the same and $X^\beta \subset X^\gamma$ by Corollary 9.4.3 (i), the second part follows from Theorem 9.2.8 $\qquad \square$.

Theorem 9.4.6 fails to hold for FK spaces, in general.

Example **9.4.8** *The space ω is an FK space, and $\omega^\alpha = \omega^\beta = \omega^\gamma = \phi$, but ϕ has no Fréchet metric.*

9.5 Relations between different kinds of duals

In this section, we establish some useful relations between different kinds of duals.
We will frequently use the following generalization of the Hahn–Banach theorem.

Theorem 9.5.1 (Hahn–Banach theorem) ([209, 3.0.1] or [208, Theorem 7.2.11]) *Let X be a subspace of a linear topological space Y and f be a linear functional on X which is continuous in the relative topology of Y. Then f can be extended to a continuous linear functional on Y.*

Proof. The proof is left to the reader. $\qquad \square$

There is a very close relation between the β–dual and the continuous dual of an FK space.

Theorem 9.5.2 *Let $X \supset \phi$ be an FK space. Then $X^\beta \subset X'$; this means that there is a linear one-to-one map $T : X^\beta \to X'$. If X has AK then T is onto.*

Proof. We define the map T by $Ta = f_a$ for all $a \in X^\beta$, where f_a is the functional with $f_a(x) = \sum_{k=0}^\infty a_k x_k$ for all $x \in X$, and observe that $Ta = f_a \in X'$ for all $a \in X^\beta$ by Theorem 9.2.6. Obviously T is linear. Furthermore, if $Ta = 0$ then $f_a(x) = \sum_{k=0}^\infty a_k x_k = 0$ for all $x \in X$, in particular, $f_a(e^{(k)}) = a_k = 0$ for all k, that is, $a = 0$. Thus $Ta = 0$ implies $a = 0$, and consequently T is one–to–one.
Now we assume that X has AK. Let $f \in X'$ be given. We define the sequence a by $a_k = f(e^{(k)})$ for $k = 0, 1, \ldots$. Let $x \in X$ be given. Then $x = \sum_{k=0}^\infty x_k e^{(k)}$, since X has AK, and $f \in X'$ implies

$$f(x) = f\left(\sum_{k=0}^\infty x_k e^{(k)}\right) = \sum_{k=0}^\infty x_k f(e^{(k)}) = \sum_{k=0}^\infty x_k a_k,$$

hence $a \in X^\beta$ and $Ta = f$. This shows that the map T is onto. $\square$

Example 9.5.3 *The continuous duals c_0^*, ℓ_1^* and ℓ_p^* for $1 < p < \infty$ are norm isomorphic with $\ell_1 = c_0^\beta$, $\ell_\infty = \ell_1^\beta$ and ℓ_q where $q = p/(p-1)$, respectively. If $X = \ell_\infty$ or $x \in c$, then we have $\|a\|_X^* = \|a\|_1$ for all $a \in X^\beta$.*

Another dual space is frequently used in the theory of sequence spaces.

Definition 9.5.4 *Let $X \supset \phi$ be an FK space. Then the set*

$$X^f = \{(f(e^{(n)}))_{n=0}^\infty : f \in X'\}$$

is called the functional dual of X.

The following results hold for the functional dual; the first statement in Part (b) is analogous to that in Corollary 9.4.3 (ii).

Theorem 9.5.5 ([209, Theorems 7.2.4 and 7.2.6]) *Let $X, Y \supset \phi$ be FK spaces. Then we have*

(a) $X^f = (\mathrm{cl}_X(\phi))^f$;

(b) $X \subset Y$ *implies* $X^f \supset Y^f$; *if X is closed in Y then $X^f = Y^f$.*

Proof.

(a) We write $Z = \mathrm{cl}_X(\phi)$.
First, we assume that $a \in X^f$, that is, $a_n = f(e^{(n)})$ $(n = 0, 1, \ldots)$ for some $f \in X'$. We write $g = f|_Z$ for the restriction of f to Z. Then we have $a_n = g(e^{(n)})$ for all $n = 0, 1, \ldots$, $g \in Z'$ and so $a \in Z^f$. Thus we have shown that

$$X^f \subset Z^f. \tag{*}$$

Conversely, let $a \in Z^f$, then $a_n = g(e^{(n)})$ $(n = 0, 1, \ldots)$ for some $g \in Z'$. By Theorem 9.5.1, g can be extended to $f \in X'$, and we have $a_n = f(e^{(n)})$ for $n = 0, 1, \ldots$, hence $a \in X^f$. Thus we have also shown that

$$Z^f \subset X^f; \tag{**}$$

and the inclusions in (*) and (**) yield $X^f = Z^f$.
This completes the proof of Part (a).

(b) We assume $a \in Y^f$. Then $a_n = f(e^{(n)})$ $(n = 0, 1, \dots)$ for some $f \in Y'$. Since $X \subset Y$, we have $g = f|_X \in X'$ by Theorem 9.2.8, and $a_n = g(e^{(n)})$ $(n = 0, 1, \dots)$, so $a \in X^f$. Thus we have shown that $Y^f \subset X^f$.

If X is closed in Y, then the FK topologies are the same by Theorem 9.2.8, and we obtain $X^f = (\mathrm{cl}_X(\phi))^f = (\mathrm{cl}_Y(\phi))^f = Y^f$ from Part (a). □

It might be expected from $X \subset X^{\dagger\dagger}$ in Corollary 9.4.3 (iii) that X is contained in X^{ff}; but this is not the case, in general, as the following Example 9.5.6 will show. We will, however, see later in Theorem 9.5.11 that $X \subset X^{ff}$ for BK spaces with AD.

Example **9.5.6** ([209, Example 7.2.5]) *Let* $X = c_0 \oplus z$ *where* z *is an unbounded sequence.* *Then* X *is a BK space,* $X^f = \ell_1$ *and* $X^{ff} = \ell_\infty$, *so* $X \not\subset X^{ff}$.

The next results establish some relations between the functional, $\beta-$ and $\gamma-$duals.

Theorem 9.5.7 ([209, Theorem 7.2.7]) *Let* $X \supset \phi$ *be an FK space.*

(a) *Then we have* $X^\beta \subset X^\gamma \subset X^f$.

(b) *If* X *has AK, then* $X^\beta = X^f$.

(c) *If* X *has AD then* $X^\beta = X^\gamma$.

Proof.

(i) First we show that
$$X^\beta \subset X^f. \tag{9.20}$$
Let $a \in X^\beta$. We define the linear functional f by $f(x) = \sum_{k=0}^{\infty} a_k x_k$ for all $x \in X$. Then $f \in X'$ by Theorem 9.5.2, and we have $f(e^{(n)}) = a_n$ for all n, hence $a \in X^f$. Thus we have shown the inclusion in (9.20).

(b) Now we prove Part (b).
We assume that X has AK, and $a \in X^f$, hence there is $f \in X'$ such that $a_k = f(e^{(k)})$ for $k = 0, 1, \dots$. Let $x \in X$ be given. Then $x = \sum_{k=0}^{\infty} x_k e^{(k)}$, since X has AK, and it follows from $f \in X'$ that
$$f(x) = f\left(\sum_{k=0}^{\infty} x_k e^{(k)}\right) = \sum_{k=0}^{\infty} x_k f(e^{(k)}) = \sum_{k=0}^{\infty} x_k a_k,$$
hence $a \in X^\beta$. Thus we have shown that $X^f \subset X^\beta$.
Together with (9.20), this yields $X^\beta = X^f$.
This completes the proof of Part (b).

(c) Now we prove Part (c).
We assume that X has AD and $a \in X^\gamma$, and define the linear functionals f_n on X for $n = 0, 1, \dots$ by
$$f_n(x) = \sum_{k=0}^{n} a_k x_k \quad (x \in X).$$
Since X is an FK space, we have $f_n \in X'$ for all n. Furthermore, $a \in X^\gamma$ implies that the sequence $(f_n)_{n=0}^{\infty}$ is pointwise bounded, hence equicontinuous by the uniform boundedness principle, Theorem 3.7.4. Since $\lim_{n\to\infty} f_n(x)$ exists for each $x \in \phi$ and X has AD, the limits $\lim_{n\to\infty} f_n(x)$ must exist for all $x \in X$ by the convergence lemma, Theorem 3.7.6, hence $a \in X^\beta$. Thus we have shown that $X^\gamma \subset X^\beta$.
We also have $X^\beta \subset X^\gamma$ by Corollary 9.4.3 (i), hence $X^\beta = X^\gamma$.
This completes the proof of Part (c).

(a) Finally we prove Part (a).
     First we observe that $\mathrm{cl}_X(\phi) \subset X$ implies $X^\gamma \subset (\mathrm{cl}_X(\phi))^\gamma$ by Proposition 9.4.2 (ii). Furthermore, we have

$$\mathrm{cl}_X(\phi))^\gamma = (\mathrm{cl}_X(\phi))^\beta \subset (\mathrm{cl}_X(\phi))^f = X^f$$

by Part (c), (9.20) and Theorem 9.5.5 (a). Thus we have shown that $X^\gamma \subset X^f$. This completes the proof of Part (a).   □

The next result gives a relation between the functional and continuous duals of an FK space.

Theorem 9.5.8 ([209, Theorems 7.2.10 and 7.2.12]) *Let $X \supset \phi$ be an FK space.*

(a) *Then the map $q : X' \to X^f$ given by $q(f) = (f(e^{(k)}))_{k=0}^\infty$ is onto. Moreover, if $T : X^\beta \to X'$ denotes the map of Theorem 9.5.2, then $q(Ta) = a$ for all $a \in X^\beta$.*

(b) *Then $X^f = X'$, that is, the map q of Part (a) is one to one, if and only if X has AD.*

     Proof.

(a) Let $a \in X^f$ be given. Then there is $f \in X'$ such that $a_k = f(e^{(k)})$ for all k, and so $q(f) = (f(e^{(k)}))_{k=0}^\infty = a$. This shows that q is onto.
     Now let $a \in X^\beta$ be given. We put $f = Ta \in X'$ and obtain $q(Ta) = q(f) = (f(e^{(k)}))_{k=0}^\infty = ((Ta)(e^{(k)}))_{k=0}^\infty = (a_k)_{k=0}^\infty = a$.

(b) First we assume that X has AD. Then $q(f) = 0$ implies $f = 0$ on ϕ, hence $f = 0$, since X has AD. This shows that q is one–to–one.
     Conversely we assume that X does not have AD. By Theorem 9.5.1, there exists an $f \in X'$ with $f \neq 0$ and $f = 0$ on ϕ. Then we have $q(f) = 0$, and q is not one to one.□

Example 9.5.9 ([209, Example 7.2.11]) *We have $c^\beta = c^f = \ell_1$. The map T of Theorem 9.5.2 is not onto. We consider $\lim \in X'$. If there were $a \in X^f$ with $\lim a = \sum_{k=0}^\infty a_k x_k$ then it would follow that $a_k = \lim e^{(k)} = 0$, hence $\lim x = 0$ for all $x \in c$, contradicting $\lim e = 1$. Also the map q of Theorem 9.5.8 is not onto, since $q(\lim) = 0$.*

If X and Y are BK spaces, then we know from Theorem 9.4.6 and Corollary 9.4.7 that the multiplier spaces $M(X,Y)$ and the α–, β– and γ–duals are also BK spaces. Now we are going to show that an analogous statement is true for the functional dual of a BK space.

Theorem 9.5.10 ([209, Theorem 7.2.14]) *Let $X \supset \phi$ be a BK space. Then X^f is a BK space.*

Proof. Let the map $q : X^* \to X^f$ be defined as in Theorem 9.5.8. Then q is linear and onto by Theorem 9.5.8, hence a quotient map in the sense of Definition 2.12.1. Let X^f have the quotient topology $\mathcal{Q}_q$.

(i) First we show that $(X^f, \mathcal{Q}_q)$ is a Banach space.
     Since

$$N(q) = \{f \in X^* : f|_\phi = 0\}$$

     is closed in X, the set $\{0\}$ is closed in Y by Theorem 2.12.11, hence the quotient topology $\mathcal{Q}_q$ of Y is separated by Theorem 2.12.9 (1) and Definition 2.12.10. Since X^* is a Banach space by Corollary 4.4.5, Corollary 3.3.9 implies that Y is a Banach space.

(ii) Now we show that the coordinates in Y are continuous.
Let P_n denote the nth coordinate on X^f. We write $g_n = P_n \circ q : X^* \to \mathbb{C}$ for all n. It follows that

$$|g_n(f)| = |P_n(q(f))| = |f(e^{(n)})| \leq \|f\| \cdot \|e^{(n)}\| \text{ for all } f \in X^*,$$

hence $g_n \in X^{**}$, and so $P_n \in (X^f)^*$ by Theorem 2.12.6. $\qquad\square$

Now we are going to show that if X is a BK space with AD then $X \subset X^{ff}$, the analogue of $X \subset X^{\dagger\dagger}$ in Corollary 9.4.3 (iii).

Theorem 9.5.11 ([209, Theorem 7.2.15]) *Let $X \supset \phi$ be a BK space.*
(a) Then $X^{ff} \supset \mathrm{cl}_X(\phi)$.
(b) If X has AD, then $X \subset X^{ff}$.

Proof.

(i) First we have to show that $\phi \subset X^f$ in order for X^{ff} to be meaningful.
Since X is a BK space, we have $P_k \in X'$ for all k where $P_k(x) = x_k$ $(x \in X)$. It follows by Theorem 9.5.8 (a) that, for each k, $q(P_k) = (P_k(e^{(n)}))_{n=0}^{\infty} = e^{(k)}$, hence $\phi \in X^f$.
This completes the proof of Part (i).

(b) Now we prove Part (b) of the theorem.
We assume that X has AD and have to show that $X \subset X^{ff}$.
Let $x \in X$ be given. We define the functional $f : X' \to \mathbb{C}$ by $f(\psi) = \psi(x)$ for all $\psi \in X'$. Then we have
$$|f(\psi)| = |\psi(x)| \leq \|\psi\|\|x\|,$$
and consequently $f \in X''$. Let $q : X' \to X^f$ be the map of Theorem 9.5.8 (a) which is an isomorphism (onto, Definition 1.6.1 (d)) by Theorem 9.5.8 (b), since X has AD. Thus the inverse map $q^{-1} : X^f \to X'$ exists. We define the map $g : X^f \to \mathbb{C}$ by $g(b) = \psi(x)$ $(b \in X^f)$ where $x = q^{-1}(b)$. It follows that
$$|g(b)| = |\psi(x)| \leq \|\psi\| \cdot \|x\| = \|\psi\| \cdot \|q^{-1}(b)\| \leq \|\psi\| \cdot \|q^{-1}\| \cdot \|b\|$$
and the open mapping theorem, Theorem 3.5.3, yields $\|q^{-1}\| < \infty$. Thus we have $g \in (X^f)'$, and also $g(b) = g(q(\psi)) = f(\psi) = \psi(x)$ for all $\psi \in X'$. Finally, it follows for $\psi = P_k$ that
$$x_k = P_k(x) = g(q(P_k)) = g(e^{(k)}) \text{ for all } k,$$
hence $x \in X^{ff}$. Thus we have shown that $X \subset X^{ff}$.
This completes the proof of Part (b) of the theorem.

(a) Finally, we prove Part (a) of the theorem.
We put $Y = \mathrm{cl}_X(\phi)$. Then Y has AD and $Y \subset Y^{ff}$ by what we have just shown. But $Y \subset X$ implies $Y^{ff} \subset X^{ff}$ by Theorem 9.5.5 (b). $\qquad\square$

The condition that X has AD is not necessary for $X \subset X^{ff}$, in general.

Example **9.5.12** ([209, Example 7.2.16]) *Let $X = c_0 \oplus z$ with $z \in \ell_\infty$. Then we have $X^{ff} = \ell_1^f = \ell_\infty \supset X$, but X does not have AD.*

9.6   Properties of the transposes of matrices

Our next results connect mapping properties of a matrix A with those of its transpose A^T. The results are related to, but different from, the adjoint map.

Theorem 9.6.1 ([209, Theorem 8.3.8]) *Let $X \supset \phi$ be an FK space and Y be any set of sequences.*

(a) If $A \in (X, Y)$ then $A^T \in (Y^\beta, X^f)$.

(b) If X and Y are BK spaces and Y^β has AD then we have $A^T \in (Y^\beta, \mathrm{cl}_{X^f}(X^\beta))$.

  Proof.

(a) First we prove Part (a).
Let $A \in (X, Y)$ and $z \in Y^\beta$ be given. We define the functional $f : X \to \mathbb{C}$ by

$$f(x) = \sum_{n=0}^{\infty} z_n(A_n x) \text{ for all } x \in X.$$

Since X is an FK space, $Ax \in Y$ by assumption and $z \in Y^\beta$, we have $f \in X'$ by Theorem 9.5.2. We also have

$$f(e^{(k)}) = \sum_{n=0}^{\infty} z_n a_{nk} = A_k^T z \text{ for all } k,$$

hence $A^T z \in X^f$. Thus we have shown that $A^T \in (Y^\beta, X^f)$.
This completes Part (a) of the proof.

(b) Now we prove Part (b).
We assume that X and Y are BK spaces and Y has AD. Then we have $X^\beta \subset X^f$ by Theorem 9.5.7 (a) and (c), and X^f is a BK space by Theorem 9.5.10. Also $\mathrm{cl}_{X^f}(X^\beta)$ is a closed subspace of X^f. Since $A \in (X, Y)$, we have $A_n = (a_{nk})_{k=0}^{\infty} \in X^\beta$ for all n, but

$$(A^T)_k e^{(n)} = \sum_{j=0}^{\infty} a_{jk} e_j^{(n)} = a_{nk} \text{ for all } n \text{ and } k,$$

that is,

$$\left(A^T e^{(k)}\right)_{k=0}^{\infty} = A_n \in X^\beta \text{ for all } n$$

which means that $A^T \in (X^\beta)^f$.
This and $A \in (Y^\beta, X^f)$ imply $A^T \in (Y^\beta, \mathrm{cl}_{X^f}(X^\beta))$ by Theorem 9.3.5.
This completes Part (b) of the proof.   □

One may be misled to assuming that $A \in (X, Y)$ implies $A^T \in (Y^\beta, X^\beta)$. The next example shows that this is not true, in general, even if X and Y are BK spaces.

Example **9.6.2** ([209, Note on pp. 123/124]) *Let $X = \ell_1 \oplus e$, $Y = \ell_1$ and $A_n x = x_n - x_{n-1}$ for $n = 0, 1, \dots$. Then we obviously have $A \in (X, Y)$, but $A^T \notin (Y^\beta, X^\beta) = (\ell_\infty, cs)$, since*

$$X^\beta = cs, \tag{*}$$

as we will show below, and since, for $x = ((-1)^n)_{n=0}^{\infty}$,

$$A^T x = 2x \notin cs.$$

We show (*).
First we assume $a \in X^{\beta}$. Then $\sum_{k=0}^{\infty} a_k x_k$ converges for all $x = x^{(1)} + \lambda \cdot e \in X$, where $x^{(1)} \in \ell_1$ and $\lambda \in \mathbb{C}$, in particular, if $x^{(1)} = 0$ and $\lambda = 1$ then, for $x = e \in X$, the series $\sum_{k=0}^{\infty} a_k e_k = \sum_{k=0}^{\infty} a_k$ converges, hence $a \in cs$. Thus we have shown that

$$X^{\beta} \subset cs. \tag{**}$$

Conversely, we assume $a \in cs$. Let $x \in X$ be given. Then we have $x = x^{(1)} + \lambda \cdot e$ for some sequence $x^{(1)} \in \ell_1$ and some scalar $\lambda \in \mathbb{C}$, and so the series

$$\sum_{k=0}^{\infty} a_k x_k = \sum_{k=0}^{\infty} a_k \left(x_k^{(1)} + \lambda \cdot e \right) = \sum_{k=0}^{\infty} a_k x_k^{(1)} + \lambda \sum_{k=0}^{\infty} a_k$$

converges since $a \in cs \subset \ell_{\infty} = \ell_1^{\beta}$. So we have $a \in X^{\beta}$. Thus we have shown that

$$X^{\beta} \subset cs. \tag{***}$$

Now (*) follows from (**) and (***).

The following result, however, holds for BK spaces with AK.

Theorem 9.6.3 ([209, Theorem 8.3.9]) *Let X and Y be BK spaces with AK and $Y = Z^{\beta}$. Then we have*

(a) $(X, Y) = (X^{\beta\beta}, Y)$;

(b) $A \in (X, Y)$ *if and only if* $A^T \in (Z, X^{\beta})$.

Proof. Since X is a BK space with AK, X^{β} is a BK space by Corollary 9.4.7, and $X^{\beta} = X^f$ by Theorem 9.5.7 (b).

(b) First we prove Part (b).

  (i) We show that $A \in (X, Y)$ implies $A^T \in (Z, Y^{\beta})$.
     First we assume $A \in (X, Y)$. As $Z^{\beta\beta} \supset Z$ by Corollary 9.4.3 (iii) and $X^{\beta} = X^f$, it follows by Theorem 9.6.1 that

$$A^T \in (Y^{\beta}, X^f) = (Y^{\beta}, X^{\beta}) = (Z^{\beta\beta}, X^{\beta}) \subset (Z, X^{\beta}).$$

  This completes Part (i) of the proof.

  (ii) Now we show that $A^T \in (Z, Y^{\beta})$ implies $A \in (X, Y)$.
     We assume $A^T \in (Z, X^{\beta})$. As $Z^f = Z^{\beta}$ by Theorem 9.5.7 (b) and $X^{\beta\beta} \supset X$ by Corollary 9.4.3 (iii), it follows by Theorem 9.6.1 that

$$A = A^{TT} = (X^{\beta\beta}, Z^f) = (X^{\beta\beta}, Z^{\beta}) \subset (X, Y).$$

  This completes Part (ii) of the proof.

Thus we have proved Part (b).

(a) To prove Part (a), we first observe that $X \subset X^{\beta\beta}$ implies

$$(X^{\beta\beta}, Y) \subset (X, Y).$$

Conversely, we assume $A \in (X, Y)$. Then we have $A^T \in (Z, X^\beta)$ by Part (b), and Theorem 9.6.1 and $Z^f = Z^\beta = Y$ imply

$$A = A^{TT} \in (X^{\beta\beta}, Z^f) \in (X^{\beta\beta}, Z^\beta) = (X^{\beta\beta}, Y).$$

Thus we also have

$$(X, Y) \subset (X^{\beta\beta}, Y).$$

This completes the proof of Part (a).   □

9.7   Matrix transformations between the classical sequence spaces

We apply the results of the previous sections to give necessary and sufficient conditions on the entries of a matrix A to be in any of the classes (X, Y) where X and Y are any of the classical sequence spaces ℓ_p $(1 \le p \le \infty)$, c_0 and c with the exceptions of (ℓ_p, ℓ_r) where both $p, r \ne 1, \infty$ (the characterizations are not known, in general, except for (ℓ_2, ℓ_2)), and of (ℓ_∞, c) (Schur's theorem), and (ℓ_∞, c_0) (no functional analytic proof seems to be known).

The class (ℓ_2, ℓ_2) was characterized by Crone ([49] or [175, pp. 111–115]).

For the sake of completeness, we characterize the classes (ℓ_∞, c_0) and (ℓ_∞, c). The fact that no functional analytic proof seems to be known for these two cases may be attributed to the fact that ℓ_∞ has no Schauder basis by Example 3.4.3 (d). Instead the classical method of the gliding hump has to be used in the proof of the characterizations. Matrices in (ℓ_∞, c) are called *coercive*.

We need the following result.

Lemma 9.7.1 *If* $\sum_{k=0}^\infty |a_{nk}| < \infty$ *for each* n *and* $\sum_{k=0}^\infty |a_{nk}| \to 0$ $(n \to \infty)$ *then* $\sum_{k=0}^\infty |a_{nk}|$ *is uniformly convergent in* n.

Proof. Let $\varepsilon > 0$ be given. Since $\sum_{k=0}^\infty |a_{nk}| \to 0$ $(n \to \infty)$, there is $N \in \mathbb{N}_0$ such that $\sum_{k=0}^\infty |a_{nk}| < \varepsilon$ for all $n > N$. Furthermore, since $\sum_{k=0}^\infty |a_{nk}| < \infty$ for each n with $0 \le n \le N$, there is an integer $m(n)$ such that $\sum_{k=m(n)}^\infty |a_{nk}| < \varepsilon$. We choose $m = \max_{0 \le m \le N} m(n)$. Then we obtain $\sum_{k=m}^\infty |a_{nk}| < \varepsilon$ for all n, and so $\sum_{k=0}^\infty |a_{nk}|$ is uniformly convergent in n.□

Now we are going to prove Schur's famous theorem which gives the characterizations of the classes (ℓ_∞, c) and (ℓ_∞, c_0).

Theorem 9.7.2 (Schur, 1920) ([189]) *We have*

(a) $A \in (\ell_\infty, c)$ *if and only if*

$$\sum_{k=0}^\infty |a_{nk}| \quad \text{converges uniformly in } n \qquad (9.21)$$

and

$$\lim_{n \to \infty} a_{nk} = \alpha_k \quad \text{for each } k; \qquad (9.22)$$

(b) $A \in (\ell_\infty, c_0)$ *if and only if the condition in (9.21) holds and*

$$\lim_{n \to \infty} a_{nk} = 0 \text{ for each } k. \qquad (9.23)$$

Proof.

(a) We show Part (a).

(i) First we show the sufficiency of the conditions in (9.21) and (9.22). We assume that the conditions in (9.21) and (9.22) are satisfied. First we show that

$$\sup_n \sum_{k=0}^\infty |a_{nk}| < \infty. \qquad (*)$$

By (9.21), there is $k_0 \in \mathbb{N}_0$ such that

$$\sum_{k=k_0+1}^\infty |a_{nk}| < 1 \text{ for all } n = 0, 1, \dots.$$

Also (9.22) implies $(a_{nk})_{n=0}^\infty \in c \subset \ell_\infty$ for every $k \in \mathbb{N}_0$. Thus, for every $k \in \mathbb{N}_0$, there is a constant M_k such that $|a_{nk}| < M_k$ for all $n = 0, 1, \dots$. We put $M = 1 + \sum_{k=0}^{k_0} M_k$ and obtain

$$\sum_{k=0}^\infty |a_{nk}| \leq \sum_{k=0}^{k_0} |a_{nk}| + \sum_{k=k_0+1}^\infty |a_{nk}| \leq M \text{ for all } n,$$

that is, $(*)$ holds.

Now $(*)$ and (9.22) imply $(\alpha_k)_{k=0}^\infty \in \ell_1$. To see this, let $m \in \mathbb{N}_0$ be given. Then we have

$$\sum_{k=0}^m |\alpha_k| = \lim_{n \to \infty} \sum_{k=0}^m |a_{nk}| \leq \sup_n \sum_{k=0}^\infty |a_{nk}| < \infty,$$

and so $(\alpha_k)_{k=0}^\infty \in \ell_1$, since $m \in \mathbb{N}_0$ was arbitrary. Therefore the series $\sum_{k=0}^\infty \alpha_k x_k$ converges for each $x \in \ell_\infty$. Furthermore, $x \in \ell_\infty$ and (9.21) together imply that $A_n x$ is absolutely and uniformly convergent in n, since

$$\sum_{k=0}^\infty |a_{nk} x_k| \leq \left(\sum_{k=0}^\infty |a_{nk}| \right) \sup_k |x_k| < \infty.$$

Therefore, we have that

$$\lim_{n \to \infty} A_n x = \sum_{k=0}^\infty \left(\lim_{n \to \infty} a_{nk} x_k \right) = \sum_{k=0}^\infty \alpha_k x_k,$$

exists, hence $Ax \in c$. This shows the sufficiency of the conditions, and completes Part (i) of the proof.

(ii) Now we show the necessity of the conditions in (9.21) and (9.22). We assume $A \in (\ell_\infty, c)$. It follows from $e^{(k)} \in \ell_\infty$ $(k = 0, 1, \dots)$ that, for each k, there exists a complex number α_k such that (9.22) holds. Also $c \subset \ell_\infty$ implies $(\ell_\infty, c) \subset (\ell_\infty, \ell_\infty)$ and so $(*)$ holds by Theorem 9.3.3 (c) and Example 9.5.3. We put $b_{nk} = a_{nk} - \alpha_k$

$(n, k = 0, 1, \dots)$. Then (*) and (9.22) imply $\alpha = (\alpha_k)_{k=0}^{\infty} \in \ell_1$. So we obtain $B \in (\ell_\infty, c)$. We are going to show that this implies

$$\lim_{n \to \infty} \sum_{k=0}^{\infty} |b_{nk}| = 0. \tag{9.24}$$

Then it will follow by Lemma 9.7.1 that $\sum_{k=0}^{\infty} |b_{nk}|$ converges uniformly in n whence $\sum_{k=0}^{\infty} |a_{nk}| = \sum_{k=0}^{\infty} |b_{nk} + \alpha_k|$ converges uniformly in n, which is the condition in (9.21).

To show that (9.24) must hold, we assume that it is not satisfied and construct a sequence $x \in \ell_\infty$ with $B(x) \notin c$ which is a contradiction to $B \in (\ell_\infty, c)$.

If $\sum_{k=0}^{\infty} |b_{nk}| \not\to 0 \ (n \to \infty)$ then there is a positive real c such that

$$\limsup_{n \to \infty} \sum_{k=0}^{\infty} |b_{nk}| = c,$$

hence, for some subsequence $(n_j)_{j=0}^{\infty}$,

$$\lim_{j \to \infty} \sum_{k=0}^{\infty} |b_{n_j,k}| = c.$$

We omit the indices j, that is, we assume without loss of generality

$$\lim_{n \to \infty} \sum_{k=0}^{\infty} |b_{nk}| = c. \tag{9.25}$$

It follows from (9.22) that

$$\lim_{n \to \infty} b_{nk} = 0. \tag{9.26}$$

By (9.25) and (9.26), there is an integer $n(1)$ such that

$$\left| \sum_{k=0}^{\infty} |b_{n(1),k}| - c \right| < \frac{c}{10} \text{ and } |b_{n(1),0}| < \frac{c}{10}.$$

Since

$$\|B_{n(1)}\|_1 = \sum_{k=0}^{\infty} |b_{n(1),k}| < \infty,$$

we may choose an integer $k(2) > 0$ such that

$$\sum_{k=k(2)+1}^{\infty} |b_{n(1),k}| < \frac{c}{10},$$

and it follows that

$$\left| \sum_{k=0}^{k(2)} |b_{n(1),k}| - c \right| \leq \left| \|B_{n(1)}\|_1 - c \right| + \sum_{k=k(2)+1}^{\infty} |b_{n(1),k}| + |b_{n(1),0}| < \frac{3c}{10}.$$

Now we choose an integer $n(2) > n(1)$ such that

$$\sum_{k=0}^{k(2)} |b_{n(2),k}| < \frac{c}{10} \text{ and } \left| \sum_{k=0}^{\infty} |b_{n(2),k}| 1 - c \right| < \frac{c}{10},$$

and an integer $k(3) > k(2)$ such that

$$\sum_{k=k(3)+1}^{\infty} |b_{n(2),k}| < \frac{c}{10}.$$

Again it follows that

$$\left| \sum_{k=k(2)+1}^{k(3)} |b_{n(2),k}| - c \right| < \frac{c}{10}.$$

Continuing in this way, we can determine sequences $(n(r))_{r=1}^{\infty}$ and $(k(r))_{r=1}^{\infty}$ of integers with $n(1) < n(2) < \dots$ and $0 = k(1) < k(2) < \dots$ such that

$$\begin{cases} \sum_{k=0}^{k(r)} |b_{n(r),k}| < \frac{c}{10}, \quad \sum_{k=k(r+1)+1}^{\infty} |b_{n(r),k}| < \frac{c}{10} \\ \text{and} \left| \sum_{k=k(r)+1}^{k(r+1)} |b_{n(r),k}| - c \right| < \frac{3c}{10}. \end{cases} \tag{9.27}$$

Now we define the sequence x by

$$x_k = \begin{cases} 0 & (k = 0) \\ (-1)^r \mathrm{sgn}(b_{n(r),k}) & (k(r)+1 \le k \le k(r+1)) \\ & (r = 0, 1, \dots). \end{cases}$$

Then we obviously have $x \in \ell_{\infty}$ and $\sup_k |x_k| \le 1$, and we conclude from (9.27) that

$$|B_{n(r)}(x) - (-1)^r c| \le \sum_{k=0}^{k(r)} |b_{n(r),k}||x_k| + \sum_{k=k(r+1)+1}^{\infty} |b_{n(r),k}||x_k| +$$

$$+ \left| \sum_{k=k(r)+1}^{k(r+1)} b_{n(r),k} x_k - (-1)^r c \right|$$

$$\le \sum_{k=0}^{k(r)} |b_{n(r),k}| + \sum_{k=k(r+1)+1}^{\infty} |b_{n(r),k}| +$$

$$+ \left| (-1)^r \left(\sum_{k=k(r)+1}^{k(r+1)} |b_{n(r),k}| - c \right) \right|$$

$$< \frac{c}{10} + \frac{c}{10} + \frac{3c}{10} = \frac{c}{2}.$$

Consequently, the sequence $(B_n(x))_{n=0}^{\infty}$ is not a Cauchy sequence and so not convergent. Thus if (9.24) is false then there is a sequence $x \in \ell_{\infty}$ such that $(B_n(x))_{n=0}^{\infty}$ is not convergent, which is a contradiction to $B(x) \in c$ for all $x \in \ell_{\infty}$. Therefore (9.24) must hold. This completes the proof of the necessity of the conditions, that is, of Part (ii).

(b) Part (b) is proved in exactly the same way as Part (a) by putting $\alpha_k = 0$.

$\square$

In the following theorem we write $\sup_N$ and $\sup_K$ for the suprema taken over all finite subsets N and K of $\mathbb{N}_0$, respectively.

Theorem 9.7.3 *Let $1 < p, r < \infty$, $q = p/(p-1)$ and $s = r/(r-1)$. Then the necessary and sufficient conditions for $A \in (X, Y)$ can be read from the following table*

From	ℓ_∞	c_0	c	ℓ_1	ℓ_p
To ℓ_∞	1.	2.	3.	4.	5.
c_0	6.	7.	8.	9.	10.
c	11.	12.	13.	14.	15.
ℓ_1	16.	17.	18.	19.	20.
ℓ_r	21.	22.	23.	24.	unknown

where

1., 2., 3. (1.1^*) $\displaystyle\sup_n \sum_{k=0}^{\infty} |a_{nk}| < \infty$

4. (4.1^*) $\displaystyle\sup_{n,k} |a_{nk}| < \infty$

5. (5.1^*) $\displaystyle\sup_n \sum_{k=0}^{\infty} |a_{nk}|^q < \infty$

6. (6.1^*) $\displaystyle\lim_{n\to\infty} \sum_{k=0}^{\infty} |a_{nk}| = 0$

7. (1.1^*) *and* (7.1^*), *where* (7.1^*) $\displaystyle\lim_{n\to\infty} a_{nk} = 0$ *for every k*

8. (1.1^*), (7.1^*) *and* $(8.1)^*$, *where* (8.1^*) $\displaystyle\lim_{n\to\infty} \sum_{k=0}^{\infty} a_{nk} = 0$

9. (4.1^*) *and* (7.1^*)

10. (5.1^*) *and* (7.1^*)

11. (11.1^*) *and* (11.2^*), *where*

     (11.1^*) $\displaystyle\sum_{k=0}^{\infty} |a_{nk}|$ *converges uniformly in n*

     (11.2^*) $\displaystyle\lim_{n\to\infty} a_{nk} = \alpha_k$ *exists for every k*

12. (1.1^*) *and* (11.2^*)

13. $(1.1)^*$, (11.2^*) *and* (13.1^*), *where* (13.1^*) $\displaystyle\lim_{n\to\infty} \sum_{k=0}^{\infty} a_{nk} = \alpha$ *exists*

14. (4.1^*) *and* (11.2^*)

15. (5.1^*) *and* (11.2^*)

16., 17., 18. (16.1^*), *where* (16.1^*) $\displaystyle\sup_N \left(\sum_{k=0}^{\infty} \left| \sum_{n\in N} a_{nk} \right| \right) < \infty$

19. (19.1^*) $\displaystyle\sup_k \sum_{n=0}^{\infty} |a_{nk}| < \infty$

20. (20.1^*) $\sup_N \left(\sum\limits_{k=0}^{\infty} \left| \sum\limits_{n \in N} a_{nk} \right|^q \right) < \infty$

21., 22., 23. (21.1^*) $\sup_K \left(\sum\limits_{n=0}^{\infty} \left| \sum\limits_{k \in K} a_{nk} \right|^r \right) < \infty$

24. (24.1^*) $\sup\limits_{k} \sum\limits_{n=0}^{\infty} |a_{nk}|^r < \infty.$

Proof. Condition (1.1^*) in **1** and **2**, and Conditions (4.1^*) and (5.1^*) in **4** and **5** follow from Theorem 9.3.3 (c) and Example 9.5.3. Since $c_0 \subset c \subset \ell_\infty$, we also obtain Condition (1.1^*) in **3**.
Theorem 9.3.5 and the conditions in **2**, **4** and **5** yield those in **7** and **12**, **9** and **14**, and in **10** and **15**. Now the conditions in **8** and **13** follow from those in **7** and **8** by Theorem 9.3.6. The conditions in **6** and **11** are those given in Schur's theorem, Theorem 9.7.2 (b) and (a). Conditions (16.1^*) in **16** and **17**, and (20.1^*) in **20** follow from Theorem 9.3.3 (d) and Example 9.5.3. Since $c_0 \subset c \subset \ell_\infty$, we also obtain Condition (16.1^*) in **18**. Furthermore, Condition (19.1^*) in **19** follows from Theorem 9.6.3 with $Z = c_0$, hence $Y = c_0^\beta = \ell_1$, and **1**.
Conditions (21.1^*) in **21** and **22**, and (24.1^*) in **24** follow from Theorem 9.6.3 and those in **20** and **1**. Since $c_0 \subset c \subset \ell_\infty$, we also obtain Condition (21.1^*) in **23**. □

Remark 9.7.4 *By Example 9.3.4, we obtain $A \in (\ell_1, \ell_1)$ if and only if*

$$\sup_{k} \sum_{n=0}^{\infty} |a_{nk}| < \infty. \tag{9.28}$$

*The condition in (9.28) can also be obtained by applying Theorem 9.6.3 (b) with $X = \ell_1$, $Z = c_0$ and $Y = Z^\beta = \ell_1$, and using the condition in (1.1^*) of Theorem 9.7.3 **1**.*
We would obtain from Theorem 9.3.3 (d) and Example 9.5.3 that $A \in (\ell_1, \ell_1)$ if and only if

$$\sup_N \left(\sup_k \left| \sum_{n \in N} a_{nk} \right| \right) < \infty. \tag{9.29}$$

It is easy to see that Condition (16.1^*) of Theorem 9.7.3 and (9.29) are equivalent.

Remark 9.7.5 *The results of Theorem 9.7.3 can be found in [209] and [194]; many references to the original proofs can be found in [194]. We give the following list of references for the parts of Theorem 9.7.3*

Parts	in [209]	in [194]
1., 2. and **3.**	Example 8.4.5A, p. 129	**1.** (1.1)
4.	Example 8.4.1A, p. 126	**6.** (6.1)
5.	Example 8.4.5D, p. 129	**5.** (5.1)
6.	Theorem 1.7.19, p. 17	**21.** (21.1)
7.	Example 8.4.5A, p. 129	**23.** (1.1) and (11.2)
8.	Example 8.4.5A, p. 129	**22.** (1.1), (11.2) and (22.1)
9.	Example 8.4.1A, p. 126	**28.** (6.2) and (11.2)
10.	Example 8.4.6D, p. 129	**27.** (5.1) and (11.2)

11.	Theorem 1.17.8, p.15	10. (10.1) and (10.4)
12.	Example 8.4.5A, p. 129	12. (1.1) and (10.1)
13.	Example 8.4.5A, p. 129 or	11. (1.1), (10.1)and (11.1)
	Theorem 1.3.6, p. 6	
14.	Example 8.4.1A, p. 126	17. (6.1) and (10.1)
15.	Example 8.4.5D, p. 129	16. (5.1) and (10.1)
16., 17. and 18.	Example 8.4.9A, p. 130	72. (72.2)
19.	Example 8.4.1D, p. 126	77. (77.1)
20.	Example 8.4.1D, p. 126	76. (76.1)
21., 22. and 23.	Example 8.4.8A, p. 131	63. (63.1)
24.	Example 8.4.1D, p. 126	68. (68.1).

The conditions for the class (ℓ_∞, c_0) in [209, Theorem 1.7.19, p. 17] are that

$$\sum_{k=0}^{\infty} |a_{nk}| \text{ converges uniformly in } n \tag{i}$$

and (7.1^*) in Theorem 9.7.3 7. Two more sets of equivalent conditions for the class (ℓ_∞, c) are given in [209, Theorem 1.17.8, p.15], they are

$$\begin{cases} \textbf{11. } (11.2^*), \sum_{k=0}^{\infty} |a_{nk}| < \infty \text{ for all } n, \\ \sum_{k=0}^{\infty} |\alpha_k| < \infty \text{ with } \alpha_k \text{ from } \textbf{11. } (11.2^*) \text{ and} \\ \lim_{n\to\infty} \sum_{k=0}^{\infty} |a_{nk} - \alpha_k| = 0 \end{cases} \tag{ii}$$

or

$$\begin{cases} \textbf{11. } (11.2^*), \text{ and} \\ \lim_{n\to\infty} \sum_{k=0}^{\infty} |a_{nk}| = \sum_{k=0}^{\infty} |\alpha_k|, \text{ both series being convergent.} \end{cases} \tag{iii}$$

Alternative equivalent conditions for the classes (ℓ_∞, ℓ_1), (c_0, ℓ_1) and (c, ℓ_1) are given in [209, Example 8.4.9A, p.132], namely

$$\sup_K \sum_{n=0}^{\infty} \left| \sum_{k \in K} a_{nk} \right| < \infty, \tag{iv}$$

and in [194, 72.], namely

$$\sup_{N,K} \left| \sum_{n \in N} \sum_{k \in K} a_{nk} \right| < \infty, \tag{72.1}$$

where the supremum is taken over all finite subsets N and K of $\mathbb{N}_0$, or (72.3) which is (iv), or

$$\sum_{n=0}^{\infty} \left| \sum_{k \in K_*} a_{nk} \right| \text{ converges for all } K_* \in \mathbb{N}_0. \tag{72.4}$$

If we apply Theorem 9.6.3 with $X = Z = c_0$ and $Y = c_0^\beta = \ell_1$ then we have $A \in (c_0, \ell_\infty)$ if and only if $A^T \in (c_0, \ell_\infty)$, that is, the conditions are symmetric in n and k, and (iv) follows in the same way as $\textbf{16 } (16.1^*)$. It can easily be shown that the other conditions given here are equivalent.

An alternative condition for the class (ℓ_1, ℓ_p) is given in [209, Example 8.4.1D, p. 126], namely

$$\sup_k \sum_{n=0}^{\infty} |a_{nk}|^p < \infty, \tag{v}$$

which can be obtained by applying Theorem 9.6.3 with $X = \ell_1$, $Z = \ell_q$ and $Y = \ell_p$ and then using 5 (5.1*) with A and p replaced by A^T and q.

Finally, an alternative condition for the classes (c_0, ℓ_r), (c, ℓ_r) and (ℓ_∞, ℓ_r) is given in [194, 63.], namely

$$\sum_{n=0}^{\infty} \left| \sum_{k \in K_*} a_{nk} \right| \quad converges \ for \ all \ K_* \subset \mathbb{N}_0. \tag{63.1}$$

Part (b) of the next result gives the characterization of regular matrices mentioned at the beginning of this section.

Theorem 9.7.6 *(a) Let $A \in (c, c)$ and $x \in c$. Then putting*

$$\chi = \chi(A) = \lim_{n \to \infty} \sum_{k=0}^{\infty} a_{nk} - \sum_{k=0}^{\infty} \alpha_k, \ where \ \alpha_k = \lim_{n \to \infty} a_{nk} \ for \ each \ k, \tag{9.30}$$

we have

$$\lim_{n \to \infty} A_n x = \chi \cdot \lim_{k \to \infty} x_k + \sum_{k=0}^{\infty} \alpha_k x_k. \tag{9.31}$$

(b) A matrix A is regular if and only if the condition in (1.1) of Theorem 9.7.3 **1** holds and*

$$\lim_{n \to \infty} a_{nk} = 0 \ for \ all \ k; \tag{9.32}$$

and

$$\lim_{n \to \infty} \sum_{k=0}^{\infty} a_{nk} = 1. \tag{9.33}$$

Proof.

(a) If $A \in (c, c)$ then the conditions in (1.1*), (11.2*) and (13.1*) of Theorem 9.7.3 **13** hold, so $\chi(A)$ is defined. We write $\|A\| = \sup_n \sum_{k=0}^{\infty} |a_{nk}|$, so the condition in (1.1*) of Theorem 9.7.3 **13** can be rewritten as

$$\|A\| < \infty. \tag{9.34}$$

We know from the proof of Theorem 9.7.2 that $(\alpha_k)_{k=0}^{\infty} \in \ell_1$, and so $\alpha(x) = \sum_{k=0}^{\infty} \alpha_k x_k$ converges for all $x \in c$, and

$$\sum_{k=0}^{\infty} |\alpha_k| \leq \|A\|. \tag{9.35}$$

(i) We show that

$$\lim_{n \to \infty} A_n x = \sum_{k=0}^{\infty} \alpha_k x_k \ for \ all \ x \in c_0. \tag{*}$$

Let $x \in c_0$ and $\varepsilon > 0$ be given. We choose an integer $K = K_\varepsilon$ such that

$$|x_k| < \frac{\varepsilon}{4\|A\| + 1} \ for \ all \ k > K_\varepsilon, \tag{9.36}$$

and, by (11.2*) in Theorem 9.7.3 **13**, we choose an integer $N = N_\varepsilon$ such that

$$\sum_{k=0}^{K} |a_{nk} - \alpha_k||x_k| < \frac{\varepsilon}{2} \text{ for all } n > N. \tag{9.37}$$

Let $n > N$ be given. Then (9.37), (9.36), (9.34) and (9.35) imply

$$\left| A_n x - \sum_{k=0}^{\infty} \alpha_k x_k \right| \leq$$

$$\sum_{k=0}^{K} |a_{nk} - \alpha_k||x_k| + \sum_{k=K+1}^{\infty} (|a_{nk}| + |\alpha_k|)|x_k| <$$

$$\frac{\varepsilon}{2} + \frac{\varepsilon}{4\|A\| + 1} \left(\sum_{k=0}^{\infty} |a_{nk}| + \sum_{k=0}^{\infty} |\alpha_k| \right) \leq$$

$$\frac{\varepsilon}{2} + \frac{\varepsilon}{4\|A\| + 1} (\|A\| + \|A\|) \leq \varepsilon.$$

Hence (*) is valid.

Now if $x \in c \setminus c_0$, then we have $\xi = \lim_{k \to \infty} x_k \neq 0$. We consider the sequence $x' = x - \xi e$. Then $x' \in c_0$, and (*) implies $\lim_{n \to \infty} A_n x' = \sum_{k=0}^{\infty} \alpha_k x_k'$. We conclude from (13.1*) in Theorem 9.7.3 **13** and the linearity of A on ω_A that

$$A_n x = A_n x' + \xi A_n e \to \sum_{k=0}^{\infty} \alpha_k x_k' + \xi \lim_A e =$$

$$\xi(\lim_{n \to \infty} A_n e - \sum_{k=0}^{\infty} \alpha_k) + \sum_{k=0}^{\infty} \alpha_k x_k \ (n \to \infty). \tag{9.38}$$

(Note that $\sum_{k=0}^{\infty} \alpha_k$ is convergent, since $(\alpha_k)_{k=0}^{\infty} \in \ell_1 \subset cs$.)

(b) Part (b) is an immediate consequence of Part (a) and Theorem 9.7.3 **13**. □

As an immediate consequence of Theorems 9.7.6 and 9.7.2 we obtain the following famous result due to *Steinhaus*.

Theorem 9.7.7 (Steinhaus) ([193, 132])
A regular matrix cannot sum all bounded sequences.

Proof. If A were a regular matrix that sums all bounded sequences, then we would have by (9.30), (9.32), (9.33) and (9.21)

$$1 = \chi(A) = \lim_{n \to \infty} \sum_{k=0}^{\infty} a_{nk} - \sum_{k=0}^{\infty} \alpha_k = 0,$$

a contradiction. □

An interesting application of Schur's theorem, Theorem 9.7.2, is to show that *strong* and *weak convergence* coincide in ℓ_1.

We recall that a sequence (x_n) in a normed space $(X, \|\cdot\|)$ is said to be *weakly convergent* to a limit $x \in X$ if $f(x_n) \to f(x)$ $(n \to \infty)$ for all $f \in X^*$; it is said to be *strongly convergent* to a limit $x \in X$ if $\|x_n - x\| \to 0$ $(n \to \infty)$. Since

$$|f(x_n) - f(x)| = |f(x_n - x)| \leq \|f\|^* \cdot \|x_n - x\| \text{ for all } f \in X^*,$$

strong convergence implies weak convergence. The converse implication is not true in general. To see this, we consider the sequence $(e^{(n)})_{n=0}^{\infty}$ in $(\ell_2, \|\cdot\|_2)$. Let $f \in \ell_2^*$ be given, then, by Theorem 4.7.1, there is a sequence $a = (a_k)_{k=0}^{\infty} \in \ell_2$ such that $f(x) = \sum_{k=0}^{\infty} a_k x_k$ for all $x \in \ell_2$. It follows that $f(e^{(n)}) = a_n \to 0$ $(n \to \infty)$, and so the sequence $(e^{(n)})_{n=0}^{\infty}$ is weakly convergent to zero. But on the other hand, we have $\|e^{(m)} - e^{(n)}\|_2 = \sqrt{2}$ for all $m \neq n$, hence $(e^{(n)})_{n=0}^{\infty}$ is not a Cauchy sequence, and so not convergent in ℓ_2. Therefore the sequence $(e^{(n)})_{n=0}^{\infty}$ is not strongly convergent.

The following well–known result, however, holds in ℓ_1.

Theorem 9.7.8 *Strong and weak convergence of sequences coincide in ℓ_1.*

Proof. We assume that the sequence $(x^{(n)})_{n=0}^{\infty}$ is weakly convergent to x in ℓ_1, that is

$$f(x^{(n)}) - f(x) \to 0 \ (n \to \infty) \text{ for each } f \in \ell_1^*.$$

By Theorem 4.7.1, to every $f \in \ell_1^*$, there corresponds a sequence $a \in \ell_\infty$ such that

$$f(y) = \sum_{k=0}^{\infty} a_k y_k \text{ for all } y \in \ell_1.$$

We define the matrix $B = (b_{nk})_{n,k=0}^{\infty}$ by $b_{nk} = x_k^{(n)} - x_k$ $(n, k = 0, 1, \dots)$. Then we have for all $a \in \ell_\infty$

$$f(x^{(n)}) - f(x) = f(x^{(n)} - x) = \sum_{k=0}^{\infty} a_k(x_k^{(n)} - x_k)$$

$$= \sum_{k=0}^{\infty} b_{nk} a_k \to 0 \ (n \to \infty),$$

that is, $B \in (\ell_\infty, c_0)$. It follows by (6.1^*) in Theorem 9.7.3 **6** that

$$\lim_{n \to \infty} \|x^{(n)} - x\|_1 = \lim_{n \to \infty} \left(\sum_{k=0}^{\infty} |x_k^{(n)} - x_k| \right) = \lim_{n \to \infty} \sum_{k=0}^{\infty} |b_{nk}| = 0.$$

So the sequence $(x^{(n)})_{n=0}^{\infty}$ is strongly convergent to x in ℓ_1. $\qquad \square$

9.8 Compact matrix operators

In this section, we establish some identities or estimates for the Hausdorff measure of noncompactness of matrix operators from BK spaces with AK into the spaces c_0, c and ℓ_1. This is achieved by applying Theorems 9.3.3, 7.9.3 and 7.11.4. The identities or estimates and Theorem 7.11.5 yield the characterizations of the corresponding classes of compact matrix operators.

We need the following concept and result. A norm $\|\cdot\|$ on a sequence space X is said to be *monotonous*, if $x, \tilde{x} \in X$ with $|x_k| \leq |\tilde{x}_k|$ for all k implies $\|x\| \leq \|\tilde{x}\|$. Part (a) of the following lemma generalizes Example 7.9.6.

Lemma 9.8.1 *(a) Let X be a monotonous BK space with AK, $\mathcal{P}_n : X \to X$ be the* projector onto the linear span of $\{e^{(1)}, e^{(2)}, \ldots, e^{(n)}\}$ and $\mathcal{R}_n = I - \mathcal{P}_n$, where I is the identity on X. Then we have

$$\lim_{n\to\infty} \left(\sup_{x\in Q} \|\mathcal{R}_n(x)\| \right) \text{ exists for all } Q \in \mathcal{M}_X \tag{9.39}$$

and

$$a = \lim_{n\to\infty} \|\mathcal{R}_n\| = 1. \tag{9.40}$$

(b) Let $\mathcal{P}_n : c \to c$ be the projector onto the linear span of $\{e, e^{(1)}, e^2, \ldots, e^{(n)}\}$ and $\mathcal{R}_n = I - \mathcal{P}_n$. Then (9.39) holds and

$$a = \lim_{n\to\infty} \|\mathcal{R}_n\| = 2. \tag{9.41}$$

Proof. It is clear that

$$\mu_n(Q) = \sup_{x\in Q} \|\mathcal{R}_n(x)\| < \infty \text{ for all } n \text{ and all } Q \in \mathcal{M}_X \text{ or } Q \in \mathcal{M}_c.$$

(a) Let $x \in \mathcal{M}_X$ be given. Since X is a monotonous BK space with AK, we obtain for all n and all $x \in \mathcal{M}_X$

$$\|\mathcal{R}_n(x)\| = \|x - x^{[n]}\| \ge \|x - x^{[n+1]}\| = \|\mathcal{R}_{n+1}(x)\|. \tag{9.42}$$

Let $n \in \mathbb{N}_0$ and $\varepsilon > 0$ be given. Then there exists a sequence $x^{(0)} \in Q$ such that $\|\mathcal{R}_n(x^{(0)})\| \ge \mu_{n+1}(Q) - \varepsilon$ and it follows from (9.42) that

$$\mu_n(Q) \ge \|\mathcal{R}_n(x^{(0)})\| \ge \|\mathcal{R}_{n+1}(x^{(0)})\| \ge \mu_{n+1}(Q) - \varepsilon. \tag{9.43}$$

Since $n \in \mathbb{N}_0$ and $\varepsilon > 0$ were arbitrary, we have $\mu_n(Q) \ge \mu_{n+1}(Q) \ge 0$ for all n, and so $\lim_{n\to\infty} \mu_n(Q)$ exists. This shows that the statement in (9.39) holds. Furthermore, since the norm $\|\cdot\|$ is monotonous, we have $\|\mathcal{R}_n(x)\| = \|x - x^{[n]}\| \le \|x\|$ for all $x \in X$ and all n, hence

$$\|\mathcal{R}_n\| \le 1 \text{ for all } n. \tag{9.44}$$

To prove the converse inequality, given $n \in \mathbb{N}_0$, we have $\|\mathcal{R}_n(e^{(n+1)})\| = \|e^{(n+1)}\| \ne 0$, and consequently

$$\|\mathcal{R}_n\| \ge 1 \text{ for all } n. \tag{9.45}$$

Finally, (9.44) and (9.45) imply (9.43).

(b) This is Example 7.9.7. □

Example 9.8.2 *The Hausdorff measure of noncompactness of any operator $L \in \mathcal{B}(\ell_1)$ is given by*

$$\|L\|_\chi = \lim_{m\to\infty} \left(\sup_k \sum_{n=m}^\infty |a_{nk}| \right), \tag{9.46}$$

where $A = (a_{nk})_{n,k=0}^\infty$ is the matrix that represents L (Theorem 9.3.3 (b)).

Proof. We write $\overline{B}_{\ell_1}$ for the closed unit ball in ℓ_1 and $A^{<m>}$ for the matrix with the rows $A_n^{<m>} = 0$ for $n \le m$ and $A_n^{<m>} = A_n$ for $n \ge m+1$. Then we obviously have $(\mathcal{R}_m \circ L)(x) = A^{<m>}(x)$ for all $x \in \ell_1$ and we obtain by (7.63) in Theorem 7.11.4, Remark 7.9.5, (9.39) and (9.40) in Lemma 9.8.1, and (9.14) in Example 9.3.4

$$\|L\|_\chi = \chi\left(L(\overline{B}_{\ell_1})\right) = \lim_{m\to\infty} \left(\sup_{\|x\|=1} \|(\mathcal{R}_m \circ L)(x)\| \right) = \lim_{m\to\infty} \|A^{<m>}\|$$

$$= \lim_{m \to \infty} \left(\sup_k \sum_{n=m+1}^{\infty} |a_{nk}| \right),$$

that is, (9.46) holds.

Remark 9.8.3 *It follows from (9.46) and (7.68) in Theorem 7.11.5 that $L \in \mathcal{K}(\ell_1)$ if and only if*

$$\lim_{m \to \infty} \left(\sup_k \sum_{n=m}^{\infty} |a_{nk}| \right) = 0. \qquad (9.47)$$

Obviously, we have $I \in \mathcal{B}(\ell_1) \setminus \mathcal{K}(\ell_1)$.

For each $m \in \mathbb{N}_0$, we denote by N_m the set of all integers greater than or equal to m. Part (a) of the next result is taken from [55].

Theorem 9.8.4 *Let X and Y be BK spaces, X have AK, $L \in \mathcal{B}(X, Y)$ and $A \in (X, Y)$ be the matrix that represents L (Theorem 9.3.3 (b)).*

(a) ([55, Theorem 3.4]) *If $Y = c$, then we have*

$$\frac{1}{2} \cdot \lim_{m \to \infty} \left(\sup_{n \geq m} \|A_n - (\alpha_k)_{k=0}^{\infty}\|_X^* \right) \leq \|L\|_X \leq$$

$$\lim_{m \to \infty} \left(\sup_{n \geq m} \|A_n - (\alpha_k)_{k=0}^{\infty}\|_X^* \right), \quad (9.48)$$

where

$$\alpha_k = \lim_{n \to \infty} a_{nk} \text{ for all } k. \qquad (9.49)$$

(b) *If $Y = c_0$, then we have*

$$\|L\|_X = \lim_{m \to \infty} \left(\sup_{n \geq m} \|A_n\|_X^* \right). \qquad (9.50)$$

(c) *If $Y = \ell_1$, then we have*

$$\lim_{m \to \infty} \left(\sup_{N_m} \left\| \sum_{n \in N_m} A_n \right\|_X^* \right) \leq \|L\|_X \leq 4 \cdot \lim_{m \to \infty} \left(\sup_{N_m} \left\| \sum_{n \in N_m} A_n \right\|_X^* \right), \quad (9.51)$$

where the supremum taken over all finite subsets N_m of integers $\geq m$.

Proof. We write $\| \cdot \| = \| \cdot \|_X^*$ and $\|A\| = \|A\|_{(X, \ell_\infty)} = \sup_n \|A_n\|$, for short.

(a) Let $A = (a_{nk})_{n,k=0}^{\infty} \in (X, c)$. Then $\|A\| < \infty$, α_k in (9.49) exists for all k and $\|L\| = \|A\|$ by Theorems 9.3.3 (c), 9.3.5 and (9.11) in Theorem 9.3.3 (c).

(i) We show that $(\alpha_k)_{k=0}^{\infty} \in X^\beta$.
Let $x \in X$ be given. Since X has AK, there exists a positive constant C such that $\|x^{[m]}\| \leq C\|x\|$ for all $m \in \mathbb{N}_0$ and it follows that

$$\left| \sum_{k=0}^{m} a_{nk} x_k \right| = \left| A_n x^{[m]} \right| \leq C\|A_n\|^* \|x\| \leq C\|A\| \cdot \|x\| \text{ for all } n \text{ and } m,$$

hence by (9.49)

$$\left|\sum_{k=0}^{m}\alpha_k x_k\right| = \lim_{n\to\infty}\left|\sum_{k=0}^{m}a_{nk}x_k\right| \le C\|A\|\cdot\|x\| \text{ for all } m. \tag{9.52}$$

This implies $(\alpha_k x_k)_{k=0}^{\infty} \in bs$, and since $x \in X$ was arbitrary, we conclude $(\alpha_k)_{k=0}^{\infty} \in X^{\gamma}$, and $X^{\gamma} = X^{\beta}$ by Theorem 9.5.7 (c), since X has AK and so AD.

Also $(\alpha_k)_{k=0}^{\infty} \in X^{\beta}$ implies $\|(\alpha_k)_{k=0}^{\infty}\|^* < \infty$ by Theorem 9.5.2.

(ii) Now we show that

$$\lim_{n\to\infty} A_n x = \sum_{k=0}^{\infty}\alpha_k x_k \text{ for all } x \in X. \tag{9.53}$$

Let $x \in X$ and $\varepsilon > 0$ be given. Since X has AK, there exists $k_0 \in \mathbb{N}_0$ such that

$$\|x - x^{[k_0]}\| \le \frac{\varepsilon}{2(M+1)}, \text{ where } M = \|A\| + \|(\alpha_k)_{k=0}^{\infty}\|^*. \tag{9.54}$$

It also follows from (9.49) that there exists $n_0 \in \mathbb{N}_0$ such that

$$\left|\sum_{k=0}^{k_0}(a_{nk}-\alpha_k)x_k\right| < \frac{\varepsilon}{2} \text{ for all } n \ge n_0. \tag{9.55}$$

Let $n \ge n_0$ be given. Then it follows from (9.54) and (9.55) that

$$\left|A_n x - \sum_{k=0}^{\infty}\alpha_k x_k\right| \le \left|\sum_{k=0}^{k_0}(a_{nk}-\alpha_k)x_k\right| + \left|\sum_{k=k_0+1}^{\infty}(a_{nk}-\alpha_k)x_k\right|$$

$$< \frac{\varepsilon}{2} + \|A_n - (\alpha_k)_{k=0}^{\infty}\|^*\|x - x^{[k_0]}\| < \frac{\varepsilon}{2} + \frac{\varepsilon}{2} = \varepsilon.$$

Thus we have shown that (9.53).

(iii) Now we show the inequalities in (9.48).

Let $y = (y_n)_{n=0}^{\infty} \in c$ be given. Then, by Example 3.4.3 (c), y has a unique representation $y = \eta e + \sum_{n=0}^{\infty}(y_n - \eta)e^{(n)}$, where $\eta = \lim_{n\to\infty} y_n$. We obtain $\mathcal{R}_m y = \sum_{n=m+1}^{\infty}(y_n - \eta)e^{(n)}$ for all m. Writing $y_n = A_n x$ for $n = 0, 1, \ldots$ and $B = (b_{nk})_{n,k=0}^{\infty}$ for the matrix with $b_{nk} = a_{nk} - \alpha_k$ for all n and k, we obtain from (9.53)

$$\|\mathcal{R}_m(Ax)\| = \sup_{n\ge m+1} |y_n - \eta| = \sup_{n\ge m+1}\left|A_n x - \sum_{k=0}^{\infty}\alpha_k x_k\right| = \sup_{n\ge m+1} |B_n x|,$$

whence

$$\sup_{x\in\overline{B}_X}\|\mathcal{R}_m(Ax)\| = \sup_{n\ge m+1}\|B_n\|^* \text{ for all } m.$$

Now the inequalities in (9.48) follow from (7.63) in Theorem 7.11.4, (9.39) and (9.41) in Lemma 9.8.1, and (7.56) in Theorem 7.9.3.

Thus we have shown Part (a).

(b)   Part (b) follows from (a) with $\alpha_k = 0$ and $a = \lim_{m\to\infty} \|\mathcal{R}_m\| = 1$.

(c)   It follows from Theorem 9.3.3 (d) and the fact that $\mathcal{R}_m$ is given by the matrix $A^{<m+1>}$ that

$$\sup_{N_{m+1}} \left\| \sum_{n\in N_{m+1}} A_n \right\| = \sup_N \left\| \sum_{n\in N} A_n^{<m+1>} \right\|_X^* \leq \|\mathcal{R}_m \circ L\|_X^* =$$

$$\|A^{<m+1>}\|_{(X,\ell_1)} \leq 4 \cdot \sup_N \left\| \sum_{n\in N} A_n^{<m+1>} \right\|_X^* = 4 \cdot \sup_{N_{m+1}} \left\| \sum_{n\in N_{m+1}} A_n \right\|. \quad (9.56)$$

Now the inequalities in (9.51) follow from (7.63) in Theorem 7.11.4, (9.39) and (9.40) in Lemma 9.8.1, and Remark 7.9.9. $\qquad\square$

9.9   The class $\mathcal{K}(c)$

The vital assumption in Theorem 9.8.4 is that the initial space has AK. Hence it cannot be applied in the case of $\mathcal{B}(c)$.

The next result gives the representation of the general operator $L \in \mathcal{B}(c)$, a formula for its norm $\|L\|$, an estimate for its Hausdorff measure of noncompactness $\|L\|_\chi$, and a formula for the limit of the sequence $L(x)$.

Theorem 9.9.1 *We have $L \in \mathcal{B}(c)$ if and only if there exist a sequence $b \in \ell_\infty$ and a matrix $A = (a_{nk})_{n,k=0}^\infty \in (c_0, c)$ such that*

$$L(x) = b\xi + Ax \text{ for all } x \in c, \text{ where } \xi = \lim_{k\to\infty} x_k, \quad (9.57)$$

$$a_{nk} = L_n(e^{(k)}), \quad b_n = L_n(e) - \sum_{k=0}^\infty a_{nk} \text{ for all } n \text{ and } k, \quad (9.58)$$

$$\beta = \lim_{n\to\infty} \left(b_n + \sum_{k=0}^\infty a_{nk} \right) \text{ exists}, \quad (9.59)$$

and

$$\|L\| = \sup_n \left(|b_n| + \sum_{k=0}^\infty |a_{nk}| \right). \quad (9.60)$$

Moreover, if $L \in \mathcal{B}(c)$, then we have

$$\frac{1}{2} \cdot \limsup_{n\to\infty} \left(\left| b_n - \beta + \sum_{k=0}^\infty \alpha_k \right| + \left\| \tilde{A}_n \right\|_1 \right)$$

$$\leq \|L\|_\chi \leq \limsup_{n\to\infty} \left(\left| b_n - \beta + \sum_{k=0}^\infty \alpha_k \right| + \left\| \tilde{A}_n \right\|_1 \right), \quad (9.61)$$

where

$$\alpha_k = \lim_{n\to\infty} a_{nk} \text{ for all } k \in \mathbb{N}_0 \quad (9.62)$$

and $\tilde{A} = (\tilde{a}_{nk})_{n,k=0}^{\infty}$ *is the matrix with* $\tilde{a}_{nk} = a_{nk} - \alpha_k$ *for all* n *and* k; *we also have*

$$\eta = \lim_{n \to \infty} L_n(x) = \xi\beta + \sum_{k=0}^{\infty} \alpha_k(x_k - \xi) = \left(\beta - \sum_{k=0}^{\infty} \alpha_k\right)\xi + \sum_{k=0}^{\infty} \alpha_k x_k$$

for all $x \in c$. (9.63)

Proof.

(i) First we assume that $L \in \mathcal{B}(c)$ and show that L has the given representation and satisfies (9.60).

We assume $L \in \mathcal{B}(c)$.

We write $L_n = P_n \circ L$ for $n = 0, 1, \dots$ where $P_n : c \to \mathbb{C}$ is the nth coordinate with $P_n(x) = x_n$ for all $x = (x_k)_{k=0}^{\infty} \in c$. Since c is a BK space, we have $L_n \in c^*$ for all $n \in \mathbb{N}$, that is, by (4.55) (with a and b replaced by b_n and A_n, respectively)

$$L_n(x) = b_n \cdot \lim_{k \to \infty} x_k + \sum_{k=0}^{\infty} a_{nk} x_k \text{ for all } x \in c \tag{9.64}$$

with b_n and a_{nk} from (9.58); we also have by (4.51)

$$\|L_n\| = |b_n| + \sum_{k=0}^{\infty} |a_{nk}| \text{ for all } n. \tag{9.65}$$

Now (9.64) yields the representation of the operator L in (9.57).

Furthermore, since $L(x^{(0)}) = Ax^{(0)}$ for all $x^{(0)} \in c_0$, we have $A \in (c_0, c)$ and so

$$\|A\| = \sup_n \sum_{k=0}^{\infty} |a_{nk}| < \infty$$

by (1.1*) in Theorem 9.7.3 **12**. Also $L(e) = b + Ae$ by (9.57), and so $L(e) \in c$ yields (9.59), and we obtain

$$\|b\|_\infty \le \|L(e)\|_\infty + \|A\| < \infty,$$

that is, $b \in \ell_\infty$. Consequently, we have

$$\sup_n \|L_n\| = \sup_n \left(|b_n| + \sum_{k=0}^{\infty} |a_{nk}|\right) < \infty.$$

Since $|\lim_{k \to \infty} x_k| \le \sup_k |x_k| = \|x\|_\infty$ for all $x \in c$, we obtain by (9.59)

$$\|L(x)\|_\infty = \sup_n \left| b_n \cdot \lim_{k \to \infty} x_k + \sum_{k=0}^{\infty} a_{nk} x_k \right|$$

$$\le \left[\sup_n \left(|b_n| + \sum_{k=0}^{\infty} |a_{nk}|\right)\right] \cdot \|x\|_\infty = \sup_n \|L_n\| \cdot \|x\|_\infty,$$

hence $\|L\| \le \sup_n \|L_n\|$. We also have

$$|L_n(x)| \le \|L(x)\|_\infty \le \|L\| \text{ for all } x \in S_c \text{ and all } n,$$

that is, $\sup_n \|L_n\| \le \|L\|$, and we have shown (9.60).

This completes Part (i) of the proof.

(ii) Now we show that if L has the given representation then $L \in \mathcal{B}(c)$.
We assume $A \in (c_0, c)$ and $b \in \ell_\infty$ and that the conditions in (9.57), (9.59) and (9.60) are satisfied. Then we obtain from $A \in (c_0, c)$ by (1.1*) in Theorem 9.7.3 **12** that $\|A\| < \infty$, and this and $b \in \ell_\infty$ imply $\|L\| < \infty$ by (9.60), hence $L \in \mathcal{B}(c, \ell_\infty)$. Finally, let $x \in c$ be given and $\xi = \lim_{k \to \infty} x_k$. Then $x - \xi \cdot e \in c_0$, so by (9.57)

$$L_n(x) = b_n \cdot \xi + \sum_{k=0}^{\infty} a_{nk} x_k = \left(b_n + \sum_{k=0}^{\infty} a_{nk} \right) \cdot \xi + A_n(x - \xi \cdot e) \text{ for all } n,$$

and it follows from (9.59) and $A \in (c_0, c)$ that $\lim_{n \to \infty} L_n(x)$ exists. Since $x \in c$ was arbitrary, we have $L \in \mathcal{B}(c)$.
This completes Part (ii) of the proof.

(iii) Now we show that if $L \in \mathcal{B}(c)$, then $\|L\|_\chi$ satisfies the inequalities in (9.61).
We assume $L \in \mathcal{B}(c)$.
Let $x \in c$ be given, $\xi = \lim_{k \to \infty} x_k$ and $y = L(x)$. Then we have by Part (i) of the proof

$$y = b \cdot \xi + Ax, \text{ where } A \in (c_0, c) \text{ and } b \in \ell_\infty,$$

and we note that the limits α_k in (9.62) exist for all k by (11.2*) in Theorem 9.7.3 **12**, and $(\alpha_k)_{k=0}^{\infty} \in \ell_1$ by Part (a) (i) of the proof of Theorem 9.8.4. So we can write

$$y_n = b_n \cdot \xi + A_n x = \xi \cdot \left(b_n + \sum_{k=0}^{\infty} a_{nk} \right) + A_n(x - \xi \cdot e) \text{ for all } n \in \mathbb{N}. \tag{9.66}$$

Since $A \in (c_0, c)$, it follows from (9.53) in Part (a) (ii) of the proof of Theorem 9.8.4 that

$$\lim_{n \to \infty} A_n(x - \xi \cdot e) = \sum_{k=0}^{\infty} \alpha_k(x_k - \xi) = \sum_{k=0}^{\infty} \alpha_k x_k - \xi \sum_{k=0}^{\infty} \alpha_k. \tag{9.67}$$

So we obtain by (9.66), (9.59) and (9.67)

$$\eta = \lim_{n \to \infty} y_n = \lim_{n \to \infty} \left[\xi \left(b_n \cdot \sum_{k=0}^{\infty} a_{nk} \right) + A_n(x - \xi \cdot e) \right] \tag{9.68}$$

$$= \xi \cdot \lim_{n \to \infty} \left(b_n + \sum_{k=0}^{\infty} a_{nk} \right) + \lim_{n \to \infty} A_n(x - \xi \cdot e)$$

$$= \xi \cdot \beta + \sum_{k=0}^{\infty} \alpha_k x_k - \xi \sum_{k=0}^{\infty} \alpha_k = \xi \left(\beta - \sum_{k=0}^{\infty} \alpha_k \right) + \sum_{k=0}^{\infty} \alpha_k x_k,$$

that is, we have shown (9.63).
For each m, we have

$$\mathcal{R}_m(y) = \sum_{n=m+1}^{\infty} (y_n - \eta) e^{(n)} \text{ for } y \in c \text{ and } \eta = \lim_{n \to \infty} y_n.$$

Writing

$$f_n^{(m)}(x) = (\mathcal{R}_m(L(x)))_n \,,$$

we obtain for $n \geq m + 1$ by (9.66) and (9.68)

$$f_n^{(m)}(x) = y_n - \eta = \xi \cdot b_n + A_n(x) - \left[\xi \left(\beta - \sum_{k=0}^{\infty} \alpha_k \right) + \sum_{k=0}^{\infty} \alpha_k x_k \right]$$

$$= \xi \cdot \left(b_n - \beta + \sum_{k=0}^{\infty} \alpha_k \right) + \sum_{k=0}^{\infty} (a_{nk} - \alpha_k) x_k.$$

Since $f_n^{(m)} \in c^*$, we have by (4.51)

$$\left\| f_n^{(m)} \right\| = \left| b_n - \beta + \sum_{k=0}^{\infty} \alpha_k \right| + \sum_{k=0}^{\infty} |a_{nk} - \alpha_k|,$$

and it follows that

$$\sup_{x \in S_c} \| \mathcal{R}_m(L(x)) \|_\infty = \sup_{n \geq m+1} \| f_n^{(m)} \|$$

$$= \sup_{n \geq m+1} \left(\left| b_n - \beta + \sum_{k=0}^{\infty} \alpha_k \right| + \sum_{k=0}^{\infty} |a_{nk} - \alpha_k| \right).$$

Now the inequalities in (9.61) follow from (7.63) in Theorem 7.11.4, (9.39) and (9.41) in Lemma 9.8.1, and (7.58) in Remark 7.9.4. □

We obtain the following result for $L \in \mathcal{B}(c, c_0)$ as in Theorem 9.9.1 with $\beta = \alpha_k = 0$ for all k and by replacing the factor $1/2$ in (9.61) by 1.

Corollary 9.9.2 *We have $L \in \mathcal{B}(c, c_0)$ if and only if there exist a sequence $b \in \ell_\infty$ and a matrix $A = (a_{nk})_{n,k=0}^{\infty} \in (c_0, c)$ such that (9.57) holds, where the sequence b and the matrix A are given by (9.58), and the norm of $\|L\|$ satisfies (9.60).*
Moreover, if $L \in \mathcal{B}(c, c_0)$, then we have

$$\|L\|_\chi = \limsup_{n \to \infty} (|b_n| + \|A_n\|_1). \tag{9.69}$$

Now we give the estimates or identities for the Hausdorff measures of noncompactness of bounded linear operators between the spaces c_0 and c.

Theorem 9.9.3 *Let $L \in \mathcal{B}(X, Y)$ where X and Y are any of the spaces c_0 and c. Then the estimates or identities for the Hausdorff measure of the operators L can be obtained from the following table*

From	c_0	c
To c_0	1.	2.
c	3.	4.

where

1. $\displaystyle \|L\|_\chi = \limsup_{n \to \infty} \left(\sum_{k=0}^{\infty} |a_{nk}| \right)$

2. $\displaystyle \|L\|_\chi = \limsup_{n \to \infty} \left(|b_n| + \sum_{k=0}^{\infty} |a_{nk}| \right)$

3. $\displaystyle \frac{1}{2} \limsup_{n \to \infty} \left(\sum_{k=0}^{\infty} |a_{nk} - \alpha_k| \right) \leq \|L\|_\chi \leq \limsup_{n \to \infty} \left(\sum_{k=0}^{\infty} |a_{nk} - \alpha_k| \right)$

4. $\displaystyle \frac{1}{2} \limsup_{n \to \infty} \left(\left| b_n - \beta + \sum_{k=0}^{\infty} \alpha_k \right| + \sum_{k=0}^{\infty} |a_{nk} - \alpha_k| \right) \leq \|L\|_\chi$

$$\leq \limsup_{n \to \infty} \left(\left| b_n - \beta + \sum_{k=0}^{\infty} \alpha_k \right| + \sum_{k=0}^{\infty} |a_{nk} - \alpha_k| \right).$$

Proof **1** and **3**. The conditions follow from Theorem 9.8.4 (b) for **1** and (a) for **3**, and (7.58) in Remark 7.9.4 and the fact that $\| \cdot \|_X^* = \| \cdot \|_1$ for $X = c_0$ and $X = c$ by Example 9.5.3.

2 and **4**. The conditions follow from (9.69) in Corollary 9.9.2 and from (9.62) in Theorem 9.9.1 (b). $\qquad \square$

Applying (7.68) in Theorem 7.11.5 we obtain the following characterizations for compact operators between the spaces c_0 and c.

Corollary 9.9.4 *Let $L \in \mathcal{B}(X,Y)$ where X and Y are any of the spaces c_0 and c. Then the necessary and sufficient conditions for $L \in \mathcal{K}(X,Y)$ can be obtained from the following table*

From	c_0	c
To c_0	1.	2.
c	3.	4.

where

$$
\textbf{1.} \quad \|L\|_\chi = \lim_{n\to\infty} \left(\sum_{k=0}^{\infty} |a_{nk}| \right) = 0
$$

$$
\textbf{2.} \quad \|L\|_\chi = \lim_{n\to\infty} \left(|b_n| + \sum_{k=0}^{\infty} |a_{nk}| \right) = 0
$$

$$
\textbf{3.} \quad \lim_{n\to\infty} \left(\sum_{k=0}^{\infty} |a_{nk} - \alpha_k| \right) = 0
$$

$$
\textbf{4.} \quad \lim_{n\to\infty} \left(\left| b_n - \beta + \sum_{k=0}^{\infty} \alpha_k \right| + \sum_{k=0}^{\infty} |a_{nk} - \alpha_k| \right).
$$

The following example will show that a regular matrix cannot be compact; this is a well–known result by *Cohen* and *Dunford* ([45]).

Example 9.9.5 *Let $A \in (c,c)$. It follows from the condition in* **4** *in Corollary 9.9.4 with $b_n = 0$ for all n and $\beta = \lim_{n\to\infty} \sum_{k=0}^{\infty} a_{nk}$ that L_A is compact if and only if*

$$
\lim_{n\to\infty} \left(\left| \sum_{k=0}^{\infty} \alpha_k - \beta \right| + \sum_{k=0}^{\infty} |a_{nk} - \alpha_k| \right) = 0. \tag{9.70}
$$

If A is regular, then $\alpha_k = 0$ for $k = 0, 1, \ldots$ and $\beta = 1$ by (9.32) and (9.33) in Theorem 9.7.6 (b), and so

$$
\left| \sum_{k=0}^{\infty} \alpha_k - \beta \right| + \sum_{k=0}^{\infty} |a_{nk} - \alpha_k| = 1 + \sum_{k=0}^{\infty} |a_{nk}| \geq 1 \text{ for all } n.
$$

Thus a regular matrix cannot satisfy (9.70) and consequently cannot be compact.

9.10   Compact matrix operators between the classical sequence spaces

In this section we characterize the known classes of compact matrix operators between the classical sequence spaces. The characterizations are achieved in most cases by the application of our results on the Hausdorff measure of noncompactness. A notable exception is the characterization of the class $\mathcal{K}(\ell_1, \ell_\infty)$ ([179, Theorem 5]) which uses a completely different approach.

We need some more results. The following result is well known (cf. for instance [179, Theorem 2 (c)]; the proof in [179] is totally different from our proof.

Lemma 9.10.1 *Let X be a BK space, Y be a BK space with AK, and $A \in (X, Y)$. Then $L_A \in \mathcal{K}(X, Y)$ if and only if*

$$\lim_{n \to \infty} \|L_{A^{<n>}}\| = 0, \tag{9.71}$$

where the matrices $A^{<n>}$ for $n = 0, 1, \ldots$ are obtained by replacing the rows $A_0, A_1, \ldots, A_n$ of A by the sequence $(0, 0, \ldots)$.

Proof. We assume $A \in (X, Y)$. Since Y is a BK space with AK, we obtain for the operator L_A with $L_A(x) = Ax$ for all $x \in X$ by (7.63) in Theorem 7.11.4, (7.55) and (7.56) in Theorem 7.9.3

$$\frac{1}{a} \cdot \limsup_{n \to \infty} \left(\sup_{x \in \overline{B}_X} \|(\mathcal{R}_n \circ L_A)(x)\| \right) \le \|L_A\|_\chi = \chi(L(\overline{B}_X)) \le$$

$$\limsup_{n \to \infty} \left(\sup_{x \in \overline{B}_X} \|(\mathcal{R}_n \circ L_A)(x)\| \right) = \limsup_{n \to \infty} \|\mathcal{R}_n \circ L_A\| = \limsup_{n \to \infty} \|L_{A^{<m>}}\|.$$

Now the statement of the lemma follows by applying (7.68) in Theorem 7.11.5.   □

Lemma 9.10.2 *Let X and Z be BK spaces with AK and $Y = Z^\beta$, and $\| \cdot \|_{X^\beta} = \| \cdot \|_X^*$ and $\| \cdot \|_{Z^\beta} = \| \cdot \|_Z^*$. If $A \in (X, Y)$, then we have $\|L_A\| = \|L_{A^T}\|$.*

Proof. We note that $A \in (X, Y)$ if and only if $A^T \in (Z, X^\beta)$ by Theorem 9.6.3 (b).

(i)   First we show that

$$\|L_{A^T}\| \le \|L_A\|. \tag{9.72}$$

Let $A \in (X, Y)$ and $z \in Z$. Since $Ax^{[m]} \in Y = Z^\beta$ for all $m \in \mathbb{N}_0$ and all $x \in X$, we obtain

$$\sum_{n=0}^\infty z_n A_n x^{[m]} = \sum_{n=0}^\infty z_n \sum_{k=0}^\infty a_{nk} x_k^{[m]} = \sum_{n=0}^\infty z_n \sum_{k=0}^m a_{nk} x_k$$

$$= \sum_{k=0}^m x_k \sum_{n=0}^\infty a_{nk} z_n = \sum_{k=0}^m (A_k^T z) x_k,$$

and $A^T z \in X^\beta$ implies

$$\sum_{k=0}^\infty (A_k^T z) x_k = \lim_{m \to \infty} \sum_{k=0}^m (A_k^T z) x_k = \lim_{m \to \infty} \sum_{n=0}^\infty z_n A_n x^{[m]}.$$

Since X has AK, we have

$$\|L_{A^T}\| = \sup_{\|z\|=1} \|A^T z\|_{X^\beta} = \sup_{\|z\|=1} \|A^T z\|_X^* = \sup_{\|z\|=1} \sup_{\|x\|=1} \left| \sum_{k=0}^\infty (A_k^T z) x_k \right|$$

$$= \sup_{\|z\|=1} \sup_{\|x\|=1} \lim_{m\to\infty} \left| \sum_{n=0}^\infty z_n A_n x^{[m]} \right| \le \sup_{\|x\|=1} \lim_{m\to\infty} \|L_A x^{[m]}\|$$

$$= \sup_{\|x\|=1} \|L_A(x)\| = \|L_A\|.$$

Thus we have shown (9.72).

(ii) Now we show that
$$\|L_A\| \le \|L_{A^T}\|. \tag{9.73}$$

Let $A^T \in (Z, X^\beta)$ and $x \in X$. Since $Ax \in (X, Y) = (X, Z^\beta)$, we obtain for all $z \in Z$,

$$\sum_{n=0}^\infty z_n A_n x = \sum_{n=0}^\infty z_n \sum_{k=0}^\infty a_{nk} x_k = \lim_{m\to\infty} \sum_{n=0}^m z_n \sum_{k=0}^\infty a_{nk} x_k$$

$$= \lim_{m\to\infty} \sum_{k=0}^\infty x_k \sum_{n=0}^m a_{nk} z_n = \lim_{m\to\infty} \sum_{k=0}^\infty (A_k^T z^{[m]}) x_k.$$

Since $Y = Z^\beta$ and Z has AK, we have

$$\|L_A\| = \sup_{\|x\|=1} \|L_A x\| = \sup_{\|x\|=1} \|Ax\|_Z^* = \sup_{\|x\|=1} \sup_{\|z\|=1} \left| \sum_{n=0}^\infty z_n A_n x \right|$$

$$= \sup_{\|x\|=1} \sup_{\|z\|=1} \lim_{m\to\infty} \left| \sum_{k=0}^\infty (A_k^T z^{[m]}) x_k \right| = \sup_{\|z\|=1} \lim_{m\to\infty} \|A^T z^{[m]}\|$$

$$= \sup_{\|z\|=1} \|A^T z\| = \|L_{A^T}\|.$$

Thus we have shown (9.73). $\square$

Remark 9.10.3 *If $X = \ell_\infty$ or $X = c$, Z is a BK space with AK, $Y = Z^\beta$ and $\|\cdot\|_{Z^\beta} = \|\cdot\|_Z^*$, then $(\ell_\infty, Y) = (c_0^{\beta\beta}, Y) = (c_0, Y)$ by Theorem 9.6.3 (a), and since $c_0 \subset c \subset \ell_\infty$, we also have $(c, Y) = (c_0, Y)$, and Theorem 9.6.3 (b) and the fact that $X^\beta = c_0^\beta$ imply that $A \in (X, Y)$ if and only if $A^T \in (Z, X^\beta)$.*
Furthermore, since $\sup_m \|x^{[m]}\|_\infty = \|x^{[m]}\|_\infty$ for all $m \in \mathbb{N}_0$ and all $x \in X$ when $X = c$ or $X = \ell_\infty$, we also obtain $\|L_A\| = \|L_{A^T}\|$, if we modify the proof of (9.72) to

$$\|L_{A^T}\| \le \sup_{\|x\|_\infty=1} \sup_m \|L_A(x^{[m]})\| \le \sup_{\|x\|_\infty=1} \sup_m \|L_A\| \cdot \|x^{[m]}\|_\infty$$

$$= \sup_{\|x\|_\infty=1} \|L_A\| \cdot \|x\|_\infty = \|L_A\|.$$

We note that if X and Y are any of the classical sequence spaces with the properties of Lemma 9.10.2 or Remark 9.10.3, then the following result can be used which we give without proof.

Lemma 9.10.4 ([179, Theorem 3]) *Let X, Y and Z be BK spaces with the properties in Lemma 9.10.2 or Remark 9.10.3 and $A \in (X, Y)$. Then we have $L_A \in \mathcal{K}(X, Y)$ if and only if $L_{A^T} \in \mathcal{K}(Z, X^\beta)$.*

Lemma 9.10.5 ([179, Corollary p. 84]) *Let X, Y and Z be BK spaces with the properties in Lemma 9.10.2 or Remark 9.10.3, $A \in (X,Y)$ and X^β have AK. Then we have $L_A \in \mathcal{K}(X,Y)$ if and only if*

$$\lim_{n \to \infty} \|L_{A^{>n<}}\| = 0, \tag{9.74}$$

where the matrices $A^{>n<}$ for $n = 0,1,\dots$ are obtained by replacing the columns $A^0, A^1, \dots, A^n$ of A by the sequence $(0,0,\dots)$.

Proof. If $A \in (X,Y)$, then we have by Lemma 9.10.4 that $L_A \in \mathcal{K}(X,Y)$ if and only if $L_B \in \mathcal{K}(Z,X^\beta)$, where $B = A^T$, and since X^β has AK and trivially $B^{<n>} = A^{>n<}$, (9.74) follows from (9.71), Lemma 9.10.2 and Remark 9.10.3. $\qquad\square$

Lemma 9.10.6 *Let $1 \leq p < \infty$ and $A \in (\ell_1, \ell_p)$. Then we have*

$$\|L_A\| = \|A\|_{(\ell_1,\ell_p)} = \sup_k \left(\sum_{n=0}^{\infty} |a_{nk}|^p \right)^{1/p}. \tag{9.75}$$

Proof. We assume $A \in (\ell_1, \ell_q)$. Then $L_A \in \mathcal{B}(\ell_1, \ell_q)$ by Theorem 9.3.3 (a), and we obtain by Minkowski's inequality for all $x \in \ell_1$

$$\|L_A(x)\|_p = \left(\sum_{n=0}^{\infty} |A_n x|^p \right)^{1/p} \leq \left(\sum_{n=0}^{\infty} \left(\sum_{k=0}^{\infty} |a_{nk} x_k| \right)^p \right)^{1/p}$$

$$\leq \sum_{k=0}^{\infty} \left(|x_k| \left(\sum_{n=0}^{\infty} |a_{nk}|^p \right)^{1/p} \right) \leq \|A\|_{(\ell_1,\ell_p)} \cdot \|x\|_1,$$

whence

$$\|L_A\| \leq \|A\|_{(\ell_1,\ell_p)}. \tag{9.76}$$

We also have for each k

$$\|L(e^{(k)})\|_p = \left(\sum_{n=0}^{\infty} |A_n e^{(k)}|^p \right)^{1/p} = \left(\sum_{n=0}^{\infty} |a_{nk}|^p \right)^{1/p} \leq \|L_A\| \cdot \|e^{(k)}\|_1 = \|L_A\|,$$

hence $\|A\|_{(\ell_1,\ell_p)} \leq \|L_A\|$ and the last inequality and (9.76) together imply (9.75). $\qquad\square$

Remark 9.10.7 *We already know the case $p = 1$ of Lemma 9.10.6 from (9.14) in Example 9.3.4.*

Now we give the known characterizations of the classes of compact matrix operators between the classical sequence spaces.

Theorem 9.10.8 *Let $1 < p, r < \infty$, $q = p/(p-1)$, $s = r/(r-1)$, N_m and K_m denote any finite subsets of nonnegative integers greater or equal to m, and $A \in (X,Y)$. Then the necessary and sufficient conditions for $L_A \in \mathcal{K}(X,Y)$ can be read from the following table:*

From	ℓ_∞	c_0	c	ℓ_1	ℓ_p
To ℓ_∞	1.	2.	3.	4.	5.
c_0	6.	7.	8.	9.	10.
c	11.	12.	13.	14.	15.
ℓ_1	16.	17.	18.	19.	20.
ℓ_r	21.	22.	23.	24.	unknown

where

1., 2., 3. (1.1^+) $\displaystyle\lim_{m\to\infty}\left(\sup_n \sum_{k=m}^{\infty}|a_{nk}|\right)=0$

4. (4.1^+) $\begin{cases} \displaystyle\lim_{m\to\infty}\left(\sup_{0\le n\le m}|a_{n,k_1}-a_{n,k_2}|\right)=\sup_{n\ge 0}|a_{n,k_1}-a_{n,k_2}| \\ \text{\textit{uniformly in} } k_1 \text{ \textit{and} } k_2 \text{ \textit{for} } 0\le k_1,k_2<\infty \end{cases}$

5. (5.1^+) $\displaystyle\lim_{m\to\infty}\left(\sup_n \sum_{k=m}^{\infty}|a_{nk}|^q\right)^{1/q}=0$

6., 7., 8. (6.1^+) $\displaystyle\lim_{m\to\infty}\left(\sum_{k=0}^{\infty}|a_{nk}|\right)=0$

9. (9.1^+) $\displaystyle\lim_{m\to\infty}\left(\sup_{n\ge m;k\ge 0}|a_{nk}|\right)=0$

10. (10.1^+) $\displaystyle\lim_{m\to\infty}\left(\sup_{n\ge m}\left(\sum_{k=0}^{\infty}|a_{nk}|^q\right)^{1/q}\right)=0$

11., 12. (11.1^+) $\begin{cases} \displaystyle\lim_{m\to\infty}\left(\sup_{n\ge m}\left(\sum_{k=0}^{\infty}|a_{nk}-\alpha_k|\right)\right)=0, \\ \text{\textit{where} } \alpha_k=\lim_{n\to\infty}a_{nk} \text{ \textit{for all} } k \end{cases}$

13. (13.1^+) $\begin{cases} \displaystyle\lim_{n\to\infty}\left(\left|\sum_{k=0}^{\infty}\alpha_k-\beta\right|+\sum_{k=0}^{\infty}|a_{nk}-\alpha_k|\right)=0, \\ \text{\textit{where} } \beta=\lim_{n\to\infty}\sum_{k=0}^{\infty}a_{nk} \end{cases}$

14. (14.1^+) $\displaystyle\lim_{m\to\infty}\left(\sup_{n\ge m;k\ge 0}|a_{nk}-\alpha_k|\right)=0$

15. (15.1^+) $\displaystyle\lim_{m\to\infty}\left(\sup_{n\ge m}\left(\sum_{k=0}^{\infty}|a_{nk}-\alpha_k|^q\right)^{1/q}\right)=0$

16., 17., 18. (16.1^+) $\displaystyle\lim_{m\to\infty}\left(\sup_{N_m}\left(\sum_{k=0}^{\infty}\left|\sum_{n\in N_m}a_{nk}\right|\right)\right)=0$

19. (19.1^+) $\displaystyle\lim_{m\to\infty}\left(\sup_k \sum_{n=m}^{\infty}|a_{nk}|\right)=0$

20. (20.1^+) $\displaystyle\lim_{m\to\infty}\left(\sup_{N_m}\left(\sum_{k=0}^{\infty}\left|\sum_{n\in N_m}a_{nk}\right|^q\right)^{1/q}\right)=0$

21., 22., 23. (21.1^+) $\displaystyle\lim_{m\to\infty}\left(\sup_{K_m}\left(\sum_{n=0}^{\infty}\left|\sum_{k\in K_m}a_{nk}\right|^r\right)^{1/r}\right)=0$

24. (24.1^+) $\displaystyle\lim_{m\to\infty}\left(\sup_k\left(\sum_{n=m}^{\infty}|a_{nk}|^r\right)^{1/r}\right)=0.$

Proof. **1, 2** and **3.** Since $X^\beta=\ell_1$ for $X=\ell_\infty,c_0,c$, (1.1^+) follows from (9.74) in Lemma

9.10.5, (9.10) and (9.11) in Theorem 9.3.3 (c) and the fact that $\|\cdot\|_{\infty}^{*} = \|\cdot\|_1$ by Example 9.5.3.

4. This is [179, Theorem 5].

5. We obtain (5.1^{+}) by a similar argument as in the proof of (1.1^{+}), using the fact that $\|\cdot\|_p^{*} = \|\cdot\|_q$ by Example 9.5.3.

6, 7 and **8.** Condition (6.1^{+}) in **6** follows by an analogous argument as that in the proof of **1** of Theorem 9.9.3 and by (7.63) in Theorem 7.11.4; Condition (6.1^{+}) in **7** is **1** in Corollary 9.9.4; and Condition (6.1^{+}) in **8** follows from (9.70) in Example 9.9.5 with $\beta = \alpha_k = 0$ for all k.

9 and **10.** Conditions (9.1^{+}) in **9** and (10.1^{+}) in **10** follow by (9.50) in Theorem 9.8.4 (b), since $\|\cdot\|_{\ell_1}^{*} = \|\cdot\|_{\infty}$ and $\|\cdot\|_{\ell_p}^{*} = \|\cdot\|_q$ by Example 9.5.3.

11. Condition (11.1^{+}) in **11** follows by an argument analogous to that in the proof of **3** of Theorem 9.9.3 and by (7.63) in Theorem 7.11.4 and Condition (11.1^{+}) in **12** is **3** in Corollary 9.9.4.

12. This follows by (9.48) in Theorem 9.8.4 (a) and since $\|\cdot\|_{c_0}^{*} = \|\cdot\|_1$ by Example 9.5.3.

13. This follows from Example 9.9.5.

14 and **15.** These parts follow by (9.48) in Theorem 9.8.4 (a), and since $\|\cdot\|_{\ell_1}^{*} = \|\cdot\|_{\infty}$ and $\|\cdot\|_{\ell_p}^{*} = \|\cdot\|_q$ by Example 9.5.3.

16, 17 and **18.** Since ℓ_1 has AK, these parts follow from (9.71) in Lemma 9.10.1, and (9.12) and (9.13) in Theorem 9.3.3 (d).

19. This is (9.47) in Remark 9.8.3.

20. Condition (20.1^{+}) follows similarly as condition (16.1^{+}).

21, 22 and **23.** It follows by Lemma 9.10.4 that $L_A \in \mathcal{K}(X, \ell_r)$ if and only if $L_{A^T} \in \mathcal{K}(\ell_s, \ell_1)$ for $X = \ell_{\infty}, c_0, c$. We put $B = A^T$. Since ℓ_1 has AK, we obtain by (9.71) in Lemma 9.10.1 and (9.12) and (9.13) in Theorem 9.3.3 (d)

$$\|L_{(A^T)<m-1>}\| = \|L_{B<m-1>}\| = \sup_N \left(\sum_{k=0}^{\infty} \left| \sum_{n \in N} b_{nk}^{<m-1>} \right|^r \right)^{1/r}$$

$$= \sup_{N_m} \left(\sum_{k=0}^{\infty} \left| \sum_{n \in N_m} b_{nk} \right|^r \right)^{1/r} = \sup_{N_m} \left(\sum_{k=0}^{\infty} \left| \sum_{n \in N_m} a_{kn} \right|^r \right)^{1/r}$$

$$= \sup_{K_m} \left(\sum_{n=0}^{\infty} \left| \sum_{k \in K_m} a_{nk} \right|^r \right)^{1/r}.$$

This immediately yields Condition (21.1^{+}).

24. This follows from (9.71) in Lemma 9.10.1 and (9.75) in Lemma 9.10.6. $\quad\square$

Pitt's famous theorem [155] states that in some cases every bounded matrix operator between certain sequence spaces is compact. Here we give the following version of Pitt's theorem as stated in [67, p. 175]. A simple proof can be found in [52].

Theorem 9.10.9 (Pitt) *Let $1 \le q < p \le \infty$, and put $X_p = \ell_p$ if $p < \infty$ and $X_{\infty} = c_0$. Then we have $\mathcal{B}(X_p, \ell_q) \subset \mathcal{K}(X_p, \ell_q)$.*

Remark 9.10.10 *Sargent [179, (a)–(g), p.85] characterized several classes of compact matrix operators between the classical sequence spaces without the use of the Hausdorff measure of noncompactness. The conditions for*

- *$L_A \in \mathcal{K}(\ell_{\infty}, \ell_{\infty}), \mathcal{K}(c_0, \ell_{\infty}), \mathcal{K}(c, \ell_{\infty})$ $((1.1^{+})$ in **1–3** of Theorem 9.10.8) can be found in [179, (b) $(p = \infty)$, (f), and (g), p. 85];*

- $L_A \in \mathcal{K}(\ell_p, \ell_\infty)$ for $1 < p < \infty$ ((5.1^+) in **5** of Theorem 9.10.8) can be found in [179, (b), p. 85];

- $L_A \in \mathcal{K}(\ell_1, \ell_1)$, $L_A \in \mathcal{K}(\ell_1, \ell_r)$ for $1 < r < \infty$ ((19.1^+) in **19** and (24.1^+) in **24** of Theorem 9.10.8) can be found in [179, (a), p. 85].

The remaining conditions in [179, p. 85] can be obtained by the use of Pitt's theorem, Theorem 9.10.9.

- We have $\mathcal{K}(c_0, \ell_r) = \mathcal{B}(c_0, \ell_r)$ for $1 \leq r < \infty$ by Pitt's theorem, Theorem 9.10.9, and [179, (9) in (e), p. 85] follows from (16.1^+) in **17** of Theorem 9.7.3 for $r = 1$ and from (21.1^+) in **22** of Theorem 9.7.3 for $1 < r < \infty$. Furthermore, since $\mathcal{B}(c_0, \ell_r) = \mathcal{B}(c, \ell_r) = \mathcal{B}(\ell_\infty, \ell_r)$, we also obtain [179, (9) in (d) and (g)] for the characterizations of $\mathcal{K}(c, \ell_r)$ and $\mathcal{K}(\ell_\infty, \ell_r)$.
  Compare these conditions to those in (16.1^+) in **16–18** and (21.1^+) in **21–23** of Theorem 9.10.8.

- We have $\mathcal{K}(\ell_p, \ell_1) = \mathcal{B}(\ell_p, \ell_1)$ for $1 < p \leq \infty$ by Pitt's theorem, Theorem 9.10.9, and the condition in [179, (c), p.85] follows from (16.1^+) in **16** of Theorem 9.7.3 for $p = \infty$ and (20.1^+) in **16** of Theorem 9.7.3 for $1 < p < \infty$.
  Compare these conditions with the corresponding ones in Theorem 9.10.8.

Chapter 10

Fixed Point Theory

Chapter 10 is different from the previous chapters in the sense that it deals with nonlinear functional analysis, perhaps the most important part of which is fixed point theory with its applications in the solution of nonlinear differential and integral equations that appear in physics, chemistry, biology and economics. It includes many of the most important results in fixed point theory, starting with Banach's classical contraction principle for which several different proofs are presented, as well as various corollaries and examples. Furthermore, it presents results by Edelstein and Rakotch, the concept of nonlinear contraction by Boyd and Wong, and theorems by Meir–Keeler, Kannan, Chatterje and Zamfirescu. It also deals with results related to the concepts of Ćirić's generalized contractions and quasi–contractions, and establishes the theorems by Reich, Hardy–Rogers, Caristi, and Bollenbacher and Hicks. The chapter closes with studies of the Mann iteration and fixed point theorems by Mann, Reinermann, Franks and Marzec for real functions on compact intervals of the real line.

10.1   Introduction

Fixed point theory is a major branch of nonlinear functional analysis because of its wide applicability. Numerous questions in physics, chemistry, biology, and economics lead to various nonlinear differential and integral equations.

The classical Banach contraction principle [9] is one of the most useful results in metric fixed point theory. Due to its applications in mathematics and other related disciplines, this principle has been generalized in many directions. Extensions of Banach's contraction principle have been obtained either by generalizing the distance properties of the underlying domain or by modifying the contractive condition on the mappings.

We refer the reader for further studies on fixed point theory to the monograph [114].

This chapter deals with a great number of the most important results in fixed point theory, starting with Banach's classical contraction principle, Theorem 10.2.2, for which several different proofs are presented, as well as various corollaries and examples. Furthermore, it presents results by Edelstein and Rakotch, Theorems 10.3.1, 10.3.2, 10.3.3, 10.4.2 and10.4.4; the concept of nonlinear contraction by Boyd and Wong in Definition 10.5.1; and theorems by Meir–Keeler, Theorem 10.6.2, Kannan, Theorem 10.7.1, Chatterjee, Theorem 10.7.6, and Zamfirescu, Theorem 10.7.7. It also deals with results related to the concepts of Ćirić's generalized contractions and quasi–contractions in Definitions 10.8.1 and 10.10.1, and establishes the theorems by Reich, Theorem 10.9.1, Hardy–Rogers, Theorem 10.9.4, Caristi, Theorem 10.11.1, and Bollenbacher and Hicks, Theorem 10.12.2. The chapter closes with studies of the Mann iteration and fixed point theorems by Mann, Theorem 10.13.2, Reinermann, Theorem 10.14.3, and Franks and Marzec for real functions on compact intervals of the real line, Theorem 10.14.4.

10.2   Banach contraction principle

In this section we are going to study the famous *Banach fixed point theorem*, also commonly referred to as the *Banach contraction principle*. This principle of 1922 marks the beginning of fixed point theory in metric spaces.

Definition 10.2.1 Let (X, d) be a metric space. A mapping $f : X \to X$ is called a *contraction* if there exists some $q \in [0, 1)$ such that

$$d(f(x), f(y)) \leq q \cdot d(x, y), \text{ for all } x, y \in X. \tag{10.1}$$

We shall also use the term *q–contraction* for contraction.

We observe that every contraction is a continuous mapping. The following theorem shows the existence and uniqueness of a fixed point of an arbitrary contraction on a complete metric space. It is important to mention that there exists a continuous mapping without a fixed point property.

Theorem 10.2.2 (Banach contraction principle)
If (X, d) is a complete metric space and $f : X \to X$ is a contraction, then the mapping f has a unique fixed point in $z \in X$.

Proof. Let $x_0 \in X$ be arbitrary. We define a sequence (x_n) in X such that $x_n = f(x_{n-1})$ for $n \in \mathbb{N}$.

(i)   First we prove that (x_n) is a Cauchy sequence, hence convergent, since X is a complete metric space X.
We obtain for any $n \in \mathbb{N}$,

$$d(x_n, x_{n+1}) = d(f(x_{n-1}), f(x_n)) \leq q \cdot d(x_{n-1}, x_n)$$
$$\leq \cdots \leq q^n \cdot d(x_0, x_1),$$

and therefore, if $m > n$,

$$d(x_n, x_m) \leq \sum_{k=n}^{m-1} d(x_k, x_{k+1}) \leq \sum_{k=n}^{m-1} q^k d(x_0, x_1) \leq \frac{q^n}{1-q} d(x_0, x_1). \tag{10.2}$$

Since $0 \leq q < 1$, it follows that $\lim_{n,m \to \infty} d(x_n, x_m) = 0$, hence (x_n) is a Cauchy sequence. Moreover, X is a complete metric space, and so there exists some $z \in X$ such that $\lim_{n \to \infty} x_n = z$.

(ii)   Now we show that $f(z) = z$ by estimating $d(x_n, f(z))$ for $n \in \mathbb{N}$.
Since

$$0 \leq d(x_n, f(z)) = d(f(x_{n-1}), f(z)) \leq q \cdot d(x_{n-1}, z),$$

we obtain $\lim_{n \to \infty} d(x_n, f(z)) = 0$, and by the uniqueness of the limit of any convergent sequence in a metric space, we conclude that $f(z) = z$.

(iii) It remains to prove that such z is uniquely determined.

We assume $f(y) = y$ for some $y \in X$ with $y \neq z$, then

$$d(z, y) = d(f(z), f(y)) \leq q \cdot d(z, y)$$

and $(1 - q)d(z, y) \leq 0$, which contradicts our assumption because $0 < 1 - q \leq 1$. $\quad\square$

Corollary 10.2.3 *Let $f : X \to X$ be a q–contraction on a complete metric space X and $z \in X$ be the fixed point of the function f. Then we have*

(1) the sequence $(f^n(x))$ converges for each $x \in X$ and converges to z;

(2) $d(x, z) \leq \dfrac{1}{1 - q} \cdot d(x, f(x))$;

(3) $d(f^n(x), z) \leq \dfrac{q^n}{1 - q} \cdot d(x, f(x))$;

(4) $d(f^{n+1}(x), z) \leq q \cdot d(f(x), x)$;

(5) $d(f^{n+1}(x), z) \leq \dfrac{q}{1 - q} \cdot d(f^n(x), f^{n+1}(x))$.

Proof. We only prove the inequalities in (2) and (3); the proofs of the other conditions are straightforward.

(2) The inequality in (2) follows from

$$d(x, z) = \lim_{n \to \infty} d(x, f^n(x)) \leq \lim_{n \to \infty} \sum_{k=0}^{n-1} d(f^k(x), f^{k+1}(x))$$

$$= \sum_{k=0}^{\infty} d(f^k(x), f^{k+1}(x)) \leq \sum_{k=0}^{\infty} q^k d(x, f(x))$$

$$= \frac{1}{1 - q} \cdot d(x, f(x)).$$

(3) It follows from $f(z) = z$ that $d(x, a)f^n(z) = z$, and we have from the first part of the proof

$$d(f^n(x), z) = d(f^n(x), f^n(z)) \leq q^n d(x, z) \leq \frac{q^n}{1 - q} \cdot d(x, f(x)).$$

$\quad\square$

Remark 10.2.4 *There exist various approaches to the Banach fixed point theorem, but the proof above gives a method of how to find the fixed point for a contraction f. It is also known as Picard's iteration method or* fixed point iteration. *It is based on the idea of defining a sequence of successive iterations. We start with any $x_0 \in X$ and define $x_n = f(x_{n-1})$ for $n \in \mathbb{N}$. The proof presented above guarantees the existence of a limit $\lim_{n \to \infty} x_n = z \in X$ such that $f(z) = z$. If we let $m \to \infty$ in (10.2), then*

$$d(x_n, z) \leq \frac{q^n}{1 - q} d(x_0, x_1),$$

and this is an estimate for the error made by approximating the solution z by the n–th iteration x_n.

We now present a few different proofs of Theorem 10.2.2.

Proof of Theorem 10.2.2 (Joseph and Kwack [103]). Let $c = \inf\{d(x, f(x)) : x \in X\}$. If $c > 0$, then $c/q > c$ and there exists $x \in X$ such that

$$d(f(x), f(f(x))) \leq q \cdot d(x, f(x)) < c,$$

which is a contradiction. Hence we must have $c = 0$. Let (x_n) be a sequence in X such that $d(x_n, f(x_n)) \to 0$ as $n \to \infty$. We show that (x_n) is a Cauchy sequence, since

$$d(x_n, x_m) \leq d(x_n, f(x_n)) + d(f(x_n), f(x_m)) + d(f(x_m), x_m)$$

implies

$$(1 - q)d(x_n, x_m) \leq d(x_n, f(x_n)) + d(x_m, f(x_m)).$$

Hence there exists $z \in X$ such that $\lim_{n \to \infty} x_n = z$, and $\lim_{n \to \infty} d(x_n, f(x_n)) = 0$ implies $\lim f(x_n) = z$. It follows from $d(f(x_n), f(z)) \leq qd(x_n, z)$ that $\lim_{n \to \infty} f(x_n) = f(z)$, hence $f(z) = z$. The uniqueness of the fixed point of the function f follows from the contractive condition of f. □

Proof of Theorem 10.2.2 (Palais [149]). Let $x_1, x_2 \in X$. Then we have

$$d(x_1, x_2) \leq d(x_1, f(x_1)) + d(f(x_1), f(x_2)) + d(f(x_2), x_2),$$

that is,

$$(1 - q)d(x_1, x_2) \leq d(x_1, f(x_1)) + d(f(x_2), x_2).$$

Hence we obtain the *fundamental contraction inequality*

$$d(x_1, x_2) \leq \frac{1}{1 - q} \cdot [d(x_1, f(x_1)) + d(x_2, f(x_2))], \text{ for all } x_1, x_2 \in X. \qquad (10.3)$$

If x_1 and x_2 are fixed points of the function f, then it follows from (10.3) that $x_1 = x_2$, that is, the contraction can have at most one fixed point.

Let $x \in X$, $n, m \in \mathbb{N}$, and $x_1 = f^n(x)$ and $x_2 = f^m(x)$. We obtain from (10.3) that

$$d(f^n(x), f^m(x)) \leq \frac{1}{1 - q} \cdot [d(f^n(x), f(f^n(x))) + d(f^m(x), f(f^m(x)))] \qquad (10.4)$$

$$\leq \frac{q^n + q^m}{1 - q} \cdot d(x, f(x)). \qquad (10.5)$$

Since $0 \leq q < 1$, it follows that $\lim_{n \to \infty} q^n = 0$, hence $d(f^n(x), f^m(x)) \to 0$ as $n \to \infty$ and $m \to \infty$. Therefore the Cauchy sequence $(f^n(x))$ converges, that is, there exists $z \in X$ such that $\lim_n f^n(x) = z$. Because of the continuity of the function f, we have $f(z) = f(\lim_n f^n(x)) = \lim_n f(f^n(x)) = z$. We note that letting $m \to \infty$ in (10.4), we obtain

$$d(f^n(x), z) \leq \frac{q^n}{1 - q} \cdot d(x, f(x)). \qquad □$$

Proof of Theorem 10.2.2 (Boyd and Wong [29]). We define $\varphi(x) = d(x, f(x))$ for $x \in X$. Since f is a contraction, the function $\varphi : X \to \mathbb{R}$ is continuous and $\varphi(f^n(x)) \to 0$ as $n \to \infty$, for each $x \in X$. We put

$$C_m = \left\{ x \in X : \varphi(x) \leq \frac{1}{m} \right\}.$$

It follows from the conditions above that C_m is a closed and nonempty subset of X for each $m = 1, 2, \ldots$. Now we estimate the diameter of the set C_m. Let $x, y \in C_m$. Then we have

$$d(x, y) \leq d(x, f(x)) + d(f(x), f(y)) + d(f(y), y) \leq \frac{2}{m} + qd(x, y),$$

hence

$$\text{diam}(C_m) \leq \frac{2}{m(1-q)}.$$

Since each C_m is a closed, nonempty subset of X, $C_1 \supset C_2 \supset C_3 \supset \ldots$ and $\text{diam}(C_m) \to 0$ as $m \to \infty$, it follows by Cantor's intersection theorem that $\bigcap_m C_m = \{\xi\}$.
Since $f(C_m) \subset C_m$ for each m, it follows that ξ is a fixed point of the function f, and clearly the fixed point is unique. (We note $f(\{\xi\}) = f(\bigcap_m C_m) \subset \bigcap_m f(C_m) \subset \bigcap_m C_m = \{\xi\}$.)
We have for each $x \in X$

$$d(f^n(x), \xi) = d(f^n(x), f^n(\xi)) \leq q^n d(x, \xi) \to 0 \ (n \to \infty).$$

Since

$$d(x, \xi) \leq d(x, f(x)) + d(f(x), f(\xi)) \leq d(x, f(x)) + qd(x, \xi),$$

it follows that

$$d(x, \xi) \leq \frac{d(x, f(x))}{1-q}.$$

Hence we again have the estimate

$$d(f^n(x), \xi) \leq \frac{q^n}{1-q} \cdot d(x, f(x)). \qquad \square$$

Corollary 10.2.5 *Let S be a closed subset of a complete metric space (X, d) and $f : S \to S$ be a contraction. For an arbitrary point $x_0 \in S$, the iterative sequence $x_n = f(x_{n-1})$ $(n \in \mathbb{N})$ converges to the fixed point of the mapping f.*

The following example will show that the statement in Corollary 10.2.5 does not hold without the assumption that the set S is closed, in general.

Example 10.2.6 *Let d be the natural metric on $\mathbb{R}$ defined by $d(x, y) = |x-y|$ for all $x, y \in \mathbb{R}$, and $S = B_0(1) = \{x \in \mathbb{R} : |x| < 1\}$. Then the mapping*

$$f : S \to S \text{ with } f(x) = \frac{x+1}{2},$$

is a contraction without a fixed point in S.

Banach's fixed point theorem has wide and diverse applications, for instance, in solving various kinds of equations, inclusions, etc.

Example 10.2.7 *If X is a Banach space, $A, B \in \mathcal{B}(X)$, A is an invertible operator and $\|B-A\| \cdot \|A^{-1}\|^{-1} < 1$, then the invertibility of B follows from Banach's fixed point theorem.*

Proof. It is sufficient to show that, for any $y \in X$, the equation $Bx = y$ has a unique solution $x \in X$.
We choose an arbitrary point y in X. If $Bx_0 = y$ for some $x_0 \in X$, then

$$y = Bx_0 = (B - A)x_0 + Ax_0 \text{ and } A^{-1}y = A^{-1}(B - A)x_0 + x_0.$$

We put $z = A^{-1}y$ and $C = A^{-1}(B - A)$. Then we have $x_0 = z - Cx_0$.
The idea is to show that the function $f : X \to X$ defined by $f(x) = z - Cx$ for $x \in X$ is a contraction and x_0 is its fixed point.
The following inequalities hold for all $x, y \in X$

$$\|f(x) - f(y)\| = \|C(x - y)\| \leq \|A^{-1}\| \cdot \|B - A\| \cdot \|x - y\|.$$

Since $\|A^{-1}\| \cdot \|B - A\| < 1$, f is a contraction and x_0 is the unique fixed point of f. Based on a few elements of an iterative sequence $(f^n(x))$,

$$z - Cx, \ z - C(z - Cx) = z - Cx + C^2x, \ z - Cx + C^2x - C^3x, \ldots.$$

We may assume, and then easily prove that, because of $\|C\| < 1$, this sequence converges to $z - Cx + C^2x - C^3x + \cdots$.

We observe that if $A = I$ and $\|C\| < 1$, then $I - C + C^2 - C^3 + \cdots$ is an inverse of $I + C$. $\square$

The following corollary shows a relation between f^n and f in the case when f^n is a contraction.

Corollary 10.2.8 (Bryant [34]) *If (X, d) is a complete metric space and $f : X \to X$ is a mapping such that f^n is a contraction for some $n \geq 1$, then f has a unique fixed point in X.*

Proof. By Banach's fixed point theorem, there exists a unique $z \in X$ such that $f^n(z) = z$. Since $f^n(f(z)) = f(f^n(z)) = f(z)$, it follows that $f(z) = z$. Every fixed point of f is, at the same time, a fixed point of f^n, thus z is the unique fixed point of f. $\square$

As observed in [34], the mapping f mentioned in Corollary 10.2.8 need not be continuous as in Theorem 10.2.2.

Example 10.2.9 (Bryant [34]) *We define $f : [0, 2] \to [0, 2]$ by $f(x) = 1$ for $x \in [0, 1)$, and $f(x) = 2$ for $x \in [1, 2]$. Then $f^2(x) = 2$ for $x \in [0, 2]$ and $f^2 : [0, 2] \to [0, 2]$ is a contraction although f is not continuous.*

Since the proof of Banach's contraction principle is based on an iterative sequence for a point $x \in X$, the next reasonable step in the research was to check local properties and modify this result.

Theorem 10.2.10 *Let (X, d) be a complete metric space and $B_r(x_0)$ denote, as usual, the open ball in X for some $x_0 \in X$ and $r > 0$. Also let $f : B_r(x_0) \to X$ be a contraction, that is,*

$$d(f(x), f(y)) \leq q \cdot d(x, y) \text{ for some } q \in [0, 1) \text{ and all } x, y \in B_r(x_0) \qquad (10.6)$$

and

$$d(f(x_0), x_0) < (1 - q)r. \qquad (10.7)$$

Then the mapping f has a unique fixed point in $B_r(x_0)$.

Proof. We choose $r_0 \in [0, r)$ such that (10.7) holds. Then we show that $f : \overline{B}_{r_0}(x_0) \to \overline{B}_{r_0}(x_0)$, where $\overline{B}_{r_0}(x_0)$ is the closure of $B_{r_0}(x_0)$. This follows, since, for any $x \in \overline{B}_{r_0}(x_0)$,

$$d(f(x), x_0) \leq d(f(x), f(x_0)) + d(f(x_0), x_0)$$
$$\leq q \cdot d(x, x_0) + (1 - q)r_0.$$

Hence f has a unique fixed point $z \in \overline{B}_{r_0}(x_0)$. It easily follows from (10.6) that z is the unique fixed point of f in $B_r(x_0)$. $\square$

10.3 Edelstein's results

For a function $f : X \to X$ on a complete metric space (X, d) which satisfies the condition

$$d(f(x), f(y)) < \lambda \, d(x, y) \text{ for all } x, y \in X \text{ with } x \neq y, \qquad (10.8)$$

where $0 \leq \lambda < 1$, the Banach contraction principle yields the existence and uniqueness of fixed points.

If we take $\lambda = 1$ in the condition in (10.8) then we obtain a contractive map, that is, a map which satisfies the condition

$$d(f(x), f(y)) < d(x, y) \text{ for all } x, y \in X \text{ with } x \neq y. \qquad (10.9)$$

In 1962, Edelstein [62] published a paper in which he studied the fixed points of contractive maps using the next condition and assumption.

The condition in (10.9) together with the assumption of the existence of $x \in X$ such that the iterative sequence $(f^n(x))$ contains a convergent subsequence $(f^{n_k}(x))$ in X, that is,

$$\text{there exists } x \in X \text{ such that } \{f^n(x)\} \supset \{f^{n_k}(x)\} \text{ with } \lim_{k \to \infty} f^{n_k}(x) \in X, \qquad (10.10)$$

provides the existence of a fixed point of f.

Theorem 10.3.1 (Edelstein [62]) *Let X be a metric space and $f : X \to X$ be a contractive map that satisfies the condition in (10.10). Then $u = \lim_{k \to \infty} f^{n_k} x$ is the unique fixed point of f.*

Proof. Let $\Delta = \{(x, x) : x \in X\}$, $Y = (X \times X) \setminus \Delta$, and $r : Y \to \mathbb{R}$ be the map defined by

$$r(x, y) = \frac{d(f(x), f(y))}{d(x, y)}. \qquad (10.11)$$

The function r is continuous on Y, and there exists a neighborhood U of points $(u, f(u))$ such that $(x, y) \in U$ implies

$$0 \leq r(x, y) < R < 1. \qquad (10.12)$$

Let $B_1 = B_\rho^1(u)$ and $B_2 = B_\rho^2(f(u))$ be the open balls with centres in u and $f(u)$, and radius ρ such that

$$\rho < \frac{1}{3} d(u, f(u)) \qquad (10.13)$$

and $B_1 \times B_2 \subset U$.

It follows from (10.10) that there exists a natural number N such that $k > N$ implies $f^{n_k}(x) \in B_1$, and (10.9) implies $f^{n_k+1}(x) \in B_2$.

For $k > N$, (10.13) implies

$$d(f^{n_k}(x), f^{n_k+1}(x)) > \rho, \qquad (10.14)$$

and it follows from (10.11) and (10.12) that

$$d(f^{n_k+1}(x), f^{n_k+2}(x)) < R d(f^{n_k}(x), f^{n_k+1}(x)). \qquad (10.15)$$

Hence, (10.15) implies for $l > j > N$ that

$$d(f^{n_l}(x), f^{n_l+1}(x)) \leq d(f^{n_{l-1}+1}(x), f^{n_{l-1}+2}(x))$$

$$< Rd(f^{n_{l-1}}(x), f^{n_{l-1}+1}(x)) \leq \cdots$$
$$< R^{l-j}d(f^{n_j}(x), f^{n_j+1}(x)) \to 0 \; (l \to \infty),$$

which is a contradiction to (10.14). Thus we must have $f(u) = u$.
We assume that $v \neq u$ is also a fixed point of the function f. Then we have

$$d(f(u), f(v)) = d(u, v),$$

which is a contradiction to (10.9).   □

The condition in (10.10) is always satisfied for a compact space. Therefore we have:

Theorem 10.3.2 (Edelstein [62]) *Let (X, d) be a compact metric space and $f : X \to X$ be a map. We assume*

$$d(f(x), f(y)) < d(x, y) \text{ for all } x, y \in X \text{ with } x \neq y.$$

Then the function f has a unique fixed point.

We obtain the following result on the iteration sequence from Theorem 10.3.1.

Theorem 10.3.3 (Edelstein [62])
We assume that the conditions of Theorem 10.3.1 are satisfied. If the sequence $(f^n(p))$ for $p \in X$ contains a convergent subsequence $(f^{n_k}(p))$ then its limit $u = \lim_{n \to \infty} f^n(p)$ in X exists and u is a fixed point of f.

Proof. By Theorem 10.3.1, we have $u = \lim_{k \to \infty} f^{n_k}(p)$. For given $\delta > 0$, there exists $n_0 \in \mathbb{N}$ such that $k > n_0$ implies $d(u, f^{n_k}(p)) < \delta$. If $m = n_k + l > n_k$, then we have

$$d(u, f^m(p)) = d(f^l(u), f^{n_k+l}(p)) < d(u, f^{n_k}(p)) < \delta. \qquad \square$$

10.4   Rakotch's results

The problem of defining a family of functions $F = \{\alpha(x, y)\}$ which satisfy the conditions $0 \leq \alpha < 1$ and $\sup \alpha(x, y) = 1$ such that Banach's fixed point theorem is satisfied when the constant α is replaced by $\alpha(x, y) \in F$ was suggested by *H. Hanani*, and *Rakotch* published a result related to this problem in 1962 [157]. In this section, we present some results from the mentioned paper.

Definition 10.4.1 ([157, Definition 2]) Let (X, d) be a metric space. We denote by F_1 the family of all functions $\alpha(x, y)$ which satisfy the following conditions:

(1) $\alpha(x, y) = \alpha(d(x, y))$, that is, α depends only on the distance of x and y;

(2) $0 \leq \alpha(d(x, y)) < 1$ for all $x \neq y$;

(3) $\alpha(d)$ is a monotone decreasing function of d.

Theorem 10.4.2 ([157, Theorem 1]) *Let (X, d) be a metric space, $f : X \to X$ be a contractive map, $M \subset X$ and $x_0 \in M$ such that*

$$d(x, x_0) - d(f(x), f(x_0)) \geq 2d(x_0, f(x_0)) \text{ for all } x \in X \setminus M, \qquad (10.16)$$

and let $f(M)$ be a subset of a compact subset of X. Then there exists a unique fixed point of f.

Proof. We assume $f(x_0) \neq x_0$ and put $x_n = f^n(x_0)$ for $n = 1, 2, \ldots$, that is,

$$x_{n+1} = f(x_n) \text{ for } n = 0, 1, \ldots. \tag{10.17}$$

By Edelstein's theorem (Theorem 10.3.1), it suffices to show that $x_n \in M$ for each n. Since f is a contractive map, the sequence $(d(x_n, x_{n+1}))$ is not increasing. Hence $f(x_0) \neq x_0$ implies

$$d(x_n, x_{n+1}) < d(x_0, x_1) \text{ for } n = 1, 2, \ldots. \tag{10.18}$$

We obtain from the triangle inequality

$$d(x_0, x_n) \leq d(x_0, x_1) + d(x_1, x_{n+1}) + d(x_n, x_{n+1}).$$

Now (10.17) and (10.18) yield

$$d(x_0, x_n) - d(f(x_0), f(x_n)) < 2d(x_0, f(x_0)),$$

and (10.16) implies $x_n \in M$ for all n. $\qquad\square$

Corollary 10.4.3 ([157, Corollary, p. 460]) *Let f be a contractive map for which there exists a point $x_0 \in X$ such that for all $x \in X$*

$$d(f(x), f(x_0)) \leq \alpha(x, x_0)d(x, x_0), \tag{10.19}$$

where $\alpha(x, y) = \alpha(d(x, y)) \in F_1$. If $B_r(x_0)$ is the open ball in X, where

$$r = \frac{2d(x_0, f(x_0))}{1 - \alpha(2d(x_0, f(x_0)))},$$

and $f(B_r(x_0))$ is a subset of a compact subset of X, then the function f has a unique fixed point.

Proof. If we put $M = B(x_0, r)$ in Theorem 10.4.2, then by (10.19), the monotony of $\alpha(d)$ and $r \geq 2d(x_0, fx_0)$, the condition $d(x, x_0) \geq r$ implies

$$\begin{aligned}
d(x, x_0) - d(f(x), f(x_0)) &\geq d(x, x_0) - \alpha(d(x, x_0))d(x, x_0) \\
&= [1 - \alpha(d(x, x_0))]d(x, x_0) \geq [1 - \alpha(r)]r \\
&\geq [1 - \alpha(2d(x_0, f(x_0)))]r = 2d(x_0, f(x_0)),
\end{aligned}$$

that is, we have (10.16). $\qquad\square$

Theorem 10.4.4 ([157, Theorem 2]) *Let $f : X \to X$ be a contractive map on a complete metric space. We assume that there exist $M \subset X$ and a point $x_0 \in M$ such that*

$$d(x, x_0) - d(f(x), f(x_0)) \geq 2d(x_0, f(x_0)) \text{ for each } x \in X \setminus M, \tag{10.20}$$

$$d(f(x), f(y)) \leq \alpha(x, y)d(x, y) \text{ for all } x, y \in M, \tag{10.21}$$

where

$$\alpha(x, y) = \alpha(d(x, y)) \in F_1.$$

Then the function f has a unique fixed point.

Proof. We assume $f(x_0) \neq x_0$ and define the sequence (x_n) by $x_n = f^n(x_0)$ for $n = 1, 2, \ldots$. As in Theorem 10.4.2, we have by (10.20)

$$d(x_n, x_{n+1}) < d(x_0, x_1) \text{ for } n = 1, 2, \ldots \tag{10.22}$$

and $x_n \in M$ for each n.

We are going to prove that the sequence (x_n) is bounded. It follows from (10.21) and the definition of the sequence (x_n) that

$$d(x_1, x_{n+1}) = d(f(x_0), f(x_n)) \leq \alpha(d(x_0, x_n))d(x_0, x_n), \tag{10.23}$$

and, by the triangle inequality, we have

$$d(x_0, x_n) \leq d(x_0, x_1) + d(x_1, x_{n+1}) + d(x_n, x_{n+1}).$$

Hence (10.22) and (10.23) imply

$$[1 - \alpha(d(x_0, x_n))]d(x_0, x_n) < 2d(x_0, x_1).$$

If $d(x_0, x_n) \geq d_0$ for some given d_0, then we have by the monotony of α

$$\alpha(d(x_0, x_n)) \leq \alpha(d_0).$$

So we obtain

$$d(x_0, x_n) < \frac{2d(x_0, x_1)}{1 - \alpha(d(x_0, x_n))} \leq \frac{2d(x_0, x_1)}{1 - \alpha(d_0)} = C.$$

Hence we have for $R = \max\{d_0, C\}$

$$d(x_0, x_n) \leq R \text{ for } n = 1, 2, \ldots, \tag{10.24}$$

that is, the sequence (x_n) is bounded.

Let $p > 0$ be an arbitrary natural number. It follows from (10.21) that

$$d(x_{k+1}, x_{k+p+1}) \leq \alpha(x_k, x_{k+p})d(x_k, x_{k+p}),$$

that is,

$$d(x_n, x_{n+p}) \leq d(x_0, x_p) \prod_{k=0}^{n-1} \alpha(x_k, x_{k+p}).$$

Now (10.24) implies

$$d(x_n, x_{n+p}) \leq R \prod_{k=0}^{n-1} \alpha(x_k, x_{k+p}). \tag{10.25}$$

We prove that (x_n) is a Cauchy sequence. It is enough to show that, for every $\varepsilon > 0$, there exists N which depends only on ε (and not on p) such that, for all $p > 0$, we have $d(x_N, x_{N+p}) < \varepsilon$ (since the sequence $(d(x_n, x_{n+p}))$ is not increasing).

If $d(x_k, x_{k+p}) \geq \varepsilon$ for $k = 0, 1, \ldots, n - 1$, then we obtain from (10.21) (because of the monotony of the function α)

$$\alpha(x_k, x_{k+p}) = \alpha(d(x_k, x_{k+p})) \leq \alpha(\varepsilon),$$

and then (10.25) implies

$$d(x_n, x_{n+p}) \leq R[\alpha(\varepsilon)]^n.$$

Since $\alpha(\varepsilon) < 1$ and $[\alpha(\varepsilon)]^n \to 0$ as $n \to \infty$, there exists a natural number N, independent of p, such that $d(x_N, x_{N+p}) < \varepsilon$ for each $p > 0$. Hence (x_n) is a Cauchy sequence.

Since X is a complete metric space, there exists $u \in X$ such $u = \lim_{n \to \infty} x_n$. Because of the continuity of the function f, u is a fixed point of f. $\qquad \square$

In particular, if $M = X$, we obtain the next corollary.

Corollary 10.4.5 ([157, Corollary, p. 463]) *Let (X, d) be a complete metric space and*

$$d(f(x), f(y)) \leq \alpha(x, y)d(x, y) \text{ for all } x, y \in X, \qquad (10.26)$$

where $\alpha(x, y) \in F_1$. Then the function f has a unique fixed point.

Remark 10.4.6 ([157, Remark 4]) *The preceding corollary and Theorem 10.4.4 are generalizations of Banach's fixed point theorem.*

10.5   Boyd and Wong's nonlinear contraction

Here we present some results by Boyd and Wong [30] in 1969. In this paper, Boyd and Wong studied fixed points for maps of the kind introduced in the next definition.

Definition 10.5.1 ([30, p. 458]) *Let (X, d) be a metric space. A map $f : X \to X$ which satisfies the condition*

$$d(f(x), f(y)) \leq \Psi(d(x, y)) \text{ for all } x, y \in X, \qquad (10.27)$$

where Ψ is a function defined on the closure of the range of d, is called a Ψ contraction. We denote the image of d by P and the closure of P by $\overline{P}$. Hence $P = \{d(x, y) : x, y \in X\}$.

Rakotch [157] proved that if $\Psi(t) = \alpha(t)t$, where α is a decreasing function with $\alpha(t) < 1$ for all $t > 0$, then the map f satisfying (10.27) has a unique fixed point u. It can be shown that if $\Psi(t) = \alpha(t)t$ and α is an increasing function with $\alpha(t) < 1$ for all $t \geq 0$, then the conclusion of Banach's theorem holds true. Boyd and Wong proved that it is enough to assume that $\Psi(t) < t$ for all $t > 0$ and Ψ is semicontinuous, and if a metric space is convex, then the last condition can be omitted.

We recall that a function $\varphi : X \to E$ $(E \subset \mathbb{R})$ is said to be *upper semi–continuous from the right at $t_0 \in X$* if $t_n \to t_0+$ implies $\limsup_{n \to \infty} \varphi(t_n) \leq \varphi(t_0)$. A function $\varphi : X \mapsto E$ $(E \subset \mathbb{R})$ is said to be *upper semi–continuous from the right on X* if it is upper semi–continuous from the right at every $t \in X$.

Theorem 10.5.2 ([30, Theorem]) *Let (X, d) be a complete metric space and $f : X \to X$ be a map satisfying (10.27), where $\Psi : \overline{P} \mapsto [0, \infty)$ is upper semi–continuous from the right on $\overline{P}$ and satisfies $\Psi(t) < t$ for all $t \in \overline{P} \setminus \{0\}$. Then the function f has a unique fixed point x_0 and $f^n(x) \to x_0$ $(n \to \infty)$ for each $x \in X$.*

Proof. Let $x \in X$ and

$$c_n = d(f^n(x), f^{n-1}(x)) \text{ for } n = 1, 2, \dots. \qquad (10.28)$$

Then, because of (10.27), the sequence (c_n) is monotone decreasing. We put $\lim_{n \to \infty} c_n = c \geq 0$, and prove $c = 0$. If $c > 0$, then we have

$$c_{n+1} \leq \Psi(c_n), \qquad (10.29)$$

hence

$$c \leq \limsup_{t \to c+} \Psi(t) \leq \Psi(c) < c, \qquad (10.30)$$

which is a contradiction.

We are going to prove that $(f^n(x))$ is a Cauchy sequence for each $x \in X$. Then the limit point of this sequence is the unique fixed point of the function f. We assume that $(f^n(x))$ is not a Cauchy sequence. Then there exist $\varepsilon > 0$ and sequences $(m(k))$ and $(n(k))$ of natural numbers with $m(k) > n(k) \geq k$ such that

$$d_k = d(f^{m(k)}(x), f^{n(k)}(x)) \geq \varepsilon \text{ for all } k = 1, 2, \ldots. \tag{10.31}$$

We may assume that

$$d(f^{m(k)-1}(x), f^{n(k)}(x)) < \varepsilon, \tag{10.32}$$

and choose $m(k)$ as the smallest integer greater than $n(k)$ which satisfies (10.31). It follows from (10.28) that

$$d_k \leq d(f^{m(k)}(x), f^{m(k)-1}(x)) + d(f^{m(k)-1}(x), f^{n(k)}(x)) \leq c_m + \varepsilon \leq c_k + \varepsilon. \tag{10.33}$$

Hence $d_k \to \varepsilon$ as $k \to \infty$. Since

$$d_k = d(f^{m(k)}(x), f^{n(k)}(x)) \leq d(f^{m(k)}(x), f^{m(k)+1}(x)) +$$
$$+ d(f^{m(k)+1}(x), f^{n(k)+1}(x)) + d(f^{n(k)+1}(x), f^{n(k)}(x))$$
$$\leq 2c_k + \Psi(d(f^{m(k)}(x), f^{n(k)}(x))) = 2c_k + \Psi(d_k), \tag{10.34}$$

letting $k \to \infty$ in (10.34), we obtain $\varepsilon \leq \Psi(\varepsilon)$. This is a contradiction, because we have $\Psi(\varepsilon) < \varepsilon$ for $\varepsilon > 0$. $\qquad\square$

The following example will show that the condition of the continuity of the function Ψ in Theorem 10.5.2 cannot be dropped, in general.

Example 10.5.3 ([30, Remark 1]) *Let* $X = \{x_n = n\sqrt{2} + 2^n : n = 0, \pm 1, \pm 2, \ldots\}$ *have the metric* $d(x, y) = |x - y|$. *Then* X *is a closed subset of the real numbers, and so complete. We assume that for each* $p \in P$ ($p \neq 0$), *there exists a unique pair* (x_n, x_m) *such that* $p = d(x_n, x_m)$. *We assume that*

$$d(x_j, x_k) = d(x_m, x_n) \text{ for some integers } j, k, m, n \text{ with } j > k \text{ and } m > n.$$

Then we obtain

$$-(m - n - j + k)\sqrt{2} = 2^j - 2^k - 2^m + 2^n. \tag{10.35}$$

Since the left-hand side in (10.35) is irrational or equal to zero and the right-hand side is rational, it follows that both sides are equal to zero. Hence we have for $m - n = j - k = s$

$$2^{n+s} - 2^n = 2^{k+s} - 2^k, \tag{10.36}$$

which is only possible for $n = k$. *We define the functions* f *by* $f(x_n) = x_{n-1}$ *and* Ψ *on* P *by*

$$\Psi(p) = |x_{n-1} - x_{m-1}| \text{ if } p = |x_n - x_m|. \tag{10.37}$$

We put $\Psi(p) = 0$ *for* $p \in \overline{P} \setminus P$.
Then we have $\Psi(t) < t$ *for all* $t \in \overline{P} \setminus \{0\}$ *and*

$$d(f(x), f(y)) = \Psi(d(x, y)), \tag{10.38}$$

but the function f *has no fixed points.*

Theorem 10.5.2 shows that it is not possible to extend the function Ψ from the set P to the set $\overline{P}$ such that it is upper semi–continuous from the right with $\Psi(t) < t$ for $t \in \overline{P} \setminus \{0\}$. This can directly be seen for the point $\sqrt{2} \in \overline{P} \setminus P$.

If the condition $\Psi(t) < t$ is replaced by $\Psi(t_0) = t_0$ for some value t_0, then Theorem 10.5.2 does not hold. This is shown in the next example.

Example **10.5.4** ([30, Remark 2]) *Let* $X = (-\infty, -1] \cup [1, \infty)$ *and* $d(x, y) = |x - y|$ *for all* $x, y \in X$. *Also let*

$$f_1(x) = \begin{cases} \dfrac{1}{2}(x+1) & \text{for } x \geq 1 \\ \dfrac{1}{2}(x-1) & \text{for } x \leq -1. \end{cases} \quad \text{and} \quad f_2(x) = -f_1(x).$$

Now the functions f_1 *and* f_2 *satisfy (10.27), if we define*

$$\Psi(t) = \begin{cases} \dfrac{1}{2}t & \text{for } t < 2 \\ \dfrac{1}{2}t + 1 & \text{for } t \geq 2. \end{cases}$$

We know that the function Ψ *satisfies all the conditions in Theorem 10.5.2, but* $\Psi(2) = 2$. *The function* f_1 *has two fixed points* -1 *and* 1 *and the function* f_2 *has no fixed points.*

Theorem 10.5.2 is a generalization of Rakotch's theorem. This is shown in the next example.

Example **10.5.5** ([30, Remark 3]) *Let* $X = [0, 1] \cup \{2, 3, 4 \ldots\}$ *be the complete metric space with its metric* d *defined by*

$$d(x, y) = \begin{cases} |x - y| & \text{if } x, y \in [0, 1] \\ x + y & \text{if at least one of } x, y \notin [0, 1]. \end{cases}$$

We define the function $f : X \to X$ *by*

$$f(x) = \begin{cases} x - \dfrac{1}{2}x^2 & \text{for } x \in [0, 1] \\ x - 1 & \text{for } x = 2, 3, \ldots. \end{cases}$$

If $x, y \in [0, 1]$ *for* $x - y = t > 0$, *then we have*

$$d(f(x), f(y)) = (x - y)\left(1 - \frac{1}{2}(x + y)\right) \leq t\left(1 - \frac{1}{2}t\right),$$

and if $x \in \{2, 3, 4, \ldots\}$ *and* $x > y$, *then we have*

$$d(f(x), f(y)) = f(x) + f(y) < x - 1 + y = d(x, y) - 1.$$

We define the function Ψ *by*

$$\Psi(t) = \begin{cases} t - \dfrac{1}{2}t^2 & \text{for } 0 \leq t \leq 1 \\ t - 1 & \text{for } 1 < t < \infty. \end{cases}$$

The function Ψ is upper semi–continuous from the right on the set $[0, \infty)$, $\Psi(t) < t$ for all $t > 0$, and the condition in (10.27) is satisfied.
Since

$$\lim_{n \to \infty} \frac{d(f(n), 0)}{d(n, 0)} = 1,$$

there is no decreasing function α with $\alpha(t) < 1$ for all $t > 0$ which satisfies (10.26). Furthermore, since

$$\lim_{x \to 0} \frac{d(f(x), 0)}{d(x, 0)} = 1,$$

there is no increasing function α with $\alpha(t) < 1$ for all $t > 0$ which satisfies (10.26).

10.6   Theorem of Meir-Keeler

In 1969, Meir and Keeler [138] proved a very interesting theorem and showed that the conclusion of Banach's fixed point theorem can be extended to a more general class of contractions. In this section, we present some results of the paper mentioned.

Definition 10.6.1 Let (X, d) be a metric space. The function $f : X \mapsto X$ is said to be a *weakly uniformly strict contraction*, or a *Meir–Keeler contraction (MK contraction)* if, for every $\varepsilon > 0$, there exists $\delta > 0$ such that

$$\varepsilon \le d(x, y) < \varepsilon + \delta \text{ implies } d(f(x), f(y)) < \varepsilon. \tag{10.39}$$

Theorem 10.6.2 (Meir and Keeler [138]) *Let (X, d) be a complete metric space and $f : X \to X$ be a function. If (10.39) is satisfied, then f has a unique fixed point z. Moreover, we have for each $x \in X$*

$$\lim_{n \to \infty} f^n(x) = z. \tag{10.40}$$

Proof. First we note that (10.39) implies that f is a contractive map, that is,

$$x \ne y \text{ implies } d(f(x), f(y)) < d(x, y). \tag{10.41}$$

Hence f is a continuous function and has at most one fixed point.
We note that if $(f^n(x))$ is a Cauchy sequence for each $x \in X$, then the function f has a unique fixed point, and (10.40) is satisfied. This follows from the following consideration. Since X is a complete space, every Cauchy sequence $(f^n(x))$ has a limit $z(x)$. The continuity of f implies

$$f(z(x)) = f\left(\lim_{n \to \infty} f^n(x)\right) = \lim_{n \to \infty} f^{n+1}(x) = z(x).$$

Hence $z(x)$ is the unique fixed point of f.
The proof of the theorem will be complete if we show that the sequence $(f^n(x)) = (x_n)$ of iterations is a Cauchy sequence for each $x \in X$. Let $x \in X$ and $c_n = d(x_n, x_{n+1})$ for $n = 1, 2, \ldots$. It follows from (10.41) that (c_n) is a decreasing sequence. If $\lim_{n \to \infty} c_n = \varepsilon > 0$, then the implication in (10.39) is not true for c_{m+1}, where c_m is chosen such that $c_m < \varepsilon + \delta$. This implies $\lim_{n \to \infty} c_n = 0$.
We assume that there exists a sequence (x_n) which is not a Cauchy sequence. Then there exists $2\varepsilon > 0$ such that, for each $m_0 \in \mathbb{N}$, there exist $n, m \in \mathbb{N}$ with $n, m > m_0$ and $d(x_m, x_n) > 2\varepsilon$. It follows from (10.39) that there exists $\delta > 0$ such that

$$\varepsilon \le d(x, y) < \varepsilon + \delta \text{ implies } d(f(x), f(y)) < \varepsilon. \tag{10.42}$$

The implication in (10.42) remains true if we replace δ by $\delta' = \min\{\delta, \varepsilon\}$. Let $m_0 \in \mathbb{N}$ be such that $c_{m_0} < \delta'/3$, and let $m, n > m_0$ be such that $m < n$ and $d(x_m, x_n) > 2\varepsilon$. We prove that there exists $j \in \{m, m+1, \ldots, n\}$ such that

$$\varepsilon + \frac{2\delta'}{3} < d(x_m, x_j) < \varepsilon + \delta'. \tag{10.43}$$

To prove (10.43), we note that $d(x_{n-1}, x_n) < \delta/3$. Since $d(x_m, x_n) > 2\varepsilon$ and $d(x_m, x_n) \le d(x_m, x_{n-1}) + d(x_{n-1}, x_n)$, it follows that

$$\varepsilon + \frac{2\delta'}{3} < d(x_m, x_{n-1}). \tag{10.44}$$

Let k be the smallest natural number in $\{m, m+1, \ldots, n\}$; (clearly $m < k \le n-1$) such that

$$\varepsilon + \frac{2\delta'}{3} < d(x_m, x_k) \tag{10.45}$$

holds. We prove $d(x_m, x_k) < \varepsilon + \delta'$. If we assume that this is not true, then we have

$$\varepsilon + \delta' \le d(x_m, x_k) \le d(x_m, x_{k-1}) + d(x_{k-1}, x_k) < d(x_m, x_{k-1}) + \frac{\delta'}{3},$$

that is,

$$\varepsilon + \frac{2\delta'}{3} < d(x_m, x_{k-1}). \tag{10.46}$$

This is a contradiction to the minimality condition of k in the inequality in (10.45). Therefore the inequality in (10.43) must hold.
Now

$$d(x_m, x_k) \le d(x_m, x_{m+1}) + d(x_{m+1}, x_{k+1}) + d(x_{k+1}, x_k),$$

(10.42) and (10.43) imply

$$d(x_m, x_j) \le c_m + \varepsilon + c_k < \frac{\delta'}{3} + \varepsilon + \frac{\delta'}{3}.$$

This is a contradiction to (10.43). Hence (x_n) is a Cauchy sequence. □

It is well known that the Meir-Keeler theorem generalizes Banach's contraction principle [9] and Edelstein's theorem [62].

Theorem 10.6.3 (Banach [9]) *Let (X, d) be a complete metric space and $f : X \to X$ be a contraction, that is, there exists $q \in [0, 1)$ such that*

$$d(f(x), f(y)) \le q \cdot d(x, y) \text{ for all } x, y \in X. \tag{10.47}$$

Then f has a unique fixed point.

Proof. Let $\varepsilon > 0$ and $\delta = (1/q - 1)\varepsilon$. Then it follows from $d(x, y) < \varepsilon + \delta$ and $x \ne y$ that $d(f(x), f(y)) \le qd(x, y) < q\varepsilon + q\delta = \varepsilon$. Hence the function f satisfies (10.39) and the proof follows from Theorem 10.6.2. □

Theorem 10.6.4 (Edelstein [62]) *Let (X, d) be a compact metric space and $f : X \to X$ be a map. We assume that*

$$d(f(x), f(y)) < d(x, y) \text{ for all } x, y \in X \text{ with } x \ne y.$$

Then the function f has a unique fixed point.

Proof (Suzuki [196]). We assume that the function f does not satisfy the condition in (10.39). Then there exist $\varepsilon > 0$ and sequences (x_n) and (y_n) in X such that

$$d(x_n, y_n) < \epsilon + \frac{1}{n} \text{ and } d(f(x_n), f(y_n)) \geq \epsilon. \qquad (10.48)$$

Since X is a compact set, there exist subsequences (x_{n_k}) and (y_{m_k}) of the sequences (x_n) and (y_n), which converge to some $x_0 \in X$ and some $y_0 \in X$, respectively. The continuity of the function f implies

$$d(x_0, y_0) \leq \varepsilon \leq d(f(x_0), f(y_0)) < d(x_0, y_0).$$

This is a contradiction, and consequently the function f must satisfy the condition in (10.39). Now the proof follows from Theorem 10.6.2. $\square$

Rakotch [157], and Boyd and Wong [30] assumed that, among other conditions, the following inequalities are satisfied:

$$d(f(x), f(y)) \leq \psi(d(x, y)) \text{ and } \psi(t) < t \text{ for all } t \neq 0. \qquad (10.49)$$

The next example shows that the Meir–Keeler theorem holds even if the condition in (10.49) is not satisfied.

Example **10.6.5** *Let* $X = [0,1] \cup \{3, 4, 6, 7, \ldots, 3n, 3n+1, \ldots\}$ *be endowed with the Euclidean metric and the function* f *be defined by*

$$f(x) = \begin{cases} \dfrac{x}{2} & \text{for } 0 \leq x \leq 1 \\ 0 & \text{for } x = 3n \\ 1 - \dfrac{1}{n+2} & \text{for } x = 3n+1. \end{cases}$$

Then the function f *satisfies (10.39), and it follows from*

$$d(f(x), f(y)) \leq \psi(d(x, y)) \text{ for all } x, y \in X \qquad (10.50)$$

that $\psi(1) = 1$.

10.7   Theorems by Kannan, Chatterjee, and Zamfirescu

In 1968, Kannan [109] proved the following fixed point theorem which is independent from Banach's fixed point theorem.

Theorem 10.7.1 (Kannan [109]) *If* (X, d) *is a complete metric space,* $0 \leq q < 1/2$ *and* $f : X \to X$ *is a map such that*

$$d(f(x), f(y)) \leq q[d(x, f(x)) + d(y, f(y))] \text{ for all } x, y \in X, \qquad (10.51)$$

then f *has a unique fixed point, that is, there exists one and only one* $z \in X$ *such that* $f(z) = z$.

Proof. (Joseph and Kwack [103]) Let

$$c = \inf\{d(x, f(x)) : x \in X\}.$$

Then we have $c \geq 0$. If $c > 0$, then $c(1-q)/q > c$ implies the existence of $x \in X$ such that $d(x, f(x)) < c(1-q)/q$. Now we have

$$d(f(x), f^2(x)) \leq \frac{q}{1-q} d(x, f(x)) < c,$$

which is a contradiction, and so $c = 0$. Hence there exists a sequence (x_n) in X such that $\lim_n d(x_n, f(x_n)) = 0$. It follows from

$$d(x_m, x_n) \leq d(x_m, f(x_m)) + d(f(x_m), f(x_n)) + d(x_n, f(x_n))$$
$$\leq (1+q)[d(x_m, f(x_m)) + d(x_n, f(x_n))],$$

that (x_n) is a Cauchy sequence. So there exists $z \in X$ such that $\lim_{n \to \infty} x_n = z$. It follows that $\lim_{n \to \infty} f(x_n) = z$.
We prove $f(z) = z$. It follows from

$$d(z, f(z)) \leq d(z, f(x_n)) + d(f(x_n), f(z))$$
$$\leq d(z, f(x_n)) + q[d(x_n, f(x_n)) + d(z, f(z))],$$

as $n \to \infty$ that

$$d(z, f(z)) \leq q d(z, f(z)),$$

and so $z = f(z)$. Now (10.51) implies that the map f has a unique fixed point. $\square$

Remark 10.7.2 *Banach's condition (10.1) and Kannan's condition (10.51) are independent. The condition in (10.1) implies the continuity of the map f, but this is not the case for the condition in (10.51). This follows from the following example.*

Example 10.7.3 *Let $X = [0,1]$ and $f(x)$ be defined by*

$$f(x) = \begin{cases} \dfrac{x}{4} & \text{for } x \in [0, 1/2) \\ \dfrac{x}{5} & \text{for } x \in [1/2, 1]. \end{cases}$$

The map f is discontinuous at the point $x = 1/2$ and so the condition in (10.1) is not satisfied, but the condition in (10.51) is satisfied for $q = 4/9$.

Example 10.7.4 *Let $X = [0,1]$ and $f(x) = x/3$ for $x \in [0,1]$. Clearly, the condition in (10.1) is satisfied, but the condition in (10.51) is not satisfied (we may take $x = 1/3$ and $y = 0$).*

Remark 10.7.5 *We note that Kannan's fixed point theorem, Theorem 10.7.1, is not an extension of Banach's contraction principle, Theorem 10.2.2. It is also known that a metric space X is complete if and only if every Kannan mapping has a fixed point, while there exists a metric space X such that X is not complete and every contractive mapping on X has a fixed point ([46, 195]).*

In 1972, Chatterjee [41] proved the following variant of Kannan's fixed point theorem.

Theorem 10.7.6 *If (X, d) is a complete metric space, $0 \leq q < 1/2$ and $f : X \to X$ is a map which satisfies the condition*

$$d(f(x), f(y)) \leq q[d(x, f(y)) + d(y, f(x))] \text{ for all } x, y \in X,$$

then the function f has a unique fixed point.

Proof (Fisher [68]). Let $x \in X$. Then we have

$$\begin{aligned}
d(f^n(x), f^{n+1}(x)) &\leq q[d(f^{n-1}(x), f^{n+1}(x)) + d(f^n(x), f^n(x))] \\
&= qd(f^{n-1}(x), f^{n+1}(x)) \\
&\leq q[d(f^{n-1}(x), f^n(x)) + d(f^n(x), f^{n+1}(x))],
\end{aligned}$$

hence

$$\begin{aligned}
d(f^n(x), f^{n+1}(x)) &\leq \frac{q}{1-q} d(f^{n-1}(x), f^n(x)) \\
&\leq \left(\frac{q}{1-q}\right)^2 d(f^{n-2}(x), f^{n-1}(x)) \\
&\leq \left(\frac{q}{1-q}\right)^n d(x, f(x)).
\end{aligned}$$

So we obtain

$$\begin{aligned}
d(f^n(x), f^{n+r}(x)) &\leq d(f^n(x), f^{n+1}(x)) + \cdots + d(f^{n+r-1}x, f^{n+r}(x)) \\
&\leq \left[\left(\frac{q}{1-q}\right)^n + \cdots + \left(\frac{q}{1-q}\right)^{n+r-1}\right] d(x, f(x)) \\
&\leq \left(\frac{q}{1-q}\right)^n \frac{1}{1-q} d(x, f(x)).
\end{aligned}$$

Since $q(1-q)^{-1} < 1$, it follows that $(f^n(x))$ is a Cauchy sequence in X. Since X is a complete metric space, there exists $z \in X$ such that $z = \lim_n f^n(x)$.
Now we have

$$\begin{aligned}
d(z, f(z)) &\leq d(z, f^n(x)) + d(f^n(x), f(z)) \\
&\leq d(z, f^n(x)) + q[d(f^{n-1}(x), f(z)) + d(f^n(x), z)].
\end{aligned}$$

Letting $n \to \infty$, we obtain

$$d(z, f(z)) \leq qd(z, f(z)),$$

and since $q < 1/2$, we have

$$f(z) = z.$$

Hence z is a fixed point of the function f.
We assume that the function f has one more fixed points $z' \in X$. Then we have

$$\begin{aligned}
d(z, z') &= d(f(z), f(z')) \\
&\leq q[d(z, f(z')) + d(z', f(z))] \\
&= 2qd(z, z').
\end{aligned}$$

Since $q < 1/2$, it follows that $z = z'$, that is, the fixed point of the function f is unique. $\square$

In 1972, Zamfirescu [214] combined the results of Banach, Kannan and Chatterjee.

Theorem 10.7.7 (Zamfirescu [214]) *Let (X, d) be a complete metric space and $f : X \to X$ be a map for which there exist real numbers $0 \le \alpha < 1$, $0 \le \beta < 1/2$ and $\gamma < 1/2$ such that, for each $x, y \in X$, at least one of the following conditions is satisfied:*

$$(z_1) \qquad d(f(x), f(y)) \le \alpha d(x, y);$$
$$(z_2) \qquad d(f(x), f(y)) \le \beta[d(x, f(x)) + d(y, f(y))];$$
$$(z_3) \qquad d(f(x), f(y)) \le \gamma[d(x, f(y)) + d(y, f(x))].$$

Then the function f has a unique fixed point.

Proof. Let $x, y \in X$. Then at least one of the conditions (z_1), (z_2) or (z_3) is satisfied. If (z_2) is satisfied, then we have

$$d(f(x), f(y)) \le \beta[d(x, f(x)) + d(y, f(y))]$$
$$\le \beta\{d(x, f(x)) + [d(y, x) + d(x, f(x)) + d(f(x), f(y))]\}.$$

This implies

$$(1 - \beta)d(f(x), f(y)) \le 2\beta d(x, f(x)) + \beta d(x, y),$$

that is,

$$d(f(x), f(y)) \le \frac{2\beta}{1 - \beta} d(x, f(x)) + \frac{\beta}{1 - \beta} d(x, y).$$

Similarly, if (z_3) is satisfied, we get the following estimate

$$d(f(x), f(y)) \le \gamma[d(x, f(y)) + d(y, f(x))] \le$$
$$\le \gamma[d(x, f(x)) + d(f(x), f(y)) + d(y, x) + d(x, f(x))] \le$$
$$\le \gamma[2d(x, f(x)) + d(f(x), f(y)) + d(x, y)].$$

Hence we have

$$d(f(x), f(y)) \le \frac{2\gamma}{1 - \gamma} d(x, f(x)) + \frac{\gamma}{1 - \gamma} d(x, y).$$

We put

$$\lambda = \max\left\{\alpha, \frac{\beta}{1 - \beta}, \frac{\gamma}{1 - \gamma}\right\}.$$

Then we have $0 \le \lambda < 1$, and if (z_2) or (z_3) is satisfied for each $x, y \in X$, then

$$d(f(x), f(y)) \le 2\lambda \cdot d(x, f(x)) + \lambda \cdot d(x, y). \qquad (10.52)$$

In a similar way, it can be shown that if (z_2) or (z_3) is satisfied, then

$$d(f(x), f(y)) \le 2\lambda \cdot d(x, f(y)) + \lambda \cdot d(x, y). \qquad (10.53)$$

Obviously, (10.52) and (10.53) follow from (z_1).
It follows from (10.52) that the function f has at least one fixed point. Now we prove the existence of a fixed point of f. Let $x_0 \in X$ and

$$x_n = f^n(x_0) \text{ for } n = 1, 2, \ldots$$

be the Picard iteration of f.
If $x = x_n$ and $y = x_{n-1}$ are two successive approximations, then it follows from (10.53) that

$$d(x_{n+1}, x_n) \le \lambda \cdot d(x_n, x_{n-1}).$$

So $(x_n)_{n=0}^{\infty}$ is a Cauchy sequence, and consequently convergent. Let $z \in X$ be its limit. Then we have

$$\lim_{n \to \infty} d(x_{n+1}, x_n) = 0.$$

By the triangle inequality and (10.52), it follows that

$$d(z, f(z)) \leq d(z, x_{n+1}) + d(f(x_n), f(z))$$
$$\leq d(z, x_{n+1}) + \lambda \cdot d(z, x_n) + 2\lambda d(x_n, f(x_n)),$$

and letting $n \to \infty$, we obtain $d(z, f(z)) = 0$, hence $f(z) = z$.   □

Remark 10.7.8 ([172]) *If a function f satisfies the condition in Theorem 10.7.6, we write $f \in (Z)$, in particular, if f satisfies one of the conditions in (z_i) for $i = 1, 2, 3$ in this theorem, then we write $f \in (Z; z_i)$ for $i = 1, 2, 3$.*

We consider the conditions (Z'): there exist nonnegative functions a, b and c satisfying the following condition

$$\sup_{x, y \in X} (a(x, y) + 2b(x, y) + 2c(x, y)) \leq \lambda < 1,$$

such that, for each $x, y \in X$,

$$d(f(x), f(y)) \leq a(x, y)d(x, y) + b(x, y)(d(x, f(x)) + d(y, f(y)))$$
$$+ c(x, y)(d(x, f(y)) + d(y, f(x)));$$

and (Z''): There exists a constant h with $0 \leq h < 1$ such that, for all $x, y \in X$,

$$d(f(x), f(y)) \leq h \max\left\{ d(x, y), \frac{d(x, f(x)) + d(y, f(y))}{2}, \right.$$
$$\left. \frac{d(x, f(y)) + d(y, f(x))}{2} \right\}. \tag{10.54}$$

It can be proved ([172]) that the conditions in (Z), (Z') and (Z'') are equivalent.

We show that (Z) implies (Z').

If the function f and $x, y \in X$ satisfy $(Z; z_1)$, then we define $a(x, y) = \alpha$ and $b = c = 0$. If for $x, y \in X$, for which the function f satisfies $(Z; z_2)$, we define $b(x, y) = \beta$ and $a = c = 0$, and similarly, in the case of $(Z; z_3)$, we define $c(x, y) = \gamma$ and $a = b = 0$.

We show that (Z') implies (Z'').

We put

$$M(x, y) = \max\left\{ d(x, y), \frac{d(x, f(x)) + d(y, f(y))}{2}, \right.$$
$$\left. \frac{d(x, f(y)) + d(y, f(x))}{2} \right\}. \tag{10.55}$$

Let $f \in (Z')$. Then we have

$$d(f(x), f(y)) \leq [a(x, y) + 2b(x, y) + 2c(x, y)]M(x, y) \leq \lambda M(x, y),$$

and $f \in (Z'')$.

We show that (Z'') implies (Z).

For each $x, y \in X$, for which $M(x, y) = d(x, y)$, the function f satisfies $(Z; z_1)$ with $\alpha = h$. If $M(x, y) = [d(x, f(x)) + d(y, f(y))]/2$, then the function f satisfies $(Z; z_2)$ with $\beta = h/2$, and the function f satisfies $(Z; z_3)$ with $\gamma = h/2$, if $M(x, y) = [d(x, f(y)) + d(y, f(x))]/2$.   □

10.8 Ćirić's generalized contraction

In [42], *Ćirić* generalized the well-known contractive condition and introduced a concept of a generalized contraction defined as follows.

Definition 10.8.1 (Ćirić [42]) Let (X, d) be a metric space. A mapping $f : X \to X$ is a λ-generalized contraction if, for all $x, y \in X$, there exist some nonnegative numbers $q(x, y)$, $r(x, y)$, $s(x, y)$ and $t(x, y)$ such that

$$\sup_{x,y \in X} \{q(x, y) + r(x, y) + s(x, y) + 2t(x, y)\} = \lambda < 1,$$

and for all $x, y \in X$,

$$d(f(x), f(y)) \leq q(x, y)d(x, y) + r(x, y)d(x, f(x)) + s(x, y)d(y, f(y))$$
$$+ t(x, y)(d(x, f(y)) + d(y, f(x))). \tag{10.56}$$

Obviously, this condition is equivalent to the fact that there exists a constant $h \in (0, 1)$ such that, for all $x, y \in X$,

$$d(f(x), f(y)) \leq h \max \left\{ d(x, y), \ d(x, f(x)), d(y, f(y)), \right.$$
$$\left. \frac{d(x, f(y)) + d(y, f(x))}{2} \right\}. \tag{10.57}$$

The next example shows that (10.56) indeed generalizes (10.1).

Example **10.8.2** *Let* $X = [0, 2] \subseteq \mathbb{R}$ *and*

$$f(x) = \begin{cases} \dfrac{x}{9} & \text{for } 0 \leq x \leq 1 \\ \dfrac{x}{10} & \text{for } 1 < x \leq 2. \end{cases}$$

The map f *does not satisfy (10.1) since, for* $x = 999/1000$ *and* $y = 1001/1000$,

$$d(f(x), f(y)) = \frac{981}{90000} > 5 \cdot \frac{180}{90000} = 5d(x, y).$$

But (10.56) holds for $q(x, y) = 1/10$, $r(x, y) = s(x, y) = 1/4$ *and* $t(x, y) = 1/6$ *for all* $x, y \in X$.

Example **10.8.3** *Let* $X = [0, 10] \subset \mathbb{R}$ *and* $f(x) = 3/4$ *for each* $x \in X$. *For* $x = 0$ *and* $y = 8$, *the function* f *satisfies (10.51) with* $q < 3$. *But the condition in (10.56) is satisfied on all of* X *with* $q(x, y) = 3/4$ *and* $r(x, y) = s(x, y) = t(x, y) = 1/20$.

Definition 10.8.4 Let (X, d) be a metric space, $f : X \to X$ be a map, and $x \in X$. An f-orbit of the element x is the set $O(x; f)$ defined by

$$O(x; f) = \{f^n(x) : n \in \mathbb{N}_0\}.$$

If f is given, then the usual notation is $O(x)$. Furthermore, for all $n \in \mathbb{N}$, we define the set

$$O(x, n) = \{x, f(x), f^2(x), \ldots, f^n(x)\}.$$

The space X is said to be an f-*orbitally complete metric space* if any Cauchy sequence in $O(x; f)$ for $x \in X$ converges in X.

Obviously, every complete metric space is f–orbitally complete, but the converse implication does not hold, in general. It is clear from the proof of Banach's theorem that it is enough to assume that (X, d) is f–orbitally complete instead of complete. The same remark applies for λ–generalized contractions, as is stated in the following theorem.

Theorem 10.8.5 (Ćirić [42]) *If $f : X \to X$ is a λ–generalized contraction on an f–orbitally complete metric space X, then, for any $x \in X$, the iterative sequence $(f^n(x))$ converges to the unique fixed point z of f, and*

$$d(f^n(x), z) \leq \frac{\lambda^n}{1 - \lambda} \cdot d(x, f(x)).$$

Proof. For an arbitrary $x \in X$, we define the sequence (x_n) by $x_0 = x$ and $x_n = f(x_{n-1})$ for $n \in \mathbb{N}$. Then we obtain from (10.56)

$$
\begin{aligned}
d(x_n, x_{n+1}) &= d(f(x_{n-1}), f(x_n)) \leq q(x_{n-1}, x_n)d(x_{n-1}, x_n) \\
&\quad + r(x_{n-1}, x_n)d(x_{n-1}, f(x_{n-1})) + s(x_{n-1}, x_n)d(x_n, f(x_n)) \\
&\quad + t(x_{n-1}, x_n)(d(x_{n-1}, f(x_n)) + d(x_n, f(x_{n-1}))) \\
&= q(x_{n-1}, x_n)d(x_{n-1}, x_n) + r(x_{n-1}, x_n)d(x_{n-1}, x_n) \\
&\quad + s(x_{n-1}, x_n)d(x_n, x_{n+1}) + t(x_{n-1}, x_n)d(x_{n-1}, x_{n+1}),
\end{aligned}
$$

and moreover

$$
\begin{aligned}
d(x_n, x_{n+1}) &\leq (q(x_{n-1}, x_n) + r(x_{n-1}, x_n))d(x_{n-1}, x_n) \\
&\quad + s(x_{n-1}, x_n)d(x_n, x_{n+1}) \\
&\quad + t(x_{n-1}, x_n)(d(x_{n-1}, x_n) + d(x_n, x_{n+1})).
\end{aligned}
$$

So we have

$$d(x_n, x_{n+1}) \leq \frac{q(x_{n-1}, x_n) + r(x_{n-1}, x_n) + t(x_{n-1}, x_n)}{1 - s(x_{n-1}, x_n) - t(x_{n-1}, x_n)} d(x_{n-1}, x_n). \tag{10.58}$$

Because of

$$q(x, y) + r(x, y) + t(x, y) + \lambda s(x, y) + \lambda t(x, y) \leq \lambda,$$

we get

$$\frac{q(x, y) + r(x, y) + t(x, y)}{1 - s(x, y) - t(x, y)} \leq \lambda \text{ for all } x, y \in X$$

and, combined with (10.58), it follows that

$$d(x_n, x_{n+1}) \leq \lambda d(x_{n-1}, x_n). \tag{10.59}$$

We remark that (10.59) allows us to consider f as a contraction under special assumptions, and

$$d(x_n, x_{n+1}) \leq \lambda d(x_{n-1}, x_n) \leq \cdots \leq \lambda^n d(x, f(x)).$$

Obviously, we have for all $m \geq n$

$$d(x_n, x_m) \leq \sum_{k=n}^{m-1} d(x_k, x_{k+1}) \leq \sum_{k=n}^{m-1} \lambda^k d(x, f(x)),$$

hence

$$d(x_n, x_{n+p}) \leq \frac{\lambda^n}{1 - \lambda} d(x, f(x)) \tag{10.60}$$

implies that (x_n) is a Cauchy sequence in $O(x)$. Let $z \in X$ denote its limit. It remains to show that $f(z) = z$ by estimating $d(f(z), f(x_n))$.

$$d(f(z), f(x_n)) \leq q(z, x_n)d(z, x_n) + r(z, x_n)(d(z, x_{n+1}) + d(x_{n+1}, f(z)))$$
$$+ s(z, x_n)d(x_n, x_{n+1}) + t(z, x_n)(d(z, x_{n+1}) + d(f(z), x_n))$$
$$\leq \lambda d(z, x_n) + (r(z, x_n) + t(z, x_n))d(z, x_{n+1})$$
$$+ r(z, x_n)d(f(x_n), f(z)) + s(z, x_n)d(x_n, x_{n+1})$$
$$+ t(z, x_n)(d(f(z), f(x_n)) + d(f(x_n), x_n))$$
$$\leq d(z, x_n) + \lambda d(z, x_{n+1})$$
$$+ (r(z, x_n) + t(z, x_n))d(f(z), f(x_n)) + \lambda d(x_n, x_{n+1})$$
$$\leq \lambda(d(z, x_n) + d(z, x_{n+1}) + d(x_n, x_{n+1})) + \lambda d(f(z), f(x_n)).$$

Thus we have

$$d(f(z), f(x_n)) \leq \frac{\lambda}{1 - \lambda} [d(z, x_n) + d(z, x_{n+1}) + d(x_n, x_{n+1})],$$

that is, z is a fixed point of the function f. The uniqueness easily follows from (10.56) and the estimation inequality is implied by (10.60). □

Remark 10.8.6 *The contractive condition (10.56) for generalized contractions implies many others, thus Theorem 10.8.5 has numerous consequences among which we will state two theorems analogous to Corollaries 10.2.8 and 10.2.10 of Banach's theorem.*

Theorem 10.8.7 *If $f : X \to X$ is a map of an f-orbitally complete metric space (X, d) such that, for some $k \in \mathbb{N}$, f^k is a λ-generalized contraction for all $x \in X$, then the iterative sequence $(f^n(x))$ converges to a unique fixed point z of f, and*

$$d(f^n(x), z) \leq (\lambda')^n \rho(x, f(x)), \text{ where } \lambda' = \lambda^{1/k}$$
$$\text{and } \rho(x, f(x)) = \max\{\lambda^{-1}d(f^r(x), f^{r+k}(x)) : r = 0, 1, \ldots, k - 1\}.$$

Proof. The existence of a unique fixed point directly follows from Theorem 10.8.5. It remains to estimate $d(f^n(x), z)$ for each $n \in \mathbb{N}$. Since $n = mk + r$ for $m = [n/k]$ and $0 \leq r < k$, we have

$$d(f^n(x), z) = d(f^{mk}(f^r(x)), z) \leq \frac{\lambda^m}{1 - \lambda}d(f^r(x), f^k(f^r(x)))$$
$$= (\lambda^{1/k})^{mk+r-r}d(f^r(x), f^{k+r}(x))$$
$$\leq (\lambda^{1/k})^{mk+r-k}d(f^r(x), f^{r+k}(x))$$
$$= (\lambda^{1/k})^n \lambda^{-1}d(f^r(x), f^{r+k}(x)),$$

hence

$$d(f^n(x), z) \leq (\lambda^{1/k})^n \max\{\lambda^{-1}d(f^r(x), f^{r+k}(x)) : r = 0, 1, \ldots k - 1\}. \quad □$$

As in the case of Theorem 10.2.10, we may consider some local properties of Theorem 10.8.5.

Theorem 10.8.8 *Let $f : B \to X$ be a map of an f-orbitally complete metric space (X, d), where $B = B_r(x_0) = \{x \in X : d(x_0, x) \leq r\}$ for some $x_0 \in X$ and $r > 0$. If f is a λ-generalized contraction on B and*

$$d(x_0, f(x_0)) \leq (1 - \lambda) \cdot r, \tag{10.61}$$

then the sequence $(f^n(x_0))$ converges to a unique fixed point z of f in B and

$$d(f^n(x_0), z) \leq \lambda^n \cdot r \text{ for } \lambda = \sup_{x, y \in B} [q(x, y) + r(x, y) + 2t(x, y)].$$

Proof. It is clear that $x_n \in B$ for all $n \in \mathbb{N}$, due to (10.61) and mathematical induction. Analogously as in the proof of Theorem 10.8.5, it follows that $(f^n(x_0))$ is a Cauchy sequence in B and its limit is a fixed point of f. Inequality (10.56) guarantees uniqueness. □

10.9   The Reich and Hardy–Rogers theorems

In 1971, Reich [168] proved the following theorem which generalizes Banach's and Kannan's theorems. (We note that for $a = b = 0$, we obtain Banach's theorem, Theorem 10.2.2, and for $a = b$ and $c = 0$, we obtain Kannan's theorem, Theorem 10.7.1.)

Theorem 10.9.1 (Reich [168]) *Let (X, d) be a complete metric space and $f : X \to X$ be a map for which there exists nonnegative numbers a, b and c with $a + b + c < 1$ such that for all $x, y \in X$,*

$$d(f(x), f(y)) \le ad(x, f(x)) + bd(y, f(y)) + cd(x, y). \qquad (10.62)$$

Then the map f has a unique fixed point.

Proof. Let $x \in X$. We consider the sequence $(f^n(x))$. If we put $x = f^n(x)$ and $y = f^{n-1}(x)$ in (10.62), then we have for all $n \ge 1$

$$d(f(f^n(x)), f(f^{n-1}(x))) \le$$
$$ad(f^n(x), f(f^n(x))) + bd(f^{n-1}(x), f(f^{n-1}(x))) + cd(f^n(x), f^{n-1}(x)).$$

Hence, we obtain
$$d(f^{n+1}(x)), f^n(x))) \le pd(f^n(x), f^{n-1}(x)),$$

where $0 \le p = (b + c)/(1 - a) < 1$. It follows that

$$d(f^{n+1}(x)), f^n(x))) \le p^n d(x, f(x)),$$

and for every $m > n$,

$$d(f^m(x)), f^n(x))) \le \frac{p^n}{1 - p} \cdot d(x, f(x)).$$

Thus $(f^n(x))$ is a Cauchy sequence, and there exists $z \in X$ with $z = \lim_{n \to \infty} f^n(x)$. We are going to show that $f(z) = z$. It suffices to show that $\lim_{n \to \infty} f^{n+1}(x) = f(z)$. When we choose $x = f^n(x)$ and $y = z$ in (10.62), then we have for all $n \ge 1$

$$d(f^{n+1}(x)), f(z)) \le ad(f^n(x), f^{n+1}(x)) + bd(z, f(z)) + cd(f^n(x), z)$$
$$\le ad(f^n(x), f^{n+1}(x)) + bd(f^{n+1}(x), f(z)) + bd(f^{n+1}(x), z) + cd(f^n(x), z)$$
$$\le ap^n d(x, f(x)) + bd(f^{n+1}(x), f(z)) + bd(f^{n+1}(x), z) + cd(f^n(x), z).$$

Thus we obtain for $n \to \infty$

$$d(f^{n+1}(x)), f(z)) \le \frac{ap^n d(x, f(x)) + bd(f^{n+1}(x), z) + cd(f^n(x), z)}{1 - b} \to 0.$$

We are going to show that the map f has a unique fixed point. If we assume that $x, y \in X$ with $x \ne y$ are fixed points of the map f, then we have

$$d(x, y) = d(f(x), f(y)) \le ad(x, f(x)) + bd(y, f(y)) + cd(x, y) = cd(x, y),$$

which implies $x = y$. □

Example **10.9.2** *Let* $X = [0,1]$ *have the natural metric and the map* $f : X \to X$ *be defined by* $f(x) = x/3$ *for* $0 \leq x < 1$ *and* $f(1) = 1/6$. *Then the map* f *does not satisfy Banach's condition, since it is not continuous; neither does it satisfy Kannan's condition, since*

$$d(f(0), f(1/3)) = \frac{1}{2}\left[d(0, f(0)) + d(1/3, f(1/3))\right].$$

But the map f *satisfies the condition in (10.62), for instance, for* $a = 1/6$, $b = 1/9$ *and* $c = 1/3$.

Corollary 10.9.3 (Reich [168]) *Let* (X, d) *be a complete metric space and* $f_n : X \to X$ *for* $n = 1, 2, \ldots$ *be a sequence of maps satisfying the condition in (10.62) with the same constants* a, b *and* c *and with the fixed points* $z_n \in X$. *We define the map* $f : X \to X$ *by* $f(x) = \lim_{n \to \infty} f_n(x)$ *for* $x \in X$. *Then the map* f *has a unique fixed point* $z \in X$ *and* $z = \lim_{n \to \infty} z_n$.

Proof. Since the metric d is a continuous function, it follows that the function f satisfies the condition in (10.62), and therefore has a unique fixed point $u \in X$. We note that

$$d(z_n, z) = d(f_n(z_n), f(z)) \leq d(f_n(z_n), f_n(z)) + d(f_n(z), f(z))$$
$$\leq ad(z_n, f_n(z_n)) + bd(z, f_n(z)) + cd(z_n, z) + d(f_n(z), f(z)).$$

Hence we have

$$d(z_n, z) \leq \frac{(b+1)d(f_n(z), f(z))}{1-c} \to 0 \ (n \to \infty). \qquad \square$$

Hardy and Rogers [83] improved some of Reich's results [168] including the following theorem.

Theorem 10.9.4 (Hardy and Rogers [83])
Let (X, d) *be a metric space and* $f : X \to X$ *be a map such that for all* $x, y \in X$,

$$d(f(x), f(y)) \leq ad(x, f(x)) + bd(y, f(y))$$
$$+ cd(x, f(y)) + ed(y, f(x)) + hd(x, y), \qquad (10.63)$$

where $a, b, c, e, h \geq 0$ *and* $\alpha = a + b + c + e + h$.

(i) *If* (X, d) *is a complete metric space and* $\alpha < 1$, *then the map* f *has a unique fixed point.*

(ii) *If* (X, d) *is compact,* f *is continuous and the condition in (10.63) is replaced by*

$$d(f(x), f(y)) < ad(x, f(x)) + bd(y, f(y))$$
$$+ cd(x, f(y)) + ed(y, f(x)) + hd(x, y), \qquad (10.64)$$

for all $x \neq y$, *and* $\alpha = 1$, *then* f *has a unique fixed point.*

The following lemma is essential in the proof of this theorem, but for the reader's convenience, we state it separately.

Lemma 10.9.5 *We assume that (10.63) is satisfied and* $\alpha < 1$. *Then there exists* $\beta < 1$ *such that*
$$d(f(x), f^2(x)) \leq \beta d(x, f(x)). \qquad (10.65)$$

If $\alpha = 1$ *and (10.64) is satisfied, then*

$$x \neq f(x) \ \text{implies} \ d(f(x), f^2(x)) \leq \beta d(x, f(x)). \qquad (10.66)$$

Proof. In the first case, for $\alpha < 1$, we put $y = f(x)$, and observe

$$d(f(x), f^2(x)) \leq \frac{a+h}{1-b} \cdot d(x, f(x)) + \frac{c}{1-b} \cdot d(x, f^2(x)), \tag{10.67}$$

which, along with $d(f(x), f^2(x)) \geq d(f^2(x), x) - d(f(x), x)$ and (10.67), leads to

$$d(f^2(x), x) - d(f(x), x) \leq \frac{a+h}{1-b} \cdot d(x, f(x)) + \frac{c}{1-b} \cdot d(x, f^2(x)),$$

that is,

$$d(f^2(x), x) \leq \frac{1+a+h-b}{1-b-c} \cdot d(x, f(x)). \tag{10.68}$$

Now, inserting (10.68) in (10.67), we obtain

$$d(f(x), f^2(x)) \leq \frac{a+c+h}{1-b-c} \cdot d(x, f(x)), \tag{10.69}$$

and replacing a and c by b and e (which is permitted because of the symmetry of the metric d), we get

$$d(f(x), f^2(x)) \leq \frac{b+e+h}{1-a-e} \cdot d(x, f(x)).$$

If we put

$$\beta = \min\left\{\frac{a+c+h}{1-b-c}, \frac{b+e+h}{1-a-e}\right\},$$

then (10.65) is satisfied.

The remainder of the lemma is shown analogously. $\qquad\square$

Proof of Theorem 10.9.4. To prove Part (i), we first observe that, by (10.65), for all $m > n$,

$$d(f^m(x), f^n(x)) \leq d(f^m(x), f^{m-1}(x)) + \cdots + d(f^{n+1}(x), f^n(x))$$
$$\leq \beta^n(1 + \beta + \cdots + \beta^{m-n})d(x, f(x))$$
$$\leq \frac{\beta^n}{1-\beta} \cdot d(x, f(x)).$$

Hence $(f^n(x))$ is a Cauchy sequence and $z \in X$ is its limit. It remains to show that $f(z) = z$. This follows directly from $\lim_{n\to\infty} f^{n+1}(x) = f(z)$.
The following inequality holds by (10.63)

$$d(z, f(z)) \leq d(f^{n+1}(x), f(z)) + d(f^{n+1}(x), z)$$
$$\leq ad(f^n(x), f^{n+1}(x)) + bd(z, f(z))$$
$$+ cd(f^n(x), f(z)) + (e+1)d(f^{n+1}(x), z) + hd(f^n(x), z). \tag{10.70}$$

Letting $n \to \infty$ in (10.70), we obtain

$$d(z, f(z)) \leq (b+c)d(z, f(z)),$$

and $b + c < 1$ implies $z = f(z)$. The uniqueness clearly follows from (10.63). We note that, under the assumptions in (ii), there is some $y \in X$ such that

$$\inf\{d(x, f(x)) : x \in X\} = d(y, f(y)).$$

Because of (10.66), it follows that $y = f(y)$. The uniqueness is shown as previously discussed. $\qquad\square$

10.10   Ćirić's quasi-contraction

In 1971, Ćirić [43] used a concept of generalized contraction to replace the linear combination of distances in (10.56) by their maximum, and defined a new class of contractive mappings called *quasi–contractions*.

Definition 10.10.1 (Ćirić [43]) A map $f : X \to X$ of a metric space (X, d) is a quasi–contraction if there exists some λ with $0 < \lambda < 1$ such that

$$d(f(x), f(y)) \leq \lambda \cdot \max\{d(x, y), d(x, f(x)), d(y, f(y)), d(x, f(y)), d(y, f(x))\} \qquad (10.71)$$

for all $x, y \in X$.

Obviously, if a mapping f satisfies condition (10.1), then (10.71) also holds. Ćirić [43] presented an example which shows that the converse implication is not true, in general.

The following example due to Rhoades generalizes Ćirić's example.

Example **10.10.2 (Rhoades)** ([172]) *Let* $f(x) = 0$ *for all* $0 \leq x < 1$ *and* $f(1) = 1/2$. *Then the function* f *satisfies (10.71) but not (10.56). We note that*

$$d\left(f\left(\frac{1}{2}\right), f(1)\right) = \frac{1}{2} = \frac{d\left(\frac{1}{2}, f(1)\right) + d\left(1, f(\frac{1}{2})\right)}{2},$$

$$d\left(\frac{1}{2}, 1\right) = d\left(\frac{1}{2}, f\left(\frac{1}{2}\right)\right) = d(1, f(1)) = \frac{1}{2},$$

$$d(f(x), f(y)) = 0 \text{ for all } x \neq y \text{ and } x, y \neq 1,$$

$$d(f(x), f(1)) = \frac{1}{2} \leq \frac{3}{4} \cdot d(1, f(x)) = \frac{3}{4} \text{ for } x \neq 1.$$

Theorem 10.10.3 (Ćirić [43]) *If* $f : X \to X$ *is a quasi–contraction on an* f*–orbitally complete metric space* (X, d)*, then* f *has a unique fixed point* z *in* X*, and the iterative sequence* $(f^n(x))$ *converges to* z *for any* $x \in X$*. Moreover, we have*

$$d(f^n(x), z) \leq \frac{\lambda^n}{1 - \lambda} d(x, f(x)).$$

Proof. We put $\alpha(x, n) = \text{diam}(O(x, n))$, and $\alpha(x) = \text{diam}(O(x))$ where diam denotes a diameter of a set.

Then we have

$$\alpha(f(x), n - 1) = \text{diam}(\{f(x), f^2(x), \ldots, f^n(x)\}) \leq \lambda \alpha(x, n). \qquad (10.72)$$

Obviously, if $\alpha(f(x), n - 1) = d(f^j(x), f^k(x))$ for $1 \leq j < k \leq n$, then (10.71) yields

$$\begin{aligned}
\alpha(f(x), n - 1) &= d(f(f^{j-1}(x)), f(f^{k-1}(x))) \\
&\leq \lambda \max\{d(f^{j-1}(x), f^{k-1}(x)), d(f^{j-1}(x), f^j(x)), d(f^{k-1}(x), f^k(x)), \\
&\qquad\qquad d(f^{j-1}(x), f^k(x)), d(f^{k-1}(x), f^j(x))\} \\
&\leq \lambda \text{diam}(\{f^{j-1}(x), f^j(x), \ldots, f^k(x)\}) \\
&\leq \lambda \text{diam}(\{x, f(x), \ldots, f^n(x)\}) \\
&= \lambda \alpha(x, n),
\end{aligned}$$

and (10.72) holds.
Furthermore, we obtain from (10.72),

$$\alpha(x, n) = d(x, f^k(x)) \text{ for some } k \leq n. \tag{10.73}$$

It follows from (10.72), (10.73) and the triangle inequality that

$$\begin{aligned}
\alpha(x, n) = d(x, f^k(x)) &\leq d(x, f(x)) + d(f(x), f^k(x)) \\
&\leq d(x, f(x)) + \alpha(f(x), n-1) \\
&\leq d(x, f(x)) + \lambda \alpha(x, n),
\end{aligned}$$

and

$$\alpha(x, n) \leq \frac{1}{1-\lambda} \cdot d(x, f(x)). \tag{10.74}$$

Since $\lim_{n\to\infty} \alpha(x, n) = \alpha(x)$, (10.74) implies

$$\alpha(x) \leq \frac{1}{1-\lambda} \cdot d(x, f(x)), \tag{10.75}$$

so the f–orbit of x has a finite diameter.
We write $\beta_n(x)$ for the diameter of $\alpha(f^n(x))$.
The sequence $(\beta_n(x))$ is non–increasing and bounded, so there exists $\lim_{n\to\infty} \beta_n(x) = \beta(x)$ and $\beta(x) \leq \beta_n(x)$ for all $n \in \mathbb{N}$.
Letting $n \to \infty$ in (10.72), we obtain

$$\alpha(f(x)) \leq \lambda \alpha(x), \tag{10.76}$$

hence

$$\beta_{n+1}(x) = \alpha(f(f^n(x))) \leq \lambda \alpha(f^n(x)) = \lambda \beta_n(x) \ (n \in \mathbb{N})$$

and

$$\beta(x) \leq \lambda \beta(x),$$

so $\beta(x) = 0$ and $(f^n(x))$ is a Cauchy sequence in X.
Let $z = \lim_{n\to\infty} f^n(x)$. Because of (10.71), we have

$$d(f(z), f(f^n(x))) \leq$$
$$\lambda \max \left\{ d(z, f^n(x)), d(z, f(z)), d(f^n(x), f^{n+1}(x)), d(z, f^{n+1}(x)), d(f^n(x), f(z)) \right\},$$

hence

$$d(f(z), z) \leq \lambda \, d(z, f(z)),$$

that is, $f(z) = z$. The uniqueness also follows from (10.71).
We obtain from (10.76), $\alpha(f^n(x)) \leq \lambda^n \alpha(x)$ and combined with (10.75)

$$\alpha(f^n(x)) \leq \frac{\lambda^n}{1-\lambda} d(x, f(x)).$$

If $n, m \in \mathbb{N}$ and $m \geq n$, then

$$d(f^n(x), f^m(x)) \leq \alpha(f^n(x)) \leq \frac{\lambda^n}{1-\lambda} d(x, f(x)),$$

and when $m \to \infty$, then

$$d(f^n(x), z) \leq \frac{\lambda^n}{1-\lambda} d(x, f(x)). \qquad \square$$

10.11   Caristi's theorem

There are many extensions of Banach's contraction principle, one of the most studied ones is that by Caristi [39], 1976. Caristi's theorem [39] may be motivated by the following consideration. If (X, d) is a metric space and $T : X \to X$ is a contraction with a Lipschitz constant $k \in [0, 1)$, then we have

$$
\begin{aligned}
d(x, T(x)) &= \frac{1}{1-k} \cdot d(x, T(x)) - \frac{k}{1-k} \cdot d(x, T(x)) \\
&\leq \frac{1}{1-k} \cdot d(x, T(x)) - \frac{1}{1-k} \cdot d(T(x), T(T(x))) \\
&= \phi(x) - \phi(T(x)),
\end{aligned}
$$

for all $x \in X$, where $\phi(x) = (1 - k)^{-1} d(x, T(x))$.

It is well known that Caristi's theorem (or the Caristi–Kirk, or the Caristi–Kirk–Browder theorem) is equivalent to Ekeland's variation principle [61] which is very important because of its numerous applications. The original proof of the Caristi–Kirk theorem is rather complicated and, in the literature, there are several different proofs of that theorem.

We mention that the map $\varphi : X \to \mathbb{E}$ ($\mathbb{E} \subset \mathbb{R}$) is lower semicontinuous at $x \in X$ if, for every sequence (x_n), it follows from $\lim_{n \to \infty} x_n = x$ that $\varphi(x) \leq \liminf_{n \to \infty} \varphi(x_n)$. The map $\varphi : X \to \mathbb{E}$ is lower semicontinuous on X if it is lower semicontinuous at every $x \in X$.

Theorem 10.11.1 (Caristi [39]) *Let (X, d) be a complete metric space, $T : X \to X$ and $\phi : X \to [0, \infty)$ be lower semicontinuous such that*

$$
d(x, T(x)) \leq \phi(x) - \phi(T(x)) \text{ for all } x \in X. \tag{10.77}
$$

Then T has a fixed point.

Proof (*Ćirić [44]*). For each $x \in X$, we put

$$
\begin{aligned}
P(x) &= \{y \in X : d(x, y) \leq \phi(x) - \phi(y)\}, \\
\alpha(x) &= \inf \{\phi(y) : y \in P(x)\}.
\end{aligned}
$$

Since $x \in P(x)$, $P(x)$ is a nonempty set and $0 \leq \alpha(x) \leq \phi(x)$.

Let $x \in X$. We define the sequence (x_n) in X such that $x_1 = x$, and if $x_1, x_2, \dots, x_n$ are already defined then we define $x_{n+1} \in P(x_n)$ such that $\phi(x_{n+1}) \leq \alpha(x_n) + 1/n$. Hence the sequence (x_n) satisfies the following conditions:

$$
\begin{aligned}
d(x_n, x_{n+1}) &\leq \phi(x_n) - \phi(x_{n+1}); \\
\alpha(x_n) &\leq \phi(x_{n+1}) \leq \alpha(x_n) + 1/n. \tag{10.78}
\end{aligned}
$$

Since $(\phi(x_n))$ is a decreasing sequence of real numbers, there exists $\alpha \geq 0$ such that

$$
\alpha = \lim_{n \to \infty} \phi(x_n) = \lim_{n \to \infty} \alpha(x_n). \tag{10.79}
$$

Let $k \in \mathbb{N}$. It follows from (10.78) that there exists N_k such that $\phi(x_n) < \alpha + 1/k$ for every $n \geq N_k$. Hence the monotonicity of the sequence $(\phi(x_n))$ for $m \geq n \geq N_k$ implies $\alpha \leq \phi(x_m) \leq \phi(x_n < \alpha + 1/k$, that is,

$$
\phi(x_n) - \phi(x_m) < 1/k \text{ for each } m \geq n \geq N_k. \tag{10.80}
$$

We have from the triangle inequality and the inequality in (10.78)

$$d(x_n, x_m) \leq \sum_{s=n}^{m-1} d(x_s, x_{s+1}) \leq \phi(x_n) - \phi(x_m). \tag{10.81}$$

Now (10.80) implies

$$d(x_n, x_m) < 1/k \text{ for each } m \geq n \geq N_k.$$

Since (x_n) is a Cauchy sequence and X is a complete metric space the sequence converges to some $z \in X$.

Since ϕ is lower semicontinuous, we obtain from (10.81) that

$$\phi(z) \leq \liminf_{m\to\infty} \phi(x_m) \leq \liminf_{m\to\infty} [\phi(x_n) - d(x_n, x_m)] = \phi(x_n) - d(x_n, z),$$

and so

$$d(x_n, z) \leq \phi(x_n) - \phi(z).$$

Hence we have $z \in P(x_n)$ for all $n \in \mathbb{N}$ and $\alpha(x_n) \leq \phi(z)$. Now (10.79) implies $\alpha \leq \phi(z)$. On the other hand, since ϕ is lower semicontinuous, (10.79) implies $\phi(z) \leq \liminf_{n\to\infty} \phi(x_n) = \alpha$. Hence we have $\phi(z) = \alpha$.

Since $z \in P(x_n)$ for each $n \in \mathbb{N}$, (10.77) implies $Tz \in P(z)$, that is,

$$\begin{aligned} d(x_n, Tz) &\leq d(x_n, z) + d(z, Tz) \\ &\leq \phi(x_n) - \phi(z) + \phi(z) - \phi(Tz) \\ &= \phi(x_n) - \phi(Tz). \end{aligned}$$

Hence we have $Tz \in P(x_n)$ for each $n \in \mathbb{N}$. It follows that

$$\phi(Tz) \geq \alpha(x_n) \text{ for each } n \in \mathbb{N}.$$

Now (10.79) implies

$$\phi(Tz) \geq \alpha.$$

Since (10.77) implies $\phi(Tz) \leq \phi(z)$ and $\phi(z) = \alpha$, we have

$$\phi(z) = \alpha \leq \phi(Tz) \leq \phi(z),$$

and so $\phi(Tz) = \phi(z)$. Now (10.77) implies

$$d(z, Tz) \leq \phi(z) - \phi(Tz) = 0,$$

that is, $Tz = z$. $\square$

Theorem 10.11.2 (Ekeland [61], 1972) *Let $\phi : X \to \mathbb{R}$ be an upper semicontinuous function on the complete metric space (X, d). If ϕ is bounded above then there exists $z \in X$ such that*

$$\phi(z) < \phi(x) + d(z, x) \text{ for } x \in X \text{ with } x \neq z. \tag{10.82}$$

Proof (Ćirić [44]). We are going to show that z from the proof of Theorem 10.11.1 is the desired point. Using the same notations for $x \neq z$ we have to prove $x \notin P(z)$. We suppose that this is not the case, that is, for some $v \neq z$, we have $v \in P(z)$. Then $0 < d(z, v) \leq \phi(z) - \phi(v)$ implies $\phi(v) < \phi(z) = \alpha$.

Since

$$d(x_n, v) \leq d(x_n, z) + d(z, v)$$

$$\leq \phi(x_n) - \phi(z) + \phi(z) - \phi(v)$$
$$= \phi(x_n) - \phi(v),$$

it follows that $v \in P(x_n)$. Hence we have

$$\alpha(x_n) \leq \phi(v) \text{ for all } n \in \mathbb{N}.$$

We obtain for $n \to \infty$

$$\alpha \leq \phi(v),$$

which is a contradiction to $\phi(v) < \alpha = \phi(z)$. Hence we have $x \notin P(z)$ for $x \in X$ with $x \neq z$, and so

$$x \neq z \text{ implies } d(z, x) > \phi(z) - \phi(x). \qquad \square$$

Remark 10.11.3 *Theorems 10.11.2 and 10.11.1 are equivalent.*

Proof (Brézis and Browder [31]). By Theorem 10.11.2 there exists $z \in X$ which satisfies the condition in (10.82). It follows that $Tz = z$, for $Tz \neq z$ would imply $\phi(Tz) - \phi(z) > -d(z, Tz)$, which contradicts (10.77).
Now we show that Theorem 10.11.1 implies Theorem 10.11.2. Indeed, if we assume that the conclusion of Theorem 10.11.2 is not true, then, for each $x \in X$, there exists $y \in X$ with $y \neq x$ such that $\phi(y) - \phi(x) \leq -d(x, y)$. Hence we may define a map $T : X \to X$ which satisfies (10.77), but does not have a fixed point. $\square$

We are going to present a proof of Caristi's theorem given by Kirk and Saliga [116]. First we prove a result by Brézis and Browder [31], the well–known *Brézis–Browder [31] principle of ordering.*
Let $(X, \preceq)$ be a partially ordered set. We denote $S(x) = \{y \in X : x \preceq y\}$ for $x \in X$. A sequence (x_n) in X is said to be *increasing* if $x_n \preceq x_{n+1}$ for each $n \in \mathbb{N}$.

Theorem 10.11.4 (Brézis and Browder [31]) *Let $(X, \preceq)$ be a partially ordered set and the function $\phi : X \to \mathbb{R}$ satisfy the following conditions:*

(1) $x \preceq y$ implies $\phi(x) \leq \phi(y)$;

(2)   for every increasing sequence (x_n) in X with $\phi(x_n) \leq C < \infty$ for each $n \in \mathbb{N}$, there exists $y \in X$ such that $x_n \preceq y$ for each $n \in \mathbb{N}$;

(3)   for each $x \in X$ there exists $u \in X$ such that $x \preceq u$ and $\phi(x) < \phi(u)$.

Then $\phi(S(x))$ is a bounded set for each $x \in X$.

Proof. For $a \in X$, let

$$p(a) = \sup_{b \in S(a)} \phi(b).$$

We are going to show that $p(x) = +\infty$ for each $x \in X$. We assume that $p(x) < \infty$ for some $x \in X$. We define a sequence (x_n) by induction such that $x_1 = x$, $x_{n+1} \in S(x_n)$ and $p(x_n) \leq \phi(x_{n+1}) + (1/n)$ for each $n \in \mathbb{N}$. Since $\phi(x_{n+1}) \leq p(x) < \infty$, the condition in (2) implies that there exists $y \in X$ such that $x_n \preceq y$ for each n. It follows from the condition in (3) that there exists $u \in X$ such that $y \preceq u$ and $\phi(y) < \phi(u)$. Since $x_n \preceq u$, we have $\phi(u) \leq p(x_n)$ for all n. Furthermore, we have $x_{n+1} \preceq y$, so $\phi(x_{n+1}) \leq \phi(y)$, and consequently

$$\phi(u) \leq p(x_n) \leq \phi(x_{n+1}) + (1/n) \leq \phi(y) + (1/n) \text{ for all } n \in \mathbb{N},$$

hence $\phi(u) \leq \phi(y)$, which is a contradiction. $\square$

Theorem 10.11.5 *Let $(X, \preceq)$ be a partially ordered set, $x \in X$ and $S(x) = \{y \in X : x \preceq y\}$. We assume that the map $\psi : X \to \mathbb{R}$ satisfies the following conditions:*

(a) $x \preceq y$ *with* $x \neq y$ *implies* $\psi(x) < \psi(y)$;

(b) *for each increasing sequence (x_n) in X, for which $\psi(x_n) \leq C < \infty$ for each $n \in \mathbb{N}$, there exists $y \in X$ such that $x_n \preceq y$ for each $n \in \mathbb{N}$;*

(c) *for each $x \in X$, the set $\psi(S(x))$ is bounded above.*

Then, for each $x \in X$ there exists $x' \in S(x)$ such that x' is maximal in X, that is, $\{x'\} = S(x')$.

Proof. We apply Theorem 10.11.4 to the set $X = S(x)$; since the conditions in (1) and (2) of Theorem 10.11.4 are satisfied, and the conclusion of the theorem does not hold, it follows that the condition in (3) is not satisfied for some $x' \in S(x)$. Hence we have $S(x') = \{x'\}$.□

We remark that the map $\varphi : X \to \mathbb{R}$ is lower semicontinuous from above if $x_n \in X$ for $n = 1, 2, \ldots$, $\lim_{n \to \infty} x_n = x$ and $(\varphi(x_n)) \downarrow r$ imply $\varphi(x) \leq r$.

Theorem 10.11.6 (Kirk and Saliga [116])

We assume that (X, d) is a complete metric space and $T : X \to X$ is an arbitrary map such that we have for each $x \in X$

$$d(x, T(x)) \leq \varphi(x) - \varphi(T(x)), \tag{10.83}$$

where the map $\varphi : X \to \mathbb{R}$ is bounded above and lower semicontinuous. Then the map T has a fixed point in X.

Proof. We introduce *Brøndsted's partial order* $\preceq$ on X as follows: For each $x, y \in X$, we have

$$x \preceq y \text{ if and only if } d(x, y) \leq \varphi(x) - \varphi(y),$$

and let $\psi = -\varphi$. Then the condition in (a) of Theorem 10.11.5 is satisfied, and the condition in (c) follows from the fact that the map φ is bounded below. To show the condition in (b), we assume that (x_n) is an increasing sequence in $(X, \preceq)$ such that $\psi(x_n) \leq C < \infty$ for each n. Then $(\varphi(x_n))$ is a decreasing sequence in $\mathbb{R}$, and there exists $r \in \mathbb{R}$ such that $\lim_{n \to \infty} \varphi(x_n) = r$. Since $(\varphi(x_n))$ is a decreasing sequence, we have for each $m > n$

$$\lim_{n,m \to \infty} d(x_n, x_m) \leq \lim_{n,m \to \infty} [\varphi(x_n) - \varphi(x_m)] = 0.$$

Hence (x_n) is a Cauchy sequence in X. It follows that there exists $x \in X$ such that $\lim_{n \to \infty} x_n = x$. From $\varphi(x_n) \downarrow r$ and $\varphi(x) \leq r$, it follows that

$$d(x_n, x) \leq \lim_m d(x_n, x_m) \leq \lim_m [\varphi(x_n) - \varphi(x_m)]$$
$$= \varphi(x_n) - r \leq \varphi(x_n) - \varphi(x).$$

Hence x is an upper bound for the sequence (x_n) in $(X, \preceq)$ and so we have proved the condition in (b). Now it follows by Theorem 10.11.5 that $(X, \preceq)$ has a maximal element x'. Since (10.83) implies $x' \preceq T(x')$, we have $T(x') = x'$. □

Siegel [191] proved in 1977 in an original way, a generalized version of Caristi's theorem. Here we present some of his results [191].

Let (X, d) be a complete metric space, $\phi : X \to \mathbb{R}^+$, the set of nonnegative real numbers, and $g : X \to X$ be a not necessarily continuous map such that $d(x, g(x)) \leq \phi(x) - \phi(g(x))$ for all $x \in X$.

If a sequence of functions f_i for $i \leq 1 < \infty$ is given, then we define the product

$$\prod_{k=1}^{\infty} f_k x = \lim_{k \to \infty} f_k f_{k-1} \cdots f_1 x,$$

if the limit exists, and call it the *countable decomposition of the given sequence of functions.*

Definition 10.11.7 Let $\Phi = \{f : f : X \to X \text{ and } d(x, f(x)) \leq \phi(x) - \phi(f(x))\}$. We put $\Phi_g = \{f : f \in \Phi \text{ and } \phi(f) \leq \phi(g)\}$.

Lemma 10.11.8 *Let ϕ be an upper semicontinuous function, and (x_i) be a sequence in X such that $d(x_i, x_{i+1}) \leq \phi(x_i) - \phi(x_{i+1})$ for each i. Then there exists $\overline{x} \in X$ such that $\overline{x} = \lim_{i \to \infty} x_i$ and $d(x_i, \overline{x}) \leq \phi(x_i) - \phi(\overline{x})$ for each i.*

Proof. Since the sequence $(\phi(x_i))_i$ is not increasing and bounded below by zero, and since $d(x_i, x_j) \leq \phi(x_i) - \phi(x_j)$ for $i \leq j$, (x_i) is a Cauchy sequence in X. Let $\overline{x} = \lim x_i$. Since ϕ is an upper semicontinuous function, it follows that

$$d(x_i, \overline{x}) = \lim_{j \to \infty} d(x_i, x_j) \leq \phi(x_i) - \lim_{j \to \infty} \phi(x_j) \leq \phi(x_i) - \phi(\overline{x}). \qquad \square$$

Lemma 10.11.9 *The sets Φ and Φ_g are closed by the composition of functions and if ϕ is an upper semicontinuous function then the sets Φ and Φ_g are closed by the countable composition of sequences of functions.*

Proof. We prove that the sets Φ and Φ_g are closed by the composition of functions. If $f_1, f_2 \in \Phi$, then we have

$$\begin{aligned} d(x, f_2 f_1(x)) &\leq d(x, f_1(x)) + d(f_1(x), f_2 f_1(x)) \\ &\leq (\phi(x) \phi(f_1(x))) + (\phi(f_1(x)) - \phi(f_2 f_1(x))) \\ &= \phi(x) - \phi(f_2 f_1(x)). \end{aligned}$$

Hence we have $f_2 f_1 \in \Phi$. If $f_1 \in \Phi_g$, then $\phi(f_1(x)) - \phi(f_2(f_1(x))) \geq 0$ implies $\phi(f_2 f_1) \leq \phi(g)$, and so $f_2 f_1 \in \Phi_g$.
The remainder of the proof follows from the fact that, for each $x \in X$, the sequence $(x_i) = (f_i f_{i-1} \cdots f_1)(x)$ satisfies the conditions of Lemma 10.11.8. $\qquad \square$

Definition 10.11.10 Let E be a subset of a complete metric space (X, d) and $\phi : X \to \mathbb{R}^+$. Then we introduce the following notations:

(1) $r(E) = \inf_{x \in E}(\phi(x))$.

(2) Let $\Phi' \subseteq \Phi$. For each $x \in X$, we put $S_x = \{fx : f \in \Phi'\}$.

Lemma 10.11.11 *We have $\mathrm{diam}(S_x) \leq 2(\phi(x) - r(S_x))$.*

Proof. We have

$$\begin{aligned} d(f_1(x), f_2(x)) &\leq d(x, f_1(x)) + d(x, f_2(x)) \\ &\leq \phi(x) - \phi(f_1(x)) + \phi(x) - \phi(f_2(x)) \\ &\leq 2(\phi(x) - r(S_x)). \square \end{aligned}$$

The main result of Siegel's paper [191] is the following theorem.

Theorem 10.11.12 (Siegel [191], 1977). *Let $\Phi' \subseteq \Phi$ be sets of functions closed by the composition of functions. Also let $x_0 \in X$.*

(a) *If the set Φ' is closed for the composition of a countable sequence of functions, then there exists $\overline{f} \in \Phi'$ such that $\overline{x} = \overline{f}(x_0)$ and $g(\overline{x}) = \overline{x}$ for all $g \in \Phi'$.*

(b) *If the elements of Φ' are continuous functions, then there exists a sequence of functions $f_i \in \Phi'$ and $\overline{x} = \lim_{i \to \infty} f_i f_{i-1} \cdots f_1(x_0)$ such that $g(\overline{x}) = \overline{x}$ for each $g \in \Phi'$.*

Proof. Let (ε_i) be a sequence of positive real numbers converging to zero and $\varepsilon > 0$. Then there exists $f_1 \in \Phi'$ such that $\phi(f_1(x_0)) - r(S_{x_0}) < \varepsilon/2$. We put $x_1 = f_1(x_0)$. Since the set Φ' is closed under the composition of functions it follows that $S_{x_1} \subseteq S_{x_0}$ and

$$\mathrm{diam}(S_{x_1}) \leq 2(\phi(x_1) - r(S_{x_1})) \leq 2(\phi(f_1(x_0)) - r(S_{x_0})) < \varepsilon_1.$$

Continuing in this way, we obtain a sequence of function f_i such that $x_i = f_i(x_{i-1})$, $S_{x_{i+1}} \subseteq S_{x_i}$ and $\mathrm{diam}(S_{x_i}) < \varepsilon_i$.

We know from the condition in (a) that there exists $\overline{f} = \prod_{i=1}^{\infty} f_i \in \Phi'$. Let $\overline{x} = \overline{f}(x_0)$. Since $\overline{x} = \prod_{j=i+1}^{\infty} f_j(x_i)$, it follows that $\overline{x} \in S_{x_i}$ for all i. On the other hand, $\lim_{i \to \infty} \mathrm{diam}(S_{x_i}) = 0$ implies $\overline{x} = \cap_{i=0}^{\infty} S_{x_i}$.

Now we prove that $g(\overline{x}) = \overline{x}$ for each $g \in \Phi'$. This is a consequence of the fact that $g(\overline{x}) \in S_{x_i}$ for each i, and because of $g(\overline{x}) = g(\prod_{j=i+1}^{\infty} f_j(x_i))$.

We know from the condition in (b) that there exists

$$\overline{x} = \lim_{i \to \infty} f_i f_{i-1} \cdots f_1(x_0) = \lim_{i \to \infty} x_i.$$

Since $(x_j)_{j>i} \subseteq S_i$ for each i, it follows that $\overline{x} \in \overline{S_i}$, where $\overline{S_i}$ is the closure of S_i. Since $\mathrm{diam}(\overline{S_i}) = \mathrm{diam}(S_i)$ it follows that $\overline{x} = \cap_{i=0}^{\infty} \overline{S_i}$.

We are going to show that $g(\overline{x}) = \overline{x}$ for each $g \in \Phi'$. We note that $g(x_i) \in S_{x_i}$ for each i. Since g is a continuous function, for each $\varepsilon > 0$, there exists an i_0 such that

$$\{x \in X : d(g(\overline{x}), x) < \varepsilon\} \bigcap S_{x_i} \neq \emptyset \text{ for all } i > i_0.$$

Hence if $i > i_0$, it follows that $d(g(\overline{x}), \overline{x}) < \varepsilon + \varepsilon_i$. Now $\varepsilon_i \to 0$ implies $d(g(\overline{x}), \overline{x}) \leq \varepsilon$, and since ε is arbitrary, we have $g(\overline{x}) = \overline{x}$. $\qquad\square$

Remark 10.11.13 *In the previous theorem, in the condition in (b), we may take $\Phi' = \{g^n\}$, the set of continuous functions and their finite iterations. Then $\overline{x} = \lim_{n \to \infty} g^n(x_0)$ as in Banach's contraction theorem.*

10.12   A theorem by Bollenbacher and Hicks

The following result is related to Caristi's theorem 10.11.1.

Theorem 10.12.1 (Eisenfeld and Lakshmikantham [64])
Let (X, d) be a metric space and $f : X \to X$ be a map. Then there exists a map $\phi : X \mapsto [0, \infty)$ for which

$$d(x, f(x)) \leq \phi(x) - \phi(f(x)) \text{ for } x \in X, \tag{10.84}$$

if and only if the series

$$\sum_{n=0}^{\infty} d(f^n(x), f^{n+1}(x)) \qquad (10.85)$$

converges for each $x \in X$.

Proof. We assume that the condition in (10.84) is satisfied. We show that the series in (10.85) converges. This follows from the fact that, for each $n \in \mathbb{N}$,

$$\sum_{k=0}^{n} d(f^k(x), f^{k+1}(x)) = d(x, f(x)) + \cdots + d(f^{n-1}(x), f^n(x))$$
$$\leq (\phi(x) - \phi(fx)) + \cdots + (\phi(f^{n-1}(x)) - \phi(f^n(x)))$$
$$= \phi(x) - \phi(f^n(x)) \leq \phi(x).$$

If the series (10.85) converges for each $x \in X$, then we define a map $\phi : X \to [0, \infty)$ by

$$\phi(x) = \sum_{k=0}^{\infty} d(f^k(x), f^{k+1}(x)) \text{ for all } x \in X.$$

Clearly this map ϕ satisfies the condition in (10.84). $\qquad \square$

Let $x \in X$ and $O(x, \infty) = \{x, f(x), f^2(x), \ldots\}$ be the orbit of x. The map $G : X \to [0, \infty)$ is said to be f–orbitally lower semicontinuous at x if, for each sequence (x_n) in $O(x, \infty)$, it follows from $\lim_{n \to \infty} x_n = u$ that $G(u) \leq \liminf_{n \to \infty} G(x_n)$.

We note that if the condition in (10.84) is satisfied for each $y \in (x, \infty)$, then the series (10.85) converges for x, since the sequence of partial sums is nondecreasing and bounded by $\phi(x)$.

In 1988, Bollenbacher and Hicks [25] proved the following very interesting theorem, the corollaries of which include many generalizations of Banach's fixed point theorem.

Theorem 10.12.2 (Bollenbacher and Hicks) ([25, Theorem 3]) *Let (X, d) be a metric space, and $f : X \to X$ and $\phi : X \to [0, \infty)$. We assume that there exists x such that*

$$d(y, f(y)) \leq \phi(y) - \phi(f(y)) \text{ for each } y \in O(x, \infty),$$

and that each Cauchy sequence in $O(x, \infty)$ converges to some point in X. Then we have:

(1) $\lim_{n \to \infty} f^n(x) = z$ *exists;*

(2) $d(f^n(x), z) \leq \phi(f^n(x))$;

(3) $f(z) = z$ *if and only if $G(x) = d(x, f(x))$ is f–orbitally lower semicontinuous at x;*

(4) $d(f^n(x), x) \leq \phi(x)$ *and $d(z, x) \leq \phi(x)$.*

Proof.

(1) First we prove the statement in (1).
    It follows from Theorem 10.12.1 that the series

$$\sum_{k=0}^{\infty} d(f^k(x), f^{k+1}(x))$$

converges.

We prove that $(f^n(x))$ is a Cauchy sequence. This follows since, for each $m > n$,

$$d(f^n(x), f^m(x)) \le d(f^n(x), f^{n+1}(x)) + \cdots + d(f^{m-1}(x), f^m(x))$$

$$= \sum_{k=n}^{m-1} d(f^k(x), f^{k+1}(x)),$$

and from the fact that the series $\sum_{k=n}^{\infty} d(f^k(x), f^{k+1}(x))$ converges. Hence there exist $z \in X$ such that the statement in (1) is satisfied.

(2) Now we prove the inequality in (2).
The inequality in (2) follows from

$$0 \le d(f^n(x), f^m(x)) \le \sum_{k=n}^{m-1} d(f^k(x), f^{k+1}(x))$$

$$\le \sum_{k=n}^{m-1} [\phi(f^k(x)) - \phi(f^{k+1}(x))] = \phi(f^n(x)) - \phi(f^m(x)) \le \phi(f^n(x)),$$

as $m \to \infty$.

(3) Now we prove the statement in (3).
We assume that $x_n = f^n(x) \to z$ as $(n \to \infty)$. If G is f–orbitally lower semicontinuous at x, then we have

$$0 \le d(z, f(z)) = G(z) \le \liminf_{n \to \infty} G(x_n) = \liminf_{n \to \infty} d(f^n(x), f^{n+1}(x)) = 0,$$

and so $f(z) = z$.
Now we assume $f(z) = z$ and that (x_n) is a sequence in $O(x, \infty)$ such that $\lim_{n \to \infty} x_n = z$. Then we have

$$G(z) = d(z, f(z)) = 0 \le \liminf_{n \to \infty} d(x_n, f(x_n)) = \liminf_{n \to \infty} G(x_n),$$

and so G is an f–orbitally lower semicontinuous function at x.

(4) Finally, we prove the inequalities in (4).
The inequalities in (4) follow from

$$d(x, f^n(x)) \le d(x, f(x)) + d(f(x), f^2(x)) + \cdots + d(f^{n-1}(x), f^n(x))$$

$$\le [\phi(x) - \phi(f(x))] + [\phi(f(x)) - \phi(f^2(x))] + \dots$$

$$+ [\phi(f^{n-1}(x)) - \phi(f^n(x))]$$

$$= \phi(x) - \phi(f^n(x)) \le \phi(x),$$

and since as $n \to \infty$, we also get $d(x, z) \le \phi(x)$. $\qquad \square$

Corollary 10.12.3 ([93, Theorem, p. 327] or [25, Corollary, p. 899]) *Let (X, d) be a complete metric space and $0 < k < 1$. We assume that, for $f : X \to X$, there exists x such that*

$$d(f(y), f^2(y)) \le kd(y, f(y)) \text{ for each } y \in O(x, \infty). \qquad (10.86)$$

Then we have

(1) $\lim_{n \to \infty} f^n(x) = z$ *exists;*

(2) $d(f^n(x), z) \le k^n(1-k)^{-1}d(x, f(x))$;

(3) $f(z) = z$ if and only if $G(x) = d(x, f(x))$ is an f-orbitally lower semicontinuous function at x;

(4) $d(f^n(x), x) \le (1-k)^{-1}d(x, f(x))$ and $d(z, x) \le (1-k)^{-1}d(x, f(x))$.

Proof.

(i) First we show the statements in (1), (3) and (4).
Let $\phi(y) = (1-k)^{-1}d(y, f(y))$ for all $y \in O(x, \infty)$. If we take $y = f^n(x)$ in (10.86), then we get

$$d(f^{n+1}(x), f^{n+2}(x)) \le kd(f^n(x), f^{n+1}(x)),$$

and so

$$d(f^n(x), f^{n+1}(x)) - kd(f^n(x), f^{n+1}(x)) \le d(f^n(x), f^{n+1}(x)) - d(f^{n+1}(x), f^{n+2}(x)).$$

Hence we have

$$d(f^n(x), f^{n+1}(x)) \le \frac{1}{(1-k)} \cdot [d(f^n(x), f^{n+1}(x)) - d(f^{n+1}(x), f^{n+2}(x))],$$

that is,

$$d(y, f(y)) \le \phi(y) - \phi(f(y)).$$

Now the statements in (1), (3) and (4) follow immediately from Theorem 10.12.2.

(ii) Finally, we prove the inequality in (2).
We remark that (10.86) implies $d(f^n(x), f^{n+1}(x)) \le k^n d(x, fx)$, and Theorem 10.12.2 implies

$$d(f^n(x), z) \le \phi(f^n(x)) = \frac{1}{1-k} \cdot d(f^n(x), f^{n+1}(x)) \le \frac{k^n}{1-k} \cdot d(x, f(x)),$$

hence (2). □

Remark 10.12.4 ([25, Remarks, p. 900]) *We remark that it is not necessary for ϕ to be a lower semicontinuous function, but it is enough that the condition in (10.84) is satisfied only on $O(x, \infty)$ for some x. Furthermore, it can be easier to check that G is a lower semicontinuous function than to check this for the function ϕ. Even if ϕ is an lower semicontinuous function and (10.84) is satisfied for each $x \in X$, it is not necessary in Caristi's theorem that $f(z) = z$, but only $f(x_0) = x_0$ for some x_0 in X.*

Example **10.12.5** ([25, Example 1]) *Let $X = [0, 1]$ and $\phi(x) = x$ for all $x \in X$. We define the map f by*

$$f(x) = \begin{cases} 0 & \text{for } x \in \left[0, \frac{1}{2}\right] \\ \dfrac{x}{2} + \dfrac{1}{4} & \text{for } x \in \left(\frac{1}{2}, 1\right]. \end{cases}$$

For each $x \in [0, 1/2]$, we have $d(x, f(x)) = d(x, 0) = x$ and $\phi(x) - \phi(f(x)) = \phi(x) - 0 = x - 0 = x$. If $x \in (1/2, 1]$, then $d(x, f(x)) = x/2 - 1/4 = \phi(x) - \phi(f(x))$. Hence we have $d(x, f(x)) = \phi(x) - \phi(f(x))$ for all $x \in X$. We note that 0 is the only fixed point of the function f. If $x > 1/2$, then $\lim f^n(x) = 1/2 \ne f(1/2) = 0$.

Example **10.12.6** ([25, Example 2]) *Let* $X = \{(x,y) : 0 \leq x, y \leq 1\}$, *$d$ be the usual metric on X and $f(x,y) = (x,0)$ for all $(x,y) \in X$. Then $f(f(p)) = f(p)$ for all $p \in X$ and $0 = d(f(p), f^2(p)) \leq (1/2)d(p, f(p))$. As in Corollary 10.12.3, let $\phi(p) = 2d(p, f(p))$ and $d(p, f(p)) \leq \phi(p) - \phi(f(p))$. This example shows that, even if both maps f and ϕ are continuous, then f may have more fixed points than one.*

Example **10.12.7** ([25, Example 3]) *We define the map $f : [-1,1] \to [-1,1]$ by*

$$f(x) = \begin{cases} -1 & \text{for } x < 0 \\ \dfrac{x}{4} & \text{for } x \geq 0. \end{cases}$$

We note that $d(f(x), f^2(x)) \leq (1/4)d(x, f(x))$ for all $x \in [-1,1]$. As in Corollary 10.12.3, let $\phi(x) = (4/3)d(x, f(x))$ for all $x \in [-1,1]$. If $x < 0$, then we have $\lim_{n\to\infty} f^n(x) = -1 = f(-1)$, and if $x > 0$, then we have $\lim_{n\to\infty} f^n(x) = 0 = f(0)$. Hence 0 and -1 are the only fixed points of the map f. In this example, f and ϕ are discontinuous functions, $\phi(x) = (4/3)d(x, f(x))$ is an upper semicontinuous function and $d(x, f(x)) \leq \phi(x) - \phi(f(x))$.

10.13 Mann iteration

The continuous function $f : [0,1] \to [0,1]$ with $f(x) = -x$ for $x \in [0,1]$ has a unique fixed point 0, but the Picard iteration sequence $(f^n(x_0))$ diverges for all initial values $x_0 \neq 0$. This was noted by Reinermann [169].

The *Mann iterations* are more general than the Picard iterations, that is, the Picard iterations are special cases of the Mann iterations which Mann introduced in his paper [136] in 1953.

Let E be a convex compact subset of a Banach space X, and $f : E \to E$ be a continuous map. By Schauder's fixed point theorem [181], Theorem 7.3.8, there exists at least one fixed point of the function f, that is, there exists at least one $p \in E$ such that $f(p) = p$.

In 1953, Mann ([136]) studied the problem of constructing a sequence (x_n) in E which converges to a fixed point of f. Usually an arbitrary initial value $x_1 \in E$ is chosen, and then the sequence of successive iterations (x_n) of x_1 defined by

$$x_{n+1} = f(x_n) \text{ for } n = 1, 2, \ldots$$

is considered. If this sequence converges, then its limit is a fixed point of the function f.

Definition 10.13.1 (Dotson [57]) Let X be a linear space, C be a convex subset of X, $f : C \to C$ be a map and $x_1 \in C$. We assume that the infinite matrix $A = (a_{nj})$ satisfies the conditions

(A_1) \qquad $a_{nj} \geq 0$ for all $j \leq n$ and $a_{nj} = 0$ for $j > n$;

(A_2) \qquad $\sum_{j=1}^{n} a_{nj} = 1$ for each $n \geq 1$;

(A_3) \qquad $\lim_{n\to\infty} a_{nj} = 0$ for each $j \geq 1$.

We define the sequence (x_n) by $x_{n+1} = f(v_n)$, where

$$v_n = \sum_{j=1}^{n} a_{nj} x_j.$$

The sequence (x_n) is called the *Mann iterative sequence*, or simply, *Mann iteration*, and usually denoted by $M(x_1, A, f)$.

Hence the matrix A in Definition 10.13.1 has the following form

$$
A = \begin{bmatrix}
1 & 0 & 0 & \cdots & 0 & 0 \\
a_{21} & a_{22} & 0 & \cdots & 0 & 0 \\
\cdot & \cdot & \cdot & \cdot & \cdot & \cdot \\
a_{n1} & a_{n2} & \cdots & a_{nn} & 0 & 0 \\
\cdot & \cdot & \cdot & \cdot & & \cdot
\end{bmatrix}.
$$

Theorem 10.13.2 (Mann) ([136, Theorem 1]) *If one of the sequences (x_n) or (v_n) is convergent, then they both converge. In this case, they converge to the same limit point which is a fixed point of the function f.*

Proof. Let $\lim_{n \to \infty} x_n = p$. Since A is a regular matrix, it follows that $\lim_{n \to \infty} v_n = p$. The continuity of the function f implies $\lim_{n \to \infty} f(v_n) = f(p)$, and from $f(v_n) = x_{n+1}$, it follows that $f(p) = p$. If we assume $\lim_{n \to \infty} v_n = q$, then $\lim_{n \to \infty} x_{n+1} = f(q)$, and the regularity of the matrix A implies $\lim_{n \to \infty} v_n = f(q)$. Hence we have $f(q) = q$. $\square$

Example **10.13.3** ([136, pp. 508/509]) *Let A be the Cesàro matrix of order 1, that is,*

$$
A = \begin{bmatrix}
1 & 0 & 0 & 0 & \cdots \\
\dfrac{1}{2} & \dfrac{1}{2} & 0 & 0 & \cdots \\
\dfrac{1}{3} & \dfrac{1}{3} & \dfrac{1}{3} & 0 & \cdots \\
\cdot & \cdot & \cdot & \cdot & \cdot \cdot \cdot \\
\dfrac{1}{n} & \dfrac{1}{n} & \dfrac{1}{n} & \cdots & \dfrac{1}{n} & 0 & 0 & \cdots \\
\cdot & \cdot & \cdot & \cdot & \cdot & \cdot \cdot \cdot
\end{bmatrix}.
$$

The matrix A satisfies all the assumptions for a matrix in this section. In this case, the Mann method $M(x_1, A, f)$ is usually referred to as the mean value method, *where the initial value is $x_1 \in X$ and*

$$
x_{n+1} = f(v_n) \quad \text{and} \quad v_n = \frac{1}{n} \sum_{k=1}^{n} x_k \quad \text{for all } n = 1, 2, \ldots.
$$

We note

$$
v_{n+1} - v_n = \frac{n \sum_{k=1}^{n+1} x_k - (n+1) \sum_{k=1}^{n} x_k}{(n+1)n} = \frac{f(v_n) - v_n}{n+1}. \tag{10.87}
$$

In many special problems, the iterative method $M(x_1, A, f)$ converges even when the method $T^n x_1$ diverges.

Example **10.13.4** ([136, p. 509]) *Let $X = \{x \in \mathbb{R}^2 : \|x\| \leq 1\}$, where $\|\cdot\|$ is the Euclidean norm. Furthermore, let A be the Cesàro matrix of order 1 and the function $f : X \to X$ be the rotation about the origin by the angle $\pi/4$. Then the Picard iteration $f^n(x_1)$ does not converge for any $x_1 \in X \setminus \{0\}$. Using the Mann method $M(x_1, A, f)$, the sequences (x_n) and (v_n) always converge (on a spiral) to the origin, independently of the choice of the initial value x_1.*

Definition 10.13.5 (Dotson) ([57, Definition]) The Mann iterative method $M(x_1, A, f)$ is called the *normal Mann iterative method* if the matrix $A = (a_{nj})$, besides the conditions (A_1), (A_2) and (A_3), also satisfies the next two conditions

(A_4) $\qquad a_{n+1,j} = (1 - a_{n+1,n+1})a_{nj}$ for $(j = 1, 2, \ldots, n; n = 1, 2, \ldots)$;

(A_5) $\qquad$ either $a_{nn} = 1$ for all n, or $a_{nn} < 1$ for all $n > 1$.

In his paper [57], Dotson proved the following theorem.

Theorem 10.13.6 (Dotson) ([57, Theorem 2]) *The following statements are true:*

(a) *The Mann method $M(x_1, A, f)$ is normal if and only if the matrix $A = (a_{nj})$ satisfies the conditions in (A_1), (A_2), (A_4), (A_5) and (A_3'), where*

$$\sum_{n=1}^{\infty} a_{nn} \text{ is a divergent series.} \qquad (A_3')$$

(b) *The matrices $A = (a_{nj})$ (except for the identity matrix) in all normal Mann methods $M(x_1, A, f)$ are constructed as follows:*
Let $0 \le c_n < 1$ for all $n = 1, 2, \ldots$ and the series $\sum_{n=1}^{\infty} c_n$ be divergent. Then the matrix $A = (a_{nj})$ is defined by

$$\begin{cases} a_{11} = 1, \ a_{1j} = 0 \text{ for } j > 1; \\ a_{n+1,n+1} = c_n \text{ for } n = 1, 2, \ldots \\ a_{n+1,j} = a_{jj} \prod_{i=j}^{n}(1 - c_i) \text{ for } j = 1, 2, \ldots, n \\ a_{n+1,j} = 0 \text{ for } j > n+1 \text{ and } n = 1, 2, \ldots \end{cases}$$

(c) *The sequence (v_n) in the normal Mann method $M(x_1, A, f)$ satisfies*

$$v_{n+1} = (1 - c_n)v_n + c_n f(v_n) \text{ for } n = 1, 2, \ldots,$$

where

$$c_n = a_{n+1,n+1} \text{ for all } n.$$

Proof.

(a) The statement in (a) follows from the following well–known result on infinite products, namely, that if $0 \le c_n < 1$ for all n, then $\lim_{n \to \infty} \prod_{k=1}^{n}(1 - c_k) = 0$ if and only if the series $\sum_{k=1}^{\infty} c_k$ diverges.

(b) Now we show the statement in (b).
We note that if the matrix A satisfies the conditions in $(A1)$–$(A5)$, then it satisfies the condition in (b).
It can also be proved that if the matrix A satisfies the conditions in (b), where $c_n = a_{n+1,n+1}$ for all $n \in \mathbb{N}$, then it satisfies the conditions in $(A1)$–$(A5)$.

(c) The proof of (c) follows if we use the condition in $(A4)$ and the definitions of the sequences (v_n) and (x_n) in Mann's method $M(x_1, A, f)$. $\qquad \square$

Example **10.13.7 (Dotson)** ([57, pp. 66/67]) *For each λ with $0 \le \lambda < 1$, let the infinite matrix $A_\lambda = (a_{nj})$ be defined by*

$$\begin{cases} a_{n1} = \lambda^{n-1}, \\ a_{nj} = \lambda^{n-j}(1 - \lambda) \text{ for } j = 2, 3, \ldots, n, \\ a_{nj} = 0 \text{ for } j > n \text{ and } n = 1, 2, 3, \ldots, \end{cases}$$

where, for $\lambda = 0$, *we put* $a_{nn} = 1$ *for all* n. *Hence* A_0 *is the infinite identity matrix. It can be shown that for each* λ *with* $0 \leq \lambda < 1$, $M(x_1, A_\lambda, f)$ *is a normal Mann method with* $c_n = a_{n+1,n+1} = 1 - \lambda$ *for all* $n = 1, 2, 3 \ldots$. *Hence the sequence* (v_n) *in the normal Mann method* $M(x_1, A_\lambda, f)$ *is defined by*

$$v_{n+1} = \lambda v_n + (1 - \lambda) f(v_n) \text{ for all } n.$$

Let $g_\lambda = \lambda I + (1 - \lambda) f$ *(where* I *is the identity map). Hence we have*

$$v_{n+1} = g_\lambda(v_n) = g_\lambda^n(v_1) = g_\lambda^n(x_1) \text{ for all } n.$$

We note that $g_0 = f$ *and, in this case, the sequence* (v_n) *is obtained by Picard's iteration* $(f^n(x_1))$. *The sequence* $(g_{1/2}^n(x_1))$ *of Picard's iterations of the map* $g_{1/2} = (1/2)(I + f)$ *was studied by Krasnoselski [119] and Edelstein [63], and the sequence* $(g_\lambda^n(x_1))$ *of Picard's iterations of the map* g_λ *for* $0 < \lambda < 1$ *which was studied by Schäfer [180], Browder and Petryshyn [33], and Opial [148].*

In the literature, mainly the normal Mann iterative method is studied.

10.14   Continuous functions on compact real intervals

Now we consider the case when the Banach space is the real line $\mathbb{R}$, and the convex compact set E is a closed interval and A is the Cesàro matrix of order 1.

Theorem 10.14.1 (Mann) *([136, Theorem 4]) Let* $f : [a, b] \to [a, b]$ *be a continuous map which has a unique fixed point* $p \in [a, b]$ *and* A *be the Cesàro matrix of order* 1. *Then Mann's sequence* $M(x_1, A, f)$ *converges to* p *for each* $x_1 \in [a, b]$.

Proof. It follows from (10.87) that $v_{n+1} - v_n \to 0$ as $n \to \infty$. Since f is a continuous function and p is the unique fixed point of f, it follows that $f(x) - x > 0$ for $x < p$ and $f(x) - x < 0$ for $x > p$. Hence, for each $\delta > 0$, there exists $\varepsilon > 0$ such that $|x - p| \geq \delta$ implies $|f(x) - x| \geq \varepsilon$. It follows from (10.87) that

$$v_{n+1} = v_1 + \sum_{k=1}^{n} \frac{f(v_k) - v_k}{k + 1}.$$

Now from our previous considerations, we have $\lim_{n \to \infty} v_n = p$, and by Theorem 10.13.2, we obtain $\lim_{n \to \infty} x_n = p$. $\qquad \square$

In higher dimensional spaces, results similar to that of Theorem 10.14.1 have not been obtained.

Remark 10.14.2 *Reinermann [169, p. 210] defined a summability matrix* A *as follows*

$$a_{nk} = \begin{cases} c_k \prod_{j=k+1}^{n} (1 - c_j) & \text{for } k < n \\ c_n & \text{for } k = n \\ 0 & \text{for } k > n, \end{cases} \tag{10.88}$$

where the real sequence (c_n) *satisfies the following conditions*

(i) $c_0 = 1$,

(ii) $0 < c_n < 1$ *for* $n \geq 1$,

(iii) $\sum_{k=0}^{\infty} c_k$ *diverges.*

It can be proved that A *is a regular matrix, and satisfies the following conditions*

$$0 \leq a_{nk} \leq 1 \ \textit{for} \ n, k = 0, 1, 2, \ldots,$$

$$\sum_{k=0}^{n} a_{nk} = 1 \ \textit{for} \ n = 0, 1, 2, \ldots.$$

Reinermann also considered the condition $c_n = 1$, *since he included the identity matrix in his considerations. Since the identity matrix is of no special interest, in all interesting applications, it is assumed that* $c_n < 1$. *Then he considered the iterative scheme* $z_0 = x_0 \in E$ *and* $x_{n+1} = \sum_{k=0}^{n} a_{nk} f(x_k)$, *which can be written as*

$$x_{n+1} = (1 - c_n)x_n + c_n f(x_n). \tag{10.89}$$

It is well known by Brouwer's fixed point theorem, Theorem 7.3.5, that a continuous map from $[a, b]$ to $[a, b]$ has at least one fixed point. Reinermann proved the following result.

Theorem 10.14.3 (Reinermann) ([169, Satz]) *Let* $a, b \in \mathbb{R}$, $a < b$, $E = [a, b]$ *and* $f : E \to E$ *be a continuous map with at most one fixed point. If the matrix* A *is defined by* *(10.88) and the sequence* (c_n) *satisfies the conditions in (i)–(iii) of Remark 10.14.2 and* $\lim c_n = 0$, *then the iterative scheme (10.89), for* $x_0 \in [a, b]$, *converges to the fixed point of* f.

Proof. Without loss of generality, we may assume $a = 0$ and $b = 1$. By Brouwer's fixed point theorem and our assumption, there exists a unique fixed point $x \in [0, 1]$ of the function f. Now we have

$$\text{for all } y \in [0, 1] \text{ with } y < x \text{ it follows that } f(y) - y > 0; \tag{10.90}$$

$$\text{for all } y \in [0, 1] \text{ with } (y > x) \text{ it follows that } f(y) - y < 0. \tag{10.91}$$

If $x = 0$, then we obviously have (10.90). If $x > 0$ and if there exists $y_1 \in [0, 1]$ with $y_1 < x$ such that $f(y_1) - y_1 \leq 0$, then $f(0) - 0 = f(0) \geq 0$ implies that there exists $z \in [0, y_1]$ such that $f(z) = z$. Now $z \neq x$, which is a contradiction to the uniqueness of the fixed point. The case (10.91) is proved analogously.
There are two alternatives, **I** and **II** for the sequence (x_n):
I. There exists $n_1 \in \mathbb{N}$ such that $x_{n_1} = x$.
Then $x_n = x$ for all $n \geq n_1$ and the theorem is proved.
II. For each $n \in \mathbb{N}$, we have $x_n \neq x$.
In this case, we have the following three possibilities:

1. There exists $n_0 \in \mathbb{N}$ such that $x_n < x$ for all $n > n_0$. Then we have

    $$x_{n+1} - x_n = c_n(f(x_n) - x_n),$$

    and (10.90) implies that (x_n) is a monotone increasing sequence; so the sequence converges, since $x_n \leq 1$ for all n. By Theorem 10.13.2, and since the function f has only one fixed point $x \in [0, 1]$, it follows that $\lim_n x_n = x$.

2. There exists $m_0 \in \mathbb{N}$ such that $x_n > x$ for all $n \geq m_0$. In this case, it follows by (10.91) that $\lim_{n \to \infty} x_n = x$, as in Case **1**.

3. We assume that possibilities **1** and **2** are not true. Let $\varepsilon > 0$ be given. We choose $n_0 \in \mathbb{N}$ such that

$$|x_{n+1} - x_n| < \varepsilon \text{ for all } n \geq n_0.$$

This is possible, since

$$|x_{n+1} - x_n| \leq 2c_n \text{ and } \lim_{n \to \infty} c_n = 0.$$

We are going to prove that there exists $n_1 \in \mathbb{N}$ with $n_1 \geq n_0$ such that $|x_{n_1} - x| < \varepsilon$, that is,

$$\text{there exists } n_1 \geq n_0 \text{ such that } -\varepsilon < x_{n_1} - x < \varepsilon. \tag{10.92}$$

If (10.92) is not true, then

$$x_n \leq x - \varepsilon \text{ or } x_n \geq x + \varepsilon \text{ for each } n \geq n_0.$$

Now, if $x_{n_0} \leq x - \varepsilon$, then $x_n \leq x - \varepsilon$ for all $n \geq n_0$ (because of $|x_{n+1} - x_n| < \varepsilon$), that is, the condition in **1** is satisfied. Analogously, if $x_{n_0} \geq x + \varepsilon$, then $x_n \geq x + \varepsilon$ for all $n \geq n_0$ (again because of $|x_{n+1} - x_n| < \varepsilon$), that is, the condition in **2** is satisfied. Hence, in all cases, the conditions in **1** or **2** are satisfied. So we have shown (10.92). We are going to prove that we have $|x_n - x| < \varepsilon$ for all $n \geq n_1$. This is true for $n = n_1$. If $n \geq n_1$ and if $|x_n - x| < \varepsilon$, then we have the following possibilities **A** and **B**:

A. $x - \varepsilon < x_n < x$. Then we have (a) or (b) for x_{n+1}:

   (a) $x_{n+1} < x$. In this case, $x_{n+1} - x_n = c_n(f(x_n) - x_n)$ and (10.90) imply $x_{n+1} - x_n > 0$, hence we have

$$|x_{n+1} - x| = x - x_{n+1} < x - x_n = |x_n - x| < \varepsilon.$$

   (b) $x_{n+1} > x$. Now we have

$$|x_{n+1} - x| = x_{n+1} - x < x_{n+1} - x_n = |x_{n+1} - x_n| < \varepsilon.$$

B. $x < x_n < x + \varepsilon$. Now (10.91) implies the conclusion as in **A**, that is, $|x_{n+1} - x| < \varepsilon$. Hence $|x_{n+1} - x| < \varepsilon$. It follows by mathematical induction that $|x_n - x| < \varepsilon$ for all $n \geq n_1$, thus $\lim_n x_n = x$. $\qquad\square$

We note that if we put $c_n := 1/(n+1)$ for all n, then Theorem 10.14.3 implies Theorem 10.14.1.

In 1971, Franks and Marzec [69] showed that the condition of the uniqueness of the fixed point p in Theorem 10.14.1 is not necessary.

Theorem 10.14.4 (Franks and Marzec) ([69, Theorem]) *Let $f : [0,1] \to [0,1]$ be a continuous function. Then the iterative sequence*

$$x_{n+1} = f(\tilde{x}_n) \text{ for } n = 1, 2, \ldots \tag{10.93}$$

$$\tilde{x}_n = \sum_{k=1}^{n} \frac{x_k}{n} \text{ for } n = 1, 2, \ldots, \tag{10.94}$$

$$\tilde{x}_1 = x_1 \in [0,1], \tag{10.95}$$

converges to a fixed point of the function f in the interval $[0,1]$.

Proof. It follows from (10.93) and (10.94) that

$$\tilde{x}_{n+1} = \frac{f(\tilde{x}_n) - \tilde{x}_n}{n+1} + \tilde{x}_n \text{ for } n = 1, 2, \dots. \tag{10.96}$$

Since $\tilde{x}_n$ and $f(\tilde{x}_n) \in [0,1]$ for all n, we have

$$|\tilde{x}_{n+1} - \tilde{x}_n| \leq \frac{1}{n+1} \text{ for } n = 1, 2, \dots. \tag{10.97}$$

It suffices to prove that this sequence is convergent and its limit $\xi \in [0,1]$ is a fixed point of the function f.

1. We prove that the sequence $(\tilde{x}_n)$ is convergent. The sequence $(\tilde{x}_n)$ is in $[0,1]$, and so has at least one accumulation point. We assume that the sequence $(\tilde{x}_n)$ has two distinct accumulation points ξ_1 and ξ_2 with $\xi_1 < \xi_2$.

    a. We are going to show that we have, from the assumption above, $f(x) = x$ for all $x \in (\xi_1, \xi_2)$. Let $x^* \in (\xi_1, \xi_2)$. If $f(x^*) > x^*$, then, since f is a continuous function, there exists $\delta \in (0, (x^* - \xi_1)/2)$ such that $|x - x^*| < \delta$ implies $f(x) > x$. Hence $|\tilde{x}_n - x^*| < \delta$ implies $f(\tilde{x}_n) > \tilde{x}_n$. Thus we obtain from (10.96) that

    $$|\tilde{x}_n - x^*| < \delta \text{ implies } \tilde{x}_{n+1} > \tilde{x}_n. \tag{10.98}$$

    By (10.97), there exists N such that

    $$|\tilde{x}_{n+1} - \tilde{x}_n| < \delta \text{ for } n = N, N+1, \dots. \tag{10.99}$$

    Since $\xi_2 > x^*$ is an accumulation point of the sequence $(\tilde{x}_n)$, we can choose N such that $\tilde{x}_N > \tilde{x}^*$. It follows from (10.98) and (10.99) that

    $$\tilde{x}_n > x^* - \delta > \xi_1 \text{ for } n = N, N+1, \dots.$$

    Thus ξ_1 is not an accumulation point of the sequence $(\tilde{x}_n)$, which contradicts our assumption.
    If $f(x^*) < x^*$, then, similarly as above, we obtain that ξ_2 is not an accumulation point of the sequence $(\tilde{x}_n)$, which again is a contradiction. Hence $f(x^*) = x^*$ for each $x^* \in (\xi_1, \xi_2)$.

    b. Let us prove that ξ_1 and ξ_2 are not accumulation points of the sequence $(\tilde{x}_n)$. We note that

    $$\tilde{x}_n \notin (\xi_1, \xi_2) \text{ for } n = 1, 2, \dots. \tag{10.100}$$

    If $f(\tilde{x}_n) = \tilde{x}_n$, then (10.96) implies $\tilde{x}_m = \tilde{x}_n$ for all $m > n$. So neither ξ_1 nor ξ_2 can be an accumulation point of the sequence $(\tilde{x}_n)$. Furthermore, (10.97) and (10.100) imply that there exists a natural number M such that $\tilde{x}_M \geq \xi_2$ for all $n > M$. Hence ξ is not an accumulation point of the sequence $(\tilde{x}_n)$. It follows from $\tilde{x}_M \leq \xi_1$ that $\tilde{x}_n < \xi_1 < \xi_2$ for all $n > M$. Hence ξ_2 is not an accumulation point of the sequence $(\tilde{x}_n)$. Consequently the sequence $(\tilde{x}_n)$ cannot have two distinct accumulation points, and so this sequence is convergent. We put $\lim_n \tilde{x}_n = \xi \in [0,1]$.

2. We show that $f(\xi) = \xi$. We assume $f(\xi) > \xi$. Let

$$\varepsilon = \frac{f(\xi) - \xi}{2} > 0.$$

Since the sequence $(\tilde{x}_n)$ converges to ξ and the function f is continuous, there exists a natural number N such that $f(\tilde{x}_n) - \tilde{x}_n > \varepsilon$ for each $n > N$. It follows from (10.96) that

$$\tilde{x}_{n+1} - \tilde{x}_n = \frac{f(\tilde{x}_n) - \tilde{x}_n}{n+1} > \frac{\varepsilon}{n+1}.$$

Hence we have

$$\lim_{m \to \infty} (\tilde{x}_{N+m} - \tilde{x}_N) = \lim_{m \to \infty} \sum_{n=N}^{m-1} (\tilde{x}_{n+1} - \tilde{x}_n)$$

$$\geq \lim_{m \to \infty} \sum_{n=N}^{m-1} \frac{\varepsilon}{n+1} = \infty.$$

So $\tilde{x}_n \to \infty$ as $n \to \infty$, which contradicts the fact that $\tilde{x}_m \in [0,1]$ for all m. If $f(\xi) < \xi$, then it can be shown that $\tilde{x}_n \to -\infty$ as $n \to \infty$, which again is a contradiction. So we have $f(\xi) = \xi$. $\qquad\square$

We close with the following remark.

Remark 10.14.5 *Rhoades ([171] and [170]) among other things, generalized many results presented in this section.*

Bibliography

[1] A. Aasma, H. Dutta, and P. N. Natarajan. *An Introductory Course in Summability Theory*. Wiley, 2017.

[2] P. Aiena. *Fredholm and Local Spectral Theory, with Applications to Multipliers*. Kluwer Academic Publishers, 2004.

[3] P. Aiena. *Semi–Fredholm Operators, Perturbation Theory and Localized Svep*. Merida, Venezuela, 2007.

[4] R. R. Akhmerov, M. I. Kamenskii, A. S. Potapov, A. E. Rodkina, and B. N. Sadovskii. *Measures of Noncompactness and Condensing Operators*. Birkhäuser Verlag, Basel, 1992.

[5] G. Alexits. *Konvergenzprobleme der Orthogonalreihen*. Verlag der Ungarischen Akademie der Wissenschaften, Budapest, 1960.

[6] A. Ambrosetti. Un teorema di esistenza per le equazioni differenziali negli spazi di Banach. *Rend. Sem. Univ. Padova*, 39:349–360, 1967.

[7] C. Apostol. The reduced minimum modulus. *Michigan Math. J.*, 32:279–294, 1985.

[8] F. V. Atkinson. On relatively regular operators. *Acta. Sci. Math. Szeged*, 15:38–56, 1953.

[9] S. Banach. Sur les opérations dans les ensembles abstraits et leur applications aux équations intégrales. *Fund. Math.*, 3:133–181, 1922.

[10] S. Banach. Sur les functionnelles linéaires. *Studia Math.*, 1:211–216, 1929.

[11] S. Banach and H. Steinhaus. Sur le principe de la condensation de singularités. *Fundamenta Math.*, 9:50–61, 1927.

[12] J. Banás and K. Goebel. *Measures of Noncompactness in Banach Spaces*, volume 60 of *Lecture Notes in Pure and Applied Mathematics*. Marcel Dekker, New York and Basel, 1980.

[13] J. Banaś and M. Mursaleen. *Sequence Spaces and Measures of Noncompactness with Applications to Differential and Integral Equations*. Springer, New Delhi, Heidelberg, New York, Dordrecht, London, 2014.

[14] J. Banaś and B. Rzepka. An application of a measure of noncompactness in the study of asymptotic stability. *Appl. Math. Letters*, 16:1–6, 2003.

[15] B. A. Barnes, G. J. Murphy, M. R. F. Smyth, and T. T. West. *Riesz and Fredholm Theory in Banach Algebras*. Pitman, Boston, 1982.

[16] A. Ben-Israel and T. N. E. Greville. *Generalized Inverses: Theory and Applications*. Wiley-Interscience, 1974.

[17] A. Ben-Israel and T. N. E. Greville. *Generalized Inverses: Theory and Applications.* Springer Verlag, New York, Second edition, 2003.

[18] C. Benitez. Orthogonality in normed linear spaces: A classification of the different concepts and some open problems. *Revista Matematica de la Universidad Complutense de Madrid,* 2, número suplementario:53–57, 1989.

[19] A. A. Bennett. Newton's method in general analysis. *Proc. Nat. Acad. Sci. U.S.A.,* 2:592–598, 1916.

[20] S. Berberian. *Lectures in Functional Analysis and Operator Theory.* Springer Verlag, New York, Heidelberg, Berlin, 1974.

[21] S. K. Berberian. The Weyl spectrum of an operator. *Indiana Univ. Math. J,* 20:529–544, 1970.

[22] A. Bielecki. Une remarque sur la méthode de Banach–Cacciopoli–Tikhonov dans la théorie de l'équation $s = f(x, y, z, p, q)$. *Bull. Akad. Polon. Sci., Cl III,* 4:265–268, 1956.

[23] A. Bielecki. Une remarque sur la méthode de Banach–Cacciopoli–Tikhonov dans la théorie des équations differentiélles ordinaires. *Bull. Akad. Polon. Sci., Cl III,* 4:261–264, 1956.

[24] G. Birkhoff. Orthogonality in linear metric spaces. *Duke Math. J.,* 1:169–172, 1935.

[25] A. Bollenbacher and T. L. Hicks. A fixed point theorem revisited. *Proc. Amer. Math. Soc.,* 102:898–900, 1988.

[26] F. F. Bonsall and J. Duncan. *Complete Normed Algebras.* Springer–Verlag, Berlin, Heidelberg, New York, 1973.

[27] J. Boos. *Classical and Modern Methods in Summability.* Oxford University Press, Oxford, 2000.

[28] K. Borsuk. Drei Sätze über die n–dimensionale euklidische Sphäre. *Fundamenta Mathematica,* 20:177–190, 1932.

[29] D. W. Boyd and J. S. W. Wong. Another proof of contraction mapping theorem. *Canad. Math. Bull.,* 11:605–606, 1968.

[30] D. W. Boyd and J. S. W. Wong. On nonlinear contractions. *Proc. Amer. Math. Soc.,* 20:458–464, 1969.

[31] H. Brézis and F. E. Browder. A general principle on ordered sets in nonlinear functional analysis. *Advances in Mathematics,* 21:355–364, 1976.

[32] F. E. Browder. On the spectral theory of elliptic differential operators. *Math. Ann.,* 142:22–130, 1961.

[33] F. E. Browder and W. V. Petryshyn. The solution by iteration of linear functional equations in Banach spaces. *Bull. Amer. Math. Soc.,* 72:571–575, 1966.

[34] V. W. Bryant. A remark on a fixed point theorem for iterated mappings. *Amer. Math. Monthly,* 75:399–400, 1968.

[35] D. Buckholtz. Inverting the difference of Hilbert space projections. *Amer. Math. Monthly,* 104:60–61, 1997.

[36] S. R. Caradus. *Operator Theory of the Pseudo–Inverse*, volume 38. Queen's Papers in Pure and Applied Mathematics, Queen's University, Kingston, Ontario, 1974.

[37] S. R. Caradus. *Generalized Inverses and Operator Theory*, volume 50. Queen's Papers in Pure and Applied Mathematics, Queen's University, Kingston, Ontario, 1978.

[38] S. R. Caradus, W. E. Pfaffenberger, and B. Yood. *Calkin Algebras and Algebras of Operators on Banach Spaces*. Marcel Dekker, New York, 1974.

[39] J. Caristi. Fixed point theorems for mappings satisfying inwardness conditions. *Trans. Amer. Math. Soc.*, 215:241–251, 1976.

[40] A. Cellina. On the nonexistence of solutions of differential equations in nonreflexive spaces. *Bull. Amer. Math. Soc.*, 78, no. 6:1069–1072, 1972.

[41] S. K. Chatterjee. Fixed point theorems. *C. R. Acad. Bulgare Sci.*, 15:727–730, 1972.

[42] Lj. B. Ćirić. Generalized contractions and fixed point theorems. *Publ. Inst. Math.*, 12(26):19–26, 1971.

[43] Lj. B. Ćirić. A generalization of Banach's contraction principle. *Proc. Amer. Math. Soc.*, 45:267–273, 1974.

[44] Lj. B. Ćirić. *Some Recent Results in Metrical Fixed Point Theory*. University of Belgrade, Belgrade, 2003.

[45] L. W. Cohen and N. Dunford. Transformations on sequence spaces. *Duke Mathematical Journal*, 3, no. 4:689–701, 1937.

[46] E. H. Connell. Properties of fixed point spaces. *Proc. Amer. Math. Soc.*, 10:974–979, 1959.

[47] J. B. Conway. *A Course in Functional Analysis*. Springer–Verlag, New York, Heidelberg and Tokyo, 1985.

[48] R. C. Cooke. *Infinite Matrices*. MacMillan and Co. Ltd, London, 1950.

[49] L. Crone. A characterization of matrix mappings on ℓ^2. *Math. Z.*, 123:315–317, 1971.

[50] J. Daneš. On the Istrăţescu's measure of noncompactness. *Bull. Math. Soc. Sci. Math. R. S. Roumanie*, 16:403–406, 1972.

[51] G. Darbo. Punti uniti in transformazioni a condominio non compatto. *Rend. Sem. Math. Univ. Padova*, 24:84–92, 1955.

[52] S. Delpech. A short proof of Pitt's compactness theorem. *Proc. American Math. Soc.*, 137, No. 4:1371–1372, 2009.

[53] C. A. Desoer and B. H. Whalen. A note on pseudoinverses. *SIAM J.*, 11:442–447, 1963.

[54] J Diuedonné. Deux examples singuliérs d'équations différentielles. *Acta Sci. Math. (Szeged)*, 12, Leopoldo Fejér et Frederico Riesz, LXX annos natis dedicatus, pars B:38–40, 1950.

[55] I. Djolović and E. Malkowsky. A note on compact operators on matrix domains. *J. Math. Anal. Appl.*, 340:291–303, 2008.

[56] C. T. J. Dodson and P. C. Parker. *A User's Guide to Algebraic Topology.* Kluwer, Dordrecht, The Netherlands, 1997.

[57] W. G. Dotson. On the Mann iterative methods. *Trans. Amer. Math. Soc.*, 149:65–73, 1970.

[58] H. R. Dowson. *Spectral Theory of Linear Operators.* Academic Press, London and New York, 1978.

[59] M. P. Drazin. Pseudoinverse in associative rings and semigroups. *Amer. Math. Monthly*, 65:506–514, 1958.

[60] J. Dugundji. *Topology.* Allyn and Bacon, Boston, London, Sydney and Toronto, 1966.

[61] M. Edelstein. An extension of Banach's contraction principle. *Proc. Amer. Math. Soc.*, 12:7–10, 1961.

[62] M. Edelstein. On fixed and periodic points under contractive mappings. *J. London Math. Soc.*, 37:74–79, 1962.

[63] M. Edelstein. A remark on a theorem of M. A. Krasnoselskii. *Amer. Math. Monthly*, 73:509–510, 1966.

[64] J. Eisenfeld and V. Lakshmikantham. Fixed point theorems on closed sets through abstract cones. *Appl. Math Comput.*, 3:155–166, 1977.

[65] P. A. Enflo. A counterexample to the approximation problem in Banach spaces. *Acta Math.*, 130:309–317, 1973.

[66] J. Banaś et al., editor. *Advances in Nonlinear Analysis via the Concept of Measure of Noncompactness.* Springer Nature Nature, 2017.

[67] M. Fabian, P. Hababa, P. Hájek, V. Montesinos Santalucia, J. Pelant, and V. Zizler. *Functional Analysis and Infinite Dimensional Geometry.* CMS Books in Mathematics. Springer Verlag, 2001.

[68] B. Fisher. A fixed point theorem. *Mathematics Magazine*, 48:223–225, 1975.

[69] R. L. Franks and R. P. Marzec. A theorem on mean–value iterations. *Proc. Amer. Math. Soc.*, 30(20):324–326, 1971.

[70] G. Freud. *Orthogonale Polynome.* Birkhäuser Verlag, Basel, 1969.

[71] I. M. Gelfand. Normierte Ringe. *Mat. Sborn. (N. S.)*, 9(51):3–24, 1941.

[72] A. N. Godunov. On the Peano theorem in Banach spaces. *Funkc. Analiz. Prilož.*, 9, 1:56–60, 1975.

[73] K. Goebel and W. Rzymowski. An existence theorem for the equations $x' = f(t, x)$ in Banach space. *Bull. Acad. Polon. Sci., Sér. Math. Astronom. Phys.*, pages 367–370, 1970.

[74] I. C. Gohberg and M. S. Krein. The basic propositions on defect numbers, root numbers and indices pf linear operators. *Amer. Math. Soc. Transl.*, Ser. 2(13):185–264, 1960.

[75] S. Goldberg. *Unbounded Linear Operators with Applications.* McGraw–Hill, New York, 1966.

[76] S. Grabiner. Ascent, descent, and compact perturbations. *Proc. Amer. Math. Soc.*, 71:79–80, 1978.

[77] S. Grabiner. Uniform ascent and descent of bounded operators. *J. Math. Soc. Japan*, 34(29):317–337, 1982.

[78] C. W. Groetsch. *Generalized Inverses of Linear Operators: Representation and Approximation*. Marcel Dekker, Inc., New York and Basel, 1977.

[79] K. Gustafson and J. Weidman. On the essential spectrum. *J. Math. Appl.*, 25:121–127, 1969.

[80] O. Hadžić. Some properties of measures of noncompactness in paranormed spaces. *Proc. Amer. Math. Soc.*, 102:843–849, 1988.

[81] H. Hahn. Über Folgen linearer Operationen. *Monatsh. Math. Phys.*, 32:3–88, 1922.

[82] P. R. Halmos. *A Hilbert Space Problem Book*. Van Nostrand Co., Princeton, New Jersey, 1967.

[83] G. E. Hardy and T. D. Rogers. A generalization of a fixed point theorem of Reich. *Canad. Math. Bull.*, 16:201–206, 1973.

[84] G. H. Hardy. *Divergent Series*. Oxford University Press, Oxford, 1973.

[85] G. H. Hardy and W. W. Rogosinski. *Fourier Series*. Cambridge University Press, Cambridge, 1950.

[86] R. Harte. A quantitative Schauder theorem. *Math. Z.*, 185:243–245, 1984.

[87] R. Harte. Regular boundary elements. *Proc. Amer. Math. Soc.*, 99:328–330, 1987.

[88] R. Harte. *Invertibility and Singularity for Bounded Linear Operators*. Marcel Dekker Inc., New York and Basel, 1988.

[89] R. Harte. A note on generalized inverse functions. *Proc. Amer. Math. Soc.*, 104:551–552, 1988.

[90] R. Harte. Polar decomposition and the Moore–Penrose inverse. *Panamerican Mathematical Journal*, 2, no. 4:71–76, 1992.

[91] R. Harte and M. Mbekhta. On generalized inverses in c^*–algebras. *Studia Math.*, 103:71–77, 1992.

[92] H. G. Heuser. *Functional Analysis*. John Wiley and Sons, Chichester, New York, Brisbane, Toronto, Singapore, 1982.

[93] T. L. Hicks and B. E. Rhoades. A Banach type fixed point theorem. *Math. Japon.*, 24:327–330, 1979.

[94] S. H. Hochwald and B. B. Morrel. Some consequences of left invertibility. *Proc. Amer. Math. Soc.*, 100:109–110, 1987.

[95] V. Istrăţescu. On a measure of noncompactness. *Bull. Math. Soc. Sci. Math. R. S. Roumanie (N. S.)*, 16:195–197, 1972.

[96] V. Istrăţescu. *Fixed Point Theory: An Introduction*. Reidel Publishing Company, Dordrecht, Boston and London, 1981.

[97] S. Izumino. Convergence of generalized inverses and spline projectors. *J. Approx. Theory*, 38:269–278, 1983.

[98] D. Jackson. *Fourier Series and Orthogonal Polynomials*. Menasha: The Carus Mathematical Monographs, 1941.

[99] R. C. James. Orthogonality in normed linear spaces. *Duke Math. J.*, 37:291–392, 1945.

[100] R. C. James. A nonreflexive Banach space isometric with its second conjugate. *Proc. Nat. Acad. Sci. U.S.A.*, 37:174–177, 1951.

[101] A. Jeribi. *Spectral Theory and Applications of Linear Operators and Block Operator Matrices*. Springer Verlag, Heidelberg, New York, Dordrecht, London, 2015.

[102] P. Jordan and J. von Neumann. On inner product in linear metric spaces. *Ann. of Math.*, 36(2):719–723, 1935.

[103] J. E. Joseph and M. H. Kwack. Alternative approaches to proofs of contraction mapping fixed point theorems. *Missouri J. Math Sci.*, 11:167–175, 1999.

[104] M. A. Kaashoek. Stability theorems for closed linear operator. *Indag. Math.*, 27:452–466, 1965.

[105] M. A. Kaashoek. Ascent, descent and defect, a note on a paper by A. E. Taylor. *Math. Annalen*, 172:105–115, 1967.

[106] M. A. Kaashoek and D. C. Lay. Ascent, descent, and commuting perturbations. *Trans. Amer. Math. Soc.*, 186:35–47, 1972.

[107] S. W. Kaczmarz and H. Steinhaus. *Theorie der Orthogonalreihen*. Warszawa–Lwów, 1935.

[108] P. K. Kamthan and M. Gupta. *Sequence Spaces and Series*. Marcel Dekker, New York, 1981.

[109] R. Kannan. Some results on fixed points, *Bull. Cal. Math.*, 60:71–76, 1969.

[110] I. Kaplansky. Regular Banach algebras. *J. Indian Math. Soc. (N. S.,* 12:57–62, 1948.

[111] T. Kato. Perturbation theory for nullity, deficiency and other quantities of linear operators. *J. Analyse Math.*, 6:261–322, 1958.

[112] T. Kato. *Perturbation Theory for Linear Operators*. Springer–Verlag, Berlin, 1966.

[113] R. Khalil. Orthogonality in Banach spaces. *Math. J. Toyama Univ.*, 15:185–205, 1990.

[114] M. Khamsi and W. Kirk. *An Introduction to Metric Spaces and Fixed Point Theory*. John Wiley & Sons, Inc., New York, 2000.

[115] C. F. King. A note on Drazin inverses. *Pacific J. Math.*, 70:383–390, 1977.

[116] W. A. Kirk and L. M. Saliga. The Brézis–Browder order principle and extensions of Caristi's theorem. *Nonlinear Analysis*, 47:2765–2778, 2001.

[117] J. J. Koliha. Range projections of idempotents in C^*–algebras. *Demonstratio Mathematica*, 34(1):91–103, 2001.

[118] J. J. Koliha and V. Rakočević. Fredholm properties of the difference of orthogonal projections in a Hilbert space. *Integr. Equ. Oper. Theory*, 52:125–134, 2005.

[119] M. A. Krasnoselski. Two remarks on the method of successive approximations. *Uspehi Math. Nauk (N. S.)*, 10(1):123–127, 1955.

[120] H. Kroh and P. Volkman. Störungssätze für Semifredholmoperatoren. *Math. Z.*, 148:295–297, 1976.

[121] K. Kuratowski. Sur les espaces complets. *Fund. Math.*, 15:301–309, 1930.

[122] K. Kuratowski. *Topologie I*, volume 20. Monografie Matematyczne, Warsaw, 1958.

[123] S. Kurepa. Generalized inverse of an operator with a closed range. *Glasnik Matematički*, 3(23):207–214, 1968.

[124] S. Kurepa. Quadratic functionals contained on an algebraic basic set. *Glasnik Matematički*, 6(26):265–275, 1971.

[125] S. Kurepa. On biomorphisms and quadratic forms on groups. *Aequationes Mathematicae*, 9:30–45, 1973.

[126] J. P. Labrousse. Une charactérisation topologique des générateurs infinitésimaux de semigroupes analytiques et de contractions sur un espace de Hilbert. *Atti della Accademia Nazionale dei Lincei*, 52:631–633, 1972.

[127] J.-Ph. Labrousse. Les opérateurs quasi–Fredholm: Une généralisation des opérateurs semi–Fredholm. *Rend. Circ. Math. Palermo*, (2) XXIX:161–258, 1980.

[128] J.-Ph. Labrousse and M-Mbekhta. Les opérateurs points de continuite pour la conorme et l'inverse de Moore–Penrose. *Houston Journal of Mathematics*, 18:7–23, 1992.

[129] L. E. Labuschagne and J. Swart. An elementary proof of a classical semi–Fredholm perturbation theorem. *Rev. Roumaine Math. Pures Appl.*, 41:357–361, 1996.

[130] A. Lebow and M. Schechter. Semigroups of operators and measures of noncompactness. *J. Funct. Anal.*, 7:1–26, 1971.

[131] V. E. Ljance. Some properties of idempotent operators. *Teor. i Prikl. Mat. Lvov*, 1:16–22, 1959.

[132] I. J. Maddox. On theorems of Steinhaus type. *J. London Math, Soc.*, 42:239–244, 1967.

[133] I. J. Maddox. *Elements of Functional Analysis*. Cambridge University Press, Cambridge, 1971.

[134] E. Malkowsky and V. Rakočević. *An Introduction into the Theory of Sequence Spaces and Measures of Noncompactness*, volume 9(17) of *Zbornik radova, Matematčki institut SANU*, pages 143–234. Mathematical Institute of SANU, Belgrade, 2000.

[135] E. Malkowsky and V. Rakočević. *Advances in Nonlinear Analysis via the Concept of Measure of Noncompactness*, chapter titled On some results using measures of noncompactness, pages 127–180. Springer Verlag, 2017.

[136] W. R. Mann. Mean value methods in iteration. *Proc. Amer. Math. Soc.*, 4:506–510, 1953.

[137] M. Mbekhta. Résolvant généralisé et théorie spectrale. *J. Operator Theory*, 21:69–105, 1989.

[138] A. Meir and E. Keeler. A theorem on contraction mappings. *J. Math. Anal. Appl.*, 28:326–329, 1969.

[139] P. M. Miličić. Sur les suites orthonormaux dans les espaces normés. *Mat. Vesnik*, 33:95–102, 1981.

[140] P. M. Miličić. Sur la g–orthogonalité dans des espaces normés. *Mat. Vesnik*, 39:325–334, 1987.

[141] V. Müller. On the regular spectrum. *J. Operator Theory*, 31:363–380, 1994.

[142] V. Müller. *Spectral Theory of Linear Operators and Spectral System in Banach Algebras*. Birkhäuser Varlag, Basel – Boston – Berlin, 2001.

[143] G. J. Murphy and T. T. West. Decomposition of index–zero Fredholm operators. *Proc. R. Ir. Acad.*, 81A(1):49–54, 1981.

[144] M. Z. Nashed. *Generalized Inverses and Applications*. Academic Press, New York, 1976.

[145] M. Z. Nashed. Inner, outer, and generalized inverses in Banach and Hilbert spaces. *Numer. Funct. Anal. and Optimatiz.*, 9:261–325, 1987.

[146] S. Nikolsky. Linear equations in normed linear spaces. *Bulletin de l'Académie des Sciences de l'URSS, Série mathématique*, 7:147–166, 1943.

[147] R. D. Nussbaum. The radius of the essential spectrum. *Duke Math. J.*, 38:473–478, 1970.

[148] Z. Opial. Weak convergence of the sequence of successive approximations for nonexpansive mappings. *Bull. Amer. Math. Soc.*, 73:591–597, 1966.

[149] R. S. Palais. A simple proof of the Banach contraction principle. *J. Fixed Point Theory Appl.*, 2:221–223, 2007.

[150] R. Penrose. On best approximate solutions of linear matrix equations. *Proc. Cambridge Philos. Soc.*, 52:17–19, 1956.

[151] A. Peyerimhof. *Lectures on Summability*. Springer Verlag, Heidelberg, Berlin, New York, 1969. Lecture Notes in Mathematics 107.

[152] A. Peyerimhoff. Über ein Lemma von Herrn Chow. *J. London Math. Soc.*, 32:33–36, 1957.

[153] W. Pfaffenberger. On the ideals of strictly singular and inessential operators. *Proc. Amer. Math. Soc.*, 25(3):603–607, 1970.

[154] A. Pietsch. *Operator Ideals*. North–Holland, Amsterdam, New York, Oxford, 1980.

[155] H. R. Pitt. A note on bilinear forms. *J. London Math. Soc.*, s1-11, no. 3:174–180, 1936.

[156] V. Pták. Extremal operators and oblique projections. *Časopis Pěst. Mat.*, 110:343–350, 1985.

[157] E. Rakotch. A note on contractive mappings. *Proc. Amer. Math. Soc.*, 13:459–465, 1962.

[158] V. Rakočević. On one subset of M. Schechter's essential spectrum. *Mat. Vesnik*, 33:389–391, 1981.

[159] V. Rakočević. On the essential approximate point spectrum II. *Mat. Vesnik*, 36:89–97, 1984.

[160] V. Rakočević. Approximate point spectrum and commuting compact perturbations. *Glasgow Math. J.*, 28:193–198, 1986.

[161] V. Rakočević. Moore–Penrose inverse in Banach algebras. *Proc. R. Ir. Acad.*, 88A:57–60, 1988.

[162] V. Rakočević. A note on regular elements in Calkin algebras. *Collect. Math*, 43:37–42, 1992.

[163] V. Rakočević. Generalized spectrum and commuting compact perturbations. *Proc. Edinburgh Math. Soc.*, 36:197 209, 1993.

[164] V. Rakočević. Semi–Fredholm operators with finite ascent or descent and perturbations. *Proc. Amer. Math. Soc.*, 123:3823–3825, 1995.

[165] V. Rakočević. Semi–Browder operators and perturbations. *Studia Math.*, 122:131–137, 1996.

[166] V. Rakočević. On the norm of idempotent operators in a Hilbert space. *Amer. Math. Monthly*, 107:748–750, 2000.

[167] V. Rakočević and J. Zemánek. Lower s-numbers and their asymptotic behaviour. *Studia Math.*, 91:231–239, 1988.

[168] S. Reich. Some remarks concerning contraction mappings. *Canad. Math. Bull.*, 14:121–124, 1971.

[169] J. Reinermann. Über Toeplitzsche Iterationsverfahren und einige ihrer Anwendungen in der konstruktiven Fixpunktheorie. *Studia Math.*, 32:209–227, 1969.

[170] B. E. Rhoades. *Constructive and Computational Methods for Differential and Integral Equations*, volume 430 of *Lecture Notes in Mathematics*, chapter titled Fixed point iterations using infinite matrices II, pages 390–395. Springer–Verlag, New York, Berlin, 1974.

[171] B. E. Rhoades. Fixed point iterations using infinite matrices. *Trans. Amer. Math. Soc.*, 196:161–176, 1974.

[172] B. E. Rhoades. A comparison of various definitions of contractive mappings. *Trans. Amer. Math. Soc.*, 26:257–290, 1977.

[173] C. E. Rickart. *General Theory of Banach Algebras*. Van Nostrand, Princeton, N.J., 1960.

[174] F. Riesz. Über lineare Funktionalgleichungen. *Acta Math.*, 41:71–98, 1918.

[175] W. H. Ruckle. *Sequence Spaces*. Pitman, Boston, London, Melbourne, 1981. Research Notes in Mathematics 49.

[176] W. Rudin. *Functional Analysis*. McGraw–Hill Book Company, New York, 1973.

[177] W. Rzymowski. On the existence of solution of the equation $x' = f(t, x)$ in Banach space. *Bull. Acad. Polon. Sci., Sér. Math. Astronom. Phys.*, 1971:295–299, 1971.

[178] N. Salinas. Operators with essentially disconnected spectrum. *Acta Sci. Math. (Szeged)*, 33:193–205, 1972.

[179] W. L. C. Sargent. On compact matrix transformations between sectionally bounded *BK*–spaces. *J. London Math. Soc.*, 41:79–87, 1966.

[180] H. Schäfer. Über die Methode sukzessiver Approximation. *JBer. Deutsch. Math. Verein.*, 59:131–140, 1957.

[181] J. Schauder. Der Fixpunktsatz in Funktionalräumen. *Studia Math.*, 2:171–180, 1930.

[182] J. Schauder. Über lineare, vollstetige Funktionaloperatoren. *Studia Math.*, 2:183–196, 1930.

[183] M. Schechter. *Fredholm Operators and the Essential Spectrum*. New York University, Courant Institute of Mathematical Sciences, New York, 1965.

[184] M. Schechter. Invariance of essential spectrum. *Bull. Amer. Math. Soc.*, 71:365–367, 1965.

[185] M. Schechter. Basis theory of Fredholm operators. *Annali della Scuola Normale Superiore di Pisa, Classe di Scienze*, 21(2):261–280, 1967.

[186] M. Schechter. On the invariance of the essential spectrum of an arbitrary operator II. *Ricerche Mat.*, 16:3–26, 1967.

[187] M. Schechter. *Principles of Functional Analysis*. Academic Press, New York, 1971.

[188] M. Schechter. *Principles of Functional Analysis*, volume 36. American Mathematical Society, Providence, Rhode Island, USA, 2 edition, 2002.

[189] I. Schur. Über lineare Transformationen in der Theorie der unendlichen Reihen. *J. Reine Angew. Math.*, 151:79–111, 1920.

[190] M. Ó. Searcóid. The continuity of the semi–Fredholm index. *IMS Bulletin*, 29:13–18, 1992.

[191] V. M. Sehgal. A new proof of Caristi's fixed point theorem. *Proc. Amer. Math. Soc.*, 66:54–56, 1977.

[192] A. D. Sokal. A really simple proof of the uniform boundedness theorem. *Amer. Math. Monthly*, 118:450–452, 2011.

[193] H. Steinhaus. Remarks on the generalisation of the idea of limit. *Prace mat. fiz.*, 22:113–119, 1911.

[194] M. Stieglitz and H. Tietz. Matrixtransformationen in Folgenräumen. Eine Ergebnisübersicht. *Math. Z.*, 154:1–16, 1977.

[195] P. V. Subrahmanyam. Completeness and fixed points. *Monathsh. Math.*, 80(4):325–330, 1975.

[196] T. Suzuki. Some notes on Meir–Keeler contractions and *L*–functions. *Bull. Kyushu Inst. Tech. Pure Appl. Math.*, 53:1–13, 2006.

[197] S. Szufla. Measure of non compactness and ordinary differential equations in Banach spaces. *Bull. Acad. Polon. Sci., Sér. Math. Astronom. Phys.*, 19, 9, 1971.

[198] A. E. Taylor. Theorems on ascent, descent, nullity and defect of linear operators. *Math. Ann.*, 63:18–49, 1966.

[199] A. E. Taylor and D. C. Lay. *Introduction to Functional Analysis*. John Wiley & Sons, Inc., New York, Chichester, Brisbane, Toronto, 2 edition, 1980.

[200] O. Toeplitz. Über allgemeine Mittelbildungen. *Prace. Mat. Fiz.*, 22:113–119, 1911.

[201] J. M. Ayerbe Toledano, T. Dominguez Benavides, and G. Lopez Acedo. *Measures of Noncompactness in Metric Fixed Point Theory*, volume 99 of *Operator Theory Advances and Applications*. Birkhäuser Verlag, Basel, Boston, Berlin, 1997.

[202] F. G. Tricomi. *Vorlesungen über Orthogonalreihen*, volume 76 of *Grundlehren der mathematischen Wissenschaften in Einzeldarstellungen*. Springer–Verlag, Berlin, Heidelberg, New York, second edition, 1970.

[203] I. Vidav. On idempotent operators in a Hilbert space. *Publ. Inst. Math. (Beograd)*, 4(18):157–163, 1964.

[204] T. T. West. A Riesz-Schauder theorem for semi–Fredholm operators. *Proc. Roy. Irish Acad. Sect. A*, 87:137–146, 1987.

[205] T. T. West. Removing the jump–Kato's decomposition. *Rocky Mountain J. Math.*, Vol 20, No 2:603–612, 1990.

[206] N. Wiener. Certain iterative characteristics of bilinear operations. *Bull. Soc. Math. France*, 50:119–134, 1922.

[207] A. Wilansky. *Functional Analysis*. Blaisdell Publishing Company, New York, 1964.

[208] A. Wilansky. *Modern Methods in Topological Vector Spaces*. McGraw Hill, New York, 1978.

[209] A. Wilansky. *Summability through Functional Analysis*, volume 85. North–Holland, Amsterdam, 1984. Mathematical Studies.

[210] K-W. Yang. Index of Fredholm operators. *Proc. Amer. Math. Soc.*, 41:329–330, 1973.

[211] B. Yood. Properties of linear transformations preserved under addition of a completely continuous transformation. *Duke Math. J.*, 18:599–612, 1951.

[212] J. Yorke. A continuous differential equation in Hilbert space without existence. *Funkc. Ekvac.*, 13:19–21, 1970.

[213] K. Yosida. *Functional Analysis*. Classics in Mathematics. Springer Verlag, Berlin, Heidelberg, New York, 6th edition, 1980.

[214] T. Zamfirescu. Fixed point theorems in metric spaces. *Archiv der Mathematik*, 23:292–298, 1972.

[215] K. Zeller. Abschnittskonvergenz in FK–Räumen. *Math. Z.*, 55:55–70, 1951.

[216] K. Zeller. Allgemeine Eigenschaften von Limitierungsverfahren. *Math. Z.*, 53:463–487, 1951.

[217] K. Zeller. Matrixtransformationen von Folgenräumen. *Univ. Rend. Mat.*, 12:340–346, 1954.

[218] K. Zeller and W. Beekmann. *Theorie der Limitierungsverfahren*. Springer Verlag, Heidelberg, Berlin, New York, 1968.

[219] J. Zemánek. The stability radius of a semi–Fredholm operator. *Integral Equations and Operator Theory*, 8:137–144, 1985.

[220] J. Zemánek. The reduced minimum modulus and the spectrum. *Integral Equations and Operator Theory*, 12:449–454, 1989.

[221] A. Zygmund. *Trigonometric Series I–II*. Cambridge University Press, Cambridge, 1968.

Russian Bibliography

[AKP] Р. Р. Ахмеров, М. И. Каменский, А. С. Потапов и др., Меры некомпактности и уплотняющие операторы, *Новосибирск, Наука, 1986*

[Atk] Ф. В. Аткинсон, Нормальная разрешимость линейных уравнений в нормированных пространствах, *Мат. Сборник,* **28**, (1951) 3–14

[GGM1] Л. С. Гольденштейн, И. Ц. Гохберг и А. С. Маркус, Исследование некоторых свойств линейных ограниченных операторов в связи с их q-нормой, *Уч. зап. Кишиневского гос. ун-та,* **29**, (1957), 29–36

[GGM2] Л. С. Гольденштейн, А. С. Маркус, О мере некомпактности ограниченных множеств и линейных операторов, *В кн.: Исследование по алгебре и математическому анализу, Кишинев: Картя Молдавеняске* (1965) 45–54.

[Nikol] Д. Н. Никольская, Критерий устойчвости точечного спектра при полне непрерывных возмущениях, *Математические Заметки* **18, 4**, (1975) 601–617

[E–K] В. М. Ени и Г. Т. Карауш, Об одном характеристическом свойстве конечомерных пространств, *Бельцкий гос. пед. институт, Ученые записки* **1**, (1958) 19–20

Index

Milton Keynes UK
Ingram Content Group UK Ltd.
UKHW051942071024
449327UK00026B/2123